The Largest Locomotive in the World.

Weight of Engine, 230,000 lbs. Total Weight of Engine and Tender, 334,000 lbs. Complete Description Given on Page 470.

Latest Development of the American Freight Locomotive.
Cylinders, 21x34 inches; Driving Wheels, 55 inches; Weight of Engine, 212,750 lbs.
Complete Description Given on Page 473.

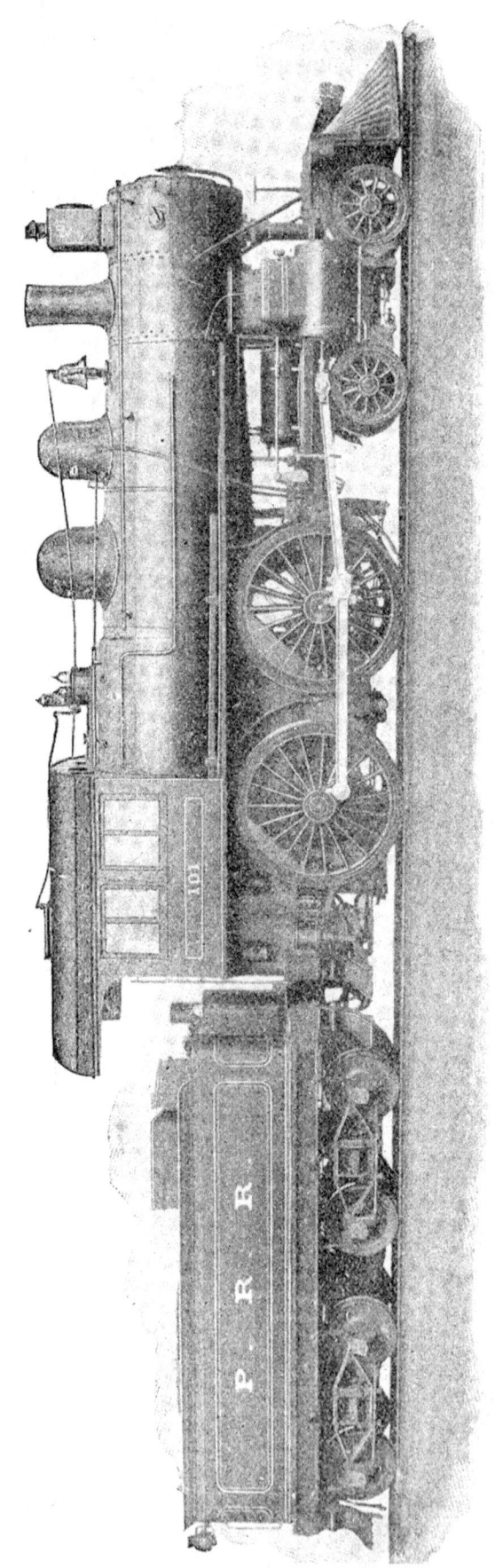

Latest Development of the American Passenger Locomotive.
Cylinders, 18½x26 inches; Driving Wheels, 80 inches; Total Weight of Locomotive in Working Order, 134,500 lbs. Complete description Given on Page 480.

THE
LOCOMOTIVE UP TO DATE.

BY
CHAS. McSHANE,
AUTHOR OF
One Thousand Pointers For Machinists and Engineers

ILLUSTRATED.

Forty-fourth Thousand.

GRIFFIN & WINTERS,
NEW YORK LIFE BUILDING,
CHICAGO, ILL.
1900.

Copyrighted by
CHAS. McSHANE,
1899.

©2008-2010 Periscope Film LLC
All Rights Reserved
ISBN #978-1-935700-21-0
www.PeriscopeFilm.com

PREFACE.

It is unnecessary for me to point out in detail what I regard as the most valuable features of this work, as my judgment may be warped by the circumstances that I have found certain subjects more interesting than others. Indeed, in my judgment, every work must stand or fall upon its own merits, and nothing that an author can say in reference to it can avail to change the ultimate verdict of those who subject it to the decisive practical tests that they are called upon to apply in the discharge of their duties.

I desire to express my sincere thanks to the Railway Public for the generous reception accorded my former efforts, and trust that in this work I may be deemed, in some degree at least, to have made good my pledges to the gentlemen who requested the preparation by me of a more comprehensive work on the same subject.

The number of contributors, inventors, locomotive works and mechanical journals who have rendered assistance are too numerous to admit of my thanking them by name, but I beg here to renew to them all the acknowledgments which have already been made to each in person.

C. McS.

Chicago, January 21, 1899.

TABLE OF CONTENTS.

History of the Locomotive	13
Slide Valves	19
Balanced Slide Valves	64
Link Motion	97
The Shifting Link	99
Errors of the Link Motion	131
Locomotive Valve Setting	146
Rules for Valve Setting	172
Other Valve Gears	190
The Steam Indicator	212
Injectors and Boiler Checks	230
Lubricators	269
Steam and Air Gauges	280
Metallic Packings	292
General Information	298
Locating Blows and Pounds	313
Compound Locomotives	326
Breakdowns	401
Combustion	421
Incrustation	445
Modern Locomotives	470
Extraordinary Fast Runs	504
General Machine Shop Work (Erecting Department)	512
Shoes and Wedges	512
Quartering Driving Wheels	534
Modern Counter-Balancing	537
Miscellaneous Points, Tables, Etc.	556
General Machine Work	563
Metric System	578
Compressed Air	583
Air Brake	609

BRIEF HISTORY OF THE LOCOMOTIVE.

The first self-moving steam engine of which there is any record was built by Mr. Nicholas Cagnot in France in 1769. It was, of course, of very crude form, being mounted on a carriage and run upon the public highway, but it was from this insignificant looking little engine that we can trace locomotive construction and development down to the present monsters of the rail. It is true that Isaac Newton has received credit for being the original inventor of the steam engine, but the boiler constructed by him in 1680, and called an engine, cannot properly be so termed because it failed to move, and therefore failed to develop any power. It consisted of a spherical boiler mounted on a carriage and the intended propulsion was through the force of escaping steam against the atmosphere; naturally it proved a complete failure.

To Mr. James Watt, more than any other man, is due the honor of first controlling and utilizing steam for power and perfecting the steam engine, although Newcomer and others had used steam for lifting water, etc., long before Watt's time. To Trevithick of England is due the honor of first applying the steam engine to rails, or tramways. His engine bore his name and was first run on the Merthyr Tydvil Tramway in South Wales, February 1, 1804. It was a pronounced success, although it appeared that Mr. Trevithick had built a machine which he could not control, as the engine run off the track and was badly wrecked the first day. The presumption is, however, that it was due to his inexperience as an engine driver. Many other locomotives were soon afterward built in England, several of which have since become famous. American engineers were not sleeping, however, while all this experimenting was taking place across the water. Mr. Nathan Reed, of Salem, Mass., built an engine as early as 1790, it being the first locomotive ever built in America; like the Cagnot engine it was also mounted on a carriage and run on the public highway.

The first locomotive to run on rails in America was the "Tom Thumb," built by Mr. Peter Cooper, of New York, in 1829. The "Stourbridge Lion" being imported from England the same year;

it was the first locomotive to cross the ocean. It was about this time that locomotive construction actively began on both sides of the water. The most important factor in the success of the locomotive was found to be the mechanism employed to distribute and control the steam, which was called the valve gear.

Many different forms of valve gear were in use in those early days. What is known as "Hook Motion" became the standard form of valve gear in this country, and it remained in use for many years after the invention of the shifting link. Those who never saw hook motions can gratify their curiosity by inspecting "Old Ironsides," which is located in the Field Columbian Museum at Chicago, Ill.

It is believed by many that Mr. William T. James of New York, a most ingenious mechanic who invented the double eccentrics, also invented the link, as he used a link of crude form as early as 1831; but the Stephenson link, which is at present the standard form of link used in this country, was invented in 1842 by Mr. William Howe, an employe of Robert Stephenson & Co., of New Castle, England. Several other valve gears have since been invented but each in turn failed to demonstrate its superiority over the shifting link which is used in one or other of its many forms on a large majority of locomotives. In the early days of locomotive construction many locomotives were imported to this country, but the tide of importation has turned and our builders are now shipping locomotives to all parts of the world. This is due to the fact that we build the fastest, most powerful and best locomotives in the world; the American locomotive being noted for its simplicity, convenience, speed and power. The largest locomotive ever built, up to the present time, is illustrated on the first page of this book. It was built by the Pittsburg Locomotive & Car Works of Pittsburg, Pa., for the Union Railway of Pittsburg, Pa., and completed during October, 1898. In working order the engine and tender weighs 334,000 pounds.

The smallest locomotive built for actual service of which there is any record, was built in Belgium and only weighed 2,420 pounds. The cylinders were $3\frac{1}{8} \times 6\frac{1}{4}$ inches, and diameter of driving wheels $15\frac{1}{4}$ inches. This locomotive, together with a car and one mile of portable track, was presented to the Sultan of Morocco, by the King of Belgium to be used in the gardens of the palace. This imperial toy had necessarily to be carried in pieces from the port of landing to the Capital by the primitive mode of freight transportation, the pack-saddle, the heaviest parts, the boiler and lower frame, weighing only 660 pounds.

This work does not treat upon electricity, but a history of the locomotive would be incomplete without reference to the electric locomotive now used on the B. & O. Ry. Therefore, we have shown a view of the engine, giving its principal dimensions and performances on page 503. While it is probable that electric locomotives will eventually supersede the steam engine, the development of electricity as applied to railways is yet in a crude form, and it will require many years to retire the steam locomotive from all kinds of service.

WHAT THE LOCOMOTIVE HAS DONE FOR AMERICA.*

In the course of three-quarters of a century, a vast wilderness on the American continent has been changed from gloomy, untrodden forests, dismal swamps and pathless prairies into the abode of a high civilization. Prosperous states teeming with populous towns, fertile farms, blooming gardens and comfortable homes have arisen from regions where formerly savage men and wild animals united to maintain sterile desolation. The most potent factor in the beneficent change effected has been the construction of railroads.

There are numerous navigable rivers and lakes furrowing the great continent, but geographically they are far apart, and there is no means of reaching vast regions except by land transportation.

Long before a railroad was built anywhere, American engineers and public men perceived the possibilities of the steam engine as a means of accelerating land travel, and this century was a very few years old when projects began to be agitated in different states, to construct railways or tramways, on which the steam engine could be employed as motive power.

More than a century has passed since Oliver Eames, the inventor of the high-speed, high-pressure steam engine, predicted that his engine would some day enable passengers to leave Washington after breakfast, dine in Philadelphia, and sup in New York. The distance is 229 miles, and it is now traversed by numerous express trains on the Pennsylvania and Baltimore & Ohio railroads in about five hours.

When Eames' prediction was made, it took a week to travel over the atrocious roads that separated New York from Washington, and it seemed absurd to suppose that the journey could be made in a day; but Americans have always been very cordial in their welcome of improvements. They have always reposed great faith in the men who showed themselves capable of inventing or improving mechanical devices to lighten human toil or lessen animal drudgery. While the natural pessimist might say that the promised mechanical revolution smacked of the miraculous or bordered upon the ridiculous, the mass of the people were ardent believers that Oliver Evans' prediction would come true, and their confidence inspired hope in others.

Before the first decade of this century closed there were steamboats plying on the Hudson, and their success brought new

*Extract from article by Mr. Angus Sinclair in "Pall Mall Magazine."

confidence to those who hoped to see the empty inland regions supplied with the means of transportation, that would encourage settlers to establish homes in the wilderness which constituted the greater portion of the United States territory. There was, therefore, little proselyting to do in favor of railroad construction.

When the era of railroad construction began, the aim at first was to connect industrial centers, or to connect inland waterways with those of the seaboard. The joining of the leading cities by rail made the most rapid progress, and this had not been done to a great extent when the demand for high-speed trains arose.

The average American has always been in a hurry. He wants to do two days' work between each rising and setting of the sun. Any ordeal that keeps him idle or inert is particularly galling. No matter what improvements in the acceleration of transportation might be made, the mind of the traveler anticipates them. The canal boat was an improvement on walking, but the passenger was watching early and late, to see that the towing mules received due inspiration from the driver's whip to make the best possible time. When the steamboat pushed canal travel to the rear, the American traveler was ready to see his hogs thrown into the furnace if it would add a few revolutions to the propelling-wheel. When railroads began to make better time than any other method of transportation, the busy man soon got to consider it intolerable that he should be kept one hour in a train when the run might have been made in fifty minutes.

The American locomotive, which was worked out on native lines, and would not have been greatly retarded in development had no railways been constructed in other countries, went through a remarkable brief period of evolution. Many people, all over the world, believe that the locomotive was invented and perfected by George Stephenson, and that the machine emerged from his hand a perfect engine. That is not true even regarding Great Britain. During the first decade after the opening of the Liverpool & Manchester Railway, a host of inventors labored to design a better locomotive than Stephenson had built, and many curious mistakes were made.

A story is told of Napier, a famous Scotch engineer, who had been invited to witness a test of a locomotive designed by an ingenious individual shortly after the "Rocket" had won its first triumph. The inventor wished to interest capitalists in his engine, and tried to obtain Napier's endorsement. He succeeded in bringing Napier into the presence of the capitalists, but when the attempt was made to have the engineer testify in favor of the engine, nothing was forthcoming but a succession of protesting grunts. Losing patience, the inventor exclaimed, "Well, you must admit that you saw the engine running." "You may call that running," was the reply; "all I saw was you fellows shovin' her."

We had the "shovin'" period in America, but it did not last long. Colonel Long, one of the most eminent pioneer civil engineers, designed a locomotive for the Baltimore & Ohio Railroad when competitive designs were still in order. In a speech made years afterwards, he admitted that his locomotive was not a success. "On the trial trip," he remarked, "it took seven hours to run four miles, and we were moving all the time."

This was our early evolution period, and it brought forth some engines that were fearfully and wonderfully made; but they served a useful purpose, since their designs stood forth as dreadful examples of what not to do.

The first form of successful locomotive consisted of a strong rectangular frame which carried the boilers and cylinders, and had fittings to keep the axles of two pairs of wheels in a parallel position. Although a crude machine, weighing little more than a modern fire-engine, it possessed all the essential elements of a modern locomotive. It was light, but the permanent way of our early railroads was relatively lighter. On this account, the first radical change in the pioneer locomotive was made. The purpose of the improvement was, to distribute the weight of the engine over a longer base, which was done by carrying the front end of the locomotive on a four-wheel truck, or "bogie," as it is called in Europe, and the back end upon the driving wheels. A single pair of driving wheels was for a short time popular; but experience soon demonstrated that two pairs coupled gave superior service. That constitutes what is now called the "American" locomotive, which is the representative type on this continent, and far outnumbers all other forms combined.

For the first twenty years of railway history, the train speed was very moderate, but at the end of that time an agitation arose for trains to be run at the ambitious speed of a mile a minute. The men in charge of the motive power of several railroads were ordered to build locomotives that would maintain this speed, and a variety of engines, with single drivers about 7 feet in diameter, were put into service. They could attain the required speed with a light train, where a long run could be made without checking speed; but that condition existed on very few railroads, and the big-wheeled engines soon fell into disrepute. They were too slow in making the numerous starts required.

Besides, the time for a mile-a-minute speed had not arrived. For the first fifty years of American railroad history, there were scarcely any stations or junctions protected by fixed signals. There were no continuous train brakes in use, and for a considerable part of that period there was no reliable system for regulating the movements of trains on the single track, which was almost universal. "Running by the smoke or headlight" was a common practice. By that practice, the safety of the train depended on the care and vigilance of the engineer, who avoided collisions by watching in daytime for the smoke of engines coming in the other direction, and at night kept a keen watch for the glare of an approaching headlight. The locomotive engineers, as our engine drivers are called, became wonderfully skillful in avoiding accidents, in early days of crude practice and appliances, and their successors are equally efficient under changed circumstances. The numerous responsibilities, the sudden calls to meet emergencies coolly and courageously, develop all the higher attributes of manhood. Under whatever name you find the men who run locomotives, they are reliable and trustworthy; no matter in what clime or country they may be met, close acquaintance will prove them, as a class, to be as manly and self-reliant as any other portion of the population.

The general application of the Westinghouse air brake to our passenger equipment prepared the way for our modern express trains. This began in 1873, and the merits of the invention were so quickly appreciated that the only restraint to its introduction was the limit of manufacturing facilities.

The traveling public understand that the Westinghouse air brake has promoted the safety of railway travel; but it requires the man who has run locomotives with defective means of stopping to fully realize the value of a good brake. When running at high speed in the anti-brake days, the writer used to feel that the driver was like the man who pulled the trigger of a rifle to send the ball into space. He could start it into speed, but he must wait for air friction, or the hitting of some object to stop it.

A first-class locomotive does not differ very much when found in England, the Continent of Europe, in India, or in America; but the special claim made in favor of American locomotives is that they are less complicated than anything to be found elsewhere. Then men who have been most influential in designing our motive power have nearly all run locomotives in their time, and they thoroughly appreciate the advantages of simplicity of parts, and providing every convenience for the men operating the machines. On all our modern locomotives, the engineer can reach every appliance used in operating the engine without moving from his seat or turning round. With this convenience, he need never divert his attention from watching for signals. This is strongly in contrast with many locomotives to be found in foreign countries. English locomotives are much more convenient than they used to be, but they do not compare favorably with American engines in this respect.

LOCOMOTIVE SLIDE VALVES.

The slide valves upon a locomotive which control the distribution of steam, receive their motion from eccentrics fastened to the main shaft or axle. An eccentric is a wheel or disc, having its axis placed out of its center and used for obtaining a reciprocating or alternate motion from a circular one or vice versa. The crank motion of the eccentrics being transmitted to the valves by the use of eccentric straps, blades, rocker arms, valve stems, etc.

The movement of the valve will be fully explained later on, but we shall first study the construction of the valve itself.

For many years the plain slide valve was considered the best form of valve for locomotive service. But of recent years, with the gradual and continued enlargement of the locomotive and the introduction of higher steam pressures the increased friction between the valves and their seats and the stresses imposed upon all the parts of the valve gear, together with the enormous power required of the engineer to reverse the engine, became a very serious problem, and the necessity for some means to reduce this increased friction became clearly apparent.

Reversing cylinders were suggested, but it was found that the friction could be reduced to a minimum by balancing the valve, or in other words, by removing the steam pressure from the top of the valve by mechanical means, and as a result innumerable forms of balance valves have been invented. A few of the best forms of these valves are in general use at the present time. While the various methods of balancing the valve are too numerous to mention, few efforts have been made to change the face of the valve, which in most cases remains the same as the plain slide valve.

A supplementary port added to the plain slide valve and known as the "Allen Ported Valve" has been in use for a number of years, and is admitted to be a decided improvement to the plain slide valve, especially for high speeds. A few other forms of valves having supplementary ports and double openings are at present in use and promise good results.

We shall first confine our investigations to the plain slide valve, and afterwards to balanced valves.

THE PLAIN SLIDE VALVE.

INVENTION OF.

The slide valve in a crude form, was invented by Matthew Murray, of Leeds, England, toward the end of the eighteenth century. It was subsequently improved by James Watt, but the long D slide valve which we use at the present day, is credited to Murdock, an assistant of Watt. It came into general use with the introduction of the locomotive, although Oliver Eames, of Philadelphia, appears to have perceived its actual value, for he applied it to engines of his own build years before the locomotive era. But it was upon the locomotive that it clearly demonstrated its real value; its simplicity of construction and its durability together with the high speed at which it could be worked at once commended it to the designers of locomotives in those days, and although repeated efforts have been made to displace it, it is still employed in one or other of its many forms on a great majority of locomotives.

ELEMENTARY PRINCIPLES.

All slide valves must be capable of fulfilling the three following conditions, and if a slide valve cannot do this the engine will not work satisfactorily.

1st. Steam must be admitted into the cylinder at one end only at the same time.

2nd. It should permit the steam to escape from one end of the cylinder, at least as soon as it is admitted into the other end.

3rd. It should cover the steam ports so as not to allow steam to escape from the steam chest into the exhaust ports.

CONSTRUCTION OF THE PLAIN SLIDE VALVE.

Unless the reader has a thorough knowledge of what is meant by the technical terms, lead, lap, cut-off, compression, release, etc., he could not follow an explanation of the construction of the slide valve. We will, therefore, first explain what is meant by those terms, and afterward explain the construction of the valve.

MEANING OF LEAD, LAP, CUT-OFF, COMPRESSION, ETC.

The term "outside lap" or "inside lap" implies the amount on each side of the valve. The two outside edges of the valve are called the steam edges, and the two inside are called the exhaust edges.

"*Outside lap,*" frequently called steam lap, is that portion of the valve which overlaps the steam ports, when the valve stands central upon the valve seat, it is that part of the valve marked L and indicated by the space between lines A and B in Fig. 1.

"*Inside lap*" of a valve, sometimes called exhaust lap, exhaust cover, and inside cover, is that portion of the valve which over-

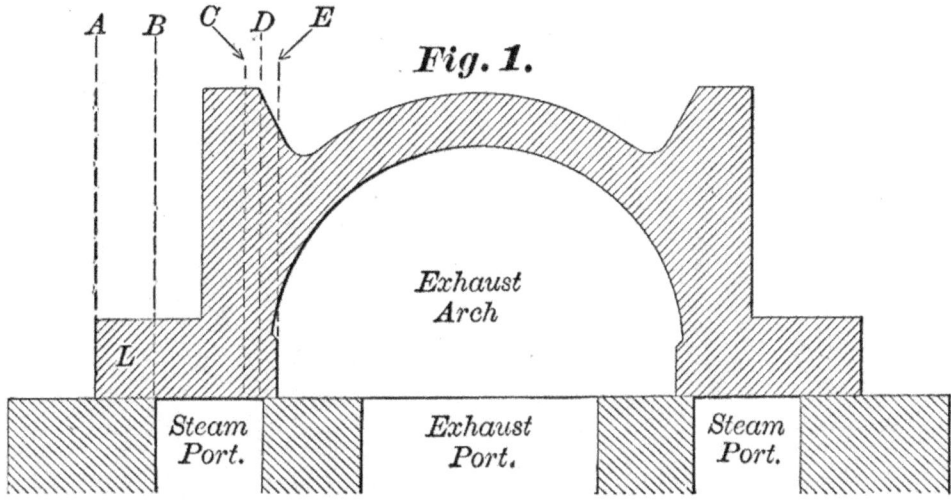

Fig. 1.

laps the two bridges of the valve seat, when the valve stands central upon the seat; as shown in Fig. 1, and indicated by the space between lines D and E.

"*Inside clearance,*" sometimes called negative exhaust lap, inside lead, or exhaust lead, is no portion of the valve; but is the space between the inside edges of the exhaust arch and the

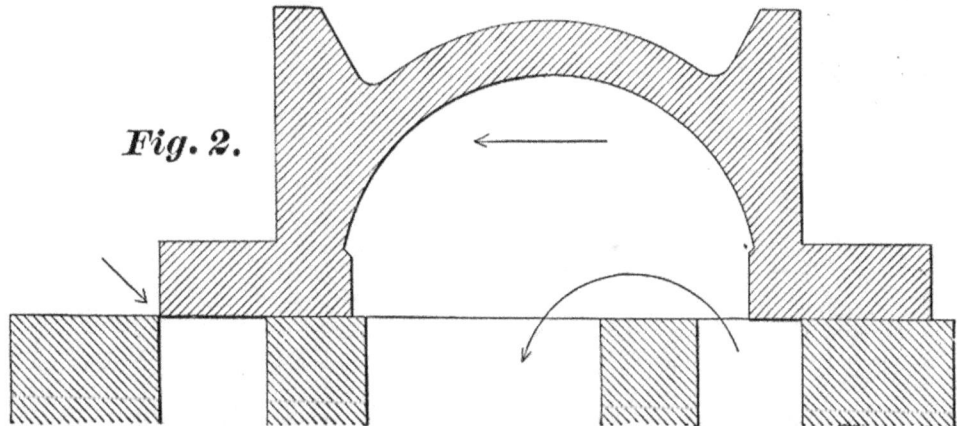

Fig. 2.

bridges when the valve stands central upon the valve seat. As indicated by the space between lines C and D in Fig. 1. The term inside clearance means that amount on each side.

"*Cut-off*" means the cutting off of live steam before the piston has completed its stroke, and thereby utilizing the expansive force of steam. The point of cut-off is reached when the steam edge of the valve completely closes the steam port, as shown in Fig. 2.

"*Compression*" means the cutting off of the exhaust steam before the piston has completed its stroke, to be compressed by the advancing piston, and its pressure increased to arrest the motion of the reciprocating parts. The point at which com-

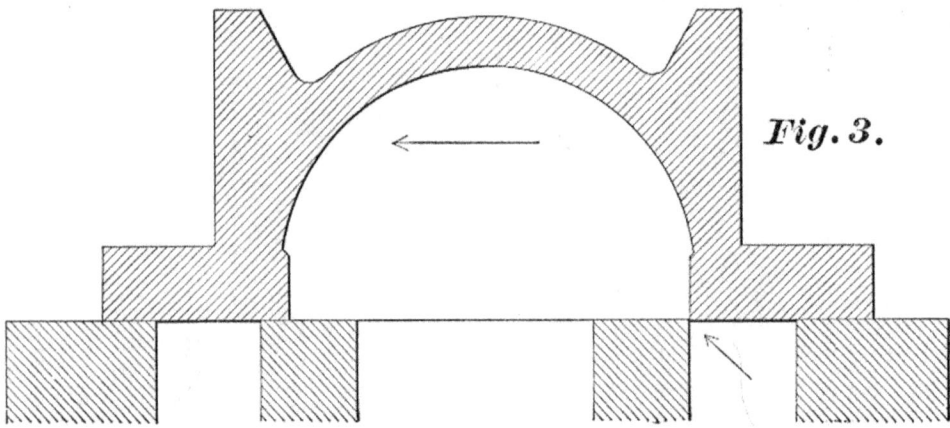

Fig. 3.

pression begins, is reached, when the inside, or exhaust edge of the valve, has completely closed the steam port and thereby cut-off the exhaust steam, as shown in Fig. 3.

"*Release or Exhaust*" means the release of the expanded steam from the cylinder, this point is reached when the inside or exhaust edge of the valve opens the port and permits the

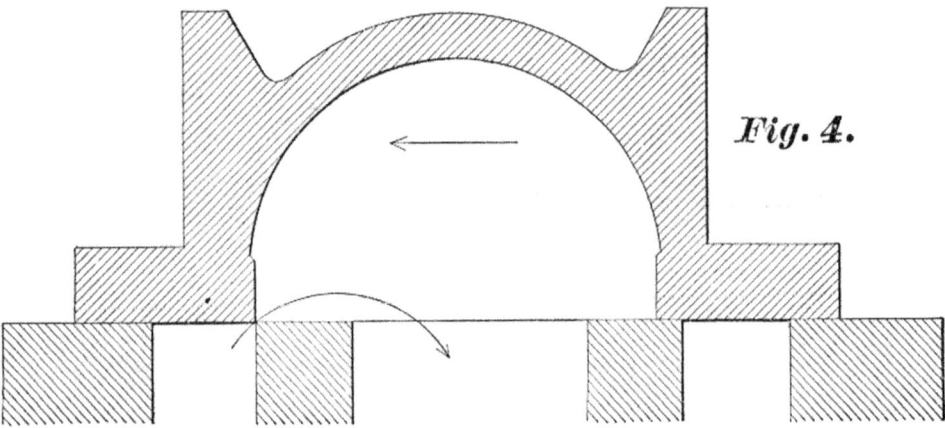

Fig. 4.

steam to escape, as shown in Fig. 4, it is at this point the engine exhausts or puffs.

"*Expansion*" means the expanding of the steam encased in the cylinder, and its time or duration lasts from the point of cut-off, Fig. 2, to the point of release or exhaust, Fig. 4. Therefore the space the valve travels during expansion equals the total of the outside and inside lap of the valve.

"*Lead,*" sometimes called steam lead, is no portion of the

valve, it means the width of the opening of the steam port to admit steam into the cylinder when the piston is at the beginning of its stroke. It is indicated by the letter L in Fig. 5.

"*Over travel*" is the distance the steam edge of the valve travels after the steam port is wide open, as indicated by space between lines A and B in Fig. 6.

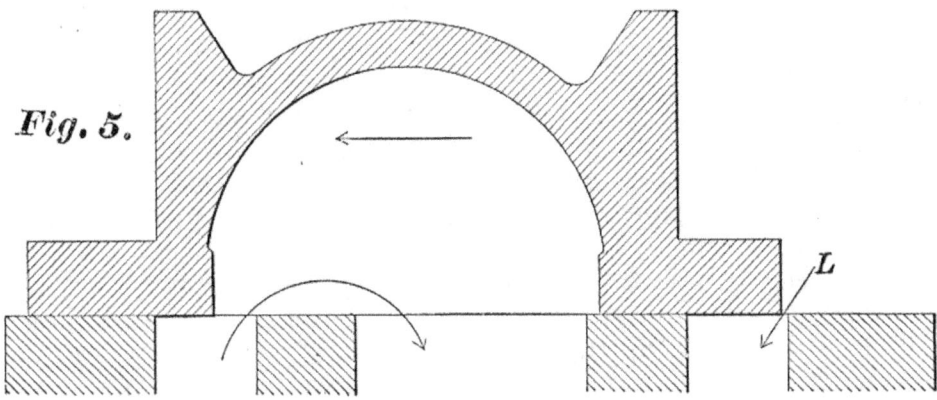

Fig. 5.

"*Travel*," stroke, or throw of the valve is the linear distance through which any part of it travels.

"*Clearance*" is all the waste space between the valve and the piston when the piston is at the beginning of a stroke.

"*Seal*" is an overlapping of the steam edges of the valve to prevent leakage.

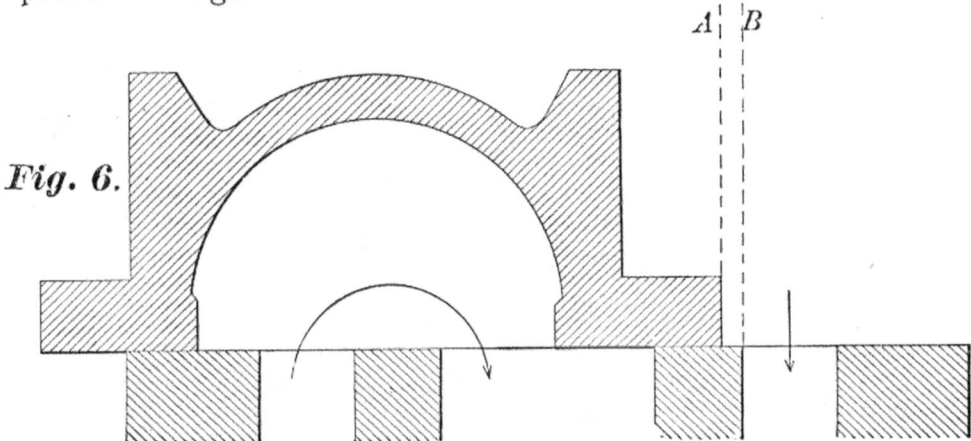

Fig. 6.

Now that the reader understands the technical definition of the terms lap, lead, etc., a further explanation of the reasons why the valve is given these functions becomes necessary.

THE EFFECT OF LAP.

When only one slide valve is used for the whole distribution of steam in one cylinder, as in locomotives, and the valve has no lap, we may justly name the form of such a valve a primitive one, because valves without lap, or only a trifling amount, about 1-16 of an inch, were used in locomotives years ago, when the great necessity for an early and liberal exhaustion was not so well understood as at present, the chief aim then being to secure a timely and free admission of steam. Such valves, as we have stated before, will admit steam during the whole length of the stroke, or, in other words, follow full stroke, and release the steam in one end of the cylinder at the same moment, or nearly so, that the steam is admitted into the other end; this is certainly no profitable way of using steam, for the following reasons: The process of exhausting steam requires *time,* and therefore the release of steam should begin in one end of the cylinder some time before steam is admitted into the other end, or, we may say, the steam which is pushing the piston ahead should be released before the end of the stroke has been reached. This cannot be accomplished with a valve having no lap, and, consequently, when such a valve is used, there will not be sufficient time for the exhausting of steam, thus causing considerable back pressure in the cylinder. In order to secure an early exhaust, lap was introduced; first $\frac{3}{8}$ of an inch lap was adopted, then $\frac{1}{2}$ of an inch. But it soon became apparent that working the steam expansively (a result of lap, besides gaining an early exhaust) additional economy in fuel was obtained, hence the lap was again increased until it became $\frac{7}{8}$ of an inch, and, in some cases, 1 inch, and even more than this. At the present time the lap of a valve in ordinary locomotives with 17x24 inches, or 18x24 inch cylinders is $\frac{7}{8}$ to 1 inch, and, in a few cases slightly exceeding this. From these remarks we may justly conclude, that in these days, the purpose of giving lap to the valve, is to cause it to cut off steam at certain parts of the stroke of the piston, so that during the remaining portion of the stroke the piston is moved by the expansion of the steam. When steam is used in this manner, it is said to be used expansively.

THE EFFECT OF LEAD.

The valve is given lead in order that the steamport will have a greater opening at the beginning of the piston's stroke. Where (the advocates of an early admission claim) it is mostly needed, it also permits of an earlier cut-off, increases compression, and helps to fill the waste volume of clearance. (See index for Lead.)

CONSTRUCTION OF THE VALVE SEAT, AND VALVE.

As the construction of a valve depends entirely upon the proportions of the valve seat, before we enter upon a more thorough study of the valve we will first call the reader's attention to a few things to be considered when designing a valve, and afterward the various effects of lead, lap, travel, etc. We will then study the relative positions of the valve and piston and explain the correct manner to design a valve seat, the valve face and parts.

AREA OF STEAM PORTS.

The area of the steam port depends largely upon the speed and other requirements of the engine, and the dimensions of its other parts. Its area is next in importance to the cut-off, and it is considered the *base* from which all other dimensions are derived when proportioning a valve face and its seat. The higher the speed required the larger the port is made in order to secure a free admission and release. The proportions given in the following tables have been found to give good results. To find the proper area for a steam port multiply the area of the piston in square inches by the number opposite to the given piston speed.

The average piston speed for locomotives varies from 600 to 800 feet per minute.

Speed of Piston in feet per minute.	Number to multiply by.
100	0.02
200	0.04
300	0.06
400	0.07
500	0.09
600	0.10
700	0.12
800	0.14
900	0.15
1000	0.17

Another rule for determining the area of the steam ports for locomotives is given as follows: Multiply the square of the diameter of the cylinder by .078. The ports are usually made a length equal to the diameter of the cylinder, but the longer the port can be made the better the results it will give; as it gives a greater opening for admission and release, reduces the travel necessary for a full port opening and diminishes the area on the back of the valve, thereby requiring less power to move the valve. The steam ports in American locomotives are much larger than those used in English locomotives with the same size cylinders, and the advisability of reducing our port area has received considerable attention from American engineers during the past few months.

THE BRIDGES.

For the same reason the steam ports are made narrow, viz., to reduce the pressure required to move the valve. The bridges should also be made as narrow as possible, but they must be made strong enough to resist the highest pressure, therefore their proper width is considered equal to the thickness of the walls of the cylinder. Although they are usually made a little wider, yet the face may be beveled without materially affecting its strength and it must be remembered that a reduction of $\frac{1}{8}$ inch in its width, will reduce the width of the valve $\frac{1}{4}$ inch, thereby decreasing the area on top. The over-travel must be considered, and a sufficient surface left when the valve is at extreme travel, to make a steam tight joint; $\frac{1}{4}$ inch is considered sufficient. The wear must also be considered, too narrow a bridge would not maintain a steam tight joint. The width of the bridge is usually less than the steam port, on American locomotives they vary from 15-16 to $1\frac{1}{4}$ inch.

EXHAUST PORT.

The exhaust port should be more than twice as wide as the steam port, especially with over travel unless inside clearance is used, as you will see by referring to Fig. 6; otherwise it would cramp or choke the exhaust. But it should not be made too wide, as it will add unnecessarily to the size of the valve, and hence to the pressure upon it, which adds to the friction, wear and tear, of all the valve gear; further than this the size of the exhaust port or cavity has no influence upon the valve. The rule for finding the width of the exhaust port is as follows: Add the width of one steam port to one-half the travel of the valve, and from that amount subtract the width of one bridge. Another rule for determining the area of the exhaust port is to multiply the square of the diameter of the cylinder by .178.

LONGITUDINAL WIDTH OF VALVE SEAT.

Except when Allen or special valves are used, the width of the valve seat is not particular; but if possible, it should be made wide enough to permit of a surface for the valve equal to the width of one bridge when the valve is at extreme travel as shown in Fig. 6; unless that would permit of a shoulder being worn on the seat when engine is hooked up in working notch, which should be avoided.

EFFECT OF LEAD LAP, ETC.

If the valve has neither inside lap, nor inside clearance, the exhaust arch should be the width of both bridges and exhaust port. If the valve has no outside lap there would be no cut-off or expansion. If the valve had no inside lap, compression at one end and release at the other would be simultaneous. If the valve had no lap or lead the eccentric should be at right angle with the crank pin.

The more *"outside lap"* the valve has, the greater the throw required and the later the admission of steam takes place, it also hastens the cut off and prolongs expansion, and necessarily shortens the period the port is open. Outside lap has no effect on compression or exhaust.

"*Inside lap*" prolongs the period of expansion, hastens compression and thereby increases it. It retards and tends to choke the exhaust, but has no effect upon steam admission or point of cut-off.

"*Inside clearance*" or negative inside lap, delays compression, but hastens the exhaust release, thereby making a quicker engine, but has no effect upon the cut-off or point of admission. With inside clearance the point of compression and release as shown in Fig. 7 would be reversed, release taking place before compression. The evil effects of inside clearance in connecting the opposite ends of the cylinder can be overcome by adding an equal amount to the exhaust edge of the valve lip.

The least "*travel*" that will give a full port opening equals twice the outside lap of the valve, plus twice the steam port width. One-half the travel of the valve should always be less than the width of the lap the steam port and the bridge added together. In order to keep the steam port wide open during any portion of the stroke the travel must be greater than the sum of the outside lap and the width of both steam ports, this is usually done on the locomotive. The more travel the valve has the longer the steam port will remain open therefore the freer the steam admission.

"*Over travel*" tends to choke the exhaust, increases the sharpness of the cut-off, retards compression and gives a later release. In order to secure sufficient port openings with an early cut-off it is necessary to give over travel at other points. When the cut-off occurs too late by reason of over travel you can remedy the evil effects by increasing the outside lap. And delayed compression may be neutralized by increasing the inside lap, if the exhaust takes place too late cut out the inside lap, if there is none, give the valve inside clearance.

"*Lead*" increases as the cut-off is made earlier; this is done by bringing the reverse lever nearer the center notch, and is caused by the radius of the link (as explained in Link Motion). Increased lead hastens every operation of the valve. The greater the speed the more lead is required to permit of smooth running. (See Rule 31 for Valve Setting.)

"*Clearance*" is given to prevent the piston from striking either cylinder head in the event of lost motion in the main rod; it also helps to prevent bursting the cylinder when there is water in it. Clearance lessens the actual expansion rate owing to its waste space, but it also economizes on live steam, no engine can be constructed without some waste space between the valve and piston.

The *"Angularity"* of the connecting rod increases the lead in front and decreases it behind. It retards the cut-off and exhaust in front and hastens each behind. This evil is overcome by back-setting the saddle pin (see Link Motion, and Angularity of main rod).

MERITS OF THE SLIDE VALVE.

A slide valve, its seat and parts should be so proportioned that steam be admitted in sufficient volume at the beginning of the piston stroke, that the cut-off takes place at the earliest point at which the engine can develop required power, that release occurs at the latest point consistent with the speed required and before admission at the other end. That the exhaust closure be at that point at which compression shall be sufficient to arrest the motion of the reciprocating parts, and it may be nearly, or quite equal to the initial boiler pressure.

RELATIVE POSITIONS OF THE VALVE AND PISTON.

It is now time the reader should familiarize himself with the constantly changing positions of the valve and piston. But as we shall see later the motion of the piston is not symmetrical which is wholly due to the varying angularity of the main rod, so we shall first study the different positions of the crank pin during the various operations of the valve, and after we know the relative positions of the valve and crank pin we will study the relative positions of the crank pin and piston. In order to illustrate this subject clearly we have adopted four diagrams, most of the parts being represented by their center lines and center points only, in order to make them as plain as possible the dimensions used for these illustrations are 18x24 inch cylinders; steam ports, $1\frac{1}{4}$ inches; exhaust port, $2\frac{3}{4}$ inches; bridges, $1\frac{1}{8}$ inches; outside lap, 13-16 of an inch; inside lap, 1-16 of an inch; travel of valve, 5 inches. The diagrams representing full gear; the small arrows of these diagrams indicate the direction the pin is moving, and the larger dotted circle represents the path of the center of the crank pin, and the small dotted circle the path of the center of the eccentric.

Fig. 7 shows the valve at the point of lead opening, which was more clearly shown in Fig. 5. Now we find the crank pin is slightly above the forward dead center and almost at the beginning of a stroke, and the engine is beginning to take steam in the forward end of the cylinder.

Fig. 8 shows the valve at the point of cut-off as was shown by Fig. 2. We find the crank pin has traveled about three-fourths of its stroke and during that time the forward steam port remained open for the free admission of steam. At this point live steam is cut-off and the steam in the cylinder begins to expand.

THE LOCOMOTIVE UP TO DATE. 29

Fig. 9 shows the point of compression, which was also shown by Fig. 3. At this point the exhaust edge of the valve closes the back steam port which you will notice has been open to the exhaust prior to the beginning of this stroke. The unexhausted steam that yet remains in the cylinder must now be compressed by the advancing piston until the piston has completed its stroke

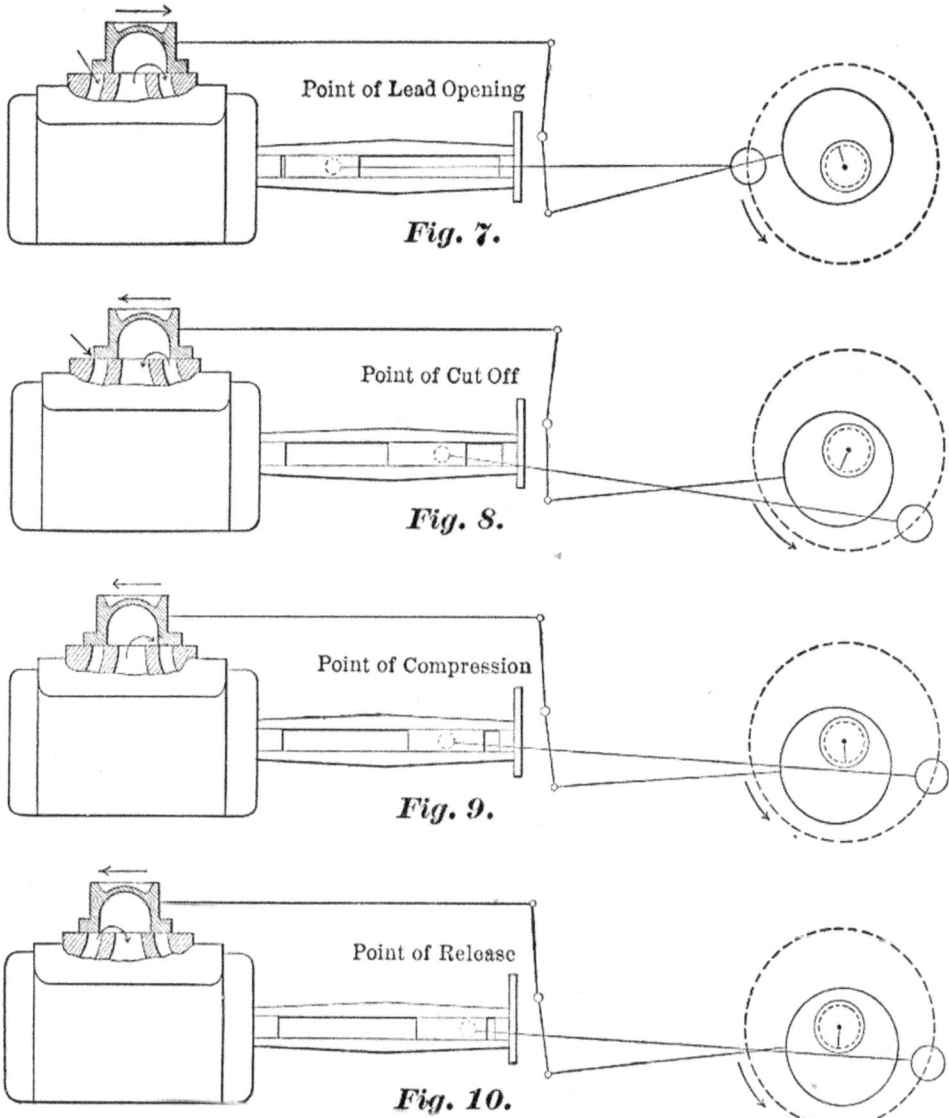

Fig. 7.

Fig. 8.

Fig. 9.

Fig. 10.

which will not be until the piston has reached the back dead center; but we find the crank pin is yet some distance below the back center, yet closer to it than it was at the point of cut-off.

Fig. 10 shows the valve at the point of release, as was shown in Fig. 4. At this point we find the exhaust edge of the valve releases the steam from the forward end of the cylinder where we have seen it was admitted during three-fourths of this same stroke, until the valve had reached the point of cut-off where expansion began. Therefore we find that expansion lasts only from the point of cut-off to the point of release and as the crank pin has not yet reached the back dead center, we find that expansion lasts during the very small portion of the piston's stroke. Now you will notice that while the crank pin continues in the same rotary motion the motion of the valve has been reversed (it was reversed before the point of cut-off was reached); and it is now about ready to take steam at the back steam port, which it will do slightly before the crank pin reaches the back center as it did for this stroke before the pin reached the forward center. Then each operation of the valve will be repeated in the return stroke as they were in this stroke. The reader who has carefully studied the different positions of the crank pin in the four preceding diagrams will readily understand the construction of Fig. 11, which combines all the positions of the crank pin shown in the preceding cuts and also the positions for the return stroke. The dotted circle represents the path of the center of the crank pin.

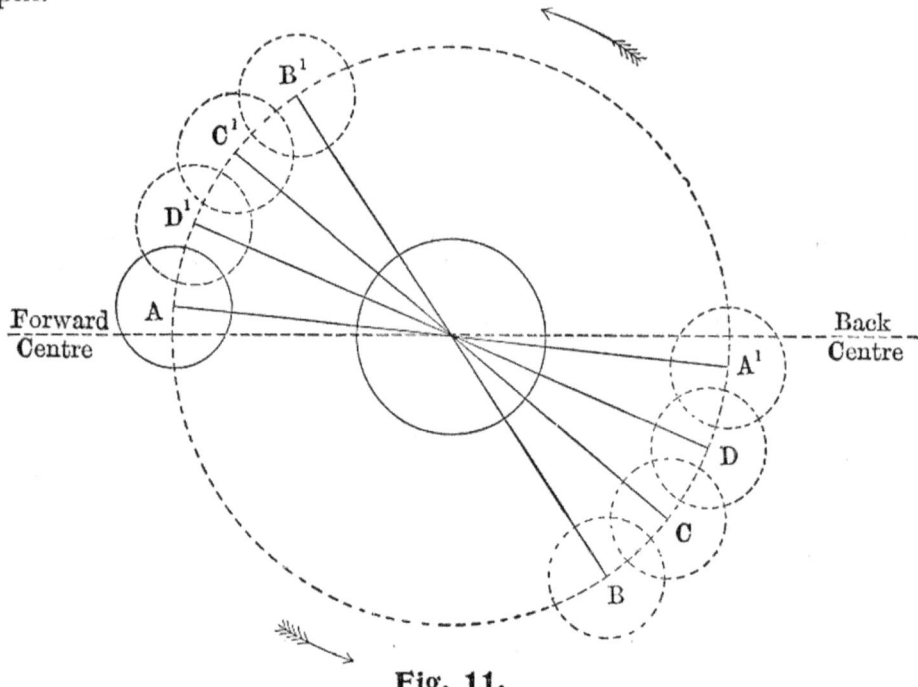

Fig. 11.

A indicates the point of admission as shown by Fig. 7.
B indicates the point of cut-off as shown by Fig. 8.
C indicates the point of compression as shown by Fig. 9.
D indicates the point of release as shown by Fig. 10.
A^1 indicates the point of admission for the return stroke.
B^1 indicates the point of cut-off for the return stroke.
C^1 indicates the point of compression for the return stroke.
D^1 indicates the point of release for the return stroke.

Now while admission must always precede every other operation of the valve, you will notice by Fig. 11, that it is the last operation in each stroke and takes place slightly before the beginning of each stroke; this is caused by giving the engine lead and the distance the crank pin will be from each dead center at the point of lead opening will be in proportion to the amount of lead given. If the valve has inside clearance instead of inside lap the points B and C and B^1 and C^1 would be reversed; while if the valve was line and line inside, release at one end and compression at the other would be simultaneous. As the reverse lever is drawn closer to the center notch each operation of the valve takes place earlier in the stroke.

RELATION BETWEEN MOTION OF CRANK-PIN AND MOTION OF PISTON.*

Now, since the aim of giving lap to a valve is to cause it to cut off steam at designated parts of the stroke of the piston, it will be necessary first to study the existing relation between the motion of the crank-pin and the motion of the piston.

In order to illustrate this subject plainly, we have adopted in Fig. 12 a shorter length for the connecting-rod, than is used in locomotives.

The circumference of the circle $A\ B\ M\ D$, drawn from the center of the axle, and with a radius equal to the distance between the center of axle and that of the crank-pin, represents the path of the latter. We will assume that the motion of the crank-pin is uniform, that is, that it will pass through equal spaces in equal times. The direction in which the crank-pin moves is indicated by the arrow marked 1, and the direction in which the piston moves is indicated by arrow 2.

In order to trace the motion of the piston it it not necessary to show the piston in our illustration, because the connection between the crosshead pin P and the piston is rigid; hence, if we know the motion of one of these we also know the motion of the other; they are alike.

The line $A\ C$ represents the line of motion of the center of cross-head pin P, consequently no matter what position the crank may occupy, the center P will always be found in the line $A\ C$. The semi-circumference $A\ B\ D$ will be the path of the center of the crank-pin P during one stroke of the piston; the point A will be the position of the crank at the beginning of the stroke, and B the position of the same at the end of the stroke. The semi-circumference $A\ B\ D$ is divided into 12 equal parts, although any other number would serve our purpose as well. The

---

*This construction of the slide-valve is from the pen of Mr. J. G. A. Meyer, M. E., whose authenticity on the subject is so well known that his name needs no introduction here.

distance between the centers D and P represents the length of the connecting-rod.

From the point A as a center, and with a radius equal to $D\ P$ (the length of the connecting-rod) an arc has been drawn, cutting the line $A\ C$ in the point a; this point is the position of the center P of the crosshead pin, when the center of the crank is at A. Once more, from the point 1 on the semi-circumference as a center and with the radius $D\ P$, another arc has been drawn cutting the line $A\ C$ in the point $1p$, and this point indicates the position of the crosshead pin when the crank-pin is at the point 1. In a similar manner the points $2p$, $3p$, $4p$, etc., have been obtained, and these points indicate the various positions of cross-head pin when the crank-pin is in the corresponding positions as 2, 3, 4, etc.

Now notice the fact that the spaces from A to 1 and from 1 to 2, etc., in the semi-circumference $A\ B\ D$ are all equal, and the crank-pin moves through each of these spaces in equal times, that is, if it requires one second to move from A to 1, it will also require one second to move from 1 to 2. The corresponding spaces from a to $1p$ and from $1p$ to $2p$, etc., on the line $A\ C$ are not equal, and yet, the crosshead pin must move through these spaces in equal times; if it requires one second to move from a to $1p$, it will also require one second to move from $1p$ to $2p$. But this last space is greater than the first. Here, then, we see that the crosshead pin, and therefore the piston, has a variable motion, that is, the piston will, at the commencement of its stroke, move comparatively slow, and increase in speed as it approaches the center of the stroke, and when the piston is moving away from the center of stroke, its speed is constantly decreasing. This variable motion of the piston is mostly caused by changing its rectilinear motion into a uniform rotary motion, and partly by the angle formed by the center line $D\ P$ of the connecting-rod and the line $A\ C$, an angle which is constantly changing during the stroke. Also notice that the distance from a to $1p$ nearest one end of the stroke is smaller than the distance from b to $11p$ nearest the other end of the stroke, and if we compare the next space $1p$ to $2p$ with the space $11p$ to $10p$, we again find that the former is smaller than the latter, and by further comparison we find that all the spaces from a to $6p$ are smaller than the corresponding spaces from b to $6p$, and consequently when the crank-pin is at point 6, which is the center of the path of the crank-pin during one stroke, the crosshead pin P will be at $6p$ and not in the center of its stroke. Thus we see that the motion of the piston is not symmetrical, and this is wholly due to the varying angularity of the connecting-rod during the stroke. If we make the connecting-rod longer, but leave the stroke the same, the difference between the spaces b to $11p$ and a to $1p$ will be less, and the same can be said of the other spaces. Again, if we consider the length of the connecting-rod to be infinite, then the difference between the spaces nearest the ends of the stroke will vanish, and the same result is true for the other spaces. Hence, when the length of the connecting-rod is assumed to be infinite the motion of the piston will be symmetrical, but still remain variable, in fact the piston will have the same motion as that shown in Fig. 13. In this figure we have dispensed with the connecting-rod, and in its place extended the piston-rod, and to its end a slotted crosshead is attached in which the crank-pin is to work. Although such mechanism is never used in a locomotive, yet with its aid we can establish a simple method for finding the position of the piston when that of the crank is known. In this figure, as in Fig. 12, the circumference $A\ B\ D\ M$ will represent the path of the center of the crank-pin, and from

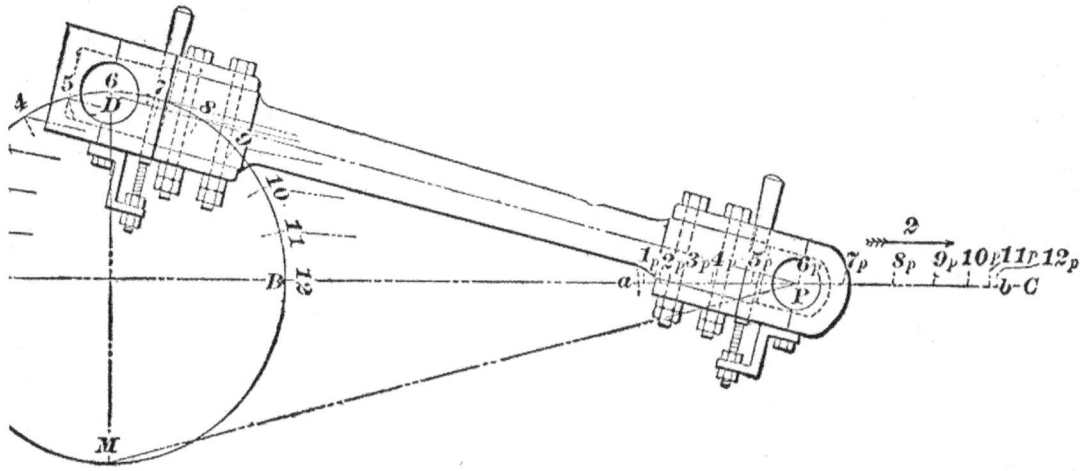

Fig. 12.

the nature of this mechanism it must be evident that at whatever point in the circumference $A\ B\ D\ M$ the crank-pin center may be located, the center line $i\ h$, of the slotted crosshead will always stand perpendicular to the line $A\ C$, and also pass through the center of crank-pin.

In Fig. 13, when the crank-pin is at A, the piston will be at the commencement of its stroke. During the time the crank-pin travels from A to point 8 the piston will travel through a portion of its stroke equal to the length $A\ E$, which is the distance between the dotted line $i\ h$ and the full line $i\ h$. If now we assume the points 1, 2, 3, etc., in the semi-circumference $A\ B\ D$ to be the various positions of the crank-pin during one stroke, and then drawn through these points lines perpendicular to the line $A\ C$, cutting the latter in the points $1p$, $2p$, $3p$, we obtain corresponding points for the position of the piston in the cylinder. Thus, for instance, when the crank-pin is at point 1 the piston will then have moved from the commencement of its stroke through a distance equal to $A\ 1p$ and when the crank-pin is at point 2, the piston will then have traveled from *A* to 2*p*, and so on.

From the foregoing, we can establish a simple method, as shown in Fig. 14, for finding the position of the piston when that of the crank is known. The diameter, *A B*, represents the stroke of the piston, and the semi-circumference *A B D* represents the path of the center of the crank-pin during one stroke. For convenience, we may divide the diameter into an equal number of parts, each division indicating one inch of the stroke. In this particular case (Fig. 14), we have assumed the stroke to be 24 inches; hence the diameter has been divided into 24 equal parts.

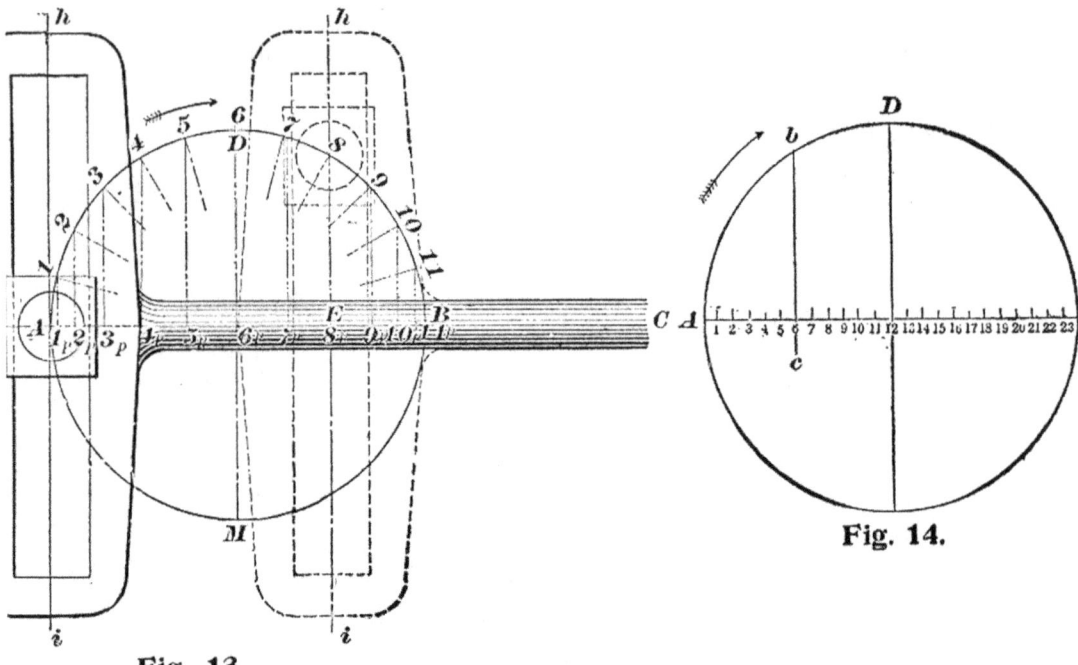

Fig. 13.

Fig. 14.

Let the arrow indicate the direction in which the crank is to turn, and *A* the beginning of the stroke; then, to find the distance through which the piston must travel from the commencement of its stroke during the time that the crank travels from *A* to *b*, we simply draw through the point *b* a straight line *b c* perpendicular to *A B*; the distance between the line *b c* and the point *A* will be that portion of the stroke through which the piston has traveled, when crank-pin has reached the point *b*. In our figure we notice that the line *b c* intersects *A B* in the point 6; hence the piston has traveled six inches from the commencement of the stroke.

If this method of finding the position of the piston, when that of the crank is known, is thoroughly understood, then the solutions of the following problems relating to lap of the slide valve will be comparatively easy:

PROBLEMS RELATING TO LAP OF THE SLIDE VALVE.

To find the point of cut-off when the lap and travel of the valve are given, the valve to have no lead.

Example 18.—Lap of valve is one inch; travel, 5 inches; no lead; stroke of piston, 24 inches. At what part of the stroke will the steam be cut off?

We must first find the center c, Fig. 15, of the circle $a\ b\ m$, whose circumference represents the path of the center of eccentric, and this is found, as the reader will remember, by placing the valve in a central position, as shown in dotted lines in this figure. Then the edge c of the valve will be the center of the circle. The valve drawn in full lines shows its position at the commencement of the stroke of piston. Through the edge c_2 draw the line $i\ h$ perpendicular to the line $A\ B$; the line $i\ h$ will intersect the circumference $a\ b\ m$ in the point y, and this point

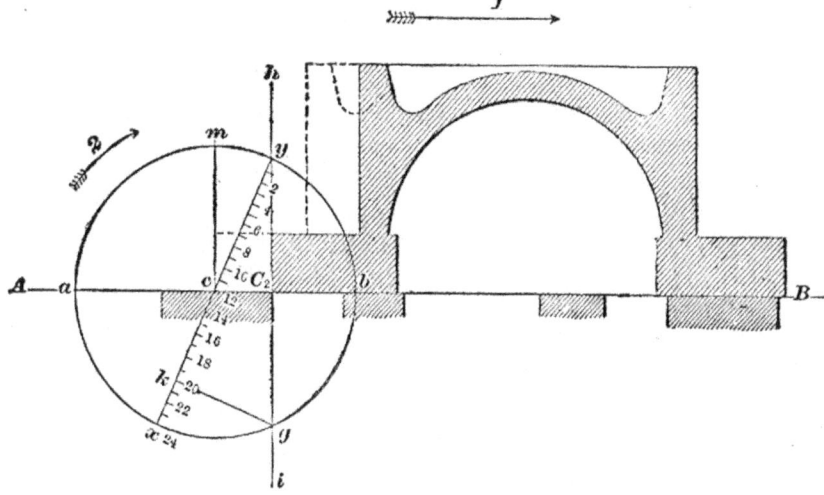

Fig. 15.

will be the center of eccentric when the piston is at the beginning of its stroke. Now, assume that the circumference $a\ b\ m$ also represents, on a small scale, the path of the center of the crankpin; then the diameter $y\ x$ of this circle will represent the length of the stroke of the piston; the position of this diameter is found by drawing a straight line through the point y (the center of the eccentric when the piston is at one end of its stroke) and the center c. Also assume that the point y represents the center of the crank-pin when the piston is at the beginning of its stroke. To make the construction as plain as possible, divide the diameter $y\ x$ into 24 equal parts, each representing one inch of the stroke of piston, and for convenience number the divisions as shown. The arrow marked 1, shows the direction in which the valve must travel, and arrow 2 indicates the direction in which the center y must travel. Now it must be evident, because the points y_1 and C_2 will always be in the same line, that during the time the center y of the eccentric travels through the arc $y\ g$, the valve not only opens the steam port, but, as the circumference $a\ b\ m$ indicates, travels a little beyond the port, and then closes the same, or, in short, during the time the center of eccentric travels from y to g, the port has been fully opened and closed; and the moment that the center of eccentric reaches the point g, the admission of steam into the cylinder is stopped. We have assumed that the point y also represents the position of the center of crank-pin at the beginning of the stroke; and, since the crank and eccentric are fastened to the same shaft, it follows that during the time the center of eccentric travels from y to g the crank-pin will move through the same arc, and when the steam is cut off the crank-pin will be at the point g. Therefore, through the point g draw a straight line $g\ k$ perpendicular to the line $y\ x$; the line $g\ k$ will intersect the line $y\ x$ in the point k, and this point coincides with the mark 20; hence steam will be cut off when the piston has traveled 20 inches from the beginning of its stroke.

LEAD WILL EFFECT THE POINT OF CUT-OFF.

In Fig. 15 the valve had no lead; if, now, in that figure, we change the angular advance of the eccentric so that the valve will have lead, as shown in Fig. 16, then the point of cut-off will also be changed. How to find the point of cut-off when the valve has lead, is shown in Fig. 16.

Example 19.—The lap of valve is 1 inch, its travel 5 inches; lead $\frac{1}{4}$ of an inch (this large amount of lead has been chosen for the sake of clearness in the figure); stroke of piston, 24 inches; at what part of the stroke will the steam be cut off?

On the line $A\ B$, Fig. 16, lay off the exhaust and steam ports; also on this line find the center c of the circle $a\ b\ m$ in a manner similar to that followed in the last construction, namely, by placing the valve in a central position, as shown by the dotted lines, and marked D, and then adopting the edge c of the valve as the center of the circle $a\ b\ m$; or, to use fewer words, we may say from the outside of the edge s of the steam port, lay off on the line $A\ B$ a point c whose distance from the edge s will be equal to the lap, that is, 1 inch. From c as a center, and with a radius of $2\frac{1}{2}$ inches (equal $\frac{1}{2}$ of the travel), describe the circle $a\ b\ m$, whose circumference will represent the path of the center of eccentric. The lead of the valve in a locomotive is generally 1-32 and sometimes as much as 1-16 of an inch, when the value is in full gear, but for the sake of distinctness we have adopted in this construction a lead of $\frac{1}{4}$ of an inch. Draw the section of the valve, as shown in full lines, in a position that it will occupy when the piston is at the beginning of its stroke, and consequently the distance between the edge c_2 of the valve and the edge s of the steam port will, in this case, be $\frac{1}{4}$ of an inch. Through c_2 draw a straight line perpendicular to $A\,B$, intersecting the circumference $a\,b\,m$ in the point y; this point will be the center of the eccentric when the piston is at the beginning of its stroke, and since it is assumed that the circumference $a\,b\,m$ also represents the path of the center of the crank-pin, the point y will also be the position of the same when the piston is at the commencement of its stroke. Through the points y and c draw a straight line $y\,x$, to represent the stroke of the piston, and divide it into 24 equal parts. Through the point s draw a straight line perpendicular to $A\,B$, intersecting the circumference $a\,b\,m$ in the point g, and through g draw a straight line perpendicular to $y\,x$, and intersecting the latter in the point k; this point will be the point of cut-off, and since the distance between the point k and 19 is about $\frac{1}{8}$ of the space from 19 to 20,

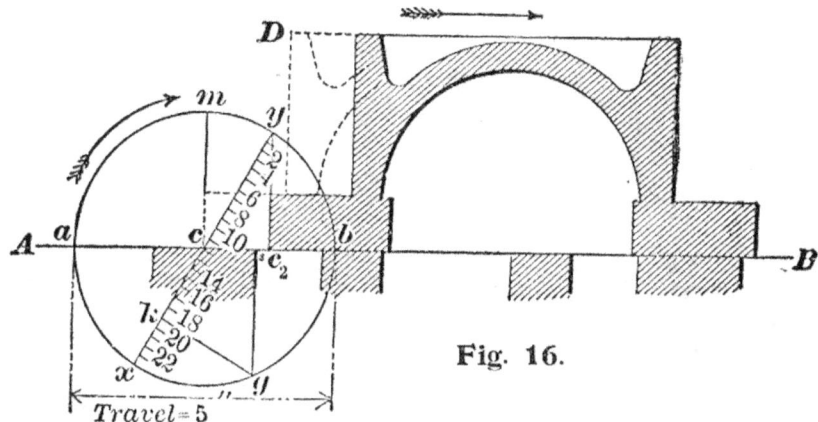

Fig. 16.

we conclude that the piston has traveled $19\frac{1}{8}$ inches from the beginning of its stroke when the admission of steam into the cylinder is suppressed.

Here we see that when a valve has no lead, as in Fig. 15, the admission of steam into the cylinder will cease when the piston has traveled 20 inches; and when the angular advance of the eccentric is changed, as in Fig. 16, so that the valve has $\frac{1}{4}$ of an inch lead, the point of cut-off will be at $19\frac{1}{8}$ inches from the beginning of the stroke, a difference of $\frac{7}{8}$ of an inch between the point of cut-off in Fig. 15 and that in Fig. 16. But the lead in locomotive valves in full gear is only about 1-32 of an inch, which will affect the point of cut-off so very little that we need not notice its effect upon the period of admission, and, therefore, lead will not be taken into consideration in the following examples.

THE TRAVEL OF THE VALVE WILL AFFECT THE POINT OF CUT-OFF.

Fig. 17 represents the same valve and ports as shown in Fig. 15, but the travel of the valve in Fig. 17 has been increased to $5\frac{3}{4}$

inches. The point of cut-off k has been obtained by the same method as that employed in Figs. 15 and 16, and we find that this point k coincides with point 21. Now, notice the change caused by an increase of travel. When the travel of the valve is 5 inches, as shown in Fig. 15, the admission of steam into the cylinder will cease when the piston has traveled 20 inches from the commencement of its stroke, and when the travel of the same valve is increased $\frac{3}{4}$ of an inch, as shown in Fig. 17, the admission of the steam will not be suppressed until the piston has traveled 21 inches. Here we notice a difference of 1 inch between the two points of cut-off. But it must be remembered that when the travel of a valve for a new engine is to be found or established, the point of cut-off does not enter the question; we simply assign such a travel to the valve that steam ports

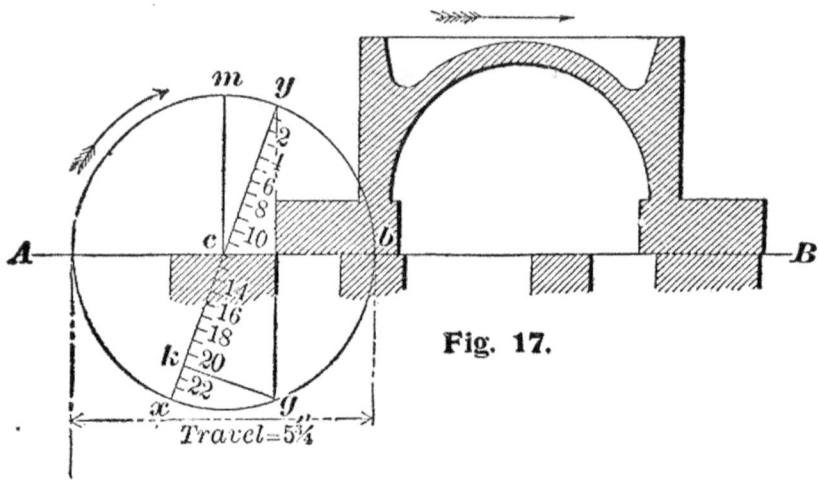

Fig. 17.

will be fully opened, or slightly more, when the valve is in full gear. The point of cut-off is regulated by the lap and position of the eccentric.

In order to find the point of cut-off it is not necessary to make a drawing of the valve, as has been done in Fig. 15. The only reason for doing so was to present the method of finding the point of cut-off to the beginner in as plain a manner as possible. In order to show how such problems can be solved without the section of the valve, and, consequently, with less labor, another example, similar to Example 18, is introduced.

Example 20.—Lap of valve is $1\frac{3}{8}$ inches; travel, $5\frac{1}{2}$ inches; stroke of piston, 24 inches; width of steam port, $1\frac{1}{4}$ inch; find the point of cut-off.

Fig. 18. Draw any straight line, as $A B$; anywhere on this line mark off $1\frac{1}{4}$ inch, equal to the width of the steam port. From the edge s of the steam port lay off on the line $A B$ a point c, the distance between the points s and c being $1\frac{3}{8}$ inches; that is, equal to the amount of lap. From c as a center, and with a radius equal to half the travel, namely, 2¾ inches, draw a circle, *a b m;* the circumference of this circle will represent the path of the center of the eccentric, and also that of the crank-pin. Through *s* draw a straight line *i h* perpendicular to *A B;* this line *i h* will intersect the circumference *a b m* in the points *y* and *g*. Through the points *y* and *c* draw a straight line *y x;* the diameter *y x* will represent the stroke of the piston. Divide *y x* into 24 equal parts; through the point *g* draw a straight line *g k* perpendicular to *y x*, and intersecting *y x* in the point *k*, and this point is the point of cut-off. Since *k* coincides with the point 18, it follows that the piston had traveled 18 inches from the beginning of its stroke when the flow of the steam into the cylinder ceased.

Now we may reverse the order of this construction and thus find the amount of lap required to cut off steam at a given portion of the stroke.

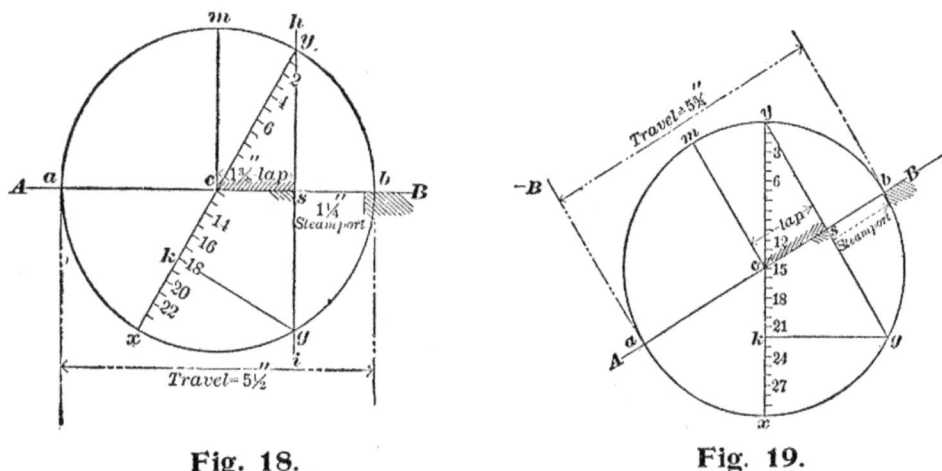

Fig. 18.   **Fig. 19.**

Example 21.—Travel of valve is 5¾ inches; stroke of piston, 30 inches; steam to be cut off when the piston has traveled 22 inches from the beginning of the stroke; width of steam port, 1⅜ inch; find the lap.

Fig. 19. Draw a circle *a b m* whose diameter is equal to the travel of the valve, viz., 5¾ inches. Through the center *c* draw the diameter *y x*. In this figure we have drawn the line *y x* vertically, which was done for the sake of convenience; any other position for this line will answer the purpose equally well. The circumference *a b m* represents the path of the center of the eccentric, also that of the crank-pin; the diameter *y x* will represent the stroke of the piston, and, therefore, is divided into 30 equal parts. The steam is to be cut off when the piston has traveled 22 inches from the beginning of the stroke, therefore, through the point 22 draw a straight line *g k* perpendicular to *y x*, the line *g k* intersecting the circumference *a b m* in the point *g*. Join the points *y* and *g* by a straight line. Find the center *s* of the line *y g*, and through *s* and perpendicular to the line *y g*, draw the line *A B*; if the latter line is drawn accurately it will always pass through the center *c*. The distance between the points *s* and *c* will be the amount of lap required, and in this example it is 1 7-16 inch.

It sometimes occurs, in designing a new locomotive, and often in designing stationary or marine engines, that only the width of steam port and point of cut-off is known, and the lap and

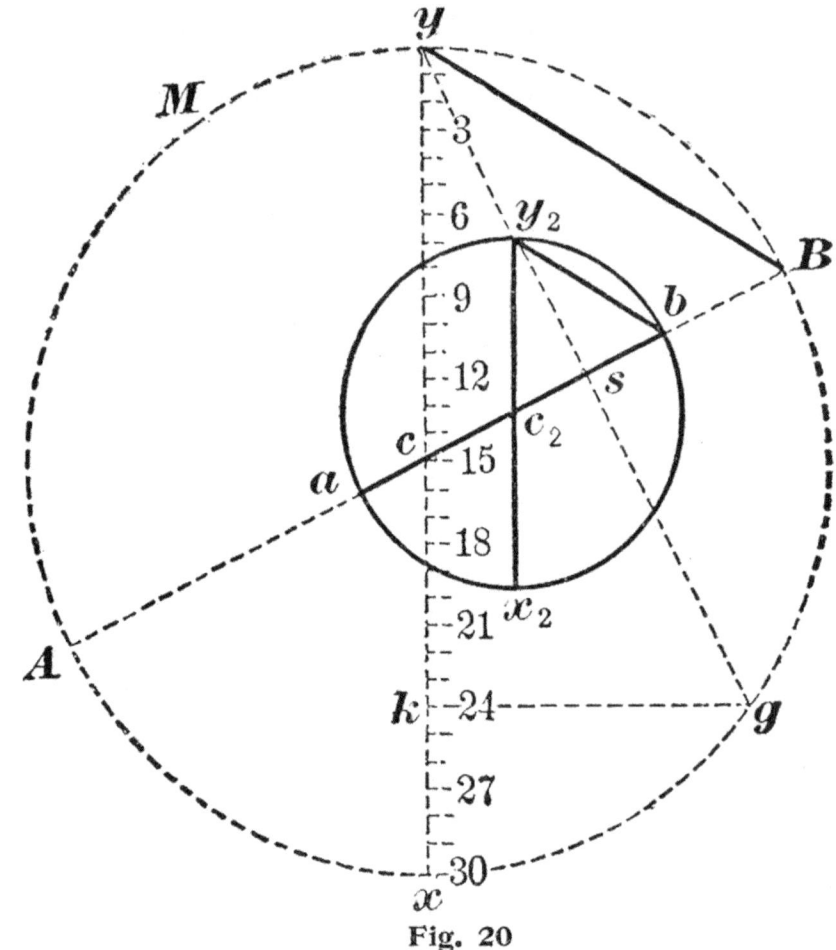

Fig. 20

travel of the valve is not known. In such cases both of these can be at once determined by the following method:

Example 22.—The width of the steam port is 2 inches; the stroke of piston, 30 inches; steam to be cut off when the piston has traveled 24 inches from the beginning of its stroke; find the lap and travel of the valve.

Fig. 20. Draw any circle, as *A B M*, whose diameter is larger than what the travel of the valve is expected to be. Through the center *c* draw the diameter *y x*, and, since the stroke of piston is 30 inches divide *y x* into 30 equal parts. Steam is to be cut off when the piston has traveled 24 inches; therefore, through point 24 draw a straight line gk perpendicular to the diameter yx, intersecting the circumference ABM in the point g. Join the points y and g by a straight line; through the center s of the line yg draw a line AB perpendicular to yg. So far, this construction is precisely similar to that shown in Fig. 19, and in order to distinguish this part of the construction from that which is to follow, we have used dotted lines; for the rest full lines will be used. It will also be noticed by comparing Fig. 20 with Fig. 19 that, if the diameter AB had been the correct travel of valve, then cs would have been the correct amount of lap. But we commenced this construction with a travel that we know to be too long; hence, to find the correct travel and lap, we must proceed as follows: Join the points B and y. From s toward B, lay off on the line AB a point b; the distance between, the points s and b must be equal to the width of the steam port plus the amount that the valve is to travel beyond the steam port, which, in this example, is assumed to be $\frac{1}{8}$ of an inch. Therefore the distance from s to b must be $2\frac{1}{8}$ inches. Through b draw a straight line by_2 parallel to By, intersecting the line yg in the point y_2. Through the point y_2 draw a straight line $y_2 x_2$ parallel to the line yx, and intersecting the line AB in the point c_2. From c_2 as a center, and with a radius equal to $c_2 b$, or $c_2 y_2$, describe a circle $a b y_2$. Then ab will be the travel of the valve, which, in this case, is $7\frac{5}{8}$ inches, and the distance from c_2 to s will be the lap, which, in this example, is 1 11-16 inch.

PRACTICAL CONSTRUCTION OF THE SLIDE VALVE.

It should be obvious, and, therefore, almost needless to remark here, that the foregoing graphical methods employed in the solutions of the problems relating to the slide valve are applicable to everyday practice, the writer believes that these methods are the simplest and best to adopt for ordinary use, and without these it would be difficult to construct a valve capable of performing the duty assigned to it. Of course, when a graphical method is employed, great accuracy in drawing the lines is necessary.

We will give a practical example, in which one of the objects aimed at, is to show the application of one of the foregoing methods to ordinary practice.

Example 23.—The width of the steam ports is $1\frac{1}{4}$ inch; length of the same 14 inches; thickness of bridges $1\frac{1}{8}$ inch; width of exhaust port $2\frac{1}{2}$ inches; travel of valve $4\frac{3}{4}$ inches; stroke 24 inches; steam to be cut off when the piston has traveled $20\frac{3}{4}$ inches from the beginning of its stroke; the edges of the exhaust cavity are to cover the steam ports, and not more, when the valve stands in a central position; construct the valve.

Fig. 21. Draw a straight line AB to represent the valve-seat through any point in AB; draw another line DC perpendicular to AB; the line DC is to represent the center of exhaust port and the center of valve. Draw the exhaust port, bridges and steam ports as shown.

The question now arises: How long shall we make the valve? Or, in other words, what shall be the distance between the outside edges of the valve c and c_2? If the valve had to admit steam during the whole stroke of the piston, or as the practical man would say, "follow full stroke," then the distance between the edges c and c_2 would be equal to the sum of twice the width of one steam port plus twice the width of one bridge plus the width of the exhaust port, hence we would have $2\frac{1}{2}+2\frac{1}{4}+2\frac{1}{2}=7\frac{1}{4}$ inches for the length of the valve. But, according to the conditions given in the example, the valve must cut off steam when the piston has traveled $20\frac{3}{4}$ inches, therefore the valve must have lap,

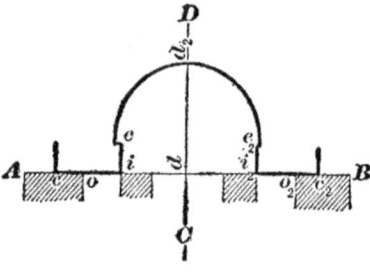

Fig. 21.

and the amount of lap that is necessary for this purpose must be determined by the method shown in Fig. 19, and given in connection with Example 21. Following this method, we find that the required lap is $\frac{7}{8}$ of an inch, therefore the total length of the valve will be $7\frac{1}{4}+(\frac{7}{8}\times 2)=9$ inches; or, we may say, that the distance between the edges c and c_2 must be equal to twice the width of one steam port plus twice the width of one bridge plus the width of the exhaust port plus twice the lap, consequently we have $2\frac{1}{2}+2\frac{1}{4}+2\frac{1}{2}+1\frac{3}{4}=9$ inches for the length of the valve. Through the points c and c_2 (each point being placed $4\frac{1}{2}$ inches from the center line CD), draw lines perpendicular to AB; these lines will represent the outside surfaces containing the edges c and c_2. These surfaces must be square with the surface AB, because, if they are not so, but are such as shown in Fig. 24, the distance between the edges c and c_2 will decrease as the valve wears, and when this occurs, the valve will not cut off the steam at the proper time. Now, in regard to the cavity of the valve. One of the conditions given in our example is, that the edges of the cavity must cover the steam ports, and no more, when the valve stands in a central position, therefore the inner edges i and i_2 of the valve must be $4\frac{3}{4}$ inches apart, which is equal to twice the width of one bridge plus the width of the exhaust port; consequently, when the valve stands midway of its travel, the inner edges of the valve (being $4\frac{3}{4}$ inches apart), the inner edges of the steam ports coincide. Through the points i and i_2 (each being placed $2\frac{3}{8}$ inches from the center line CD), draw the straight lines $i\,e$ and $i_2\,e_2$ perpendicular to AB. These lines will represent the sides of the cavity containing the inner edges i and i_2 of the valve, and these sides must be square with the surface AB; if these are otherwise, for instance, such as shown in Fig. 24, the distance between the edges i and i_2 will change as the valve wears, and then the valve will not perform its duty correctly. The depth $d\,d_2$ of the cavity is generally made from $1\frac{1}{4}$ to $1\frac{1}{2}$ times the width of the exhaust port. The writer believes that making the depth of the cavity $1\frac{1}{2}$ times the width of the

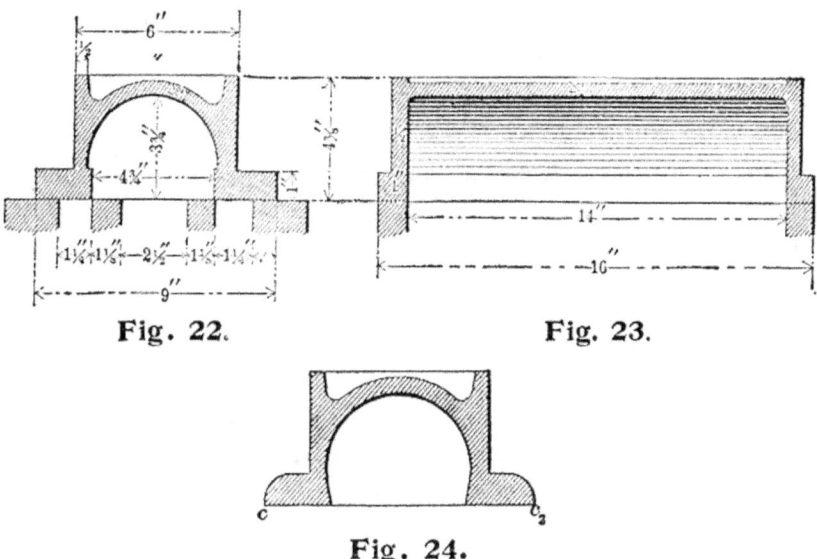

Fig. 22.   Fig. 23.

Fig. 24.

exhaust port is the best practice. In our example the width of the exhaust port is $2\frac{1}{2}$ inches, and $2\frac{1}{2}+1\frac{1}{4}=3\frac{3}{4}$ inches, which will be the distance from d to d_2, that is, the depth of the cavity. The curved surface of the cavity is generally a cylindrical surface, and when it is so, as in our example, this surface must be represented in Fig. 21 by an arc of a circle. The sides $i\,e$ and $i_2\,e_2$ must be planed, and to do this conveniently, these sides must extend a little beyond the curved surface, toward the center CD. Consequently, through the point d_2 draw an arc whose center is in the line CD, and whose radius is such that will allow the sides to project about 1-16 of an inch. Here, then, we have lines which completely represent the cavity of the valve and the valve face. If we now add to these lines the proper thickness of metal as shown in Fig. 22, this section of the valve will be complete.

Fig. 23 shows a section of the valve taken at right angles to that shown in Fig. 22. Since the ports are 14 inches long, the cavity of the valve must be 14 inches wide, as shown. The amount that the valve overlaps the ends of the steam ports must be sufficient to prevent leakage. For a valve of the size here shown, 1 inch overlap is allowed, and the thickness of metal around the cavity is generally one-half of an inch. For smaller valves the overlap at each end of the steam port is from $\frac{3}{4}$ to $\frac{7}{8}$ of an inch, and the thickness of metal around the cavity is $\frac{3}{8}$ of an inch.

The valve here shown is suitable for a locomotive cylinder 16 inches in diameter, and a piston speed of 525 feet per minute, and the dimensions here given agree with those of the valves that are at present in use.

INSIDE LAP, CLEARANCE, AND INSIDE LEAD.

Now, a few words in regard to some other terms used in connection with the slide valve.

Inside Lap.—The amount that the inside edges i and i_2 of the valve, Fig. 25, overlap the inside edges s and s_2 of the steam

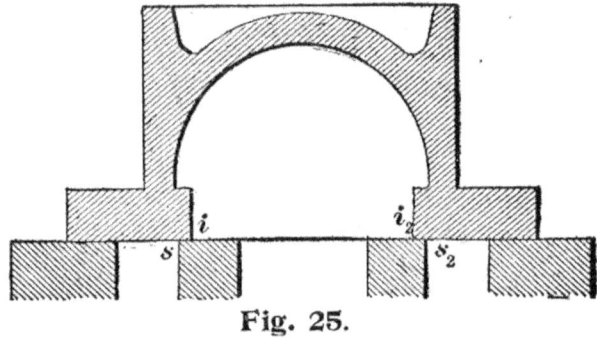

Fig. 25.

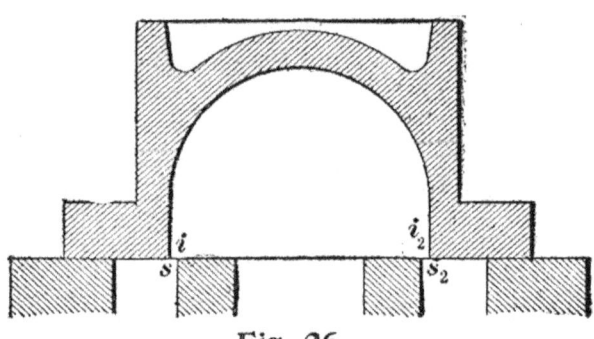

Fig. 26.

ports, when the valve stands midway of its travel, is called inside lap; thus, the distance from s to i, or from s_2 to i_2, represents the inside lap. Its purpose is to delay the release of steam.

The amount of inside lap is comparatively small, rarely exceeding $\frac{1}{8}$ of an inch, and in a number of locomotives the valves have no inside lap. Rules for determining the inside lap cannot be given, because engineers do not agree on this subject. The writer believes that for slow-running locomotives, particularly if these have to run over steep grades, a little inside lap will be beneficial. For ordinary passenger locomotives, running on comparatively level roads, no inside lap should be used.

Inside Clearance.—When the valve stands midway of its travel, as shown in Fig. 26, and its inside edges i and i_2 do not cover the steam ports, then the amount by which each edge of the valve comes short of the inner edges of the steam ports is called inside clearance; thus, the distance from i to s, or from i_2 to s_2, represents inside clearance. The purpose of inside clearance is

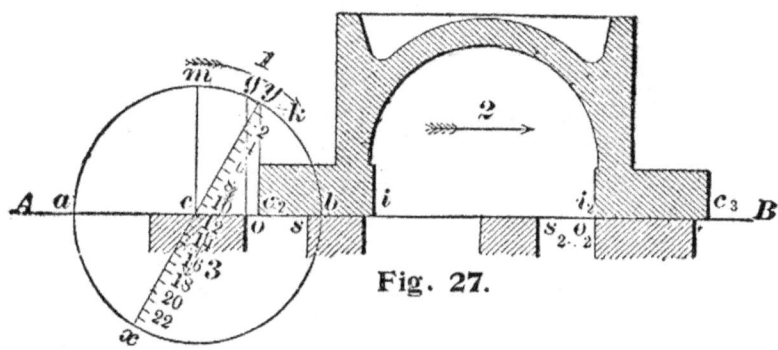

Fig. 27.

to hasten the release, and is sometimes adopted in very fast-running locomotives. It seldom exceeds 1-64 of an inch. Good judgment and great experience are required for determining the amount of clearance, and for deciding for what classes of locomotives it should be used. In ordinary passenger locomotives the valves have no inside clearance.

The width of opening of the steam port for the release of steam at the beginning of the stroke is called inside lead; thus, when the piston is at the beginning of its stroke, and the valve occupying the position as shown in Fig. 27, then the distance between the inner edges i_2 of the valve and the inner edge s_2 of the steam port is called inside lead. The simple terms "lead" and "lap" are used among engineers to designate outside lead and lap; hence, the necessity of using the terms "inside lead" and "inside lap" when such is meant.

THE EVENTS OF THE DISTRIBUTION OF STEAM.

In the distribution of steam during one revolution of the crank, four distinct events occur, namely:

1st. The admission of steam.

2d. The cutting off, or, in other words, the suppression of steam.

3d. The release of steam.

4th. The compression of steam.

The outside edges c_2 and c_3 of the valve, and the outside edges o and o_2 of the steam ports, will regulate the admission and suppression of steam; the inner edges i and i_2 of the valve and the inner edges s and s_2 of the steam ports control the release and compression of steam. The parts of the stroke of the piston during which these events will happen can be found by the following methods:

Example 24.—Travel of valve, 5 inches; lap, 1 inch; lead, ¼ of an inch; stroke of piston, 24 inches; no inside lap or clear-

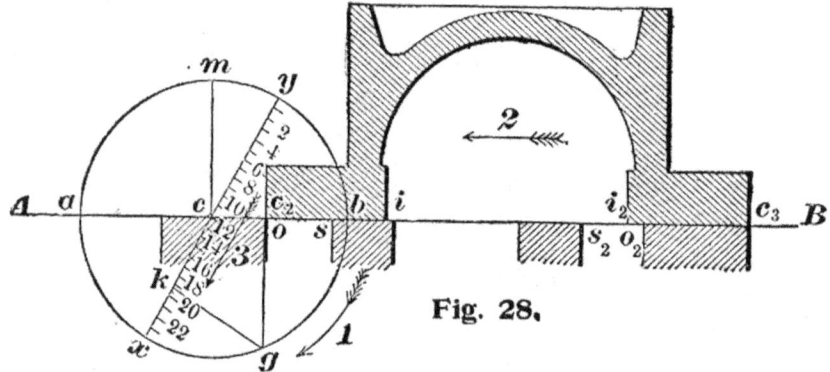

Fig. 28.

ance. Find at what part of the stroke the admission, suppression release and compression will take place.

In Figs. 27, 28 and 29 the valve occupies different positions, but the sections of the valve in these figures are exactly alike, because they represent one and the same valve. In Fig. 27 the distance between the edge c_2 of the valve and the edge o of the steam port is ¼ of an inch, which is the amount of lead given in

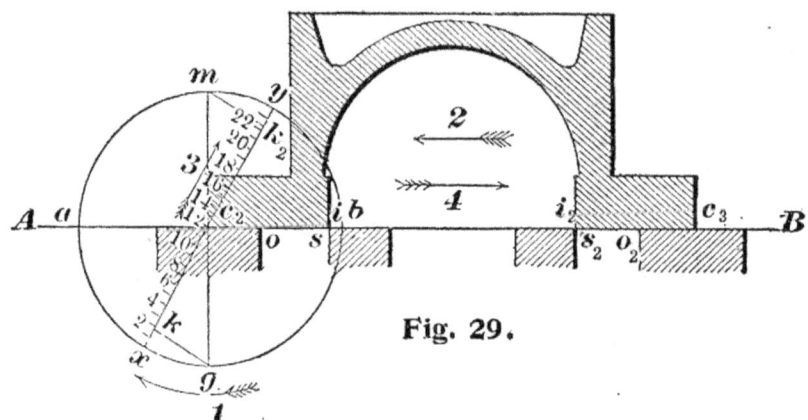

Fig. 29.

our example; hence, this position of the valve indicates that the piston is at the beginning of its stroke. In Fig. 28 the edge c_2 of the valve and the edge o of the steam port coincide, and, since the valve is moving in the direction indicated by arrow 2, the suppression commences, or, in other words, the valve is cutting off steam when it is in the position as here shown. In Fig. 29 the inside edge i of the valve coincides with the inner edge s of the steam port, and, since the valve is moving in the direction indicated by arrow 2, the release must commence when the valve arrives in the position here shown.

In Figs. 27, 28 and 29 the distances from the outside edge o of the steam port to the center c of the circle $a\ b\ m$ are equal; that is, the points c and o are one inch apart, which is the amount of lap. The diameters of the circles $a\ b\ m$ are all five inches long, which is the travel of the valve given in the example, and the circumference of each circle represents the path of the eccentric, and also the path of the center of the crank-pin. The point y in these figures represents the position of the center of eccentric when the piston is at the beginning of its stroke. The distance between the point y and m is the same in all figures, and consequently the angles formed by the lines $y\ x$ and $A\ B$ are equal.

When the valve occupies the position as represented in Fig. 27, the center of crank-pin will be in the line $A\ B$, and since the piston will then be at the beginning of its stroke, it follows that the line $A\ B$ will indicate the direction in which the piston must move. In order to compare the relative position of the piston with that of the valve with as little labor as possible we shall assume that the direction in which the piston moves is represented by the line $y\ x$, instead of the line $A\ B$; hence the point y will not only show the position of the center of the eccentric, but it will also indicate the position of the center of the crank-pin when the piston is at the commencement of its stroke. If these remarks are thoroughly understood, there will be no difficulty in comprehending that which is to follow.

Now let us trace the motions of the valve and piston and thus determine at what part of the stroke the events (previously named) will take place. When the piston is moving in the direction as indicated by the arrow marked 1, Fig. 27, the center of eccentric will move through part of the circumference $a\ b\ m$, and the valve will travel in the direction indicated by the arrow 2, and thus opening the steam port wider and wider until the end b of the travel is reached; then the valve will commence to return, and as it moves toward the center c, the steam port gradually closes, until the valve reaches the position as shown in Fig. 28, then the steam port will be closed and steam cut off. To find the position of the piston when the valve is cutting off steam, we draw through the edge c_2 of the valve, Fig. 28, a straight line $c_2\ g$, perpendicular to $A\ B$, and intersecting the circumference $a\ b\ m$ in the point g; through this point draw a line perpendicular to $y\ x$ intersecting the latter in the point k, and this point k being $19\frac{1}{8}$ inches from y indicates that the piston has traveled $19\frac{1}{8}$ inches from the beginning of its stroke before the steam is cut off, and that steam has been admitted into the cylinder during the time the piston traveled from y to k. As the piston continues to move toward the end x of the stroke the valve will move in the direction of the arrow 2, Fig. 28, and the steam port will remain closed so that no steam can enter the cylinder or escape from it, hence the steam that is now confined in the cylinder must push the piston ahead by its expansive force, but the moment that the valve reaches the position as shown in Fig. 29 the release of steam will commence. To find the corresponding position of piston we draw through the edge c_2 of the valve, Fig. 29, a line $c_2\ g$, perpendicular to $A\ B$ intersecting the circumference $a\ b\ m$ in the point g. Through this point draw a line $g\ k$, perpendicular to $y\ x$, intersecting the latter in the point k, and this point k being $22\frac{3}{8}$ inches from the beginning of the stroke indicates that the piston has traveled through this distance when the release of steam commences. Now notice, the steam is cut off when the piston has traveled $19\frac{1}{8}$ inches, and the release of steam commences when the piston has traveled $22\frac{3}{8}$ inches, consequently the steam is worked expansively during the time the piston moves $3\frac{1}{4}$ inches of its stroke. The steam port will remain open to the action of the exhaust during the time the piston completes its stroke and moves through a portion of its return stroke. In the meantime the valve will move to the end a of the travel and return, as indicated by arrow 4, and the moment that the valve again reaches the position shown in Fig. 29, the release of steam will be stopped. To find the corresponding position of the piston, draw through the edge c_2 of the valve, Fig. 29, a straight line $c_2\ m$ perpendicular to $A\ B$ intersecting the circumference $a\ b\ m$ in point m. Through this point draw a straight line $m\ k_2$ perpendicular to $y\ x$, and intersecting the latter in the points k_2. Since the distance between the points x and k_2 is $22\frac{1}{4}$ inches, it follows that the piston has moved through $22\frac{1}{4}$ inches of its return stroke, by the time that the release of steam will cease. As the valve continues its travel in the direction of arrow 4, Fig. 29, the steam port will remain closed until the edge c_2 of the valve coincides with the outer edge o of the steam port, and during this time, the steam which remained in the cylinder, is compressed, but as soon as the edge c_2 of the valve passes beyond the steam port edge o, the admission of steam into the cylinder will commence. To find the corresponding position of the piston, draw through the outer edge o of the steam port, Fig. 27, a straight line $o\ g$, perpendicular to $A\ B$ and intersecting the circumference $a\ b\ m$ in the point g; through this point draw a line $g\ k$ perpendicular to $y\ x$ intersecting the latter in the point k, and since the distance between the points x and k is $23\frac{7}{8}$ inches, we conclude that the piston has moved $23\frac{7}{8}$ inches of its return stroke before the admission of steam will begin. Here we see that steam will be admitted into the cylinder before the return stroke of the piston is completed, and that is the object of lead, as has been stated before. Notice once more the compression of steam will commence when the piston has traveled $22\frac{1}{4}$

inches of its return stroke, and will cease when the piston has traveled $23\frac{7}{8}$ inches of its return stroke, hence the steam is compressed during the time that the piston travels through $1\frac{5}{8}$ inches.

In each one of these figures the point g represents the relative position of the center of eccentric to that of the valve shown in the figure. The point g will always be found in the circumference $a\ b\ m$, and in a straight line $c_2\ g$, drawn perpendicular to $A\ B$, and the former passing through the outer edge c_2 of the valve.

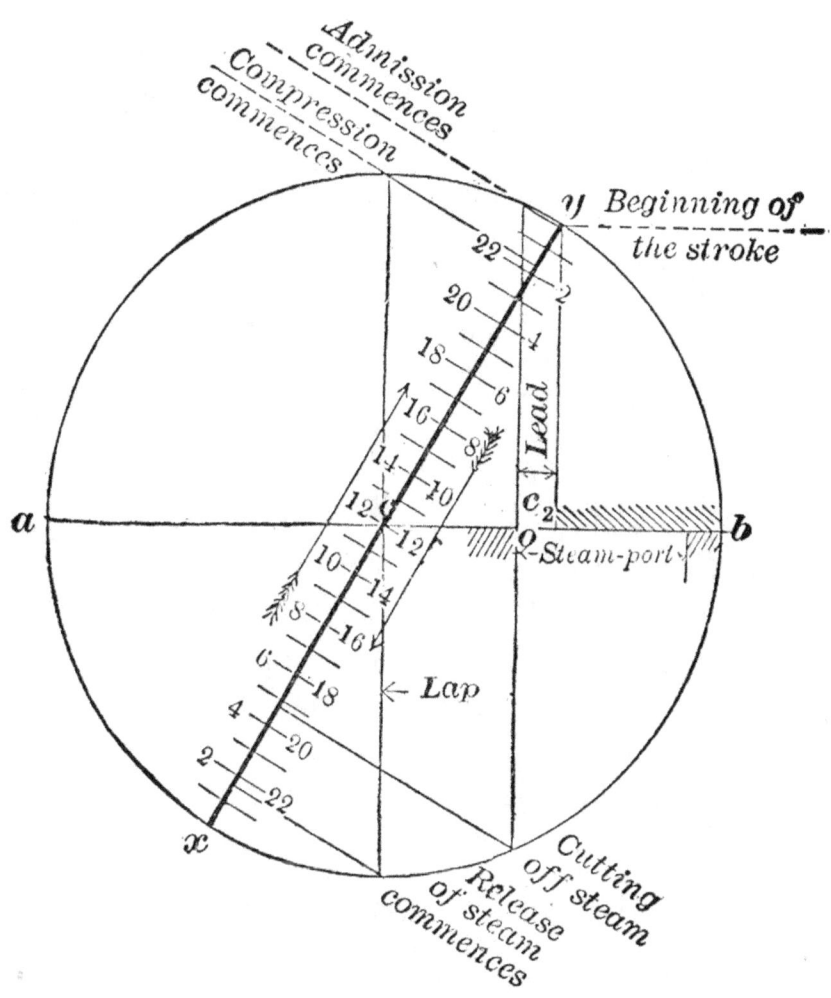

Fig. 30.

The reason why the point g should in all cases be found in the straight line $c_2\ g$ drawn through the outside edge c_2 of the valve, is this, when the center c of the circle $a\ b\ m$ is placed on the line $A\ B$ in such a position (and as has been done in these figures), so that the distance between the center c and the outside edge o of the steam port is equal to the lap, then the center g of the eccentric and the outer edge c_2 of the valve will always lie in the same straight line drawn perpendicular to $A\ B$. If the distance between center *c* and the outer edge *o* of the steam port is greater or less than the lap, then the center of the eccentric and outside edge of the valve will not lie in the same straight line drawn perpendicular to the line *A B*. Here, then, we can conceive the necessity of placing the center *c* of the circle *a b m* in the position as shown in these figures. The correctness of these remarks must be evident to the reader if the explanations in the previous examples have been understood. Again, since we have assumed that the point *g* not only represents the center of the eccentric, but also the center of the crank-pin, it follows that in order to determine how far the piston has moved from the beginning *y* of its stroke, when the crank-pin is at *g* we must draw a straight line through the point *g* perpendicular to *y x*, as has been done in these figures.

From these constructions we can form our answer to Example 24, namely:

Steam will be cut off, or, in other words, suppression will commence when the piston has traveled $19\frac{1}{2}$ inches from the beginning of its stroke, and steam will be admitted into the cylinder during the time that the piston travels through this distance. The steam will be released when the piston has traveled $22\frac{3}{8}$ inches from the beginning of its stroke, consequently the steam will be worked expansively during the time the piston travels through $3\frac{1}{4}$ inches. The release of steam will continue until the compression commences, which will occur when the piston has traveled $22\frac{1}{4}$ inches of its return stroke. The compression will cease and the admission of steam commence when the piston has traveled $23\frac{7}{8}$ inches of its return stroke.

The same answer to our example could have been obtained with less labor by a construction as shown in Fig. 30, which is nothing else but a combination of the three preceding figures; the methods of finding the different points in Fig. 30 have not been changed, and therefore an explanation in connection with this figure is unnecessary.

THE ALLEN PORTED VALVE.

The Allen ported valve, which has the supplementary port above the exhaust arch, was designed to overcome the defects of the plain slide valve. With the plain slide valve it is impossible to secure a full boiler or steam chest pressure at the beginning of a stroke, where it is most needed, without giving excessive lead which would produce a premature cut-off and impair the other operations of the valve. Besides it was found impossible to maintain a full boiler pressure during the whole period of admission, when steam was cut off short and working at a high speed. To obviate this evil, and to lessen wire drawing, the Allen valve was designed, with a supplementary port above the exhaust arch, by which steam is received from both sides of the valve at the same time to supply the same steam port, therefore giving twice the amount of opening a plain valve would have with short cut-offs. We will first call the reader's attention to Fig. 31, which shows the valve in its central position upon the valve seat. You will observe that both steam

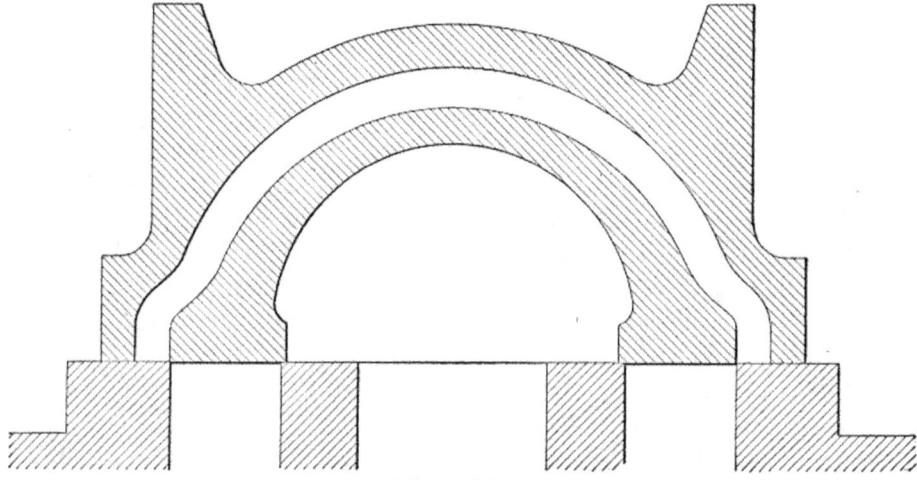

Fig. 31.

ports are completely closed, the same as with the plain slide valve, therefore the point of release or compression will not be affected unless the lead is changed, when they will take place either earlier or later in the stroke. We will next call your attention to Fig. 32. Here we find the valve moved off

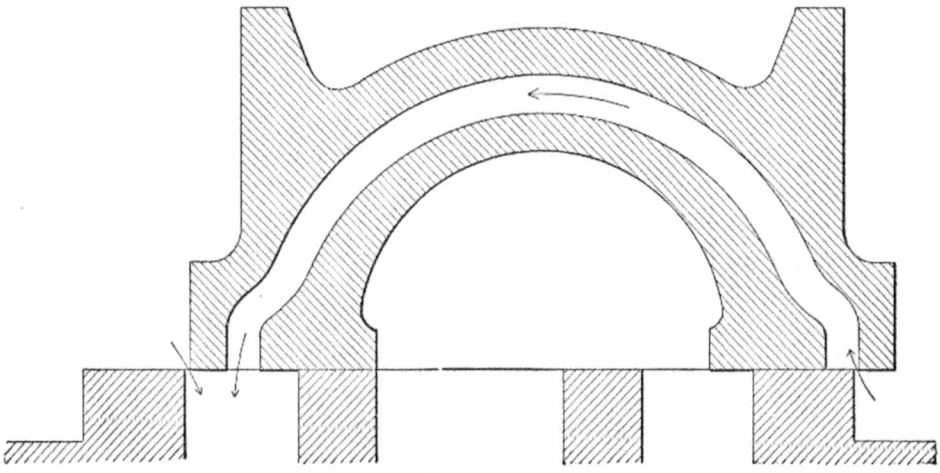

Fig. 32.

its central position and you will notice that one of the steam ports is receiving steam from each side of the valve at the same time, as indicated by the small arrows; you will also observe that the steam edge of the valve and the edge of the supplementary port open simultaneously, so they must therefore cut-off at the same time. This valve is very efficient for a high rate of speed where the travel and point of cut-off are very short. It maintains the initial pressure and the pressure during the whole period of admission is more uniform.

The wearing face of these valves being reduced they will wear very fast when not properly balanced. These valves may be used with most any form of a balance. The two forms of balance being used with these valves at present are the Richardson and the American, both of which we have shown in this book.

The Allen valve has made fair progress in American locomotive practice, but there is still considerable prejudice against it. This prejudice, we believe, is founded in want of experience, or in a mistaken view of what the capabilities of the valve are. While this conflict of opinion continues to exist, we think that some remarks made by Mr. E. M. Herr, superintendent of motive power of the Northern Pacific Railway, at a meeting of the Western Railway Club, might be studied to advantage. The discussion was on lead in the setting of locomotive valves, and in this connection he said:

"With a very long port you can give an engine less lead than with a short port, and the kind of valve used has also an effect. With an Allen ported valve you can still further reduce the lead and get the same work out of the engine. The Allen valve, in my opinion, is a very valuable device if rightly used, and I believe that many railroad men condemn it because they have not used it in just the right manner. You do not get the full advantages of an Allen valve if you give it anything like as much lead as you would a plain valve. One of the principal advantages of an Allen valve is that you can reduce the lead and still retain the mean effective pressure in the cylinder. Of course, there is another advantage with the Allen valve, and that is the more rapid admission of steam into the cylinder, and this enables a locomotive at high speeds, with an Allen valve, to very much exceed in power the same engine with a plain valve.

In making some tests on the North-western testing plant not very long ago, we showed very conclusively that at high speeds a 16x24-inch engine would develop more power with an Allen valve than a 17x24-inch engine, with practically the same size of driver, would develop with a plain valve. In fact, a 16x24-inch engine on a certain division, where the ruling grade could be approached on a good run, was put in freight service with 17-inch engines, and it did satisfactory work with the 17-inch engines, pulling over rugged parts of our division, until one day the train happened to stop at the foot of this ruling grade. That day the engine stalled, and it stalled simply because it was not as strong as a 17x24-inch engine when pulling at slow speeds. At very high speeds it was stronger."

See Rule 12 for Locomotive Valve Setting.

THE HOLT SLIDE VALVE.

The Holt slide valve with cut-off plate, designed for use upon locomotives is at present being introduced into this country. It is a product of 1895, and was invented by Mr. Holt, of England. The valve has ports on top like the ordinary cut-off valve, but the cut-off plate has no separate eccentric or other means of positive movement. It has a space of $\frac{3}{8}$ of an inch for travel, and lays loose on the top of the slide valve. The quick movement of the valve is supposed to shift the plate at each end of the stroke alternately, and thereby secure an earlier cut-off, without increasing the compression. Fuel economy is the claim made for it.

A CUT-OFF VALVE FOR LOCOMOTIVES.

This device consists of a plain slide valve having no outside lap, a loose cut-off plate encircling the valve and moved by the action of the valve. It was invented by Mr. Wm. Goodspeed, of Bloomfield, Iowa. The following is his description of the device:

"This invention is in the nature of an attachment that may be applied to the slide-valve of any ordinary engine with only slight modifications to the valve, and it is especially designed for use in connection with a link-motion for controlling the speed and direction of rotation of the engine.

My object in this invention is to provide a valve of this class of simple, strong, durable, and inexpensive construction that may be adjusted or set to automatically cut off at any desirable point without employing any valve-stem or the like in addition to the ones ordinarily used.

A further object is to provide a valve in which a comparatively large and free induction-port is provided when the valve is set to cut off at a relatively small portion of the piston's stroke to thereby tend to produce a high initial pressure and the consequent increase of steam and the increased efficiency of the engine.

A further object is to provide a valve which may be operated so as to admit steam to the cylinder during the entire stroke of the piston, so that the entire boiler-pressure may be used throughout the entire stroke of the piston when it is desired to generate an unusual great amount of power, and, further, to provide an engine of this class in which the wear upon the valve-seat will be extended over a large area, and hence the durability of the valve will be increased.

Figure 1 shows a top or plan view of the valve mounted on its seat within the steam chest. Fig. 2 is a central longitudinal vertical section on line 2 2 of Fig. 1. Fig. 3 shows a transverse sectional view through the line 3 3 of Fig. 1, and Fig. 4 shows a detail perspective view of parts of the engine and valve to illustrate certain details of construction.

Referring to the accompanying drawings, I have used the reference-numeral 10 to indicate the steam chest, having the valve-seat 11 therein provided with the induction-ports 12, leading from the valve-seat to the opposite ends of the cylinder, and also

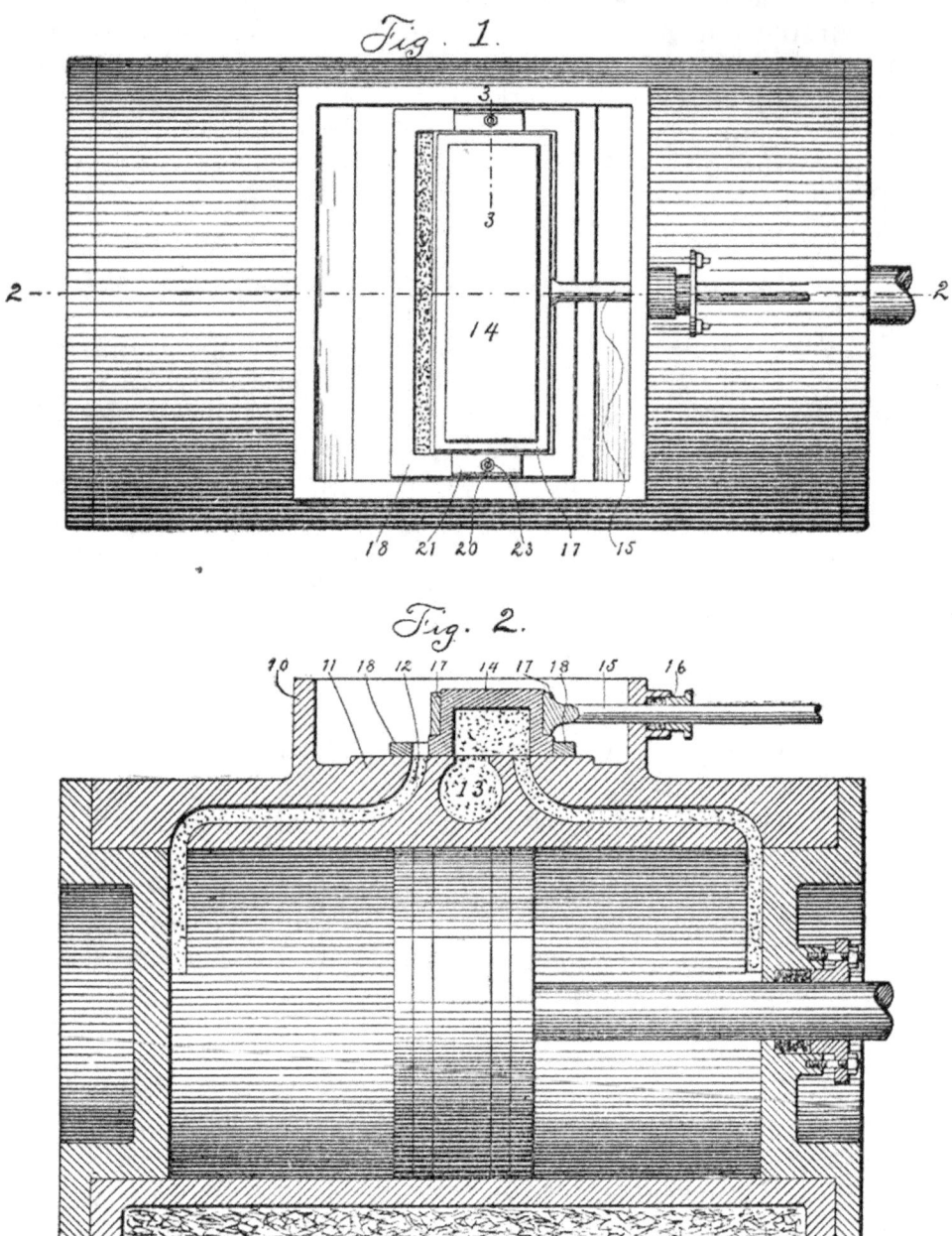

the exhaust-port 13, leading from the valve-seat at a point between the other ports to a point of discharge.

The reference-numeral 14 indicates a slide-valve which differs from the valve of ordinary construction in that it has a diminished inside or outside lap, and 15 indicates a stem passed through the packing-box 16 and having a yoke 17 on its end fixed to the valve 14.

All of the above elements are of the ordinary construction with the single exception noted.

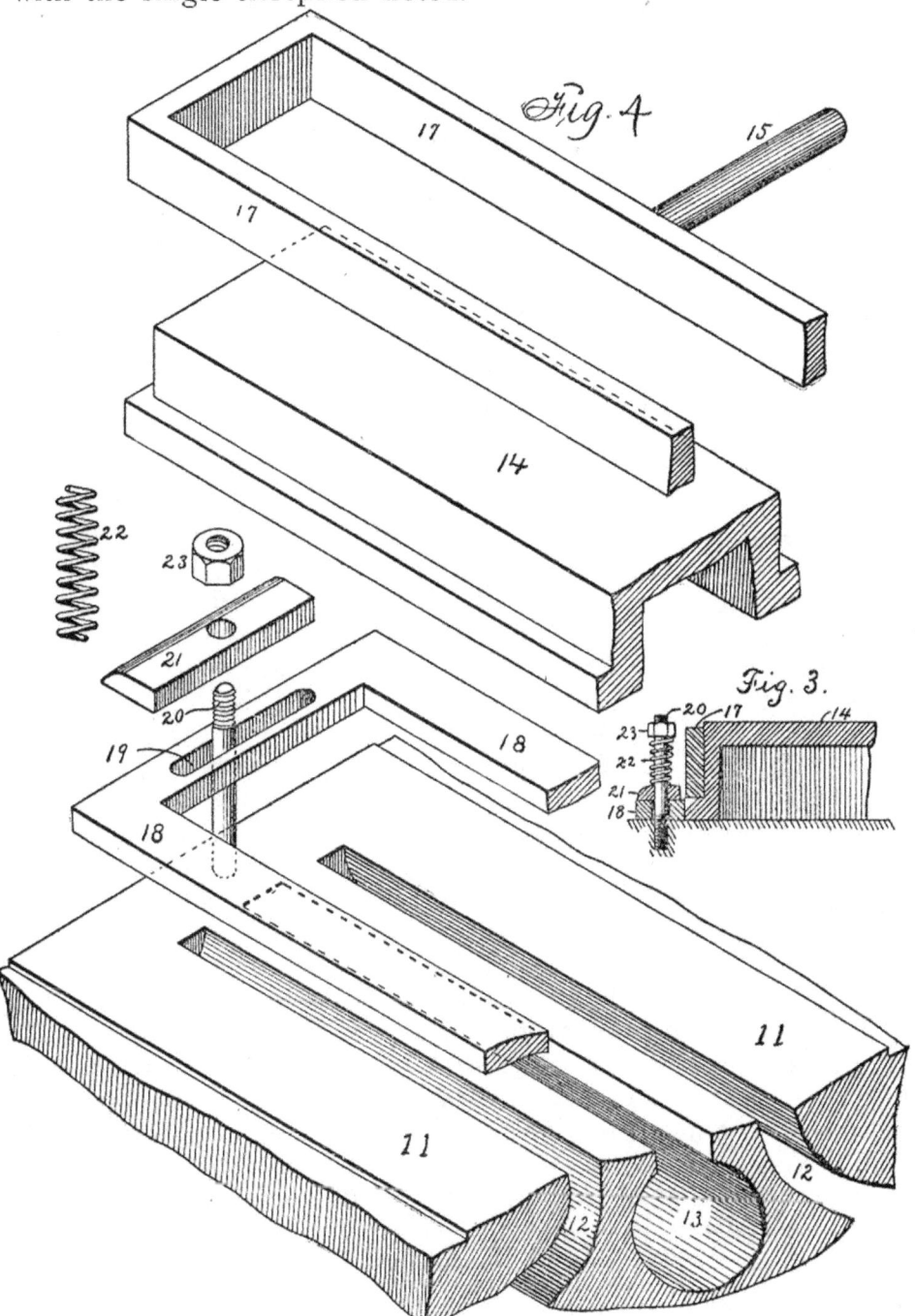

The essential novelty in the valve consists of a rectangular auxiliary cut-off 18, having a flat lower surface and having a rectangular opening in its central portion. The dimensions of this central opening are in one direction the exterior length of the valve plus the width of one of the ports 12 and in the opposite direction exactly the width of a valve. In the side pieces of the auxiliary cut-off are the longitudinal slots 19, and the reference-numeral 20 is used to indicate standards screwed into the face of the valve-seat to project upwardly through the said slots 19.

21 indicates elongated washers having openings, through which the standards 20 are passed, so that the under surfaces of the washers may engage the top surface of the auxiliary cut-off. 22 indicate expansion-springs mounted upon the standards 20, and nuts 23 are placed on the upper ends of the standards to compress the springs to thereby hold the washers 21 in frictional contact with the auxiliary cut-off 18. This arrangement is only for use when there is no steam-pressure in the chest.

In practical operation it is obvious that the valve proper will be moved a slight distance upon its seat before engaging the auxiliary cut-off, and when it does engage the cut-off will be moved slightly—that is, when the stroke of the valve is of greater length than the width of the ports leading to the cylinder.

This valve is designed for use only in connection with a link motion or other means for regulating the length of its stroke. Assuming that the link motion is set so that the valve will cut off at one-quarter stroke, and assuming the valve to be in its position for starting, that is, with a small lead opening, it is obvious that the first movement of the valve will be to open the induction-port wide. Then the auxiliary cut-off is moved by the valve so that the side of the cut-off, adjacent to the induction port, will be moved to partially cover the said induction port. Then as the valve starts on its return movement it will meet the auxiliary cut-off or that portion thereof that is partially covering the induction port at a point midway between the sides of the induction port, and hence the steam will be cut off, while the valve has only partially covered the induction port; that is, before the full stroke of the valve has been made. It is obvious that the further movement of the valve—that is, to its outer limit and part of the way back—will not open the induction port. Hence we have a one-quarter cut-off with a full opening of the induction port, and also a full opening of the exhaust port. It is to be understood in this connection, that the earlier the cut-off the greater is the length of the valve movement required.

Assuming that it is desired to work the engine to its fullest capacity, we will assume that the valve is in the same position as in the former instance—that is, with a small lead open. We will assume, further, that the link motion is set or adjusted so that the stroke imparted to the valve will be only the same as the width of the induction-port. It is obvious that in this instance the induction-port will be opened gradually until at the end of the valve's stroke it will be wide open. In this instance it will be noted that the auxiliary cut-off is not moved, and hence the friction of the valve upon its seat will be lessened, inasmuch as the area of the valve upon which the steam may press is less than with the ordinary-sized valve having considerable outside lap."

HALEY'S SLIDE VALVE.

The object of this valve is to prolong expansion and relieve excessive compression. It was especially designed for use with link motion. It is the invention of Mr. J. A. Haley, of Ft. Wayne, Ind., who explains his device as follows:

"My invention relates to improvements in slide-valves for steam engines; and its objects are to provide means whereby when the valve is operated the steam-port at one end of the cylinder or valve-seat is kept closed for a longer period than heretofore for the purpose of using the steam more expansively, while the port at the opposite end is kept open for a longer period of time for the purpose of reducing compression.

The invention relates specially to valves operated by what is known as the "link motion," and to the use of steam expansively.

I attain the objects by the mechanism illustrated in the accompanying drawings, in which—

Figure 1 is a vertical section of my improvement, showing the port-holes of the valve-seat. Fig. 2 is a bottom view of the valves shown in Fig. 1.

My invention consists in the construction and combination of a main valve constructed with shortened faces adapted to actuate independent valve-faces, two independent valve-faces adapted to supplement said shortened faces, and devices and means whereby the independent valve-faces during a portion of the travel of the main valve are not operated thereby but remain stationary and adapted to give an increase of inside lap at one part of the valve-seat with increased clearance at the other part alternately with each stroke of the main valve.

By the term, "independent valve-faces," I mean valve-faces which, during a given portion or space of the travel of the valve-rod remain stationary, not being actuated by it directly or indirectly.

The preferred form of construction of my improvement is shown in Figs. 1 and 2, and is as follows: The main valve A has portions of its faces cut out, preferably in the form of a rabbet, leaving the faces a thus shortened to rest upon the valve-seat and form with the independent valve-faces B combined valve-faces. These rabbets are cut out sufficiently deep to permit independent valve-faces B to be placed underneath them and afford room for their independent motion.

The independent valve-faces B are constructed so that when

connected together, preferably by a cross-bar H, which is also placed under a rabbet on either side of the face of the valve A there will be between the extreme length of the two faces B B and the extreme length of the rabbets on the inside of the faces of the valve A a space S. This frame consisting of the independent valve-faces B and the connecting-bars H is placed on the

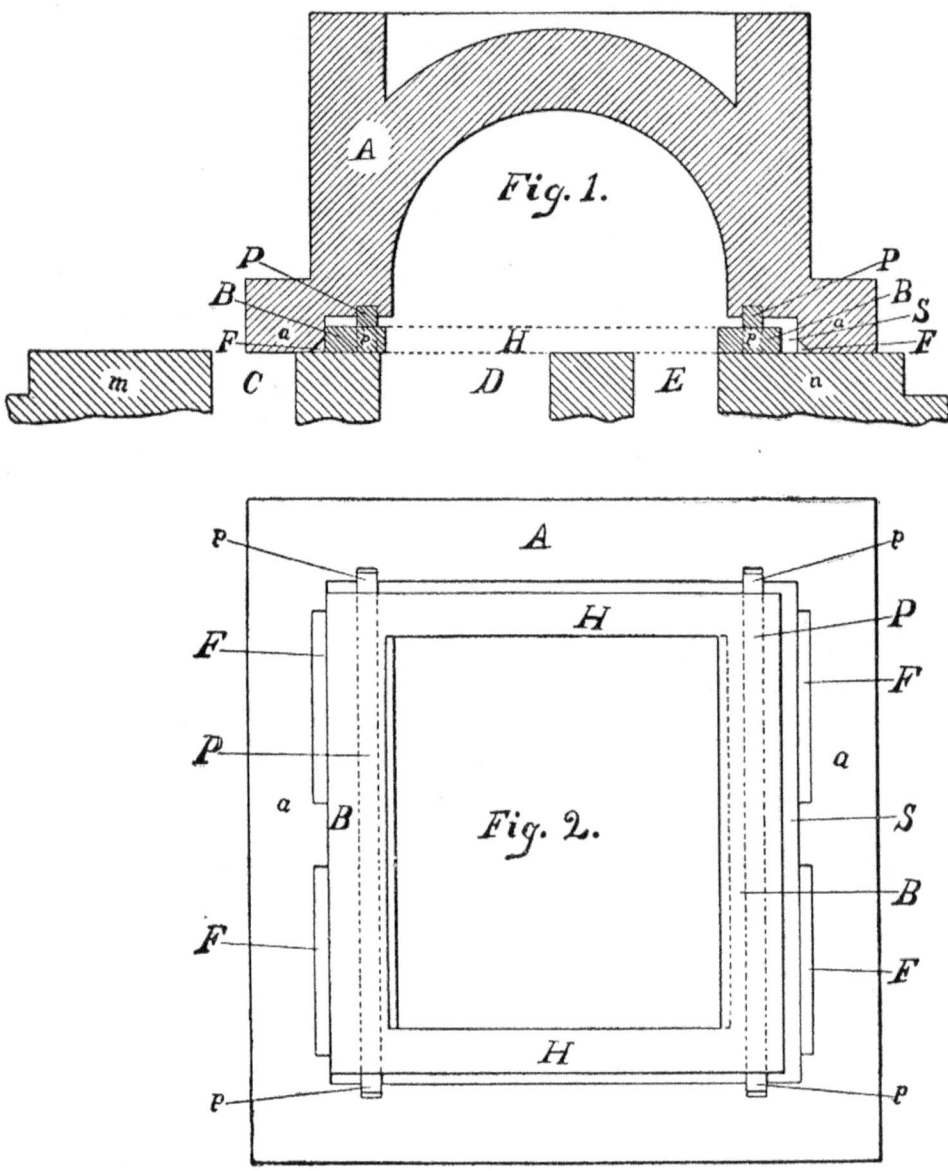

valve-seat within the rabbeted parts of the face of the main valve A, and packing P is placed above the independent faces, as shown in the drawings, and other packing *p* is placed upon the side in the side rabbets, as shown in Fig. 2, for the purpose of preventing steam from passing between the valve-faces B and *a*. The space S is for the purpose of and designed to allow a movement of the main valve A that distance or space without moving the independent valve-faces B. Its length is determined by the requirements for inside lap and inside clearance, as will appear in describing the operation. When the space S has been traversed by the valve A moving from N to M, the end of the rabbet impinges against the independent valve-face B at the port E and moves both independent valve-faces, as will be readily seen, to the end of travel of the main valve. On the return stroke the positions and movements are reversed. The faces A have chamfers F on their inside corners, as shown in the drawings. These chamfers are for the purpose of permitting the steam to operate against the ends of the independent valve faces B to aid in keeping them alternately pressed against the ends of the rabbets for the purpose of preventing any movement of the valves B, except when actuated by the faces a. This is my preferred construction, but the independent valve-faces B B are not necessarily connected together by the bars H, but could be operated between the rabbets of the face of the valve A by lugs extending down for the purpose on either side of the exhaust-cavity of the main valve, or by other projections answering the same purpose. I do not, therefore, confine myself to the particular method shown of operating the valve-faces B independently.

The operation is as follows: When the engine is first started, it is usual to use full openings of the ports. In such case my improvement has no special operation; but when the stroke of the valve is reduced or shortened for the purpose of using steam expansively then the following results take place: Say, for example, that the opening of the port C, as shown in Fig. 1, is the extreme opening for a given point of cut-off. Fig. 1 then shows that the limit of the stroke opening the port C has been reached. At this point the cylinder is therefore taking the full amount of steam at such point of cut-off, while the port E is exhausting at the other end and is open to its full extent.

The width of the exhaust cavity, measured on the valve-seat, is determined by the fixed distance apart of the inner edges of the two independent valve-faces B B, because the packings P p prevent the passage of steam between the faces B and a into and from the exhaust cavity. In the drawings such distance or width is shown, preferably, as slightly less than the distance over the inner edges of the steam ports C and E; but the spaces on the valve-seat, covered by the combined valve-faces B and a, vary during the travel of the valve. When the valve travels, say on the return stroke (See Fig. 1) the faces B remain stationary until the distance S has been traveled by the face a at the end N of the valve-seat, which then impinges against and actuates the faces B. At this point of travel, and continuing to the end of the return stroke, the combined faces B and a on the end N of the valve-seat, cover a diminished distance, while the faces B and a on the other end M cover an increased distance over the ports and valve-seats, the difference being represented by the space S. Such movment increases the inside lap of the combined valve-faces B and *a* on the end M of the valve-seat, because of the inside lengthening of the combined valve-faces, and thereby, as is well known, delays the exhaust of steam at the port C, thus permitting a greater expansion of steam in that end of the cylinder. At the same time, while so traveling, the distance covered by the combined valve-faces B and *a* at the end N has been reduced correspondingly, and the face B has remained stationary for a portion of the travel, so that the closing of the port E has been delayed, thereby giving increased clearance and reducing the compression at that end of the cylinder. This increase of the inside lap at the end M of the valve-seat and increased clearance at the opposite end N is reversed with the reverse of travel, as will readily be seen..

METZGER'S SLIDE VALVE.

The principle involved in the design of this valve has been fully explained in the description of the Haley Slide Valve, the form of construction only being different. It would be, therefore,

simply a repetition to describe the principle of this valve's action. This form of valve is the design of Mr. Jules P. Metzger, of Paterson, New Jersey.

A NEW LOCOMOTIVE VALVE AND VALVE SEAT.—C. I. & L. RY.

There is shown in the engraving illustrating this article a valve and valve seat which, while tending in a direction the limit of which has probably been tested in another design, is, we believe, quite new. It should be understood at first that the drawing shows the valve as applied to the ordinary cylinder casting A A, cored with a steam port on either side of the single exhaust port, so that it is necessary to put in a false seat, B B, 2 in. thick, and having ports suitably arranged for the openings in the valve.

This makes it necessary also that the steam chest be raised 2 in., and accordingly a filling piece 2 in. thick is placed under the steam chest. For new work the ports can, of course, be cored in the cylinder casting and then the valve seat and steam chest would be of the same height as in the ordinary design.

The valve is shown as central on the seat, and the exhaust edges of the steam ports are line and line with the inside edges of the valve, so that if it is considered that the valve is moving from right to left, the left steam port is about to open for exhaust. The double exhaust port, however, is covered ½ in. so that after the point where the interior of the valve is opened to the exhaust steam, the valve must travel ½ in. before the exhaust port begins

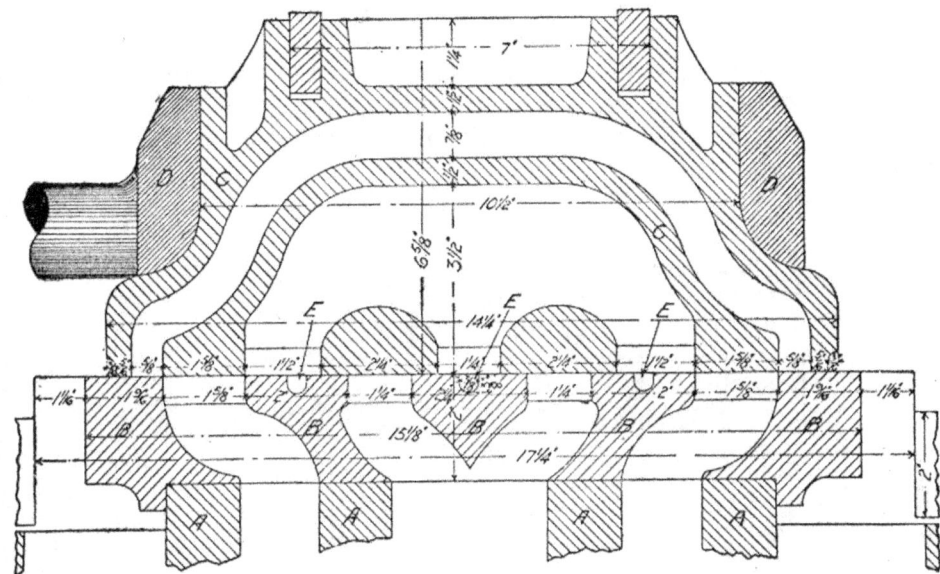

to open. The exhaust port being double, the area of the port which is uncovered by a certain movement of the valve is twice the area that would be uncovered were the port single, and a freer exhaust must therefore result. The double exhaust port and single steam port would be no better than if both ports were single were it not for the arrangement which makes it possible to open the exhaust side of the steam port ½ in. by the time the valve begins to uncover the exhaust port. The claim made for the valve is that the steam can be held in the cylinder during a longer portion of the stroke (except for the steam which expands into the inside of the valve) because when the valve begins to uncover the exhaust port the required opening is given by one-half the travel of the valve required for the single exhaust port. The grooves E E E are cut in the seat to assist lubrication; it is reasoned that they are soon filled with water or oil, or a mixture of both, and that the face of the valve is coated each time the valve passes over them.

The valve is the design of Mr. H. Watkeys, master mechanic

of the Chicago, Indianapolis & Louisville Railway, LaFayette, Ind. It has been applied to four locomotives running on that road and is giving much satisfaction.—*Railway Master Mechanic.*

FAY'S VALVE FOR REDUCING COMPRESSION.

The form of valve shown in the accompanying illustration is the invention of Mr. Henry R. Fay, of Boston, Mass. The improvement relates only to the face of the valve and its seat, and any form of balance may be used with this valve. Mr. Fay has favored us with the following information and description of his improvement:

In presenting this invention to your notice, I call your attention to its simplicity and efficiency; also to the small cost of its application, and to the fact, that there will be no maintenance expenses, as the device becomes a permanent fixture to the engine, reducing the compression, without interference with the admission and expansion of the steam.

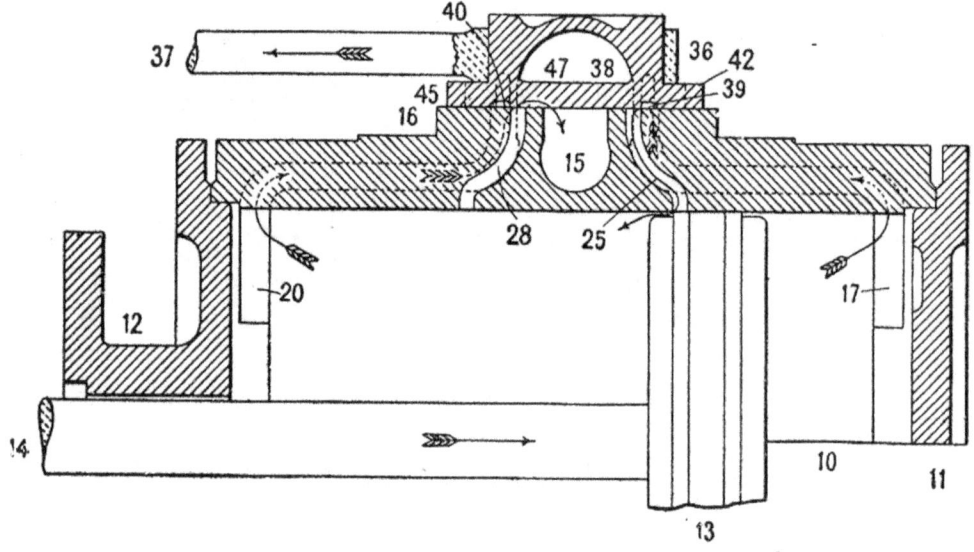

The amount of the released compression is determined by the size of the extra ports made in the cylinder, and the location of the grooves in the valves.

That excessive compression is the greatest drawback to the efficiency of our high speed locomotives cannot be questioned. It cuts down the capacity of and produces bad riding engines. These losses or defects are usually remedied to a certain extent by a wasteful distribution of steam.

This invention has been applied, during the past year, with very satisfactory results to fourteen of the Boston & Albany Railroad locomotives, seven being heavy express passenger, the others being mostly freight engines.

The advantages obtained by the use of this invention are, that smaller clearance space can be made; higher speed can be obtained, by reason of reducing excessive resistance in front of the piston when cutting off at short points of cut-off, and a saving of the general wear of engine (by giving a smooth working engine) and especially in the wear of valve and seat which has been proved in use, by reason of the valve keeping down on its seat, and not being forced off its seat by excessive compression, giving good steaming qualities, as steam cannot blow under the valve to exhaust port, and an engine will do the same work at shorter points of cut-off.

This valve can be applied to all engines using single slide valves of either piston or flat type.

This invention is of especial value to compound engines for the reason that compression, commencing at receiver pressure in high pressure cylinder, combined with the greater area of low pressure cylinder, causes excessive compression. The clearance space can be cut down, which should give economical results.

This device possesses especially valuable features when applied to the high pressure cylinder of cross compound engines, which are not shown in the drawings—an explanation of which will be furnished on application.

The device is applied to a cylinder by drilling holes (extra cylinder ports) at the end of the bridge, at an angle, as shown, from the valve seat to the bore of the cylinder, so that the piston packing will reach the first holes when the groove in the valve opens up into the admission port, after cutting off. The size of the holes in the applications that have been made are $\frac{3}{8}$ of an inch in diameter. The distance from the line of the exhaust port to the extra cylinder port in valve seat must be at least 1-64 in. greater than the width of groove in the valve. The valve has grooves (extra valve ports) $\frac{3}{8}$ in. wide and $\frac{1}{2}$ in. deep, nearly across its face. The outside edge of each groove is located at a distance from the outside edge of the valve 1-64 in. greater than the width of the admission port. Each groove is provided with the two end cavities in the sides of the valve, which are cut so as to reach the extra cylinder ports when exhaust takes place on opposite end, thereby opening communication with the other end of the cylinder and allowing the compression to pass around the piston to the other end of cylinder and exhaust as shown.

It will be seen that by placing groove in valve farther from outside edge of valve, compression release will be cut off before pre-admission takes place, thereby holding on to a certain pre-determined amount of compression if desired during the last part of the piston travel. The lugs on the corners of the valve are to prevent the uncovering of the extra cylinder ports when the valve is at its full travel.

FARRER'S VALVE.

The same principle is followed in the construction of this valve as in the Fay valve—that of reducing excessive compression. The relief ports shown in the bridges are not the full length of the bridge, but simply two small holes drilled through

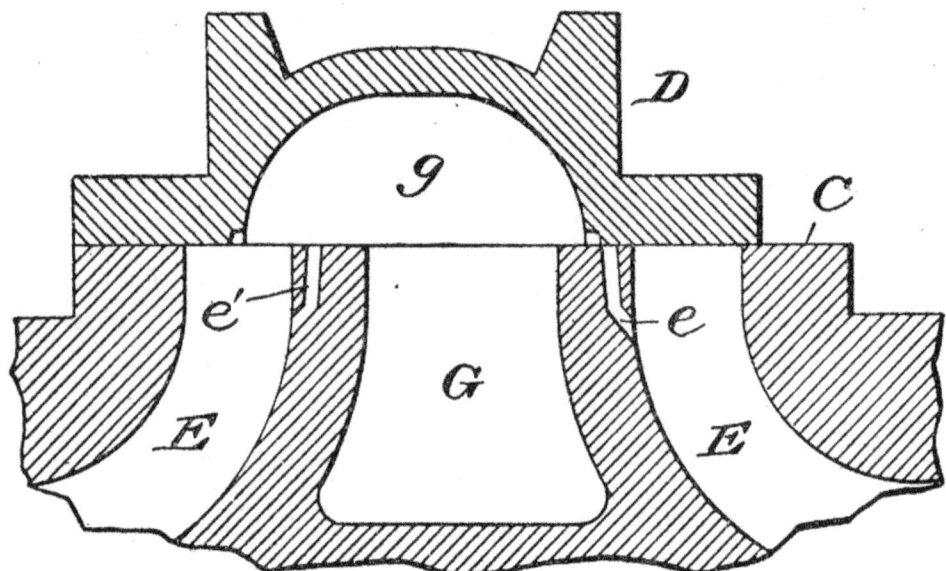

each bridge in the manner shown. The exhaust edge of the valve being provided with grooves to coincide with the relief ports, thereby permitting a portion of the air which is being compressed in the cylinder to be released and escape to the exhaust cavity, and thence to the smoke-stack. This improvement is the invention of Mr. Chas. S. Farrer, of Dunmore, Pa.

BALANCED SLIDE VALVES.

For a great many years the plain slide valve answered all the requirements for locomotive service, but with the enlargement of the locomotive and the increased steam pressures it was found almost impossible for one man to reverse an engine with such an enormous pressure on top of the large slide valves which modern locomotives require. The object of designers was, therefore, to produce a valve that would require as little power as possible to move it. To lessen this friction, what is known as the "Roller Valve" was first invented. It was a plain slide valve with rollers attached to each side of the valve. While this valve required less power to handle it, yet it failed to remove the cause of the great amount of friction between the valve and its seat, and therefore did not come into general use, although it is still in use upon some roads. But, when the balanced valve, sometimes called the equilibrium slide valve, was invented, the correct principle seems to have been followed, namely: The removal of the steam pressure from the back of the valve. These valves have been universally adopted. The various forms of these valves that have been invented are too numerous to mention, but they are all constructed upon the same principle, that of removing the pressure off the valve.

We have illustrated a few forms of these valves, varying all the way from very good to very bad. The best form of balance valve will be found to be the one which overcomes the greatest friction (within practical limits), the most simple in construction and positive in its operation and has the fewest parts which are liable to break or get out of working order.

THE GOULD BALANCE VALVES.

The Gould Balance Slide Valves are both comparatively new and we believe they deserve the careful attention of every locomotive builder or designer. The form known as the "Quick Action Valve" with the double opening deserves especial attention. Heretofore it has been the generally accepted belief among practical men, that it would be impossible to prevent the valve from lifting off its seat if all the pressure were removed from the top of the valve. But Mr. Gould holds to a contrary opinion, and after carefully studying the action of his "Quick Action Valve" the writer is inclined to agree with him.

The principles of these valves are similar to that of the piston valve, either of which can be applied to the ordinary flat valve seat. Two methods of balancing are employed, one form, the "Quick Action Valve," can be used with either a flat pressure plate or his semi-circle balance plate. It promises to be an important rival to our best forms of balance valves. For the other form of valve the balance is obtained by means of a semi-circle balance plate fitted into the steam chest lengthwise. It is not bolted to the cover like most other pressure plates, but rests on the valve seat. It has lugs which are closely fitted on the ends to prevent it from moving lengthwise; while the pressure on its back holds it down. The top of the valve is also a semi-circle in form, slightly smaller than the plate, and has balanced strips set into it which bear against the plate in the usual manner. A small port at each end permits live steam to enter between the valve and the pressure plate, which is permitted to cover sufficient area to overcome the back pressure in the cylinder, thereby obtaining almost a perfect balance. In construction the two balance plates (one for each valve) are first planed off on the edges; then the two are clamped together and bored out to the required size. The valves are finished in a similar way; first the two faces are planed off and then clamped together and turned off in a lathe, and made a little smaller than the pressure plate. When the valves require facing the same amount is taken off the bottom edges of the pressure plate, thereby retaining their original positions, and the plates automatically adjust themselves to any inequality due to the wear of the valve.

These valves are the invention of Mr. W. F. Gould, of Des Moines, Iowa, to whom we are indebted for drawings and a complete description of each of the valves, which is herewith given. The second valve described is at present being tried on the Chicago & Erie R. R. Five engines are now equipped with this valve.

DESCRIPTION OF QUICK ACTION VALVE AS SHOWN BY EXHIBIT A.

This valve is shown in a cross section, in order to show the relative positions of the ports H and D to the port leading to the cylinder (not lettered). The valve cover A is supported by four bolts to the steam chest cover, showing that the cover A can

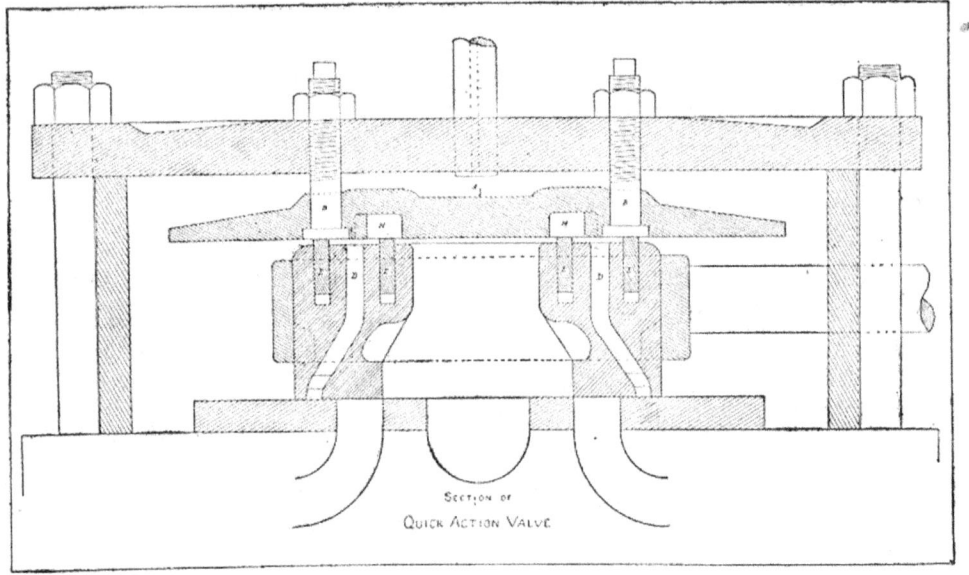

Section of Quick Action Valve

be lowered or raised at will. Letters H are ports in the base of the cover running longitudinal parallel to the steam ports. Letters D are longitudinal cavities in each end of the valve and should be the same length as the ports H, and nearly as long as the steam port which leads to the cylinder. It will be seen also that packing bars E are placed on each side of the cavity D in the top of the valve in such a manner that the distance between them is of a very little more area than the width of the steam port, for the purpose of holding the valve down when the steam port at the end of the valve is filled with steam.

In the operation of the valve, as shown herein, when the valve is pulled back so as to open the steam port, it will be observed that the packing bar nearest that side will have been also pulled back so that the steam at the top of the valve will go over the top of the packing bar E into the cavity H and down through the cavity D into the steam port, thereby forming a double opening into the cylinder. Another feature shown herein, is that the valve on its return shows the port D in communication with the steam port after the valve has cut off, thereby showing the area between the packing bar to be in communication with the steam in the steam port; this is done in order to balance the valve. It will also be observed that the outside packing bar can be placed as near the outside end of the valve as desired so as to take nearly or all the pressure off the valve.

Another feature which I wish to show is, the relative posi-

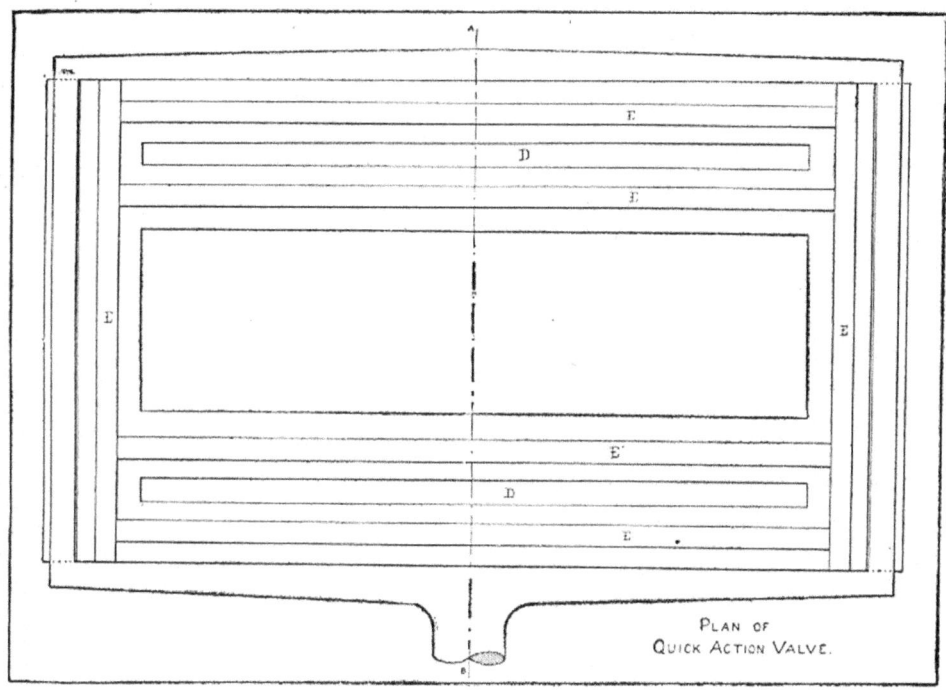

Plan of Quick Action Valve.

tion of the ports to each other when the engine is running at a high rate of speed and a comparatively short cut-off; or, in other words, using the full amount of lead. In such cases there is more or less what is known as a cushion or back pressure in the cylinder which has heretofore operated in a detrimental manner to balance valves, but in the case of this valve it is clearly shown that the port D will be in communication with the steam port and therefore entirely obviates the defects which have operated disastrously to other balance valves.

When it becomes necessary to reverse the engine at a high rate of speed it will be seen, as shown in the section of the valve, that the valve will lift and relieve the cylinder pressure as in other ordinary slide valves, for the reason that the movement of the valve will be exactly opposite to what it was when the engine was running forward.

In the plan view is shown the relative position of the end bars and the side bars E to each other, also showing the valve to have an open back so that the exhaust steam will have no effect upon the valve, but will have a tendency to keep the bottom of the cover well lubricated.

In the view of the bottom of the cover A is shown the form of the posts H with the small bars C in them. The reason they are placed diagonally is so that the packing bars E will not have

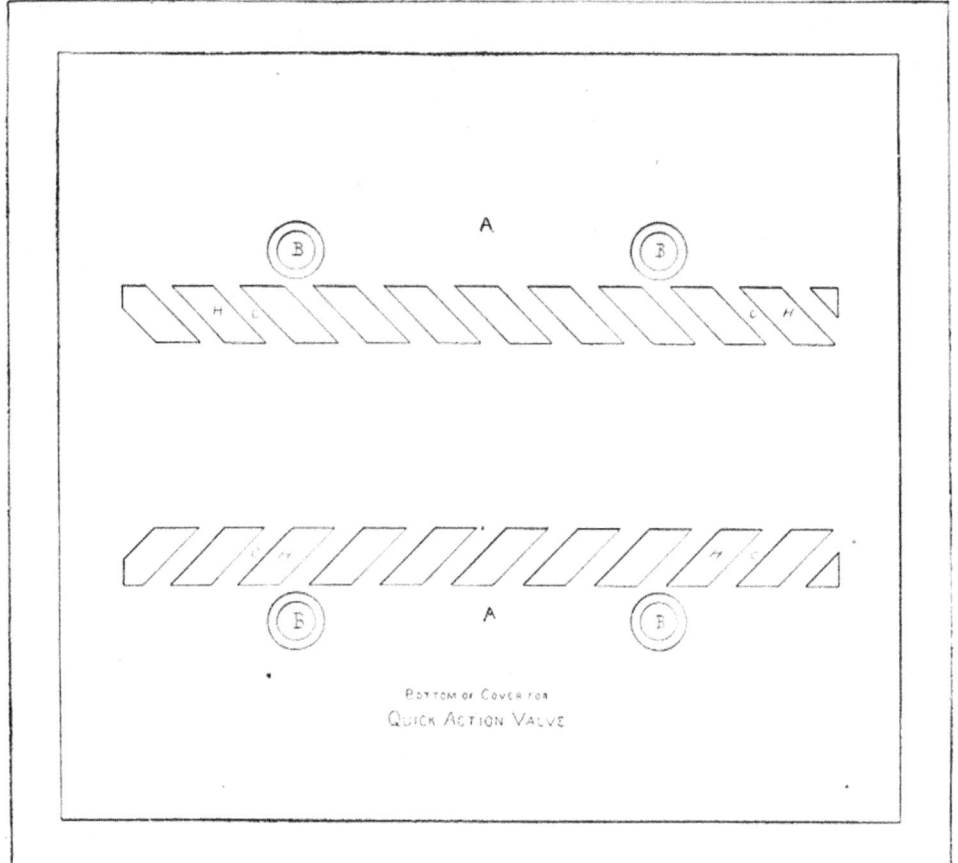

ridges worn in them while moving over the ports H. I have not shown any particular means for holding the packing bars E up against the valve cover, for the reason that I would prefer that to be done in any way most convenient to the party using them. The oil cups are screwed into the top of the steam chest cover in the usual manner and needs no further explanation. Should any further information be desired I will gladly furnish it at any time.

DESCRIPTION OF BALANCE VALVE AS SHOWN BY EXHIBIT B.

The end view of the section of the valve, as shown on line C D, shows the end of the valve, the loose valve cover, the end of the yoke, and the key A, also a representation of the valve and its cover as it appears upon the inside of the steam chest, and is described as follows: Letters h represent a loose cover, which fits over the top of the valve, letters cc represent stops, upon the inside of the steam chest cover to prevent the valve cover from having too much longitudinal movement; this movement should never exceed one-sixteenth of an inch over all. The letters ff represent spring pockets in the valve, in which small spiral springs are inclosed in order to hold the packing, which is shown by dotted lines in its place; the letter k represents a chipping strip upon the rear end of the valve, which may be faced so that the back end of the yoke d may fit it; letters n represent the sleeve, which is cast solid to each wing of the

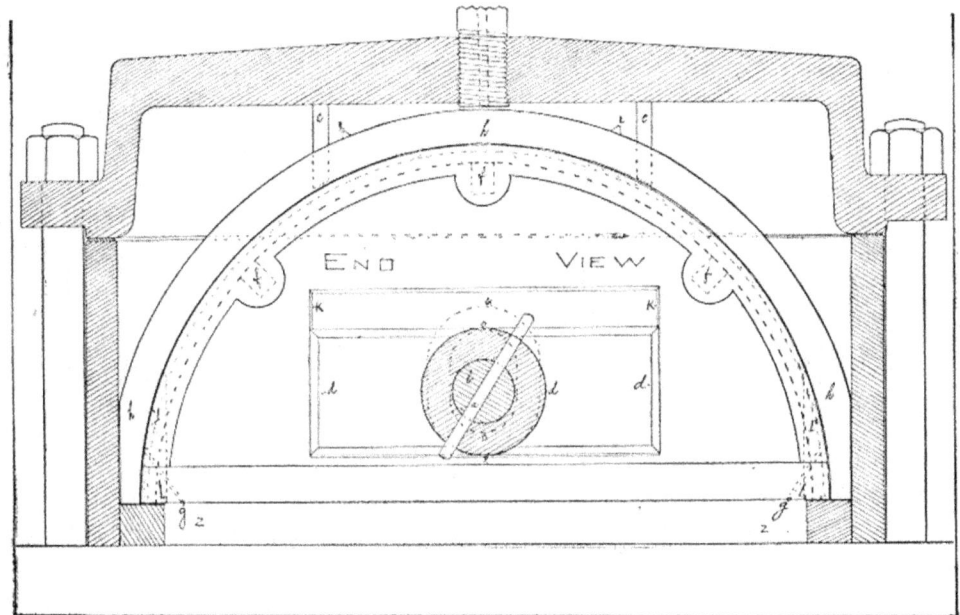

valve, and through the cavity o extends, in order to contain the valve stem b. It will be observed in all the figures that this cavity o is oblong, in order that vertical movement of the valve may not bind upon the valve stem; letters gg are two small holes drilled in the face of the valve, in order that communication may be established between the port z and the cavity x, shown in another figure; letters ee are small ribs cast upon the top of the valve cover h, so as to prevent the valve oil from running down on the sides of the cover, and also to cause the oil to come to the end of the cover, thereby more perfectly lubricating the valve.

Referring to the section on A B, it shows a side view of the valve—the packing rings are shown held up by spiral springs on one end of the valve, while the dotted lines on the opposite end of the valve shows the spring pockets; the letters X are cavities running round the whole outside faces of each end of the valve, while the packing, as shown, is placed upon each side of these cavities X. The area of the cavity X should always be a trifle more than the area of the steam port Z, so that when the steam port Z is filled with steam and communication established with cavity X by means of the hole G it will be seen that the pressure in the cavity X will equal the port pressure, and the valve be held down upon its seat. It will also be observed that the packing ring may be placed as near the outside edge of the valve as necessary to perfectly balance the same while the engine is using them.

It will be observed that there are two check stubs Y in the top of the steam chest cover. It will also be observed that they do

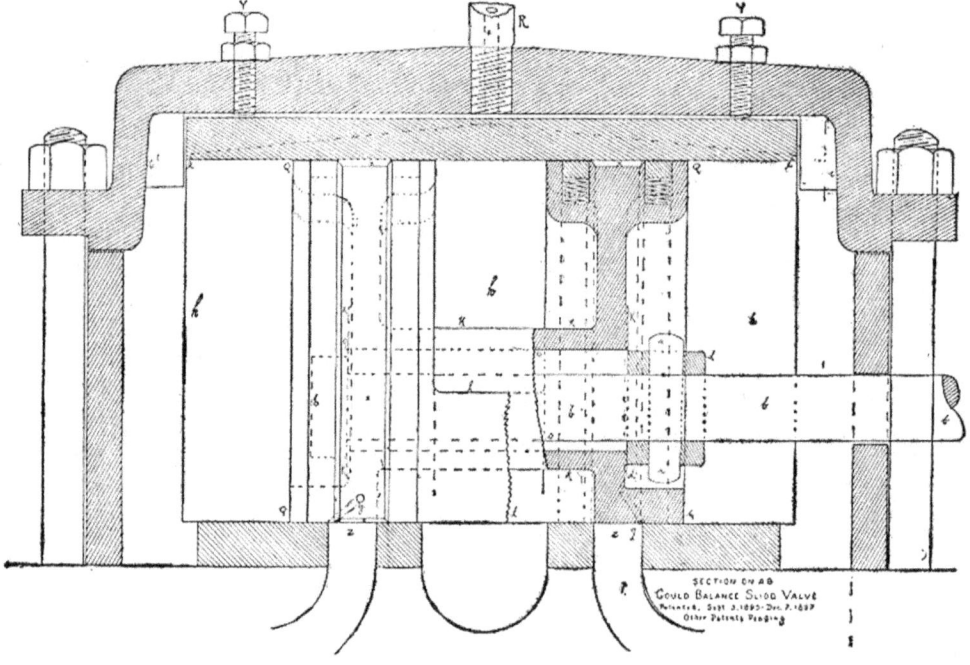

not touch the valve cover H; they are put there for the purpose of regulating the vertical movement of the valve cover, and prevent it from lifting too high from the valve; the letters L, as shown in this and the other figures, is a web which holds the bottom of the wings of the valve together in one piece; it also prevents any leaking of steam into the exhaust which might occur if the valve cover was raised from the seat. This valve, as shown, is semi-circular in form, and is constructed in the following manner: As soon as the castings come from the sand their faces are planed, and fastened together they are then put into a lathe and turned to the desired size. Grooves for the packing and the cavity X are also cut in their faces. The valve covers have their edges planed, then they are fastened together and bored out to fit the valves, thereby making one for each side of the engine. The valves and their covers should fit iron to iron when they are cold, as when the cover becomes hot it expands a little more than the valve, which leaves the valve free to move perfectly under it, should it become necessary to face the valve any amount of iron taken from the face of the valve, so as to lower it, the same amount should be planed from the bottom edge of the valve cover, thereby causing each to assume their normal position relative to each other, and as the hole O in the valve through which the valve stem passes is made oblong, it will

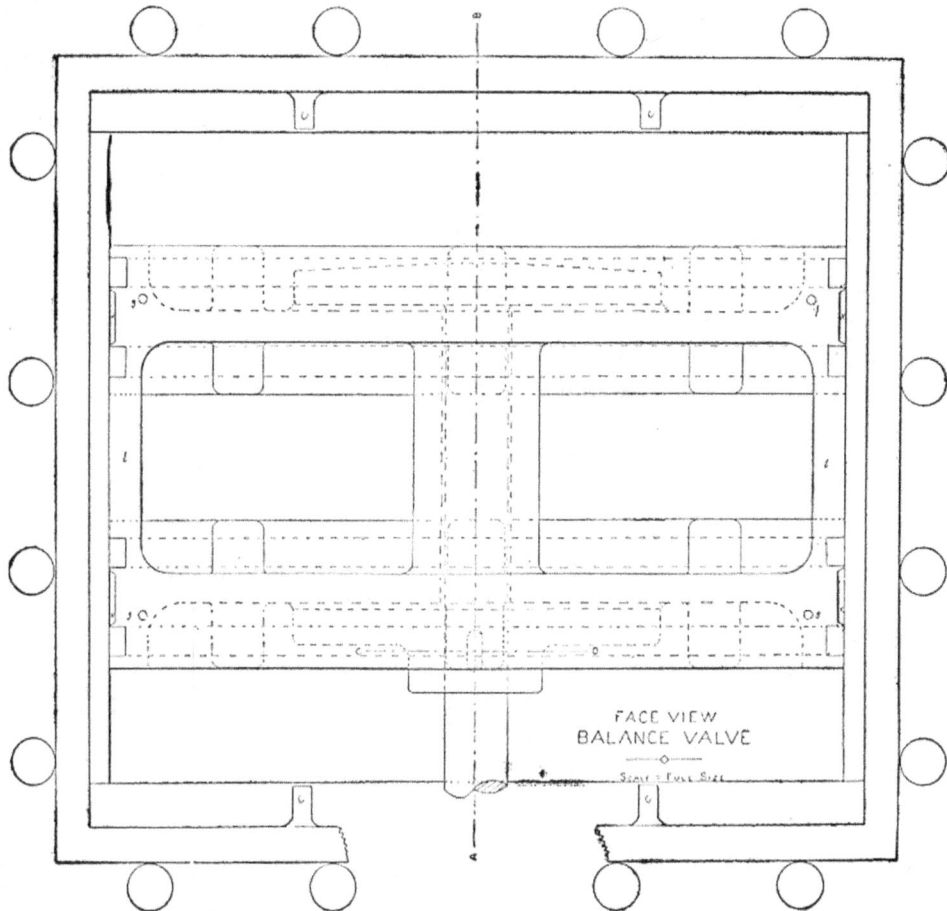

be readily seen that this will not affect the valve stem whatever. When the valve is being planed—that is, when the exhaust is being planed out, do not allow the tool to cut through the line L, as shown in the face view, but let the exhaust be of the same form as shown in the face view of the valve.

In the absence of having a definite size of steam port to work from, I cannot give the exact sizes, they would have to be governed by the size of the engine upon which the valve is used. But I will say that the packing rings should have at least $\frac{5}{8}$ face by $\frac{1}{2}$ inch deep. The valve stem should be about $1\frac{1}{2}$ inches in diameter and the stops C should not be less than ⅜ of an inch in width. The valve cover should finish 1¼ inch in thickness, so that it would never spring, and when it becomes necessary could be bored out the second time to fit the valve. The depth of the cavity X should not be less than ⅛ of an inch, and the size of the hole G should not be less than 3-16 of an inch. The size of the bolts Y should not be less than ¾ of an inch. The springs, which are designed to hold the packing, should be made of about No. 70 steel wire so that the pressure on them would be very light, because if these springs were too stiff they would cause the packing to wear too fast, or might hold the cover up from the valve when steam was not used. The key A, which is used to hold the back end of the yoke in its position, is shown full size. I have herein briefly outlined the description of the cuts here shown. Should any further information be desired it will be readily furnished.

THE PISTON VALVE.

This is a slide valve of cylindrical form which has many good qualities. It permits of a greater port area and occupies less space than the D valve, this form of valve being encased within the walls of the cylinder. It has been commonly supposed to be a perfectly balanced valve, but recent tests have proved that it is not as perfectly balanced as was generally supposed. It was found that its perfection of balance depends largely upon the width of its rings and the steam pressure, for as long as the steam exerts a pressure under the rings, holding them against the walls of the cylinder, they create an unbalanced friction equal to the area of the rings; of course much depends upon the construction of the valve, the closer the rings are fit into the grooves the better, but it is impossible to fit the rings steam tight without sticking in the grooves and the more the sides of the ring wears the greater will be the pressure under the ring and therefore the greater the friction. Owing to the construction of these valves, they being made either of one long casting or in two parts connected with a rod, or tube, when the two faces are located near the ends of the cylinder it is claimed they reduce the clearance which is so wasteful on steam, since for each stroke of the piston this unoccupied space must be filled with steam, which in no way tends to improve the engine, but rather decreased the expansive force, and increases the amount to be exhausted on the return stroke. While the correct principle for reducing the clearance seems to be followed in placing these valve faces near the ends of the cylinder it should be remembered that while the distance between the valve and cylinder is lessened the clearance space extends clear around the valve and the claim of reduced clearance for these valves has frequently been overthrown by the Indicator. But withal, this is considered one of the best forms of valves in use. Nothing is good that will not bear criticism.

The illustration we have shown is the form used with the Vauclain system of compounding (Baldwin). It is a combination piston valve, the two ends of which control the admission and exhaust of steam to and from the high pressure cylinder and the inner rings perform the same functions for the low-pressure cylinder.

The piston valve is an old affair in locomotive practice, having laid dormant for years. One of the early designs of these valves was that of Mr. Thomas S. Davis of Jersey City, in 1866. A difficulty which developed with his design and others at that time was the rapid wearing away of the valve cage at the port openings, due in part to the absence of bridge strips in the port openings, as ring-bearers for the piston. It was supposed also

that the valve piston rings needed adjustment the same as those of the steam cylinder piston. Occasionally this adjustment was faulty and, cramping the free motion of the valve, over-balanced all the advantage of the piston for the time being. The tallow then used for lubrication troubled the piston valve, as it did also the plain D valve. The cause which more than any other, however, led to the disuse of the piston valve after these early experiences was the introduction of the balance on the slide valve. This balance rendered the slide valve form acceptable at the then low pressures standard on engines.

In the meantime the piston has gained a position at the head in steam vessels, small and large, naval and commercial, at all pressures the world over. It is used in fast steamers crossing the ocean, the entire distance being passed, in some instances, without lubrication. It is also used in the best and fastest electric engines running almost continuously. The piston valve has of late years been improved in form, eliminating the features which caused objections to the old form, and with the advent of high pressures it is again gradually coming into use upon locomotives. Both the Baldwin and Brooks Locomotive Works furnish piston valves when ordered upon both simple and compound engines and in a few instances these valves are applied by railroad companies to old cylinders. But these valves should be given close investigation before a good, reliable balanced slide valve is side-tracked.

THE RICHARDSON BALANCED VALVE.*

This illustration shows a sectional view of the valve in the steam chest; the balance is secured by four rectangular packing bars fitted into the top of valve and working nicely in the grooves.

These packing bars are held lightly against the pressure plate by means of a semi-elliptic spring, until steam is admitted in chest, then steam pressure acts upon the packing bars and holds them firmly against the pressure plate and also presses them

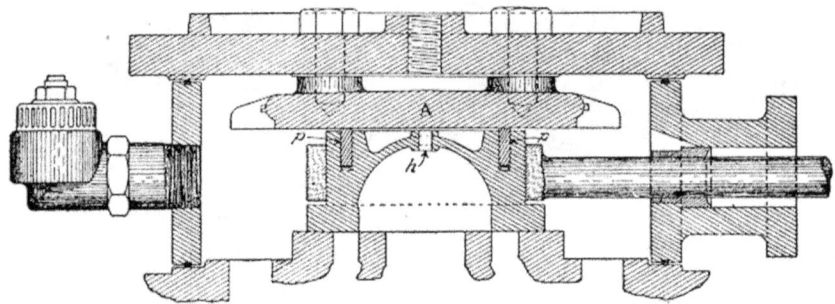

closely against the inside of their respective groove, forming thereby a steam-tight joint, and excluding the steam chest pressure from acting on the area of valve included within the packing bars.

The inventor of this form of balanced-valve, the late George W. Richardson, of Troy, N. Y., when he took up the subject, brought to his aid a ripe experience covering many years in active motor power service, and practical service has fully established the correctness of his design of balancing valves, as there are to-day three times as many locomotives running with Richardson or Allen-Richardson Balanced Valves, as all other forms of balanced valves combined.

Careful examination of the face of a worn slide-valve will show that the face wears in curved surfaces; the ends of the valve-face are worn alike in curves of the same radius, and the sides are worn in curves similar to each other but of much shorter radius than the curves at the ends. Any form of bal-

---

*These articles were prepared by Mr. M. C. Hammett, of Troy, N. Y.

ancing to successfully accomplish the desired results, and maintain tight joints for any length of time, must be constructed in conformity with the wear developed in the face of the valve, and this is just what is done in the Richardson construction, the packing bars are made in sections or separate bars, so arranged that each bar in its travel does not cross the line of travel of any other bar, thus giving each bar an independent place for wear on the balancing plate, and each bar corresponding to a portion of the valve face which wears in the same curve.

These bars are arranged relatively to each other so that they make and preserve steam-tight joints at their places of contact, and as the valve wears, each bar is free to wear uniformly throughout its entire section against the pressure plate above and maintain steam-tight joints with it.

The simplicity of construction, few parts, and reliability of this form of valve, has won the admiration of all mechanics, and been the means of the universal adoption of the valve in this country and abroad.

The use of the Richardson balancing is not restricted by any means to locomotives, but is extensively used in stationary engines and in very large marine engines, giving the best of results wherever used, rendering the handling of the engine much easier for the engineer, effecting a great saving in wear of every joint and bearing throughout the entire valve-motion

THE ALLEN-RICHARDSON BALANCED VALVE.

As shown by accompanying illustration, this is a combination of the Allen ported valve and the Richardson balancing. The Allen valve was not practical for locomotive service until suc-

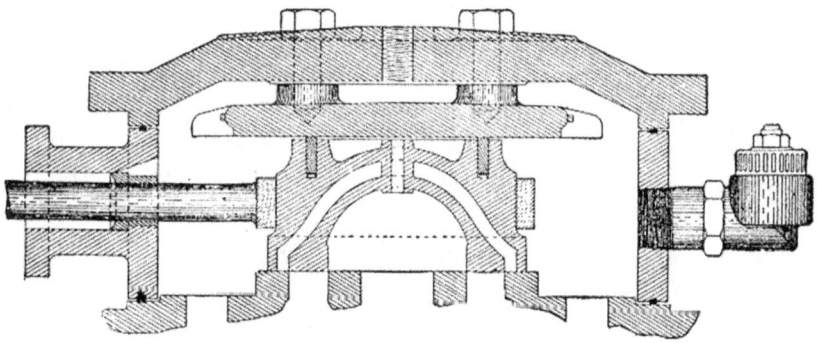

cessfully balanced by Mr. Richardson. Owing to the shell-like construction the excessive pressure when operating unbalanced, caused a springing of the valve face and very rapid wear of valve and seat, but with this form of balance, there are now thousands of our locomotives deriving the advantages of the supplemental Allen port.

THE AMERICAN BALANCE VALVE.

We have illustrated two forms of this valve, together with a sectional view of the valve in the steam chest. This design of balance was invented by Mr. W. J. Thomas of Sausilito, Cal., May 3, 1892. The improved T rings were invented June 21, 1898, by Mr. J. T. Wilson, General Manager of the American Balance Slide Valve Co., of Jersey Shore, Pennsylvania.

Experience has proved that this balance is one of the most successful of balance valves; this is due to the simplicity of construction, positive action and very large area of balance. The beveled ring is self-adjusting—no springs being required—and, therefore, not liable to get out of order.

This form of balance can be applied to almost every form of slide valve.

The American balance valve has been adopted by a great many of the principal railroads in this country, consequently we

give details of its construction, believing it will be interesting to a large number of our readers. It has also attracted the attention of foreign builders and is now in use upon many locomotives in other countries. It is claimed many thousands of these valves are in use, being used by 110 different railroads in this country and many foreign roads; also on marine and stationary engines. One of our illustrations show its application to an Allen ported valve. The claims made for this form of balance are:

1. *Self supporting*—when not under steam.
2. *Supported by steam*—when under steam.
3. *Automatic adjustment*—with or without steam.
4. *Absolute steam joints*—no waste from leakage at any time.
5. *Positive action*—impossible for ring to stick.
6. *No lateral wear*—ring moves as part of the cone itself.
7. *Permanency of cone*—cones retain original dia. and taper.
8. *Standard sizes*—rings interchangeable, old or new.
9. *Stock*—carried in stock for new work or repairs.

10. *Lathe work*—stock from the lathe is economy.
11. *Greater area of balance*—equaled by no other design.
12. *Simplicity*—always most desirable in machinery.

We herewith furnish a brief article concerning this form of balance from the pen of Mr. J. T. Wilson: "Our claims are simply an advantage in every essential feature in a balance slide valve. First of all, an absolutely steam-tight joint, not only when newly fitted, but all the time; second, greater area of balance. You will find our formula for figuring area of balance is used by no one else; they cannot use it and keep their valves on the seat; it has been tried by several ambitious railroad men. I should explain that we balance the valve in what is presumably its heaviest position, and with the steam pressure acting on the cir-

cumference of this taper ring you will observe that for the valve to lift it is necessary to force the cone up into this taper ring; and since the ring is held, by the steam chest pressure, from opening, the valve cannot lift without first overcoming the entire friction of the beveled face besides opening the ring against the chest pressure. We therefore pass over the lighter positions of the valve where a straight wall balance would allow the valve to go off its seat. It should not, however, be assumed that this taper will crowd the valve down on its seat, which would appear to be a natural conclusion to draw, from its manner of preventing the valve from leaving its seat. If the degree of taper was made great enough, 45 degrees, for instance, the action of the steam chest pressure on the circumference of the ring would, of course, wedge it in between the cone and the chest cover and exert an enormous pressure on the valve. In fact, it would not work satisfactorily at all, the friction would be too great. This, however, we do not do. We have experimented with this taper from 9 degrees to 24 degrees and have therefore found the proper degree of taper, by actual tests, with which the ring is certain to rise under all conditions and yet not crowd itself against the upper bearing more than necessary. This is demonstrated by the fact that we have rings that have run 190,000 miles with only 1-32" wear off their face. Next I would mention the facts that this form of balance is extremely simple, has no delicate parts, cannot be broken, has positive automatic adjustment, self-supporting feature of the ring, and entire absence of springs. Next we would know the cost of construction. It is cheaper to construct than any other valve in the market. It can be maintained at a very trifling cost; at about 90 per cent less than the strip balance which has been so commonly used. I might state in explanation of this that the only repairs necessary on the American balance is to put on a new ring when the old one has worn out from the top downward; new rings are 1" deep, they can easily wear $\frac{3}{8}$" and still adjust themselves; now to wear a ring $\frac{3}{8}$", assuming of course that it is made of proper metal, will require from four to eight years in continual service. When you take the old ring off the cone and place a new one from stock on the old cone, your balance is just the same as when absolutely new. This is explained by the fact that since the steam pressure on the circumference of the ring holds it firmly against the beveled face of the disc, or cone, while in operation under steam (its own tension holding it when not under steam), there is absolutely no lateral wear on the ring or disc, a new ring therefore fits an old disc at any future time. Now, since these rings are made from standard gauges, which are used on the lathe in place of a caliper or rule, and since they are all lathe work (it does not require more than twenty minutes' hand work to fasten on the L-shaped piece for covering the cut of the ring) you will therefore readily see that the expense in taking a stock ring and renewing the balance is small when compared with the work required on other designs. We would suggest that all disadvantages of other valves are removed in this valve by the taper feature of the ring; as a further illustration of this we would state that while a gauge usually frightens a workman, our gauges make him smile, because they relieve him of responsibility in the only important particular of the entire construction, i. e., with our gauge he can get the taper of the ring and the cone exactly alike. Now, in diameters we work to a line which you will readily understand is plenty accurate enough; 1-32" in the diameter of a ring either way from the sizes called for would not in anywise interfere with the service of that valve, since the ring is turned $\frac{1}{4}$" smaller than its working diameter. The ring is expanded over the cone and thus receives a tension which makes it self-supporting when not under steam; the steam, of course, on its circumference supports it when in operation. We beg to call the reader's attention to the full claims covered by the award granted us at the World's Columbian Exposition; they boiled it down to a few words, yet we believe they properly express it. And when we have the advantages, as shown in their award, we think we have about all the advantages possible. They say our award was granted us for *'Simplicity and excellence of design and large balancing area'*; these we certainly have and we can trace them all to the taper.

The latest improvement in this form of balance is the T ring; the flanges on the top of the ring give an extra width for wearing surface, which prolongs the life of the ring. The top face of the ring is provided with a small groove for oil. The L shaped joint plate forms joints both on the beveled face and at the top of the ring. The 'outside rim' or flange extending outside the taper ring is to prevent pieces of the ring from falling in the path of the valve in the event of accident to the ring.

SINGLE "DISC" BALANCED VALVE.

Always use the single balance where chest room will permit it, as one ring and disc is simpler than two.

For length of steam chest for single balance add the extreme travel of the valve to the outside diameter of disc, and to this sum add not less than $\frac{1}{2}$" for clearance—$\frac{1}{4}$" at each end of chest. If a little more clearance is desired the rims of disc may be cut $\frac{1}{8}$"—just flattened on two sides in line of valve travel. But in no case are they to be cut beyond their inside diameter.

If sufficient clearance cannot be obtained by cutting the rims $\frac{1}{8}$" each side in line of valve travel, then double balance must be used.

The ring must be protected by the disc and when figuring outside diameter of ring $\frac{1}{4}$" must be added for the joint plate and the ring must be figured when expanded on the cone until its top face is flush with top of cone, or at its greatest possible diameter.

DOUBLE "DISC" BALANCED VALVE.

When the steam chest is too short to leave clearance for the outside diameter of disc or cone of single balance at extreme travel of valve, then double balance is used. If the yoke fit (or box) of valve is large enough, two cones are cast on the valve, if the yoke fit is not large enough to cast cones on, then two discs are used. If the distance across the two discs, when they are side by side on top of valve, is greater than width of steam chest, the rims on each disc may be cut $\frac{1}{8}$" at center of valve, drawing the discs $\frac{1}{4}$" closer together; and if more clearance is necessary the rims may also be cut $\frac{1}{8}$" at ends of valve, giving $\frac{1}{4}$" more, or a total of $\frac{1}{2}$". But in no case shall the rims be cut more than $\frac{1}{8}$" or to their inside diameter.

If discs thus cut will not clear the sides of chest smaller balance must be used.

"DISC" BEARING ON VALVE.

In all cases where possible the height adjustment should be made by lowering the chest cover, or bearing plate, but when chest cover cannot be lowered the discs may be raised. When it is found necessary to raise the disc on the valve longer bolts should be used and the liners placed between the disc and the valve must be true, and large enough to give a solid bearing for disc on the valve. If found necessary to raise the disc to clear the top of valve yoke, the same rules must be observed. The bolts which fasten the disc to the valve should be steam tight on threads and steam tight under the heads, a copper washer being used under the heads, forming a bolt lock. The interior of each disc or cone is relieved to the exhaust cavity of the valve.

In "cone" balance holes are drilled through the top of valve, but in "disc" balance the relief holes pass through the bolts— $\frac{1}{4}$"—through each bolt.

SINGLE "CONE" BALANCED VALVE.

This style of valve must be cast flangeless if a valve yoke extending all around the valve (as in locomotives) is used, but need not be flangeless when made for center rod to drive the valve (as in stationary engines). In case of the locomotive yoke we recommend the yoke to be carried on the steam chest at the ends of the valve. Where old chests have rubbing strips wide enough they can be planed on top and the yoke allowed to ride on them, and in new work this can be done cheaper than to put on a front carrying horn and is more efficient than to support the yoke on the valve stem packing the valve itself. A valve need not be flangeless to thus support the yoke, it can be carried with any valve, and it insures the free upward movement of the valve at all times, which is very essential in obtaining the best results.

OUTSIDE RIMS.

The outside rim on disc or cone is merely a safeguard to the ring in case of accident—it performs no other duty. The required inside diameter of this rim must allow the ring to be expanded on the cone until the top face of the ring is flush with top of cone and still clear the $\frac{1}{8}''$ joint plate on the outside of ring. In single balance the rims may be cut $\frac{1}{8}''$ front and back, giving $\frac{1}{4}''$ more clearance, when the disc runs too close to steam chest at full travel of valve.

In double balance the rim of each disc may be cut $\frac{1}{8}''$ so as to draw the discs $\frac{1}{4}''$ closer together at center of valve, and if more clearance at sides of chest is required the rims of each disc may be cut at ends of valve also. The distance across the two discs can thus be shortened $\frac{1}{2}''$.

Note.—In no case shall the rim be cut more than $\frac{1}{8}''$. The two cones where they come together at center of valve in double balance must not be less than $\frac{3}{4}''$ apart at the bottom.

If discs thus cut will not clear the steam chest, then smaller balance must be used.

HEIGHT ADJUSTMENT.

When the valve is in position and the chest cover has been screwed down there must be $\frac{1}{8}''$ between the face of the bearing plate (sometimes called balance plate) and the top of disc or cone.

The rings are bored for this position and in this position have their proper tension. This allows the valve to lift off its seat $\frac{1}{8}''$, which it will do as soon as steam is shut off while the engine is in motion, provided it is not held down by the valve yoke.

The valve yoke must not interfere with this upward movement of the valve.

TENSION ON RING.

Rings are all bored smaller than the diameter at which they are to work; therefore when a ring is set on its proper cone it will stand higher than its working position.

The face of bearing plate must not be closer than $\frac{1}{8}''$ to top of cone after chest cover has been screwed down. In placing the cover in this position the ring is expanded over the cone until its inside diameter at bottom is the proper balancing diameter.

Owing to the natural elasticity of the ring and its expansion over the cone, a tension is placed on the ring, the action of which is (the same as the steam pressure) to close the ring on the cone, which necessarily moves upwards.

The ring is therefore self-supporting and self-adjusting. All rings are interchangeable on discs and cones of respective sizes whether standard or special.

BALANCES.

American balances are known under the following heads, one valve being balanced in each case:

Single "disc" balance, i. e., one ring and one disc.
Double "disc" balance, i. e., two rings and two discs.
Singe "cone" balance, i. e., one ring with cone cast on valve.
Double "cone" balance, i. e., two rings with cones cast on valve.

CYLINDER RELIEF.

The valve shall always be free to lift $\frac{1}{8}''$ off its seat, to allow the free passage of air from one end of the cylinder to the other, between valve and valve seat, when engine is running without steam. The tops of all American balance discs, or cones, show a polish, giving positive evidence of their contact with the bearing plate or cover, and that they therefore do float.

Our explanation is: At the first stroke of the piston, after engine has been shut off, air is compressed in one end of the cylinder while the valve is traveling a distance equal to its outside lap; at an early stage of this compression the valve is thrown off its seat and the escaping air rushes under the valve into the opposite end of the cylinder to relieve the suction which is taking place in that end; this operation is repeated so rapidly that the valve is kept floating until a slow speed has been reached.

Sufficient air is always drawn in, through the (lift valve) cylinder cocks, and exhausted through the exhaust port, to keep the current of air in the direction of the stack, but not enough to "fan the fire." This affords the most perfect cylinder relief and vacuum valves in steam chest are not a necessity.

FORMULA OF BALANCE USED.

AREA OF BALANCE FOR PLAIN VALVES.

Area of one steam port, two bridges and the exhaust port, plus 8 per cent if for single balance and plus 15 per cent if double balance.

EXAMPLE FOR SINGLE BALANCE.

S. port $1\frac{1}{4}''$+Bridge $1''$+Ex. port $2\frac{1}{2}''$+Bridge $1''=5\frac{3}{4}''\times16''$ $5\frac{3}{4}\times16=92$ sq. in.; 8 per cent of $92=7.36$; $92+7.36=99.36$. Area. Nearest dia. for $99.36=11\frac{1}{4}=$Dia. for ring.

EXAMPLE FOR DOUBLE BALANCE.

S. port $1\frac{1}{4}''$+Bridge $1''$+Ex. port $2\frac{1}{2}$+Bridge $1''=5\frac{3}{4}\times16=92$ sq. in.; 15 per cent of $92=13.80$; $92+13.80=105.80=$Area. $105.80\div2=52.90$, or area of each ring. Nearest dia. for $52.90=8\frac{1}{4}''$, dia. for each ring.

ALLEN PORTED VALVES.

For Allen valves use same formula as above; then from the area derived subtract the area of one side of the Allen port. Example: $1\frac{1}{2}''+1\frac{3}{8}''+3''+1\frac{3}{8}''=7\frac{1}{4}''\times17=123.25$ sq. in.+15 per cent$=141.73$ sqr. inches. Allen port$=\frac{1}{2}\times17=8\frac{1}{2}$; $141.73-8.50=133.23\div2=66.61$, area of each ring.

THE BARNES BALANCED VALVE.

The following description and illustration of the Barnes balanced valve is taken from *"Locomotive Engineering:"* A balanced valve having some points with a strong flavor of originality is that designed and patented by Superintendent of Motive Power Barnes, of the Wabash Railroad, and illustrated herewith. The full-sized part of a longitudinal section will perhaps give a clearer conception of the balance part of the device than will a first reference to the details. This view shows a cast-iron frame surrounding the valve, with a recess equal to the width and depth of the balance strips on its inner face.

There are four balance strips $\frac{5}{8}\times1\frac{3}{4}$ inches, carried in position by the frame, which is $1\frac{5}{8}\times2\frac{1}{2}$ inches in the over-all dimensions of its cross-section. Both the frame and strips are cored for lightness. The combination is held up to the balance plate by four helical springs, made of German silver wire, 0.125 inch diameter. It is seen that the balance strips have a very deep bearing between the valve and frame, and since the strips are carried by the frame, there must be a constant depth of support to them on the outside face, no matter how much wear takes place, nor what the lift of the springs. This condition of things tends to reduce the liability of cocked or broken strips, as has been found in service. The springs placed at the ends of the frame would seem to exert a more equable pressure on all the strips than is possible with a long spring under each individual strip, for the reason that it is a delicate undertaking to attempt

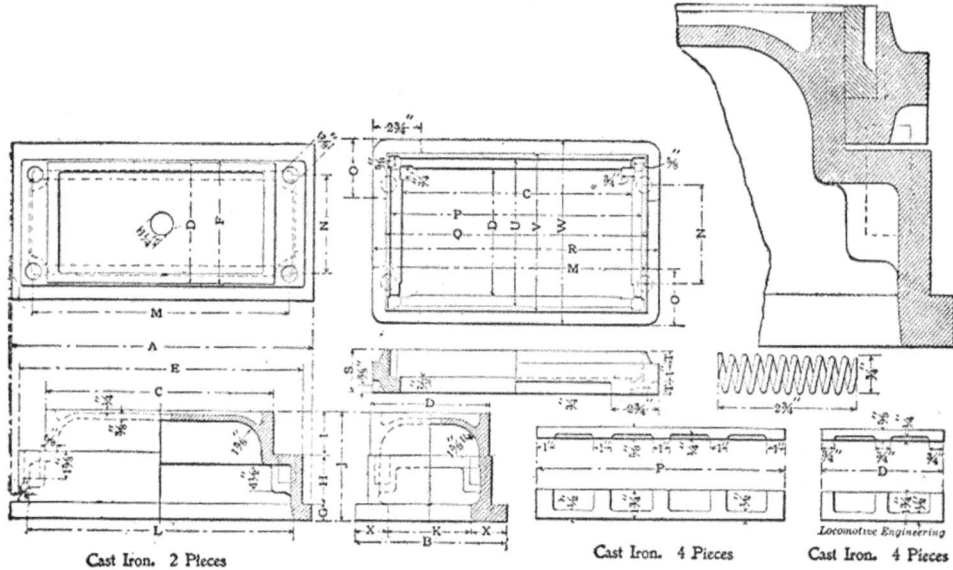

Cast Iron. 2 Pieces   Cast Iron. 4 Pieces   Cast Iron. 4 Pieces

to make four flat springs of the same degree of elasticity. On these grounds it is not unreasonable to expect a continuance of the good results so far shown by this style of valve balance. The dimensions shown are for a ten-wheeler.

THE McDONALD VALVE.

The points of advantage in this valve are as follows: That the pet cock on top of steam chest cover is always open to the atmosphere, so that should the joint leak the engineer would see

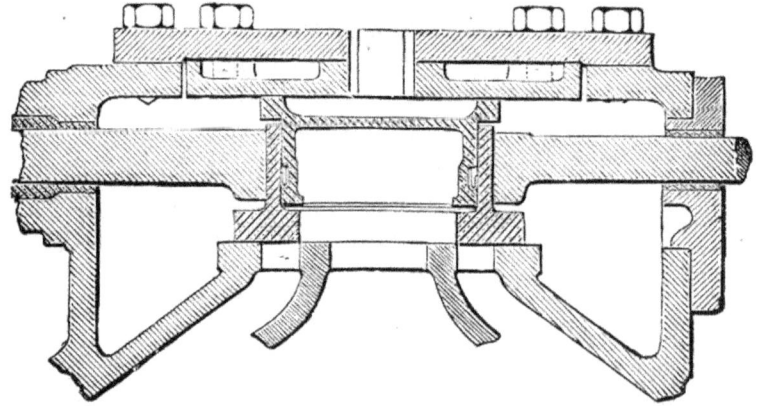

the steam escaping through the pet cock, which is sure evidence that the valve needs attention; he can then close the pet cock, and the valve will work the same as a simple D valve. The U-shaped packing strips maintain a satisfactory and tight joint, and the flat steel spring beneath the valve is of the simplest form. This valve has recently been introduced into this country. Mr. McDonald, the inventor, is a mechanical engineer at Yokohoma, Japan, and has had these valves in service some two or three years, and it is claimed they are giving excellent results.

BRIGGS BALANCED SLIDE VALVE.

The inventor, Mr. R. H. Briggs, Jr., of Amory, Miss., describes this valve as follows:

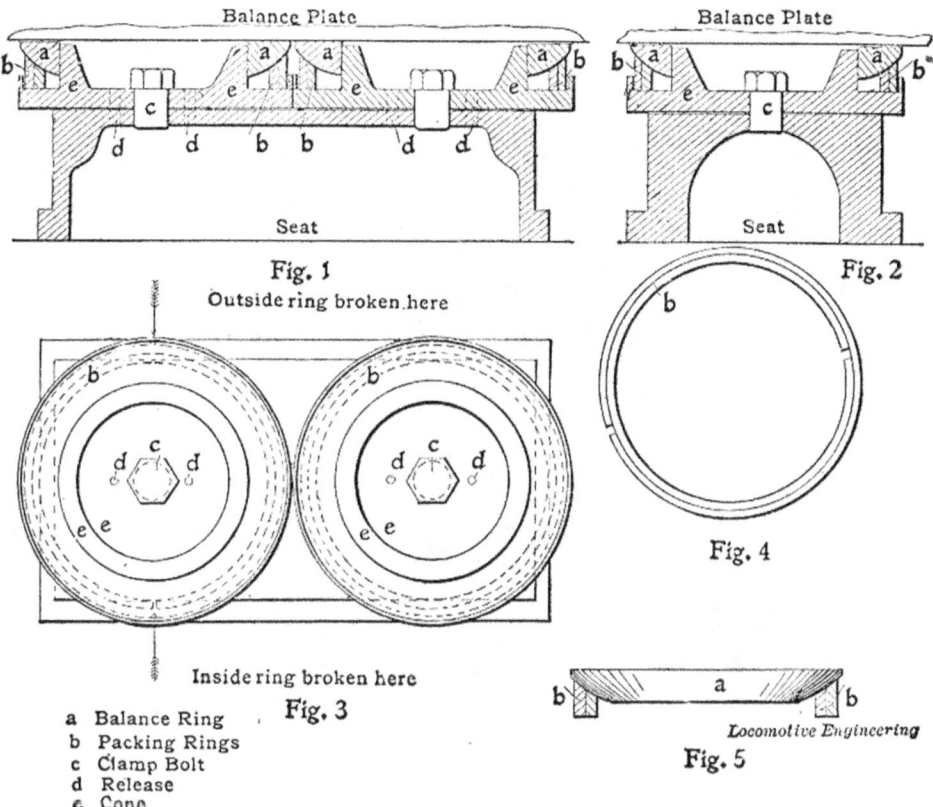

a Balance Ring
b Packing Rings
c Clamp Bolt
d Release
e Cone

BRIGGS BALANCED SLIDE VALVE.

"You can readily see that it will require very little explanation as to the operation of this improved balance. This balance consists of four parts—cone E, two packings rings B, and joint ring A. The packing rings are cut as shown in Fig. 4. To break the joints these rings are pinned together. To prevent the cuts working around, they are also ground on cone E, to prevent any leak at bottom of packing ring. They have a ball joint on top with joint ring A perfectly ground to them. Joint ring A is a nice, neat fit on cone E. Now you can very readily see that it is impossible for this valve to blow, as the greater the steam pressure on the outside of packing rings B the tighter ring A will be on the balance plate. I give these rings only 1-32" compression. I have this valve in one of our engines, and she is as tight as a corked-up bottle. Another advantage this balance has—it balances to outside diameter of outside packing ring B."

THE BROWN BALANCED SLIDE VALVE.

The object of this form of balance is to dispense with the stuffing box, the gland, and the valve stem packing, also the steam chest cover, if found desirable to do so. It is intended for either locomotive or stationary engine. It is the invention of Mr. Daniel H. Brown, of McComb, Mississippi. The following is the inventor's description of the device:

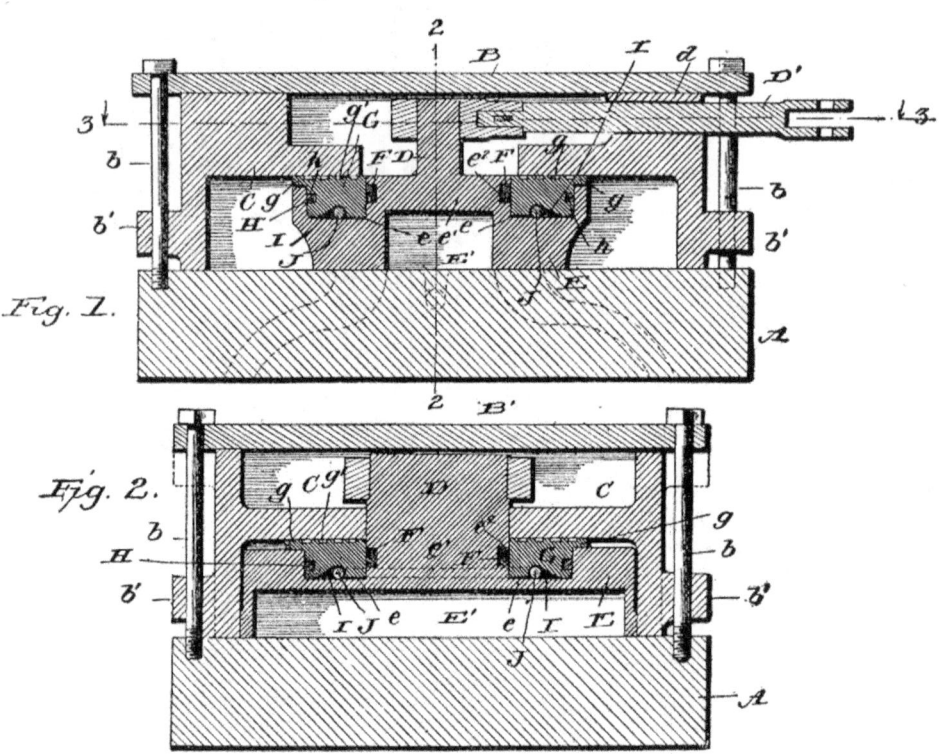

"Figure 1 is a vertical section through the valve and steam chest in the direction of movement of the valve. Fig. 2 is a vertical section taken at right angles to that of Fig. 1—on the line 2—2 of Fig. 1. Fig. 3 is a horizontal section on the line 3—3 of Fig. 1, looking in the direction of the arrows. Fig. 4 is a plan view of the valve with a portion of the same broken away.

"Referring now to the details of the drawings by letter, A designates a portion of the cylinder having the usual steam passages as indicated by dotted lines in Fig. 1; or it may be the base-piece upon which the steam chest is supported.

"B is the steam chest mounted upon the part A in the usual or any desired manner; in this instance being shown as secured in position by bolts b which engage threaded openings in the part A and pass through lugs or a flange b' on the lower portion of the steam chest as shown in Figs. 1, 2 and 3, the said bolts serving also to secure the cap or cover B' in place when one is employed.

"The steam chest is divided into two chambers or compartments by the horizontal wall or partition C which is provided centrally with a rectangular opening C' as shown best in Fig. 3. The balanced slide valve works in the lower of these compartments and the valve stem in the upper one, the valve having a vertical extension working through the opening in the diaphragm or partition to connect with the said stem as seen best in Fig. 1. D is the vertical extension and D' is the valve stem connected therewith in any suitable manner and working

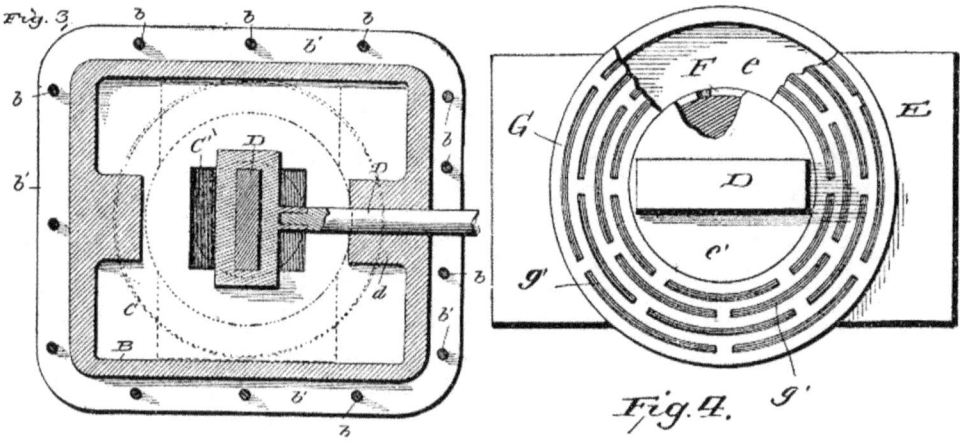

through an opening in a thickened portion d of the wall of the steam chest near the upper end thereof as seen in Figs. 1 and 3.

"E is the balanced slide valve. It has the chamber E' upon its under face and from the same projects centrally the vertical extension D above described. This valve is formed with the circular recess or chamber e and the neck e' of the valve is formed with an annular grove e^2 in which is fitted a plurality of split expansion rings F as shown in Figs. 1 and 2.

"G is an annular friction plate fitted within the chamber e as shown, having a surrounding flange g which rests upon the top face of the valve as seen best in Fig. 2 and the upper face of this friction plate is formed with arc-shaped channels g' as seen in Figs. 1, 2 and 4, while the outer periphery of the said plate is formed with an annular groove h in which is located a split expansion ring H as shown in Figs. 1 and 2.

"The under face of the friction plate or disc or ring is formed with an annular or substantially circular inclined groove I in which is located a splint ring J as shown in Figs. 1 and 2. This ring serves to seat the main ring or friction plate G.

"It will be seen that from the construction above described no packing for the valve stem is necessary and the cover to the steam chest may even be omitted in some cases. The valve reciprocates in the usual manner over the ports of the cylinder and the stem being in an independent chamber there is no leakage. The expansion rings keep the friction plate or main ring properly seated and prevent leakage therearound.

"Modifications in detail may be resorted to without departing from the spirit of the invention or sacrificing any of its advantages. The opening in the top of the steam chest may be circular as indicated by dotted lines in Fig. 3 instead of square as shown in full lines. In this case no steam chest cover will be necessary and I can use simply a round lid or cap. So also the flange or lug b' at the lower edge of the steam chest may be dispensed with and the lug provided at the upper edge as indicated by dotted lines in Fig. 2."

THE MALONE BALANCE VALVE.

This form of balance is the design of Mr. Frank P. Malone, of Evanston, Wyoming, who describes his invention in the following words:

"The object of this form of balance is to provide an improved balance ring and packing joint whereby contact between the surfaces of the balance ring and valve body will be maintained steam tight, and the contact of the balancing ring with the cover of the steam chest will also be steam tight; also to improve the construction of the slide valve and balance ring.

"Figure 1 is a plan of the valve and its packing rings. Fig. 2 is a transverse section through the steam chest and packing rings, showing parts in elevation. Fig. 3 is a section, and partial elevation, at right angles to the section Fig. 2, parts being omitted. Fig. 4 is a plan and section of the bull ring. Fig. 5 is a plan and elevation of the balance ring. Fig. 6 is a plan and elevation of one of the packing rings; and Fig. 7, an enlarged sectional detail.

"The numeral 1 indicates the steam chest, and 2 the casing thereof. These are of a common construction. So much only of the steam chest and casing are shown as are necessary to give an understanding of the invention. The cap or cover 3 of the steam chest is held down by bolts, or in other suitable and usual way. The valve 4 has its seat on the floor 5, and bears against the plate 6, which plate may be adjusted to parallelism with the valve seat, or to take up wear, by means of set screws or bolts 7 which pass through the cover of the steam chest. As shown in Figs. 1 and 3, the valve has two bull rings, 10, attached to its upper surface, by screw bolts 11, which bolts have longitudinal passages for steam. The number of rings, however, may be proportioned to the size of the valve. Each bull ring is in the form of a cup, with holes in the bottom for the passage of the retaining bolts 11. In the upturned outer flange of the bull ring there is a groove 12, in which groove two or more spring packing rings 13 are inclosed. Outside the flange of the bull ring, and inclos-

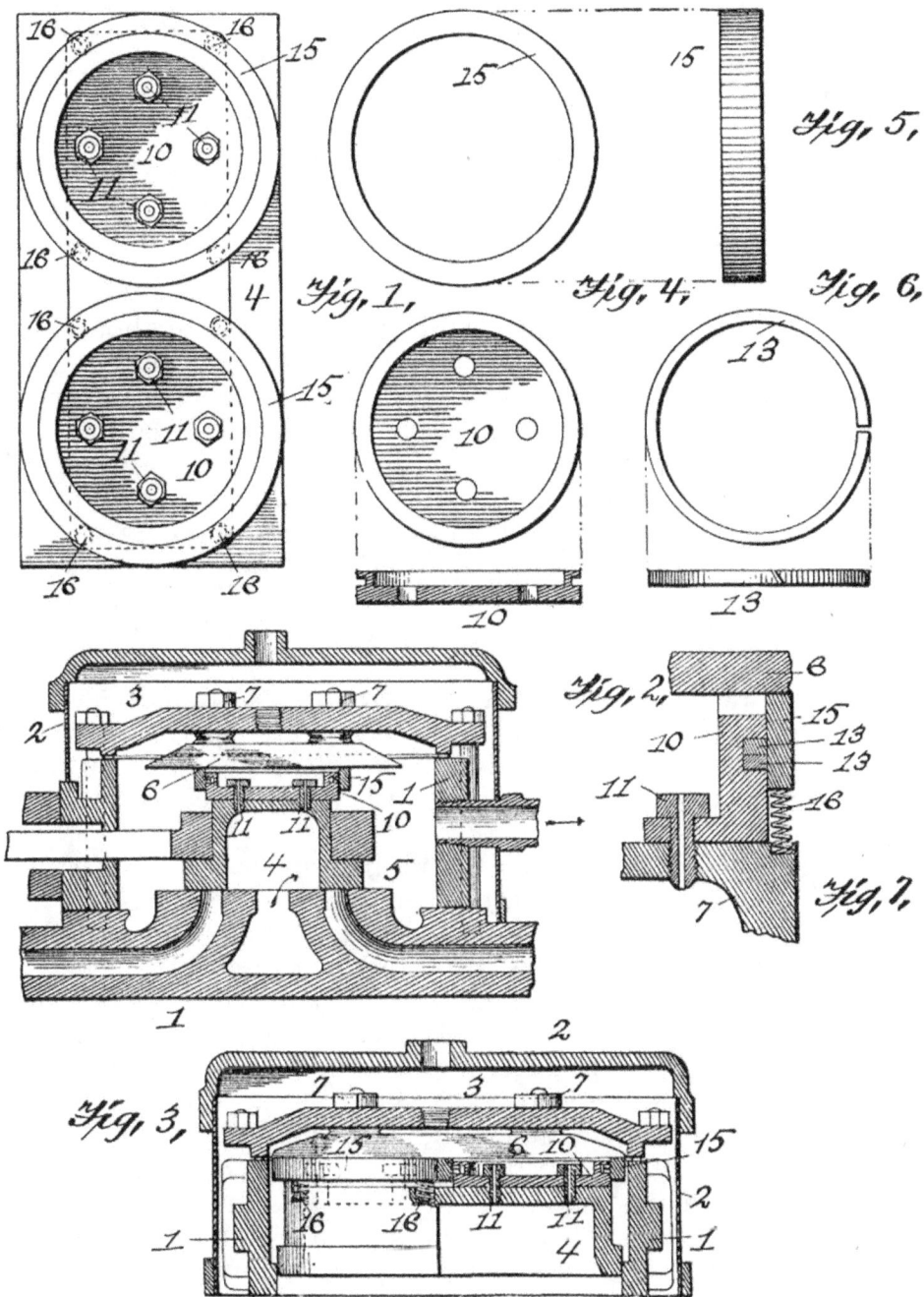

ing the packing rings I place the balance ring 15. The spring rings 13 close tightly against this packing ring, and form a steam tight joint between the packing ring and the bull ring. The balance ring is held up against the plate 6 by a number of spiral springs 16, which springs bear on the valve, and on the lower edge of the packing ring outside the bull ring. The balance ring is thus held up to its seat against the plate 6, but will yield slightly to compensate for wear, or for the lack of parallelism between the valve seat and the plate 6, or for irregularities in the valve itself. The course of the steam in the ports is as usual. The steam from below the valve passes through bolts 11 to the inside of the bull ring and balance ring, but is prevented from escaping by the tight packing rings. The area inside the balance rings should equal the pressure area below the valve, in order that the valve be balanced."

THE NICOLSON BALANCE VALVE.

The novel feature of this form of balance is the method employed to secure the balance strips and to form a steam-tight joint. We have secured the following description of this valve from the inventor, Mr. H. A. Nicholson, of Pocatello, Idaho:

"This invention relates to a union top slide valve and has for its object to provide simple means to permit the steam to pass from the exhaust chamber and lubricate the steam chest cover, to make the several parts adjustable to accommodate the various conditions, and which can be made to balance any main slide valve and can be lined up when necessary by placing liners between the parts.

"In the drawings, Figure 1 is a longitudinal sectional view of the improved device. Fig. 2 is a horizontal sectional view, on the line yy, Fig. 1, of a part of the device and a top plan view of the remaining portion of the same. Fig. 3 is a detail sectional view of a part of the device.

"Referring to the drawings, the numeral 1 designates a cast iron plate which is bolted to the top of any form of main slide valve as at 2 by five bolts 3 and 4, the bolt 4 being in the center and has drilled therein through the center of the head thereof, a hole 5 which is intersected by a smaller hole 6 through the top or head of the bolt as well as by another hole 7 also formed in the said head of the central bolt at a right angle to and intersecting the larger hole 6 and which permits the steam to pass from the exhaust chamber and lubricate the steam chest cover. The said plate 1 has four cast iron strips 8 thereon two of which are beveled as at 9 and are held to beveled sides 10 of the plate by four bolts 11 which pass through the said strips and the bearings of the plate and which have nuts 12 on their ends which may be tightened so as to regulate the tension of the strip on the plate. The said bolts 11 are encircled by coiled springs 13, arranged inside the plate and are held in place by the said nuts 12. The side strips are also held in position by cross strips 14, the side strips being protected by the outside wall of the plate through which are drilled holes 15 to give a direct steam pressure against the strips and force the latter against the beveled bearings of the plate and also against the

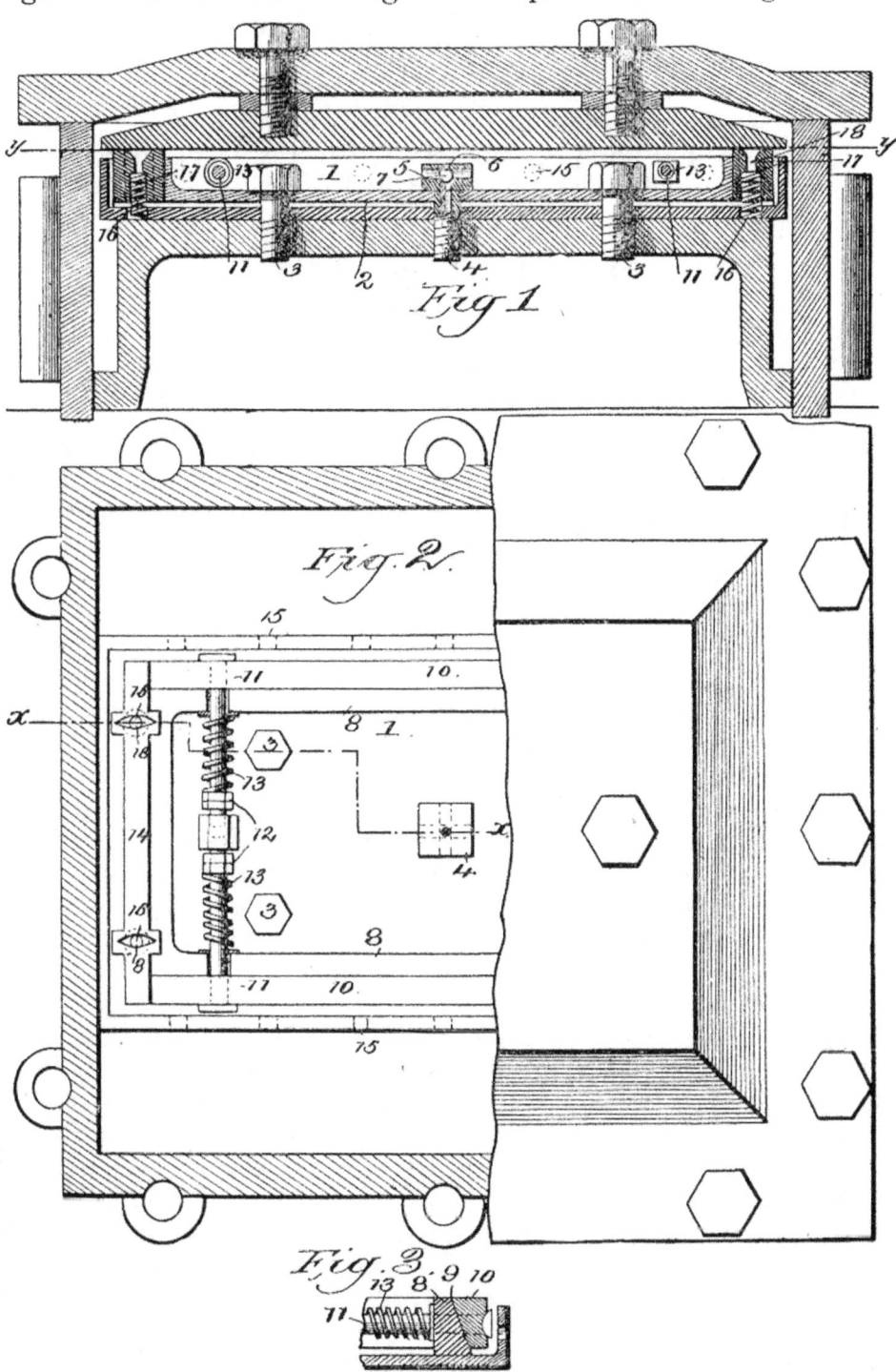

steam chest cover. The cross strips 14 are in the form of a double cross and are held against the steam chest cover by coiled steel springs 16 which are located in holes 17 drilled through the center of each cross strip and through each cross. The holes for the springs are intersected by smaller holes 18 for the purpose of lubrication. These crosses from a sufficient bearing against the steam chest cover and are also fitted in the ends of the plate so that the proper position of the strips is always maintained.

"It will be obviously apparent that many minor changes in the construction and arrangement of the several parts might be made and substituted for those shown and described without in the least departing from the nature or spirit of the invention."

SMITH'S ROLLER BALANCED VALVE.

We are indebted to "*Locomotive Engineering*" for this description of a new form of balance valve, invented by Mr. B. W. Smith, of Van Wert, O., and C. F. Hammond, of Frankfort, Ind.:

"The invention consists principally of a rocking valve provided with a cavity adapted to connect the interior of the steam chest with the cylinder port, and the latter with an exhaust chamber.

"Fig. 1 of the accompanying drawing is a sectional side elevation of the improvement.

"Fig. 2 represents the valve as incased in the steam chest.

"Fig. 3 is a sectional plan view of the same, with the valve stem and link removed, also the exhaust chamber bridge.

"Fig. 4 is a sectional view and connection link.

"The cylinder is provided with the usual inlet ports and the exhaust port; the valve seats being segmental, adapted to receive the cylindrical valves D and D^1, respectively, as shown in Fig. 1, having their ends journaled in the sides of the valve body and in the cap. The exhaust port opens into a chamber formed within the valve body and between the two valves D and D^1; the top of the chamber being closed by a transversely extending bridge forming part of the cap, and having its side edges bearing on the peripheral surfaces of the cylindrical valves D and D^1.

"The valves D and D^1 are formed in their peripheries, with cavities or recesses D^2 and D^3, respectively, arranged in such a manner that when the valves are rocked in their bearings, then the cavities alternately connect the interior of the steam chest with the corresponding cylinder port, and the ports with the exhaust chamber. In order to impart a rocking motion to the cylinder valves D and D^1, the following device is provided: On top of the valves are secured upwardly extending lugs which carry a transversely extending pin on which is pivoted a link engaging at its forward end the rounded portion of a pin carried in the lugs. The pin at valve D is formed with the squared portions fitting in vertically extending recesses arranged in the head of the valve stem.

"By reference to Figs. 1 and 2, it will be seen that the major portion of the peripheral surface of the valves D and D¹ is at all times exposed to live steam in the steam chest, so that the valves are completely counterbalanced. Since the above description appeared in *Locomotive Engineering* decided improvements in the construction of this valve are claimed. The segmental valve body has been extended upwards on the outside of each roll to a height almost equal to that of the roll, and a supplementary port has been placed in each roll, thereby permitting of a double

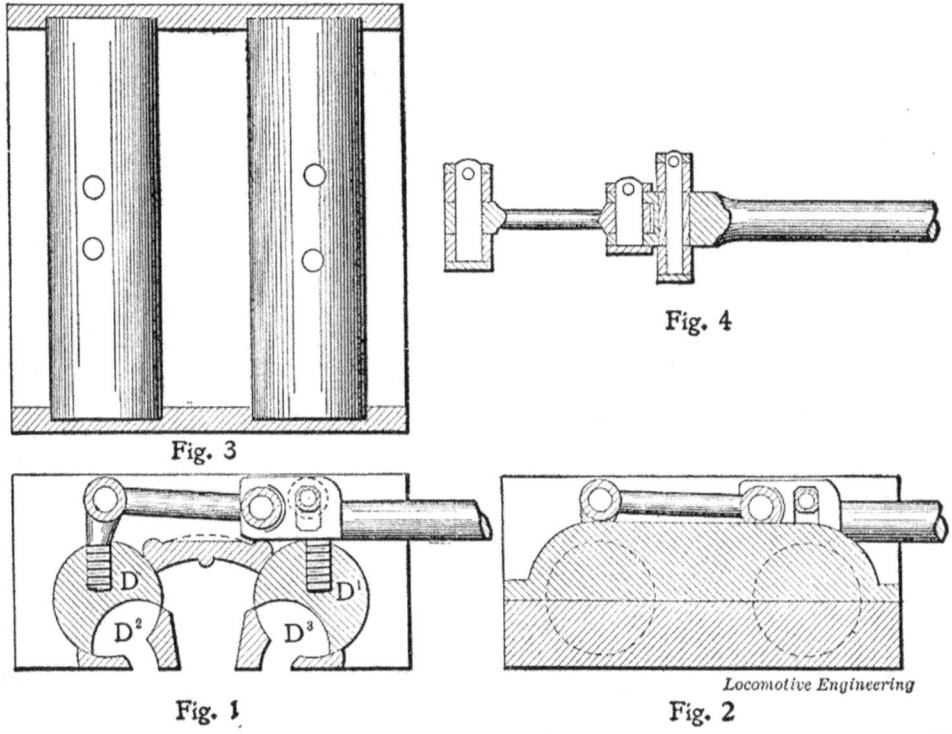

SMITH'S ROLLER BALANCED VALVE.

port opening. The blue prints were not received in time to show the improvement in this book. The claims made for this valve are: A more perfect balance, double steam port opening to cylinders and increased facilities for a free exhaust. While the first cost of this valve may be considerable, it has some very good qualities.

"There are two engines owned by the C. A. & C. R. R. now being fitted up with this valve at Mt. Vernon, Ohio, and another at the C. N. shops at Van Wert, Ohio, a five inch roll being used."

VANDEVENTER'S ROLLER BALANCE VALVE.

These figures show one form of a roller balanced valve. There are various forms of these valves in use. This design is the invention of Mr. Roland E. Vandeventer, of Mount Sterling, Ill.

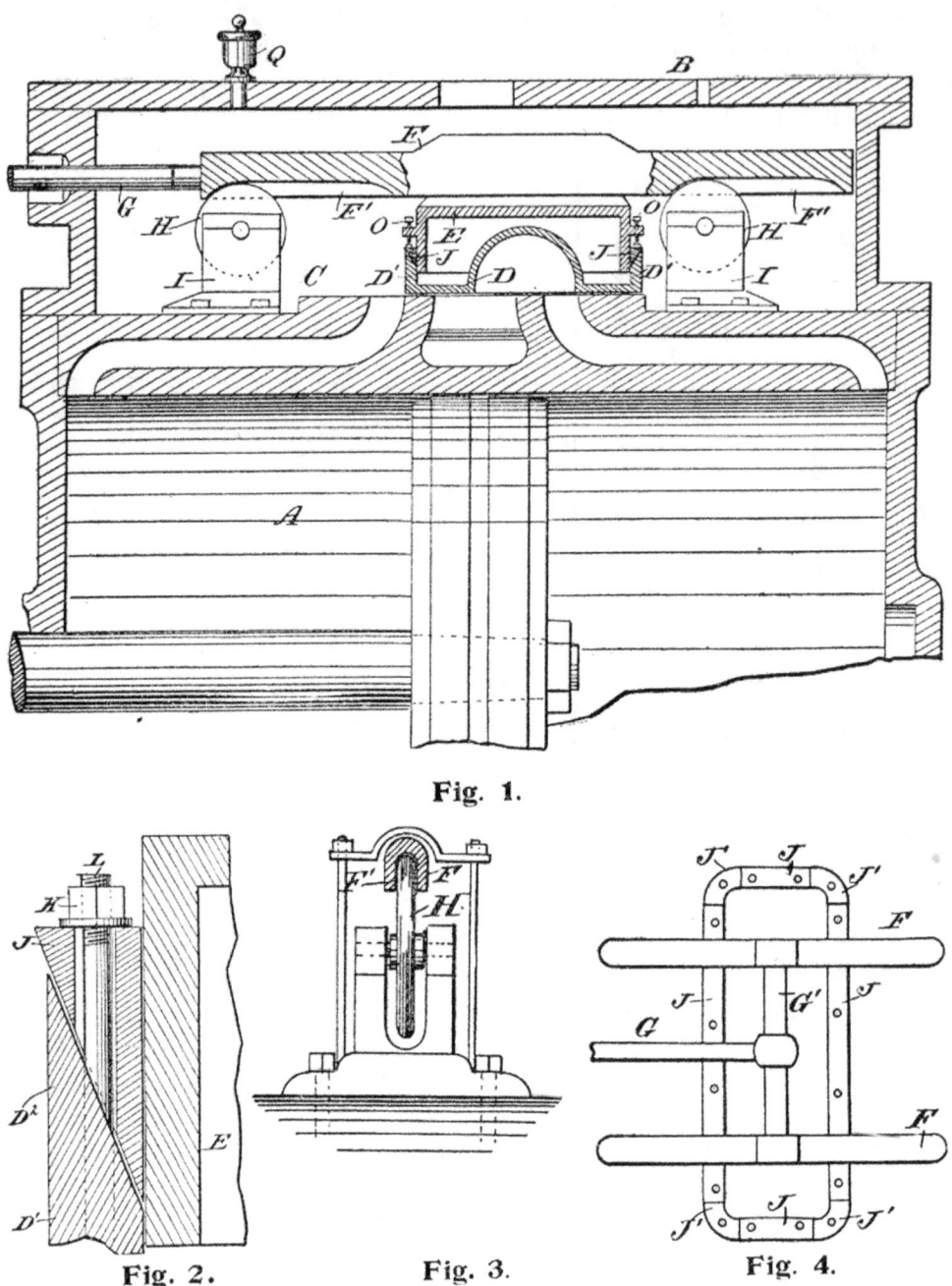

Fig. 1.

Fig. 2.　　Fig. 3.　　Fig. 4.

These valves are somewhat antiquated, therefore we do not feel justified in describing its details. We have shown enough of the valve to give the reader a fair knowledge of the principle followed in all roller valves. Fig. 1 is a sectional side elevation of the valve in the steam chest. Fig. 2 is an enlarged sectional side elevation of a part of the valve. Fig. 3 is an end elevation of the roller and part of the yoke. Fig. 4 is a plan view of the support or carrier.

THE MARGO VALVE.

The Margo balanced valve was one of the first forms of balanced valves. It had two small discs and small packing rings (similar to air pump packing rings); the steam pressure holding the disc up against the pressure plate. Much trouble was experienced with this form of valve, owing to a gumming up of its parts which would cause the disc to stick. Very few of these valves are now in use.

POWER REQUIRED TO MOVE A VALVE.

To determine the power required to move a valve multiply the area of the valve face by the steam pressure upon it less 1-3 allowed for back pressure from the steam port and exhaust port. The friction between two smooth surfaces well lubricated varies from 1-10 to 1-14 of the pressure (the weight of the valve being so slight it is seldom considered. Friction is the resistance which two contracting surfaces have to being moved one over the other and is of three kinds: Sliding, rotation, and liquid). For example: If a valve face measures 10x20 with 120 pounds' pressure, proceed as follows:

10x20=200x120=...........24,000
            8,000 less one third,
            ———
Divided by the friction, 10....16,000 1,600 lbs. power required.

This is an enormous strain on the valve gear which will cause it to wear rapidly. Of course this amount of power is not required of the engineer to reverse the engine, it is reduced according to the principles of leverage. Yet the power required of him is unnecessarily great. For the benefit of the reader we have given the proportions used by the American Balance Valve Co. for balancing their valves. See page 83.

HOLE IN THE TOP OF BALANCED VALVES.

You will notice there is always one or more small holes drilled through the top of all balanced valves. This hole is absolutely necessary, as it permits the steam which leaks through the strips, or rings, to escape through the exhaust. Without this hole the valve will lift off the seat, and the steam which would leak through the strips would practically nullify the good qualities of the balanced valve.

THE VACUUM RELIEF VALVES.

When a locomotive is running and the steam is shut off, a partial vacuum will be formed in the steam chest, causing the valve to chatter and thus ruin its mechanism, besides drawing cinders into the steam chest. To overcome this evil a vacuum relief valve is used with all balanced valves, and should be used with the plain slide valve. It admits air into the chest when the steam pressure is shut off and thereby prevents a vacuum being formed. No balanced valve must be expected to work satisfactorily without a vacuum relief valve.

LINK MOTION.

Steam has been used for heating purposes "so long that the memory of man runneth not to the contrary." But the discovery of the "power of steam" by James Watt has revolutionized the mode of travel and traffic both on land and on sea all over the world, and of all the discoveries of modern times, it is admitted to be the most valuable to mankind.

And yet the power of steam would have been of little practical benefit to the human race without the mechanical means of controlling and utilizing it. Devices of all kinds commonly called "valve gears," have been invented for this purpose; among them were many valve gears for locomotives. The best and most important of these were the various Link Motions. A link motion is a mechanical device by which the crank rotation of an engine may be reversed at the will of the engineer with only the loss of time required to overcome the momentum of the moving parts, and by which a like speed and power may be developed in a reverse direction.

It is applied principally to locomotives and marine engines where the speed and power required is variable and the motion is at one time direct, at another reverse. Many designs of Link Motion have been invented but the most important ones may be divided into four different classes:

1. The Shifting Link Motion.
2. The Stationary Link Motion.
3. The Allen Link Motion.
4. The Walschaert Link Motion.

Of these the first form, the Shifting Link Motion, is the standard link motion in this country and, excepting some slight modifications in the mode of suspension remains unchanged by the accumulated experience of half a century. Concerning the origin of this form of link motion Mr. Angus Sinclair in his work on Locomotive Running and Management says: "There is no doubt but the link was invented by Mr. Wm. T. James of New York, a most ingenious mechanic who also invented the double eccentrics. He experimented a great deal from 1830 to 1840, and while his work proved of no commercial value to him it is probable that Mr. Long, who started the Norris Locomotive

Works at Philadelphia, Pa., and introduced the double eccentrics, was indebted to Mr. James for the idea of a separate eccentric for each motion. Although the credit of inventing the shifting link was conferred upon Mr. William Howe of New Castle, England, who applied the link to the locomotives built by Messrs. Robert Stephenson & Co., and it is claimed he invented the link in practically its present form, Howe's idea was to get out an improved reversing motion. He made a sketch of the link which he explained to his employers, who were favorably impressed with his idea and permitted him to make a pattern of it and afterward gave it a trial on a locomotive constructed for the Midland Railway Company. It proved successful from the first day. Although Stephenson gave Howe the means of applying his invention, Howe failed to perceive its actual value, for it was not patented. Seeing how satisfactorily it worked Stephenson paid Howe twenty guineas ($105.00) for the device, and secured a patent in his own name.

"This is how the shifting link came to be called the 'Stephenson Link.' The credit for this invention was not extravagantly paid for."

Almost simultaneous with the appearance of the shifting link came the Stationary link motion. It was the discovery of Mr. Daniel Gooch. It accomplishes almost the same results, and has met with much favor throughout Great Britain and the continent.

The "Allen Link Motion" combines the characteristic features of both Mr. Howe's and Mr. Gooch's inventions, and in such a manner that the parts are more perfectly balanced; consequently it dispenses with the counter-weight or spring peculiar to the former of these motions.

The "Walschaert Link Motion" is extensively used in Belgium, but probably will not receive much attention from locomotive builders beyond the limits of that kingdom, unless future designers succeed in reducing the number of its connections. Unlike the three others this design of valve gear has no eccentrics but derives its motion from a crank attached to the outside of the main crank pin (which is equivalent to two eccentrics). The link being stationary, the motion is indirect in the forward gear and direct in the back gear.

Among other modern locomotive valve gears which might be mentioned here is the Joy Valve Gear, which derives its motion from the main rod, and the Lewis Valve Gear, which has no eccentrics, no links and no rocker, and derives its motion from the cross-head.

It is proposed, however, to confine our investigation to the shifting link motion only, as it is the standard valve gear in use upon our American locomotives.

THE SHIFTING LINK.

The eccentric as a means of transmitting movement to the valve motion of a steam engine, was invented by William Murdock, a Scotch engineer, and patented in 1799. Before that time the valves were operated first by hand gear, and then by rather crude attachments to the piston connections.

Although repeated efforts have been made to replace the shifting link, each one of the other motions has, in turn, failed to demonstrate its superiority over this form of valve gear. It is acknowledged (and deservedly) to be the best reversible valve gear in existence at the present day, and it is in use upon most all locomotives in this country.

The movement of a valve deriving its motion from two fixed eccentrics with a link suspended in the usual manner is extraordinarily complex. These complications arise from the distortions of the motion which are introduced by the angularity of the eccentric rods and the movement of the link itself, and by the rise and fall of the link which is commonly called "slip."

The angularity of the main rod also introduces irregularities into the motion of the piston which affects the points of cut-off and exhaust closure.

But as this valve gear has fewer parts and is less complex than other forms of valve motion, and as it is capable of an accurate adjustment which practically nullifies the irregularities in the cut-off and exhaust closure, it commends itself to all designers of locomotives.

The shifting link motion is not the most economical valve gear in use, but it is the best reversible valve gear; and as it is necessary to have a reversible gear on all locomotives, it follows that some form of slide valve must be used.

The Cam valve and the Rotary slide valve used upon the Reynolds' Corliss stationary engines are considered the most economical valves.

The general form in which the various parts of the shifting link motion are arranged is clearly shown by Fig. 1. Upon the main shaft are keyed the forward and backing eccentrics, their centers are indicated by F and B respectively. The eccentric straps are attached to the eccentric blades, and these in turn are bolted to the link near both the top and bottom; the forward eccentric blade being attached to the top of the link and the backing eccentric attached to the bottom of the link.

The slide valve is attached by its stem to the top arm of the rocker; the link block, which fits the main link and slides freely therein, is attached to the lower rocker arm.

The rocker box being rigid, the rocker, therefore, reverses

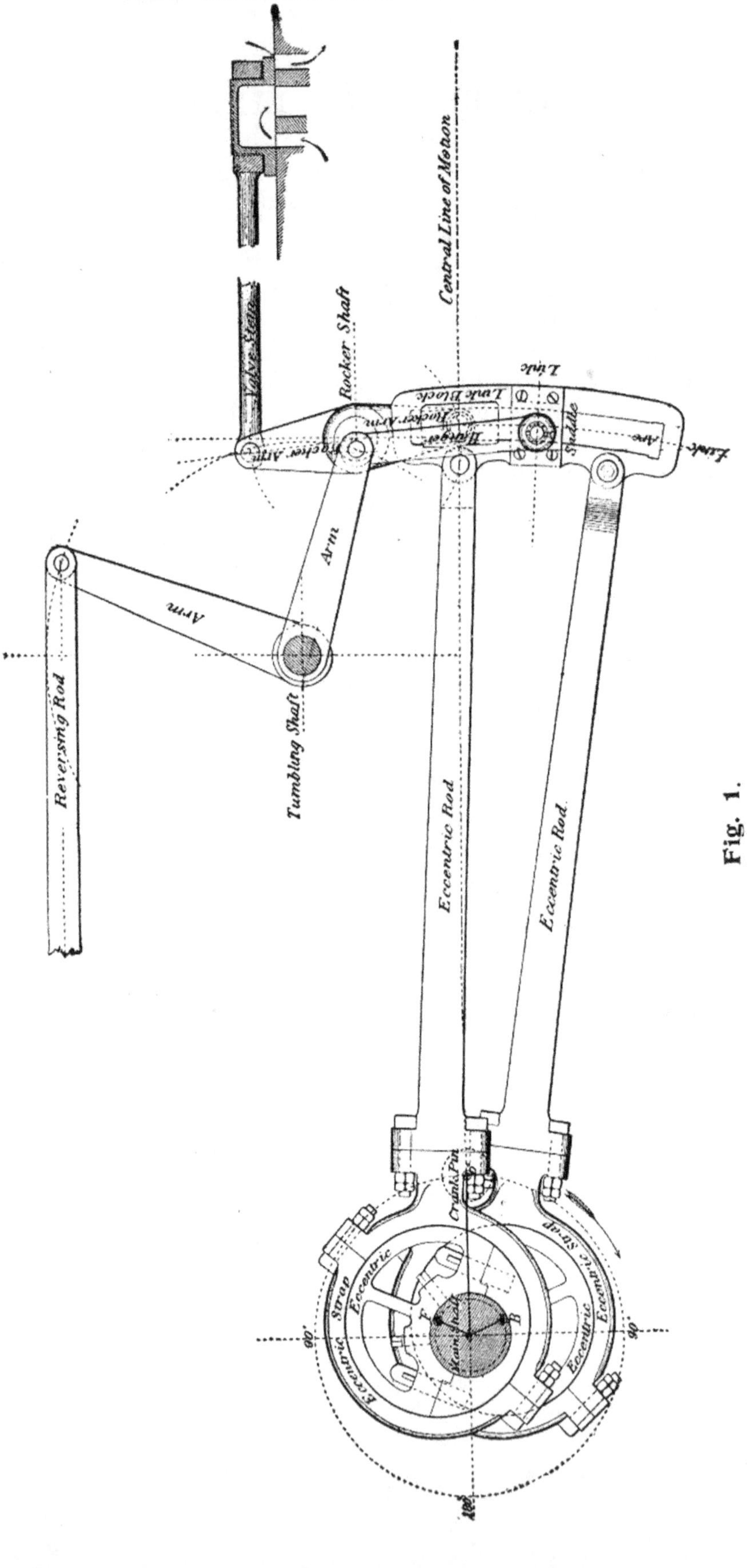

Fig. 1.

the motion of the valve. The center of the link is spanned by a plate called the saddle on which is formed the pin or stud that supports the link and eccentric rods. This saddle pin is attached to the link hanger.

The other end of the link hanger is attached to the short arm of the tumbling shaft. While the reversing rod is attached to the long arm of the tumbling shaft, both arms being rigidly secured, the shaft itself freely oscillates on properly supported bearings which are also rigid. The reversing rod is attached to the top arm of the tumbling shaft, which in turn is attached to the reverse lever (which is not shown).

In this figure the link is shown in full forward gear, thus throwing the entire influence of the forward eccentric F upon the valve motion, to the almost complete exclusion of the backing eccentric B. By drawing back the reversing rod and thereby raising the link until the pin of the other eccentric rod is in line with the pin of the lower rocker arm, the backing eccentric B will throw its entire influence upon the valve motion. It follows, therefore, that by changing the position of the reversing rod, which is accomplished by removing the reverse lever, the valve's motion may be reversed at the will of the engineer.

By intermediate suspensions of the link between either full gear and the center notch we reduce the influence of one eccentric and increase that of the other and reduce the travel of the valve.

Now that the reader has a partial knowledge of the construction of the shifting link motion, and the principles of its action, we shall proceed to a more thorough study of its various parts, together with their influence upon the motion, and also show how to lay out a link motion.

There are certain points in laying out a link motion that can be ascertained only in a technical manner, yet in ordinary shop practice, almost the entire motion work is laid out by well-established rules which, when understood, are much quicker and equally as good for all practical purposes.

Therefore, in order to show our readers wherein theory and practice differ, we shall explain first the practical methods, calling attention to those points that can be accurately determined only in a technical manner; after that we shall explain the correct technical manner for laying out a link motion, at the same time pointing out the errors of the link motion; and last, but not least, how to set a locomotive's valves.

SHOP PRACTICE.

HOW TO FIND THE TRAVEL OF A VALVE.

We shall assume that the width of the steam port is $1\frac{1}{4}''$, and that the lap is $\frac{7}{8}''$; find the travel. If the valve is to open the steam port fully and no more, for the admission of steam, then the travel can not be less than twice the sum of the width of the steam port and lap. Hence, in our example we have $1\frac{1}{4}+\frac{7}{8}=2\frac{1}{8}$ and $2\frac{1}{8}\times 2=4\frac{1}{4}''$, therefore, $4\frac{1}{4}''$ is the travel of the valve. If, however, the valve has over travel (most all locomotives have) and is to move, say $\frac{3}{8}''$ beyond the steam port, then we have $1\frac{1}{4}+\frac{3}{8}+\frac{7}{8}=2\frac{1}{2}''$ and $2\frac{1}{2}\times 2=5''$, therefore the travel of the valve is $5''$. When the rocker arms are of equal length, the throw of the eccentric is equal to the travel of the valve. You can find the travel, approximately correct, by adding together the width of one steam port, the outside lap, and one-half of the width of one bridge, and then double the entire amount. (See chapter on Locomotive Slide Valves.)

When the rocker arms are of different lengths and the throw of the eccentric is known, the travel of the valve may be determined as follows: Multiply the length of the top arm by the throw, and divide the result by the length of the lower arm.

TO FIND THE THROW OF AN ECCENTRIC.

If the eccentric is in use upon an engine or finished in the shop, you can easily determine its throw in this way: Measure the greatest distance from the axle (or the bore) to the outside face of the eccentric, then measure the least distance from the axle to the outside face. By outside face we mean the bearing on which the eccentric strap fits. The difference between the greatest and the least distance will be its throw. If the eccentric is not yet laid off and the correct throw for it is unknown it may be determined as follows: If both arms of the rocker are of equal length, the throw of the eccentric should be the same as the travel of the valve.

If the rocker arms are of unequal length the top arm being the longest, then to find the throw multiply the length of the lower arm by the travel of the valve and divide the result by the length of the top arm.

When the travel of the valve and the length of both rocker arms is known, perhaps the most simple and accurate way to determine the throw of the eccentric would be to make a sketch of the rocker arm similar to Fig. 2, using the center lines only.

We shall assume the length of the top arm is $12''$ and the length of the bottom arm $10''$ and the travel of the valve $5''$.

First, erect the perpendicular line A B; now use a pair of dividers and from a point on the line A B describe two arcs equal to the length of each rocker arm. Now locate a point on the upper arc 2½" from the line A B (which equals one-half the travel of the valve). From this point draw a line, letting it pass

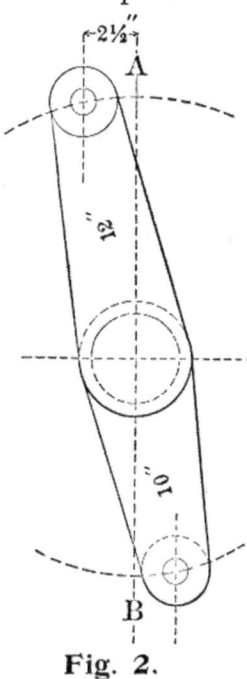

Fig. 2.

through the center of the rocker shaft and intersect the lower arc; in our figure the entire rocker is shown in this position, so measure the distance from the center of the hole in the lower rocker arm to the line A B, which will equal one-half the throw of the eccentric. Measurements should be made on each arc, but as the circular movement is so slight it has not been considered here.

HOW TO LAY OFF A NEW ECCENTRIC.

If the eccentric is in two halves, fasten both parts securely together, fit a wooden center in the hole for the shaft with a piece of tin or thin copper tacked to its face, keeping it parallel with the line R S, Fig. 3. If either side of the eccentric is planed off, keep the face of the wooden center flush with the planed side. Copper or tin may be used for a center in place of wood. If the hole for the shaft is already bored out, find its exact center, which will be at the point C, but if it has not been bored out, then draw the line A B true with planed faces of the two halves of the eccentric where they fit together. The center of the shaft should always be on the line A B. Next with a pair of hermaphrodites locate the center C from the two sides of this hole, this point represents the center of the shaft. Then erect the perpendicular line R S through the center of the rib and at right angle to the line A B, and with a pair of dividers, set to a distance equal to one-half the desired throw of the eccentric, lay off the point D, which will be the center of the eccentric; so from the point D describe another circle equal to the desired diameter of the eccentric.

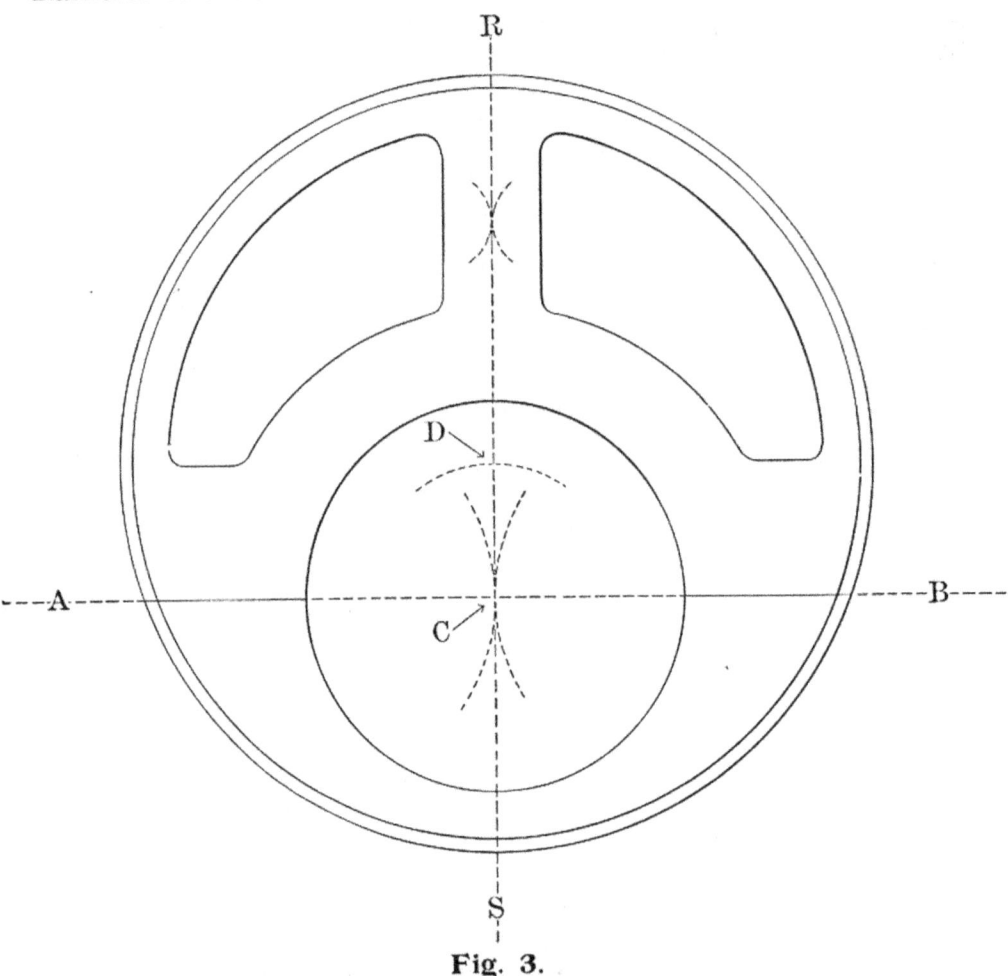

Fig. 3.

When boring out an eccentric the hole for the shaft should be bored just large enough to turn freely on the shaft when the two halves of the eccentric are securely tightened.

TO FIND THE LENGTH OF ECCENTRIC BLADES.

When speaking of the length of the eccentric blades upon a locomotive it is generally understood to mean the distance from the center of the eccentric strap to the center of the link pin hole. If the rocker has no back-set the correct length of an eccentric blade should equal the exact distance, on a horizontal line, between the center of the main driving shaft and the center of the rocker box, less the distance from the link arc to the center of the link pin hole (the center of the link block travels in the link arc). If the lower rocker arm is back-set, subtract the amount of back-set from this length.

This length is only approximately correct, as the blades are invariably adjusted a little when setting the valves, but it is near enough for all practical purposes. The correct technical length would vary only slightly. It would be a trifle longer, as the eccentric blades are crossed when the crank pin is on its back center and open when on the forward center, which tends to shorten the blade. (See Technical Points, page 124.) The effect of long and short eccentric blades is clearly shown in the chapter on "Errors of the Link Motion," page 131.

When you wish to find the length of an eccentric blade, place a straight edge across the two main shoes (see that the shoes are properly tightened), drop a plumb line through the center of the rocker-box, measure the distance from the straight edge to the line and add to this length one-half the thickness of the driving box, which is supposed to be bored out central; then substract from this sum the distance from the center of the link block to the center of the link pin hole and also subtract the backset of the lower rocker arm, if it has any backset.

Now, to find the length of the eccentric blade alone, measure the distance from the center of the eccentric strap to its butt-end or shoulder, and add to this about $\frac{1}{4}$" for liners or clearance, as the case may be; now subtract this length from the total length of the blade and strap, the remainder will be the length of the blade alone.

TO FIND THE LENGTH OF THE VALVE STEM AND YOKE.

On most all modern locomotives the cylinders and the valve seat are parallel with the wheel centers. On these engines simply plumb the outside or top arm of the rocker shaft by dropping a line through its center, the distance from this line to the center of the exhaust port will be the correct length for the valve stem from the center of the yoke to the center of the rocker pin hole.

If the valve seat is inclined, set the top arm of the rocker at right angles to the valve seat; this may be done by placing a long straight-edge on the valve seat and using a two-foot square at the rocker, and setting the center line of the rocker arm true with the square, and measure the distance from the center of the rocker arm to the center of the exhaust port; this distance will be the correct length.

To determine the length of the valve stem and yoke alone, without the valve rod, measure the distance from the center of the steam chest to the outside of the gland (when it is packed), and to this length add one-half the travel of the valve plus one inch for clearance (if space will permit add a few more inches so the gland may be repacked with the stem in any position without pinching the engine). This will be the length from the center of the valve yoke to the shoulder. The remainder of the original length will be the length of the valve rod from the center of the pin hole to its shoulder.

LAYING OFF A VALVE YOKE.

If the yoke is milled out or slotted to fit the valve, finish fitting it by hand before laying off the stem, keep the stem true with the face and sides of the valve, fit closely but do not let the yoke bind or cramp the valve. When finished the end of the stem should be permitted to raise and lower about $\frac{1}{4}$" without binding the valve. This is done on account of the circular motion of the rocker arm, and the yoke should be about 1-16" loose endways between the yoke and the valve. Now put the yoke on the valve and take it to a face plate and locate two centers from which to swing and turn the stem. Keep these two centers central both ways, and see that you have sufficient stock for the stem to finish the required size.

If the entire yoke is still in the rough take it to a face plate and block it up edgewise, use a two-foot square and surface gauge and keep the stem at right angles to the yoke, divide the stock as nearly as possible every way, then locate two centers for turning up to the required size. Then with the two-foot square lay off the yoke for the valve, making it the exact thickness of the valve longitudinally, and allow from 1-32" to 1-16" for the valve to move endwise in the yoke. Drill small holes in the centers laid off for turning, in order to retain the original centers.

FINDING THE LENGTH OF THE LINK HANGER.

On a large face plate or table lay off, full size and with correct dimensions, the tumbling shaft and rocker arm, using their center points only.

Keep the top arms of both the tumbling shaft and the rocker perpendicular, or in their correct positions; then from the center of the lower rocker arm and on a horizontal line with it, lay off the backset of the saddle stud, and from this point find the distance to the center of the short arm of the tumbling shaft, which will be the correct length of the hanger.

If one link hanger is broken and the other one is all right use the length of the good hanger.

LENGTH OF THE REACH ROD.

Plumb the reverse lever and the top arm of the tumbling shaft (if they are in position), and find the exact distance between the center of the hole in the reverse lever and the center of the hole in the tumbling shaft arm, which will be the correct length from center to center. If the reverse lever and tumbling shaft are not yet in position, put a center in the foot casting for the reverse

THE LOCOMOTIVE UP TO DATE. 107

lever and another in the tumbling shaft stand; transfer a square line from each center onto the frame and measure the distance between these two lines. If for any reason the reverse lever or tumbling shaft arm cannot be set plumb in its central position, then set either in whatever position you can for its central position and find the length as we first suggested.

TO FIND THE BACK SET OF A STANDARD ROCKER ARM.

We will assume that the rocker box is already located or that the reader knows the length of the eccentric rods and rocker arms. On a large face-plate, table or finished board draw the horizontal line A B parallel with the wheel centers (see Fig. 4), and locate the point H the correct distance from the main shaft and the right distance above the line A B; this point indicates the center of the rocker shaft. Now erect the perpendicular line C D at right angles to the valve seat, and with a pair of dividers, set to the length of each rocker arm, describe arcs from the point H equal to the length of each rocker arm. Then draw the line E F through the center of the shaft, intersecting the lower arc which represents the length of the lower rocker arm. Next, through the center H, erect the line G H at right angles to the line E F. The distance between the line C D and the line G H on the lower arc is the required backset; the line E F being the center line of motion.

If the top arm of the rocker stands plumb in its correct relation to the valve seat and the line A B bisects the center of the lower rocker arm, then the rocker will require no backset.

Remember the top and bottom arms of the rocker are entirely independent of each other as regards the backset; either one of them may require a backset and the other not have any, or they may both be backset. The top arm should be at right angles to the valve seat, and the bottom arm at right angles to the center line of motion, then each arm will travel an equal distance each way from its central position.

HOW TO FIND THE BACKSET OF A DIRECT MOTION ROCKER ARM.

The draftsman will usually determine the amount of backset for a rocker of this kind, since in doing so it is necessary to know the relative position of the center of the link block and the tumbling shaft when in their central positions. (See Technical Points, page 129.)

When these positions are known it is easy to determine the correct backset for the rocker arm. Make a full sized sketch as shown by Fig. 5. The line A B should bisect the wheel centers. The outside rocker arm should be set at right angles to the valve seat, and with a length equal to the length of the inside rocker arm describe the arc R S. Draw the line C D through

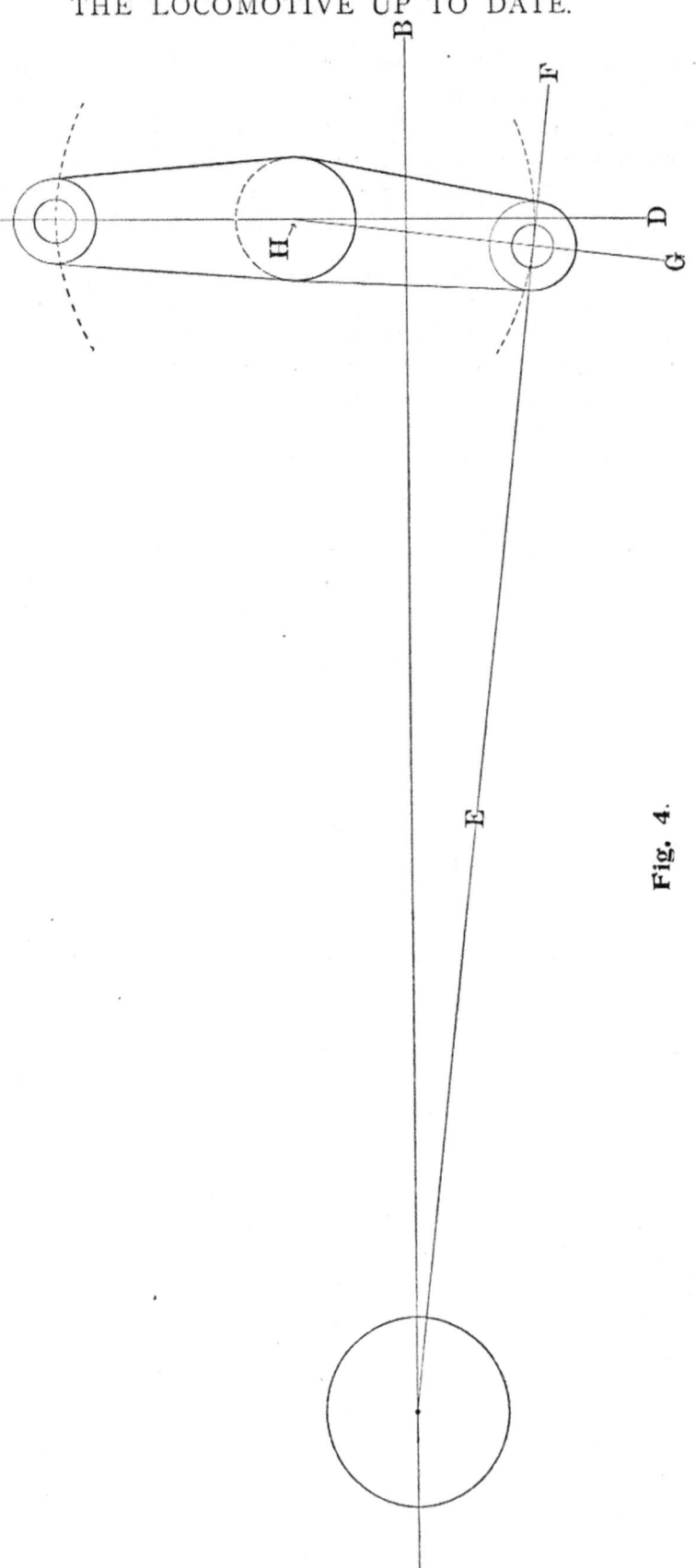

Fig. 4.

THE LOCOMOTIVE UP TO DATE.

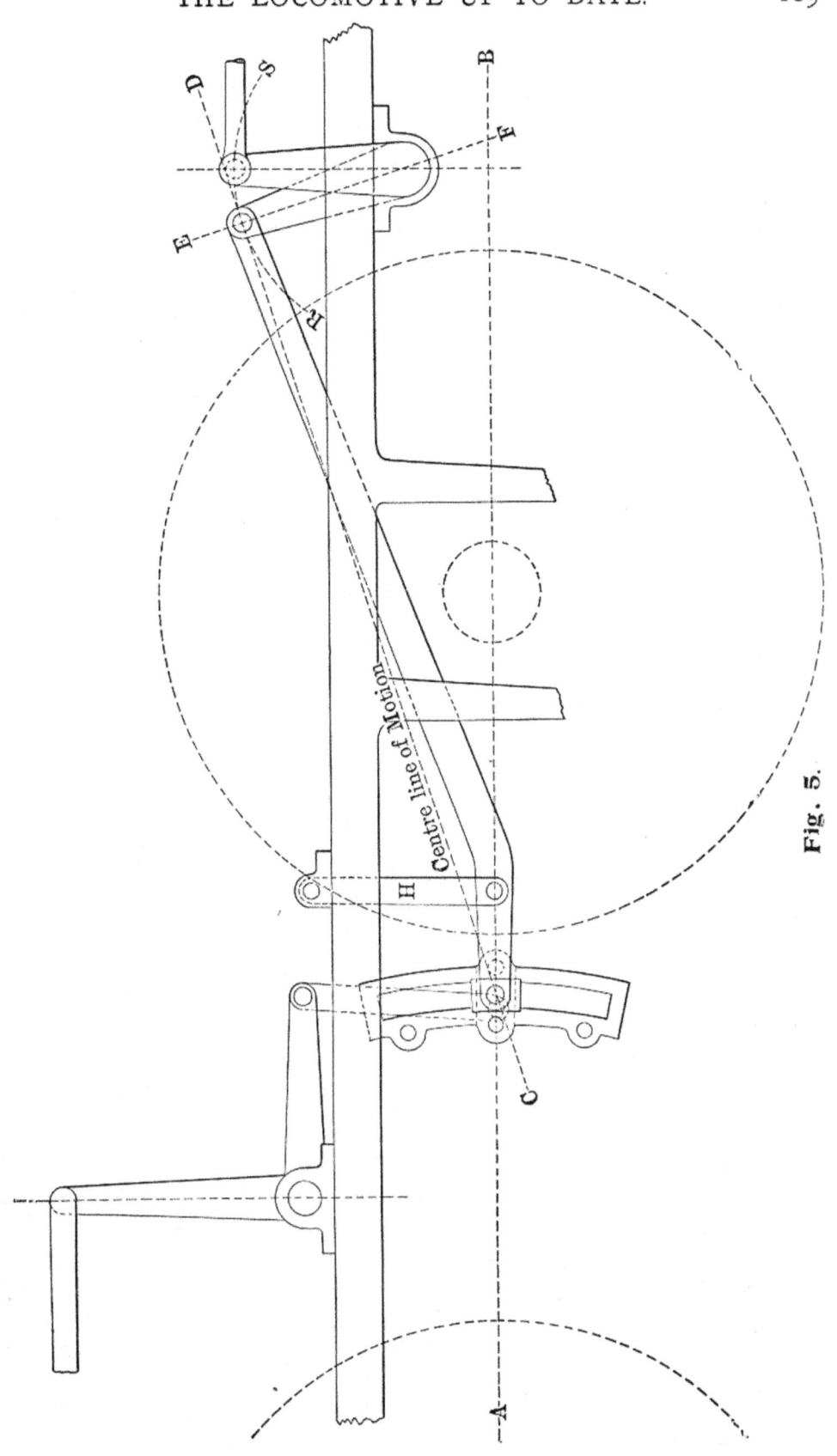

Fig. 5.

the center of the link block pin hole when the block is in its central position and let it intersect the arc R S. Then draw the line E F through the center of the rocker shaft and at right angles to the line C D. The point where the line E F intersects the line C D will be the center of the inside rocker arm, and the backset will equal the distance on the arc R S between the centers of the two arms, as is shown in the figure; the distance between the center of the link block pin hole and the center of the pin hole in the inside arm of the rocker will be the length of the extension rod. In the figure we find the link block is attached to the extension rod, while the back end of the extension rod is prevented from working up or down by the small hanger H. Excepting the slight circular movement imparted by the hanger H the center of the link block remains in a horizontal line with the center of the main shaft. We, therefore, determine the offset by the center of the link block pin hole instead of the center of the driving axle. The line C D is, therefore, the center line of motion.

LAYING OFF A ROCKER ARM.

Take the rocker to a face-plate, secure two V blocks of equal sizes and line them up with parallel strips high enough to let each arm revolve. Next place the body of the shaft on the V blocks, keeping one V block near each end; then, with a surface gauge, set to any height, scribe a line on each end of the shaft, revolving the rocker to four different positions and marking each end of the shaft for each position. From these four lines locate a center at each end of the shaft, place the arms in an upright position, and set each center true with the surface gauge; then try a two-foot square to each arm and see whether they are square with the shaft. Then place the arms in a horizontal position, keeping the two main centers true, and carry a line through the boss on each arm. Now lay off the length of each arm, and see if the rocker will true up all around. If the arm requires any backset, then with the length of the arm describe an arc on the boss and lay off the center for that arm the required distance from the central line, being careful to lay it off in the right direction. When laying off the center of each arm be careful to make all measurements at right angles to the center of the shaft. When you have the four centers correct, lay off the pinholes, making each 1-32" smaller than the small end of the taper pinhole when finished; and have a small hole drilled in each end of the shaft to retain the main centers while in the lathe. If the rocker will not true up to the correct dimensions, return it to the blacksmith then, instead of doing so after you have it half finished in the lathe. If you have plenty of stock divide it as near as possible in order to equalize the cut while in the lathe.

TO FIND THE LENGTH OF ROCKER ARMS.

This is properly the work of the designer, but when the throw of the eccentric and the travel of the valve are known it is an easy matter to determine the length of each arm, if the position of the rocker box is known. Place a straight edge across the top of the frames, and measure the distance to the center of the rocker box, then measure the distance to the center of the stuffing box. If the rocker box is above the frame the difference between these two measurements will indicate the length of the top arm, and if the rocker box is below the top of the frame measure the distance it is below the frame, and add these two measurements together, in either case add $\frac{1}{4}''$ more in order to divide the circular movement imparted to the valve stem (unless a scotch yoke is used). We shall assume that the length of the top arm is 12" and the travel of the valve 5" and the throw of the eccentric is 4". If both arms were of equal lengths we know that it would require a 5" throw to move the top arm 5". It is therefore evident that the bottom arm must be less than 12" in length. In this case it is in the proportion of 5 to 4 and may be determined thus: Multiply the length of the top arm 12" by the throw 4", which equals 48, and divide this by the travel of the valve 5", which gives 9 3-5" length of lower arm. If the throw and travel and length of the bottom arm were known, to find the length of the top arm proceed as follows: Multiply the length of the lower arm by the travel of the valve and divide the result by the throw of the eccentric.

LENGTH OF TUMBLING SHAFT ARMS, AND RELATIVE POSITION OF THE TUMBLING SHAFT AND ROCKER.

These are technical points and any attempt to determine them by plain measurement would result in a failure. (See Technical Points, page 129.)

THE QUADRANT.

The length and radius of the quadrant is properly the work of the draftsman, but they are easily determined in the shop, when the dimensions of the other parts are known. Make a sketch of the other parts, and locate the position of the reverse lever latch in each full gear. Find the distance from one extreme point to the other, and add to this the width of the reverse lever plus a little more for clearance at each end with sufficient stock for bolting both ends; this is the length of the quadrant.

To find the correct radius of the quadrant, draw the reverse lever latch up as far as it will go; then measure from the bottom of the latch to the center of the hole in the foot casting, and subtract 1-32" from this distance (clearance for the latch), the remainder will be the correct radius for the top of the quadrant. When laying off the holes in a quadrant for drilling, draw the latch up as far as possible and fasten it there. Then, if the boiler is cold, let the latch clear the quadrant 1-32" at the front end and ¼" at the back end, this will allow for the expansion of the boiler; when the boiler is warm the latch will have the same clearance at each end. There is nothing very important about a quadrant except the notches. How to lay these off will be explained later. Of course the quadrant should be set properly and securely fastened.

RADIUS OF THE LINK.

When the rocker arm has no backset, the correct radius of a link should equal the distance on a horizontal line from the center of the main shaft to the center of the rocker box. If the rocker has any backset, subtract the amount of backset from this length. The reason of this is clearly shown in our chapter entitled "Errors of the link motion." Some manufacturers of locomotives make the link radius ¾" per foot less than this length, but the distance given above is the correct length and was accepted as such by the Master Mechanics at one of their late conventions.

To find the radius, place a straight edge across the front of the main jaws (providing the front jaw is square with the top of the frame). Then drop a plumb line through the center of the rocker box, and measure the distance on a horizontal line from the straightedge to the line. Add to this length the thickness of the main shoe, plus one-half the driving box. This will be the correct length of the radius. If the main jaws both taper, find the center of the jaw (see Shoes and Wedges) and drop a line through its center. Then find the distance horizontally from this line to the other line dropped through the center of the rocker box; this length will be the correct radius. In either case if the rocker has any back-set subtract the amount of back-set from this length. This rule will not apply to valve gears having an extension rod between the link and rocker; on locomotives of this kind the radius will depend upon the relative position of the tumbling shaft and link.

LAYING OFF A NEW LINK.

Blue-prints are usually furnished for the purpose of laying off new links, but if they are not, proceed as follows: We will assume that the throw of the eccentric is 5", and that the rocker arms are the correct length to secure the required travel for the valve, and that you intend to make the link block 5" long and 3" wide. First make the straight line A A, as shown in

THE LOCOMOTIVE UP TO DATE. 113

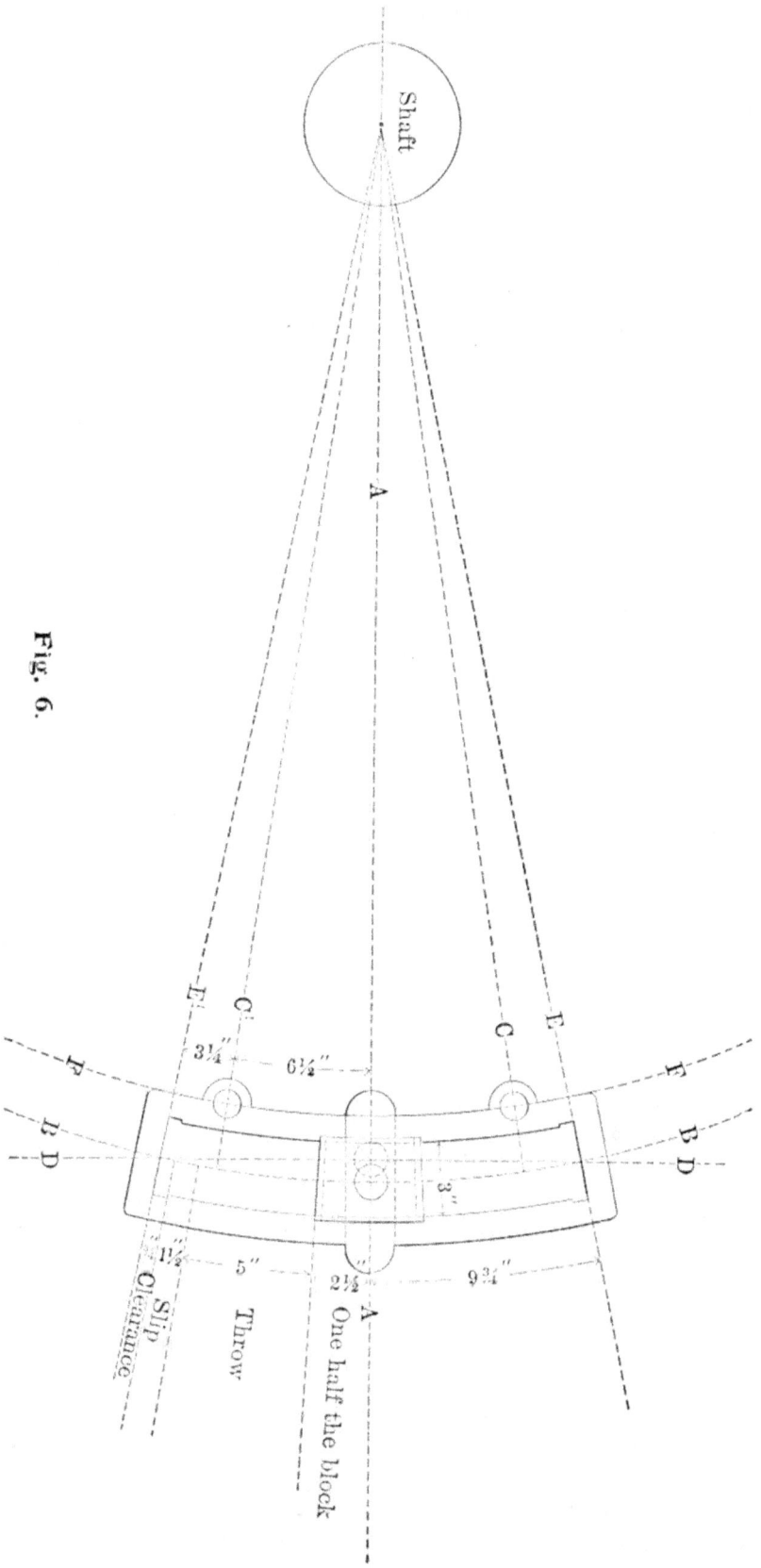

Fig. 6.

Fig. 6, through the center of the shaft and the center of the link; next with a length equal to the correct radius of the link describe the arc B B. Then add together the following amounts: $2\frac{1}{2}''$, which is one-half the length of the link block; and $5''$, which equals the throw of the eccentric; and $1\frac{1}{2}''$ for the slip of the link block (see Slip of the Link, page 115), and $\frac{3}{4}''$ for clearance between the link block and the end of the link. (In ordinary practice $1\frac{1}{2}''$ is the amount usually allowed for the slip of the link and $\frac{3}{4}''$ is considered a safe margin for clearance, although when a blue-print is furnished, you will find these amounts considerably reduced as the draftsman usually has the use of models with which he can try the maximum slip and thereby prove his work; in such a case $\frac{3}{8}''$ is considered sufficient for clearance. The slip varies on different engines, but the figures we give herewith are considered safe.) Now, by adding the above amounts, we have a total of $9\frac{3}{4}''$, which is one-half the inside length of the link. All measurements should be made on the arc B B, which is the link radius; with a flexible scale if you have one; if not, cut off a strip of tin $9\frac{3}{4}''$ long, and from the point where the line A A intersects the arc B B (which will be the center of the link block) lay off on the arc B B two additional points $9\frac{3}{4}''$ from the center; and from these two points draw the lines E and E passing through the center of the shaft. These lines indicate the inside ends of the link. Then add together $2\frac{1}{2}''$, which equals one-half the length of the link block; and $\frac{3}{4}''$, which is the clearance; this amounts to $3\frac{1}{4}''$. So, in each full gear, the center of the link block should be $3\frac{1}{4}''$ from each end of the link. Therefore from the two points where the lines E and E intersect the arc B B lay off on the arc B B two additional points $3\frac{1}{4}''$ nearer to the center of the link, and from these points draw the two lines marked C and C; the link pin holes will therefore be laid off on these two lines, which indicate full gear. Now as the link block is to be $3''$ wide, add $1\frac{1}{2}''$ to the length of the radius and describe another arc which will represent the front face of the link inside and the front face of the link block. Then subtract $1\frac{1}{2}''$ from the radius and describe another arc which will represent the back face of the link inside and the back face of the link block. Since the link block is to be $5''$ long, lay off two additional points $2\frac{1}{2}''$ from its center. From these points lay off the ends of the block, keeping them true with the center of the shaft. The link pin holes must be an equal distance from the link radius and should be as close to the radius as possible to avoid increased slip. Therefore describe the arc F F (which is called the link pin arc) as close to the arc B B as will be consistent with a proper thickness of the link. The link pin holes should be laid off at the two points where the lines C and C intersect the arc F F. Now, through the two points where the lines E and E intersect the arc B B, draw the line D D. In ordinary shop practice, when the correct position of the saddle stud is not known, the point where the lines A A and D D cross each other is used as the point of suspension. This point is only approximately correct, but near enough for all practical purposes. To find the correct point of suspension, see Technical Points, page 126. The off-set of the saddle pin is, therefore, indicated by the space between the line D D and the arc B B on the line A A.

REASON WHY A LINK SADDLE IS OFFSET.

The purpose of the offset in the link saddle is to obtain as nearly as possible an equal cut-off, and at the same time to permit the lead to be the same for each stroke, and to approach as nearly as possible a correct distribution of steam with a reciprocating engine and slide valve. It is done to overcome the inherent imperfection in the design of links, the angularity of the connecting rods, and the offset of the link pin holes.

SLIP OF THE LINK.

Link motion is especially characterized by a very important feature, that of adjustability, commonly known as the "slip of the link." In addition to the other two motions, the link block, being securely fastened to the bottom of the rocker arm, must move in the arc traversed by that arm, while the action of the eccentric rods on the link forces it to move in a sort of vertical motion during certain parts of the stroke; these two motions combined cause the link to slip on the block. This slip is caused partly by the circular movement of the lower rocker arm, thereby causing the block to slip also, but principally by the method of suspension, and the manner of attaching the eccentric blades to the link, the link pin holes being back of the link arc. The action of the link pins is similar to that of a knuckle joint between the eccentric center and the link arc through which the center of the link block must travel. The link slips most when in full gear and the slip diminishes as the block is moved toward the center of the link. By referring to "Technical Points," you will note the distortion introduced into the valve's motion by the angularity of the connecting rods and by this backset of the link pins from the link arc, and while moving the link saddle pin back tends to equalize the motion it also tends to increase the slip, which, if very great, would seriously impair the valve's motion. On marine engines they sometimes sacrifice equality of steam to a reduction of the slip; but with the long connecting rods used upon locomotives and the close proximity of the link pin holes to the end of the link, little difficulty is found in keeping the slip within practical bounds. Raising the saddle above the center of the link will also equalize the valve's motion, but in locomotive construction there are practical objections to doing this. Backsetting the link saddle pin has an effect equivalent to a lengthening of the eccentric rod during a portion of the stroke and thereby equalizes the valve's travel. (See Rule 39 for Valve Setting.) Moving the eccentric rod pin holes farther from the radius of the link, or closer together, tends to increase the slip. By referring to Fig. 6, you will notice that we allowed $1\frac{1}{2}''$ at each end of the link for the "Slip of the Link." The amount of slip varies on different locomotives, but $3''$ in the length of the link is considered a safe margin. This slip partakes of a sort of double movement during each revolution of the wheel; the block lowers and raises in the link twice during one complete revolution, as may be seen by anyone who will go under an engine and watch the link and block. The action is as follows: In full forward gear the block begins to slip down in the link immediately after the crank pin has passed the top quarter and continues to move downward until the pin has reached the forward center, at which point it begins to move upward until the pin has passed the bottom quarter; then it again begins to move downward and continues to move downward until the crank pin has reached the back center, when it again moves upward until the pin has again reached the top quarter, or starting point. In full gear backward these operations are just the reverse. The amount of slip may be ascertained by measuring the distance from the end of the link to the link block in either of these positions with the reverse lever in full gear, the difference between the greatest and least distance so obtained will indicate the amount of slip. An increase in the length of the lower rocker arm will decrease the slip, an increase in the length of the link will decrease the slip, provided the distance between the link pin holes is also increased, but it will also increase the midgear lead.

To decrease the length of the link and also the space between the link pin holes will increase the slip and reduce the midgear lead.

Mr. W. S. Auchinloss, who is a recognized authority on the design of link motions, mentions four alterations capable of reducing the slip when too great, viz.: Increase the angular advance, reduce the travel, increase the length of the link, or shorten the eccentric rods.

TECHNICAL POINTS.

In this chapter we shall endeavor to explain how to locate certain points and determine certain lengths for a shifting link motion, which could not be found or explained by plain measurement. Strict theoretical rules prohibit the use of the link templet, but as it gives the same results when properly used as a conglomeration of link arcs and circles, which only confuse and mystify the ordinary mechanic or engineer, we shall make use of the link templet to explain these subjects. Some of our drawings may appear a trifle complex, but we have made each as plain as possible for a thorough comprehension of the subject, and in order that it may be more clearly understood by all our readers we have avoided the use of Algebra and Geometry. The subjects treated in this chapter will include:

First: The angularity of the main rod, showing the crank pin at full and half stroke on a center line engine.

Second: The position of the crank pin at full and half stroke on an engine whose cylinder axis is above the wheel center.

Third: The relative position of the eccentrics to the center line of motion.

Fourth: Relative position of crank pin and eccentrics at full and half stroke.

Fifth: How to find the correct length of eccentric blades.

Sixth: How to locate the point of suspension, which indicates the position of the center of the saddle pin.

Seventh: The relative position of the tumbling (or lifting) shaft and the rocker.

Eighth: The length of the tumbling shaft arms.

ANGULARITY OF THE CONNECTING RODS.

The eccentric is in effect a crank and, being keyed onto the axle, must always remain an unvarying distance from the crank pin. It, therefore, follows that any irregularities imparted by the crank pin into the motion of the piston will also be imparted

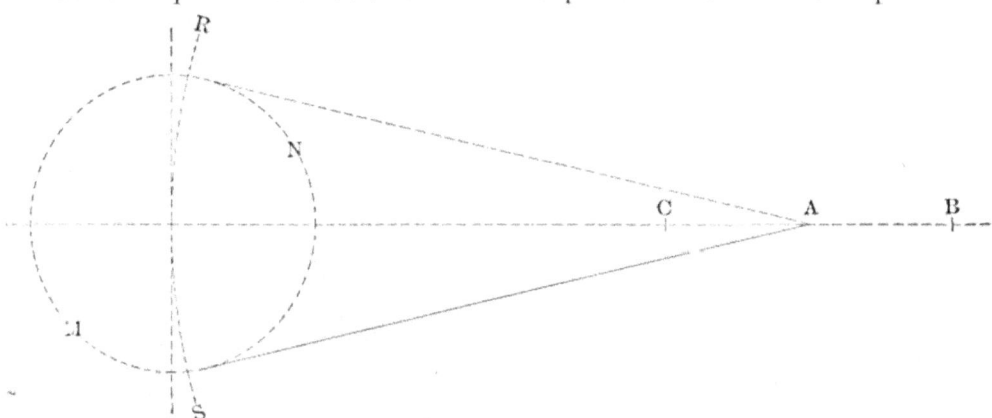

Fig. 7.

by the eccentric rods into the motion of the valve. But, since the throw of the eccentric is much less than that of the crank pin, and since the eccentric rod is proportionally longer than the main rod, it follows that the distortions in the motion of the valve are necessarily much smaller than those in the motion of the piston, and they vanish entirely when the eccentric is on either center. The point A in Fig. 7 represents the center of the cross head pin. Now, if the crank pin was on the forward dead center, it is evident that the center of the cross-head pin would be at the point B, which indicates the extreme forward travel of the center of the cross-head pin; while if the crank pin was on the back dead center, the cross-head pin would be at the point C, which indicates its extreme back travel. It now becomes necessary to locate the positions of the crank pin at half stroke. It is evident from the above brief explanation that the length of the main rod must be equal to the distance from the center of the driving axle (or shaft) to a central point between the points B and C. This point indicates the center of the cross-head pin's travel, and, therefore, the half stroke of the piston is indicated by the letter A. Now, to determine the position of the crank pin at half stroke, we must proceed as follows: From the point A, with a length equal to the length of the main rod, describe the arc R S. The two points where the arc R S intersects the crank pin circle must, therefore, indicate the positions of the crank pin at half stroke for either stroke of the piston. We find these two points are farther forward (nearer the cross-head) than the top or bottom quarter. It is, therefore, evident that the crank pin must travel farther while the cross-head pin travels from A to C than it would travel while the cross-head pin traveled from A to B.

This irregularity will be scarcely perceptible when the piston is at the beginning of its stroke, and therefore the point of lead opening, but it would seriously affect the point of cut-off which occurs at intermediate points of the piston's stroke.

Partly to overcome this imperfection of the crank motion, and in order to obtain an equal cut off, the link saddle pin is back-set, but this only approximately corrects the inherent error of crank motion.

Experiments were made by making one steam port wider than the other to overcome this defect, but that caused one exhaust to be heavier than the other and also proved injurious in other ways. To offset the effects of the angularity, valves are used upon some stationary engines which have more outside lap on one side of the valve than upon the other.

The early builders of steam engines used what is known as the slotted cross-head or Scotch yoke to overcome this defect, but it was found that setting the link saddle pin back answered the same purpose and was less expensive.

CYLINDER AXIS ABOVE THE WHEEL CENTERS.

Upon an examination of Fig. 8, we discover another form of irregularity in the piston's motion introduced by the main rod. Here we find that the line P Q represents a central line drawn through the wheel centers, and the line C B represents a line drawn through the center of the cylinder; this latter line also represents the path of the cross-head pin as explained in

Fig. 7. Now, if from the point A, as previously explained, we describe the arc R S, we find that the crank pin will be nearer the quarter in one stroke than in the other, and hence will cause a variation in the cut off for the two strokes of the piston. Now, if we examine Fig. 8 a little closer, we shall find that the error is partly overcome by the position of the crank pin at both the forward and the back dead centers. When the crosshead is at B, it follows that the position of the crank pin must be located on a line drawn from the center of the shaft to the point B; the crank pin center will therefore be located at the point G

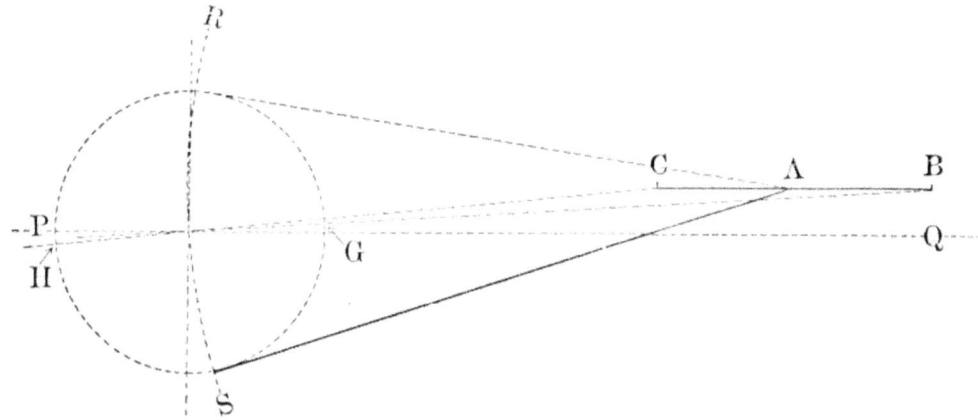

Fig. 8.

when it is on the forward dead center. Likewise a line drawn from the point C and passing through the center of the shaft will bisect the crank pin circle at the point H, which is, therefore, the position of the crank pin when on the back dead center. It will be noticed that the point G is above the horizontal line P Q, while the point H is below it, therefore a line drawn at right angles to the line P Q and passing through the center of the shaft will not indicate the correct top and bottom quarter for this kind of an engine. It will be noticed, however, that the point G is closer than the point H to the line P Q; and a line drawn from the point G to the point H would not pass through the center of the shaft. Therefore, this irregularity is not entirely overcome. When finding the length of the main rod for this kind of an engine, if the cross-head is up, the length of the rod should be found by measurements on unequal surfaces. See page 559.

Care should also be taken to get the crank pin on its correct center when putting up a main rod or finding the travel marks on these engines.

CENTER LINE OF MOTION AND ANGULAR ADVANCE.

When no extension rod is used between the link block and the rocker arm, the center line of motion will be a line drawn through the center of the main shaft and the center of the pin hole in the lower rocker arm, as shown by Fig. 4. The upper arm of the rocker being set at right angles to the valve seat. Where an extension rod is used between the link block and the rocker arm, as shown by Fig. 5, the center line of motion will extend from the center of the link block (in its central position) to the center of the pin hole in the arm of the rocker to which the extension rod is attached.

Now, we wish to call the reader's attention to a few points shown in Fig. 9. The horizontal line P Q bisects all the wheel centers, and we shall assume that in this case this line also passes through the center of the cylinder; therefore, the crank pin will be on "dead center" when its center is located on this line P Q. We have shown it on its forward center at the point G, at which point we find that the center of the lower rocker arm is below the line P Q, and that the lower rocker arm is backset. How to determine the amount of this backset has been previously explained. Now, it will be noticed, that the center line of motion does not pass through the center of the crank pin when the rocker has a backset. It will also be noticed that the short line drawn through the two eccentric centers F and B is at right angles to the center line of motion and not parallel to the perpendicular line L M, and that the eccentric centers are unequal distances from the crank pin. It, therefore, follows that the eccentrics are set by the center line of motion and not by the crank pin.

Now, if the center of the lower rocker arm was located on the line P Q, the rocker would require no backset; and this line would then be the center line of motion; and the eccentrics would then be an equal distance from the crank pin. But, again, if the engine had a straight rocker, and the line P Q was the center line of motion, it does not necessarily follow that the eccentrics are equal distances from the crank pin. If the cylinder center was above the wheel center, as shown by Fig. 8, we would again find that the eccentric centers were at unequal distances from the crank pin. It, therefore, follows that the eccentrics are always set by the center line of motion and not by the crank pin. However, the builders of modern locomotives endeavor as far as possible to build straight line engines and use straight rockers.

Another point to which we desire to call the reader's attention is the angular advance of the eccentrics. In Fig. 9 the angular advance is equal to the distance between the short line which bisects the eccentric centers and the center of the main shaft. In this case the advance is equal for each eccentric, but in some cases the angular advance of one eccentric is altered in order to improve the other motion, as will be explained later. In a case of this kind, where the angular advance of each eccentric is different, the distance between the center of the shaft and

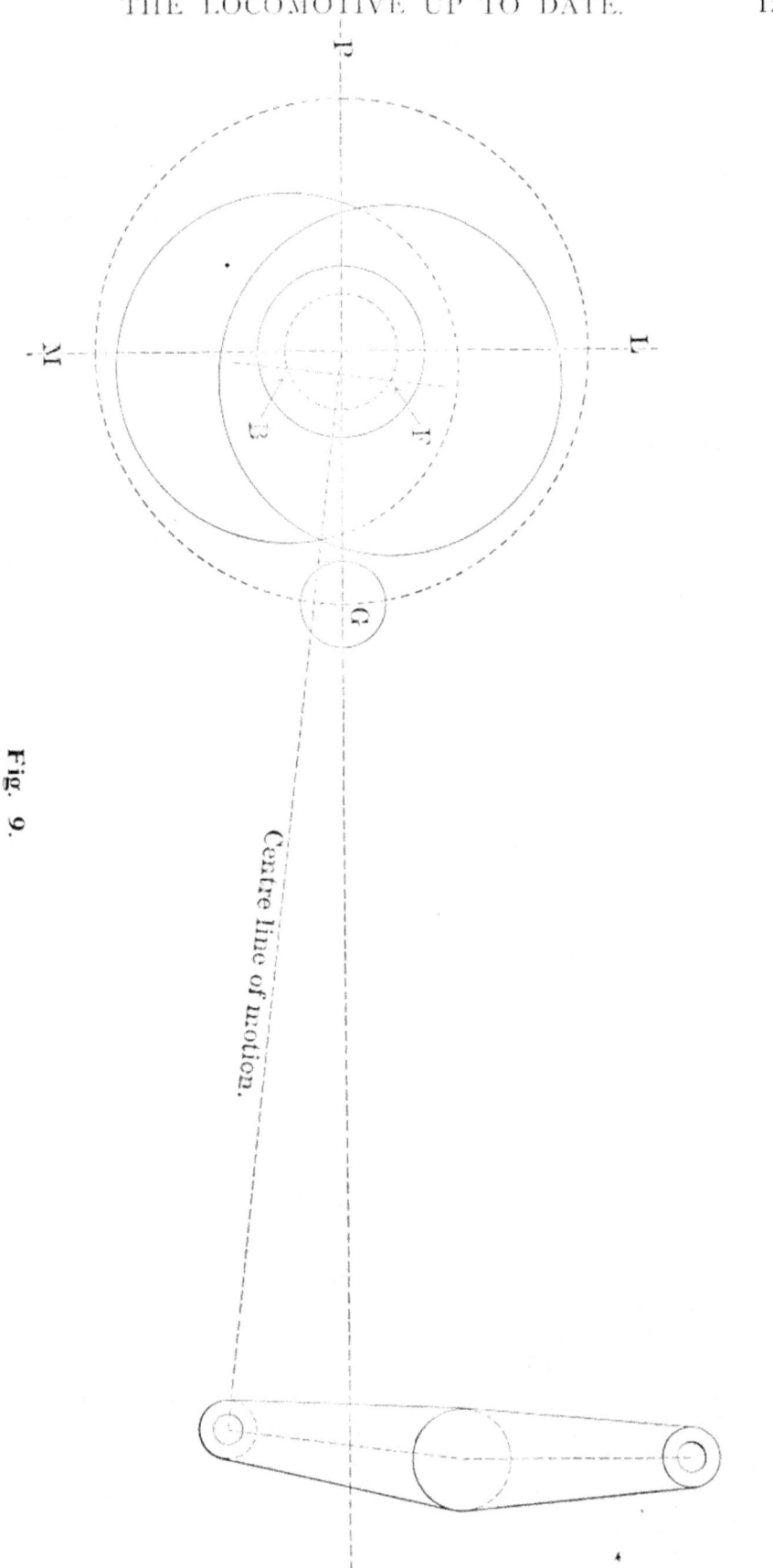

Fig. 9.

a short line drawn through the center of each eccentric, as shown by Fig. 9, would not indicate the correct angular advance of either eccentric, as such a line would not be at right angles to the center line of motion.

The angular advance of an eccentric means the distance the center of the eccentric is advanced toward the crank pin (when a rocker is used) from the center of the main shaft to a line at right angles to the line of motion and passing through the center of the eccentric.

With rocker arms of equal length the amount of angular advance should be equal to the lap of the valve plus the lead. See Rule 49, for valve setting.

RELATIVE POSITIONS OF CRANK PIN AND ECCENTRICS AT FULL AND HALF STROKE.

We have made use of the two former figures in order to call the attention of the reader to those points in the construction of

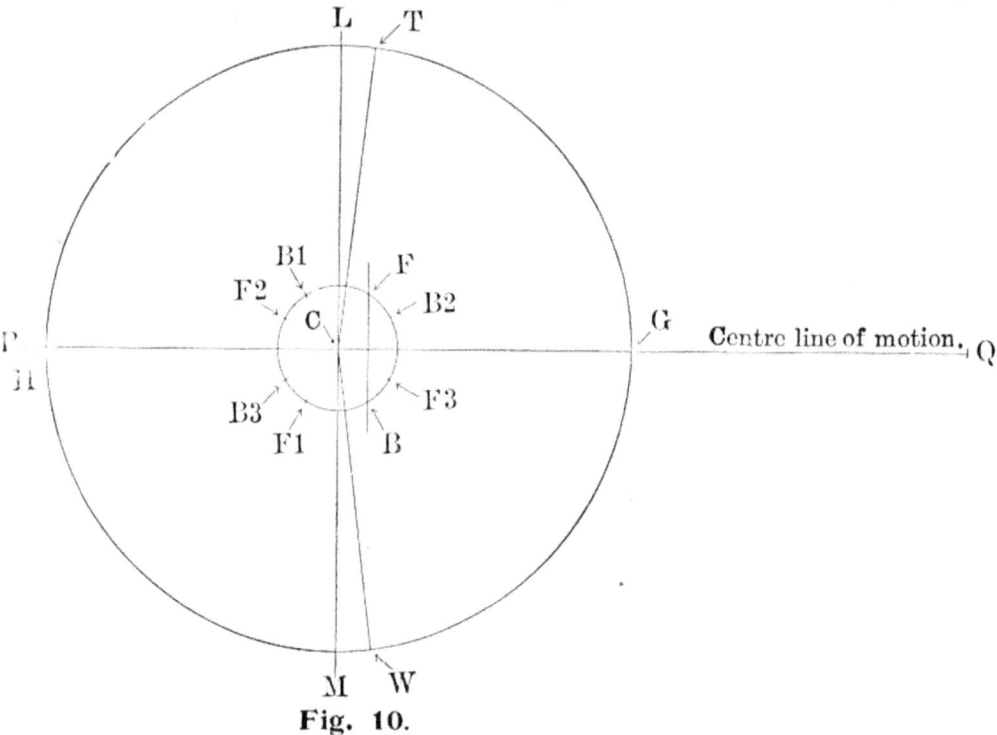

Fig. 10.

an engine which must be considered when laying off a link motion. But in order to avoid the use of complex drawings, which only tend to confuse the reader, and in order to explain the following subjects in as clear and plain a manner as possible, we shall assume that this is a "straight line" engine whose wheel centers and cylinder center are on a horizontal line, which line is also the center line of motion. This engine, of course, has a straight rocker (by straight rocker is implied one with no backset). In Fig. 10 we shall use the lines P Q and L M the same as in the former figure, but in this case the line P Q is also the center line of motion; the larger circle represents the path of the crank pin, and the small circle the path of the center of the eccentrics. Therefore the points G and H show the position of the crank pin when on forward and back center; and T and W indicate the position of crank pin at half stroke (as explained by the Angularity of the connecting rod). We shall, therefore, draw two straight lines from the points T and W to the wheel center. Next we shall locate the positions of the two eccentrics when the crank pin is at the forward center G. If the valve had no lap or lead, the eccentrics would be at right angles with the crank pin, and therefore on the line L M; but we shall assume the valve has ⅞" outside lap and 1-16" lead, therefore we must advance both eccentrics that amount toward the pin in order to give the valve the required amount of opening at the beginning of the stroke. Since the crank pin is at the forward center G, the piston must be at the forward end of the cylinder, therefore the forward steam port must be opened to admit steam. By advancing both eccentrics toward the pin, the bottom arm of the rocker is forced forward and the top arm backward, thus opening the forward port. So we make another short line parallel to the line L M and 15-16" in front of it (which is the amount of lead and lap added together); the two points where this line intersects the eccentric circle will be the centers of the two eccentrics. We shall designate them by the letters F and B, as shown, and in this case they are equal distances from a line drawn from C to G. The distance the center of each eccentric is advanced from the perpendicular line L M, which must be perpendicular to the center line of motion, is called its "angular advance." We have, therefore, located the positions of the two eccentrics in the full stroke forward. We shall now locate them in full backward stroke when the crank pin is at H (the eccentrics are always an unvarying distance from the pin, being securely fastened to the axle), therefore they will be an equal distance from a line drawn from C to H, and they are indicated by the letters F_1 and B_1. We shall next locate them at half stroke. The pin being at T, they will be equal distances from the line C T and are indicated by the letters F_2 and B_2. We will now locate them at W. They will be equal distances from the line C W and are indicated by the letters F_3 and B_3. The points marked F, F_1, F_2 and F_3 indicate the different positions of the forward or go ahead eccentric, and those marked B, B_1, B_2 and B_3 indicate the positions of the backup eccentric.

LINK TEMPLET.

We shall now proceed to make a link templet, which we shall have occasion to use later on (a templet may be made of Russian iron or tin). Fig. 11 shows the outlines of a link, the line R S being the correct radius of the link, and passes through its center; the line I J is the link pin arc and is described from the same point as the radius. The lines P Q and X Y should be parallel to the center line M N, and all three lines should be marked on the templet as shown by Fig. 12. The center of the saddle pin will be located on the line M N, at, or very near, the point where the line K L crosses it (as already explained in Shop Practice).

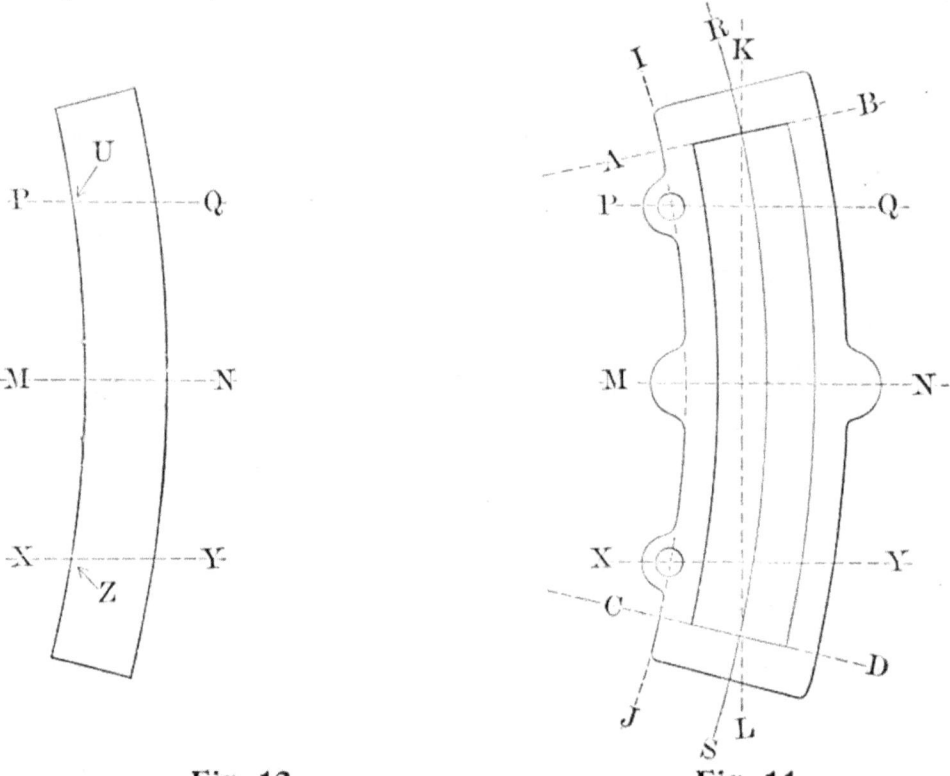

Fig. 12.   Fig. 11.

Now cut out the templet, letting it cover all the space between the lines A B, C D, I J and R S, and we have a link templet as shown by Fig. 12. The points marked U and Z indicate the centers of the link pin holes.

CORRECT LENGTH OF ECCENTRIC BLADES.

We shall now continue where we left off after locating the positions of the eccentrics for full and half stroke.

Draw the center line of motion and the perpendicular line L M the same as before. Locate the position of your eccentrics in full stroke when the crank pin is at G and H, as previously explained. Then locate your rocker arm the right distance from the center of the axle, as shown in Fig. 13, by the letters N O and parallel to the line L M. Find the exact distance between the points C and P, and from this subtract the distance from the link pin arc to the radius, as shown by Fig. 11, and the

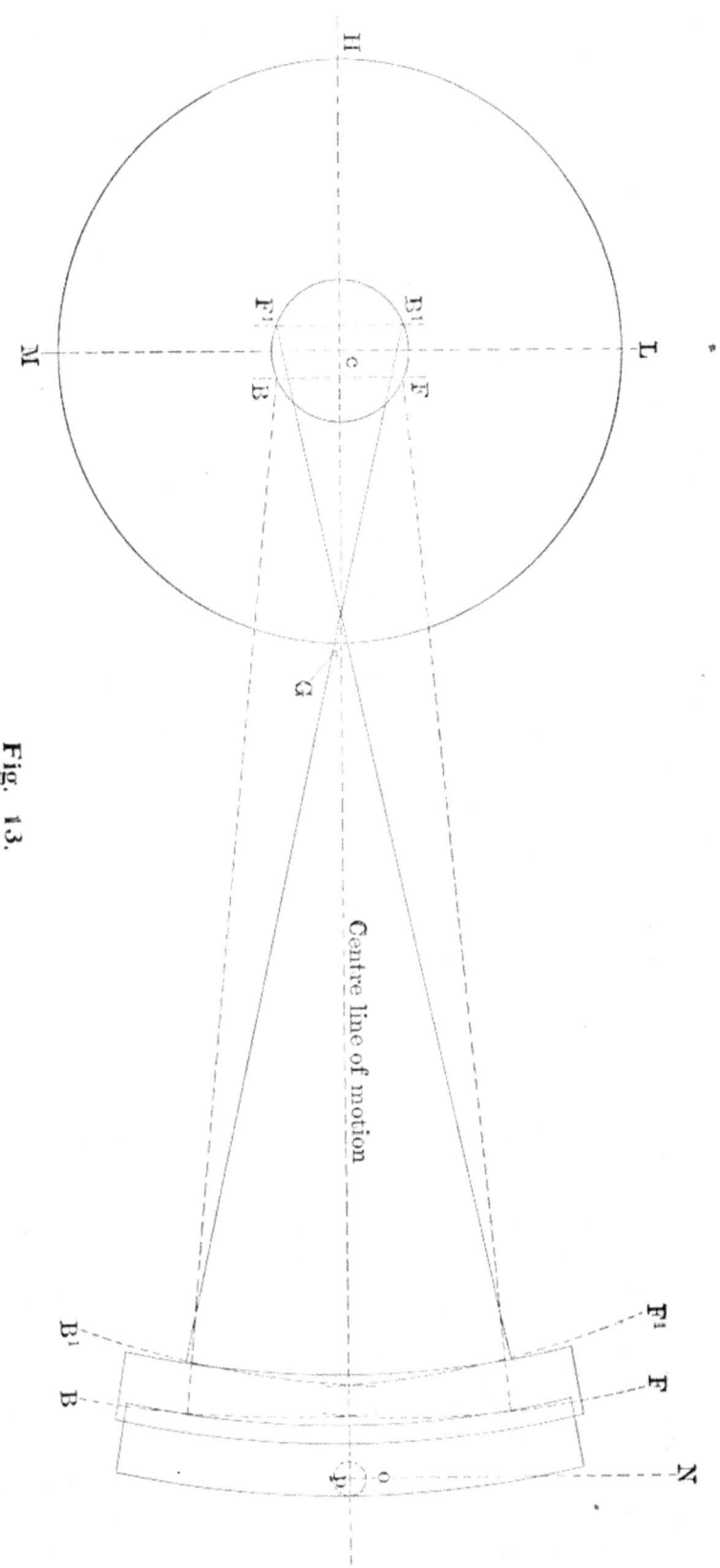

Fig. 13.

remainder will be equal to the length from the center of the eccentric strap to the center of the link pin hole in the blade. Then, with a radius equal to this length, and from the point F, describe the arc F above the center line of motion. Now, if the link is raised or lowered, this arc F indicates the path through which the link pin travels, and what is true of this one arc is also true of all that may hereafter be made. Next, from the point F1 describe the arc F1, and from B describe the arc B below the center line of motion, likewise from B1 describe the arc B1. Then place the templet on the drawing and keep the center M N on the center line of motion and let the two points marked U and Z on the templet just touch the two link arcs F and B. Then mark the shape of the templet on the drawing; this indicates the position of the link, when the crank pin is at the point marked G. Then slip the templet back until the two points marked U and Z just touch the link arcs marked F1 and B1, keeping the line M N on the center line of motion. Again mark the shape of the templet on the drawing; this indicates the position of the link when the crank pin is at H. The forward face of the link templet marked R S, in Fig. 11, indicates the center, or radius of the link; therefore, the front faces of our templet should be equal distances from the point P, which is the center of the bottom rocker arm. Now, by describing from the point P, a small circle tangent to the face of the forward templet, we find the back templet falls a little short of the circle. This is caused by the eccentric blades being crossed when the pin is on the back center, while they are not crossed when the pin is on the forward center. We must, therefore, lengthen our blades one-half the amount of the distance from our small circle to the front face of the back templet, as shown in Fig. 13. This will be the correct length of the blades and when connected up, the rocker arm will then travel an equal distance each way from its central position.

CORRECT POINT OF SUSPENSION OR POSITION OF CENTER OF SADDLE PIN.

The periods of admission and cut off may be equalized by changing the point of suspension of the link, either up or down, or horizontally. A somewhat better distribution of steam can be secured by suspending the link above its center, but in locomotive construction there are practical objections to raising it. We have already explained that the center of the saddle pin would be located on the line M N, Fig. 12, and we shall now determine its exact distance from the front face of our templet, which face is the link radius. Having already found that the inequality in the motion of the piston is greatest when the crank pin is at half stroke (the crank pin and the eccentrics being

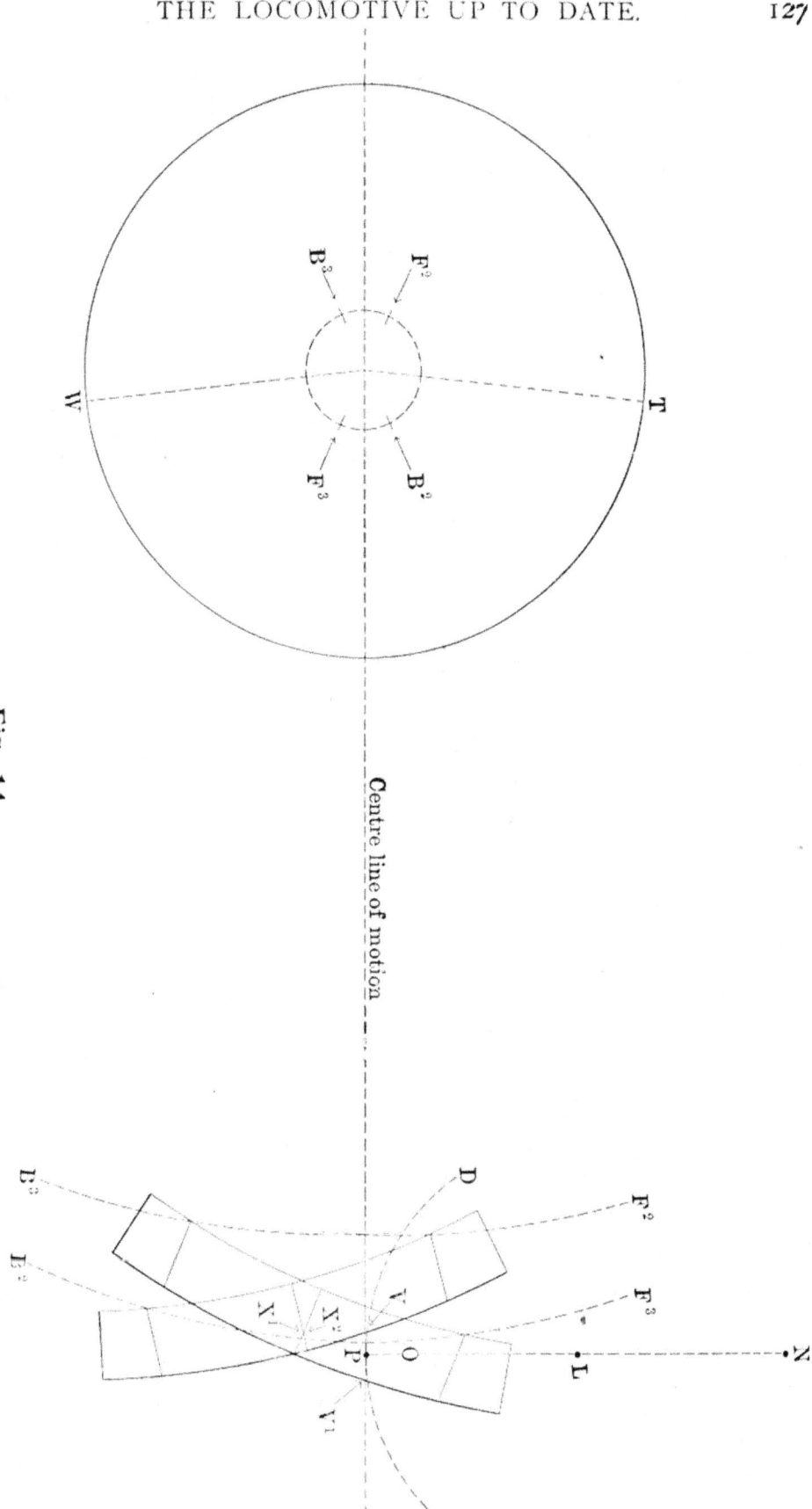

Fig. 14.

rigid), it therefore becomes necessary to locate a position for the saddle pin, so that equal portions of steam will be admitted alternately, and the cut off be equal when the crank pin is at half stroke. Therefore, draw your center line of motion and locate the positions of your eccentrics when the crank pin is at half stroke, points T and W, as previously explained, and shown in Fig. 14. Next locate your rocker arm N O, and from the center of the rocker arm L describe the arc D E now with the correct length of your eccentric rods, and from their respective points, as explained in Fig. 13, describe the four link pin arcs F, F1, B and B1. The forward arcs above the center line of motion, and the back motion arcs below it. Now, when the rocker arm stands in a vertical position, the valve stands central on the valve seat, and we know that the valve has $\frac{7}{8}$" outside lap (as previously explained); therefore the bottom of the rocker arm must move $\frac{7}{8}$" either way from its central position to open the valve, or to reach either point of cut off (assuming that the rocker arms are of equal lengths); therefore from the point P, which indicates the center of the hole in the lower rocker arm, mark off the two points, V and V1, on the arc D E, which represents the path through which the center of the lower rocker arm must travel. Now place the point U of the templet on the arc F2, and the point Z on the arc B2; then move the templet along those lines until its front face, R S, Fig. 11, touches the point marked V. The templet now represents the position of the link when the crank pin is at T, and we know that when the engine is moving forward the valve must have reached the point of cut off at the back port when the crank pin has reached this position.

Now transfer the line M N from your templet onto the drawing for future use. Again place the point U of the templet on the arc F3, and the point Z on the arc B3; then move the templet along these lines until its front face touches the point V1. The templet now represents the position of the link when the crank pin is at W, and we know that, in the forward motion, the valve must at this point be cutting off steam at the forward port; so once more transfer the line M N from the templet onto the drawing. Now, with a length equal to the distance between the lines R S and K L and on the same line M N, Fig. 11, and from the forward face of each templet, and on the line M N lay off the two points X1 and X2. If a straight line drawn through these two points is parallel with the center line of motion, then these points indicate the correct position of the center of the saddle pin. If not, then by trial locate two points that are equally distant from the front faces of the templets and that are also parallel to the line of motion; these points will be the correct position of the saddle pin. When the correct position is found, mark it on the templet for future use.

RELATIVE POSITION OF THE TUMBLING SHAFT AND ROCKER, AND LENGTH OF THE TUMBLING SHAFT ARMS.

The saddle pin is usually located in a position to obtain an equal cut off at half stroke, where the irregularities introduced by the crank pin are greatest and the tumbling (or lifting) shaft is located in a position to obtain an equal amount of lead in full stroke. Owing to the irregularities of crank motion it is impossible to obtain an equal lead, and an equal cut off at all points: If one is equal then the other will not be; this is one of the imperfections of link motion, but the difference is so slight in full gear that the cut off is considered of less importance than an equal amount of lead at the beginning of each stroke, therefore the tumbling shaft is located and the length of its arms determined to obtain the latter result. In making Fig. 15 we must combine all the foregoing problems. Make the center line of motion, locate the eccentrics at full and half stroke, locate the rocker arm N O and from its center L describe the arc D E. Now, from the position of each eccentric and with the correct length of the eccentric rods, describe each link pin arc as shown in Fig. 13, make all the forward motion arcs above the center line of motion and all the back motion arcs below it. Now locate the points X_1 and X_2 as explained in Fig. 14, which indicates the points of suspension when steam is cut off equal at half stroke. Mark these two points on your drawing, Fig. 15. Now, to determine the length of the tumbling shaft arms and the position of the tumbling shaft, we must find the points of suspension in full forward, and full backward strokes at each end of the cylinder. The valve has $\frac{7}{8}$" outside lap and 1-16" lead, therefore add the amounts together and locate the points V and V_1 on the arc D E, and 15-16" from the point marked P. Now place the templet on the drawing with the line M N below the center line of motion, place the point of templet marked U on the arc F and the point marked Z on the arc B, and move it along these lines until its front face touches the point V_1, and then mark the point X_3 on the drawing; this point will indicate the position of the saddle pin when the crank pin is at G in full gear, forward motion. Again place the templet on the drawing with the line M N below the center line of motion, place the point U on the arc F_1 and Z on the arc B_1 and move the templet until its front face touches the point V; now mark the point X_4 on the drawing. This point indicates the position of the saddle pin when the crank pin is at H in full gear forward motion. Now, we must find the points X_5 and X_6 in exactly the same manner, and using the same link arcs as before, only that the line M N of the templet must be above the center line of motion. We have not outlined the templet in these positions as it would make the drawing appear more complex, but we

130 THE LOCOMOTIVE UP TO DATE.

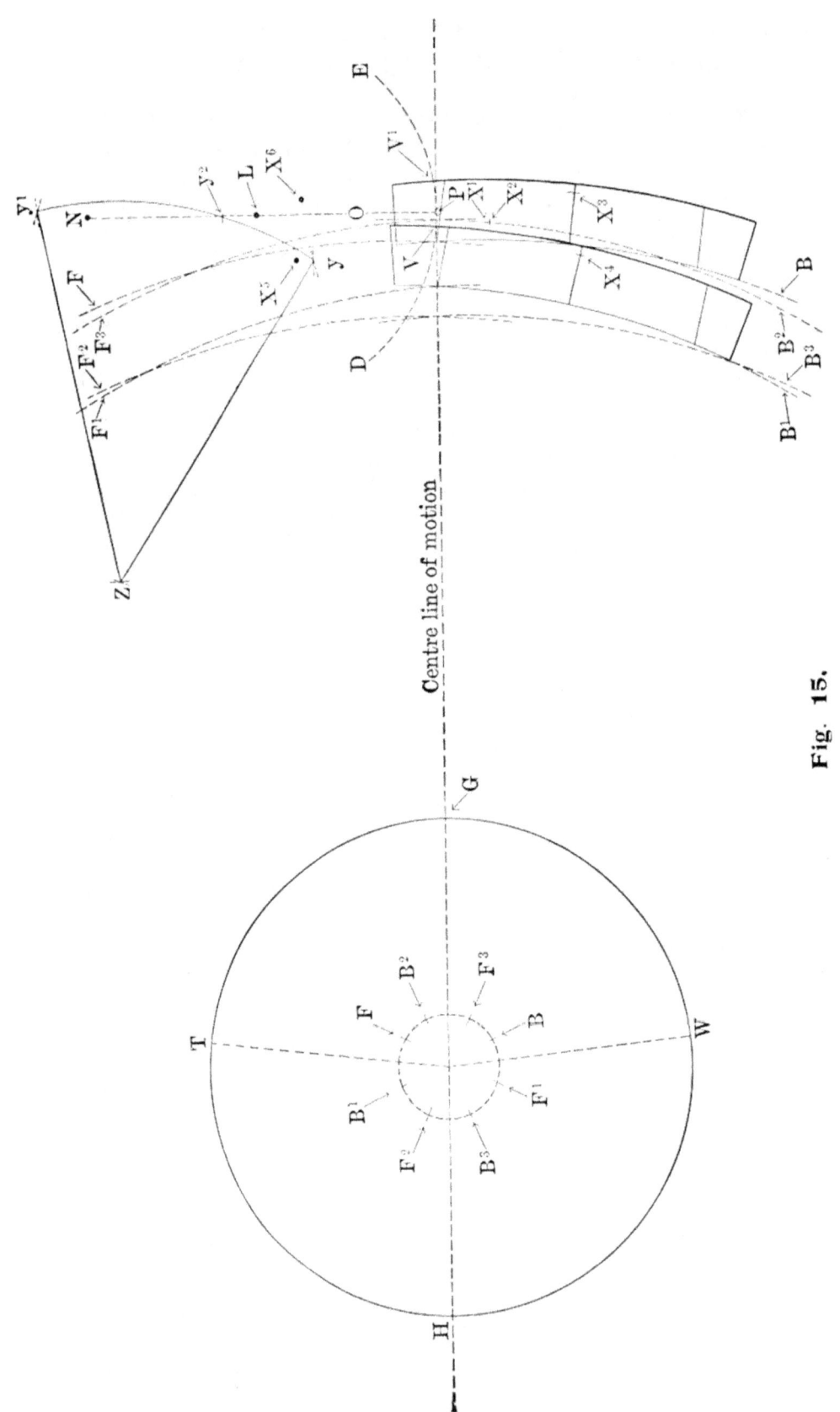

Fig. 15.

shall continue the explanation. Place the point U on the arc F and Z on the arc B, front face of the templet touching the point V1. Now mark the point X6. Again place the point U on the arc F1 and Z on the arc B1, letting the front face of the templet touch the point V. Then mark the point X5 on the drawing. These two points indicate the position of the saddle pin when the piston is at each end of the cylinder, in full gear, backward motion. Now from the two points X3, X4, with a radius equal to the length of the link hanger, find the point marked Y, and from the points X5, X6, with same radius, find the point Y1, and from the points X1, X2, with same radius, find the point Y2; now from these three points, Y, Y1, Y2, find the point Z on the drawing which indicates the position of the center of the tumbling shaft and the distance from the point Z to either of the three points, Y, Y1, Y2, equals the length of the tumbling shaft arms. The length of the top arm of the tumbling shaft is not so particular, as it is made to suit other details of the engine.

ERRORS OF THE LINK MOTION.*

As locomotives are built, there are three sources of error which tend to make cut-off, release, and compression occur at different points in the stroke for the two ends of the cylinder. These sources of error, in the order of their importance, are, the offset of the eccentric rod pins back of the link arc, the angular vibration of the eccentric rods and the angular vibration of the connecting rod. To a certain extent the latter two compensate the first, but not entirely, and to complete the compensation the hanger stud is set back of the link arc. So far as I am aware, the importance of the first two sources of error has not before been recognized.

All previous discussions of the link motion with which I am acquainted, proceed upon the assumption that the hanger stud is adjusted to correct the irregularities due to the connecting rod, although, in point of fact, the adjustment made is the direct opposite of what would be required if this were its purpose.

If a link motion be laid down with the Scotch yoke connection between the piston rod and crank pin, which obviates the error due to the connecting rod, the offset of the saddle pin necessary to obtain an equalized cut-off will be found to be greater than if the connecting rod be introduced, and if a connecting rod be shortened, this offset will be found to diminish with each shortening of the rod, until, at some very short length, the stud will be placed over the link arc. In other words, the connecting rod, instead of being a disturbing factor, as has

---

*These articles are from the pen of Mr. Frederick A. Halsey, M. E., associate editor of the *American Machinist*.

heretofore been taught, is in reality a corrective factor, since it, to a certain extent, corrects other errors, and in so far as it corrects these errors it reduces the offset of the saddle stud, the remaining offset being for the purpose of correcting the residual error due to the offset eccentric rod pins.

Before taking up the study of these errors in detail, it is advisable to examine the movement as a whole, in order to understand the means at command for accomplishing the required purpose. In this connection one fundamental fact should be kept in mind, namely, when the valve admits or shuts off steam it is displaced from its middle position by an amount equal to its lap. This will be apparent by reference to Fig. 1, in the chapter on Locomotive Slide Valves, which is a cross-section of such a valve located centrally upon the valve seat. It will be seen that both ports are covered on the outside by the lap, and that to move the valve to the admission or cut-off position for the right-hand port involves a movement to the left equal to the lap, while to move it to the admission or cut-off position for the left-hand port involves a similar movement to the right, and in all cases cut-off takes place with the center of the valve approaching the center of the seat.

THE ADJUSTMENT OF THE SADDLE STUD.

The adjustment of the residual error is accomplished by so suspending the link that it raises and falls in the course of its movement, in consequence of which the point acting upon the link block at cut-off in the rearward stroke is different from that acting at the forward stroke. Points of the link near the center give earlier cut-offs than those removed from the center, and it is obvious that by so suspending the link that the point acting upon the link block is different for the two strokes, the two points of cut-off may be altered as desired.

There are two methods by which this movement of the link can be accomplished. One, which is in universal use, consists of placing the saddle stud back of the link arc, the effect being obviously to cause the arc to rise and fall during its oscillation. The second method consists of so locating the tumbling shaft that the link hanger does not oscillate equally each side of the vertical line, but more on one side of this line than the other, the effect of which is obviously to cause the entire link to bodily rise and fall during its movement. These two methods are usually described in detail in discussions of this kind, with diagrams showing how the location of the suspension stud and of the tumbling shaft may be laid down upon the drawing board. In point of fact, however, they are not so located, and, moreover, the second method is seldom employed, as the choice of location of the reverse shaft box is usually quite restricted, and the designer is not at liberty to place it in the position which this consideration would indicate.

The location of the saddle stud is determined by trial upon the engine itself, an adjustable stud being provided, which is bolted to the link, when, by trial adjustments, the proper position is found. The link is then removed from the engine, and with the adjustable stud is taken to the link shop, where the permanent stud is made in accordance with it. In the case of a number of duplicate engines gotten out at the same time, the adjustable stud is applied to the first one only and the following engines of that lot have their permanent studs made in duplicate.

In Fig. 16 o is the central or neutral point of the movement of the lower rocker arm and pin, at which point the valve stands centrally over its seat. This point is found in the diagram by placing the link in the mid gear and the crank on the two centers successively, these positions being shown in Fig. 20. In these positions of the crank the valve and lower rocker arm pin occupy the extreme points of their travel for the mid gear and a point half way between the extreme points of the pin, that is half way between a and b of Fig. 20, locates o of Figs. 16 and 20. Measuring to the right and left of o a distance equal to the lap (the two rocker arms being supposed to be of equal length), locate points i and n, Fig. 16, at which the ports are opened or closed, according to the direction of the movement.

The valve sketches above and to the right in Fig. 16 show the valves in these positions, the upper sketch showing the valve in the act of cutting off steam for the rear port, while the lower sketch shows a similar action on the forward port. The sketch shows the link in skeleton diagram suspended in the usual manner by a hanger, which again is suspended from the reverse shaft arm. It will be seen that this hanger is not attached to the link over the center of the link arc, but at a considerable distance in the rear of this arc, and it will be seen at once from the diagram that the point i of the link at which it acts upon the link block for the forward port cut-off is farther removed from the center t of the link arc than is the point n at which it acts upon the block for the rear port cut-off. In other words, the link is nearer the full gear position for the forward port cut-off than it is for the rear port cut-off, in consequence of which the forward port cut-off is made later and the rear port cut-off earlier than they would be if the saddle stud were placed immediately over the link arc.

The positions shown in Fig. 16 may be traced through with advantage as follows: If the crank be placed upon the forward center OA, the forward eccentric will occupy the position a and the backing eccentric the position b, while if the crank occupies the back center OB, the forward eccentric will be at a^1 and the backing eccentric at b^1. If the cut-off is to be equalized

134 THE LOCOMOTIVE UP TO DATE.

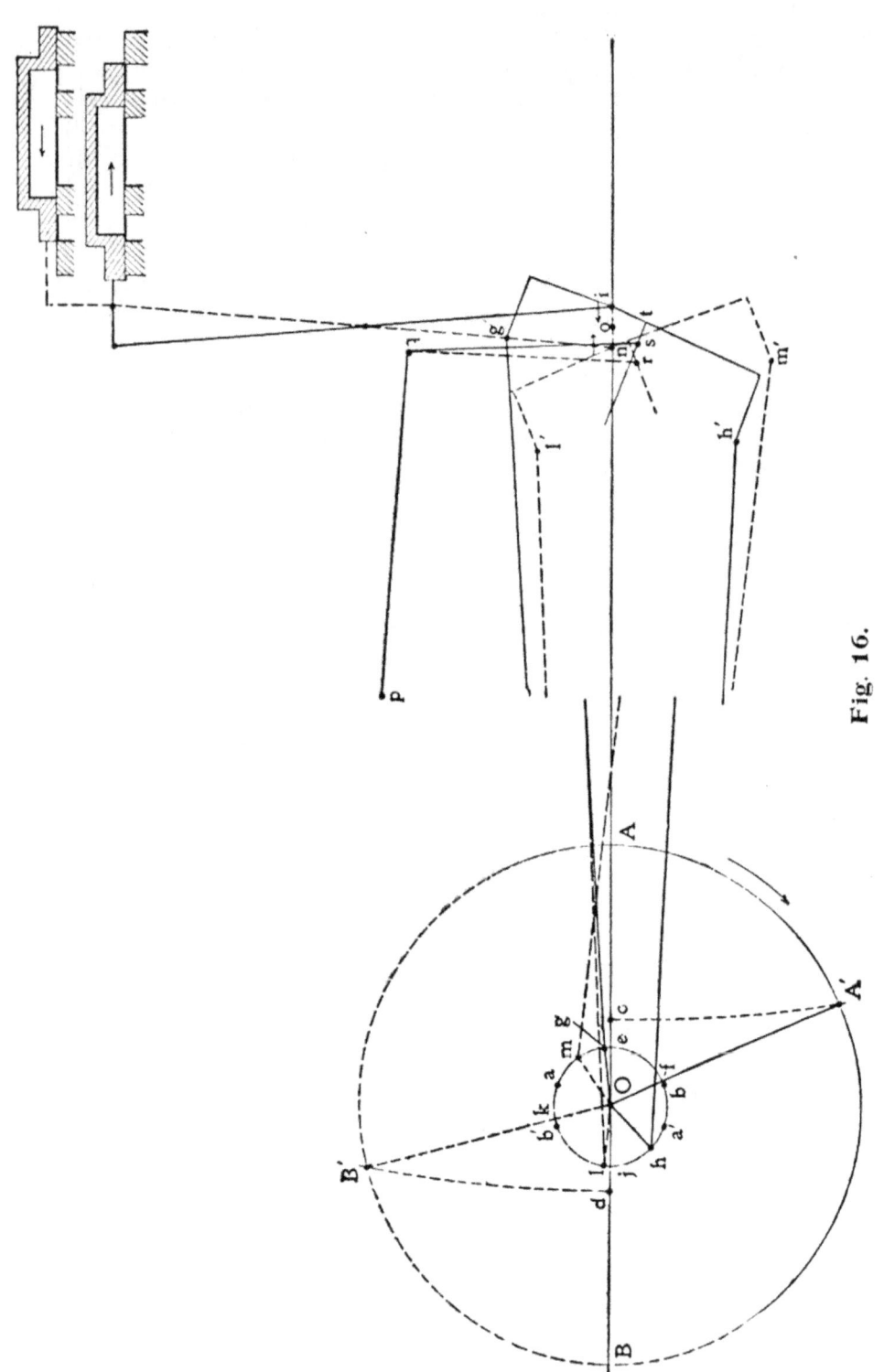

Fig. 16.

at one-third stroke, points cd may be laid down such that Ac and Bd equal one-third of the stroke.

Then with a radius equal to the length of connecting rod, arcs cA^1 and dB^1 may be drawn giving crank position OA^1 which the crank occupies at one-third stroke of the piston in the rearward motion, and OB^1, which it occupies at one-third stroke of the piston in the forward motion. Taking the distance ef in the dividers and laying it down from a and b, point g is obtained, which the forward eccentric occupies, and point h, which the backing eccentric occupies, when the crank is at OA^1. These points locate the eccentric rod pins at g^1, h^1 and give the link position shown in the full line, for cut-off at crank position OA^1. Similarly by laying off jk from a^1 and b^1, we obtain point l for the forward eccentric and m for the backing eccentric when the crank is at OB'. These points again locate the link in the dotted position for cut-off at crank position OB^1. It will be seen that if the reverse shaft arm pq be located as shown, so that the hanger swings equally each side of the vertical, points rs will occupy a horizontal straight line, and if the reverse shaft be so located that its arm is in the horizontal position when the link is raised to the mid gear, it will be raised as much above the horizontal line for one-third cut-off in the backing motion as it is here in the forward motion, when r and s will again occupy a horizontal line above the center line and the cut-off will be equalized for the backward motion as the diagram shows it to be for the forward motion, and it is in this way that the reverse shaft is located. It will be seen that this method of hanging the link introduces the element of slip by which the link rises and falls on the block. Formerly it was thought desirable to reduce this slip as much as possible, and even to be satisfied with a motion which was not perfectly equalized in order to accomplish this, but at the present time constructors do not seem to be afraid of considerable slip.

THE ERROR DUE TO THE ANGULAR VIBRATION OF THE CONNECTING ROD.

As has been explained, the offset of the saddle stud is introduced to compensate errors and bring the cut-off at the same point at both strokes. Of these errors the first in order of description, but the last in order of importance, is that due to the angular vibration of the connecting rod, the nature of which will be apparent from Fig. 17, which is a skeleton diagram of the cross-head slides, connecting rod and crank pin. It is obvious that if the connecting rod be disconnected from the crank pin and the center of the cross-head pin be placed at the center of the slide o, and the connecting rod end be then swung through an arc passing through the center of the shaft

O, this arc will cut the crank pin circle at points c d, which points do not occupy the quarter positions of the crank as they would do if the Scotch yoke were used in place of the connecting rod. The result of this error is that during the rearward movement of the cross-head from a, the crank lags behind its correct position, while during its forward movement from b, the crank runs ahead of its correct position—these errors existing at all points but being at a maximum at or near the half stroke. The study of the effect of this action upon the link is most easily made by separating it from the other error, and as there is a similar error due to the angularity of the eccentric rods, the study of the connecting rod error requires that the eccentric rod errors be gotten rid of by assuming the horizontal movements of the eccentric rod pins in the link ends to be truly the same as those of the eccentric centers. This assumption of no angular swing, on the part of the eccentric rods, involves the

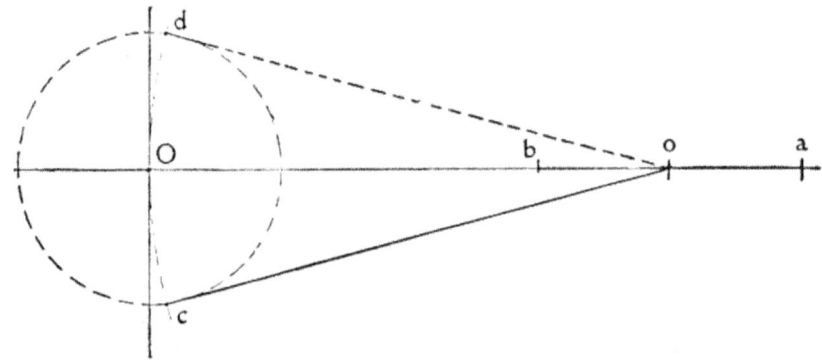

Fig. 17.

further assumption of a straight link. Similarly, the elimination of the error due to the offset eccentric rod pins, requires the location of these pins to be on the link center line. Fig. 18 has been made in accordance with these requirements, and shows the link positions for the crank positions of Fig. 16, with the errors due to the connecting rod included.

These errors in the position of the crank are obviously repeated in the eccentrics and link, that is for the rearward movement of the piston. When the crank lags behind its proper position, the eccentrics and link will do the same, and cut-off will not have occurred at the point desired. For the forward or return stroke the conditions are reversed, the crank, eccentrics and link being ahead of their correct positions, and cut-off having already occurred at the desired point of the stroke. This condition of things is shown in Fig. 18, in which the cut-off points i and n of Fig. 16 are repeated. The arrows show the direction of the motion, and it will be apparent at once that the full link has not reached the cut-off point, while the dotted line link has passed it. It is clear that these errors could be corrected by slightly raising the full line, and dropping the dotted line position, and to do this only requires that the link stud be placed outside the link center line, as shown in Fig. 19, and this adjustment of the stud will be seen to be the exact reverse of that

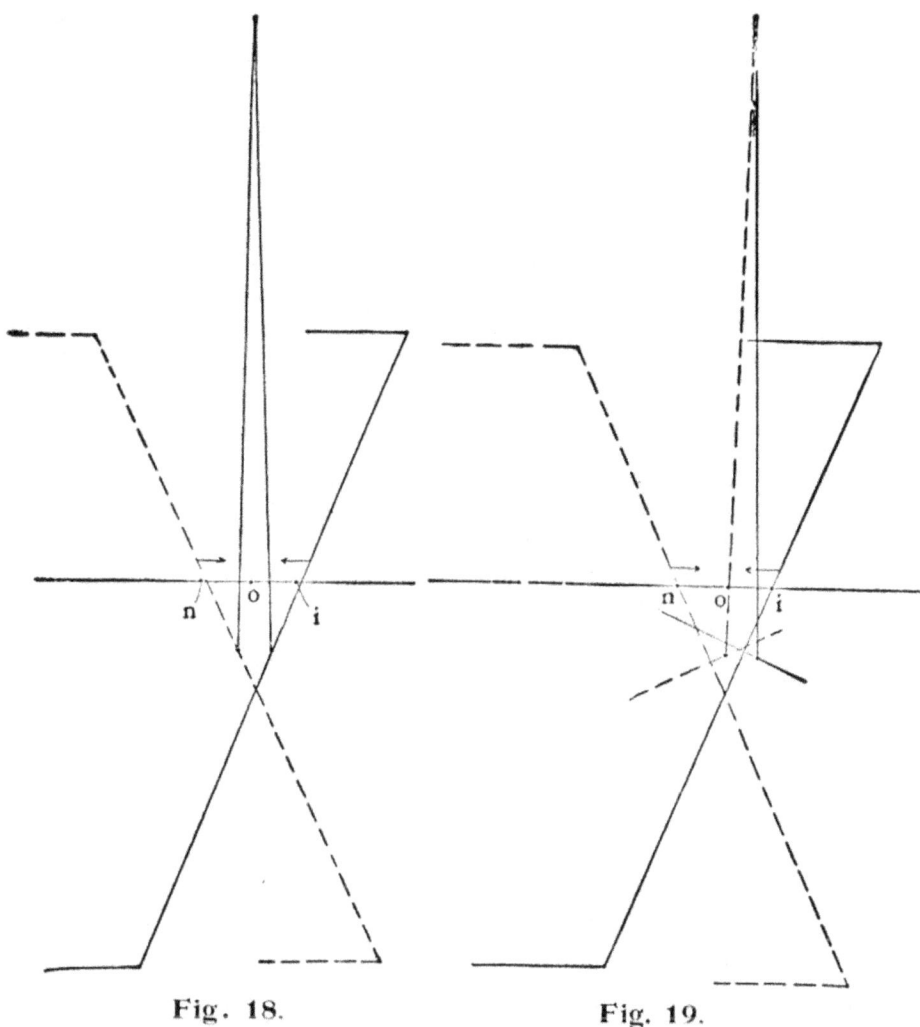

Fig. 18.   Fig. 19.

actually followed in locomotives—demonstrating the position here taken, that other errors override that due to the connecting rod.

THE ERROR DUE TO THE ANGULAR VIBRATION OF THE ECCENTRIC RODS.

It is apparent at first glance that the action between the eccentrics and link ends is in a sense similar to that between the crank and cross-head. There is, however, an important difference. Reference to Fig. 17 will recall the fact that with the connecting rod the error is zero at the centers and at its maximum near the quarter, but this is not the case with the eccentric rods, because the paths of the link ends do not pass through the

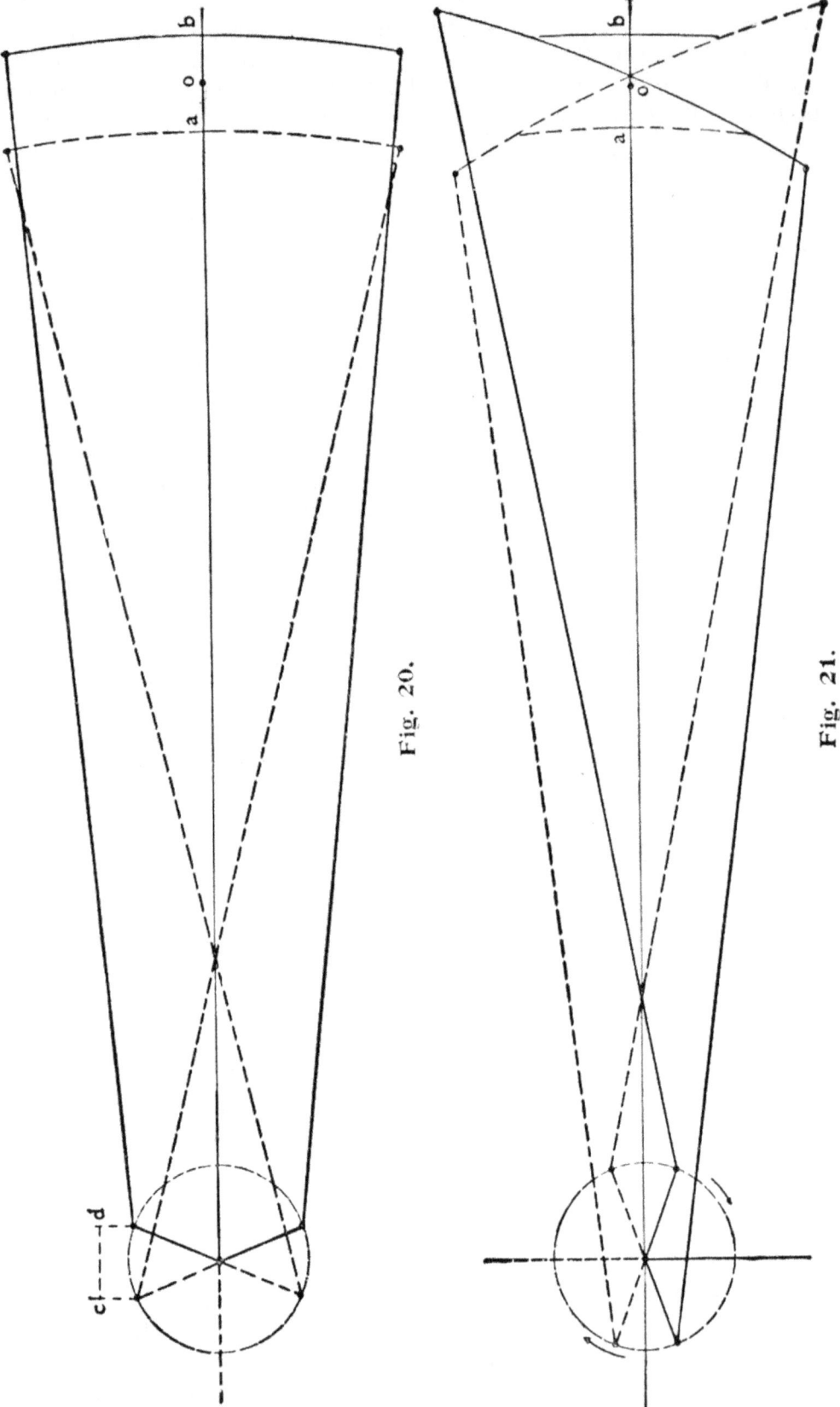

Fig. 20.

Fig. 21.

center of the shaft. Referring to Fig. 20, it will be seen that the average angle between the eccentric rods and the center line is smallest in the full line position, and that this angle increases during the entire semi-revolution and becomes a maximum at the dotted line position. In other words, the distortion, instead of increasing to a maximum for 90 degrees of rotation and then decreasing again, really increases to a maximum at 180 degrees of rotation. This increasing angle of the rods increases the

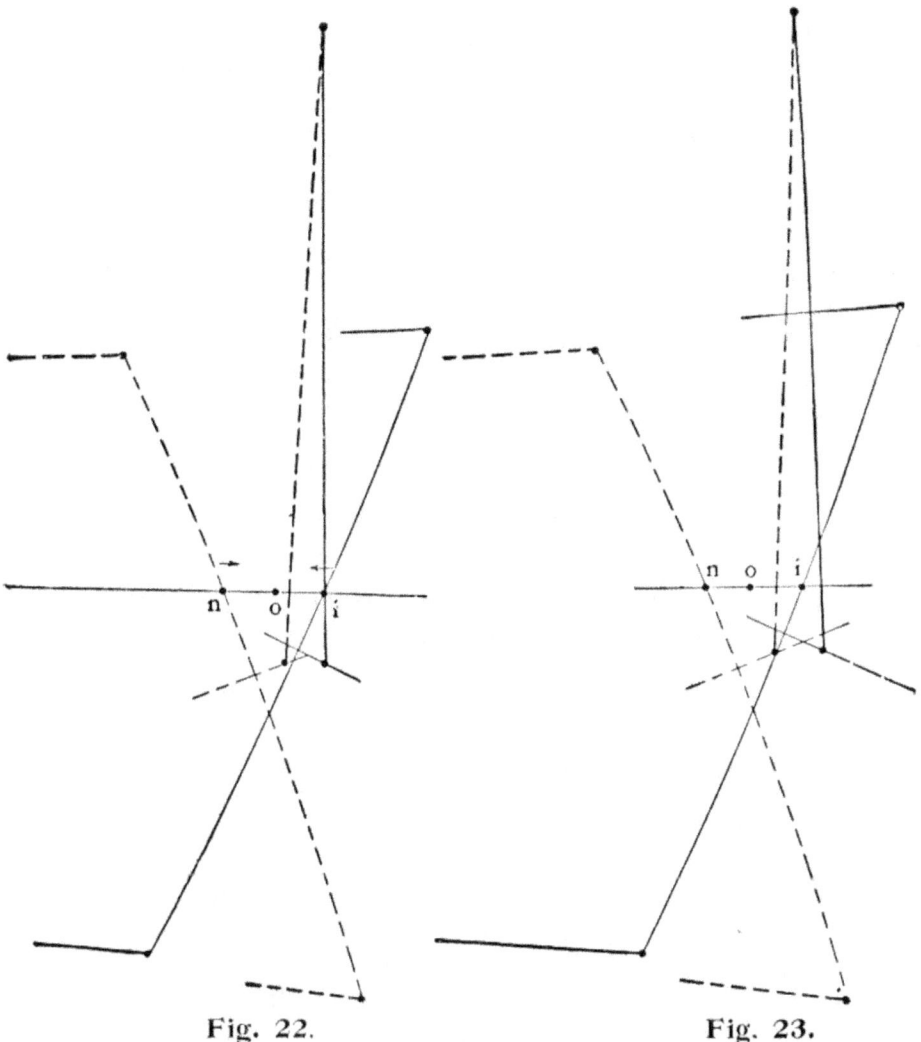

Fig. 22.   Fig. 23.

movement of the link, and whereas the stroke of a cross-head is exactly twice the length of the crank, the movement of the link center $a\ b$ is materially more than the distance $c\ d$ (in the case of this diagram nearly 50 per cent. more). Starting at b, the error in the position of the link center steadily increases during the movement toward a. The errors being greatest during the second quadrant of rotation, they increase the movement more during the second quadrant than the first. In other words, the movement of the link center is greater during the second quadrant than the first. Consequently if a diagram like Fig. 21 be laid down it will be found that the first quadrant of movement from the full line position of Fig. 20 to that of Fig. 21 leaves the link center appreciably to the right of the natural point o, and similarly a quadrant of movement from the dotted line position of Fig. 20 to that of Fig. 21 will carry the link center to the right beyond o, and the same is true of any other angle of rotation. In other words, the movement of the link center toward the left is too slow, while the movement to the right is too fast. The rocker reverses these movements on the valve, leading to too slow a movement to the right with too late a cut-off on the forward port or rearward stroke, and too fast a movement to the left with too early a cut-off on the rear port or forward stroke. These effects are obviously in the same direction as those due to the connecting rod error, and the two are, in fact, added together in an actual locomotive. The correction of this error obviously requires an adjustment of the saddle stud in the same direction as that made to correct the connecting rod error. Fig. 22 shows this in amount, this diagram having been constructed with the connecting rod error eliminated, and the offset of Fig. 22 will be seen to be greater than that of Fig. 19, for the connecting rod alone. Fig. 23 shows the offset necessary to correct the errors of both connecting and eccentric rods—the amount being approximately the sum of the offsets of Figs. 19 and 22.

THE ERROR DUE TO THE LOCATION OF THE ECCENTRIC ROD PINS BACK OF THE LINK ARC.

In the previous study of the movement of the link, the eccentric rod pins were assumed to be located in the link arc, and in previous discussions of the subject it has been tacitly assumed that the errors introduced by setting these pins back of the arc are so small as to be negligible. This, however, is by no means the case, this error being, in fact, by far the most important of the three, over correcting, as it does, both the others, and resulting in finally locating the saddle stud inside the link arc, instead of outside, where the previous errors alone would place it. This error, like the others, may be best studied by insulating it so far as possible, although it is not possible to separate it from the eccentric rod error, as will be seen. It is, however, possible to separate it from the connecting rod error. The nature of the error may be seen from Fig. 24, which shows both forms of link, one having the eccentric rod pins located in the arc, and the other having these pins located three inches back of the arc, as is customary. The eccentric rods for the former link are, of course, three inches longer than for the latter. The saddle stud is located over the center of the link arc, as is shown in the diagram, and the links are shown approximately in the positions which they would occupy for a cut-off at one-third stroke, the full line links being in position for the rearward stroke of the piston, and the dotted line links being in position for the forward stroke. The movement of the crank is supposed to be by a Scotch yoke, so that no connecting rod errors are introduced. It will be seen at once that the setting of the eccentric rod pins back of the link arc makes the lines joining the extremities of the arc and the centers of the eccentric crooked,

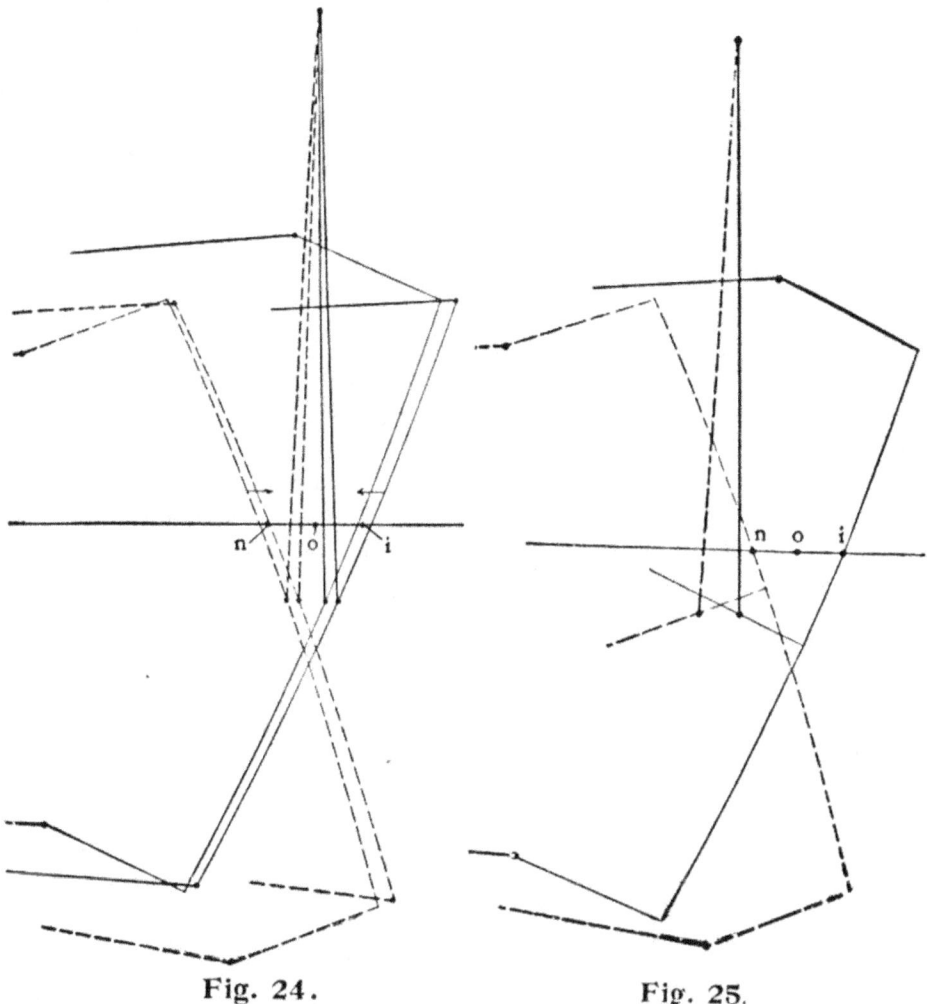

Fig. 24.   **Fig. 25.**

whereas with this pin located on the arc, this line is of course straight; consequently the effect of placing these pins back of the arc is for the positions shown, to draw the link having the offset pins nearer the shaft than the link which has the pins on the link arc. The action is that of a knuckle joint, any bending of which must draw the link toward the shaft. The rock shaft reverses this action on the valve, so that the drawing of the link toward the shaft pushes the valve away from the shaft. Pushing the valve away from the shaft quickens the cut-off for the front port or rearward stroke, and delays it for the rear port or

142   THE LOCOMOTIVE UP TO DATE.

forward stroke; that is, the effect of the offset pin is to make the cut-off too early in the rearward stroke and too late in the forward.

This effect will be seen to be the direct opposite of those produced by the connecting and eccentric rods, and it obviously calls for an adjustment of the saddle stud in the opposite direction, as shown in Fig. 25, which shows the position of the saddle stud necessary to equalize the cut-off at one-third stroke with the Scotch yoke connection, the stud being on the concave of the link, where it is in all cases located in actual engines.

THE FINAL OFFSET.

It is obvious that the final offset of the stud is the resultant of all three. The offset of the eccentric rod pins varies within

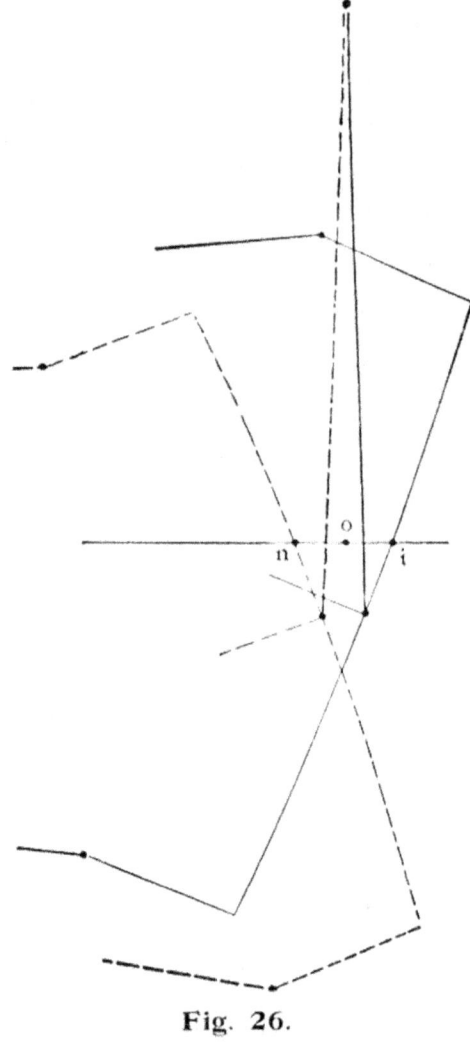

Fig. 26.

narrow limits only, but the length of the eccentric rod varies within wide limits. It is obvious that since the error of short rods alone would place the stud farther outside the link arc than long ones, they subtract more from the offset of the stud due to the eccentric rod pins than long ones, the result being that the final offset is less with short arms than with long ones; and this is found to be the case, the offset in actual engines ranging between about five-eighths of an inch and one and a quarter inches, depending on the length of the eccentric rods. The connecting rod also varies in length, but its influence is so small that the variation in its length between usual limits has but little effect on the final result. To illustrate this, Fig. 26 has been constructed, in which the connecting rod has been shortened by trial until the offset of the saddle stud disappeared, placing that stud over the link arc. With the other proportions unchanged, it was found necessary to shorten the connecting rod to a remaining length of three feet before this result was accomplished. In the diagram in which the saddle stud is located the following proportions have been used:

Stroke of piston, 24 inches.
Length of connecting rod, 91 inches.
Radius of link arc, 69 inches.
Length of link between pin centers, 12 inches.
Offset of eccentric rod pins, 3 inches.
Travel of the valve, $5\frac{1}{2}$ inches.
Lap, $\frac{7}{8}$ inch.

WARREN'S IMPROVED LINK MOTION.

The illustration herewith shown is claimed to be a correction for the errors of link motion. It was patented by Mr. W. B. Warren, General Foreman of the T. P. & W. Ry. of Peoria, Ill. Mr. Warren says it has been in use upon several locomotives on that road since 1894, giving very good results with but little wear on the blocks and face of the links. The only practical advantage claimed for this improvement is doing away with the "slip;" it is claimed the link will produce an equal cut-off in each end of the cylinder in the forward motion and an equal cut-off in the back motion when compared with the distance the wheel travels upon the rail. When a long eccentric rod is used for the forward motion (both styles being in use) a curved link hanger is used to clear the long eccentric rod coupling pin. The long eccentric rod pin is connected to the link by means of a saddle which can be located for best cut-off results and then drilled and bolted permanent. The lead is not effected by this method of coupling; as at that point the link is almost perpendicular it permits of almost a perfect admission and release.

Mr. Warren says he has never used an indicator, but has taken all points of cut-off, admission and exhaust from 6" in cut-off to full stroke, and all points show good results.

The following is Mr. Warren's complete description of this device:

Figure I is a side view, illustrative of the invention; and Fig. II is a side view of the skeleton shifting link.

Referring to the drawings, 1 represents the main driving shaft, and 2 the crank pin.

3 represents the forward motion eccentric on the shaft 1, and 4 the backward motion eccentric, which is also on the shaft 1.

5 represents the rod of eccentric 3, which is connected to the eccentric by a strap 6, and 7 represents the rod of the eccentric 4, which is connected to the eccentric by means of a strap 8.

9 represents the skeleton link, with which the rod 5 is connected at 10, and with which the rod 7 is connected at 11.

12 represents the link block, 13 the lower rocker arm, 14 the upper rocker arm, 15 the saddle pin, 16 the link lifting hanger, 17 the tumbling shaft, 18 the tumbling shaft lifting arm, 19 the tumbling shaft vertical arm, 20 the operating rod, which extends to within reach of the engineer, 21 the main valve, 22 the main valve seat, and 23 the stem of the main valve. These parts, in themselves, are all old, and require no specific explanation of their operation.

This invention resides in the manner in which the outer ends of the eccentric rods are connected to the skeleton link, with relation to the saddle pin 15, whereby a more perfect working of the parts is attained. To accomplish this object, first establish a center line A (Fig. II), from which lines B, B, C, C, and D, D, are laid off perpendicular and at right angles to the center line, and these lines are equidistant apart on the center line. Then mark off lines E, E, and F, F, parallel with the center line, and equal distances from the center line. Next locate the center of the saddle pin 15 (which is the center of suspension and axis for link 16), where the center line and the center arc line G, G, intersect with the center line C, C. Then locate the coupling pin hole 10 at the intersection of the lines E, E, and D, D, and locate the coupling pin hole 11 at the intersection of the lines B, B, and F, F, and thus have the coupling pin holes located at different distances from the main shaft and each located on a different side of the center arc G, G, and they are located at equal distances, in a perpendicular measurement from the center of the saddle pin 15, and are likewise located at equal distances in a horizontal measurement from the center of the saddle pin, thereby forming a straight line H, H, through the center of the saddle pin, the coupling pin hole 10, and the coupling pin hole 11, so that the point of connection between the eccentric rods 5 and 7, with the link 9, each travel an equal distance to or from the central line of motion at movement of the link either along the center line of motion or rotation on its center of saddle pin 15. Thus equalizing and proportioning to better advantage, the angularity of the eccentric rods, thereby producing more correct and equal cut off and exhaust of the openings in the main valve at all points of the stroke of the engine, and this is done without sacrificing the lead or admission

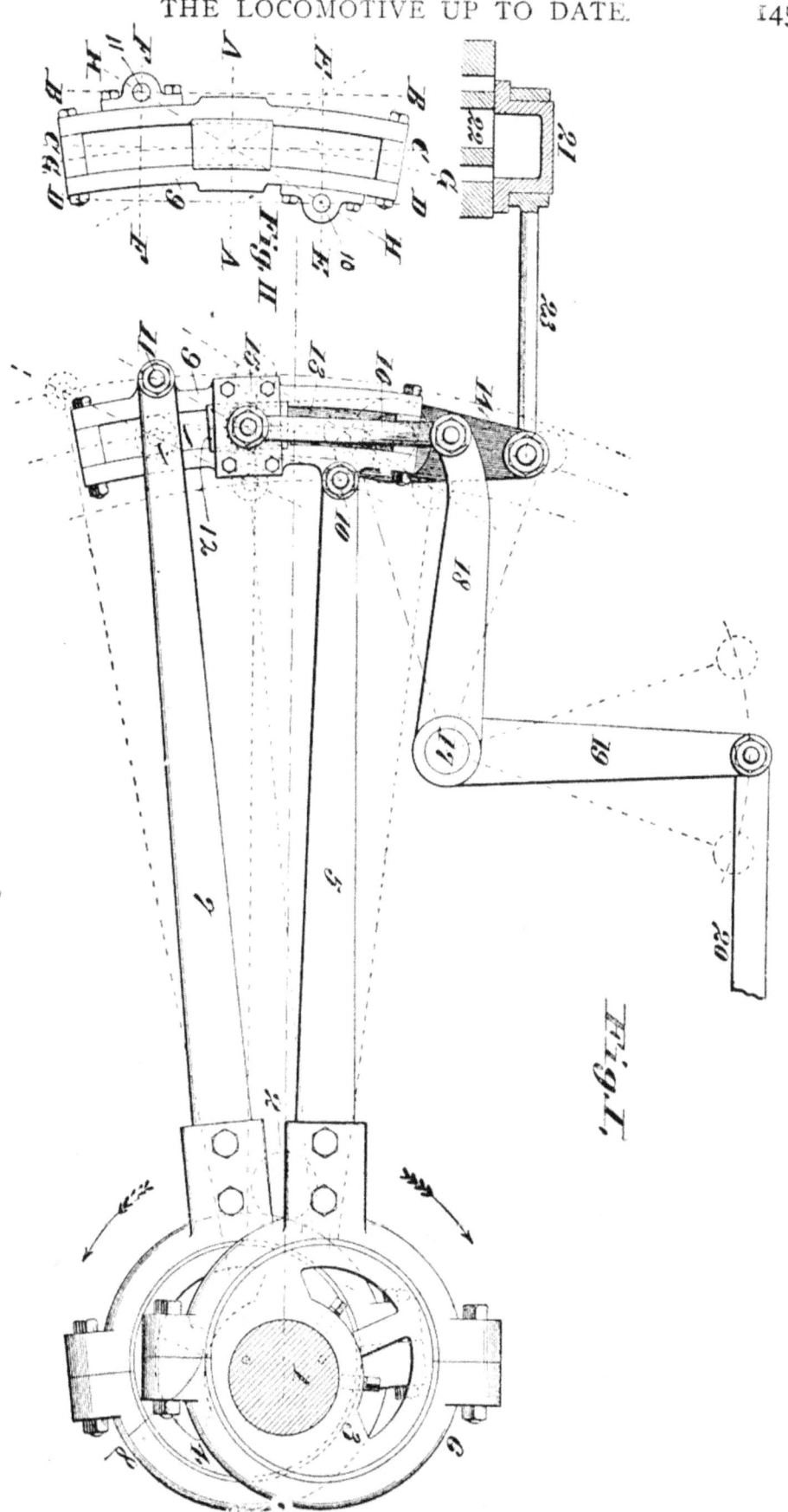

openings of the main valve. By this arrangement also, I am better able to locate the center of the saddle pin on the center arc G, G, of the link 9, thus reducing the slip of the link on its block, and reducing the strain and wear of all parts of the link motion and doing away with the customary way of equalizing the main valve cut off, and exhaust openings, which was done by locating the center of saddle pin 15 back of the center of arc G, G, and link 9, which caused slipping of the link on the block 12, resulting in wear and strain on the link motion. These advantages are attained without complication, or without the use of more parts than usually employed in link motions, and the improvement can be applied to all shifting link motions in use, and at a small cost.

It is evident that I can change the location of the holes 10 and 11, in a horizontal direction, the required amount, to correct and proportion any point of cut off and exhaust openings of the main valve, and still retain the center line of the saddle pin on the intersection of the lines, as explained.

LOCOMOTIVE VALVE SETTING.

INTRODUCTION.

This is a subject that has always been enshrouded with more or less mystery, yet there is nothing mysterious about it, and any man of ordinary intelligence who will give the subject a reasonable amount of careful study can master the principles of valve motion; and yet to understand it thoroughly, will require more diligent study than some might at first suppose. It is an object of ambition with every young mechanic to learn how to set valves; and it is also fast becoming a necessity for enginemen to learn the principles of valve motion, as each succeeding year Railway Officials are exacting of them a more thorough knowledge of the mechanism of the locomotive, as is shown by the examination questions the Firemen are required to answer before promotion. And since long continued service and familiarity with the locomotive will not teach the principles of the valve's action, it is necessary for all those who wish a thorough knowledge of this subject to secure some good book that treats upon locomotive link motion and study it at home, at the same time verifying what they learn from the book by observation and study of the engine itself. In this work we shall endeavor to explain even the minutest details and thereby make this subject sufficiently clear to be understood by anyone even though, at the start, his knowledge of the locomotive be very limited.

WHAT VALVE SETTING IMPLIES.

Locomotive valve setting means the adjustment of the valve gear so that each valve will be in such a relation to the piston that when steam is admitted to the steam chest it will also be permitted to enter one end, of at least one cylinder, and thereby set the engine in motion, and the movement of the valves must be such as to permit the driving wheels to revolve continuously in one direction until the valve's motion is reversed, and when reversed the wheels should revolve continuously in the opposite direction. This is the foundation, or first principle of valve setting, but it is by no means all that is accomplished by this process at the present day. In the early days of locomotive construction economy in the use of steam was only slightly understood, but after a few years experience it was discovered that fuel could be saved by giving lap to the valve and having the valve gear properly adjusted. This led to many experiments and to the invention and universal adoption of the Steam Indicator, an instrument for measuring and recording the pressure of steam in the cylinder during the entire stroke, thereby permitting of a correct adjustment of the valve gear; but as we have devoted a separate chapter to the use and construction of "The Steam Indicator," we believe it would only confuse the reader (who may have no knowledge of valve setting) to introduce the use of the indicator in this chapter.

Before entering upon an explanation of the method of valve setting we shall first call the reader's attention to a few new devices especially designed to facilitate locomotive valve setting.

FARRINGTON'S VALVE-SETTING MACHINE.

FOR RAISING AND REVOLVING THE DRIVING WHEELS OF LOCOMOTIVE ENGINES TO ADJUST ECCENTRICS AND VALVES.

This machine, as shown in the cut, is designed to meet a want long felt in locomotive repair shops, for some simple, cheap, effective and portable means of raising and revolving locomotive driving wheels while setting the eccentrics and valves.

The most common method, heretofore, has been to bar the engine back and forth; but this is a very slow, tedious process, involving the labor of three or four men to accomplish what one man can more easily and quickly do with this machine.

Another method has been to set wheels in the track, and by placing the forward driving wheels on these, revolve them with a lever. The objection to this is, that it is very expensive to set wheels in all the tracks where engines are liable to stand while undergoing repairs; and if there is but one pair, it adds much to the inconvenience and expense to be obliged to move engines from various parts of the shop to the place where these wheels are located.

This machine obviates all of these objections, as it is portable, and can be used in the Round House, Repair Shop, or where-

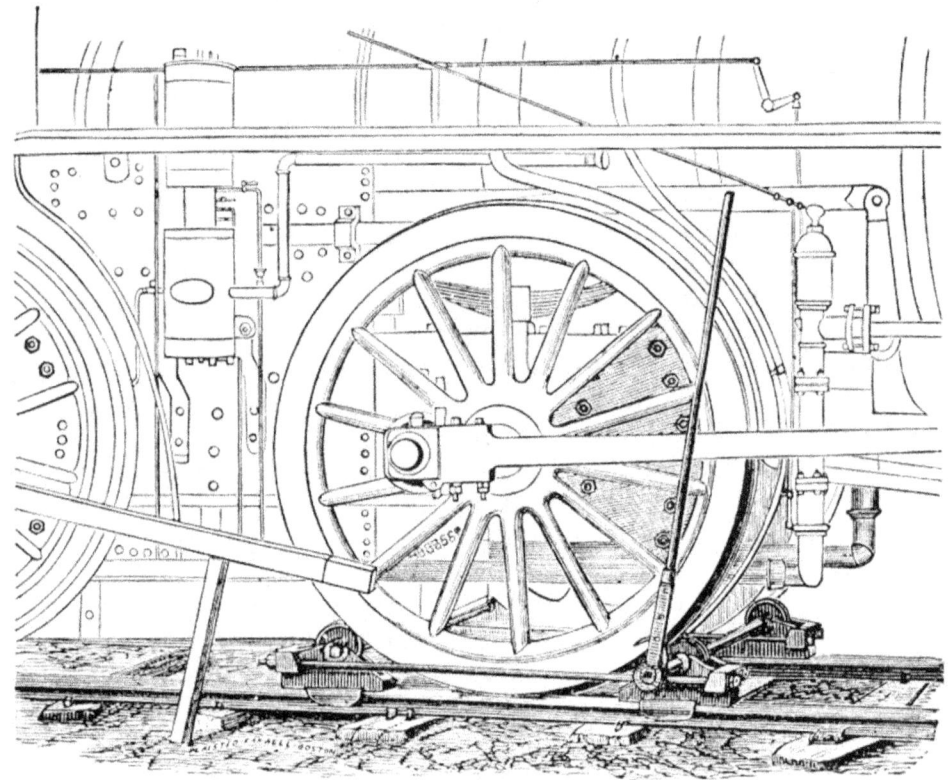

ever the engine may be; and it requires only one man to handle any part of it, or to raise and revolve the driving wheels of the heaviest locomotive.

It is confidently believed that a trial of one of these machines will satisfy any person of the great saving in time and expense effected by it, as they have met the highest commendation wherever introduced.

ROLLERS DRIVEN BY AN ENGINE.

In the Wabash shops at Springfield, Ill., a small engine that is run with air is attached to the rollers for valve setting. It performs good work rapidly, and it is claimed that it can "catch" a center every time.

ROLLERS DRIVEN BY ROPE PULLEYS.

Another device for turning the rollers when setting valves consists of worm wheels placed on each roller shaft, driven by other worms which are attached to a square shaft. At the end of the square shaft which connects the two worm wheels is placed a small lever with reversing gears. It is claimed the driving wheels may be brought within 1-16" of the center and stopped, and then by placing a monkey wrench on the square shaft and giving a slight turn the wheel may be brought to the exact center. This device is in use at the A. V. Ry. shops at Verna, Pa.

LOST MOTION ADJUSTING MACHINE.

The purpose of the device illustrated herewith is to facilitate the work of setting locomotive slide valves, by putting the valve gear under a certain amount of strain, so as to prevent lost motion or back lash, and to insure more accurate

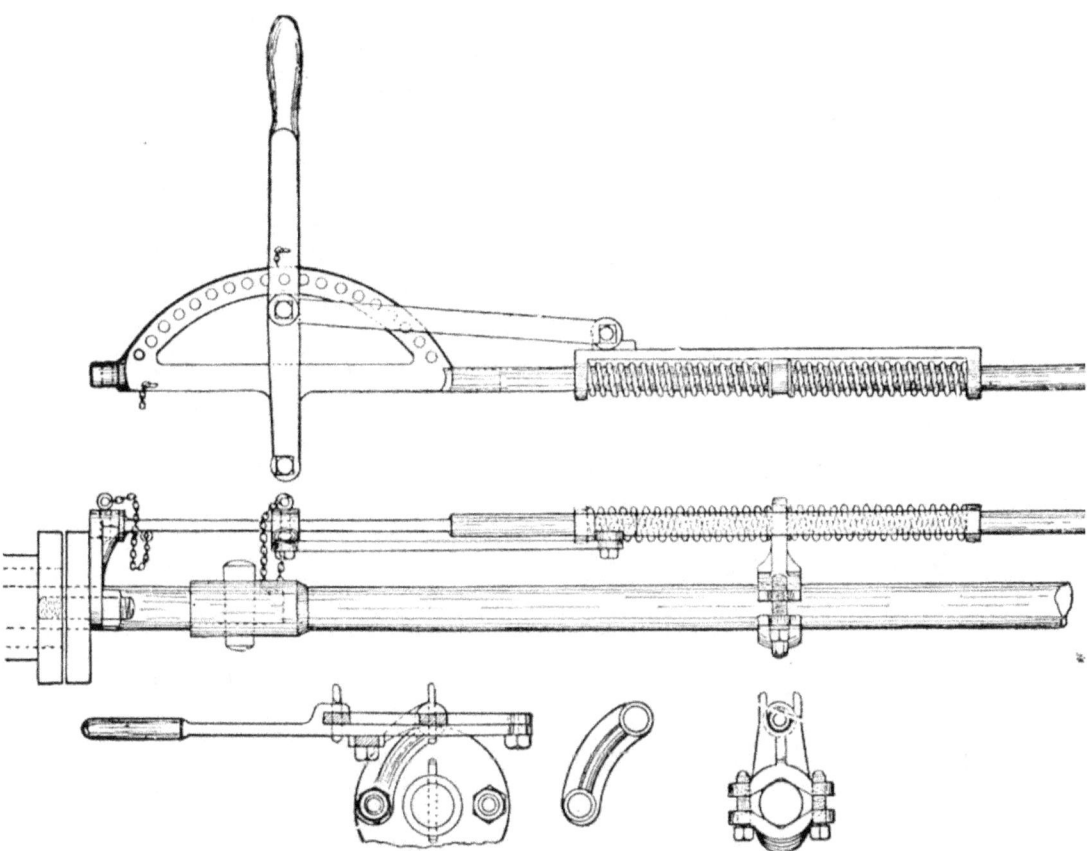

setting of the valves than if they are under no strain, as is usually the case in this work. An arm is attached to the outer stud of the valve stem gland and a clamp is bolted to the valve rod; these attachments are of malleable iron and are placed on both sides of the engine, and the malleable iron sector and the $\frac{7}{8}$" steel rod on which the two steel springs are placed are supported by these attachments on which ever side of the engine the valves are being adjusted. The springs are of No. 6 steel wire with two coils to the inch and so tempered that if compressed solid they will return to their original length. The springs bear against the upper part of the clamp and are compressed by a sliding bar with sleeve ends fitting over the $\frac{7}{8}$" rod, the bar being moved by the hand lever and connecting bar. The lever connecting bar and slide bar are of malleable iron. The clamp is so set on the valve rod as to be $26\frac{1}{2}$" from the arm when the valve is on the middle of its seat. The methods of operation are as follows: Supposing the crank to be 6" behind the back dead center and the reversing lever in full gear ahead, the lever is thrown far enough toward the cross head to get a good strain against the clamp and is then secured in position by a pin through the sector. The crank is then put on dead center and the valve rod marked with the tram; then the crank is put 6" ahead of the center and then back to the dead center with reverse lever in full back gear, the valve rod then moving against the strain of the spring; the same operations are gone through for the forward dead center, the lever then being thrown toward the other end of the sector; the object being to compress the spring in the opposite direction from which the valve rod is moving, but the spring should never be compressed solid. In setting eccentrics the spring is first compressed against the clamp so as to offer resistance as the eccentric is being turned to give the necessary lead of the valve.

ECCENTRIC STRAP AND BLADE ADJUSTER.

The accompanying illustration is self explanatory. The object of this invention is to improve the construction of the con-

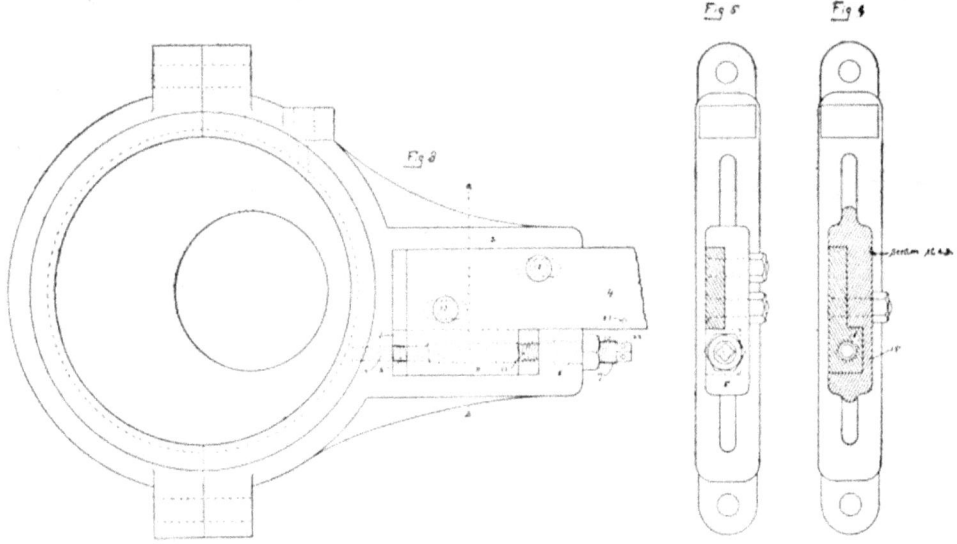

nection between the eccentric rod and the eccentric strap or yoke, and to provide simple, inexpensive and efficient means for

THE LOCOMOTIVE UP TO DATE.

THE LOCOMOTIVE UP TO DATE. 145

enabling the eccentric rod to be accurately and rapidly adjusted without necessitating its removal.

The eccentric strap has a lug cast on it, and the blade a projection as shown, each having a threaded opening to receive the longitudinal adjusting screw; this adjusting screw has a larger diameter in the strap than in the blade, but the pitch of the thread is the same. Alterations can be made quickly and accurately with this device, and it is claimed it will hold the eccentric blade in its correct position should the blade bolts work loose. It is the invention of Mr. W. N. Reazor, of Waverly, N. Y

THE INSTRUCTOR.

The illustration herewith shown is self explanatory. It is a model of the standard form of link motion in use upon American

Locomotives. One of these models would be an invaluable aid to anyone who desires to study link motion, as the interior of the steam chest and cylinder are shown, and the movement of the valve and piston can be closely followed. Some of these models have all the parts adjustable which will permit of experimenting, and the apprentice boy can "set valves" at home. Every Mechanical School and every lodge of Engineers, Firemen and Machinists should possess one of these models.

PREPARATION FOR VALVE SETTING.

First see that you are supplied with the necessary tools, which may be designated as follows: a valve tram; cross-head tram; a wheel tram; an eccentric tram; two pairs of dividers (one large and one small); hermaphrodites; a box square; a prick punch (or two); a small scale; a two-foot rule; a hand hammer and monkey wrench; necessary chisels and files; eccentric set screw wrenches and other wrenches which you may need; a piece of chalk, and don't forget your lamp. We enumerate all these tools, although an experienced man can make one tool answer for three or four; in fact the writer has often seen a "hobo" machinist start to the round house to set valves with nothing but a pair of dividers, a center punch and a two-foot rule.

In some round houses the entire engine is pinched forward and back when setting the valves, while in others the engine is moved by steam while squaring the blades, then one dead center is found on each main wheel and both eccentrics are set by it, but we will assume that we intend to do a good job and use rollers under the main wheels; so disconnect the side rods from the main crank pins, then adjust the rollers until the tread of the wheel just clears the rail sufficiently for the wheel to turn freely. Four jacks should be placed under the engine and tightened slightly to hold the weight of the engine. Now see that the valve stem keys, main-rod keys (at both ends), main wedges, eccentric set screws and blade bolts are all properly tightened and that the reverse lever latch will enter each extreme notch in the quadrant.

The next and most important thing to be considered before beginning to set the valves are the port marks. If you have the slightest doubt as to the accuracy of these marks do not proceed until you have raised the steam-chest covers and secured new port marks.

If the engine is just out of the back shop you may presume the port marks are correct, as these marks are invariably secured before the steam-chest covers are put on. At the present day most good roads have a number of valve trams made to a templet, each being an exact length; these trams are used all over the system and save much unnecessary labor in the round house. We shall first explain how to get the port marks, and afterward how to find a dead center, which is a very important part of valve setting.

HOW TO MARK PORT OPENINGS.

POINTS OF ADMISSION AND CUT-OFF.

First see that the valve stem key is properly tightened, then with nuts and old bushings clamp the steam-chest firmly to the cylinder, being careful not to cramp the copper joint on top of the steam-chest. Now examine the back end of the valve rod, and see that it divides perfectly and that it will connect with the rocker arm without cramping the valve rod or twisting the ends of the valve yoke up or down and thereby cramp the valve. If it is all right do not shove the rocker pin in with your hand and thereby permit the valve rod to spring away from the rocker arm, but tighten the nut on the pin moderately. Next see that the valve is not cocked and that its steam edges are exactly parallel with the edges of the steam ports; if not, use a small jack (or a large stud-nut) and spring the valve yoke until the edges of the valve are true with the edges of the port.

When springing the yoke or stem, be sure to have the valve moved back in the steam-chest so as not to twist that part of the valve stem which travels inside the gland. Also be very careful not to crack the stem, you may heat it with a lamp before springing or peen it a little to keep it in its proper position.

Now see if there is any lost motion between the valve and the valve yoke; if there is, use sufficient liners to take up the lost motion, placing them between the back of the valve and the valve yoke as indicated by the letter L in Fig. 27.

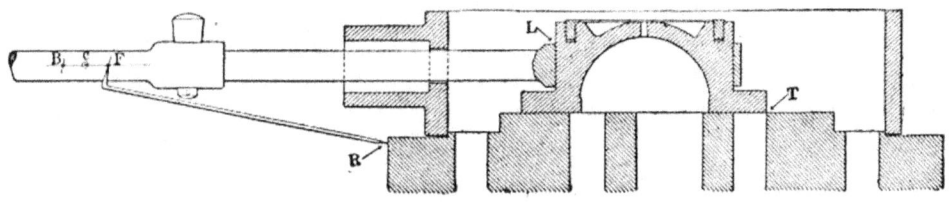

Fig. 27.

Then move the valve back until you can slip a piece of thin tin (and no more) between the forward edge of the valve and the edge of the forward steam port at the point indicated by the letter T. Now the valve is in position to mark the forward port mark on the valve rod, so with a prick punch make the small center R on the cylinder, and from this point, with the valve tram, scribe the line F on the valve rod. Then remove the liners

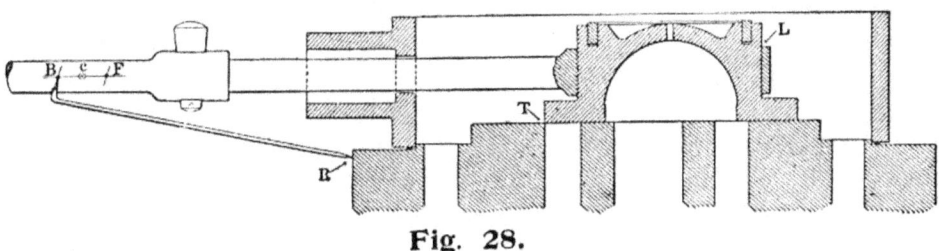

Fig. 28.

and place them at the front of the valve at the point L in Fig. 28, and move the valve forward until the back steam port begins to open, and again set the valve to the thickness of the piece of tin at the point T as shown in Fig. 28. Now, from the point R, and with the valve tram, scribe the line B on the valve rod. Then scribe a parallel line on the valve rod (if you have no box square you can easily make one out of Russia iron) and at the two points where the lines F and B cross the parallel line make two small centers; the center F is the forward port mark, and B is the back port mark.

POINTS OF RELEASE AND COMPRESSION.

Now with a small pair of dividers find the exact center between the points F and B and make another small center which is indicated by the letter c; this point represents the central position of the valve upon its seat.

The points F and B represent the points of admission and cut-off. While if the valve has neither inside-lap or inside-clearance the point c will represent the points of both release and compression. But if the valve has inside-lap or inside-clearance it would be necessary to find two additional points on the valve rod. To find these points set a small pair of dividers to a distance equal to the inside lap or inside clearance of the valve (on one side) and from the central point c describe a small circle, and the two points where the parallel line on the valve rod bisects this circle will indicate the points of release and compression. Be very careful to remember whether it is inside-lap or inside-clearance; the difference will be explained later on. Now you have finished; you have all four port marks.

The two latter points are seldom used in ordinary practice.

MEANING OF A DEAD CENTER.

By the common acceptation of the term, "dead center," we mean when the driving wheels are in such position that the centers of the crank pins and the centers of the driving axles are on a horizontal line; but this is not always the dead center. The technical definition of the term implies that the main rod be in such a position that a line drawn through its center would bisect the main driving axle, or in other words, that the center of the crank pin be on an imaginary line drawn through the center of the cross head pin and the center of the driving shaft, which point indicates the extreme travel of the cross head in either direction. Therefore, there must be two dead centers on each driving wheel, one when the crank pin is in front of the wheel center, which is called the "forward center," and one when it is behind, which is called the "back center," thereby making in all four dead centers to be found, namely: the right forward center; the right back center; the left forward center, and the left back center.

HOW TO FIND A DEAD CENTER.

There are many shops where the pinch bar is still used when setting valves, so for convenience in our explanation of this subject, and in order that all our readers may understand, we shall use the word "pinch."

We shall now proceed to find the right forward dead center. Pinch the wheel forward until the cross-head is about one inch from the extreme travel mark as shown by Fig. 29. Then use the cross-head tram, and from the point A on the forward guide block describe the arc B on the cross-head, and before moving the wheel, with the wheel tram, and from the point P describe

THE LOCOMOTIVE UP TO DATE. 155

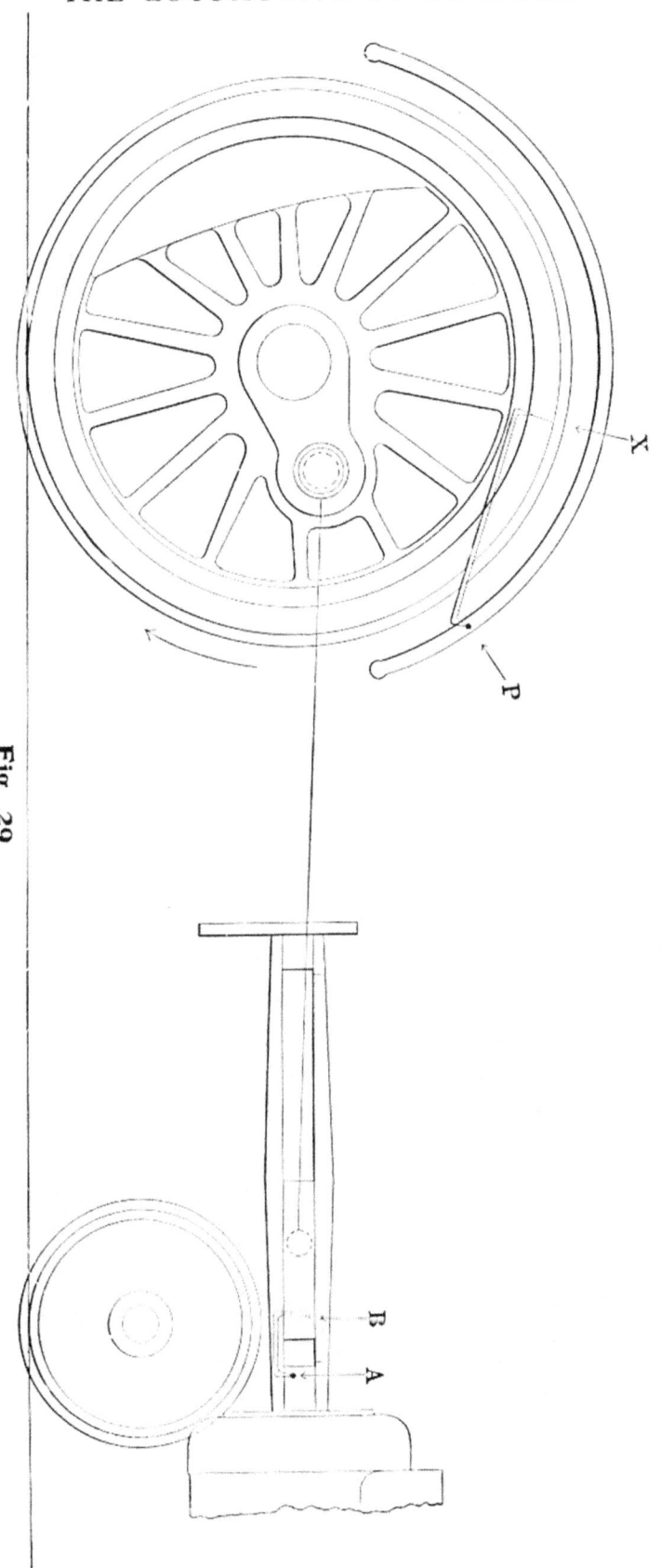

Fig. 29

156   THE LOCOMOTIVE UP TO DATE.

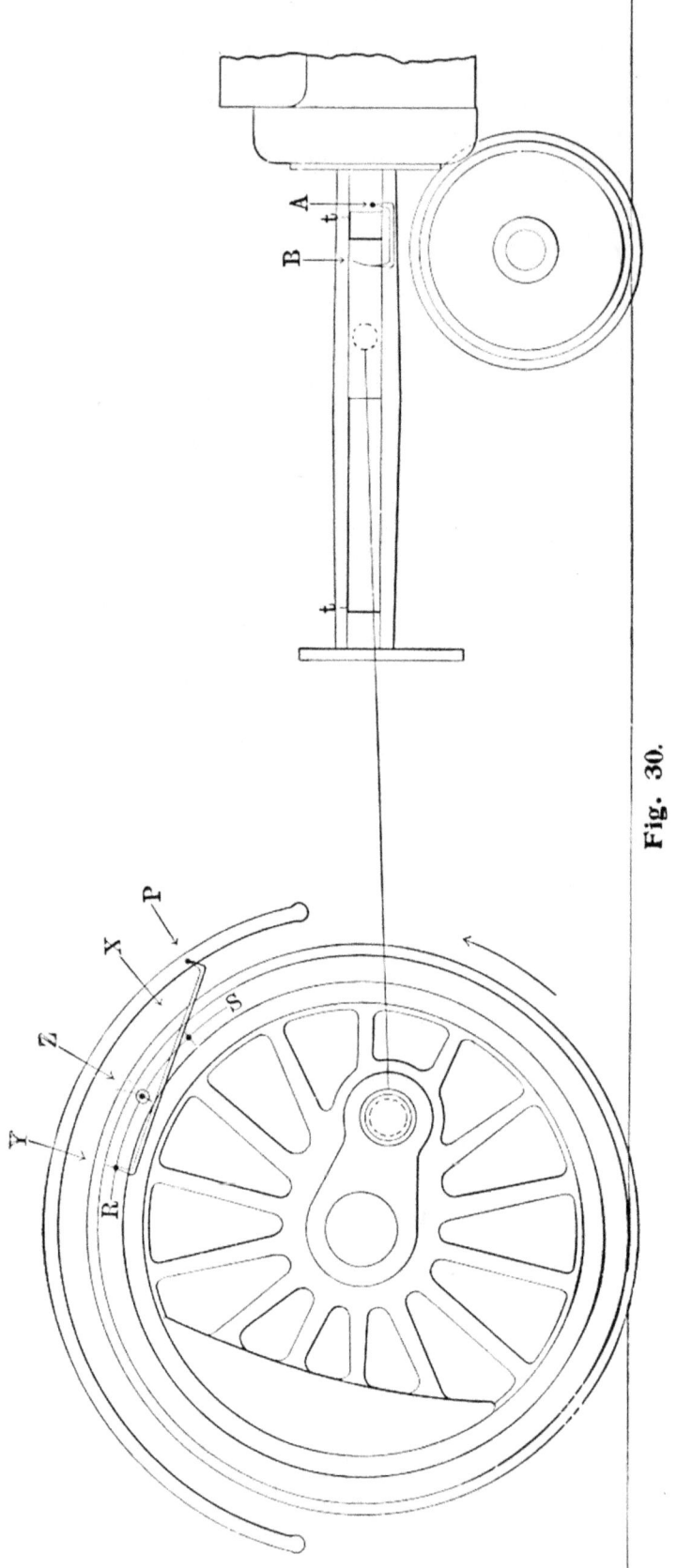

Fig. 30.

the arc X on the tire of the wheel. (The point A and P may be made at any convenient point, but if the wheel cover is used be sure it is securely fastened.)

Now turn the wheel forward, as indicated by the arrow, and as the cross-head recedes from the forward travel mark t use the cross-head tram again and catch the arc B, then stop. (See Rule 21.) Use the wheel tram again, and from the point P scribe the arc Y on the tire as shown by Fig. 30. Now, from the inside of the tire, if possible; if not, then from the outside, with a pair of hermaphrodites describe the arc R S. Next, take a small prick punch and mark two small centers at the points where the lines X and Y cross the arc R S; and from these two points find, with a pair of dividers, the exact center between them on the arc R S, which center is indicated in the illustration by the letter Z. This point is the dead center, so describe a small circle around it to keep from getting it confused with other centers.

Now that you understand the method of securing one dead center you can easily find all four, as they are all secured in exactly the same way, except that when finding the two back centers the cross-head tram should be used from the back guide block or back end of the guide.

HOW TO SET A LOCOMOTIVE'S VALVES.

Now that the reader understands how to find the four dead centers and knows what the port marks indicate, he will experience less difficulty in following us through this explanation. But as there are so many points to be considered which we can not explain as we proceed without confusing the reader, we shall frequently refer to our rules for valve setting, which are given in another chapter, a perusal of which, in advance, will greatly assist the reader in a thorough understanding of this subject. Some men use the guides and engine frame for a black-board to chalk down lead and lap at each point, but as that is unnecessary we shall not. We are now ready to begin.

TRYING THE LEAD.

You may find your first dead center at whatever point will be most convenient and thereby save unnecessary pinching, but in this explanation we shall begin at the right forward dead center. First, find the dead center; now, when you have secured this point, you will observe the crank pin is about 6" below its dead center (this distance is not always sufficient to take up all the lost motion in the valve gear). Since the wheel must turn backward as the crank pin approaches its dead center we must have the valve gear in back motion; so place the reverse lever in the back notch of the quadrant (See Rule 25). Now pinch the wheel back, and with the wheel tram catch the dead center and then stop. Then, with the valve tram, scribe a line on the valve stem, beginning at (or slightly above) the parallel line on the valve stem and let it extend considerably below the line (See Rule 23).

Now the engine is on dead center on the right side, so scribe a line on the guides at the front end of the cross head, which designates the extreme travel of the cross head (See Rule 28). Next place the reverse lever in the forward notch in the quadrant and pinch backward until you notice the valve stem move, which indicates that all the lost motion in the valve gear has been taken up (See Rule 25); then pinch forward and again catch the dead center with the wheel tram; then mark the valve stem again with the valve tram, this time above the parallel line (See Rule 23).

This gives you one dead center; go to the left side of the engine, pinch forward and find the left forward dead center and repeat the operation, marking the valve stem for each motion exactly as before, and don't forget to mark the travel; then return to the right side and get the right back dead center; again repeat the operation and then go to the left side once more; find the left back dead center and mark the valve stem and cross head the same as before. You have now finished pinching for the present; you have the engine run over once. We shall now see what changes are necessary to be made.

HOW TO MAKE ALTERATIONS.

We shall assume that the tram marks you have made on the valve stems correspond with the marks and figures we shall use to explain this subject, as shown by Fig. 31, and that you wish to give the engine 1-32" lead in full gear in both forward and back motion. We will commence by examining the two forward motion tram marks on the right side (those above the horizontal line on the right valve stem); we find it has ¼" lead at the forward port mark and 1-16" lap at the back port mark. To distinguish between lead and lap see Rule 27. Since these marks were both made while the engine was in the forward motion it follows that the length of the right forward motion eccentric blade must be altered in order to equalize these tram marks at both ends. We must now determine whether to lengthen or shorten the blade, and what amount. See Rule 29. We find that we must shorten the blade 5-32", so with a piece of chalk or pencil mark the R F eccentric blade thus (shorten 5-32"), but make no changes until you have examined all the tram marks and marked all the changes necessary to be made on the eccentric blades. Before going further, let us determine what effect the above change will make in our tram marks. When you shorten an eccentric blade you know the rocker arm will force the valve stem forward. Now we shall assume that the valve tram is set to the forward tram mark on the right valve stem, and then held stationary while this change in the length of the eccentric blade is being made. What is the result? Is not the point of the tram 5-32″ closer to the forward port mark? And if set to the back port mark, while the change was being made, does not the point of the tram extend 3-32″ back of the back port mark, since you had but 1-16″ lap at this point while the valve stem moves 5-32″? Now let us see how the port marks will come after this change is made. If we subtract 5-32″ from the ¼″ lead we find we still have 3-32″ lead at the forward port opening. Then, if we subtract the 1-16″ lap at the back point from

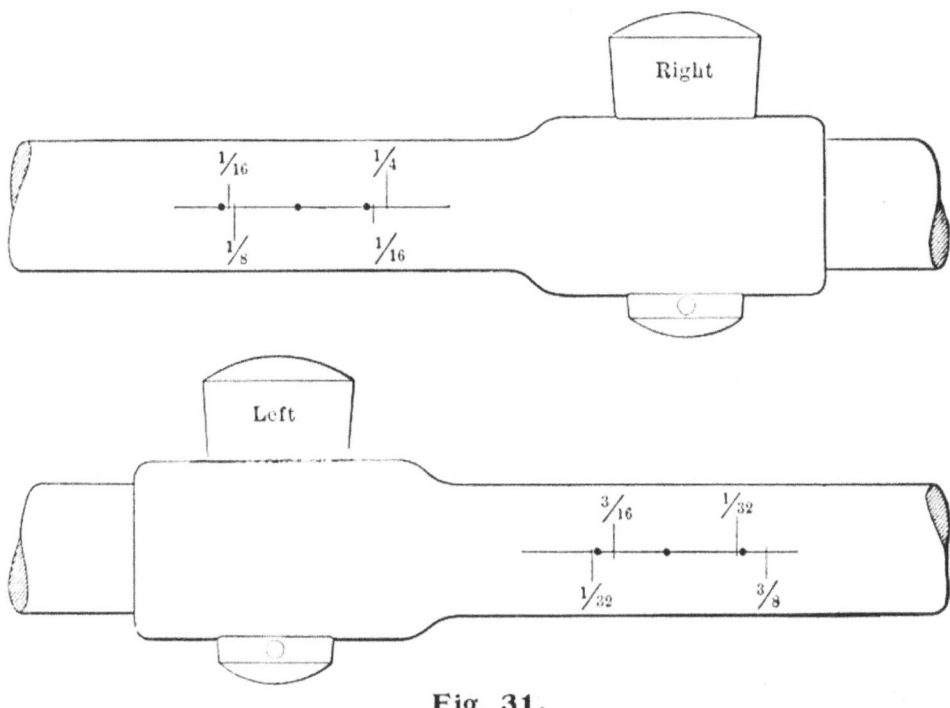

Fig. 31.

the 5-32″ of change, we find we shall have 3-32″ left, which will be lead at back port opening; therefore, we will have 3-32″ lead at both ends.

Since we have 3-32″ lead at both ends, and want the valves to have but 1-32″ lead in full gear when the valves are set, it is evident that we must reduce the lead; the amount of reduction in this case must be 1-16″. Therefore, mark on the frame or firebox with chalk, "R. F. Ecc. 1-16″ lead off." We shall next examine both back motion tram marks on the right side (those below the horizontal line). Here we find 1-16″ lead in front and ⅛″ lap behind; by observing (Rules 27 and 29) carefully we find that we must shorten the right back eccentric blade 3-32″, which will leave 1-32″ lap at both ends, but as we want 1-32″ lead at both ends, it is evident we must give the right back eccentric 1-16" more lead in order to overcome the lap and have the required amount of lead. Therefore, mark "R. B. Ecc. 1-16" lead on."

We shall next examine both of the left forward-motion tram marks. Here we find we have 3-16" lap in front and 1-32" lap behind; in this case we must lengthen the blade 5-64" (See Rule 29), which will give us 7-64" lap at both ends. Now that we want 1-32" lead instead of 7-64" lap we must give the eccentric 9-64" lead, therefore mark: "L. F. Ecc. 9-64" lead on."

Next try the two left back-motion marks. Here we find that we have 1-32" lead in front and 3/8" lead behind; therefore, according to Rule 29, we must lengthen the left back eccentric blade 11-64", which will leave 13-64" lead at both ends, but as this is more than we need we must reduce the lead on the left back eccentric; therefore, mark: "L. B. Ecc. 11-64" lead off."

Now move all the eccentric blades and eccentrics the amount each is marked (See Rules 2, 29 and 30). When you have finished this, you are ready to try the cut off.

TRYING THE CUT OFF.

A locomotive performs most of its work at early points of cut off. It is, therefore, more important to have steam perfectly equalized in the running cut off than in full gear. In other words, it is more important to have an engine "square" when hooked up in the working notch than it is to be "square" in the corner notch and "lame" or "out" when hooked up.

If link motion were a perfect valve gear it would be unnecessary to try the cut off, for if "square" in the corner notches it would necessarily be "square" in every notch; but link motion is not a perfect valve gear owing to the errors introduced by the angularity of the main rod and eccentric rods, and the off-set of the link pin holes from the link arc, but it can be made almost perfect. Off-setting the link saddle pin approximately corrects the inherent error of the link motion.

On a passenger engine you should try the cut off at about 6" and a freight engine at about 9". We shall assume that this engine is for passenger service. Pinch the engine forward until the right main crank-pin has passed the forward dead center. Now, as the cross-head recedes from the dead center, measure the distance between the forward end of the cross-head and the forward travel or "danger" mark. When this distance measures 6" stop. Place the reverse lever in the forward notch, then draw it back slowly until your valve tram is true with the right forward port mark, then stop and hook the reverse lever there, or in the nearest notch of the quadrant. Now pinch the wheel back until the right valve stem begins to move, which signifies that all the lost motion has been taken up. Then pinch the wheel forward until the valve tram is again true with the right forward port mark, then stop and again measure the distance from the cross-head to the right forward travel mark. We shall assume that the distance measures $5\frac{3}{4}''$, so chalk that amount down on the forward end of the right guide bar, thus: $5\frac{3}{4}''$.

Now go to the left side of the engine (leaving the reverse lever hooked in the same notch until you have finished). Again pinch the wheel forward until the left main crank pin has passed the forward dead center; and as the cross-head leaves the dead center, catch the left front port mark with your valve tram; then stop and measure the distance from the left cross-head to the left front travel mark. We shall assume that it measures $6\frac{1}{4}''$, so chalk that down on the front end of the left guide bar.

Now return to the right side, continuing to pinch the wheel forward until the crank pin has passed the right back dead center, then catch the right back port mark with your valve tram, then measure the distance between the back end of the right cross-head and the right back travel mark and chalk the distance on the back end of the right guide bar; we shall assume that it measures $5\frac{1}{4}''$. Again return to the left side, continuing to pinch the wheel forward until the left cross-head has passed the left back dead center, then catch the left back port mark with the valve tram and measure the distance from the back end of the left cross-head to the left back travel mark and chalk the amount on the back end of the left guide bar; we shall assume that the distance at this point measures $6\frac{1}{2}''$. Now you have finished pinching, so we shall see what alterations are necessary to equalize the cut off.

CHANGES TO MAKE IN THE CUT OFF.

By trying the cut off we have discovered that although the lead was perfect in full gear, the steam is not perfectly equalized in the running cut off. This evil is due to the inherent errors of the shifting link motion, but it should be remembered that lost motion, or an imperfect construction of any of the parts of the valve gear will seriously affect the cut off. In order to more thoroughly understand the effect of various changes that may be made to equalize the cut off, the reader should study all our rules for valve setting that refer to this subject, particular attention being called to Rules 31, 36 and 39; also read Rules 16, 17, 18, 37, 38, 40, 41 and 45.

We shall now proceed to examine the figures we made on the guides while trying the cut off. On the right side we find the engine cuts off steam at $5\frac{3}{4}''$ in front, and at $5\frac{1}{4}''$ behind; the difference being only $\frac{1}{2}''$. This amount is not considered a very serious defect in ordinary practice, so if you can locate no defect by the suggestions given in Rule 36, then we must sacrifice a perfect equalization of steam in the full gear in order to equalize the cut off; this may be accomplished by shortening the right forward eccentric blade 1-32".

When the difference in the cut off at both ends is slight, as in this case, you can tell whether to lengthen or shorten the blade by observing Rule 45. On the left side we find the engine cuts off steam at $6\frac{1}{4}$" in front, and at $6\frac{1}{2}$" behind; the difference we find to be only $\frac{1}{4}$"; according to Rule 45 we must lengthen the left forward eccentric blade 1-64". Now compare the cut off on both sides of the engine to find out which side cuts off steam quickest; to do this add the amount of cut off at both ends on each side. On the right side we have $5\frac{3}{4}$"+$5\frac{1}{4}$"=11" total amount of cut off on the right side. On the left side we have $6\frac{1}{4}$"+$6\frac{1}{2}$"= $12\frac{3}{4}$" total amount of cut off on the left side. By this process we find the right side cuts off steam earlier in the stroke than the left side; the difference between the total cut off on each side being $1\frac{3}{4}$". In this case the left side (or whichever side cuts off steam latest is called "the heavy side"). Now, in order to secure an equal distribution of steam, the cut off in both sides of the engine should be the same. It is, therefore, evident that we must make some alteration in the valve gear in order to equalize both sides. See Rules 17, 36, 37 and 40. If, after reading these rules, you fail to locate any cause for this inequality of cut off on the two sides, you must line the rocker boxes or tumbling shaft up or down. It is considered better practice to line the rocker box rather than the tumbling shaft when it can be done conveniently, for, unless you exercise due care when placing liners under the tumbling shaft stands, you may cramp the bearings; besides if the engine is new or has received a thorough overhauling and the links are properly hung it should not be done.

In the present case we shall suggest four changes, any one of which will equalize the cut off, and you may perform whichever one is most convenient. Line the left rocker-box 7-64" down or the right rocker-box 7-64" up, or line the tumbling shaft stand on the left side 5-32" up, or the right tumbling shaft stand 5-32" down. See Rules 36, 37 and 40. When you have made this alteration you have finished, and the running cut off will be equalized. In some places they try the cut off with the reverse lever in several different notches of the quadrant, and then divide the variation of cut off between full and mid-gear. The variation between this practice and the method we have explained is very slight; and while the exhaust may not be so perfect in other notches it is considered better practice to have steam perfectly equalized in the running cut off where the engine performs most of its work.

Now in order to try the exhaust opening and closure we must find additional points on the valve stems. If the valve is line and line inside, then you know that the point of opening and closure will be indicated by the dead center between the port marks, but if the valve has inside lap, it will be quite different, as the valve will cut off the exhaust before it has reached its central position, while it will not release until it has passed its central position. On the other hand if the valve had inside clearance, these conditions would be reversed. So it will be well for a novice to distinguish these points by private marks or letters when found.

To find these points use a pair of small dividers, and from the dead center between the port marks and with a radius equal to the inside lap or inside clearance, as the case may be, describe a small circle on each valve stem and make two centers where the circle crosses the horizontal line, and mark each whatever it represents, as previously explained. Now, remember the opening when the valve is moving forward will be the closure when it moves backward, and *vice versa*.

Now, when you have these marks, proceed to try each by measuring the cross-head as you did the cut-off, and mark down each point. Then compare these figures; alterations may be effected the same as in the cut-off. The process of equalizing the cut-off incidentally has the same effect upon the exhaust closure, but remember compression is of more importance than release, and compression and lead opening should be made as near perfect as possible. Indicator diagrams will expose these defects much clearer than could be explained here.

To find the maximum port opening and maximum travel of the valve, place the reverse lever in full gear and turn the wheel one complete revolution, marking the extreme travel of the valve in each direction with your valve tram; the distance between each extreme point indicates the travel of the valve, while the distance from either extreme point thus found to the port mark indicates the maximum port opening. To determine the minimum port opening and minimum valve travel place the reverse lever in the center notch and repeat the operation.

RECENT PRACTICE IN VALVE SETTING.

For many years it has been the usual practice to give a small amount of positive lead in the full gear. By many it is considered a necessity to arrest the reciprocating parts and secure a smooth running engine. The subject of lead for locomotives has received much discussion of late at the meetings of the various Railway Clubs and Conventions, many able arguments having been made pro and con; we will not attempt to discuss this subject in these pages, but as a result there is a growing tendency to reduce the full gear lead, or rather to adjust the valve gear to the running cut-off and permitting the full gears to take such lead as the length of the eccentric rods will give; the compression being considered sufficient to arrest the reciprocating parts. The amount of lead given in the running cut off varying from $\frac{1}{8}''$ to $\frac{1}{4}''$; this implies a negative lead or blind port in one or both full gears, but this was found to have no deleterious effect on the running qualities in the full gear, owing to that gear being used only at starting when the speed is slow and with the large port opening of the full gear no lead is required to properly fill the cylinder with steam. Negative lead in full gear within reasonable limits is claimed to be an advantage, giving an appreciable increase in the power of the engine when running in full gear, and a slight economy in fuel is also claimed for it.

The reduction of the lead in the running gear may be obtained by reducing the angular advance of one or both eccentrics in equal or unequal amounts. Master Mechanics who agree on the main point of a reduced running gear lead, differ in the method used to accomplish it, as may be seen by the following methods now in use:

The New York, New Haven & Hartford gives to its fast passenger engines 1-16-inch positive lead in full gear forward, and $\frac{1}{4}$-inch negative lead in the full gear back, resulting in $\frac{1}{4}$-inch positive in the running gear.

The Maine Central sets the valve line and the line for the full gear forward, and give $\frac{1}{4}$-inch negative lead in the full gear back on passenger engines.

The Illinois Central gives to passenger engines 1-32-inch positive lead in the full gear forward and then adjusts the backing eccentric to give about 3-16-inch lead in the running cut-off.

The Lake Shore & Michigan Southern gives to passenger engines 1-16-inch negative lead in the full gear forward and 9-64-inch negative lead in the full gear back, resulting in about 5-16-inch lead for the running cut-off.

The Chicago Great Western sets the valves of passenger engines line and line in both forward and back full gears, resulting in from 3-16-inch to 9-32-inch at 6-inch cut-off, and on mogul freight engines gives 3-64 negative lead in both full gears, resulting in $\frac{1}{4}$-inch lead at 6-inch cut-off.

The Chicago & North-Western recently specified for some 19x24 passenger engines 3-16-inch negative lead in the full gear forward and $\frac{1}{4}$-inch positive lead at 6-inch cut-off, obtained by adjusting the back eccentrics. These engines had Allen valves.

The diversity of practice is apparent, but doubtless many of the apparent discrepancies would disappear if the full details of the motions were known. The effect of the length of the eccentric rods is so marked on the mid-gear leads that it has doubtless had an influence on the methods followed in different cases, even when substantially the same running-gear lead has been arrived at.

Another subject which has received equally as much attention of late is "inside clearance." Valves were formerly made line and line inside and in some cases inside lap was used, but this was found to make an engine "loagy" and as a result of many experiments, most progressive Master Mechanics now give their engines from 1-16-inch to ⅛-inch inside clearance. Economy is claimed for this practice, besides making the engine smarter. Experiments were made with as much as 5-16-inch inside clearance (on each side), but this amount proved wasteful on steam.

We herewith reproduce one of the best formulas for valve setting that it has ever been our pleasure to see.

FORMULA FOR VALVE SETTING.

Valve Motion of Engine No. _____ Class _____ out of _____ Shops.
_____ 189 _____

Cylinders _____ x _____
Steam Ports _____ x _____
Exhaust Port _____ x _____
Bridges _____
Outside Lap _____
Inside Clearance _____

Throw of Eccentric _____
Travel of Valve _____
Saddle Pin Back _____
Lead Full Gear _____
Valves Set by _____

	REVERSE LEVER NOTCH.	CUT OFF FRONT END.	CUT OFF BACK END.	DIFFERENCE IN CUT OFF.	EXHAUST OPEN. FRONT.	EXHAUST OPEN BACK	DIFFERENCE IN EXHAUST OPENING.	EXHAUST CLOSES. FRONT.	EXHAUST CLOSES. BACK.	DIFFERENCE IN EXHAUST CLOSURE.	LEAD FRONT	LEAD BACK	MAXIMUM PORT OPENING
FORWARD MOTION.													
BACK MOTION.													

REMARKS: _____

MASTER MECHANIC.

LAYING OFF A NEW QUADRANT.

This is considered the most important mechanical job to be performed upon a locomotive, not owing to the intrinsic value of the quadrant itself, but on account of the great influence it has upon the valve's motion. It is by the use of a quadrant that a locomotive is enabled to develop either speed or power and perform economical work under the various conditions a locomotive must do its work.

Therefore, the man who lays off a new quadrant should have a pretty thorough knowledge of valve motion. In some shops, only the two extreme notches of the quadrant are located, and then fine notches are slotted across its entire face. In other shops the quadrant is laid off by a templet or old quadrant, and sometimes notches are slotted across the entire face, and when the two extreme points are located, stop plugs are put into the quadrant. But for this explanation we will presume the quadrant is to be laid off while setting the valves.

The two extreme notches of the quadrant are the first to be located. You may begin on either side of the engine or at either center, but as the right side is within view of the man who handles the reverse lever, for convenience we will begin on that side. Place the main crank pin on its forward dead center; now throw the reverse lever forward until the link block clears the top of the link the amount allowed at each end of the link for clearance and slip. Now clamp the reverse lever in that position temporarily. Now have the wheel turned one complete revolution and equalize the two forward motion eccentric blades, and at the same time note the full travel of the valve (errors are sometimes made in the design or construction of the valve gear, so the required travel of the valve cannot be obtained without shortening the link block or making some other alteration).

Now place the crank pin on the forward dead center and try the lead; you need not move the eccentrics yet, but mark down the amount you must move each eccentric. Now, as the point of cut off is of more importance than the travel of the valve, proceed as follows: Turn the wheel forward and measure the distance between the forward travel mark and the cross-head (See Rule 28). When the cross head has reached the required full gear cut off which, for a 24" stroke of the piston it is usually 21" or seven-eighths of the piston's stroke, then stop. Now, with your valve tram, try the forward port mark on the right side. (See Rule 26.) Now you know the amount you must move the right forward eccentric to secure the proper lead, and by referring to Rule 31 you will find that by changing the position of the eccentric on the shaft you also change the point of cut off, so set your cross head accordingly to the amount you must move the right forward eccentric, then adjust the reverse lever until the forward port mark is right with your valve tram. This will be the position of the reverse lever for the full forward gear, so scribe the quadrant for the forward notch.

Again place the crank pin on either dead center and haul the reverse lever back until the link block clears the bottom of the link the amount of the clearance and slip, then turn the wheel backward and equalize the two back motion blades. Try the travel and locate the correct point of cut off for the full backward gear exactly as you did for the forward gear. You can then move all the eccentrics the required amount and finish setting the valves. Try the full gear lead and full gear cut-off in each motion to prove your work. Now you can locate the notches for intermediate points of cut off for both gears as follows: Suppose you wish to locate a notch in the quadrant for a six-inch cut off in the forward gear. Place the reverse lever in the forward motion and turn the wheel forward, and as the crank pin passes its forward dead center measure the distance between the cross-head and the forward extreme travel mark, and when the distance measures 6" then stop and draw the reverse lever toward the center notch, and when your valve tram is right with the forward port mark then stop and mark the quadrant at that point, which will be the correct position for the reverse lever for a 6" cut off, and all other intermediate points of cut off may be located in exactly the same way, except that in the back gear the lever should be placed in the back motion and drawn toward the center, and the wheel should be turned backward. When the full gear cut off is the same for both gears, to find the center notch use a pair of dividers and locate a central point between the two extreme notches. Always work from that face of the reverse lever to which the latch is attached.

LAYING OFF ECCENTRIC KEYWAYS BEFORE SETTING THE VALVES.

In ordinary practice, the usual method of laying off keyways for the eccentrics is to first have the keyway slotted in the eccentric, and, after the valves are set and each eccentric occupies its correct position, to scribe the keyways on the shaft. There is no better or safer way to lay off keyways than this, and yet every Master Mechanic and Foreman realizes the delay and the enormous amount of work it requires to move every eccentric, drill and chip out the keyways, fit the keys and replace the eccentrics in their original position. Much of this work and delay can be avoided by having the keyways in the axle milled or drilled out before the wheels are placed under the engine, and even if one or more of the eccentric keys should require a slight offset it will cause no serious damage. While if the man that lays them off knows his business and is accurate they can be laid off perfectly. The following method may be used for any sized shaft or any throw of eccentric with a straight rocker, but if the rocker arm has any backset a little more work will be required, because the center line of motion must be located, see page 121. It is necessary to locate the center line of motion as the eccentrics are always set by it, and not by the crankpin. These points may be laid off on the end of the shaft and transferred to their proper position with a box square; or if the wheels are on the axle, a full size drawing may be made and one or more pieces of tin cut off the right length to reach from the center

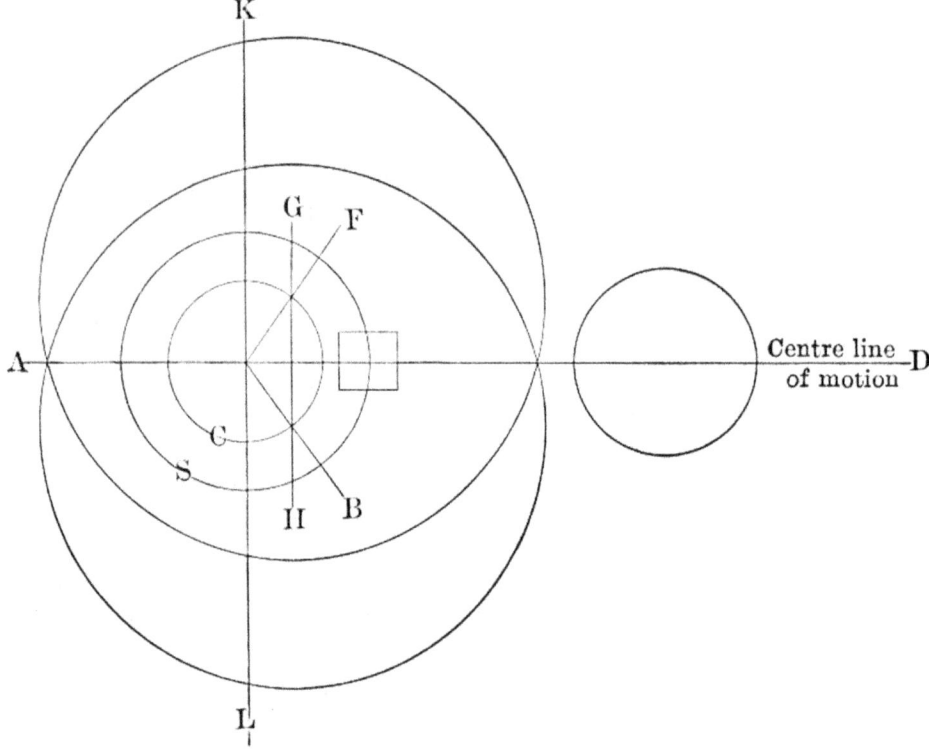

Fig. 32.

line of motion to the center of the keyway. First draw the center line of motion A D, as shown in Fig. 32; then erect the perpendicular line K L through the center of the shaft and at right angles to the center line of motion. Now, if the two arms of the rocker are the same length we know that the travel of the valve and the throw of the eccentric will also be equal. Add the amount of lead required to the outside lap of the valve; but if the valve is to have negative lead, subtract the amount from the outside lap of the valve, and make the line G H that distance from the line K L, and parallel to it.

If the arms of the rocker are of unequal lengths you must determine how far the bottom arm will move while the top arm moves the amount of the lap plus the lead and make the line

G H that distance from the line K L. The difference in the travel of each arm may easily be determined by using a pair of dividers and a scale. Now, when you have the line G H properly located, then from the center of the axle and with a length equal to one-half the throw of the eccentric, describe the circle C, which indicates the path of the center of each eccentric. The two points where the line G H intersects the circle C will indicate the positions of the center of each eccentric. From these two points and through the center of the shaft draw the two lines F and B, allowing them to extend to the outside diameter of the shaft, which is indicated by the letter S. These two lines, F and B, indicate the correct positions for the center of the keyways. So transfer them with a box square to the positions desired to locate the keyways.

EFFECT OF CHANGING MAIN DRIVERS END FOR END.*

The engravings on page 170 show the resulting effect of a change of this kind. They are made to represent the main axle and crank pin with the eccentrics keyed in their correct positions. Figs. 1 to 4, inclusive, show the eccentrics with their centers perpendicular to a center line of motion which is assumed to be coincident with the center line of cylinder; the eccentrics have an angular advance sufficient to open the steam port when the pin is on its center.

The views are all supposed to be seen from the right-hand side of the engine. Fig. 1 is the right-hand axle with the main pin on the forward center and ready to move ahead, or in the direction indicated by the arrows; the setting of the eccentrics will be seen to be correct for this movement. If this Fig. 1 is now turned end for end and brought to the left-hand side it will be seen in the position shown in Fig. 2 when viewed from the right-hand side, with the pin on the back center. While this turning of the axle has taken place, the left-hand pin has been transferred to the right-hand side, and since it was on its upper quarter prior to being moved, it follows that it would still be in the same position, as seen in Fig. 3. In order to get this pin in the same position occupied by the right-hand pin before turning took place, it must be advanced one-quarter of a turn, as shown in Fig. 4, and what was the original right-hand pin in Fig. 2 is now the left-hand pin shown in Fig. 3, since each pin must move through an angle of 90 degrees in changing position by rotation.

What effect this transfer of wheels has on the eccentrics may now be noted. In Fig. 1 the go-ahead eccentric is indicated by the letter F, and the back-up eccentric by the letter B, and it is seen by a reference to Fig. 2 that in the transfer from right to left, the eccentric marked F is now ahead of the pin instead of

---

*"*Locomotive Engineering*," February, 1898, page 83.

170 THE LOCOMOTIVE UP TO DATE.

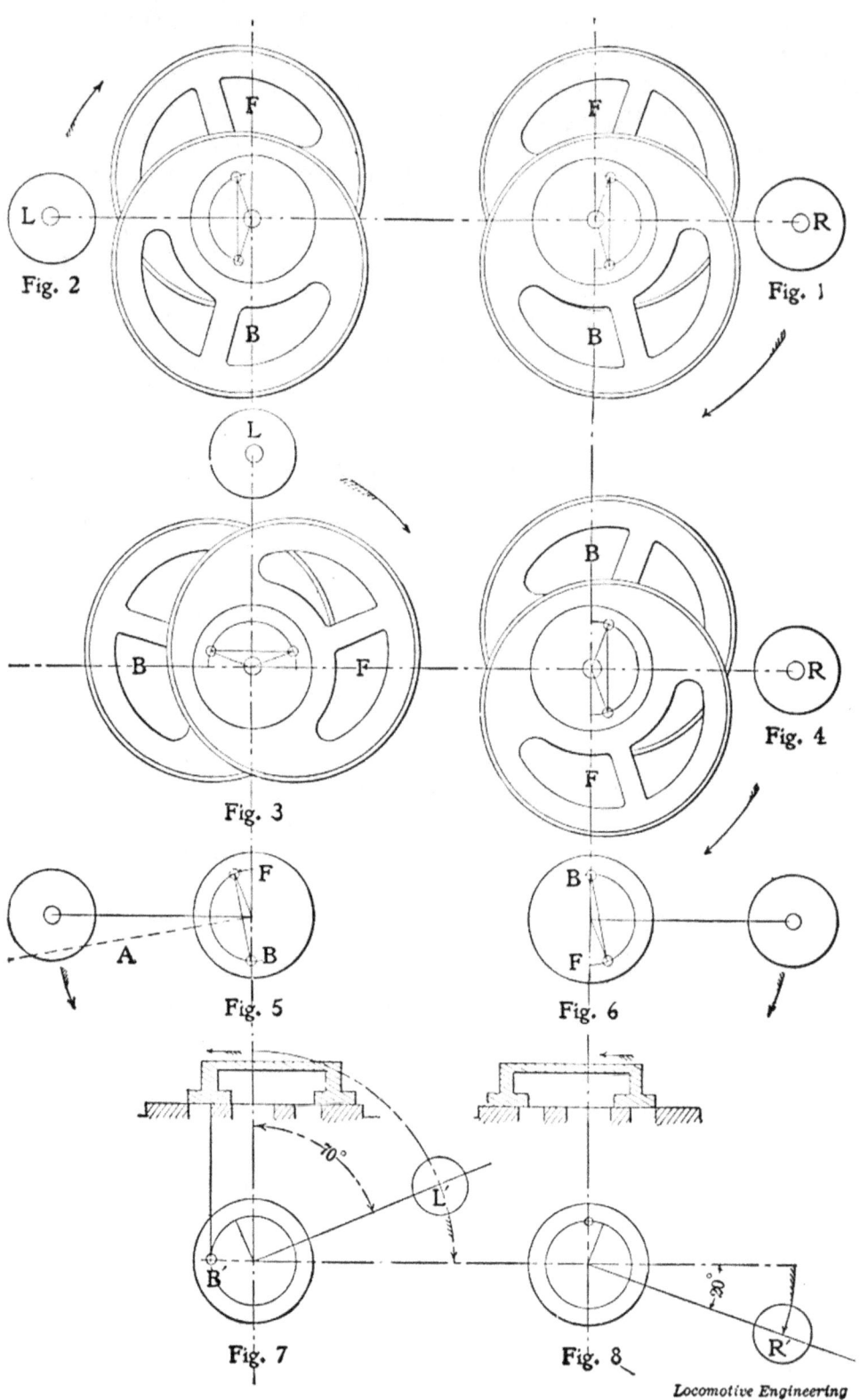

back of it, as in Fig. 1, and is so found in Figs. 3 and 4, but this reversal of eccentrics will not introduce any distortion in the valve movement when the go-ahead rod is coupled up to eccentric B, and the back-up to eccentric F, for the reason that the eccentric centers stand in the same relation to the pins as they did before the axle was turned. This is seen by comparing Figs. 1 and 4; the engine, therefore, will run just the same as before the axle was turned end for end.

What the effect will be when the center line of motion is not coincident with the center of the cylinder will next be considered. In Fig. 5 the center line of motion is shown more nearly in accord with conditions obtained in locomotive practice, but exaggerated, to make it plain. Line A shows at an angle with the center line of crank, which is assumed to be left side on forward center, and since the eccentric centers must be perpendicular to line A, it follows that the eccentrics will have a new location, as shown by their centers which are designated by the same letters as in the other figures.

On turning Fig. 5 from the left-hand side to the right, it will be in the position shown in Fig. 6, with its direction of motion changed, and eccentric centers transposed as in the former cases. The eccentric B is not the go-ahead, and since the position of its center is seen to be moved back an amount equal to the angular advance of its mate, it is plain that the lower end of rocker has been moved toward the eccentric, while the upper end of rocker, and with it the valve have been advanced over the valve seat so as to cover the port an amount equal to the lap and lead, assuming that the rocker arms are equal in length. The engine cannot, therefore, take steam on the right-hand side until after the piston has started on its return stroke.

An examination, however, of what is going on at the left hand side may be made by a reference to Fig. 7, in which the location of the go-ahead eccentric for the new conditions is shown with its center at B^1. This position causes the valve to be at the extreme forward end of its travel, and the back port is therefore wide open to admission while the pin is in its most advantageous position for work; that is, on its upper quarter, and the engine will therefore move ahead, because the pin passes through an angle of about 70 degrees before cut-off occurs, and the port, it will be seen, must remain open until the valve on the right-hand side has opened the front port to admission, as shown in Fig. 8, where the pin has traveled through an arc of about 20 degrees on its return stroke before the valve began to open the front port for admission. It will thus be seen that the engine will move in one direction continuously, even when the valve motion is badly distorted, although such an arrangement would not be conducive to economy. These engravings cover average eccentric condition, and are furnished at some length for the purpose of making it plain that such questions as these may be answered by a little mental effort, by which all ordinary valve motion problems narrow down to an easy solution. To assist to a clearer understanding of the subject than is conveyed by the engravings, it is recommended that traced copies of them be made on thin paper; the reproductions may then be turned to any position and thus prove the propositions made."

RULES FOR VALVE SETTING.

RULE 1.

RELATIVE POSITIONS OF ECCENTRICS TO CRANK-PIN.

The design of valve gear in general use upon our American locomotives, shown on page 100, has indirect motion; with this form of valve gear the forward motion or "go-ahead" eccentric should follow the crankpin when the engine moves forward, and the rearward motion or "back-up" eccentric should lead the pin.

In different forms of direct motion (See pages 109 and 191) the eccentrics should occupy reversed positions in relation to the crankpin; in other words, the forward motion or "go-ahead" eccentric should lead the crankpin when moving forward, and the rearward motion or "back-up" eccentric should follow the pin. To distinguish direct from indirect motion See Rule 44.

RULE 2.

HOW TO CHANGE THE LEAD.

On all standard engines (indirect motion), to increase the lead, move the rib (belly) of eccentric toward the crank-pin. To decrease the lead, move away from the pin. On all engines with direct motion, move the eccentric just the reverse.

RULE 3.

METHOD OF CHANGING LEAD.

To change the lead, mark the eccentric and shaft (axle) with a V chisel, or better, a three-pointed tram (few Master Mechanics will permit the use of a V chisel on the shaft). Now move the eccentric on the shaft the exact amount shown by the valve stem, providing the rocker arms are of equal lengths; if they are of different lengths move the eccentric a proportional amount. Thus, if the top arm of the rocker is 12" long and the bottom arm 9" long, by moving the eccentric 3-32" the valve stem will move $\frac{1}{8}$".

RULE 4.

TO SET AN ECCENTRIC IN CASE OF EMERGENCY.

When setting an eccentric in an emergency while out on the road, or setting temporarily in the shop, before setting the valves (assuming that there are 15 spokes in the wheel and that the valve has $\frac{7}{8}''$ outside lap and 1-32" lead is desired) place the rib (belly) of the eccentric slightly past the third spoke from

Fig. 33

the crank-pin, as shown by Fig. 33. This position will be about right on most engines; of course the exact position depends upon the lap of the valve, number of spokes, center line of motion, etc. See Rule 15.

RULE 5.

ANGULAR ADVANCE.

The "angular advance" of an eccentric means the distance the center of the eccentric is advanced toward the crank-pin from a line drawn through the center of the axle at right angles to the center line of motion. It is indicated by the space between the lines G H and K L, Fig. 32. The line A D is the center line of motion. See Rule 15. When the rocker arms are of equal length the "angular advance" is equal to the amount of lap of the valve plus the lead in full gear.

RULE 6.

TRAVEL OF VALVE IN MID-GEAR.

When the rocker arms are of equal lengths, the travel of the valve in mid-gear equals twice the angular advance plus the increase of lead in mid-gear. The increase of lead in mid-gear varies from $\frac{1}{4}''$ to $\frac{3}{8}''$, according to the radius of the link.

RULE 7.

TO FIND THE AMOUNT OF OUTSIDE LAP.

The distance between the port marks always indicates the amount of outside lap; the valve has one-half that amount on each side.

RULE 8.

TRUEING UP WORN ECCENTRICS.

When eccentrics are worn out of round very badly, remember you can reduce the size of one or all of them by trueing them up in the lathe without affecting the valve motion, providing you do not change the throw. Of course the eccentric blade will have to be lengthened so as to divide equal.

RULE 9.

THROW OF ECCENTRICS.

Keep all four eccentrics exactly the same throw; a difference in their throw would affect the travel of the valve and also the point of cut-off.

RULE 10.

STEAM PORTS.

Steam ports should all be kept exactly the same size. A variation in their size will make one exhaust heavier than the other, and sometimes break the crank-pins.

RULE 11.

AMOUNT OF LEAD TO BE GIVEN.

It is impossible to lay down any definite rule as to the proper amount of lead that should be given, although for convenience in our explanation of valve setting, the engine is given 1-32" lead all around in full gear. The amount of lead which would give the best results on one engine might not do so on another, owing to the difference in the design of different locomotives; the best practical amount for each engine can be ascertained by the use of an Indicator. The volume of the clearance space, the cut-off, and other causes influence the amount of lead given, which varies from ¼" positive lead to 3-16" negative lead. Advocates of early admission claim that the cylinder and piston have become cooled during exhaust and should be reheated as early in the stroke as practicable, while the argument in favor of a late admission is, that while the crank is at or near the dead center, pressure against the piston will have no effect to turn the shaft, but rather preventing its turning by increasing the friction on the pins and main shaft. There is reason in both arguments. But there is not much loss or gain by either early or late admissions, within practical limits, and the lead should be such as will produce the best results otherwise. There is, however, a growing tendency among the locomotive builders of the present day to reduce the full gear lead, as may be seen by referring to the dimensions of the several locomotives herewith illustrated. See Rule 13. For an explanation why an engine is given lead, see "The Slide Valve," page 24.

RULE 12.

LEAD FOR ALLEN VALVES.

A divergence of opinion obtains as to the proper amount of lead to be given Allen Valves. The best practical amount to be given each engine can be determined by the use of an Indicator. Owing to the double opening (the supplementary port being equal to the length of the steam port) the volume of steam admitted to the cylinder by an Allen Valve will be double the amount that a plain slide valve would admit with an equal amount of lead. But a decrease of lead will delay the cut off and every other operation of the valve. However, within practical limits, this variation will be so slight that it is considered good practice to give Allen valves only one-half the amount of lead given to plain slide valves. (See Article by Mr. E. M. Herr, Supt. M. P. of the Northern Pacific Ry., page 52. The Allen Valve will maintain the initial pressure better at early cut-offs than the plain slide valve, and the pressure will be more uniform during the period of admission on account of the double opening.

RULE 13.
LOST MOTION IN THE VALVE GEAR.

Lost motion in the valve gear decreases, and often nullifies the amount of lead given the engine in full gear. Therefore it is important that all lost motion should be taken up when setting the valves, otherwise an old engine may be running blind in full gear; and yet, with a valve tram apparently having the required amount of lead. This evil would be clearly shown on an Indicator diagram. See Rules 14 and 21; also see the Lost Motion Adjusting Machine, page 149.

RULE 14.
EVIL EFFECT OF INCREASED LEAD.

Lead increases as the reverse lever is drawn toward the center notch in proportion to the radius of the link. This increased lead is sometimes injurious to slow, hard-pulling engines. To remedy this, the angular advance of the back-up eccentric is decreased in order to benefit the forward gear. Many freight engines of this kind are in use on the Pennsylvania railroad that are run $\frac{3}{8}"$ blind in back gear. Decreasing the angular advance of one eccentric gives the other motion almost a constant, unvarying amount of lead between full gear and mid-gear. The Allen design of valve gear, which has a straight link, was designed to overcome this increase of lead in full gear. See page 193; also see how the eccentrics are set on various other roads. Page 164.

RULE 15.
RELATIVE DISTANCE OF ECCENTRICS FROM CRANK-PIN.

Very many believe that when the valve is given an equal amount of lead in the forward and back motion the eccentrics are an equal distance from the crank-pin, yet they are not always an equal distance even if the cut-off be equal in each extreme notch. Remember the eccentrics are set from the center line of motion and not from the crank-pin, so the bottom arm of the rocker will travel an equal distance each way from its central position (See "Technical Points," page 121). This is a technical point, but as the eccentric blades are so long and the line of motion usually so close to the crank-pin when it is on center, that for all practical purposes, set your eccentric by the pin when setting them temporarily.

RULE 16.
EQUALIZATION OF ADMISSION AND CUT-OFF.

Remember a perfect equalization of admission and cut-off for both gears is practically impossible with link motion. If the lead be perfect the cut-off will not be, and vice versa. Equalization of the back gear is usually sacrificed to benefit the forward gear. This is the inherent imperfection of link motion, but it can be made almost perfect.

RULE 17.

EQUALIZATION OF STEAM.

When the exhaust is perfect, steam is not always perfectly equalized, and it is sometimes necessary to sacrifice a perfect exhaust to a correct equalization of steam. For example, if one cylinder was considerably larger than the other on a single-expansion engine, the larger cylinder should be given an earlier cut-off, or the small cylinder a later cut-off, in order to equalize the volume of steam admitted to each cylinder. Such an inequality would be readily discovered by the use of an Indicator. An imperfect equalization of steam will impart unequal stresses upon the crank pins and, if the inequality be great, will cause the engine to jerk and sometimes break the crank-pins.

RULE 18.

PERFECT EQUALIZATION OF STEAM.

By referring to "Technical Points," we find that it is impossible to secure a perfect equalization of steam in both gears with the shifting link motion. If the lead be equal, the cut-off will not be, and vice versa. This is due to the angularity of the connecting rods and the offset of the eccentric blade pins back of the link radius. By advancing the exhaust on the back edge of the valve in proportion to the clearance, the compression may be equalized in both ends of the cylinder and the exhaust will sound perfect. This is done on some of our modern engines. Those familiar with the Indicator diagrams will readily perceive the advantage.

RULE 19.

FULL GEAR AND MID-GEAR.

Full gear means that the reverse lever is in either extreme notch in the quadrant. Mid-gear implies that the reverse lever is in the center notch of the quadrant.

RULE 20.

DEAD CENTER, QUARTER, AND EIGHTH.

By "dead center" is meant that the crank pins are parallel with the wheel centers. By "the quarter" is meant when the pin is at right angle with the dead center, either on top or bottom;

and by "eighth" is meant when the pin is half way between the dead center and the quarter.

RULE 21.

TO FIND THE DEAD CENTER.

When finding the forward dead center on the main wheel, you will notice that the main rod is pushing the cross-head when approaching the center, and pulling it after it passes the dead center, and the back center just the opposite. Therefore, if there was any lost motion in the main rod, the dead center would not be correct. To overcome this error, after having marked a line on the cross-head when approaching the dead center, pinch past the center and past the line; then back up again and catch the line with the cross-head tram. Then the strain against the cross-head will be the same each way from dead center. This rule should be strictly observed on all engines with much lost motion. Otherwise the dead center on the wheel will not be correct, and unless the dead centers are correct, it would be impossible to set the valves perfectly.

RULE 22.

DISTINCTION BETWEEN PORT MARKS AND TRAM MARKS.

The distinction between port marks and tram marks are these: While they are both made with the same tram, the port marks are the points obtained when first marking the valve stem, and they indicate the point of lead opening and the point of cut-off. While the tram marks are those marks secured while setting the valves and indicate the amount of lead or lap the engine has at given points.

RULE 23.

HOW TO SCRIBE TRAM MARKS.

Always scribe the forward motion tram marks above the parallel line on the valve stem, and the back motion tram marks below the line.

RULE 24.

TO EQUALIZE THE TRAM MARKS AND CHANGE THE LEAD.

To equalize the tram marks at each end change the length of the eccentric blade. To change the lead move the eccentric.

RULE 25.

TO TRY FORWARD AND BACK MOTION PORT MARKS.

When trying forward motion port marks always see that the reverse lever is in the forward notch, and have the engine pinched back of dead center enough to take up all lost motion (about 6"). Then pinch forward and catch the dead center; then mark the valve stem. When trying the back motion port marks, place the lever in the back notch and pinch ahead enough to take up all the lost motion; then pinch back and catch the dead center and mark the valve stem.

RULE 26.

WHICH PORT MARKS TO WORK FROM.

When the crank pin is on the forward center, always figure your valve tram marks from the front port mark; when on the back center, from the back port mark.

RULE 27.

TO DETERMINE LEAD FROM LAP.

If a tram mark comes between the port marks, it indicates so much lap (sometimes called blind); if outside, so much lead. This is the rule. But occasionally when blades and eccentrics are out bad, a tram mark may lap outside of the opposite port mark and appear as lead. Always notice which center your pin is on and which port mark you are trying. See Rule 26.

RULE 28.

MARKING EXTREME TRAVEL OF CROSS-HEAD.

Always mark the extreme travel of the cross-head on the guides while the engine is on its dead center, at each of the four points. Travel marks are sometimes called danger marks.

RULE 29.

TO DETERMINE WHETHER TO LENGTHEN OR SHORTEN AN ECCENTRIC BLADE.

Unless some positive rule is understood it requires much time for deliberation and often puzzles good mechanics to tell whether to lengthen or shorten an eccentric blade. The reader who carefully follows this explanation will experience no difficulty from this source and can tell at a glance whether to lengthen or shorten a blade, and thereby save much time and avoid mistakes. In order to more clearly explain this subject an illustration, Fig. 34, is used.

The two forward motion tram marks are shown above the parallel line and indicated by the letters F and F1, and the two back motion marks below the line and indicated by the letters B and B1. The point marked o indicates the exact center between the port marks (not the tram marks). All measurements should be made on the parallel line. First examine the two forward motion tram marks. With a small pair of dividers find the exact center between the marks F and F1; the center thus found, indicated by the letter f, is in front of the point marked o. Therefore, the eccentric blade should be shortened an amount equal to the exact distance between the point o and f. If the rocker arms are of different lengths, move the blade enough to cause the valve stem to move the amount thus indicated.

Now examine the two back motion tram marks; with the small dividers again find the exact center between the marks B and B1. The center thus found, indicated by the letter b, is back or behind the point o; therefore, the eccentric blade should

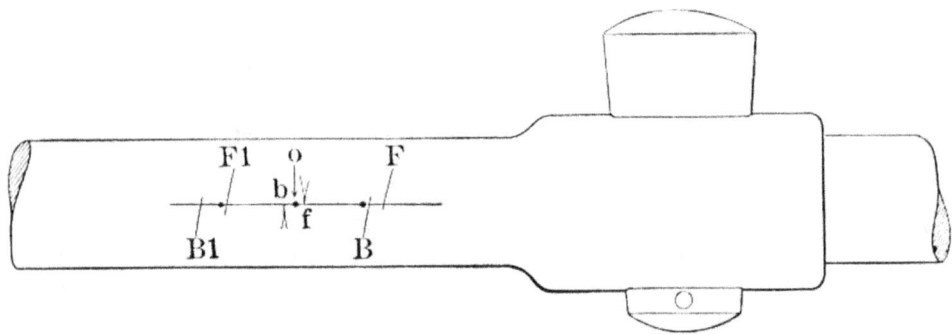

Fig. 34.

be lengthened an amount equal to the distance between the points b and o.

The rule is, *always bring these two centers together*. If the center between the tram marks is in front of the center of the port marks, *shorten the blade;* if behind it, *lengthen the blade*. On a direct motion engine the changes should be made exactly the reverse. See Rule 27. Another method largely practiced is to add the lead and lap together and divide the amount, etc. This method is often puzzling, for even good men sometimes get lead and lap mixed. The rule here given is less liable to mistakes, as it does not matter whether the marks indicate lead or lap you are safe when you work from their center.

RULE 30.

PROVING ALTERATIONS BY TRIAL.

In some shops they change the length of all eccentric blades that need alteration and then run the engine over again to prove their work, and also run the engine over again, after moving the eccentrics, before trying the cut-off. This is considered good practice but it cannot always be done on "hurry up jobs;" besides if you are accurate in making the changes it is unnecessary. See Rule 31.

RULE 31.

EFFECT OF A CHANGE OF LEAD UPON THE CUT-OFF.

If you have a "hurry up job," change the eccentric blades,

but try the cut-off before moving any of the eccentrics. You can figure this way: If the lead be increased $\frac{1}{8}"$ by moving the eccentric on the shaft, the engine will cut off steam $\frac{1}{2}"$ earlier in each stroke. And by adding together this $\frac{1}{2}"$ at both front and back points we have a total difference of 1" in the cut off on that side. Therefore, by increasing the lead $\frac{1}{8}"$ we reduce the total amount of cut off on that side 1-inch. Remember that increased lead hastens every operation of the valve (See "Lead will effect the point of cut off," page 36. It is likewise true that a decrease of lead always delays the point of cut off an equal amount. Therefore, by decreasing the lead $\frac{1}{8}"$ the cut off

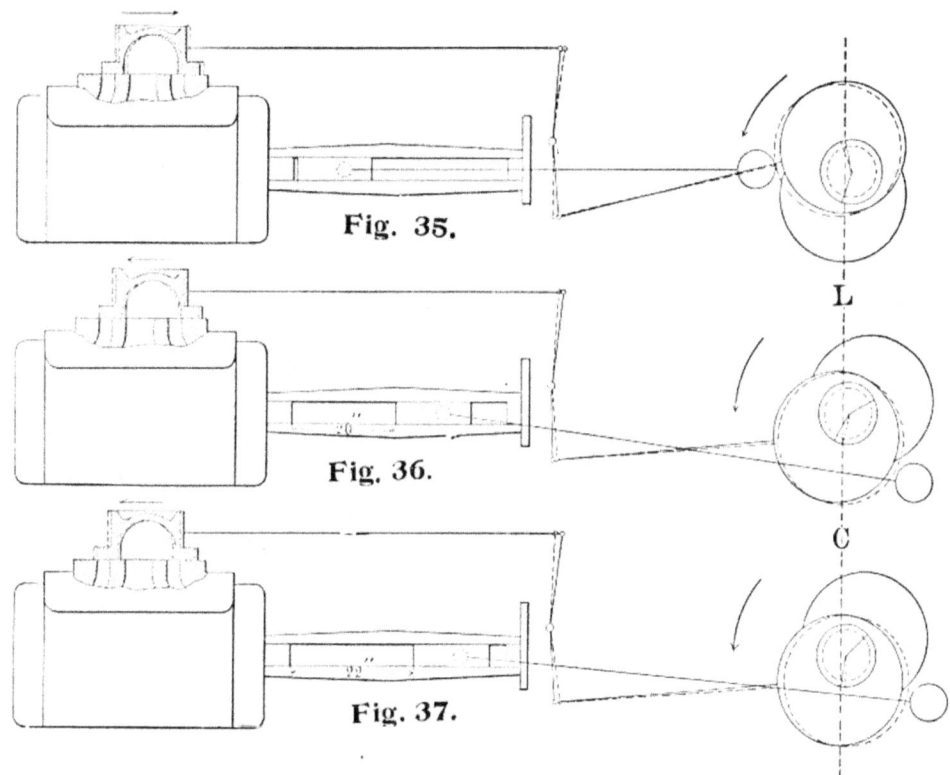

is delayed and the total amount of cut off is increased 1" on that side.

It is quite necessary that the valve setter should thoroughly understand this subject. The above explanation will become more clear by studying Figs. 35, 36 and 37.

In these three figures we have shown the valve and the eccentrics much larger than the other parts, and the increase of lead very great in order to more clearly illustrate the effect of increased lead on the point of cut-off. Some of the parts are shown by their center lines only, and the arrows indicate the motion of the crank pin in valve. The full lines of the valve, eccentrics, and rocker, indicate the various positions of these parts with only sufficient angular advance to bring the valve to the point of lead opening when the crank pin is on the dead center, as shown by Fig. 35; while the dotted lines indicate the positions those parts would occupy with an increase of lead. In Fig. 35 we find the centers of both forward and back eccentrics are on the perpendicular line L C, and the valve has reached the point of lead opening at the forward steam port. This line must therefore indicate the point of lead opening and also the point of cut off, so we have extended the line L C through all three figures. Now, if we advance the forward eccentric toward the crank pin, as shown by the dotted line in Fig. 35, we find that the valve will be opened for the admission of steam at the forward steam port; this opening of the port while the crank pin is on its center is called "lead." We shall now follow the journeyings of the crank pin and eccentric until they reach the positions shown in Fig. 36. Here we find the center of the dotted eccentric has reached the perpendicular line L C and, therefore, has reached the point of cut off, as shown by the valve. While the center of the eccentric without lead has not yet reached the line L C. Here we find, by measuring, that the distance between the travel mark and the forward end of the cross-head (Fig. 36) is 20"; therefore, with this increase of lead, we find that steam will be cut off at 20" of the piston's travel (the cross head and piston being rigid). Again suppose the crank pin and eccentrics to be moved to the positions shown by Fig. 37. Here we find the eccentric, without lead, has reached the line L C, and therefore the point of cut off, while the dotted eccentric has passed the line. Again measuring the distance between the forward travel mark and the cross head, we find it to be 22", as shown by Fig. 37. Thus proving that an increase of lead will hasten the point of cut off, while a decrease of lead will delay the point of cut off.

If the engine is cold set the cross-head to $21\frac{1}{2}$" in the forward gear and $20\frac{1}{2}$" in the back gear, and when the boiler gets warm the expansion will make the cut off equal in both gears.

RULE 32.

CONNECTING ECCENTRIC BLADES TO THE LINK.

Due care should be taken when connecting eccentric blades to the link. Remember that on all standard engines the forward motion blade should be attached to the top of the link and the back motion to the bottom of the link. See Rule 44. The eccentrics are not in the same position on all engines. On some the back motion eccentric is next to the box, and on others the forward motion eccentric is next to the driving box. When coupled up wrong the engine will move in the opposite direction from the one indicated by the position of the reverse lever. Through negligence good mechanics sometimes make this error.

RULE 33.

TO CHANGE THE VALVE STEM INSTEAD OF BOTH BLADES.

If there is a right and left thread nut on the valve stem (or rod) and if both eccentric blades on the same side need lengthening an equal amount, you can shorten the valve stem the same amount, and thereby save changing both blades; and if the blades need to be shortened then lengthen the valve stem. Of course if the rocker arms are of different lengths the change of the valve stem should be a proportioned amount. If the valve stem is the correct length do not alter it. Very few of these valve stems are now in use.

RULE 34.

HOW THE LEAD MAY BE ALTERED.

The lead of an engine can be altered by changing the length of the reach rod or link hangers, or lining the rocker arms or tumbling shaft up or down, but what is put on the two go-ahead eccentrics is taken off the back-up eccentrics, or *vice versa*.

RULE 35.

THE EFFECT OF ONE BLADE ON THE OTHER'S MOTION.

If one eccentric blade is not the right length, it will affect the other motion and cause the valves to sound "out." The effect will be greatest when the reverse lever is hooked up in mid-gear, but it will be scarcely perceptible in full gear.

RULE 36.

TO EQUALIZE THE CUT-OFF.

After the lead has been equalized in full gear it is customary to run the engine over again and see if the steam is equalized at early points of cut-off, particular attention being given the running cut-off, which varies from 4" to 6" for a passenger engine and from 6" to 9" for a freight engine. If the engine is "out" very bad in the running cut-off, locate the cause at once, before making any alterations to equalize the cut-off. Should one side of the engine cut off steam 3" or 4" quicker at one end than at the other end, and the other side of the engine cut off equal, you may be sure there is something wrong. First, examine the link saddles carefully, occasionally a saddle is fastened to the link "up side down," which would change the off-set of the saddle pin; see that the saddles are not transposed, and that the saddle is securely fastened to the link, and not working "too and fro;" sometimes a saddle becomes loose, and if the holes are oblong it will move. Next examine the back-set of the rocker arms and see whether they are the same; one arm may be bent or sprung and thereby cause the back-set of the rocker to be imperfect, but unless the arm is sprung considerably, say ¼" or more, it would not cause this defect. See that the valve stem keys and eccentric blades are tight. If you cannot locate any other cause, the defect can be remedied by changing the off set of the saddle pin. (See Rule 39.) If the difference in the cut-off, front and back, be very slight, it may be remedied by adjusting the eccentric blades, but it will be at the expense of a perfect equalization of steam in full gear. Should the running cut-off be imperfect on both sides, the cut-off taking place too early at the same end on each side, and each side at the same portion of the stroke, see if the top arm of the tumbling shaft is loose; see if the reach rod is the correct length by trying the cut-off in full gear in each motion. (See Rule 41.) If the reach rod is the correct length, this defect may be overcome by lining the rocker boxes or tumbling shaft stands up or down, or changing the length of the link hangers or "lifters."

Should the engine show 3" or 4" "heavy" on one side, in other words, carry steam that much farther on one side than on the other before cutting off, examine the arms of the tumbling shaft and see whether either arm is bent or sprung either up or down. See if the two link lifters are of an exact length, and that all the eccentrics have the same throw, and the two links the same radius, and that the off-set of the link saddles are the same. If the lap, or travel of the valves, or size of the steam ports were not the same on both sides, it would produce this defect in the cut-off. If you fail to locate the defect by the foregoing suggestions, then place a straight edge across the frames and level it by the cylinders, and try the centers in the rocker boxes, and tumbling shaft and its short arms—the centers on each side of the engine should correspond. Now you can remedy this defect by lining under the rocker box or tumbling shaft. Make any change that will raise the link or lower the link-block, on the "heavy" side of the engine, or lower the link or raise the link-block on the "light" side. (See Rules 16, 17, 18, 31, 37, 39 and 40.) If any part is right do not alter it.

RULE 37.

EFFECT OF LINING THE ROCKER BOX AND TUMBLING SHAFT.

Remember that when you place a liner under one tumbling shaft stand, you thereby raise both links, one about one-third as much as the other, and that when you line a rocker box up or down it will not affect the other side. Therefore, 1-16" liner under a rocker box will effect the cut-off as much as a 3-32" liner placed under the tumbling shaft stand.

RULE 38.

ALTERATION OF MAIN RODS.

If your main rods require alteration, figure on it when figuring on your changes for cut-off.

RULE 39.

TO CORRECT UNEQUAL CUT-OFFS BY CHANGING THE LINK SADDLE.

When an engine does not cut-off equally front and back, and you have located the cause to be in the link saddle, examine the saddle carefully and see that it has not been bolted to the link "up side down," or saddle transposed; that is, the right saddle fastened to the left link, and vice versa. If the saddles are apparently all right, then use a temporary link-saddle (See Fig. 38), if you have one; if not, use temporary bolts in the place of the link saddle bolts. Be sure to mark the original position of

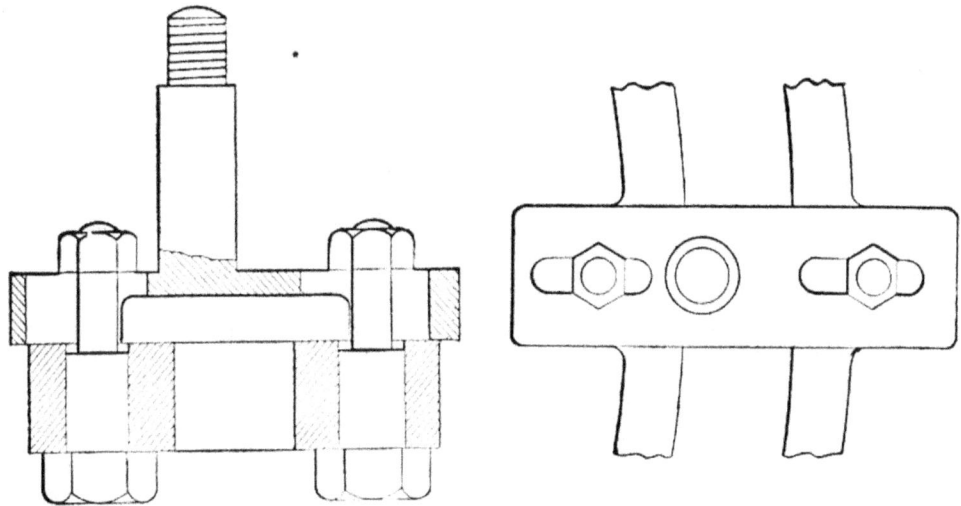

Fig. 38.

the link saddle before removing the bolts. We shall assume that the engine cuts off steam equally front and back on one side, and on the other side it cuts off at 4" in front and 7" behind. Now we must determine which way to move the saddle, whether to increase or decrease the off-set of the saddle pin. Let us first ascertain just what effect this off-set of the saddle pin has on the cut off in the forward motion.

Since the link and the link block are suspended from different points, the saddle-pin being inside the link arc, it follows that the link block pin and the link saddle pin sweep through different arcs. By referring to Fig. 39 we find that the path of the saddle pin is represented by the arc H H, and the path of the link block pin is represented by the arc R R. If we consider the position of the eccentrics, we find that the link will be lowered when the crank pin is on the forward center and slightly elevated when the crank pin is on the back center. Now we know that raising the link increases the lead and thereby hastens the cut off. It therefore follows, that this **off-set of the link saddle pin hastens the**

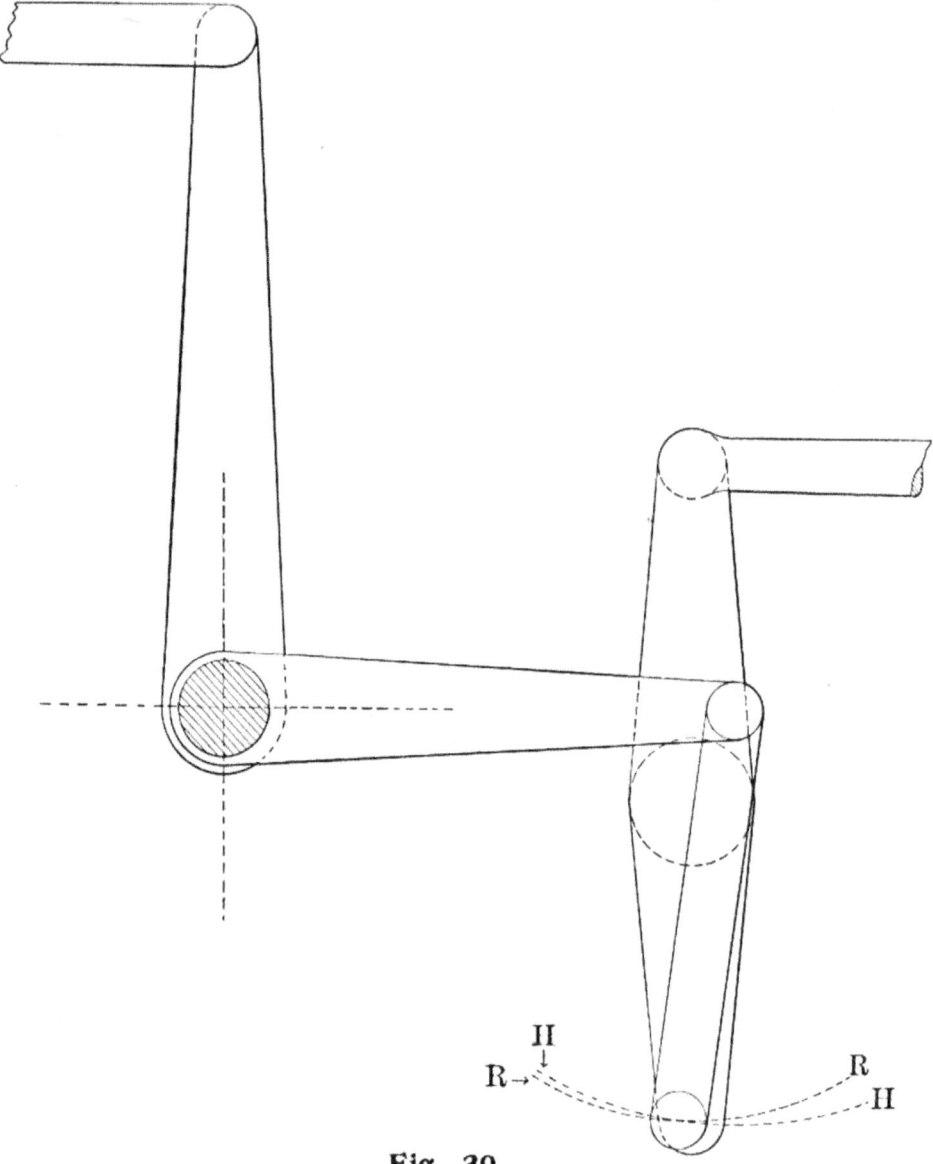

Fig. 39.

cut-off behind and delays the cut-off in front. (It has an effect equivalent to an increase in the length of the eccentric blade while the crank pin is on the forward center, and a decrease in its length when the crank pin is on the back center.) If the saddle pin had no off-set at all, the cut-off would occur earlier in the stroke in front and later behind (this imperfection is caused by the angularity of the connecting rods, the main rod and the eccentric rod, and the off-set of the eccentric rod pin holes from the link arc). Now, as the off-set of the link saddle pin tends to equalize the cut-off in both strokes, therefore it must delay the cut-off in front and hasten the cut-off behind.

As our engine cuts off at 4" in front and at 7" behind, it is apparent that we must increase the off-set; therefore, move the saddle back 9-32" from where it originally was (*3-32" for every one inch difference in the cut off*).

The rule is, *if steam be cut off earlier in the front than behind, increase the off-set. If earlier behind than in front, decrease the off-set.* After making this change, run that side of the engine over again and try the cut-off, and you will find it about right front and back. Mark the saddle when correct. If it is the original saddle that was on the link, rose-bit the holes and countersink for the heads of the bolts. If the change made be very great plug the holes in the saddle and drill new ones, or replace it with a new saddle.

The exact amount the link saddle pin should be off-set depends entirely upon the design of the engine. Engines whose cylinder centers are above the wheel center will require more off-set in the saddle than center line engines. $\frac{5}{8}$" is the distance the saddle pin is off-set on a large majority of our American engines.

RULE 40.

Lining a link, or link block, 1-16" up or down will make 1" difference in the cut-off on that side.

RULE 41.

TRYING LENGTH OF REACH ROD.

Most engines cut off steam at seven-eighths stroke in full gear. Experience has proved this point of cut-off for full gear to give the best results for the various kinds of service to which a locomotive is subject. Therefore, an engine with a 24" stroke should cut off steam at 21" in full gear.

If you have an old engine that carries her steam too far or not far enough in either gear, and you wish to try the length of the reach rod by trying the cut-off in each extreme notch of the quadrant, proceed as follows: First try the cut-off in both gears; if it carries steam too far in the forward gear (for example, if it cuts off at 22" in the forward gear, and at 20" in the back gear), place the reverse lever in the forward notch of the quadrant; now use a tram (from a running board bracket, wheel cover or anything solid) and scribe a line on the reach rod. Then turn the wheel forward, and as the crank pin recedes from either center use a rule between the end of the cross-head and the travel mark it is leaving, and when the distance measures 21" stop. Now draw the reverse lever slowly toward the center notch and catch the correct port opening with the valve stem tram;

then stop moving the lever, and again mark the reach rod with the same tram. The difference between those two lines on the reach rod is the amount it should be shortened. In cases of this kind one or more notches of the quadrant is sometimes plugged up.

RULE 42.

INSIDE LAP OR INSIDE CLEARANCE.

Remember that if the valve has no inside lap, or inside clearance, one exhaust opens and the other closes just as the valve moves either way off the center between the port marks. While if the valve has inside lap, or inside clearance, the point of release and compression will take place earlier, or later, according to the amount.

RULE 43.

RIGHT AND LEFT LEAD ENGINES.

As a rule the right crank pin leads the left pin while in forward motion. These are called Right Lead engines. When the left crank pin leads the engine it is called a Left Lead engine; there are very few of the latter kind in use in this country. There is no material difference as to which crank pin leads, the former being the established practice.

RULE 44.

TO DISTINGUISH BETWEEN DIRECT AND INDIRECT VALVE MOTION.

To distinguish a direct from an indirect valve motion is sometimes very puzzling even to those thoroughly familiar with different valve gears. The rocker arm is usually the means by which the motion imparted by the eccentric is reversed, and thereby creating indirect motion. But occasionally we see an engine with both arms of the rocker extending upward or in the same direction. See page 109. In such case the motion is not reversed; therefore, it is direct motion.

In the Walschaert Valve Gear, shown on page 195 (which is used extensively in Belgium), the crank, or single eccentric, is attached to the crank pin, and causes the motion to be direct in one gear; that is, when the engine moves forward, and indirect in the other, or when moving backward. According to the usual method of connecting the shifting link motion with the reverse lever, to reverse the arms of the tumbling shaft does not reverse the movement imparted to the valve, even though many think it does; it merely reverses the positions of the eccentric blades where they fasten to the link. The forward or "go-ahead" eccentric blade being connected to the bottom of the link.

In the Allen design of valve gear, shown on page 193, and

used upon the "Class X" Pennsylvania Railway engines, we find four arms on the tumbling shaft instead of two, thus making the link and link block each adjustable; here we find the forward eccentric blade connected to the bottom of the link, and yet, owing to the rocker arm, the motion is indirect. With some piston valves used upon single expansion engines the steam ports are so arranged that the engine takes steam from the inside edge of the valve (that is, between the two rings or ends of the valve), thus causing the steam ports to be open at a time when they would be closed if steam was taken from the outside, or if a plain slide valve were used, and vice versa. Therefore, with the same valve gear that might be used with a plain slide valve every operation of the valve will be reversed by using this type of piston valve, thereby causing the eccentrics to occupy a reverse position in relation to the crank pin.

RULE 45.

TO EQUALIZE THE RUNNING CUT-OFF.

When the difference in the cut-off in both strokes on the same side of the engine is slight, and you can locate no other defect, you should sacrifice a perfect equalization of steam in the full gear in order to equalize the steam in the running cut-off. If the cut-off shows "heavy" in front; that is, carries steam farther in front than behind, shorten the eccentric blade; if "heavy" behind, lengthen the blade. A change of 1-32" in the length of the blade will make a difference of $\frac{1}{8}$" in the cut-off at each end.

RULE 46.

GIVING LEAD WHEN BOILER IS COLD.

If the quadrant is attached to the boiler and the boiler is cold, give the two back-up eccentrics 1-32" more lead than the two go-ahead eccentrics, as the expansion of the boiler when hot will raise the links slightly and thereby equalize the lead.

RULE 47.

TO SET VALVES ON A NEW ENGINE.

After setting the valves upon a new engine or one that has received a thorough overhauling, shorten all four eccentrics' blades 1-64". When the engine settles down and the wedges are set up the blades will be the correct length and the steam will be equalized.

RULE 48.

STATIONARY LINK GIVES CONSTANT LEAD.

When the link is stationary and the link block is adjustable, the lead is constant, See Fig. 40, page 191.

RULE 49.

ANGULAR ADVANCE AND CENTER LINE OF MOTION.

So long as the angular advances of the eccentrics are laid off from a line at right angles to the center line of the link motion, the latter can be arranged at any inclination to the piston motion without affecting the action of the link.

RULE 50.

CORRECT RADIUS OF A LINK.

A link with a radius less than the length of the eccentric rods will increase mid-gear lead at the forward port and reduce the lead behind. The correct radius for a link should equal the exact length of the eccentric blades, plus the distance from link pin arc to link arc.

RULE 51.

EFFECT OF RAISING THE ENGINE.

If the valve motion of a locomotive is perfectly equalized and it becomes necessary to raise the engine at the truck center or on the driving wheels, it will have a tendency to increase the lead in the forward gear, and decrease the lead in the back gear, and increasing the lead hastens every operation of the valve, but with the long eccentric rods in general use upon locomotives the engine can be raised an inch or two, and the effect upon the valve gear will be scarcely perceptible. If the engine has settled down on her springs, after steam has been equalized, it would have an opposite effect on the valve gear.

OTHER VALVE GEARS.

THE STATIONARY LINK—INVENTED BY MR. DANIEL GOOCH.

This form of connection between the valve and eccentrics, as shown by Fig. 40, is specially applicable to those circumstances in which the former requires no rocker. The mutual relation of the parts will be clearly perceived from an examination of the figure here shown, which illustrates one of the most successful methods of suspension. The eccentrics stand in their usual location for a direct action motion. The main link is hung from a *fixed* point by a short bar called the "suspending link," and the link block connected with the valve stem through the "Radius Rod" $m\,d^1$. By means of a reversing combination the block may be carried to any point between m, the full gear forward, and n, the full gear back. But since the link arc is always struck with a radius equal to the length of the rod md^1, having its center at d^1 and d^2 in the central line of motion, when the crank occupies the zero or

THE LOCOMOTIVE UP TO DATE.

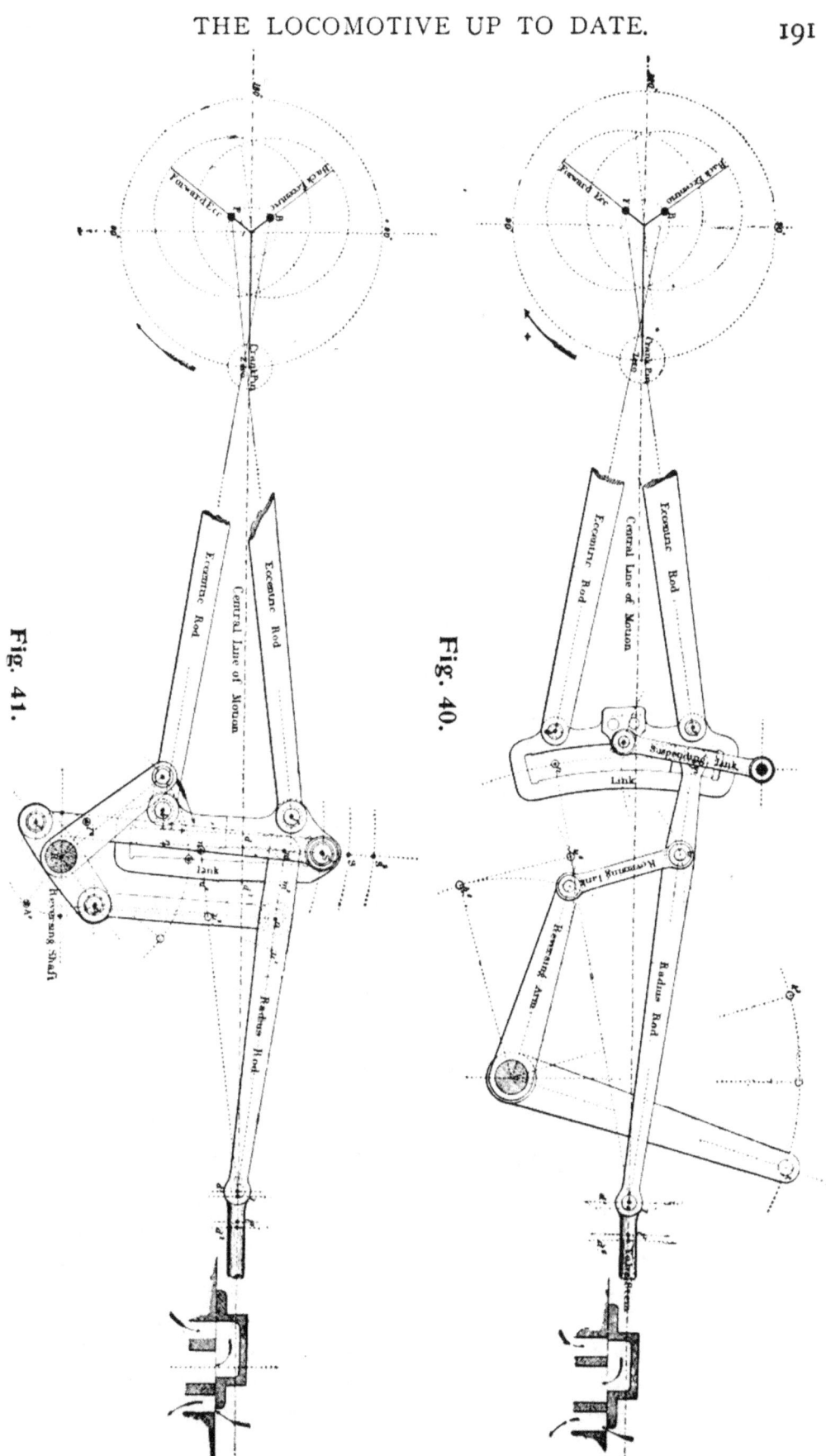

Fig. 40.

Fig. 41.

180° location, it must be evident that the block can be moved from one full gear to the other *without* altering the position of the points d^1 or a^2, consequently the lead opening will remain *constant* throughout the motion. . . . . . Now it has been invariably the custom to simply define a stationary link motion as *"one in which the lead is constant,"* leaving it to be inferred that the angular withdrawal of the crank from its zero position at the moment of pre-admission must also be a *constant* quality. Whereas, in reality, this lead angle *increases* just as much for a stationary link motion as for a shifting one. The only difference between the two is that the lead opening of the stationary link motion is more ample and the angle slightly greater, for all except the mid-gear, than with the shifting link motion. Unlike the shifting link motion, however, the lead opening is *not* dependent on the arrangement of the eccentric rods, for these may either be crossed or opened without altering the result. But for the purpose of meeting the other conditions of the motion an arrangement like the figure shown should be adopted. As a general thing more attention is paid to the equalization of the cut-off and reduction of the slip in the forward than in the back gear. For the accomplishment of this object, the center of the link should be dropped below the central line of motion, the angular advance of the backing eccentric slightly reduced and the backing eccentric rod lengthened.

The Stationary Link is seldom found in American practice, from the fact that all modern locomotives are built with steamchests on top of their cylinders, instead of at the side. On stationary engines, the link and governor are occasionally used conjointly; in such instances the stationary link will be found best adapted to the requirements of the case, because its radius rod imposes a far lighter duty upon the balls of the governor than the shifting link with its rods, hanger, and additional friction of eccentric straps. This form of link motion is largely used upon locomotives throughout Great Britain.

ALLAN LINK MOTION.

The discovery of this motion was a natural sequence to the invention of the shifting and stationary links. By it a compromise has been effected between the leading features of both motions, resulting in a more *direct* action and perfect balance of the parts together with a reduced slip of the link block. One mode of suspension, for the link in the full gear forward, appears in figure 41, in which the cross sections of the valve and its seat have been revolved for the purpose of more plainly exhibiting their relative positions. The locations of the point of suspension and attachments of the eccentric rod pins upon or back of the link arc are quite as variable for this, as for the

THE LOCOMOTIVE UP TO DATE.

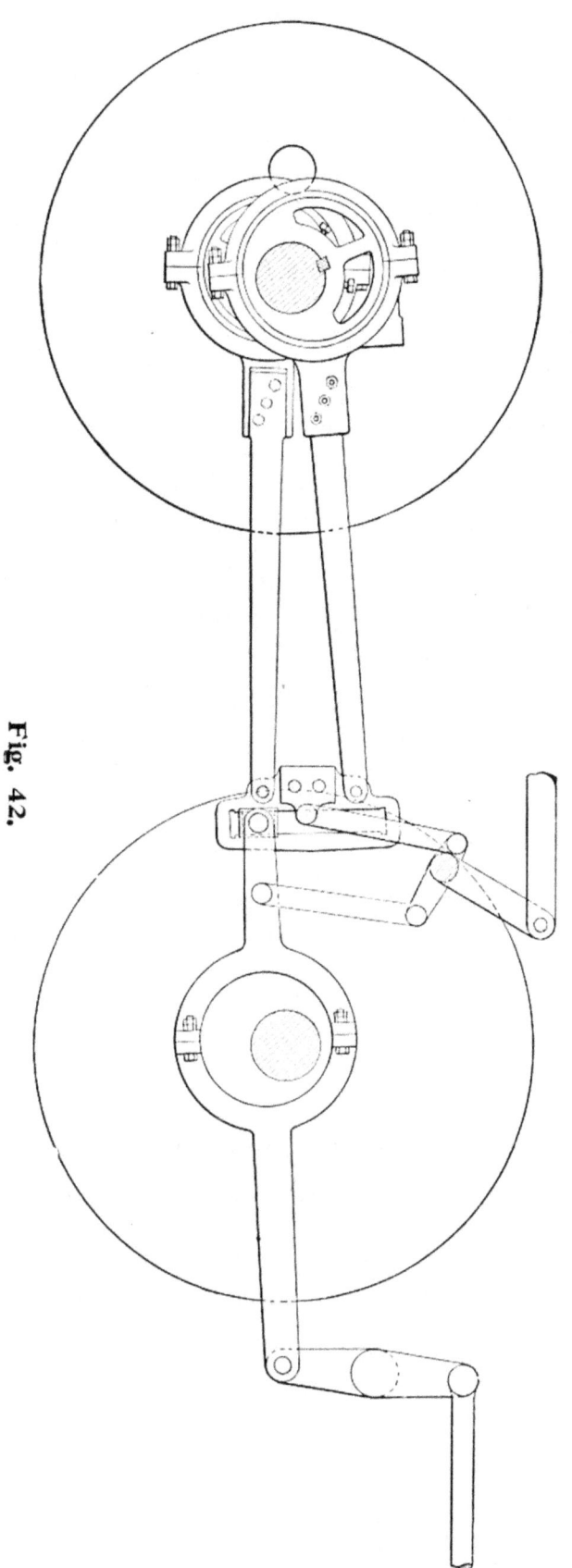

Fig. 42.

shifting link motion; and the requirements of the other details generally indicate whether the reversing shaft should be placed above or below the central line of motion.

In proportioning the parts, the main object is to move the link and radius rod (when the crank stands at the zero or 180° locations) in such a manner that the link arcs peculiar to each motion shall always be *tangent* to each other. In this case all the locations of the link block will be found in one and the same *straight* line. This peculiarity has given rise to the title, "Straight Link" motion, expressive of the form of the link.

The radius rod and main link are supported by rods from the reversing shaft arms, and the inequality in the lengths of the latter, which is essential to a proper suspension of the parts, incidentally tends to *equalize* the weights resting on the opposite sides of the reversing shaft, thus greatly facilitating a change of the motion from one full gear to the other.

Well-schemed motions of this type practically preserve the characteristic feature of the stationary link, viz., a constant lead; yet from the nature of the case they possess at times slight inequalities in one or both of the full gears. These, however, are quite insignificant for a relatively long radius rod and short travel.

This form of link motion is particularly adapted for use upon locomotives with steam chests at the side of the cylinders, better known as "inside connected" engines. Another form of the Allan link motion is shown on page 193. This form was designed for use upon the American type of locomotive, with steam chests on top of the cylinders. This form of link motion has been applied to several large Class X Freight Engines of the Pennsylvania R. R., and are at present in service on the P., Ft. W. & C. division between Ft. Wayne, Ind., and Chicago. It is claimed they are giving perfect satisfaction and showing excellent results. The straight link gives almost a constant lead, and the slip is greatly reduced.

WALSCHAERT LINK MOTION.

This motion, as shown by Fig. 43, is extensively used in Belgium and other continental countries, but probably will not receive very much attention from locomotive builders in this country, unless future designers succeed in reducing the number of its connections. Although a few engines equipped with this form of valve gear are at present in service in the United States. No eccentrics are used with this form of link motion, the valve receiving its motion from a crank attached to the outside of the main crank pin (which is equivalent to an eccentric). The link being stationary the motion is indirect in the forward gear and direct in the back gear (depending of course upon the connections). The crank, therefore, does the work of two eccentrics.

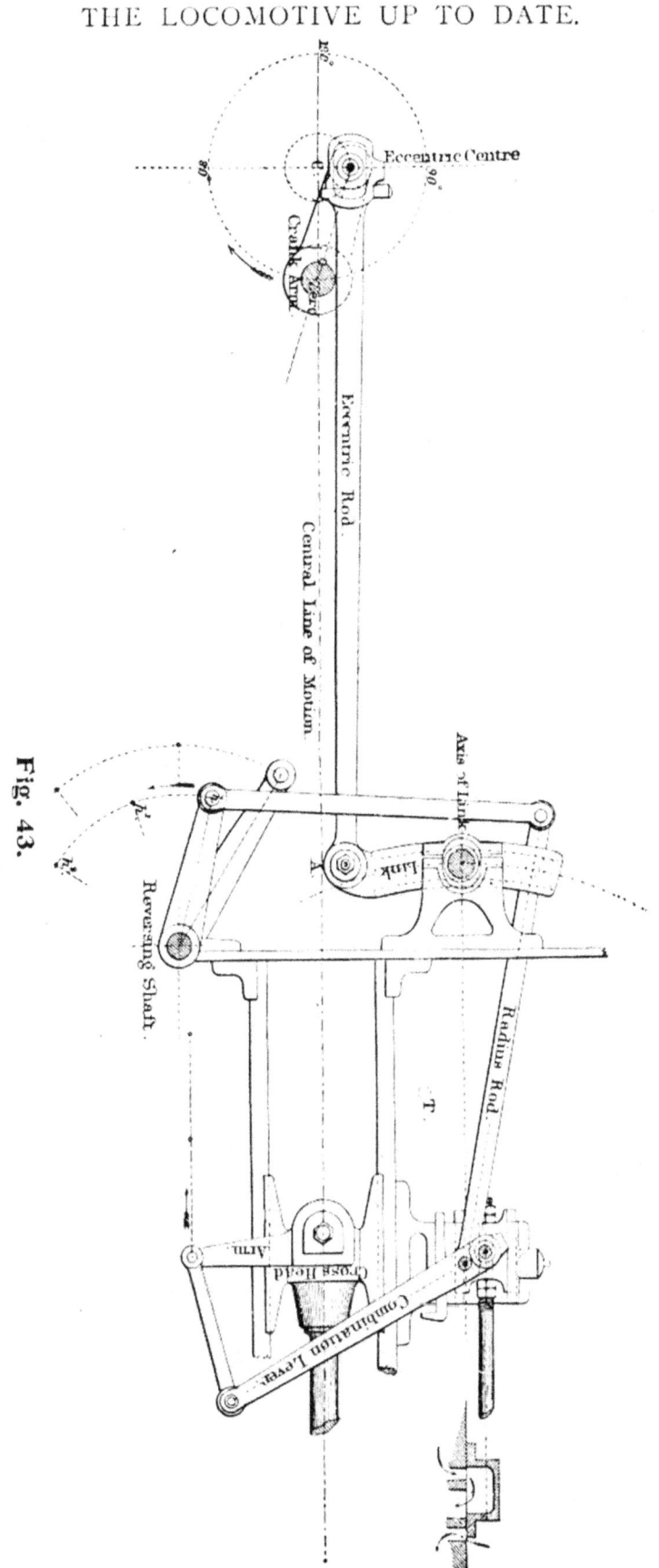

Fig. 43.

This device groups two perfectly distinct motions—the one derived from a single eccentric, the other from the cross-head of the piston rod—in such a manner that their combined effect is, when the parts are well proportioned, quite analogous to the motion obtained from the stationary link. From the nature of the connection between the cross-head and the valve-stem, the motion can be more readily applied to an outside cylinder engine than to an inside one.

The eccentric usually assumes the form of a return crank from the main crank pin. Its center then is found on a line at right angles to the crank arm. The angular advance becomes equal to zero, and hence, so far as the link will be concerned, the valve can have neither lap nor lead. The link oscillates freely about a fixed axis, and its arc has a radius equal to the length of the radius rod. This rod is moved from one full gear to the other, in the usual manner, by means of a reversing shaft with arms. From the extremity of a short arm, rigidly bolted to the cross-head pin, extends a union bar which is pinned to one end of the combination lever. By the aid of this lever, the eccentric and cross-head motions are so combined, that the latter virtually restores the angular advance discarded while locating the eccentric, and consequently enables the valve to possess both a constant lap and lead.

THE LEWIS VALVE GEAR.

This design of valve gear is a recent production and a radical departure from the old reliable link motion, dispensing with eccentrics, links, rocker arms, etc. It received considerable attention from railway officials a few years ago and attracted the general attention of practical railroad men throughout the country, but very few have ever seen this valve gear in operation, and few of these who did see it understood anything about it. It is owing to the general interest manifested in the device and the fact that it has not heretofore been illustrated, as far as we know, that we present the valve gear, together with a complete description of it by the inventor, Mr. Wallace J. Lewis, of St. Louis, Missouri.

It is an ingenious device and well worth studying. It is practically the duplex pump movement with a reversible cut-off attachment added. It is so constructed that the lap and lead movement on one side is connected by the cross-head on the same side, while increased travel to the valve is conveyed from the opposite cross-head—*the lead is constant*.

Owing to its compactness and close proximity to the steam chests it necessarily diminishes the amount of lost motion, which is often an important factor on ten and twelve wheeled engines.

This valve gear was given a trial on the Wabash, Vandalia

and Illinois Central Railways and failed to demonstrate its superiority over the shifting link; it was therefore removed, and we understand that at present it is being tried on some western road.

Mr. W C. Arp, Supt. M. P. of the Vandalia Railroad, gives the following reason for its removal:

"This gear was placed on engine 39 in the summer of 1892 and removed in the fall of 1895. The working parts of the gear were poorly constructed and gave continuous trouble in the matter of repairs. It would frequently break when the engine was in heavy service, necessitating the engine being towed back to the shop for repairs of broken parts. When this valve gear was in good condition it gave very good results."

DESCRIPTION BY MR. LEWIS.

"Figure 1 is a side elevation of the valve-gear, together with so much of the parts with which it is connected as is necessary to an understanding of the construction and operation of this device.

Fig. 2 is a plain view of the devices shown in Fig. 1; and Fig. 3 is a rear elevation of the parts shown in Fig. 1, looking in the direction of the arrow. Fig. 4 is a detail enlarged plan view of the rocker-arm 7, link 8, and sliding bar 9, and in the interengaging flanges 7^a and 8^a. Fig. 5 is a sectional view in elevation, taken on the line x x, Fig. 4, showing the link 8 and rocker-arm 7 and flange 7^a in elevation. Like symbols refer to like parts wherever they occur.

This invention relates more particularly to the construction of reversing mechanism for that class of valve gear used in conjunction with the slide valves of double engines, wherein two separate lines of reciprocating motion are applied, the one to overcome the "lap and lead" of the valve, and the other to effect the "throw" of the valve.

For purposes of illustration, the valve gear as applied to the right hand side of a locomotive engine has been selected—it being understood that it has its duplicate on the left hand side of said engine—and the two separate lines of motion have been obtained from the opposite cross-heads.

The lines of reciprocating motion generated by such engines have respectively the same rate of travel, but as the motion of the two pistons of the double engine are not synchronous, being separate in time approximately by half the period of a stroke, so the two lines of motion which are combined to operate the slide valve of each engine have the same period of complete reciprocation, but are separate in time by half of that period. The object of combining these two motions, which have the same period but are not synchronous, is to reduce a resultant reciprocation by which the slide valve may be operated with a varying rate of movement more favorable to the expansive action of steam in the cylinders than is possible with the ordinary link motion.

The function of the valve being to regulate the alternate admission and discharge of steam at each end of the cylinder the valve is accordingly set so as to be at a point beyond the middle of its travel—the required "lap and lead"—when the piston is at the beginning of its stroke, and in coming to said position the valve overcomes the "lap and lead" so that the steam may fill the clearance space at the beginning of the stroke of the piston.

In order to work effectually and economically, either with full boiler pressure or expansively, it is desirable that over the middle portion of its travel the movement of the valve should be accelerated and for the remainder of its travel the movement of the valve should be retarded, and the longer the retarded movement can be maintained, consistent with the next following rapid movement, the better will be the results.

The object of accelerating the slide valve during the middle portion of its travel is to quickly overcome the lap and lead and open the induction port (the same movement opening for the exhaust at the other end of the cylinder), and the object of retarding the movement of the slide valve during the remainder of its stroke is to obtain the full expansive energy of the steam. In order to effect this proportional and regulated acceleration and retardation of the valve travel, the two separate lines of reciprocating motion, hereinbefore referred to, are each connected at different points to the same floating lever so that a third point of said lever will yield the required varying reciprocal motion to the slide valve. The action of said two movements upon the floating lever, and thereby upon the valve, is so adjusted that when the reciprocal motion derived from the opposite side is in "mid-gear," or at the middle point of its movement, the movement derived from the near side will give the slide-valve a motion sufficient to overcome the "lap and lead." Hence it is usual to say in speaking of this class of gear that the valve receives its "lap and lead" from its own side, and its "throw" from the opposite side.

In double engines, where a reversal of travel or rotation is required, as in locomotives or marine engines, it is necessary to incorporate in the valve gearing some mechanism which, without interfering with the construction and operation of the gearing as hereinbefore specified, will accomplish the reversal of the piston when desired. To this end, the main feature of my invention, generally stated, embraces the combination with the valve, a floating lever for operating the valve, and means for actuating the floating lever at different points at the same rate of travel, though not synchronously, of a slide bar, and a movable guide

THE LOCOMOTIVE UP TO DATE.

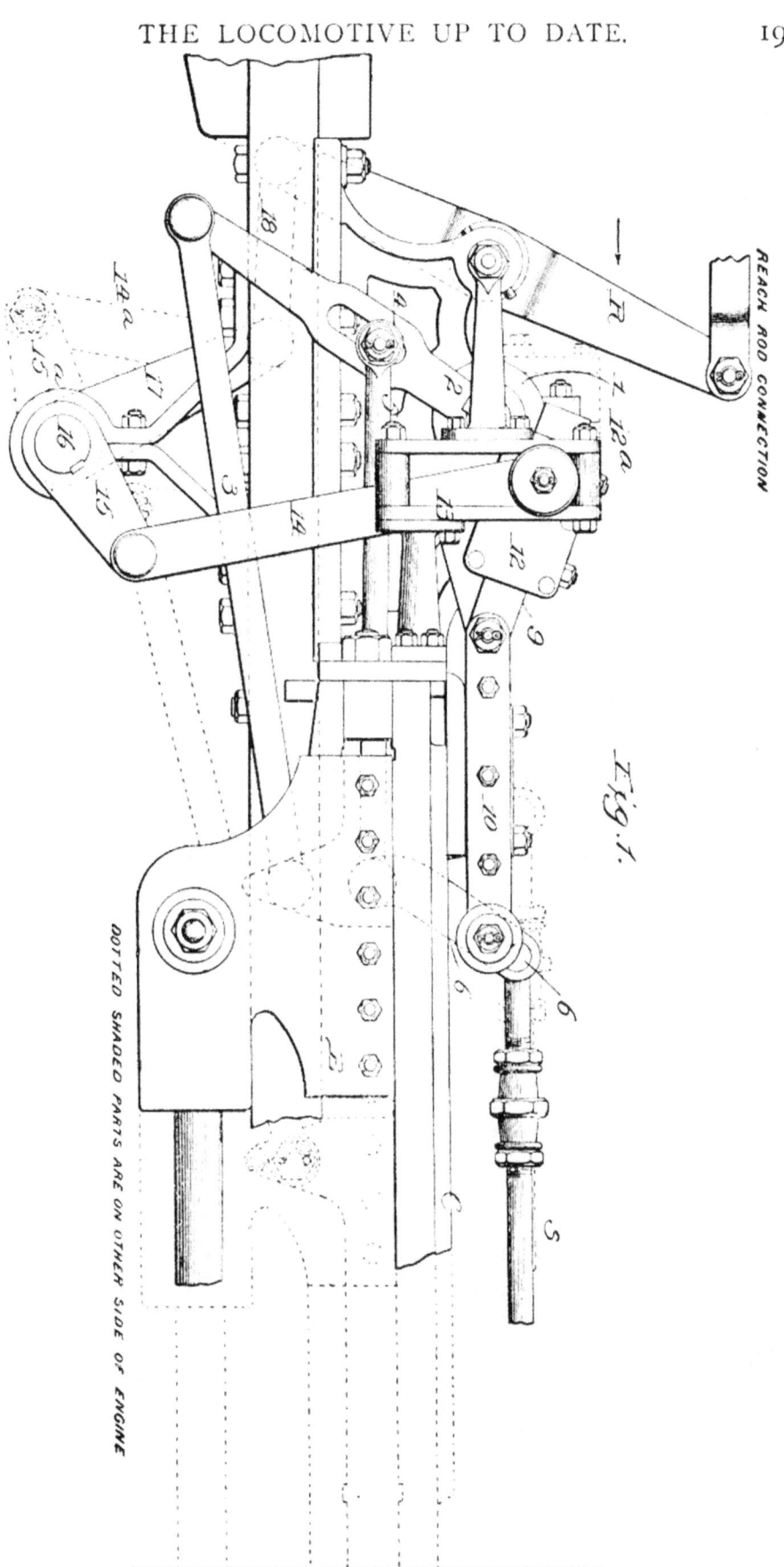

Fig. 1.

box therefor, said two latter elements being interposed between the floating lever and the means which imparts the "throw" to the valve, whereby the reversal of the engines may be easily and readily effected.

A second feature of my invention embraces an adjustable connection between the floating lever which actuates the valve, and the means for imparting the movement to overcome the "lap and lead," whereby the lead may be increased or decreased at will."

DESCRIPTION OF THE PARTS.

In the drawings A indicates a portion of one side of the frame; B, the cross-head of said side; C, the cross-head guide, and D the guide yoke which couples the cross-head guides of the opposite sides of a locomotive.

Extending transversely of the locomotive and properly journaled on its frame, are two rock shafts 1 and 1^a, the first of which—(1) receives its motion from the right hand cross-head B (or equivalent moving element), and transmits it to the opposite or left hand piston valve to effect the "throw" thereof, and the second of which (1^a) receives its motion from the left hand cross-head (or equivalent moving element) and transmits it to effect the "throw" of the right hand valve—said rock shafts thus coupling and combining the duplicate valve gear of opposite sides.

The rock shaft 1 (shown in the drawings) is provided with a lever 2, the lower end of which is connected with the cross-head B of the right hand side of the engine by means of a rod 3, and said lever 2 is provided at or near its middle with an elongated slot wherein is adjustably secured a thimble 4, which is journaled one end of a rod 5, the opposite end of which is pivoted to the lower end of a floating lever 6, the opposite end of said floating lever 6 being pivotally connected with the valve stem s of the right side or right hand valve, the right hand valve thus receiving from the right hand cross-head B (or its engine) the motion which overcomes its "lap and lead," and as the rod 5 is adjustably connected with the lever 2 it is evident that the lead of said valve may be increased or decreased by a proper adjustment between the parts 2 and 5.

It is to be understood that the parts of the valve gear thus far described have their equivalents—or counterparts—the valve gear for the valve on the left hand side of the engine, and that through said parts the second rock shaft 1^a is operated from the cross-head on the left hand side of the engine (or the opposite side as the case may be). To the near end of said rock shaft 1^a is secured a rock arm 7, connected at its outer end by a pin with the corresponding end of a link or arm 8 and thence through suitable connections with the floating lever 6 at a point between

When in a central position or "mid-gear," the center line of the pivot pin which connects arm 8 to slide link 9 coincides with the center of axial line of shaft 1ª and the nearer said center is approached the shorter will be the "cut-off."

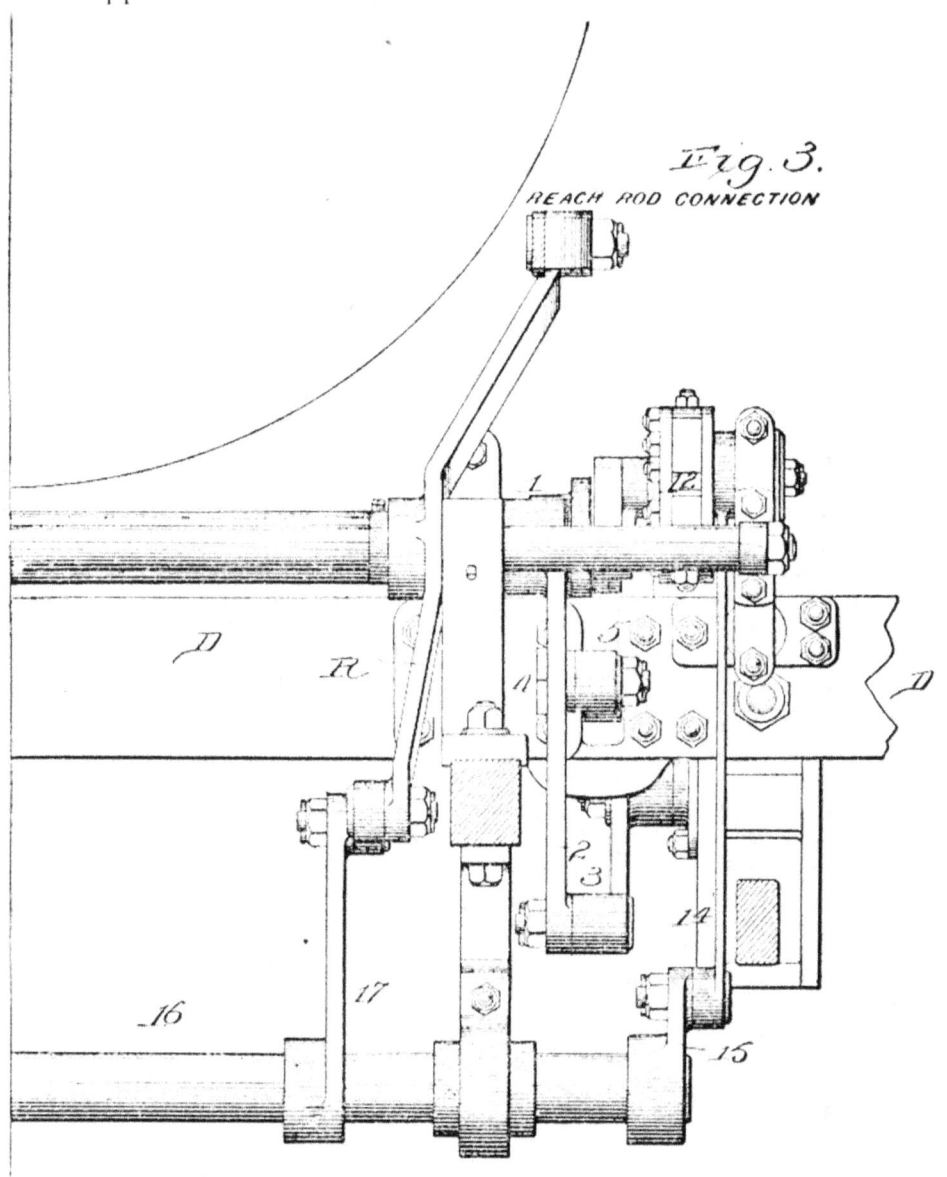

It should be noted that by throwing the pivot pin which connects the lever end of the floating lever 6 and the rod 5 out of line the admission of steam in each end of the cylinder may be equalized for all 'cut-offs.'"

THE JOY VALVE GEAR.

This valve gear is used upon several of the leading railroads in England; it is also in use upon many continental lines and in South America; at the present time strenuous efforts are

204 THE LOCOMOTIVE UP TO DATE.

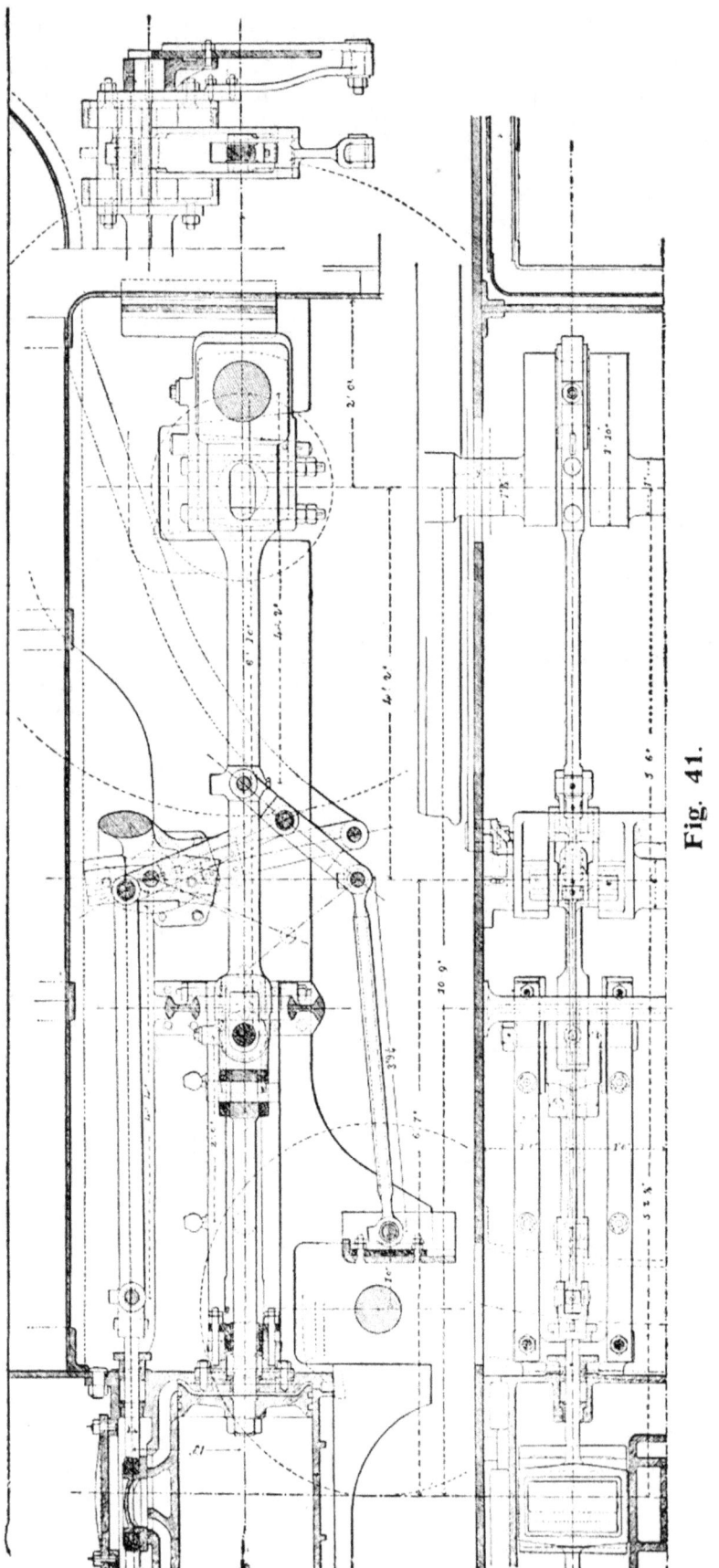

Fig. 41.

being made to introduce it into this country. Believing it will be highly interesting to our readers, we have shown two illustrations of this motion. The first one as applied to the English (inside connected) locomotives, and the other its application to American (outside cylinder) locomotives. Through the courtesy of *"The London Magazine,"* of London, England, we are enabled to give the following description of the Joy Valve Gear:

"Many advantages are claimed for the Joy Valve Gear; it is simple in construction and maintenance, the weight of the parts are less than those of the shifting link and it is generally more correct in working. The lead and cut-off are exactly equal for both ends of the cylinders, and they remain so for all grades of expansion to mid-gear.

The valve opens more rapidly than when actuated by link motions, the cut-off being prompt and the release of the exhaust quick, whilst it moves slowly during the expanding and exhausting periods. These qualities are very desirable for a locomotive slide valve, when obtained without any undue lead, compression, or too early exhaust.

The cut-off point is not limited by the throw of eccentrics, etc., but the reversing depends upon the angle to which the quadrant block guides are inclined, so that it would be only necessary to allow these to be carried over past the point usual for a full gear cut-off of say 75 per cent. to obtain a cut-off of 80 or 90 per cent.; thus the starting power of the engine can be greatly increased, and the trouble sometimes necessary of reversing to get it into a more favorable position, dispensed with.

This gear dispenses with eccentrics, and is classed among radical valve gears, in which the valve movement is taken from the connecting rod through a system of levers. It is one of the most successful and most used of them. An arrangement of it as applied to an inside cylinder engine is illustrated in Fig. 1.

The connecting rod has an enlarged boss formed in it at a suitable point, about one-third of the length of the rod measuring from the small end. This is bored out and fitted with a bush; through it a pin passes, projecting on either side to carry the forked ends of the "correcting link." This is coupled at its other end to the "anchor" link, which in turn is pinned and allowed to vibrate about a fixed point. In the illustration this point is upon a projecting lug upon the motion plate or slide bar bracket. Examples may, however, be often met with in which it is on either a frame stay or a shaft fixed across the frames, as found most suitable in the particular design of engine. The correcting link has a bearing in its central portion, to which is connected the "valve levers," these being further provided with two other bearings—one at the top end for the attachment of the valve rod end and the other close to it for the pin, which carries upon

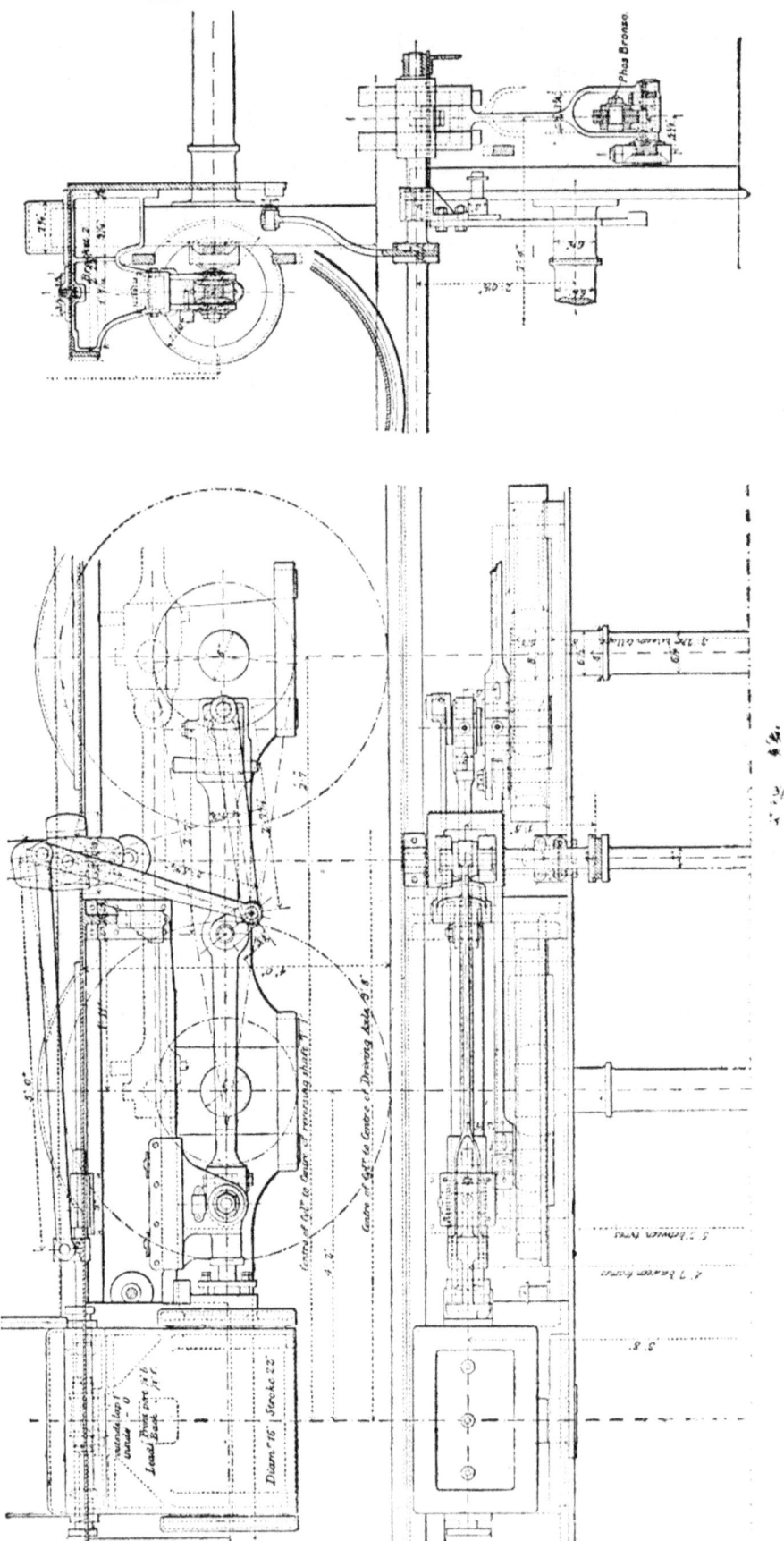

its ends the "quadrant blocks," which are fitted to sliding curved guides fixed to the reversing shaft.

It will be seen that the vibration of the connecting rod when the engine is running moves the quadrant blocks up and down in the curved guides, which compel it to take a course dependent upon the position of the guides; thus, in the sketch the guides are vertical, therefore the valve rod, and with it the valve, will have least movement horizontally, when the engine is in "mid-gear." When it is desired to run forwards, the reversing shaft has a partial revolution given to it, tilting the top of the guides over towards the cylinders; this causes the block to move in the required position and gear. When the engine is reversed to full back gear, the shaft is partially turned, so that the top of the guides lay over towards the fire box; the points of cut-off between mid-gear and full gear are settled by giving the guides more or less inclination as required.

To determine the several dimensions and positions of attachment of the various rods, etc., it is necessary to take account of all the arrangements of the engine. The position of the pin hole in the connecting rod for attachment of the correcting link is found by drawing lines from that part of the crank circle which the center of the crank pin occupies when the piston is at half stroke to the center of the small end pin, and taking a point in the length of the connecting rod which has a vertical vibration equal to at least twice the full stroke of the valve, so as to avoid too great an inclination upon the guides when the engine is put over into full forward or backward gear.

This point being fixed, mark off upon the center line of the rod the position it will occupy when the crank is upon its front and back centers respectively, and draw a vertical line—vertical to the center line of the motion—through a point exactly central between them. Now the length of the correcting link can be found; its end, which is attached to the anchor link, must be sufficiently far away to allow of the angle between its two extreme positions being less than a right angle.

The anchor link should be made as long as convenient, so as to allow the end of the correcting link to rise and fall as nearly as possible in a vertical line. It will not affect the distribution of steam which end of the engine its end is fixed, but it is usually found more convenient in locomotives to fix it forward of the crank.

Next, upon the center line of the valve spindle, which must be in the plane of vibration of the connecting rod, upon each side of the vertical line above mentioned, mark off the amount required for lap and lead. That for the front port being drawn upon the crank side of the vertical line, and that for the back port upon the cylinder side of it. Now, assuming the crank to be upon its front dead center, and the correcting link coupled to the anchor link, choose a point in the length of the correcting link, which has to be assumed, and draw a line through it to the lap and lead mark for the front center, the crank being upon that center; the point where this line crosses the vertical line will be the center line of the reversing shaft, and the center of oscillation of the curved guides, which must be concentric at each end of the stroke of the piston. The exact point in the correcting link for the attachment of the valve levers must be found so that the quadrant blocks vibrate equally on each side of the center of the quadrant block guides. It will now be seen that as the center of oscillation of the quadrant block, and the center of the reversing shaft exactly coincide with each other when the piston is at each end of its stroke, it will be possible to reverse the motion from full forward to full backward gear without communicating any movement to the valve rod, the reversing shaft simply making a partial revolution, the lead being constant in all gears. This point is important, and is made use of when the valves are being set in the shop.

The valve spindle may be of any convenient length, and is coupled up to the top of the valve levers at one end, and to a cross-head upon the valve buckle spindle at the other. Whatever length is chosen for the valve spindle will be also the length of the radius of the quadrant block guides.

It will be noticed that the lap and lead are entirely dependent upon the distance apart of the two centers of the valve levers, which are connected to the valve spindle and the quadrant blocks respectively, and can only be altered by varying this length.

The path described by the point in the connecting rod where the correcting link is connected is a horizontal ellipse when the engine is running, and the end of the valve lever to which the valve spindle is coupled describes a vertical ellipse at the same time; therefore, this particular style of motion is known as "elliptical" gear. It would be possible to couple the end of the valve levers direct to the connecting rods, and dispense with the correcting link altogether, but this would give an irregular motion to the valve, and give a different distribution of steam in the front and back strokes.

In some locomotives with outside cylinders, where it would not be possible to find room for the anchor link, as shown in Fig. 2; the correcting links are cut off at the point of attachment of the valve levers, and the anchor link, coupled to them there, and at its other end to small return cranks upon the crank pin—the stroke of which must, of course, equal the distance traveled by the other end of the anchor link.

Deviations from the above positions and proportions may be made without materially altering the correctness of the results.

The metals employed are usually wrought iron for the links, rods, pins, etc., the wearing portions being well case-hardened.

The quadrant block guides too are of wrought iron, being usually turned up and cut from a large ring of the necessary radius and case-hardened. The quadrant blocks are of bronze or cast iron; if of the latter, to prevent wear they are cast in chills and ground up to shape. The reversing shaft is of cast steel, and the bushes in the connecting rods of phosphor bronze.

TO SET THE VALVES.

To set the valves in the shop, the four dead centers of the crank are found in the same way as for link motions, as also are the cut-off points for each port. Then setting the engine upon one dead center, the rods being coupled up, reverse the motion from one full gear to the other, if the valve spindle moves, the reversing shaft has to be raised or lowered until this ceases. This being done, try the other dead center upon that side in the same way, and equalize the lead for both ends by moving the shaft backwards or forwards as found necessary. Try all dead centers in the same way, and the valves are set; the process is much simpler than that required for link motions.

THE SHELB VALVE MOTION.

This design of valve gear is a product of 1895. It is an ingenious device, and the invention of Mr. Peter J. Shelb, of Fort Worth, Texas, who has furnished us the following description of his device:

"My invention relates to a variable cut-off valve-motion, and has for its object to provide simple and efficient and at the same time readily adjustable means for varying the points at which the ports communicating with the cylinder of a locomotive are opened and closed, in accordance with the load, the desired speed, etc.

This I have accomplished in the following manner:

The eccentrics, instead of being adjustable upon the drive-shaft to vary the cut-off by the disposition of their major axes at different angles or in different relative positions, are arranged with their major axes at right angles or perpendicular to a line connecting the axes of the drive-shaft and the crank-pin, whereby the eccentrics are at their greatest throw as the crank pin approaches the dead centers.

A bell-crank lever 14, one arm of which is connected by a rod 15 to an operating or hand-lever 16, is connected by means of a link 17 to the slotted link 11, whereby the pivotal point of connection between said slotted link and the lower arm of the rocker 13 may be adjusted as desired to vary the relative movements of the eccentrics and the rocker.

A cross-head 18, which is mounted for reciprocation in guides 19 and is operated by the connecting-rod, or by other means from

14

the crank-pin, is loosely connected to the lower end of a swinging arm 20, which is preferably pivoted at its upper end to a stationary part of the framework of the engine, and this arm is connected, at an intermediate point, to one arm of an auxiliary rocking-lever 21. The fixed pivotal point of the swinging-arm 20 is shown at 22, and the connection of the arm with the crosshead is by means of a pin 23 on the latter fitting in a longitudinal slot 24 in the former. A sliding or loose connection is also pro-

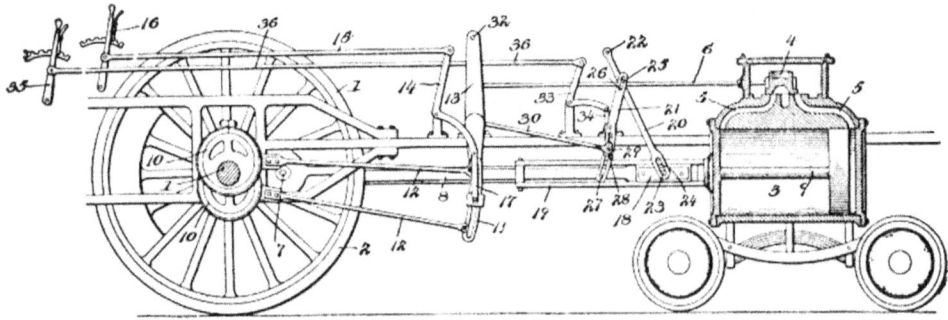

Fig. 41.

vided between the swinging-arm 20 and the upper arm of the auxiliary rocking-lever, formed by a pin 25 on the former fitting in a slot 26 in the latter.

The lower arm of the auxiliary rocking-lever is slotted as shown at 27 for the adjustment of a block 28 carrying a pivot 29 for the attachment of the rod 30 which extends to and is connected with the lower end of a link-arm 31 pivoted to the upper

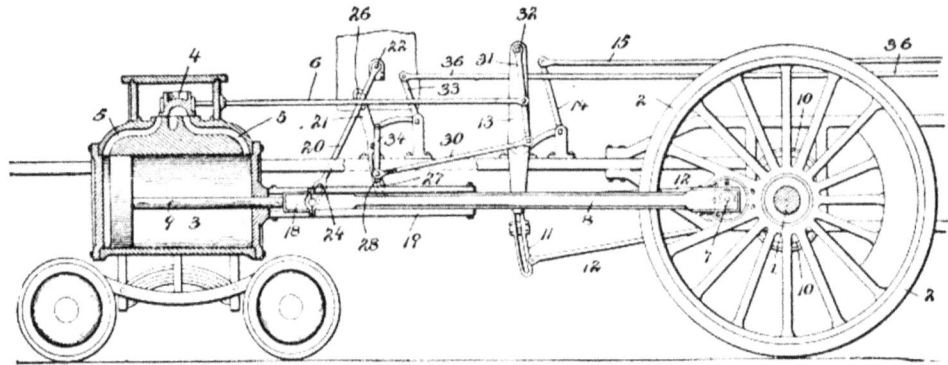

Fig. 42.

extremity of the main rocker, as shown at 32. This link arm forms the connection between the valve-stem and the rocker, and is pivotally connected, at an intermediate point to the extremity of the former. The block 28 forms the connection between the rod 30 and the lower arm of the auxiliary rocker, and as the latter is preferably pivoted to a fixed part of the framework of the engine it will be understood that by varying the position of the block in the slot the throw of the link-arm 31, and hence of the valve-stem, may be varied independently of the main rocker and connections. The adjustment of the pivotal connection of the rod 30 with the auxiliary rocker is accomplished by means of an auxiliary bell-crank lever 33, connected by a link 34 to the block, an operating lever or handle 35, and a connecting rod 36.

From the above description the operation of the improved valve-motion will be readily understood by those skilled in the art to which the invention appertains.

The adjustment of the pivotal point of opening and closing of the ports may be attained, and such opening and closing may be accomplished at any desired point of travel of the piston. It should be understood that this construction is susceptible of numerous modifications to suit different conditions under which it is employed, and it is obvious that changes in the form, proportion, and minor details of construction may be resorted to without departing from the spirit or sacrificing any of the advantages of my invention.

THE STEAM ENGINE INDICATOR.

The steam engine indicator is an instrument for measuring and recording steam pressures in the cylinder. The purpose of the instrument is the same as a recording steam gauge, with this difference, that it records the pressure at each instant of the engine stroke, and this pressure is usually a variable one in different parts of the stroke. It is a miniature of the larger cylinder whose performance it is designed to reveal. Its object is simply to put upon paper the steam pressure pushing the piston at each point in the stroke, and at the same time to put on paper what pressure is opposing the piston at each portion of the piston's

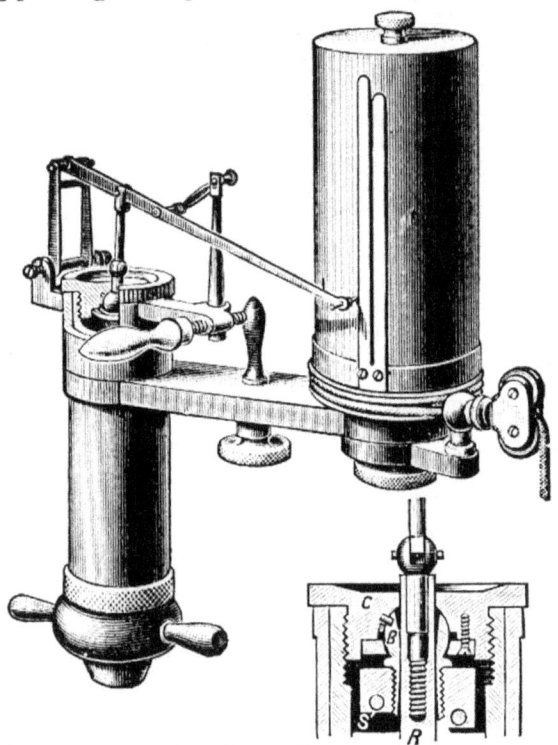

stroke. The instrument is virtually a steam gauge with a spiral spring to measure the pressures, and a pencil to record them. The indicator, although in a crude form, was invented by Mr. James Watt. It has gradually been improved, until in its present form it will point out the least distortion in a correct distribution of steam, thereby permitting of a correct adjustment of the valve gear. Therefore, its use is considered indispensable with an economical use of steam.

The degree of excellence to which steam engines of the present time have been brought is due more to the use of the indicator than to any other cause. A careful study of indicator diagrams taken under different conditions of load, pressure, etc., furnishes the only means of showing the action of steam in an engine, and of gaining a definite knowledge of the various changes of pressure that take place in the cylinder.

An indicator diagram is the result of two motions, namely: a horizontal movement of the paper in exact correspondence with the movement of the piston, and a vertical movement of the pencil in exact ratio to the pressure exerted in the cylinder of the engine. Consequently, it represents by its length, the stroke of the engine on a reduced scale, and by its height at any point, the pressure on the piston, at a corresponding point in the stroke. The shape of the diagram depends altogether upon the manner in which the steam is admitted to and released from the cylinder of the engine. The variety of shapes given from different engines, and by the same engine under different circumstances, is almost endless, and it is in the intelligent and careful measurement of these that the true value of the indicator is found, and no one at the present day can claim to be a competent engineer who has not become familiar with the use of the indicator, and skillful in turning to practical advantage the varied information which it furnishes.

A diagram shows the pressure acting on one side of the piston only, during both the forward and return stroke, whereon all the changes of pressure may be properly located, studied and measured. To show the corresponding pressures on the other side of the piston, another diagram must be taken from the other end of the cylinder. When the three-way cock is used, the diagrams from both ends are usually taken on the same paper.

Owing to the necessarily limited space we will be permitted to devote to this subject, we shall endeavor to explain only the elementary principles of the steam engine indicator for the benefit of those who have no knowledge of the indicator. For a more complete knowledge of this subject we would advise the reader to secure some special treatise on the indicator. There are many such works published. We shall briefly explain the construction and use of the indicator, and give the definitions of the technical terms used, together with its method of application and operation, and a few instructions for reading its diagrams (or cards). There are many different steam indicators upon the market, but we shall describe only one, and how to take care of the instrument. To understand the construction of one indicator implies a knowledge of all others, as the principle is the same in all; they differ only in their construction.

CONSTRUCTION AND USE OF THE STEAM INDICATOR.

The modern indicator is a very sensitive and delicate instrument, and as near perfection as any piece of mechanism can be made, and its records are universally relied upon. It is used for measuring continuously during the stroke of the piston the pressure in the cylinder, and recording that pressure on paper; the paper card being called an indicator diagram. The pencil which marks the card is connected by a lever with a small piston which moves up and down in a very small cylinder, the pencil having the same vertical movement; the travel of the pencil being three or four times greater than the piston. A small spring of known tension registers the action of the piston, and a small cock separates it from the cylinders, so when it is in communication with the cylinder it acts as a gauge, and when the steam pressure is cut off both sides of the piston are exposed to the atmosphere, the pencil thereby indicating the atmospheric pressure; the top of the piston is always exposed to the atmosphere. The paper drum which holds the cards, as may be seen by the illustration, is revolved upon its axis by a cord, which receives its motion from the cross-head, being transmitted by means of a fulcrumed lever, a cord being attached to the lever at such a point as will give the drum a proportionate travel to the stroke so diagrams of any desired size may be taken; a reducing wheel is in use which may be adjusted for any stroke. A 4" diagram is about right for locomotive practice. The height of the diagrams will be controlled by the tension of the spring used, each spring being marked. The numbers on the spring signify that a vertical movement of the pencil of one inch is accomplished by a pressure per inch in the cylinder equal to the number on the spring; but in practice springs numbered one-half as high as boiler pressure are used. An adjustable spring adapted to any speed or pressure is also in use. By recording the pressure on the piston during a complete stroke we are able to determine the expansive force of the steam and also to discover any back pressure which may be in the cylinder, and thereby determine the horse power exerted by the engine, and discover any irregularities in a correct distribution of the steam, which could not be done without the use of the indicator. It is not so difficult to learn to operate the indicator as many men believe; it is a delicate instrument, but only requires a little study, careful handling and good judgment.

METHOD OF APPLICATION TO A LOCOMOTIVE.

The use of the indicator has become so common that the cylinders of all modern engines are drilled and tapped for its attachment by the builders. When drilling holes in the cylinder for indicator pipes two holes should be drilled, one at each end of the cylinder. They should be kept as near the ends of the cylinder as possible, so the piston will not close the holes when at extreme travel. When the cylinder has been bored out or worn as large as the counter bore and the clearance becomes very small, a groove should be cut from near the end of the cylinder to the hole. The holes should be drilled on the outside of the cylinder about half way up the side for convenience, but never in the bottom of the cylinder, where there might be water. They should be tapped for a $\frac{3}{4}''$ pipe. In some cases the cylinder heads are drilled for this purpose, but the side is the best place as the pipe connection should be as short and have as few elbows as possible. (When the engine is in service, the pipes must extend above the steam chest to clear bridges, etc.) The two holes should be connected with a straight pipe, with a three way cock in the center, or T, and cock on each side of it, the indicator being attached to the T. One end of the fulcrumed lever should be securely fastened to the running board and should swing on a pivot; the lower end should be attached to the cross-head. When the drum cord is fastened to the lever, in the center of its travel, the lever should always be at perfect right angles with the cord. Pulleys should be avoided if possible.

DEFINITION OF TECHNICAL TERMS.

"*Absolute pressure*" of steam is its pressure reckoned from a vacuum; the pressure shown by the steam gauge, plus the pressure of the atmosphere.

"*Boiler pressure*" is the pressure above atmosphere; the pressure shown by a correct steam gauge.

"*Initial pressure*" is the pressure in the cylinder at the beginning of the forward stroke.

"*Terminal pressure*" (t) is the pressure that would be in the cylinder at the end of the piston's stroke if release did not take place before the end of the stroke; it can be determined by extending the expansion curve to the end of the diagram, or by dividing the pressure at the cut-off by the ratio of the expansion.

"*Mean effective pressure*" (M. E. F.) is the average pressure against the piston during its entire stroke in one direction, less the back pressure.

"*Back pressure*" is the loss in pounds per square inch required to get the steam out of the cylinder after it has done its work. On a locomotive it is shown by the distance apart of the atmospheric and counter pressure lines.

"*Total back pressure*" is the distance between the lines of counter pressure and of perfect vacuum represented in pounds.

"*Initial expansion*" is shown by the reduction of pressure in the cylinder before steam is shut off.

"*Ratio of expansion*" would be the ratio of the fall in pressure between the cut-off and the end of the stroke, providing there was no exhaust.

"*Wire drawing*" is the reduction of pressure between the

boiler and the cylinder; it often causes initial expansion. It is caused by contracted steam pipes or ports.

"*Clearance*" is all the waste space between the piston and valve, when the piston is at the end of its stroke.

"*A unit of heat*" is the heat required to increase the temperature of one pound of water one degree Fahrenheit when the temperature of the water is just above the freezing point.

"*A unit of work*" (foot pound) is one pound raised a height of one foot. One unit of heat equals 772 units of work.

"*One horse power*" (H. P.) is 33,000 pounds lifted a height of one foot in one minute; or one pound lifted 33,000 feet in one minute or an equivalent force.

"*Indicated horse power*" (I. H. P.) is the horse power shown by the indicator. It is the product of the net area of the piston, its speed in feet per minute and the mean effective pressure divided by 33,000 pounds.

"*Net horse power*" is the indicated horse power less the friction of the engine.

"*Saturated steam*," called dry steam, is steam that contains just sufficient heat to keep the water in a state of steam.

"*Superheated steam*" is steam which has an excess of heat which may be parted with without causing condensation.

"*Compression*" is the compressing of the unexhausted steam into the clearance space by the piston after exhaust closure.

"*Latent heat*" is the quality of heat expressed in heat units required to vaporize or evaporate water already heated to the temperature of the steam into which it is to be converted.

"*Sensible heat of steam*" is its heat as shown by a thermometer.

"*Piston displacement*" is the space reckoned in cubic inches swept through by the piston in a single stroke. It is found by multiplying the area of the piston in inches by its stroke in inches.

TAKING DIAGRAMS.

Diagrams should be taken from each end of the cylinder. This can be accomplished by adjusting the three-way, or other cock used, so as to place but one end of the cylinder in communication with the indicator at the same time. Each pair of diagrams should be taken under precisely the same conditions; in order to do this the operator should have an assistant in the cab who will watch the steam gauge and see that other conditions are the same. (An electrical appliance is now used whereby any number of diagrams may be taken simultaneously.) The indicator may be swung around to any desired position and the guide pulleys at the lower end of the drum may be adjusted to lead the cord fairly; adjust the cord to the proper length and see that the drum has the required travel. The paper should be so placed on the drum that the clips will hold it stretched smoothly, otherwise the diagram would be of no value. The pencil should be sharpened to a fine point and care should be taken that it does not bear too heavily on the paper—just sufficiently to make a clear mark. When everything is ready remove the pencil from contact with the paper, open the cock and let the piston and cylinder warm up a little; then move the pencil up to the paper and hold it there while the crank-pin makes a complete revolution. Then remove the pencil, close the cock, and immediately return the pencil and trace the atmospheric line. Then connect with the other end of the cylinder and repeat the operation. The conditions under which each diagram is taken should be noted on its back; speed, boiler pressure, etc. It is necessary that the operator should have the free use of both hands when taking diagrams, considering the position of the indicator and the rocking motion of a locomotive when running at a high rate of speed, an enclosed platform is usually attached to the bumper beam for the operator's safety.

An indicator should not be attached to a new engine until the engine has run a day or two, as the sand or grit which might be lodged in the steam ports or pipes would ruin the indicator.

READING DIAGRAMS.

Having taken a diagram the next thing required is to know how to read it. The only positive information the diagram conveys is the pressure in the cylinder, but having a knowledge of the various operations of the valve, much additional information is derived through the process of reasoning, the dimensions of the diagram being in a proportion to the dimensions of the cylinder. We will first refer to Fig. 1, and explain what the different lines signify. E represents the atmospheric line; this line has no connection with the conditions existing in the cylinder, it being made when communication between the cylinder and indicator is shut off and when both sides of the indicator piston are exposed to the atmosphere as previously explained. It is the base from which all other pressures are determined. F is drawn by hand and represents a perfect line of vacuum, or no pressure, and it is made a distance below E equal to a scale of the spring, to the pressure of the atmosphere, which is 14.7 pounds per square inch at the sea level. A B is called the admission line; B representing the point in the stroke at which steam is first admitted to the cylinder. A C is the steam line; beginning at A and ending at C; this line is kept up by the constant pressure until it reaches C, which is the point of cut-off, where the pressure begins to fall. C R is the expansion line; it represents the constantly decreasing pressure, after cut-off, caused by the expansion of the steam, while the piston travels a horizontal distance from C to R or from the point of cut-off to release. The exhaust line begins at R, which indicates

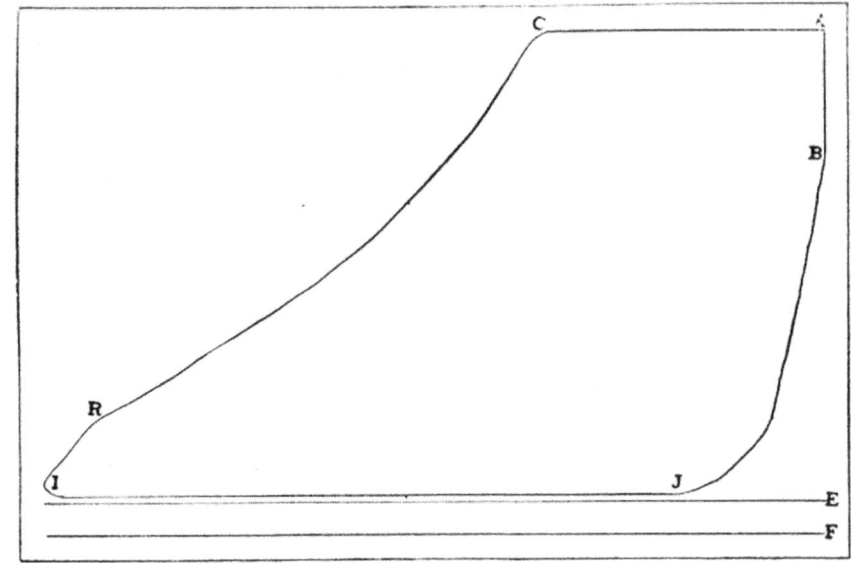

Fig. 1

the point of release or exhaust, and continues to the end of the stroke and the beginning of the return stroke. I J is the line of counter pressure; it begins at I and ends at J, where the exhaust is closed. J B is the compression line; beginning

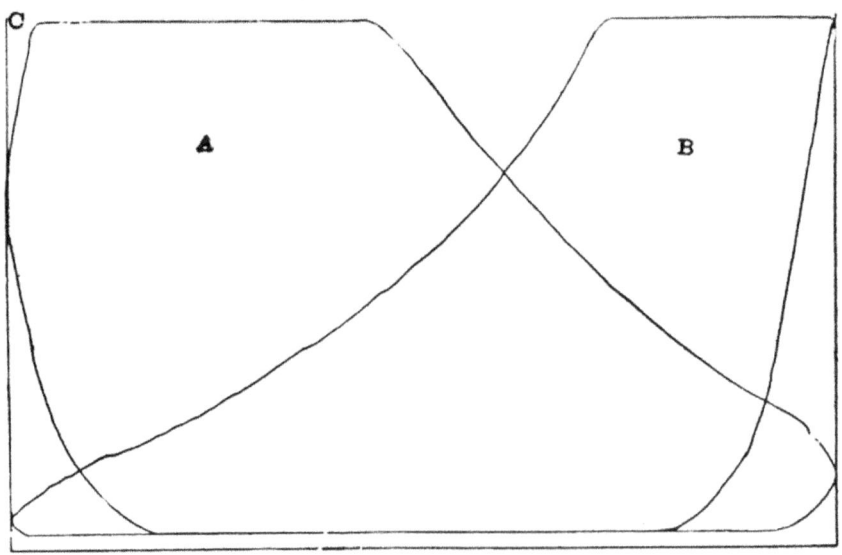

Fig. 2

at J and ending at B it represents a rise in pressure being caused by the advancing piston compressing the steam which remains in the cylinder. The space between the different points are called periods, thus: From C to R is called the period of expansion. The admission line is sometimes considered as continuing from the point of admission to the point of cut-off, and the exhaust line from the opening to the closing of the exhaust port. No matter how carefully the valves may be adjusted by

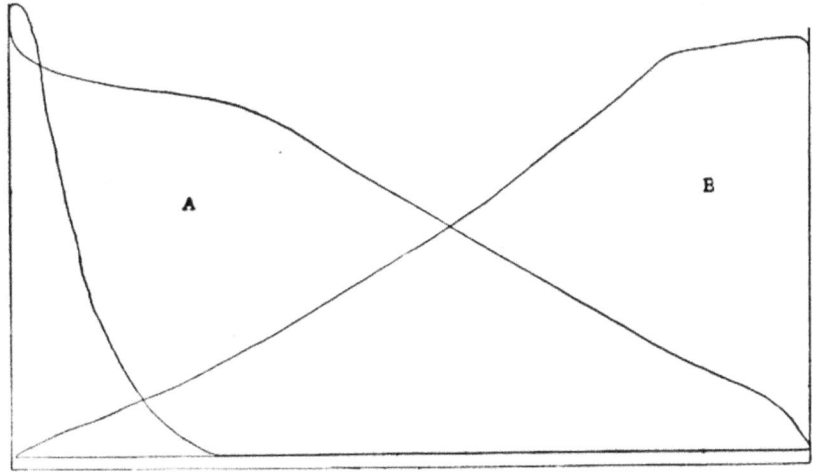

Fig. 3

use of a tram, when the indicator is applied slight irregularities will invariably appear.

The pair of diagrams shown in Fig. 2 indicate a dissorted valve gear. In Fig. 1 the admission line is vertical, showing that the steam port had sufficient opening before the beginning

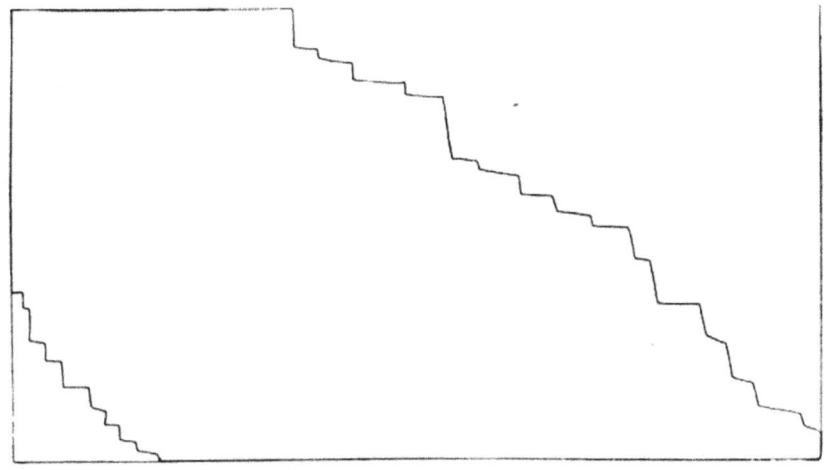

Fig. 4

of the stroke to fill the clearance space and retain its boiler pressure. (The pressure in the cylinder never quite equals boiler pressure, owing to its passage through cramped steam ports, etc.) A in Fig. 2 shows the steam line inclined in the direction of the stroke of the piston; this indicates delayed admission, as the piston should have full pressure at the beginning of its stroke. The distance the valve has moved before the piston receives full pressure may be determined by measuring the space marked C, and figuring proportionately, as you know the proportion the diagram is to the cylinder, to the stroke of

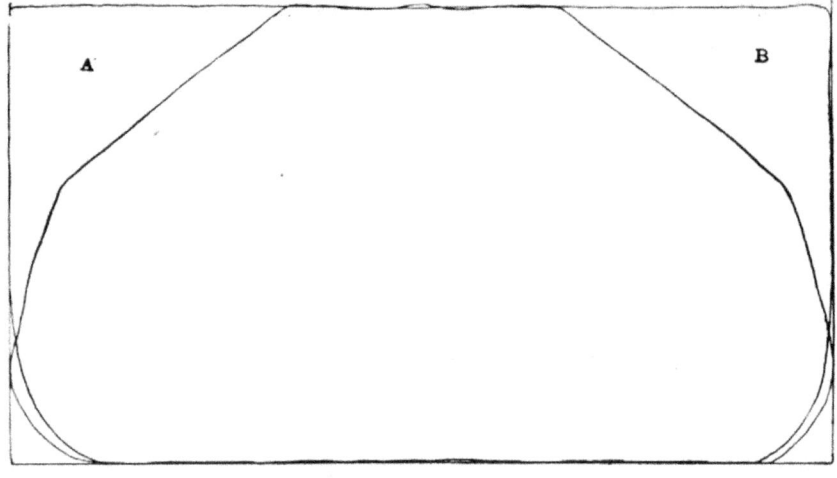

Fig. 5

the piston and to the travel of the valve. Now if B was like A the eccentric should be advanced to increase the lead, but we find B has a full pressure at the beginning of the stroke and cuts off earlier in the stroke; it therefore has excessive lead. Therefore adjust the blades until admission and cut-off are as nearly

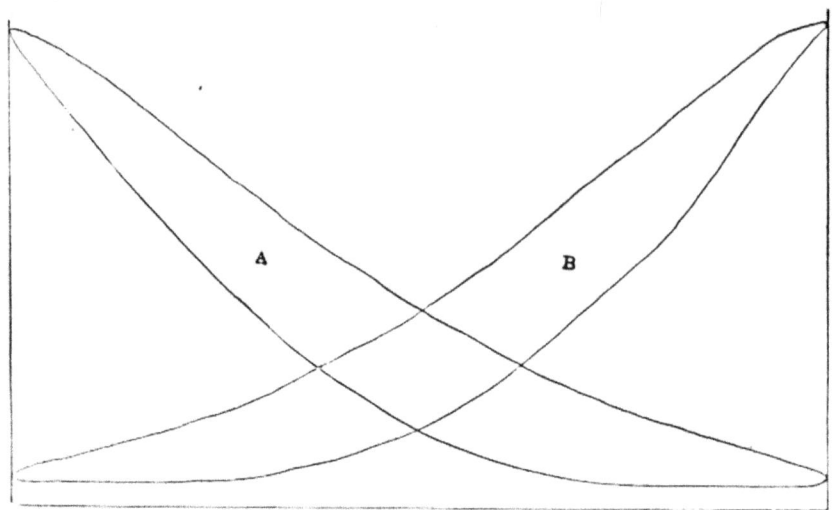

Fig. 6

equal as possible. If both diagrams were like B the lead should be reduced.

A in Fig. 3 shows too much compression; the valve closes the exhaust too early, and the pressure behind the piston is increased above steam chest pressure before the completion of the return stroke, then when the valve opens the pressure falls,

THE LOCOMOTIVE UP TO DATE.

making a loop in the diagram which sometimes causes the valve to lift off the seat. Compression for slow speed locomotives need not be very great, but B should always be higher than R as shown in Fig. 1. For a high speed, compression should be almost, if not quite, equal to steam chest pressure. B, in Fig. 3, shows no compression, the exhaust remains open until the piston has completed its stroke, which should never occur. Therefore the exhaust should be made to close later at A and earlier at B. If it is too great or too small when equalized, then the only remedy would be to change the inside lap of the valve. Back pressure will be shown by the distance apart of the atmospheric line and the line of counter pressure as shown in Fig. 1. Back pressure is largely controlled by the speed, design of the engine, adjustment of the valves, etc. Excessive back pressure may be caused by congested ports, or passages, or by late exhaust opening. An engine running at high speed in full gear will show an enormous back pressure. Water in the cylinders will also cause back pressure.

When the lines of a diagram look like the teeth of a saw as shown in Fig. 4 it is evidence that the friction in the cylinder causes the piston to stick; in this case the drum motion should be disconnected, leaving the indicator piston in communication with the engine cylinder for some time and keep it well oiled.

Fig. 5 shows a diagram taken from a freight engine with 18"x24" cylinders; steam ports 1¼x17" with ⅞" outside lap of valve and 1-16" inside lap and 1-16" lead. Engine in full gear, speed 12 miles per hour, boiler pressure 150 lbs., grade 30 feet per mile. This represents good practice, as the steam distribution is almost perfectly equalized.

Fig. 6 also represents good practice. These diagrams were taken from a passenger engine with one special car, at a speed of 60 miles per hour. The engine having 19"x24" cylinders, steam ports 1¼"x18", 13-16" outside lap, no inside lap, 1-32" lead, in full gear, reverse lever in center notch.

DESCRIPTION OF THE CROSBY STEAM ENGINE INDICATOR.

The illustration shows the design and arrangement of the parts of the Crosby Steam Engine Indicator.

Part 4 is the cylinder proper, in which the movement of the piston takes place. It is made of a special alloy, suited to the varying temperatures to which it is subjected, and secures to the piston the same freedom of movement with high pressure steam as with low; and as its bottom end is free and out of contact with all other parts, its longitudinal expansion or contraction is unimpeded and no distortion can possibly take place.

Between the parts 4 and 5 is an annular chamber, which

serves as a steam jacket; it will always be filled with steam of nearly the same temperature as that in the cylinder.

The Piston, 8, is formed from a solid piece of tool steel. Its shell is made as thin as possible consistent with proper strength. It is hardened to prevent any reduction of its area by wearing, then ground and lapped to fit (to the ten thousandth part of an inch) a cylindrical gauge of standard size. Shallow channels in its outer surface provide a steam packing, and the

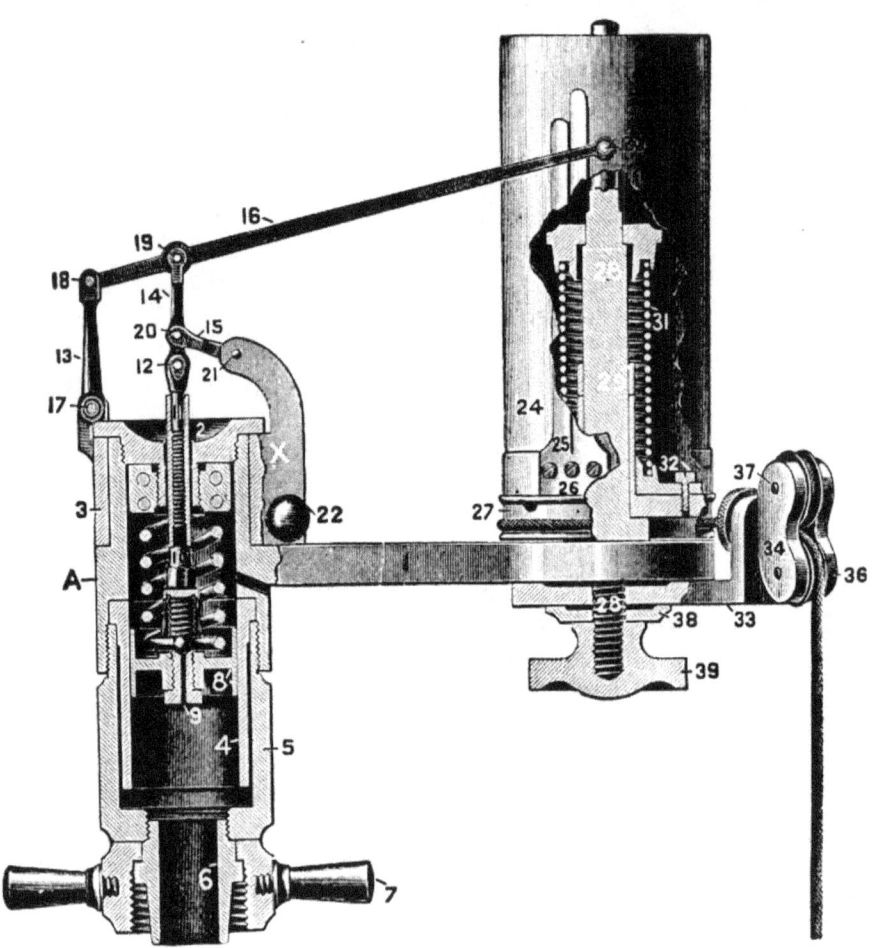

moisture and oil which they retain act as lubricants, and prevent undue leakage by the piston. The transverse web near its center supports a central socket, which projects both upward and downward; the upper part is threaded inside to receive the lower end of the piston-rod; the upper edge of this socket is formed to fit nicely into a circular channel in the under side of the shoulder of the piston-rod, when they are properly connected. It has a longitudinal slot which permits the ball bearing on the end of the spring to drop to a concave bearing in the upper end of the piston-screw 9, which is closely threaded into the lower part of the socket; the head of this screw is hexagonal and may be turned with the hollow wrench which accompanies the indicator.

The Piston-rod, 10, is of steel and is made hollow for lightness. Its lower end is threaded to screw into the upper socket of the piston. Above the threaded portion is a shoulder having in its under side a circular channel formed to receive the upper edge of the socket, when these parts are connected together. When making this connection the piston-rod should be screwed into the socket as far as it will go, that is, until the upper edge of the socket is brought firmly against the bottom of the channel in the piston-rod. This is very important, as it insures a correct alignment of the parts, and a free movement of the piston within the cylinder.

The Swivel Head, 11, is threaded on its lower half to screw into the piston-rod more or less, according to the required height of the atmospheric line on the diagram. Its head is pivoted to the piston-rod link of the pencil mechanism.

The Cap, 2, screws into the top of the cylinder and holds the sleeve and all connected parts in place. Its central hole is furnished with a hardened steel bushing which forms a durable and sure guide to the piston-rod. On its under side are two threaded portions. The lower and smaller projection is screw-threaded outside to engage with the like threads in the head of the spring and hold it firmly in place. The upper and larger projection is screw-threaded on its lower half to engage with the light threads inside the cylinder; the upper half of this larger projection—being the smooth vertical portion—is accurately fitted into a corresponding recess in the top of the cylinder, and forms thereby a guide by which all the moving parts are adjusted and kept in correct alignment, which is very important, and is impossible to secure by the use of screw threads alone.

The Sleeve, 3, surrounds the upper part of the cylinder, and supports the pencil mechanism. It turns around freely, and is held in place by the cap. The handle for adjusting the pencil point is threaded through the arm and in contact with a stop-screw in the plate 1, may be delicately adjusted to the surface of the paper on the drum. It is made of hard wood in two sections; the inner one may be used as a lock-nut to maintain the adjustment.

The Pencil Mechanism is designed to afford sufficient strength and steadiness of movement, with the utmost lightness; thereby eliminating as far as possible the effect of momentum, which is especially troublesome in high speed work. Its fundamental kinematic principle is that of the pantograph. The fulcrum of the mechanism as a whole, the point of attachment to the piston-rod, and the pencil point are always in a straight line. This gives to the pencil point a movement exactly parallel with that of the piston. The movement of the spring throughout its range bears a constant ratio to the force applied and the amount of this movement is multiplied six times at the pencil point. The pencil lever, links and pins are all made of hardened steel; the latter—slightly tapering—are ground and lapped to fit accurately, without perceptible friction or lost motion.

Springs. In order to obtain a correct diagram, the movement of the pencil of the indicator must be exactly proportional to the pressure per square inch on the piston of the steam engine at every point of the stroke; and the velocity of the surface of the drum must bear at every instant a constant ratio to the velocity of the piston.

The Piston Spring is of unique and ingenious design, being made of a single piece of the finest spring steel wire, wound from the middle into a double coil, the spiral ends of which are screwed into a brass head having four radial wings with spirally drilled holes to receive and hold them securely in place.

Adjustment is made by screwing them into the head more or less until exactly the right strength of spring is obtained, when they are there firmly fixed. This method of fastening and adjusting removes all danger of loosening coils, and obviates all necessity for grinding the wires—a practice fatal to accuracy in indicator springs.

At the bottom of the spring—in which lightness is of great importance, it being the part subject to the greatest movement—is a small steel bead, firmly attached to the wire. This takes the place of the heavy brass foot used in some indicators, and reduces the inertia and momentum at this point to a minimum, whereby a great improvement is effected. This bead has its bearing in the center of the piston, and in connection with the lower end of the piston-rod and the upper end of the piston-screw, 9 (both of which are concaved to fit), it forms a ball and socket joint which allows the spring to yield to pressure from any direction without causing the piston to bind in the cylinder, which is sure to occur when the spring and piston are rigidly united. It is of extreme importance that the spring be so designed that any lateral movement it may receive when being compressed shall not be communicated to the piston and cause errors in the diagram.

The Testing of the Spring. The rating or measurement of the springs, both in vacuum and in pressure is determined with great care and accuracy by special apparatus. The vacuum test is made by a powerful vacuum pump to which is connected a mercury column marked in inches. The pressure test is by the direct action of the steam in the cylinder of the indicator and in a mercury column, simultaneously operating with a capacity of three hundred pounds pressure per square inch. Suitable and ingenious electrical apparatus is so combined with these mercury columns that the ordinary division in inches of vacuum and in pounds pressure respectively are automatically marked on the test card on the indicator drum as the test of the spring proceeds. Each spring is tested in pressure to twice the capacity marked on the same. This method of testing pressure springs has been in use for several years and has been demonstrated to be the best system for accuracy.

The Drum Spring, 31, in the Crosby indicator is a short spiral spring. In every other make a long volute spring is used.

It is obvious from the large contact surfaces of a long volute spring that its friction would be greater than that of a short, open spiral form; also, that in a spring of each kind, for a given amount of compression—as in the movement of an indicator drum—the recoil would be greater and expended more quickly in the spiral than in the volute form.

If the conditions under which the drum spring operates be considered, it will readily be seen, that at the beginning of the stroke, when the cord has all the resistance of the drum and spring to overcome, the latter should offer less resistance than at any other time; in the beginning of the stroke in the opposite direction, however, when the spring has to overcome the inertia and friction of the drum, its energy or recoil should be greatest.

These conditions are fully met in the Crosby indicator; its drum spring being a short spiral having no friction, has a quick recoil, and is scientifically proportioned to the work it has to do. At the beginning of the forward stroke it offers to the cord only a very slight resistance, which gradually increases by compression, until at the end its maximum is reached. At the beginning of the stroke in the other direction its strength and recoil are greatest at the moment when both are most needed, and gradually decrease until the minimum is reached at the end of the stroke. Thus, by a most ingenious balancing of opposing forces, a nearly uniform stress on the cord is maintained throughout each revolution of the engine.

The Drum, 24, and its appurtenances, except the drum spring, are similar in design and function to like parts of other indicators and need not be particularly described. All the moving parts are designed to secure sufficient strength with the utmost lightness, by which the effect of inertia and momentum is reduced to the least possible amount.

The indicator is made with a drum one and one-half inches in diameter, this being the correct size for high speed work, and answering equally as well for low speeds. If, however, the indicator is to be used only for low speeds and a longer diagram is preferred, it can be furnished with a two-inch drum.

All Crosby indicators which are numbered above 3737 are changeable from right-hand to left-hand instruments if occasion requires.

HOW TO HANDLE AND TAKE CARE OF THE CROSBY INDICATOR.

The indicator is a delicate instrument, and in order to secure good results from its use, it must be handled with care and be kept in good order.

To Remove the Piston and Spring, unscrew the cap; then take hold of the sleeve and lift all the connected parts free from the cylinder. This gives access to all the parts to clean and oil them.

Never remove the pins or screws from the joints of the pencil movement, but keep them well oiled.

Important. In the under side of the shoulder of the piston-rod is a circular channel formed to receive the upper edge of the slotted socket of the piston. Whenever it is desirable to connect the piston-rod with the piston, either in the process of attaching a spring, or for the purpose of testing the freedom of movement of the piston in the cylinder without a spring, *be sure* to screw the piston-rod into the socket as far as it will go; that is, until the upper end of the socket is brought firmly against the bottom of the channel, in the piston-rod. This insures a perfectly central alignment of the parts and therefore a perfectly free movement of the piston within the cylinder.

To Attach a Spring. Hold the hollow wrench in an inverted position and insert the piston-rod until its hexagonal part engages the wrench; then with the spring inverted, insert the combined wrench and piston-rod until the bead of the spring rests in the concaved end of the latter; then invert the piston and pass the transverse wire at the bottom of the spring through the slot until the threads at the bottom of the piston-rod engage those inside the socket of the piston, and with the wrench screw it in as far as it will go; that is, until the upper edge of the socket is in contact with the bottom of the channel in the shoulder of the piston-rod. The piston screw should be loosened slightly before this last operation and afterwards set up against the bead lightly, to provide against any lost motion, yet not so as to make it rigid. Next, hold the sleeve and cap in an upright position—so that the pencil lever will drop to its lowest point—and engage the threads of the swivel head with those inside the piston-rod and screw it up until the threads on the lower projection of the cap engage those in the spring head, and continue the process until the latter is screwed firmly up against the cap. Then, letting the cap go free and holding only by the sleeve, continue to turn the piston (together with its connections) until the top of the piston-rod is flush with the shoulder on the swivel head.

The piston and its connections may now be inserted in the cylinder and the cap screwed down, which will carry all parts into their proper places.

To detach a spring simply reverse this process.

To Change the Location of the Atmospheric Line of the Diagram. First, unscrew the cap and lift the sleeve, with its connections, from the cylinder; then—holding the sleeve with the left hand—with the right hand turn the piston and connected parts towards the left, and the pencil point will be raised, or to the right and it will be lowered. One complete revolution of the piston will raise or lower the pencil point $\frac{1}{8}$ inch and this should be the guide for whatever amount of elevation or depression of the atmospheric line is needed.

To Change to a Left-Hand Instrument. If it is desired to make this change: First, remove the drum by a straight upward pull; then, with a screw driver remove the steel stop screw in the drum base, and screw it into the vacant hole marked L; next, reverse the position of the adjusting handle in the arm; also, the position of the metallic point in the pencil lever; then replace the drum and the change from right to left will be completed.

This applies to all indicators numbered above 3737.

The tension on the drum spring may be increased or diminished according to the speed of the engine on which the instrument is to be used, as follows: Remove the drum by a straight upward pull; then raise the head of the spring above the square part of the spindle and turn it to the right for more or to the left for less tension, as required; then replace the head on the spindle.

Before attaching the indicator to an engine, be sure to *blow steam freely through pipes and cock* to remove any particles of dust or grit that may have lodged in them.

After using the indicator it should be carefully wiped and oiled.

For this purpose it is not often necessary to disturb the paper drum, but the cylinder cap should be unscrewed and all the connected parts lifted out; then the piston, piston-rod and spring should be detached and all carefully wiped with cloth or tissue paper until perfectly dry, then slightly oiled with a lubricant of good quality; the inside of the cylinder should also be oiled. After this is done the piston and piston-rod should be replaced in the cylinder, but the spring should be kept on its stud in the box when the instrument is not in use. This box should be kept in a dry place, locked, and the key in the owner's pocket.

After the indicator has laid unused for any length of time the oil used at its last cleaning may have become gritty or gummy, and should be wiped off with a soft cloth or tissue paper saturated with naphtha or benzine, and freshly oiled before using it again. This keeps the instrument in prime condition and insures the best results from its use. An occasional naphtha bath is good for an indicator, as it thoroughly cleanses every part.

If any grit or other obstruction gets into the cylinder it will seriously affect the diagram and lead to bad results. It is not difficult to detect such trouble and it should be remedied at once by taking out the piston, detaching the parts and cleaning them as above described, when the disturbing cause will generally be removed.

It is essential to know whether or not the indicator is in good condition for use; especially to know that the piston has perfect freedom of motion and is unobstructed by undue friction. To test this successfully, detach the spring and afterwards replace the piston and piston-rod in their usual position, then, holding the indicator in an upright position by the cylinder in the left hand, raise the pencil arm to its highest point with the right hand and let it drop; it should freely descend to its lowest point. The inner walls of the cylinder should be frequently lubricated. The pencil should always have a smooth fine point; for this purpose a fine file is the best instrument to use.

AN IMPROVED REDUCING WHEEL.

The reducing wheel here shown is known as the Improved "Victor" Aluminum Reducing Wheel. In view of the universally conceded superiority of a good reducing wheel over the "pendulum," "pantagraph," "lazy-tongs," etc., no comments on this subject seem necessary.

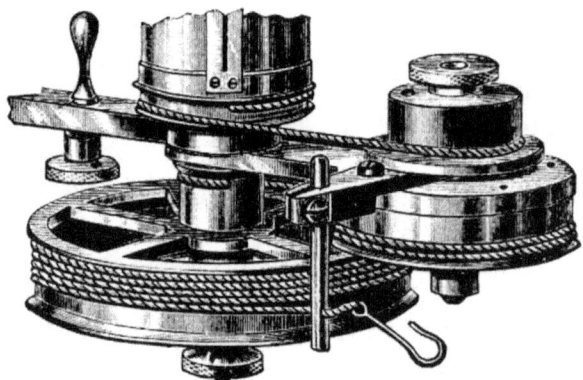

Recent improvements in the "Victor" wheel make it, we believe, nearer absolute perfection than any other. Every part is made of the material best suited to the work, and each joint is so admirably fitted that its lightness, accuracy, and durability are only equalled by the convenience and facility with which it may be applied to any indicator, stroke or speed. It has no gears, therefore no grating action. The cord wheel revolves on a polished steel spindle—not on a threaded shaft as in others. The wheel is stationary, and by means clearly shown in the cut, the guide pulley is moved across its face a distance equal to the thickness of the cord for each revolution, so that the cord will wind evenly, coil to coil, no matter in what direction it is led.

THE "LIPPINCOTT" PLANIMETER.

This instrument operates on a different principle from others. The wheel slides freely on the shaft, the reading being the distance it moves, read from a scale corresponding to indicator spring used.

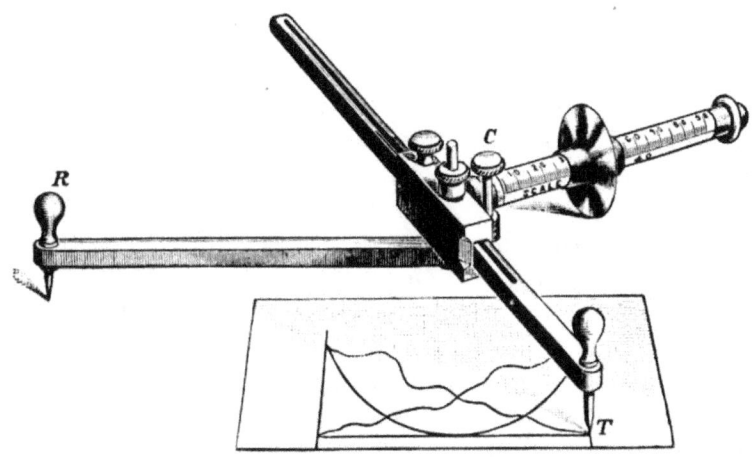

As the wheel does not scrape on the paper, the reading is not affected by the texture of the surface on which it moves. Its diameter is immaterial, therefore wear or injury to edge does not affect the accuracy.

A needle point may be protruded through the pivot screw by pressing the small plunger shown at the top, which provides a convenient method of setting the tracer bar to card length, without turning the instrument over. The scales are hermetically sealed in a glass tube, which forms a frictionless shaft upon which

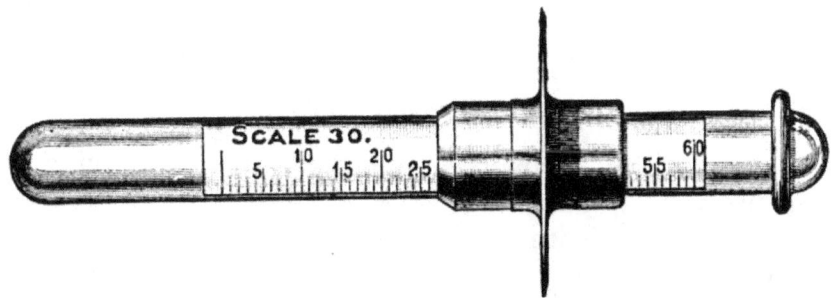

the wheel moves. This scale can never become soiled, or affected by atmospheric influences.

Three of these tubes furnished with each instrument, each containing two scales, so that *M. E. P. may be read direct for any spring*. Areas in square inches and tenths may also be read direct.

LOCOMOTIVE INJECTORS AND BOILER CHECKS.

THE INJECTOR.

The injector is without doubt the most simple, ingenious and best device ever employed for feeding water to locomotive boilers. It has almost completely replaced the old style locomotive pumps. Although on some roads a pump is still used upon one side of the engine. Pumps are considered less liable to get out of repair than injectors and are retained principally as a matter of safety and, in some cases, are only used in the event of failure of the injector to work, but there yet remains one point in favor of the pump, that is, no appreciable amount of steam is required to operate it; the plunger being attached to the cross-head it will pump water with every revolution of the wheel. But the most serious objection to the pump is the fact that it will operate only while the engine is moving. The disadvantage of such an arrangement is clearly evident. Every railroad man of considerable experience can recall the time when it was necessary to run an engine up and down the road a mile or two to "pump her up," as it was called, before the engine could start out with a train, which was a loss of much valuable time. Like the valve motion of a locomotive, a familiarity with the use of the injector does not teach the principle of its action. Realizing the importance of this subject to every man whose duties require of him the construction, care, or management of injectors, we have deemed it advisable in addition to our own knowledge to secure articles from the best known authorities on this subject.

INVENTION OF THE INJECTOR.

To Henri Jacques Giffard is due the honor of having invented the most simple apparatus for feeding boilers that has ever been devised, utilizing in a novel and ingenious way the latent power of a discharging jet of steam. His invention was patented May 8, 1858, which he named an injector. His early technical education and wonderful ingenuity well fitted him for breaking away from the old beaten paths in his experiments, and the method by which he proposed to force a continuous stream of water into a boiler was based upon purely theoretical grounds. Giffard seems to have considered the various phases of the question, for there have been few other inventions in which the underlying principles have been so thoroughly worked out by the original inventor. He designed the body of the injector and proportioned the nozzles almost as they are used to-day, and the first instrument constructed entirely fulfilled the expectation of the designer. In common with all new inventions and radical improvements, considerable difficulty was at first experienced in obtaining a fair trial of its merits; the public believing the inventor was encroaching dangerously near perpetual motion. However, the Academy of Science, of Paris, France, unsolicited, awarded to Giffard the grand mechanical prize for 1859. During the same year the injector was introduced into England by Sharp, Stewart & Co., the American patents were given to William Sellers & Co. of Philadelphia, Pa., who commenced their manufacture in 1860.

INDUCED CURRENTS.

While the theory of an injector's action is generally accepted to be that of induced currents, and is beyond doubt the principal cause of the injector's action, by reading Mr. Colvin's article on "The Theory of the Injector's Action," it will be found that the formation of the injector tubes and nozzle have much to do with the successful operation of an injector. We will first briefly state what is meant by induced currents.

It is a well-known fact that a current of any kind has a tendency to induce a movement in the same direction of any body it passes or touches; thus a passing train will draw articles in its path. For the same reason wind will cause a ripple or produce waves when passing over a body of water. Under a given pressure the velocity of escaping steam is much greater than that of the water. For example, if a boiler with 90 lbs. pressure had two holes cut in it simultaneously, one above and one below the water line, the velocity of the escaping steam would be about nine times greater than that of the water. Water being a solidity will not penetrate the atmosphere as rapidly as steam, which is only a vapor. Therefore the steam which has the greatest velocity meeting the water in the injector induces its movement, and the water which is a solidity strikes the check valve with sufficient force to raise the valve, and its momentum keeps the check valve open, but the temperature of the steam is greatly reduced before re-entering the boiler. Injectors do not always begin working when the throttle is first opened, for that reason an overflow is supplied where the water can escape until the required momentum is attained. Blowers, steam siphons, steam jets and many other instruments are operated upon the same principle.

THEORY OF THE INJECTOR'S ACTION.*

There is a wide diversity of opinion regarding the theory of the injector's working, the generally accepted theory being that

---
*By Fred H. Colvin.

the high velocity of the steam, as it issues from the boiler, strikes the column of water and, mingling with it, carries it along with it into the boiler. It forces its way into the body of water in the boiler, which, though under the same pressure as the steam which operates the injector, is a passive body and cannot resist the velocity of the inflowing water.

This is, in brief, the "velocity" theory, which is the one generally accepted by engineers and professors, and which can be enlarged upon as much as desired by the use of figures showing kinetic energy, etc., if desired.

There are a few, however, and among them the makers of well-known injectors, who do not agree with this theory and who, in the opinion of the writer, have good grounds for their opposition. They argue that, if the velocity theory was correct, injectors could be made with the steam and discharge tube of the same diameter (or with the steam tube possibly a trifle larger to allow for friction of steam before reaching the water), as the velocity would be there just the same to force the water into the boiler. In reality, an injector will not work under these conditions, but the area of the steam tube must be considerably in excess of the area of discharge tube or the injector does not do its work, and this difference in area will be found to vary as the difference between the steam pressure and the pressure the injector forces against increases.

Where an injector forces water directly into the boiler which supplies the steam, the pressure per square inch of the steam leaving the discharge tube need be but a little in excess of boiler pressure, enough to lift check valve disk and allow water to flow in.

The steam flowing into the mouth of the steam tube is condensed by the water and the water heated somewhat. This mass of water practically forms a piston against which the steam acts, and as the area (and consequently the total pressure) here is greater than the area at the throat of the discharge tube, the water is forced into the boiler.

When, however, the steam at the injector is of lower pressure (from being wire-drawn through quite a long pipe, when, as is sometimes the case, the injector is at some distance from the boiler), the area of steam tube must be larger than in the first case, to overcome the increase of pressure it must force against as compared with the lower initial pressure at steam tube.

For boiler-testing injectors, used in many railroad shops for making a pressure test with hot water, the tubes must be specially made to force against the pressure it is desired to use in testing the boiler. In some cases this is four times the initial pressure, and the writer has seen a testing injector work against about five times its initial pressure. This necessitates a steam tube having a larger area than the injector for ordinary work, the same as a steam pump requires a larger steam cylinder to force against a high pressure than a low one.

If the velocity theory was the correct one it would seem as though the pressure per square inch at the discharge tube would be practically independent of the area of steam tube, as long as the steam tube was as large as the discharge tube.

When an injector is used to force against a gauge (for testing) or into a boiler already full, the velocity seems to be lacking, the only water escaping at the overflow being little more than the amount of condensed steam. This seems, to the writer, an additional argument in favor of the "pressure" rather than the velocity theory.

Exhaust injectors are limited to low pressures, and cannot be used for much over 65 or 70 pounds, as the initial pressure is usually a little over 15 pounds. The nozzles have to be made to correspond, and it will be found that the area of the steam nozzle is greatly in excess of the discharge nozzle.

The experience of those who oppose the generally accepted theory is, that the injector's action may be likened to that of a steam pump, the forcing of water being dependent on the excess of total pressure in the steam tube over that opposing the flow of water from the discharge tube. This seems to be proved by the necessity of enlarging the area of the steam tube (and, of course, adding to the total pressure) when the pressure against discharge end begins to materially exceed the steam pressure. This can be readily illustrated by taking the outline of a steam pump and drawing the injector tubes over it, retaining the same dimensions and proportions. The diameters of the tubes are taken at their "throat," or smallest part, the tapers and curves being used to aid in the flow of water and steam; but in common practice throats only are considered in calculating pressures. The shape of the nozzle or tube affects the amount of water passing through, as can be studied in any work on hydraulics or hydro-mechanics.

VARIOUS FORMS OF INJECTORS.

Injectors may be divided into several classes, the principal ones being:

Single set of tubes,	Fixed tubes,
Double set of tubes,	Open overflow,
Adjustable and self-adjusting tubes,	Closed overflow.

And these can again be subdivided into:

Restarting,	Non-lifting.
Automatic,	Lifting,

The Giffard injector was of the open-overflow, adjustable type, as shown herewith.

The principal details have undergone many changes since then, some of which have not been improvements. Every principle for a perfect working injector is clearly shown in the patents of Giffard, which shows that the subject of injectors had been well thought out by him before the world realized that such an apparatus was possible. The exhaust injector, being in a field of its own, will not be considered here.

A closed overflow is one that is closed by hand after the injector is started and remains so until again opened by hand.

An open overflow is one that automatically opens when water or steam flows through the injector, but closes (usually by gravity, though in some cases by a light spring) when the injector gets to work. Its only use in open overflow injectors is to prevent the air being drawn in, slightly cooling the water and making a disagreeable noise.

This feature is used by some engineers to put boiler fluid into their boilers without dumping it into the water supply tank.

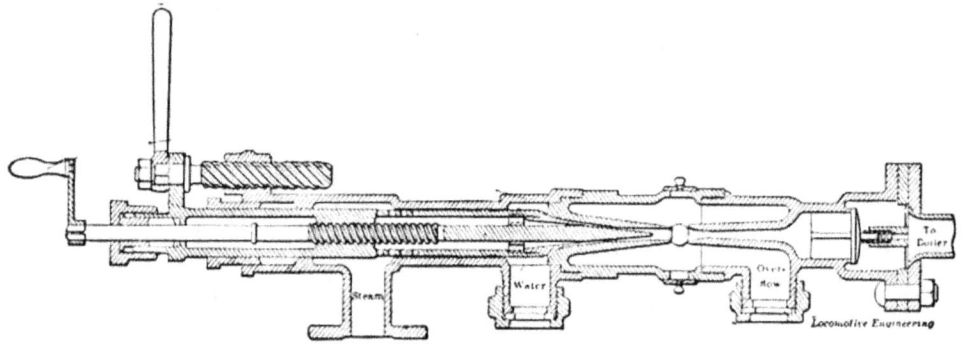

THE GIFFARD INJECTOR.

By simply removing the overflow valve disk and placing the overflow nozzle in a shallow dish holding the fluid, the fluid will be drawn into the injector and forced into the boiler without any difficulty.

The open overflow has the advantage of showing at once when the injector "breaks" or ceases to work as it should. There are cases in locomotive work, however, where the overflow is piped so that it cannot be seen by the engineer, thus entirely destroying its advantage over the closed overflow.

The use of the fixed nozzle or non-adjustable tube injector is constantly increasing, as shown by the number of restarting and other fixed tube injectors now on the market.

The fixed nozzle or tube type is somewhat different. There is no way of regulating the water supply except by the water valve in the pipe (few of the leading injectors having special valves for this purpose now), but, owing to the construction and proportion of the tubes this is seldom necessary, as they will handle quite a wide range of both water and steam without difficulty. With extreme water pressure it may be necessary to throttle it, but this seldom occurs, the combined action of the two jets, the central one and the auxiliary annular jet, giving a particularly flexible action, if such a term can be used. They will not, however, handle as wide a range as the adjustable tube injectors and cannot, from the nature of things, notwithstanding the various claims of some makers.

With this type of injector the overflow is usually situated in an overflow pipe, at any convenient point between the injector and boiler. This is often controlled by an ordinary globe valve, which must be open before starting the injector, and the outlet to this should be visible, although the sound will help greatly after becoming accustomed to it.

TO START.

Having the overflow valve open, turn on the water supply as before and open the steam valve slowly. The hot water will commence to rush out of the overflow, as this is the line of least resistance. After the steam is turned on full, begin to close the overflow valve slowly, and as the valve becomes nearly closed, the familiar hum will begin, reaching its maximum when the overflow is fully closed. The water is now going into the boiler, as there is no other outlet for it. To stop, simply shut off steam and open overflow valve ready for the next start. A "break" in the operation is indicated by a cessation of the regulation "hum," a snapping of the steam and water in the pipes, a rapid heating of pipes and injector, and, if the water supply is drawn from a tank, by the noise and possibly spraying of the water therein.

RE-STARTING INJECTORS.

These are comparatively few in number, so far as types go, although many thousands have been sold. They are not, however, so widely used in large steam plants and similar places as the regular types, but seem to be most in demand for yachts, traction and portable engines, for which they were, in fact, originally designed. Their object is to prevent a possibility of failure, by automatically starting after the water supply has been broken, from the farm engine or yacht lurching, or any similar cause. This necessitates the combination of a lifting jet, so as to lift the water to start with, and is very neatly accomplished by several makers. The writer has repeatedly tested injectors of this type by shutting off the water supply for several minutes, until both the injector and feed water were quite hot, and on opening the feed valve they would start readily. They are, however, of comparatively sensitive construction, and it has been the writer's experience that they do not meet the general requirements as well as the regular types, although he does not wish to condemn any make or antagonize any of the makers. His conclusions in this respect are borne out by the experience of many other practical engineers.

Automatic devices have been used on boilers, for starting and stopping the injector, by a float in the boiler. For this work, a restarting injector is especially adapted, but it is not a practical method in many cases.

While there are occasionally conditions in locomotive practice which are not often met with in stationary practice, still the general action is the same and the causes for failure are practically the same in each case.

Generally speaking, an injector will work when its tubes are in fair condition and it has a sufficient water supply.

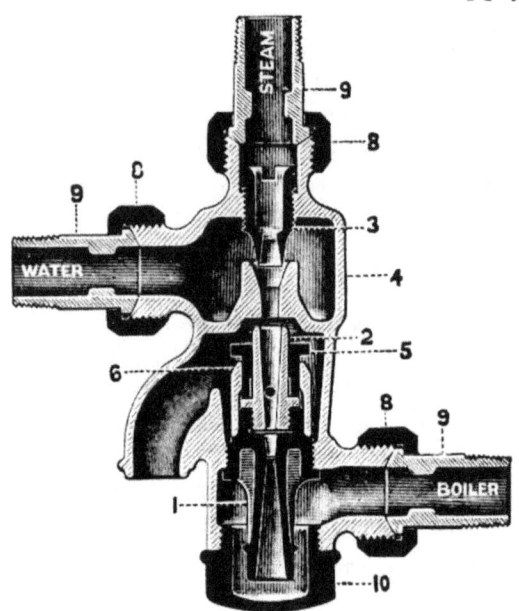

RE-STARTING INJECTOR.

The main cause of breaking is insufficient water supply from some cause or other. Briefly, and without regard to theory, the action of the injector is as follows: The water flows up the supply pipe (either by an outside pressure or by the action of a lifting jet) and through the combining tube to the overflow. The steam, entering from the steam tube, is condensed by the water in the combining tube and forces it through the tubes with sufficient pressure to overcome the pressure in the boiler, and, lifting the check valve, forces its way into the boiler.

When these conditions do not exist the injector "breaks," usually from the water supply not being sufficient to condense the steam. In this case the injector "breaks" and all the pipes become hot. Should the water be much in excess of the required quantity the injector will start as usual, but will not always force into the boiler as there is more water than can be handled by the steam.

In starting an injector having an adjustable valve in the water supply, either by moving the tubes together or by a separate valve outside, it is a good plan to give an excess of water rather than an insufficient supply, as the water supply can be readily reduced and the injector started to work without any annoying delay due to heating of pipes or water.

The act of starting (with a non-lifter) is simply opening the water valve until water shows at overflow and then opening steam valve, preferably by an easy, gradual movement.

With a lifter, the lifting jet is first opened (either separately or together with the main steam valve as in some injectors) until the water shows at overflow as before, then the main valve is opened until the steam "picks up" the water. If any water shows at overflow after injector is working, the water supply should be reduced slightly, or until the overflow ceases.

Should the steam pressure fall, other conditions remaining the same, water will again appear at overflow, there being more than is needed to condense the steam and more than it can handle. This will not cause it to break, however, but will merely cause waste of water, which is a serious question in some cases. Should the steam pressure increase materially, there will not be enough water to condense it, and steam will show at overflow, and if the rise in pressure continues the insufficient water supply will not be able to take care of the steam, and unsatisfactory working or breaking will result. The injector will also become heated and add to liability of breaking as well as delay starting again. If the tubes are adjusted to this new condition before the breaking point is reached there will be no trouble. The same difficulty will be experienced if the water supply is decreased with pressure remaining the same, which sometimes happens from the supply pipe being partially choked by foreign matter, such as waste, wood, coal, etc. The writer has seen an injector condemned when it was literally filled with small particles of coal and wood which had come in from the supply pipe. In another case the engineer, who had evidently been "seen" by a rival injector salesman, had filled the supply tank with fine coal screenings, which completely blocked the tubes. He was informed that the instrument was an injector, not a coal conveyor.

Various steam pressures require different amounts of water, and the higher the steam the more water is necessary to condense it in the mouth of the combining tube.

Stopping the injector is simply the reverse of starting.

CARE OF AN INJECTOR.

Many a fire has been killed and many an engine towed in owing to the engineer's lack of knowledge of the injector. Every man who is intrusted with the care of an injector should thoroughly understand its philosophy; if he does not he cannot expect to operate it successfully at all times. It is true, the difficulties connected with the successful operation of the injector have been materially lessened by the use of the latest improved injectors; nevertheless, every prudent engineer or thoughtful machinist makes it an object of ambition to study the principle of the injector's action and its construction, in order to enable him to locate any defect in its action. To preserve a good working injector all pipes and joints should be kept perfectly tight, and all stems well packed, as the admission of air into the injector will affect its action. The air mixing with the water has a tendency to decrease the speed of the water, by impairing its solidity and making it a semi-elastic body; such a defect can usually be detected by a bubbling sound at the check valve, the water not being a solid compact body. Small leaks neglected will eventually cause the injector to break, or not work at all. All pipes leading to the injector should be carefully examined for leaks, particularly the feed pipe above the water level. Leaks affect lifting injectors most, as it prevents the steam jet from forming a vacuum. The object of the steam jet (or primer) on all lifting injectors is to force all the air out of the injector. The atmospheric pressure on the water in the tank then forces the water up the feed pipe and into the injector, when the steam can be turned on full force. Most improved injectors have a small priming valve attached to the steam ram, so the injector must be primed before the steam valve can be fully opened.

When an injector will not prime, first, see that you have a sufficient supply of water in the tank, and if in the winter, see that the tank is not froze up to prevent the admission of air. Next see that the overflow or hose is not stopped up, or that the check is not stuck. When a check leaks or sediment gets on the seat it sometimes prevents priming. Give the check or branch pipe one or two slight taps and it may start, but don't batter it out of shape. If you have a bad check, report it. If the injector sucks air, or if the nozzles are out of line, it will not prime. If it primes all right, but breaks, it indicates that the check valve is stuck, or the injector is sucking air, or that it is not receiving sufficient water. Examine your hose; if it is all right, disconnect branch pipe from check and clean it out, or the trouble may be in the nozzle. If it primes but won't start, and steam and water both escape at the overflow, increase the steam pressure, or reduce the supply of water. (The steam throttle at the boiler should be once regulated and then marked some way to avoid further trouble with it.) In this case the check valve may not have sufficient lift. Some injectors have an adjustable combining tube which adjusts itself to the volume of steam and water passing through it; if this tube is prevented from working freely, by either grit or sand, the injector will break. Clean out the hose frequently, for cinders and dirt choke up the strainer very quickly. A frequent cause of annoyance with injectors is a leaky check valve which blows back and heats the injector so that it will not work. In such a case open the water valve and overflow and turn the primer full on and try to fill the injector with cold water, letting it run a few minutes, when the injector will probably start. Such a check should be reported at the first opportunity. Perhaps the most frequent cause of the injector refusing to work is in calcareous districts where it becomes choked with lime. This is easily discovered, as its force will gradually diminish until it refuses to operate at all, but it will have a tendency to work under low pressure. Another defect is a loose nozzle, which will work best with a high pressure of steam; this may be discovered by removing the frost plug and the main ram. When nozzles are worn too large the injector will refuse to work, and they should be replaced with new ones. In winter all injector pipes should be supplied with drain cocks at their lowest point, and kept open when not in use to prevent the pipes from freezing. Injectors should be oiled regularly, but lard oil should not be used, as it has a tendency to make the water in the boiler foam. When this occurs change the water in the boiler and tank at the first opportunity.

SIGNIFICANCE OF THE NUMBERS ON INJECTORS.

The numbers which may be seen on most all injectors indicate the exact diameter of the smallest orifice of the delivery tube, expressed in millimeters, which is equal to .03937 inches. (See Metric System, page 578.)

LOCOMOTIVE INJECTORS.*

The knowledge of the capacity of a moving jet of steam or other fluid to produce a vacuum in properly formed ducts, for the purpose of raising air, water or other fluids, and conveying them from one place to another, may be traced back to the time of Venturi, Nicholson and others. Nevertheless it must be admitted that the eminent French engineer, H. J. Giffard, was the first to conceive the idea that the kinetic energy (moving energy or momentum) of a moving mass of fluid could be utilized to overcome the static energy (resistance) of a mass of water under boiler pressure, so that the first mass of fluid would enter such boiler against the resistance of the second.

The terms "kinetic energy" and "static energy" may not be absolutely correct in a mechanical sense in this connection, but they may be used here for the purpose of designating the difference between the conditions of the two masses of fluid under consideration, which will receive notice further on.

---

*By J. A. Bischoff, of the Nathan Mfg. Co.

THE ESSENTIALLY ACTIVE PARTS OF AN INJECTOR.

1. A steam nozzle, through which the operating steam from the boiler enters the injector.

2. A combining and condensing nozzle, in which the steam and feed water meet, and in which the steam condenses and transmits its dynamic force to the water.

3. A delivery nozzle, in which the maximum velocity of the combined mixture of steam and water is attained and subsequently reduced by means of the expanding curves or tapers and increasing cross sections, to the velocity and pressure in the boiler pipe.

If these three parts, which are to be found in every injector, are looked upon as the essential components of an injector, the Marquis Mannaury d'Ectot must be considered the inventor of the injector. In 1818 he was granted a French patent for a steam jet apparatus, which was capable of raising water from a tank and delivering it into a second tank. After d'Ectot, the French engineers Pelleton and Bourdon (of steam gauge fame) published various inventions of a similar character.

On May 8, 1858, H. J. Giffard was granted his first patent for an injector, to be used as boiler feeder. The difference between his apparatus and those of former inventors, consisted mainly of a certain distance between the outlet orifice of the condensing nozzle and the inlet orifice of the delivery nozzle, which is called the overflow space and which communicates with the outer atmosphere.

In starting the injector, more water, as a rule, enters the apparatus than the injector is capable of delivering against the back pressure of the boiler, and if there were no communication with the atmosphere, the injector would "break," that is, refuse to work, and the steam blow back into the tank from which the water is taken, because the steam would naturally follow the line of least resistance. By providing for an overflow space between the condensing and delivery nozzles, the surplus water is given an opportunity to escape, until the jet of combined steam and water has attained a sufficient velocity and over pressure to open the boiler check valve and deliver the jet into the boiler. After the apparatus has been started and is in operation, the waste of water may be avoided by cutting down the supply, until no more water is seen at the overflow, or the overflow is stopped automatically, according to circumstances and the type of the instrument used.

Giffard's invention, therefore, consisted in the important discovery that, in order to deliver water through an injector into a boiler, by means of steam taken from the same boiler, the jet must first overflow into the atmosphere. Injectors made prior to Giffard were, in fact, based upon the same principle as that of Giffard's injector, but were unable to deliver against pressure.

In common with all new inventions and improvements, great difficulty was experienced in obtaining a fair trial of the merits of the injector, and in many cases the exaggerated claims of its friends interfered as much with it as the severe and condemning criticism of its enemies. The advantages, however, of the new method of boiler feeding, the simplicity and efficiency of the apparatus, and the comparatively small expense of installation and maintenance, were soon appreciated by steam users, and to-day the injector is among the most popular boiler feeding apparatus in use, and still deserves the high praise bestowed upon it in 1859 by M. Ch. Combes, Inspector General and Director of the School of Mines at Paris: "It is without doubt better than all devices hitherto used for feeding boilers, and the best that can be employed, as it is the simplest and most ingenious."

Soon after Giffard's injector had been placed on the market, the manufacture of injectors became a most important and extensive industry in the United States as well as in Europe. Hundreds of thousands of injectors have been manufactured for various purposes, and to-day there is hardly a locomotive engine running anywhere in the world which is not provided with at least one injector. Great numbers of steam vessels and many stationary plants are equipped with injectors as boiler feeders, hence it is of the utmost importance that every engineer, stationary, marine and locomotive, should thoroughly familiarize himself with the principles which underlie the operation, management and construction of injectors.

THE ACTION OF THE INJECTOR.

It is very difficult, almost impossible, to properly explain the seemingly paradoxical action of an injector, without entering into mathematical demonstrations. It will be necessary, therefore, to confine ourselves to stating the results of such mathematical deductions.

The simplest method of considering the action of the injector is to consider it from a purely mechanical point of view, as an apparatus in which the force of a jet of steam is transferred to a more slowly moving body of water, resulting in a final velocity sufficient to overcome the pressure from the boiler. We will consider, for a concrete example, a boiler containing steam at 120 lbs. pressure. According to the laws governing the flow of steam through properly built channels, the steam discharging through the minimum diameter of the steam nozzle into the atmosphere or steam of lower pressure, will reach a velocity of about 1,400 feet per second, but when it discharges into a combining tube in which there is perhaps a 20-inch vacuum (which is not unusual) the velocity will be about 3,500 feet per second. In supplying water to the injector at the rate of 13 lbs.

to 1 lb. of steam, which is a fair performance at the pressure stated, the water will receive the impulse of the moving steam, condensing the latter, and the two fluids move along together through the delivery tube with a final velocity which is very much less than the original velocity of the steam jet. This final velocity is, under the conditions named, about 170 feet per second. In order that the injector should work properly, this velocity must be greater than that with which the water in the boiler would issue from the delivery nozzle under a pressure of 120 lbs. per sq. in. This velocity is equal to that due to the head, corresponding to 120 lbs. pressure, or about 133 feet per second. It will be seen, therefore, that there is a considerable margin of available energy in favor of the moving mixture of steam and water as against the "stationary" resistance, under the same pressure, of the water in the boiler. An injector must have the proportions of steam, water and delivery areas so designed that the velocity of the moving mixture of steam and water will be greater than the velocity at which a jet of water would flow from the boiler under the same pressure. If the weight of water supplied is too great, the steam will not have power to give the water the required surplus of velocity; if there is an insufficient supply, the volume of steam will not be sufficiently reduced by condensation to pass through the nozzles, and in neither case will the injector perform its functions properly.

CAUSES WHICH PREVENT INJECTORS FROM WORKING.

In this condition there are principally two general conditions to be considered. 1. The injector refuses to lift promptly or lift at all. This may be caused by leaky joints in the suction pipe, by improperly packed water valve stems; by dirty, clogged-up strainers; clogging of the lifting steam passages in the injector; hot suction pipe, etc., etc. In connecting an injector, either when new or when it has been taken off for repairs of some kind, particular care should be exercised in testing the suction pipe for tightness, by well known methods. Particular attention should be paid to the strainer; it should be taken out, examined and cleaned before each trip of the engine. "An ounce of prevention is sometimes preferable to a pound of cure." Cheerfully undergoing a little trouble may be the cause of preventing a good deal of annoyance, anger and expense. Hot suction pipe is usually the result of leaky steam valves or leaky boiler checks. If you notice any leaks in either, have them attended to without delay. The longer repairs are put off, the more aggravated will the trouble become, and the higher will be the cost of repairs, not taking into consideration the danger of something serious occurring during a run, as a result of neglecting timely repairs.

2. The injector lifts the water, but refuses to force it into

the boiler, or forces it partly into the boiler and partly through the overflow. This may be caused by insufficient water supply as a result of improper size of the suction pipe, hose or tank valve opening, by clogged-up strainer, obstruction in the nozzles (pieces of coal, scale from steam pipe, waste, etc.), insufficient opening of the boiler-check, "sticking" of the boiler-check or of the line-check valve of the injector, by insufficient steam supply or wet steam. The pipes of an injector should never be smaller than the size called for by the injector connections, more especially the suction pipe and the clear openings of the tank valve. Sharp bends in the pipe should be avoided as much as possible. In piping an injector, the pipes, but more especially iron pipes, should be thoroughly blown out before connecting up, to remove scale and dirt from the pipes.

The "sticking" of the boiler-check or of the line-check valve is mostly caused by sediments and scale resulting from bad water. Incrustation of the nozzles by limey deposits around the points of the nozzles will also cause the injector to spray at the overflow, and result in improper action generally. The effects of bad water may be partly, if not entirely, eliminated by cleaning the injector frequently, and by placing it occasionally into an acid bath.

Injectors work best with dry steam, for which reason the steam supply pipe should be attached to the highest point of the boiler. If steam is taken from a fountain to which other steam appliances are connected, the volume or cubic contents of the fountain must be large enough to more than amply supply all appliances connected to same.

SUGGESTIONS.

Designers of locomotive engines, as a rule, do not pay that attention to the injector which its importance calls for, and beyond specifying the type and size desired, hardly any other attention at all is paid to its most desirable location and other points, with a view to assisting in the development of the best qualities and assuring reliable service. Injectors, as any other product of human endeavor, are subject to defects and failures; the very mechanical nature of the instrument calls for considerations in its arrangement and that of its accessories, which, when properly considered at the time of building the engine, would go a long way toward preventing annoyance, expense and sometimes serious inconvenience. It is perhaps within the province of this article to point out some features in this direction.

THE WICKED STRAINER.

The little conical copper strainers inside of the suction pipe should be abolished, as they are a nuisance and the cause of more trouble than they get discredit for. If a premium had been put upon designing something to readily catch and retain any dirt in such manner as to materially reduce the water supply, these strainers would undoubtedly receive first prize. The very fact that they are inside of the pipe is objectionable. The strainers should be outside of the pipe, either directly below the tank valve or well, or at the end of the suction pipe, between pipe and hose. The size of the strainer should be such that even if half filled with leaves or other matter it should still retain the full pipe capacity. It should also be so designed that it could be readily cleaned at any time and in a very few minutes. Such strainers can be obtained in the market, and their cost would be more than compensated for by avoiding troubles often caused by their absence.

PUT CLOSING VALVE IN SUCTION PIPE.

Some types of injectors are not provided with any water valve in the suction pipe. In cases where such valves are provided for in the injector, they are not considered as shut-off valves, but merely as regulators, by means of which to regulate the supply of feed water at certain pressures. For this reason some of these valves are so constructed that they will not form a tight joint when closed down. It happens then, occasionally, that in case of an accident to the boiler-check, which prevents its tight closing, and with the line-check valve of the injector leaking, the water cannot be kept in the boiler, and the engine must be side-tracked. To provide for such emergencies, it would be very advisable to place a properly sized shut-off valve in the suction pipe, between pipe and hose. With such valve at hand, and with the overflow of the injector closed, the water could not leave the boiler.

PLACE THE INJECTOR INSIDE THE CAB.

It is a very usual, but not a good practice, to place injectors outside of the cab, and run long operating rods from the injector handles into the cab. These extension rods, as a rule, are not connected up very carefully, and even if they are, the injector cannot be as readily started or as well regulated as when the operating handles are close to the injector, and therefore under better control of the operators. The result of this arrangement is a waste of time in starting the injector, and considerable waste of water through the overflow. Very often the overflow cannot be observed, and the operator judges by the sound whether the injector operates properly or not, and with some types of injectors he is compelled to "feel" the suction pipe in order to convince himself of the operative condition of the injector. All this cannot be conveniently done with the injector removed. The proper position of the injector is inside of the cab. Room can always be provided for it. The objection of the overflow splashing or the steam clouding the windows of the cab, can be overcome by providing a larger overflow pipe than is usually employed. With a large enough overflow pipe, the overflow connection can be made perfectly tight.

Most of the trouble with injectors comes from improper, slow lifting, especially when the water in the tank is low, caused by circumstances for which the injector proper is not responsible, such as bad boiler checks, leaks in the suction, obstructed strainer, etc. To reduce inconveniences from this source to a minimum, injectors should be placed as near to the water level in tank as possible. On some roads the admirable practice prevails of placing the injectors a foot or two below the highest level of the water in the tank. The ideal in this direction is the "non-lifting" injector, placed below the lowest water level in tank; not only because the source of most of the trouble with injectors, the lifting, is entirely eliminated, but because by keeping the injector submerged in water, it will be kept comparatively cool, the precipitations from bad water will not be "baked" on, and the injector will wear considerable longer without repairs.

SELLERS' INJECTORS.

We have illustrated three forms of Sellers' injectors. The first two illustrations represent their latest improved No. 10½

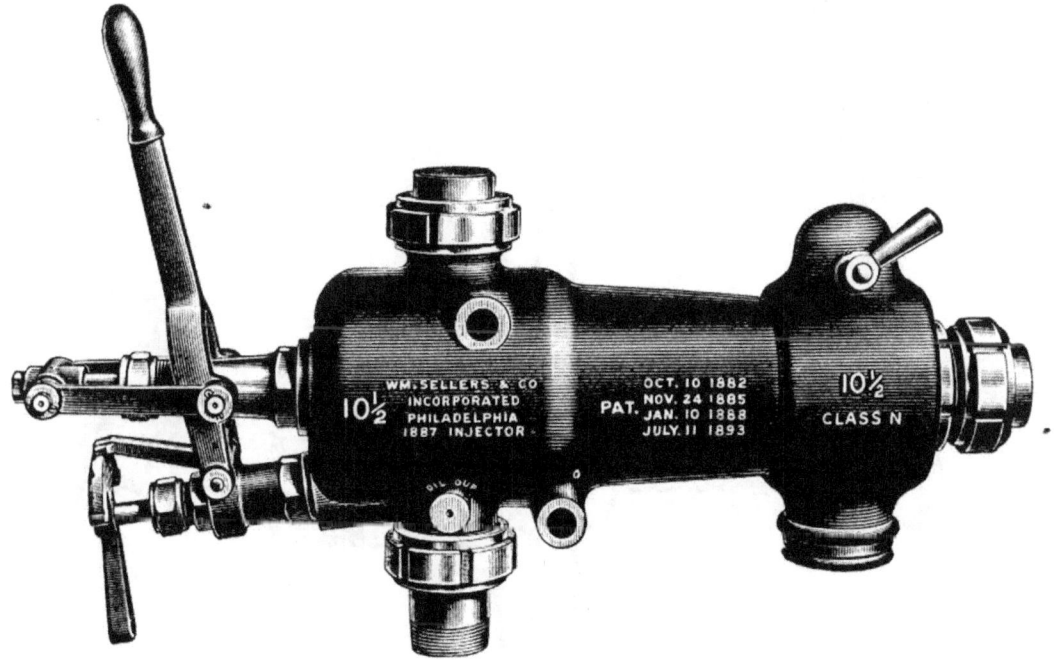

restarting injectors; that is, if the water be temporarily interrupted, the injector will start again automatically as soon as the supply is resumed. It is self adjusting, requiring no regulation of the water supply to prevent overflow above 40 lbs. steam pressure, and its construction is such that the tubes and other

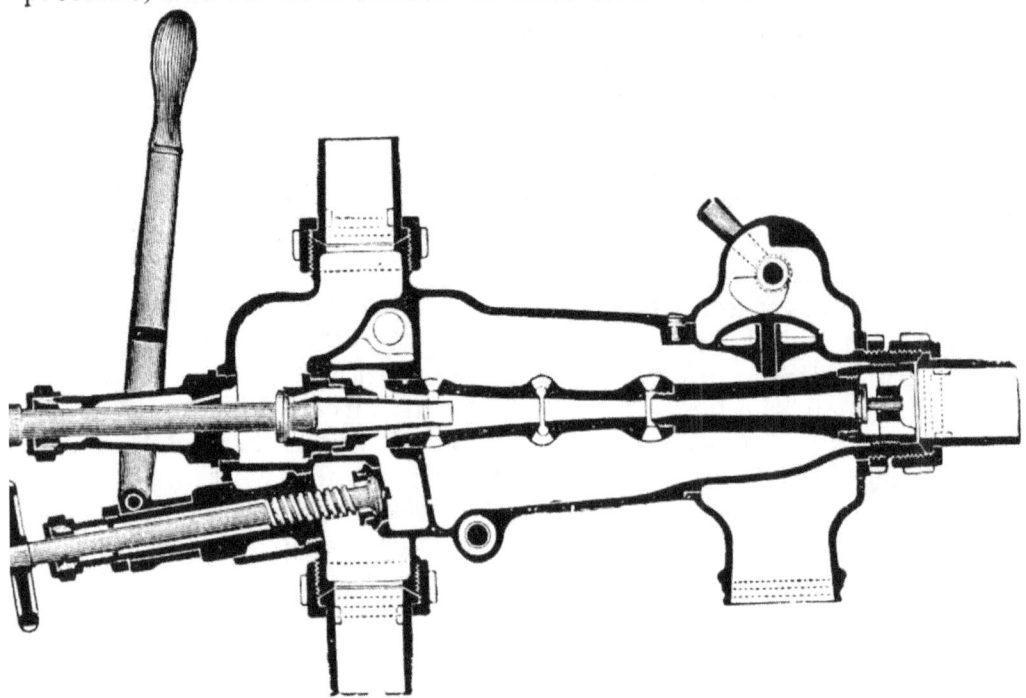

parts can be easily taken out for cleaning and repairs. In the 1893 injector the water supply valve has been altered and a grad-

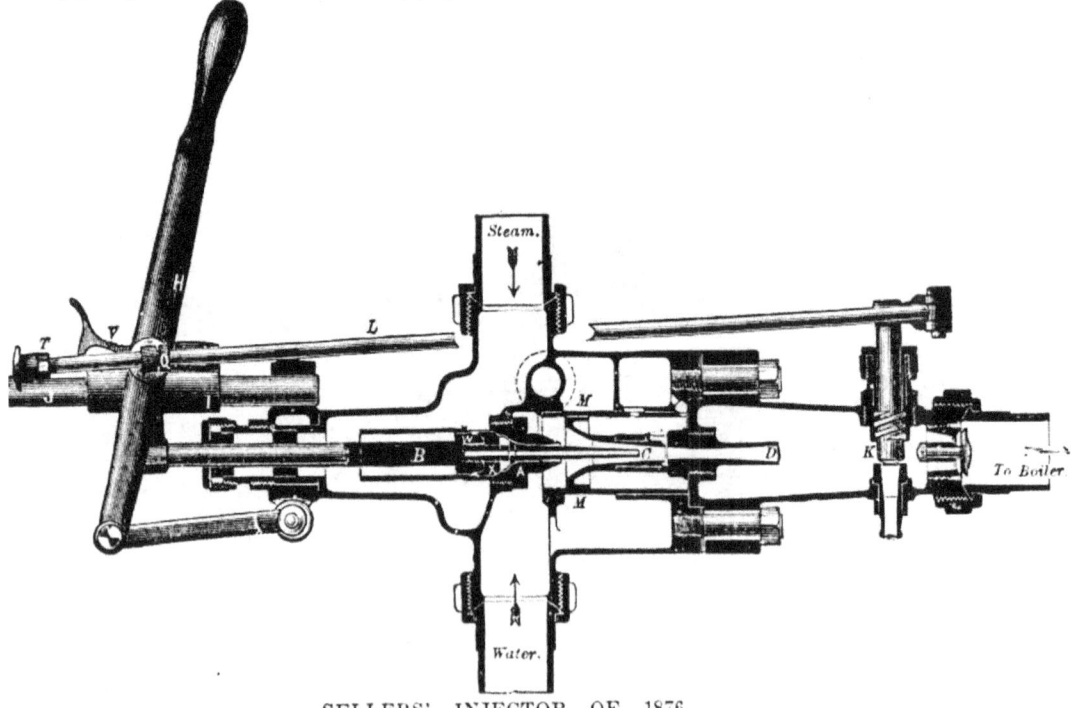

SELLERS' INJECTOR OF 1876.

uated lever and index substituted for the regulating hand wheel, the tubes and nozzles remaining unaltered.

HOW TO OPERATE.

To Start—Pull out the lever. *To Stop*—Push in the lever. Regulate for quantity with the water valve. To use as a heater, close the waste valve and draw the starting lever. When the water flows to the injector it is of course necessary to open the water valve before pulling out the lever, and to close it after pushing in the lever. In starting on high lifts and in lifting hot water, it is best to pull out the lever slowly.

As there are a great many of the old No. 6 injectors of 1876 still in use, we have shown an interior view of one of them. This is a self-adjusting injector; during its time it was adopted as a standard by the Pennsylvania Ry. Co., but it is too small to supply the large locomotive boilers in use at the present day. All the tubes of the Sellers injectors are connected together and when not corroded are easily removed all together through the delivery end. To operate the 1876 injector all that is required is the movement of the lever.

MONITOR INJECTOR OF 1888.

Since the introduction of the Monitor injector in 1880 they have been classed among the most efficient and reliable in-

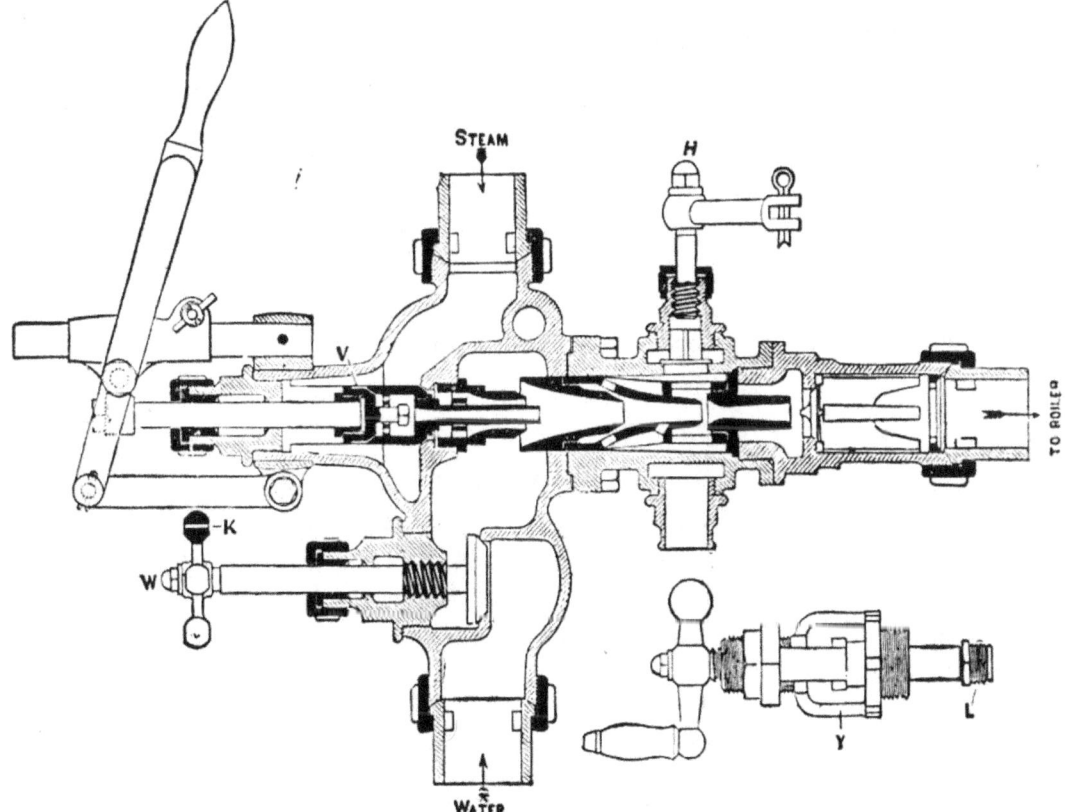

jectors in use, and they are so extensively used in this country that no introduction is necessary here. The first cut shows the latest development of the Monitor, which is a No. 9 Monitor of 1888. The latest improvement in this injector is the lever handle attachment and different form of water valve which is less liable to corrosion. The independent lifting jet, by which the Monitor has long been distinguished, is here dispensed with, and a taper lifting spindle in the steam nozzle which discharges through the combining tube is substituted. These injectors will operate under very low pressure, and the body is divided into three sections, which is very convenient for repairs.

P. R. R. STANDARD MONITOR.

The next cut shows the form of Monitor of 1888 adopted as a standard by the Pennsylvania Ry. Co., the difference being

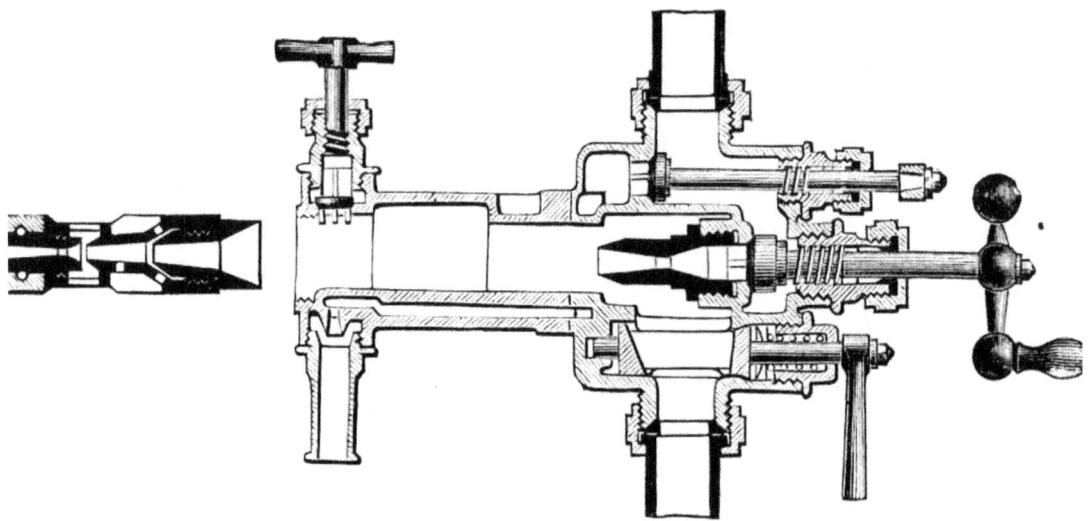

in the position of the lifting jet and in the form of nozzles, the nozzles all being attached so they may be removed through the delivery end like the Sellers nozzles.

HOW TO OPERATE.

WITH LEVER MOTION.

To Start.—Pull out the lever a short distance to lift the water; when water runs from the overflow, steadily draw back the lever until overflow ceases. Do not increase the water supply after overflow has ceased. Regulate for quantity with water valve W.

To Stop: Push in the lever.

WITH SCREW MOTION.

To Start: Open the steam valve one-quarter of a turn to lift the water. When water runs from the overflow, open steam valve until overflow ceases. Do not increase the steam supply after overflow has ceased. Regulate for quantity with water-valve W.

To Stop: Close steam valve.

To Grade Injector: Throttle water by valve W; if this is not sufficient, reduce the steam by pushing in lever handle about half way and in case of the screw motion by screwing in the steam spindle about half way.

To Use as a Heater: Close valve H and pull out lever all the way, and in case of screw motion open valve full. At all other times valve H must be kept open.

The heater cock can be worked from the cab by means of an extension rod.

The hole in the knob K of water handle W indicates the position of the water valve. One turn of the handle fully opens or entirely closes the water passage.

In either case, the knob with the hole in should be in an upright position. Intermediate positions of the knob K indicate corresponding openings in the water passage.

STANDARD MONITOR.

The next cut shows the Standard Monitor injector, which style is perhaps in most general use at the present time. A

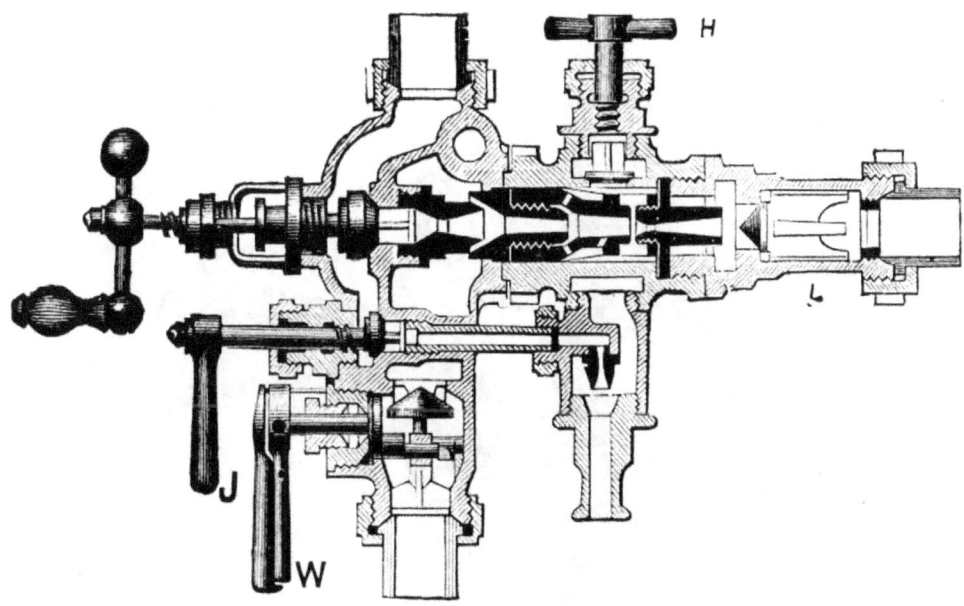

great point of convenience about these injectors is that they can be located and operated either inside or outside the cab, and the valves are all easily removed, and the body of the injector being in two sections, the nozzles are easily taken out.

HOW TO OPERATE.

To Start: Open jet J. When water appears at overflow,

open valve S until overflow ceases, then close jet J. Do not increase steam supply after overflow has ceased.

To Stop: Close valve S.

To Heat Water in Tender: Close valve H, and open valve S; but the valve H should never be closed except when the injector is to be used as a heater-cock. Regulate for quantity of water needed by valve W.

To Grade the Injector: Throttle water by valve W. If this be not sufficient, reduce the quantity of steam supplied.

A small lubricator should be attached to the oiler plug of injector to lubricate the nozzles and prevent incrustation or corrosion.

Whenever the road passes through a section of country where the water is impure or limy, the nozzles should be taken out of the injector frequently and placed over night in a bath of mineral oil, or washed in a dilution of sulphuric acid, to remove deposits from the surface.

THE METROPOLITAN 1898 LOCOMOTIVE INJECTOR.

The illustrations herewith shown represent the latest develop-

ment of the Metropolitan injector. The makers, The Hayden and Derby Mfg. Co., of New York city, have kindly furnished us the following description of this form of injector:

PRINCIPLE.

Experience has demonstrated that an injector using the double-tube principle of an independent lifting apparatus which lifts the water and in turn delivers it to the forcing apparatus which discharges it into the boiler, can be most advantageously used for feeding the modern locomotive. In fact, it is impossible to secure in any other form of injector the features which are now recognized by experienced railroad men as necessary for the proper and effective performance of the locomotive injector

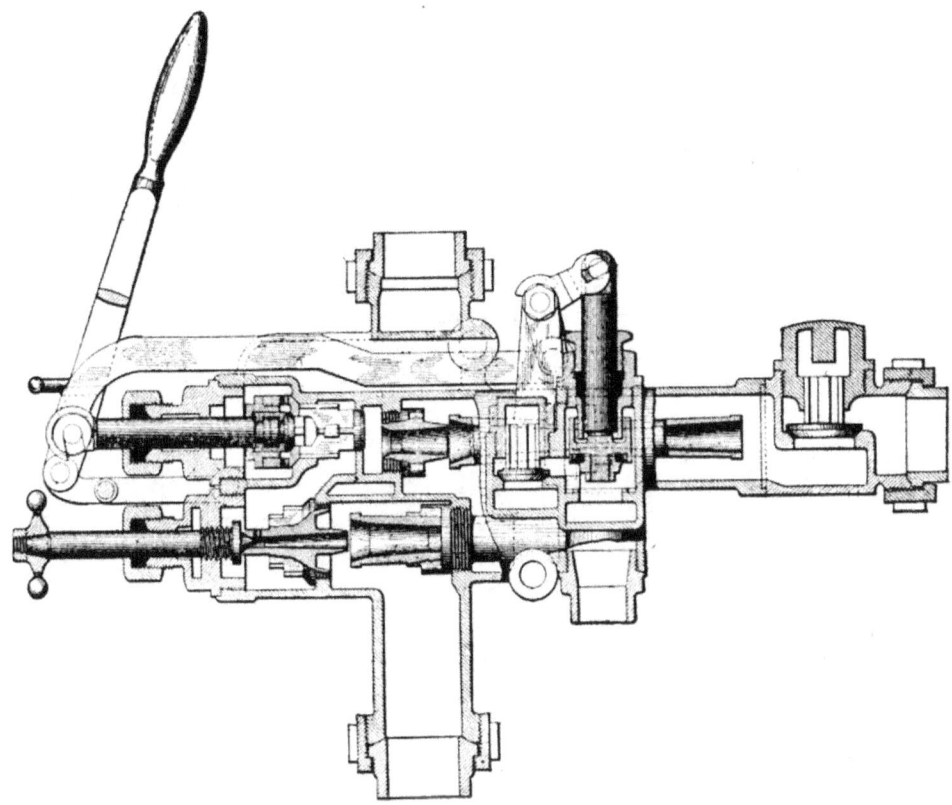

under all the conditions which exist on the modern high-pressure locomotive.

The Metropolitan "1898" Locomotive Injector is a double-tube injector, composed of a lifting set of tubes which lifts the water and delivers it to the forcing set of tubes under pressure, which in turn forces the water into the boiler.

The lifting set of tubes act as a governor to the forcing tubes, delivering the proper amount of water required for the condensation of the steam, thus enabling the injector to work without any adjustment under a great range of steam pressure, handle very hot water and admit of the capacity being regulated for light or heavy service under all conditions.

This injector will start with 30 to 35 lbs. steam pressure and without any adjustment of any kind will work at all steam pressures up to 300 lbs. In fact, at all steam pressures and under all conditions its operation is the same. When working, all the water must be forced into the boiler. It is impossible for part or all the water to waste at the overflow should the steam pressure vary.

The injector is easily handled. The lever works very freely and can be handled without care, for there is no sensitiveness whatever in starting, as is the case with most injectors, consequently any one can operate it.

The Independent Lifting Apparatus produces a strong, powerful vacuum, which enables the injector to promptly lift the water when subjected to the severe conditions of a hot suction pipe, leaking check valves, and hot water supply.

This injector will handle very hot feed water. It will start readily, taking feed water at 140° Fahr. with steam pressures up to 150 lbs., taking feed water at 135° Fahr. with a steam pressure of 175 lbs., and taking feed water at 130° Fahr. with a steam pressure of 200 lbs.

Regulation of Capacity is an important, in fact indispensable, feature of the perfect locomotive injector. With this injector the capacity can be regulated for light or heavy service under all steam pressures and with hot as well as with cold feed water. While most injectors will admit of the capacity of being regulated with low steam pressures and cold feed water, this injector is the first that admits of a successful regulation with steam pressures up to and above 200 lbs., and with the feed water heated.

CONSTRUCTION.

Durability, reliability, and low cost of maintenance depend largely upon the construction of an injector.

We have adopted such construction which experience demonstrates is the best, and have introduced several new features that will appeal to railroad men as changes in the right direction, which will effectively eliminate certain weak construction that has existed in all locomotive injectors.

The wear in the tubes is chiefly confined to the forcing combining tube. Owing to the double tube principle used in this injector a solid forcing combining tube is used; and as the wear in a combining tube is chiefly in the spill or relief holes, it will be seen that a solid tube will wear longer and be less affected by bad feed water than one having spill or relief holes. This has been demonstrated by practice.

The Horizontal Line Check Valve used in all locomotive injectors has never proved satisfactory, and it is not possible to make it so. Located in the delivery end of the injector, the area of opening around the valve is restricted, and the valve necessarily being light in construction can not be made durable and is always quickly affected by bad feed water.

We have made a radical departure in the check valve, as will be seen, and we are confident this new form of check valve will be thoroughly appreciated. We make the check-valve casing separate from the main injector casing and joined to it by a flange joint. The check valve is an upright valve made large and extra heavy, giving full opening. It is durable and always effective. By removing the cap the valve can easily be reground and examined, which examination will always disclose if the main boiler check valve is in good order. All this is done without disconnecting the injector.

Owing to the flanged joint between the check valve casing and the injector proper the injector can be removed, leaving the check valve on the delivery pipe, which permits the injector being removed even should the main boiler check leak.

The forcing tubes are removed by simply breaking the flanged joint between the injector and check-valve casing, which gives large, free opening for removing the tubes; the steam center piece never has to be removed except to regrind the steam valve.

The overflow valve is readily removed for cleaning or repairs; the connecting bar passes through a cored passage in the center of the injector and is thoroughly protected, but does not come in contact with any steam or water. By means of a handle on the connecting bar it is easily disconnected and the overflow closed, thus making the simplest form of heater without the necessity of throttling the main steam valve.

TO CONNECT AND OPERATE.

Place the injector above the level of the water in the tender within reach of the engineer. Take steam from the dome through a dry pipe; should the injector be placed outside the cab, extension fittings must be used.

To Start the Injector: Pull the lever back slightly until the resistance of the main steam valve is felt. This lifts the water. As soon as the water is lifted, pull the lever back steadily as far as it will go. The injector will then be feeding. Do not push lever in to regulate the feed; it must be pulled back as far as it will go.

To Regulate the Feed: To increase the capacity turn the wheel to the left. To decrease the capacity turn the wheel to the right.

To Use as a Heater: Close the overflow valve by disconnecting the connecting bar and pulling it back. Admit steam by pulling the lever slightly.

Note.—If the water used is limy or impure the tubes should occasionally be taken out and placed in a bath of one part muriatic acid to ten parts of water for cleaning.

LITTLE GIANT INJECTOR.

These injectors, of which we have shown a cut of the most improved pattern, are no experiments, as they have been in service for twenty-five years. They are simple in construction,

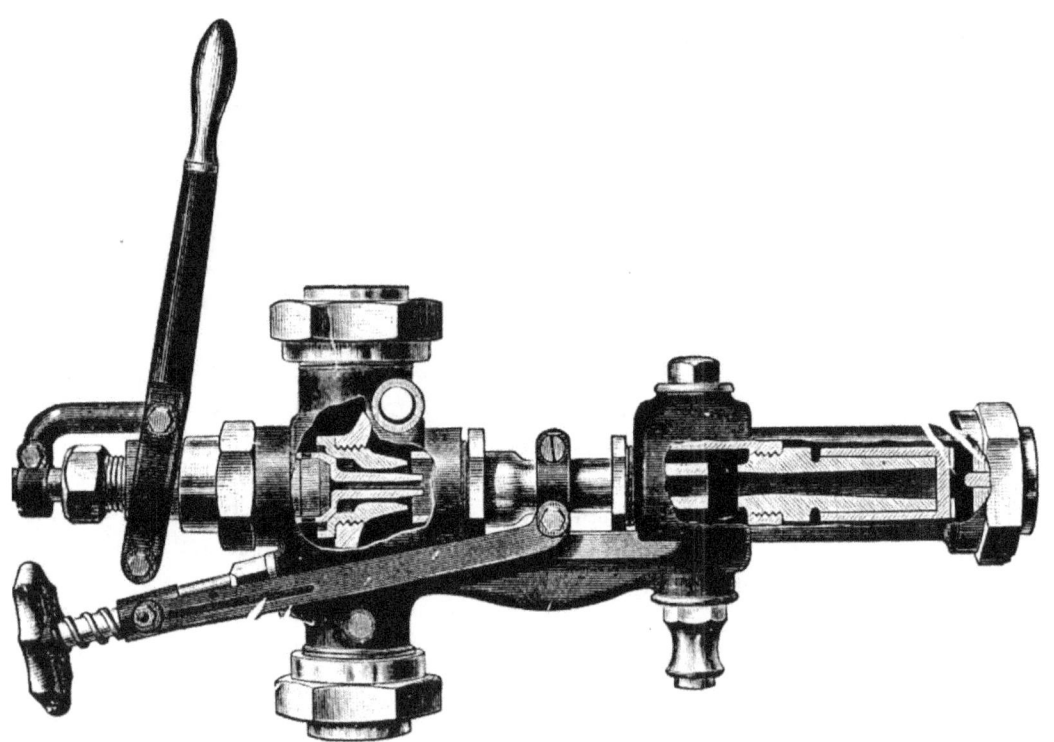

easily operated and are the only injectors made having a movable combining tube which the operator can adjust to work under different pressures of steam.

THE HANCOCK INSPIRATOR.

We have shown a sectional view of the latest Hancock locomotive inspirator of 1894, and explain its mode of action. The difference between injectors and inspirators is, that the inspirator is a double apparatus combining the lifting and forcing jets and tubes in the same instrument and operating with closed overflow, while in the injector they are independent of each other. The first Hancock inspirator was manufactured in Boston in 1874. The following is its mode of action: By a slight pull of lever No. 137, steam is admitted by valve No. 130 through main steam valve No. 126 to lifter steam nozzle No. 101, the velocity of which into No. 102 creates a vacuum, causing water to flow through No. 102, condensing the steam, and out No. 121 at the final overflow valve No. 117 in delivery chamber. By a further movement of lever No. 137 the main

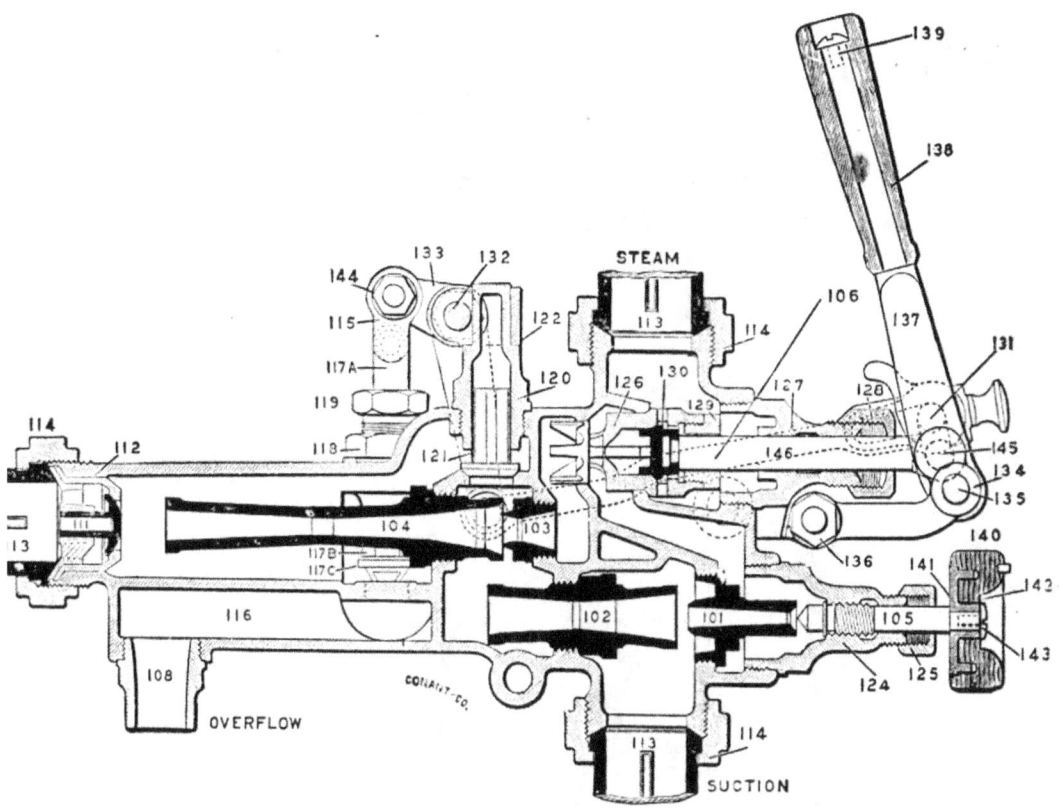

steam valve is opened—No. 126, and steam admitted to forcer steam nozzle No. 103, which takes the water from No. 121 through forcer combining tube No. 104, raising pressure in delivery chamber above boiler pressure, when line check valve No. 111 is opened, and the inspirator is at work; the combining operation takes about 30 seconds only to complete, from start to finish.

OHIO INJECTOR.

This is a thoroughly modern injector designed to meet the requirements of the present day. It is noted for its simplicity, having fewer parts than most injectors which are arranged in a convenient manner for repairs. The combining and delivery

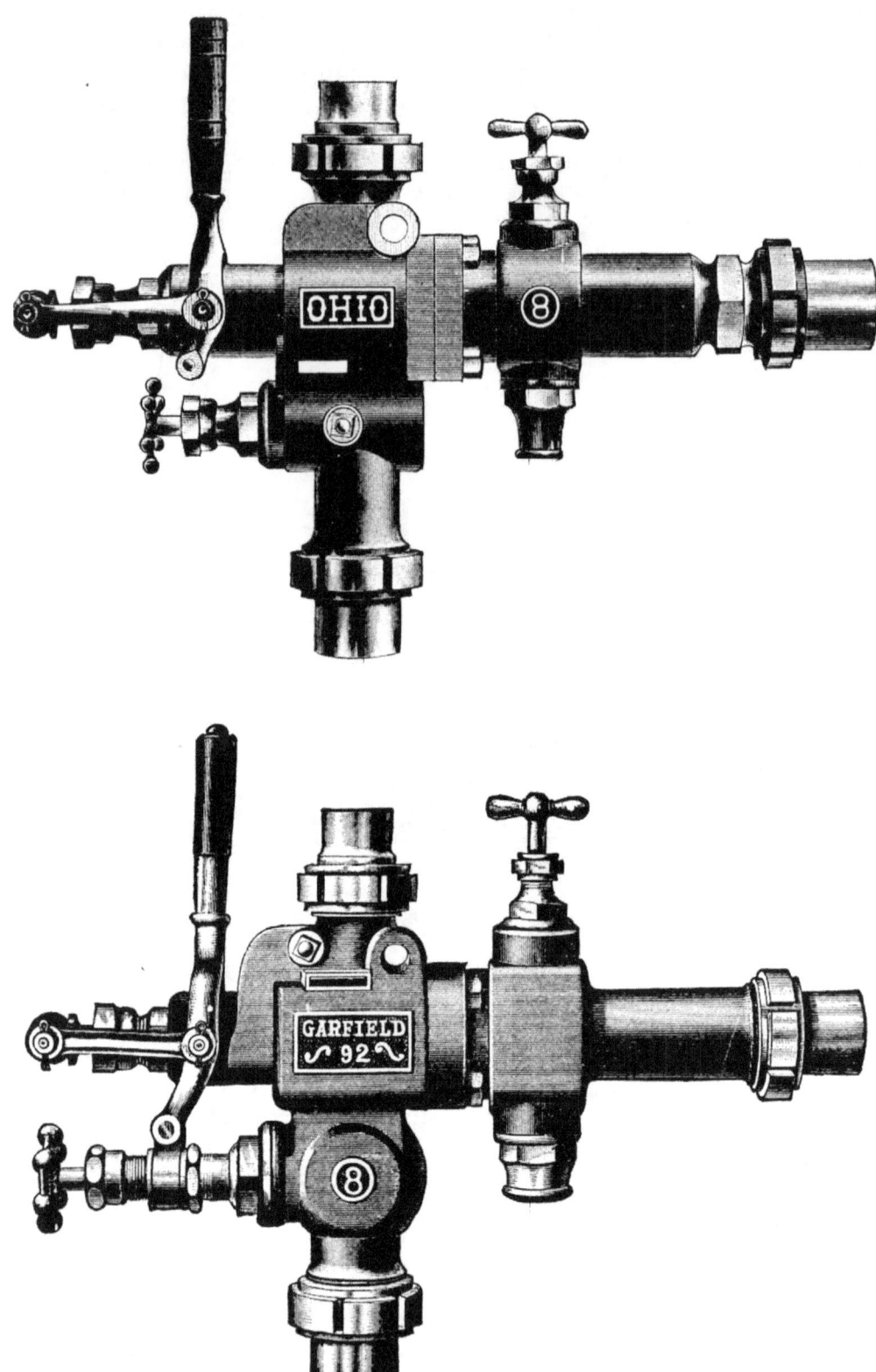

tubes are both attached to the nose of the injector so that they are easily removed and the lifting tube, instead of being screwed to the shell, is held in place by the flange joint. Quite a number of these injectors are now in use.

THE GARFIELD INJECTOR.

This is another modern injector which is in use on many of our railroads. It is constructed in a most convenient form for repairs, most of the nozzles being clamped together by the flange joint, which makes them easy to remove. It is operated by a simple movement of the lever, the supply being regulated by the water valve. Like most other injectors it is claimed to be superior to all others.

THE DODGE INJECTOR.

This is the latest type of injector, which was formerly made by the National Tube Works, and patented by W. E. Dodge, who was their superintendent for twenty-seven years. This is

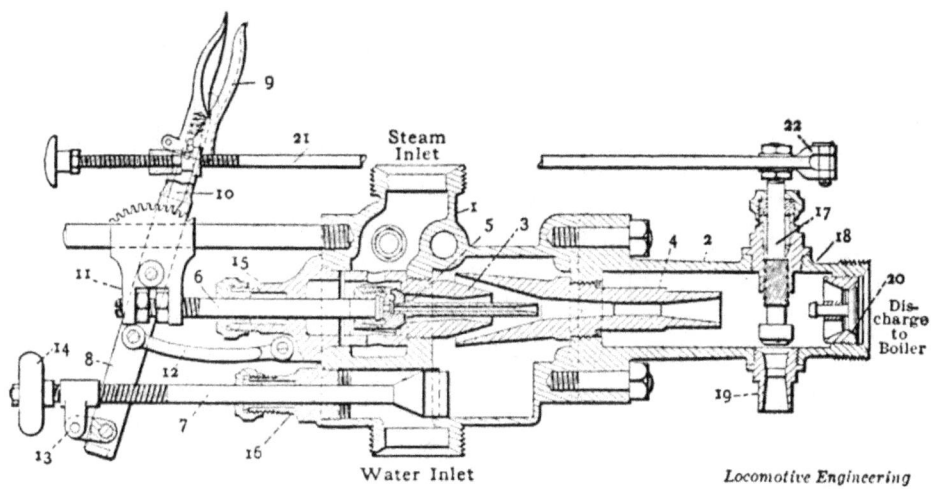

now controlled by the Fitchburg Steam Engine Co., of Fitchburg, Mass., who are now pushing it largely in railroad work.

As will be seen, this is practically a "single movement" injector, the same lever controlling both the lifting and forcing jets. It is claimed that its special design enables it to deliver water to boiler hotter (with same use of steam) than any other injector, and that it saves fuel in consequence.

It is interchangeable with any other standard make, and its ports can be removed for cleaning or renewal, without disconnecting the injector from the pipes. The range of both the water delivered and the steam pressure with which the injector can be worked is said to be equal or greater than any other. The illustration will make its workings clear to engineers and mechanics.

THE LUNKENHEIMER INJECTOR.

The Lunkenheimer injector here illustrated is an automatic, single tube machine of the fixed nozzle type. By "automatic" is meant that should the machine stop forcing (due from interruption of steam or water supply) the injector would restart without attention as soon as the supply is resumed. When the injector "breaks" from stoppage of water supply, the steam will not go down the suction pipe, thereby heating the water and rendering it too hot to work, or where supply is taken through meter injuring same, but will blow through injector into the atmosphere, thereby creating a strong draft through the machine,

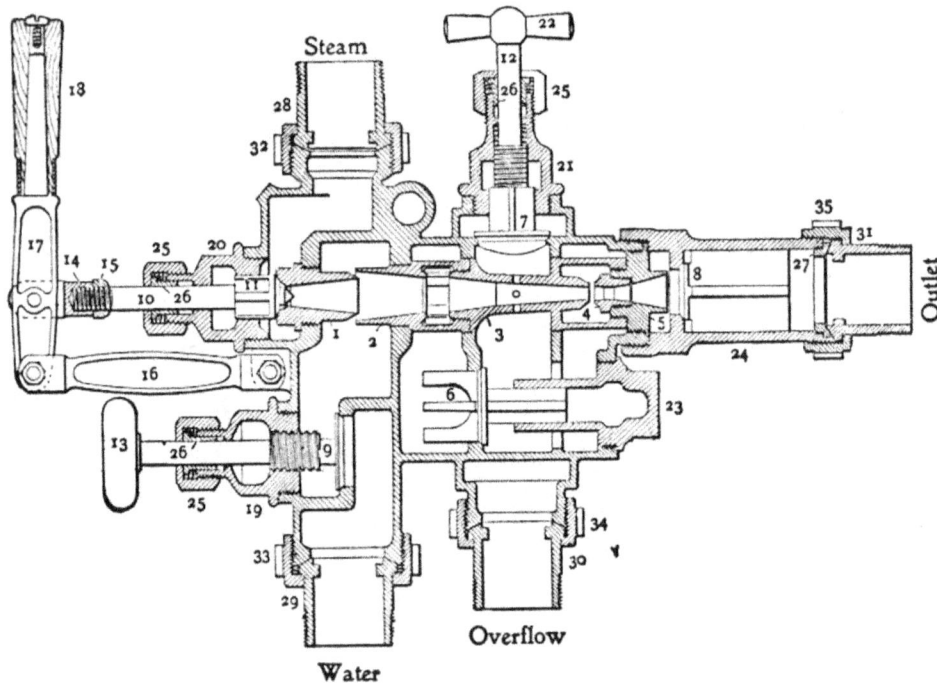

and when the water supply is resumed it will come up to the injector, which will start at once to force it into the boiler without any attention from the operator. Many injectors are claimed to be automatic in action, but few are really so, and most of those which do approach an ideal performance of this feature soon wear out, the part which makes them automatic, i. e., the check valve closing overflow chamber between water lifting and combining tubes, cuts out or scales up. In the Lunkenheimer automatic injector this valve is so made that it will not cut out, and even though it should not be perfectly tight it will not materially impair the working of the machine. It has no delicate or complicated mechanism, the several parts are large and easy of access for examination and repairs without the use of special tools to remove same. Owing to its construction the injector will start promptly at all steam pressures from 30 pounds up to 250 pounds and higher on lifts not exceeding 18 feet. It is not necessary to prime the injector in starting it, as the single movement of the lever is sufficient to admit steam and the water is promptly lifted to the injector, which will at once start to work and force it into the boiler. This injector has been put upon the market by the Lunkenheimer Company, Cincinnati, O.

THE BROWNLEY DOUBLE TUBE INJECTOR.

We present two views of this injector, which is perhaps the most simply constructed injector made, having but one valve. It is in use on the Manhattan Road of New York, where it is

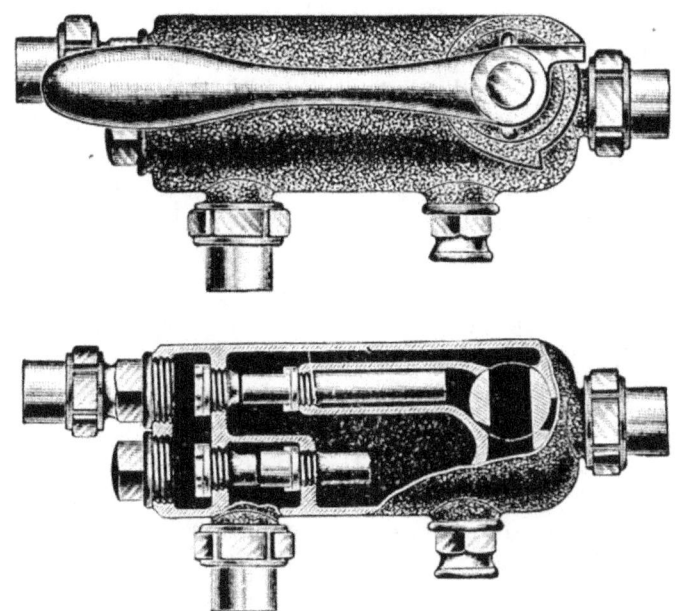

said to have given better results than any other form of lifting injector, being very powerful and at the same time economical on steam. It will work under any pressure from 15 lbs. to 350 lbs., and is operated by a simple movement of the lever.

REPAIRING INJECTORS.

The requirements to perform this work successfully are two; first, a thorough knowledge of the construction of the injector and check, and the principle of the injector's action, and secondly, the hand of a skilled workman to make the needed repairs. It is considered nice clean work in a machine shop, but every machinist is not competent to perform this work, and furthermore not every machinist who is allotted to this particular line of work performs it successfully. This is one of the many jobs in the machine shop which requires more than practice to thoroughly understand. Below we give the method of performing this work, but it requires much reflection and should be done carefully. When overhauling an injector of any make, take out the main valve (or ram), the primer cock, the water valve, and the frost plug, then take injector apart and take out the nozzle. Clean all dirt or sediment out of the shell and soak shell in benzine to remove scale from inside, then examine the nozzles carefully before putting back in; if they are cut bad or worn out larger than their standard size, turn up new ones. See that nozzles are firm when intact, and if the injector divides into two or three parts see that each joint is steam-tight, and when putting together use oil, white lead or varnish, and tighten firmly. Now examine all valves and seats and if they are cut bad face off valves, or replace them with new ones; then face seats, using a bushing to keep rosebit central. Now grind all valves and seats steam-tight, using flour emery. Now examine threads on ram and other stems; also, on all plugs and packing nuts, and if very loose, replace with new ones. Now put it together and make each joint steam-tight. Now re-pack all the stems with asbestos, if you have it; if not, then use lamp wick, or some other kind of packing. Now your injector is finished. When taking an injector apart, if some parts are very tight, warm them a little and they will readily loosen. Brass expands very quickly. If an injector is reported thus: won't work; first take the cap off check, and if it is all right, examine feed pipe and hose for holes to see if the injector sucks air. If not, then take out main valve and primer cock; if they are all right, take out frost plug and see if injector is full of lime; if you cannot locate the defect in this way then take down the injector and overhaul it, as per instructions above.

A NEW HOSE STRAINER.

The hose strainer here shown appears to be a decided improvement over the old style strainer, and something of this kind has long since been needed. This strainer permits the use of the full area of the suction pipe, and can be used for either right or left hand side of the locomotive. The screen is circular in form and rigidly attached to the cap or bonnet, which forms a receptacle for all dirt and sediment. By removing the cap or bonnet the dirt or sediment is removed with the screen which can easily be cleaned and the hose need not be disconnected. It is made in one size only, with nipple for suction pipe for 2" iron pipe or for copper pipe 2" or 2¼" outside diameter.

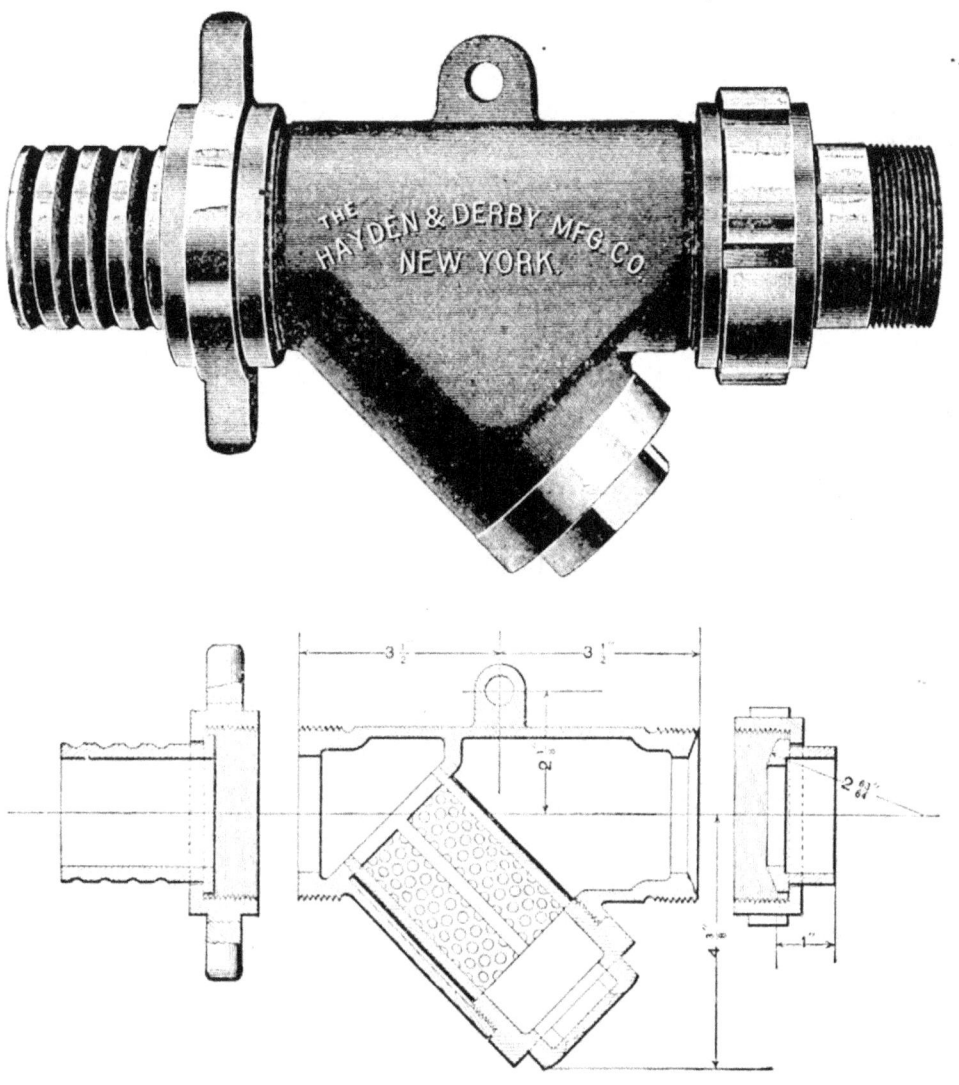

BOILER CHECKS.

Too much attention cannot be given the boiler check, for unless the check is in good working order no injector should be expected to work satisfactorily. If the check valve gets cocked, or sticks, or leaks and blows back sufficient steam to heat the injector, or if the valve has not sufficient lift, the injector will not work. So you see, while a check valve is very simple of construction, it is very important that it should be in perfect condition. Most check valves have three or four lugs extending below the valve seat. The object of these lugs is to act as guides to reseat the valve centrally upon its seat. These lugs or wings should not be too close a fit in the shell as they become corroded and may cause the valve to stick. They should be about 1-32 loose. In some of the latest improved check valves these lugs or wings are "fan" shaped or inclined from the seat, thereby creating a turbine, or revolving check valve. (See the McLeod Check Valve, page 459.) These valves give excellent results, as they prevent the formation of scale and regrind upon their own seat. The lift of a check valve should be in proportion to its own diameter and the capacity of the injector. About 5-16" is the average lift for a No. 9 injector. An extra check valve is used on some engines; it is located in the branch pipe between the injector and the boiler-check. They are intended to prevent heating the injector in case the boiler check leaks. Some of our "up-to-date" locomotives have both injectors placed on the right side, with a check valve in each branch pipe and a kind of a goose-neck is used in place of a boiler check. This s intended to relieve the fireman of all responsibility in regard to the injectors. We have shown herewith a few forms of improved check valves.

THE FOSTER SAFETY BOILER CHECK.

The cut is self-explanatory and shows the construction of this form of boiler check, which appears to be of convenient

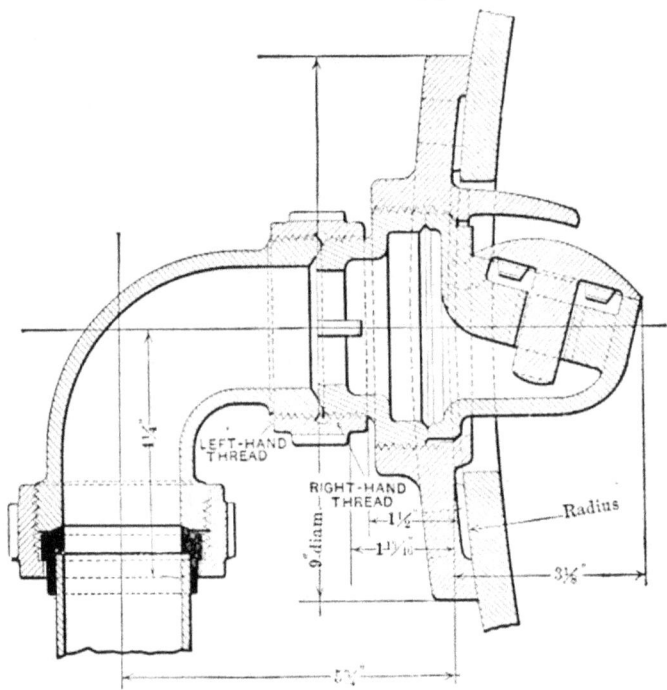

form. The flange may be attached by rivets, tap bolts or studs riveted on the inside of boiler plate. The valve chamber is readily accessible and can be removed for regrinding or other repairs without disturbing the flange. It cannot be tampered with or damaged by the engineer or fireman hammering on the valve chamber. It will be noticed this form of check has two steam tight joints, the retaining piece seating against the valve chamber and forming another seat. This valve has inwardly projecting wings which cause the valve to rotate under operation. It was invented by Mr. John M Foster, of Elizabeth, N. J.

THE HEINTZELMAN SAFETY CHECK VALVE.

This duplex safety check valve was designed by Mr. T. W. Heintzelman, Master Mechanic of the Southern Pacific Ry., Sacramento, Cal., who describes his invention as follows:

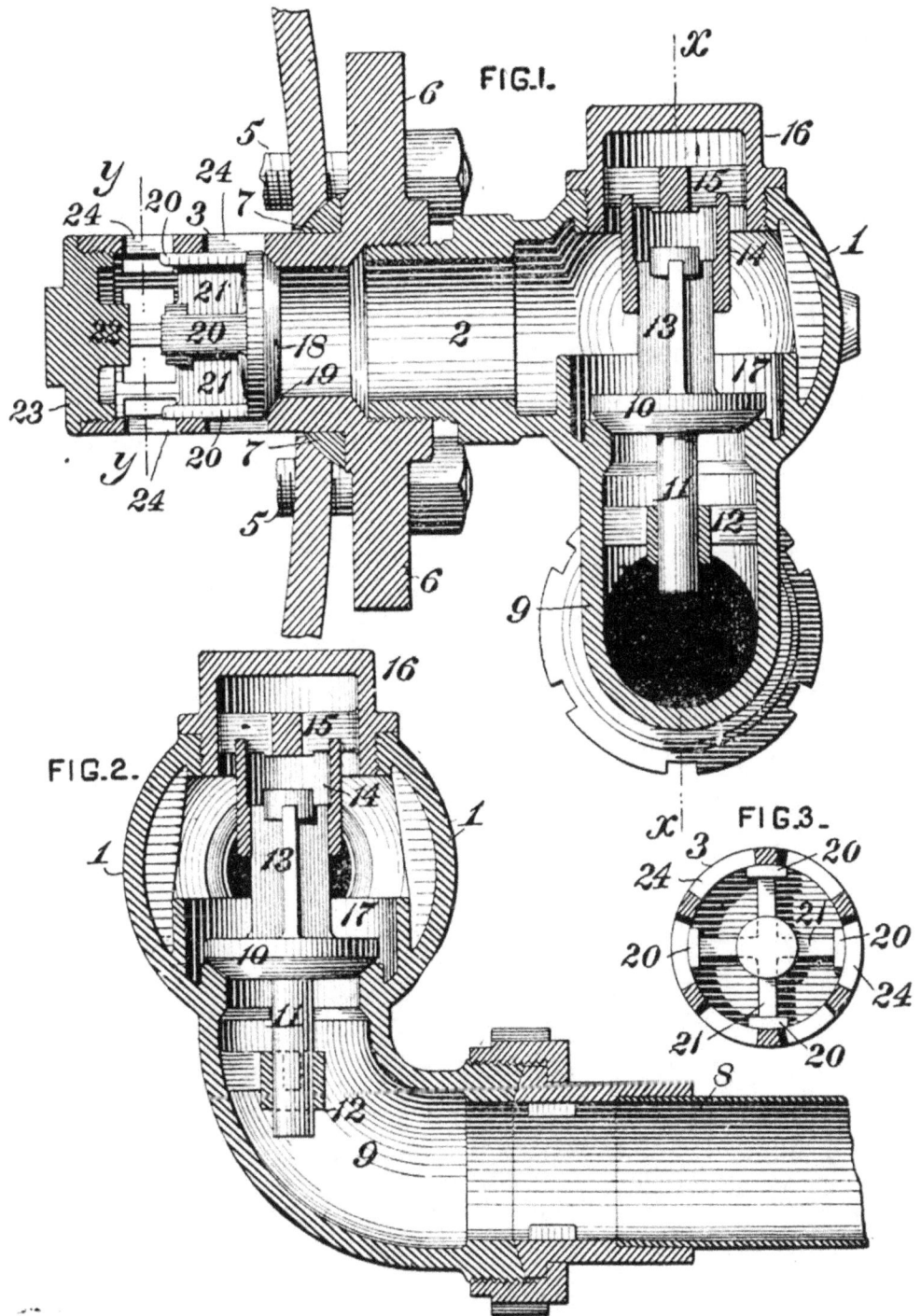

My invention relates to the type of check valves located outside of the boiler and a second or supplemental check valve having direct communication therewith, and located inside of the boiler; the construction being such that in the event of the main or outside valve being removed or broken off the inner or supplement check valve will prevent the escape of water or steam, and thereby obviate the liability to accident resulting therefrom.

My object was to produce a check valve of simple and inexpensive construction and capable of application in the ordinary manner, all of whose parts, especially such as are subject to wear in ordinary use, shall be readily accessible without the necessity of removing the entire device from the boiler. A further object was to obviate the liability of imperfect operation of the valves, by the accumulation of sediment or scale, on or around the seats.

The various parts are clearly shown in the illustration, therefore a detailed explanation of the several parts is unnecessary. It is a good, reliable and convenient form of check valve.

THE LINSTROM CHECK VALVE.

The check valve herewith shown is the invention of Mr. Chas. Linstrom, of Vicksburg, Miss., who gives the following description of it:

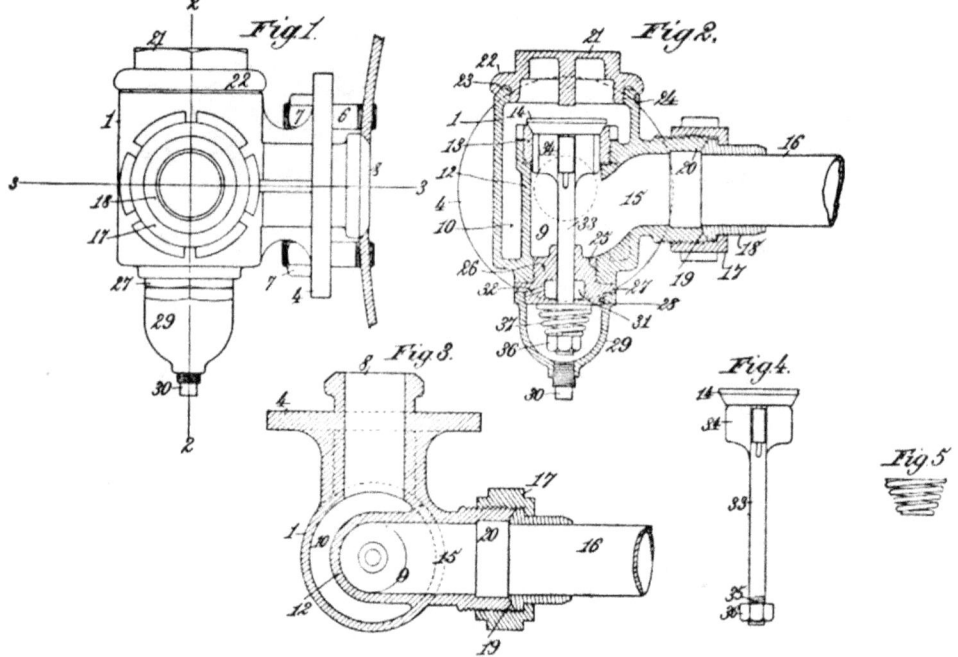

The numeral 1 indicates the valve-casing, having at one side a base-plate 4, by which the valve is secured to the shell 5 of a boiler through the medium of suitable attaching-bolts 6 and nuts 7.

The base-plate is provided with a suitable outlet 8, which opens into the boiler, so that the feed-water passing through the check-valve will flow into the boiler when the injector or the feed-pump is working.

The check-valve casing is constructed interiorly with a water-chamber 9 and a steam-chamber 10, which is cut off from the water-chamber by the dividing partition or web 12, which is constructed at its upper end portion with a screw-threaded socket, into which is screwed the ordinary brass ring 13, which constitutes the valve-seat for the disk, plate, or plug of the check-valve 14. The brass ring is in the form of an ordinary wing-nut, so that it can readily be removed and replaced.

The water-chamber 9 of the check-valve connects by an elbow-shaped passage 15 with the feed-water pipe 16, which connects in the ordinary manner with an injector or with a feed-pump. (Not necessary to illustrate.) The connection of the feed-pipe 16 with the valve-casing is preferably through the medium of a union coupling or nut 17 and a sleeve 18, having a ball-joint connection, as at 19, with the inlet 20 of the valve-casing.

The upper end of the valve-casing is constructed with an internal screw-thread, with which engages an external screw-thread on a valve-case cover 21, having a marginal flange 22, constructed at its under side with a somewhat conical bearing 23, which fits a conical bearing 24, formed on the upper end of the valve-casing and arranged at an angle of about thirty degrees to the axis of said casing, all in such manner that by screwing the valve-case cover into position the wedging action of the bearings 23 and 24 produces a perfectly steam-tight joint, while at the same time the joint is elastic, due to a slight elasticity of the flange 22 of the valve-case cover, whereby the cover will nicely and perfectly adapt itself to the valve-casing.

The lower end of the valve-casing is constructed with an internal screw-thread, into which is screwed an external screw-thread 25 on a guide-thimble 26, having a flange 27, which bears against the lower end of the valve-casing. The guide-thimble is also constructed with an external screw-threaded extension 28 to engage internal screw-threads on a cover-cap 29, having at its center a relief-plug or air-cock 30 of any desired or suitable construction. The guide-thimble is also provided internally with a chamber 31, from which a narrow channel 32 leads to the exterior of the thimble, so that the pressure of steam or water in the chamber can escape when the cover-cap 29 is removed by unscrewing it from the thimble.

The guide-thimble is constructed with a central vertical bore, through which extends a cylindrical or other suitably-shaped guide-stem 33, which depends from the check-valve disk, or plug 14.

The check-valve proper, as here shown, is composed of a winged body 34, having a beveled or inclosed disk or plate seating against a beveled or inclined valve-seat, but this valve proper may be a disk, plate, or plug of any ordinary form, shape, or configuration which will suit the conditions required.

The guide-stem 33 is formed integral with the valve disk, plate, or plug. The lower end of the guide-stem is constructed with an external screw thread 35, on which is screwed an adjusting-nut 36, and between this adjusting nut and the guide thimble 26 is arranged a special spring 37, which acts to positively and promptly reseat the valve whenever it has been opened or unseated by the working of the injector or feed pump.

The stem of the valve is guided perpendicularly by the guide thimble 26, and consequently the valve is accurately seated, and, since the guide-stem extends to the exterior of the valve casing, it is obvious that the outer end of the stem is accessible for the purpose of adjusting the tension of the spring, or for manipulation, or operating the stem and valve should the latter adhere or stick to its seat and fail to properly act when the injector or feed-pump is working. It will of course be understood that the cover-cap 29 is detached or removed from the valve-casing whenever it is desired to gain access to the outer end of the guide stem of the check-valve disk, plate, or plug.

A NEW CHECK VALVE.

Mr. Wade DeSanno, of Vallejo, California, describes the check valve here illustrated in the following words:

A is the outside valve case, B is the outside valve, C is the movable valve seat, making a joint on the lower part of the valve case at D and on the end or flange of the check pipe at E; all being held in place by the usual nut F. G is a flange (bolted to boiler in the usual way) and with an extension into the boiler, making a seat for the inside valve H, which is an ordinary four-winged valve with a round stem J, which passes through the guide K to preserve the alignment of the valve. The inside cuts no figure in general service, and only comes into service in case the outside case is carried away in a wreck, or when it is desirable to work on the outside check, when the engine is under steam. If at any time it is desired to remove the outside check valve for repairs, and there is pressure on the boiler, there is a small globe valve (not shown) screwed into the valve case at the dotted circle L. This globe valve should be opened suddenly, allowing any steam or water in the check case to escape, also telling us whether or not the inside valve is seated. The opening of the globe valve is a precautionary measure, before any attempt is made to loosen the check-pipe nut F. You will notice that the outside valve seat is not a part of the valve case, but inserted.

We don't have to unscrew the top of the valve case and try to fish out the valve, as in general practice. The set screw shown is to regulate the lift of the valve, if thought best. It will be

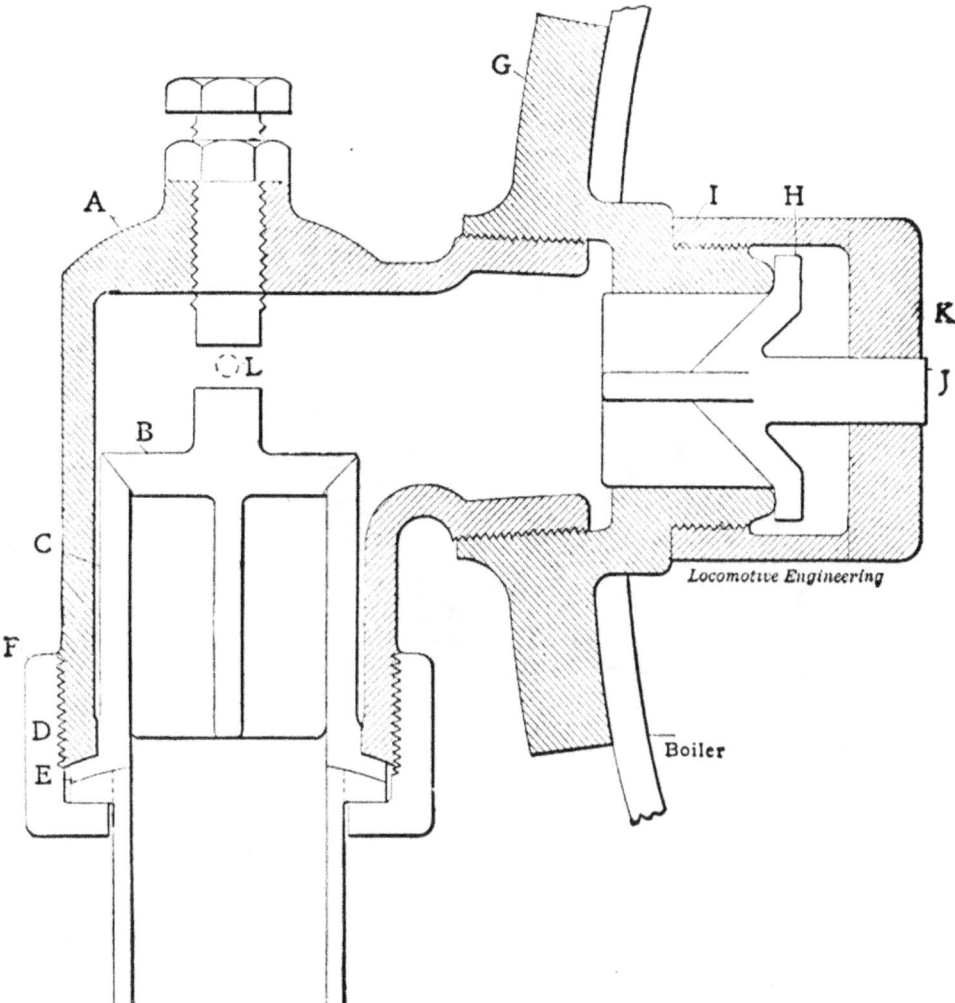

further noticed that the seat of the inside valve is rounding; it is made so that, in case of an emergency, the valve may crush through any scale that may be on the seat. No, it is not patented. Use it.

McLEOD'S TURBINE CHECK

For a complete description of this check, see the article on Boiler Cleaning by Mr. F. W. Hornish, page 450.

THE ALBIN SAFETY BOILER CHECK.

The safety check shown in the illustration has been patented by Mr. Frank Albin, of Newton, Kan., and is especially intended for use on locomotive boilers. It consists of an exterior mud pocket, which is threaded into the shell of the boiler, and receives at its outer end the injector pipe. The mud pocket is closed by a threaded cap which is perforated, and on the inner side is extended to form a valve cage in which is located a ball valve. The passage from the mud pocket to the boiler terminates in a short elbow which is screwed into the neck of the pocket and extends upward within the boiler, where it terminates in a ball valve similar to that in the pocket. The feed water, in passing through to the boiler, will deposit any solids and foreign

matter which it may contain, within the mud pocket, where it will collect and settle. It will be seen that the ball valves will prevent the return of water from the boiler, and should the mud pocket be broken off, the valve on the inside of the boiler will effectually prevent the terrible effects which ordinarily follow from the escaping water and steam in the event of collision. The inner ends of both the valve chambers are closed by spanner nuts, and the various connections are threaded, so that the device is easily taken apart for inspection. The valve in the interior of the boiler, moreover, enables the mud pocket to be opened and cleaned at any time when the boiler is under steam.—*Scientific American.*

LUBRICATORS.

The locomotive lubricator has almost entirely replaced the old style oil plugs; their universal adoption was due to the fact that they would feed oil to the cylinders under steam pressure, and could be regulated and fed continuously, besides using the oil economically. With the old style oil plugs formerly in use it was necessary to shut off steam pressure before the cylinders could be oiled. The lubricator overcame this difficulty and at

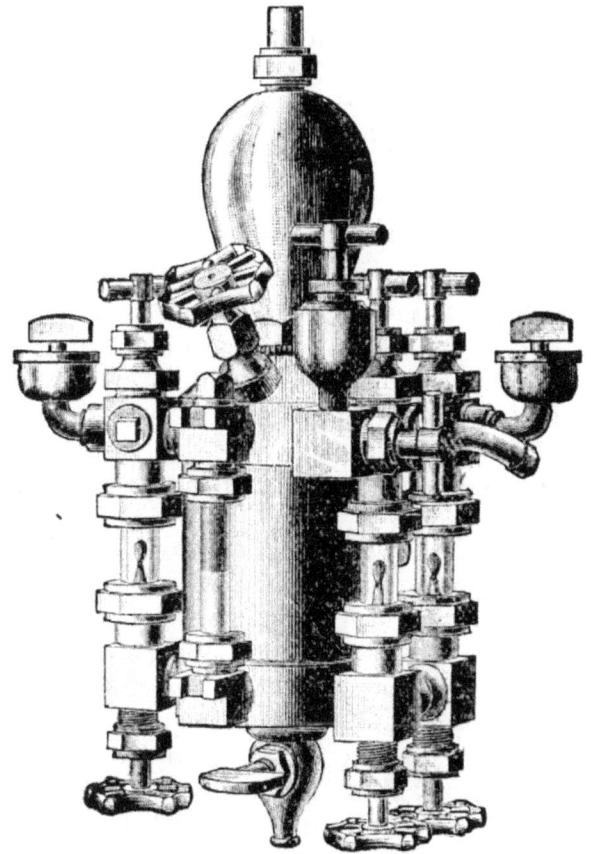

THE NATHAN.

the same time supplied a means of continuous feed which is capable of very accurate adjustment.

Lubricators are made in various forms, some supplying oil to the steam chest and cylinders only. Others to the oil pump only. But the latest designs combine both cylinder and oil pump feeds. The two lubricators in most general use are the Nathan and Detroit. The first illustration is an exterior view of the latest form of Nathan lubricator.

Mode of Action: In this lubricator as in all others the oil rises to the top of the water which is condensed steam, and then passes down through the pipes shown on the interior views, then out to the feed valves where the supply is regulated; it then passes up through the sight feed glasses which are filled with condensed steam, thence into the oil pipes to the cylinders and air pump. We have shown interior views of "The Detroit" lubricators with each part named and numbered so the reader may study their construction and become familiar with the correct names of the various parts. We also give instructions in regard to the attachment of the lubricator, how to fill it, operate it and how to overhaul or repair them when out of order.

How to Attach the Lubricator: Support lubricator with a heavy bracket (2x⅜), preferably in center of boiler head. Connect with tallow pipes, which should have a marked descent to steam chests. Remove valves from over steam chests. For steam, connect at C direct with boiler, always allowing pipe to descend gradually to the boiler to permit surplus condensation to flow back to boiler.

How to Fill: Close valves D, EE and I, and fill with clean strained oil.

How to Operate: Open steam valve M one turn for boiler pressure, then valve D, regulate feed with valves EE and I. BB, auxiliary oilers, are entirely independent of the lubricator, and are to be used the same as old cab oilers, when necessary, by simply closing the engine throttle.

TIPPETT ATTACHMENT OF DETROIT LUBRICATOR.

The accompanying illustration shows what is known as the Tippett attachment, applied by the Detroit Lubricator Co. to their sight feed lubricators, for the purpose of securing a steady feed against the back pressure of the cylinders.

Since the higher steam pressures have come into vogue there have been a great many complaints of the lubricator not feeding oil to the steam chests when engine was running with a full open throttle. The Tippett attachment consists of a pipe leading into the dry pipe of the locomotive and connecting with the two oil pipes. When the throttle is open, an extra current of steam is admitted into the oil pipes which overcomes the back pressure and creates a constant current in the oil pipes towards the cylinders as long as throttle remains open.

In the diagram showing the lubricator with the attachment connected to the locomotive, A is the auxiliary steam pipe, taking steam from the dry pipe; B is the oil pipe, C the steam pipe to the condenser of the lubricator, D an expansion joint and E the dry pipe fitting. The last named parts are shown in detail in detached views. Lubricators with this attachment are in service on several leading railroads with so satisfactory a record of service that they are being specified for attachment to new power being built. It is said that in every case in which they have been used the efficiency of the engine has been increased without in-

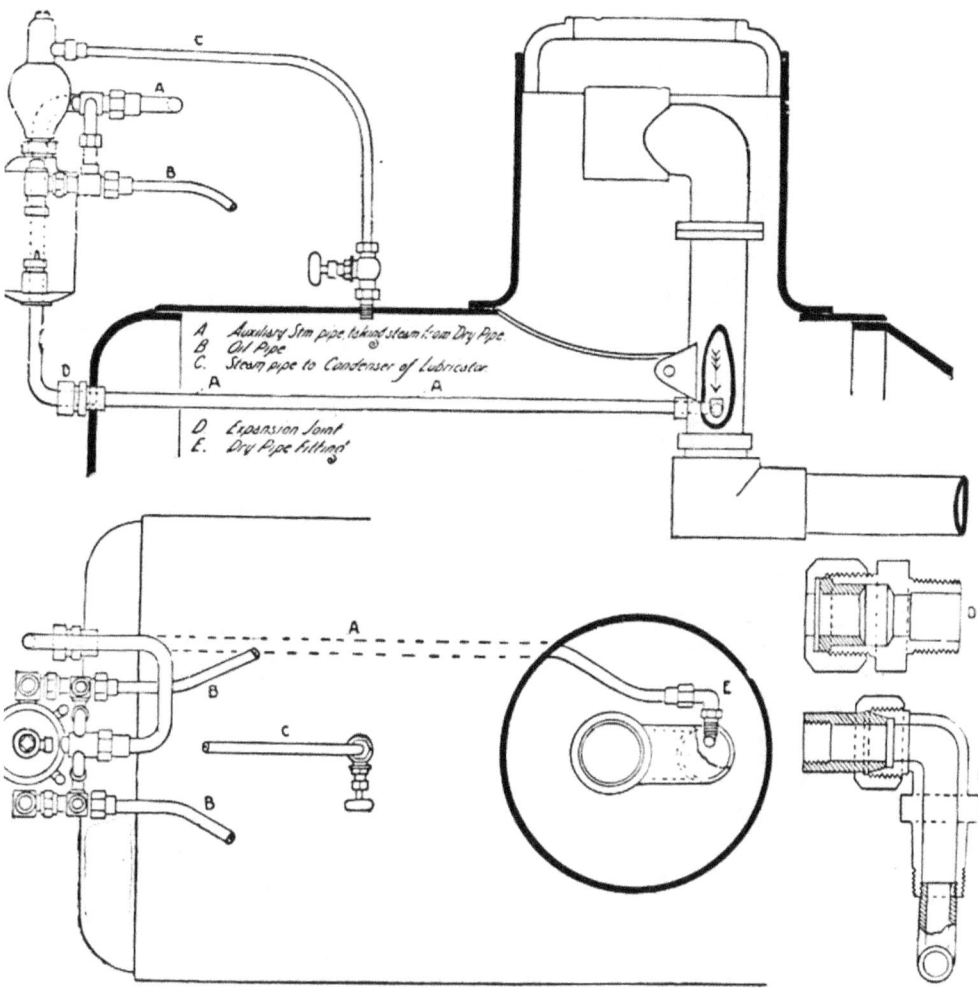

creasing the quantity of coal used, and on account of the more complete lubrication furnished to the valves the reverse lever can be operated with greater ease.

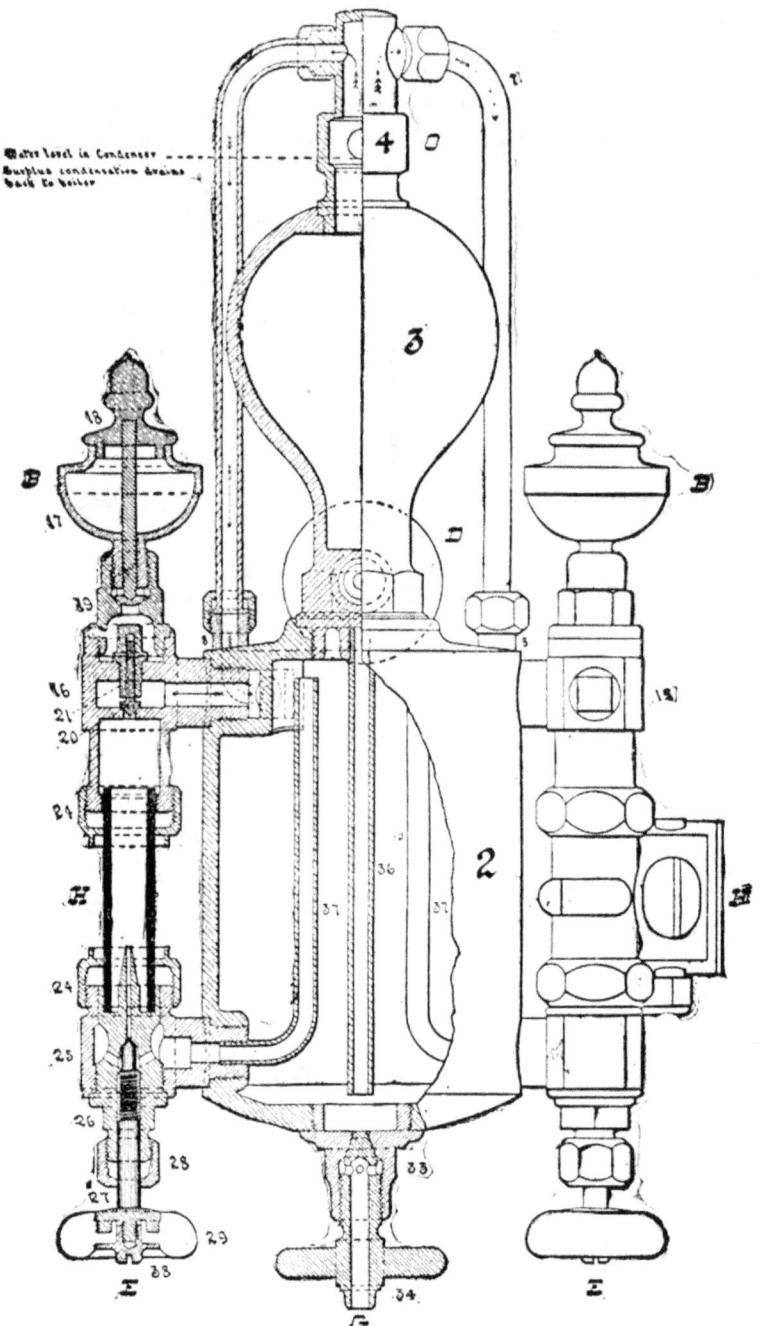

PARTS OF CYLINDER LUBRICATOR.

2—Oil Reservoir, quart size.
3—Condenser.
4—Extension Top Complete.
5—Tail Nut.
6—Tail Pipe.
7—Equalizing Tubes.
8—Elbows.
9—Water Valve Complete.
10—Water Valve Stem.
11—Water Valve Center Piece.
12—Water Valve Follower Plate
13—Water Valve Packing Nut.
14—Wood Handle.
15—Upper Feed Arm, right.
16—Upper Feed Arm, left.
17—Hand Oilers.
18—Hand Oiler Covers.
19—Hand Oiler Plugs.
20—Check Valves.

THE LOCOMOTIVE UP TO DATE.

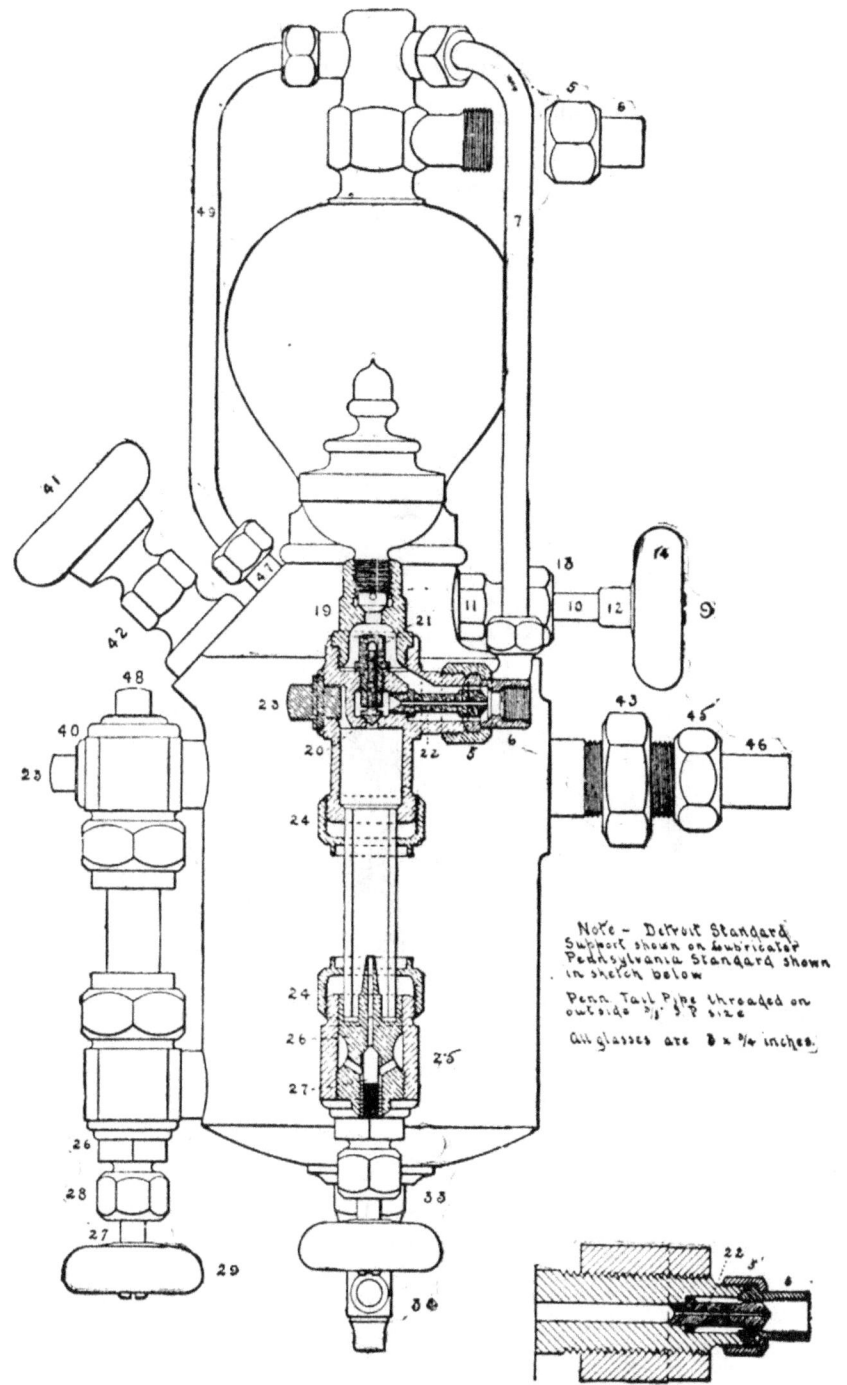

PARTS OF THE CYLINDER LUBRICATOR.

21—Check Valve Guides.
22—Nozzles.
23—Water Plugs.
24—Packing Nut for Glass.
25—Lower Feed Arm.
26—Feed Valve, Body.
27—Feed Valve, Stem.
28—Feed Valve, Packing Nut.
29—Feed Valve, Stem Handle.
30—Filler Arm.

31—Filler Plug.
32—Lower Gauge Arm.
33—Drain Valve Body.
34—Drain Valve Stem.
35—Jamb Nut.
36—Water Tube.
37—Oil Tube.
38—Handle Button.
All Glasses, 3x¾.

AIR PUMP LUBRICATOR.

The exterior view of the air pump lubricators herewith shown is of the Nathan type and the interior view represents the Detroit pattern. These lubricators should be connected to the steam pipe leading to the brake pump.

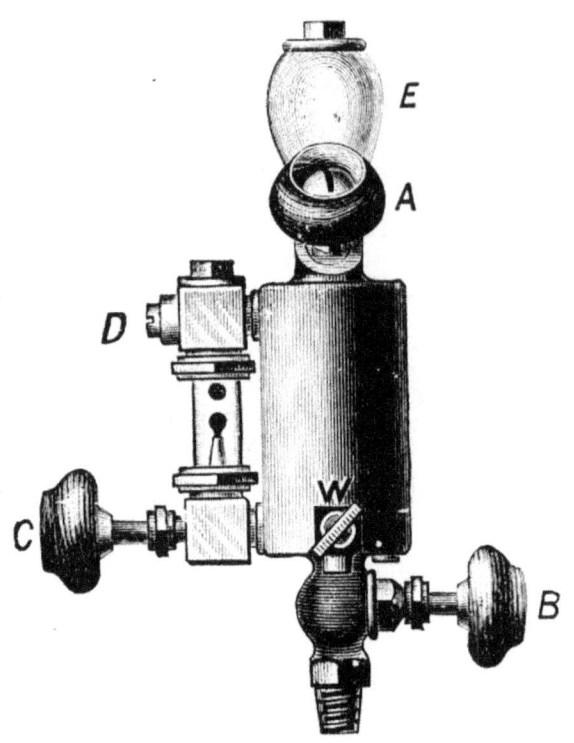

THE NATHAN.

DIRECTIONS FOR USE.

Fill the cup with clean, strained oil, through filling plug A

To Start: Open valve B, wait until glass has filled with condensed water, then regulate the feed by valve C.

To Stop: Close valves C and B.

To Renew Supply of Oil: Close valves C and B and draw off water at waste cock W; then fill the cup with oil and start again as before.

Note:—The valves (with the exception of filling plug) should be opened wide, and steam blown through them once in every two weeks at least, to cleanse them thoroughly,

PARTS.

27—Feed Valve Stem.
38—Handle Buttons.
43—Globe Valve, Center Piece.
44—Globe Valve, Stem Nut.
44—Equalizing Tube Nut.
44—Pulsating Stem Nut.
45—Globe Valve Stem.
47—¾ Nuts.
50—Feed Stem Nut.
52—Filler Plug.
57—Pulsating Valve Stem.
62—Globe Valve, Body.
63—Wood Handle.
65—Lower Feed Arm.
66—Feed Valve.
67—Upper Gauge Arm.
68—⅝ Nuts.
69—Lower Gauge Arm.
70—Drain Valve, Body.
71—Drain Valve, Center Piece.
72—Drain Valve, Stem.
110—Condenser.
111—Upper Feed Arm.
112—Support Arm.
113—Right and Left Coupling Nut.
114—Support Post.
115—Equalizing Tube.
116—Oil Reservoir.
    Gauge Glass, ⅝x3¼.
    Sight-Feed Glass, ¾x3.

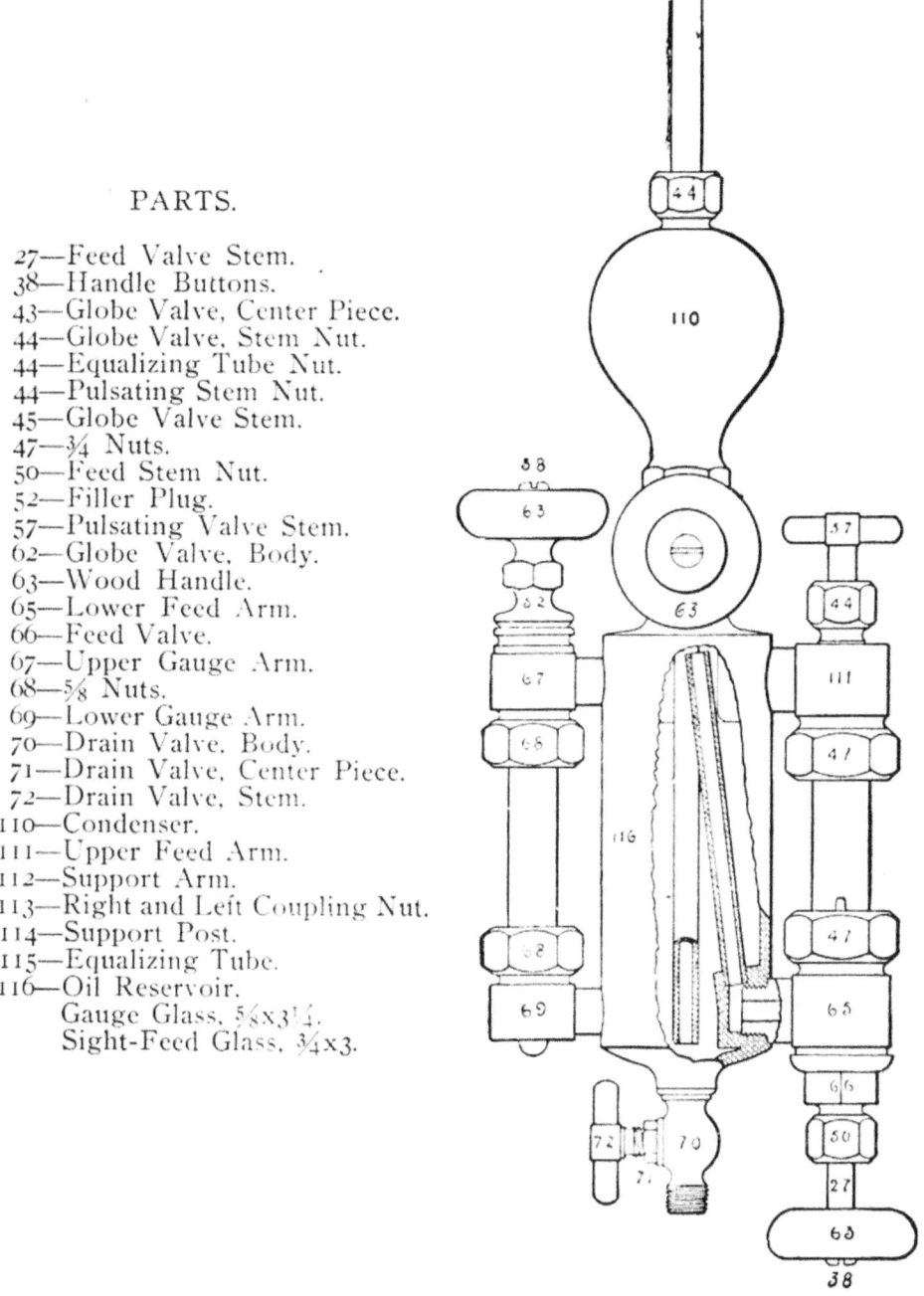

HOW TO REPAIR LUBRICATORS.

If a lubricator won't work, open the bottom cocks and turn on full head of steam and blow out. If it won't work then, take out all the glass tubes and examine small feeders and see that they are not stopped up with dirt or waste, which may have been in the oil. If they are all right, take lubricator down and screw chamber off of top and see if any of the small feed pipes inside are broke off, or stopped up; clean out good, then take out all cocks and clean them out. If threads are stripped on any of them, make new ones. Screw all of them in tight, making steam tight joints, and put up, and you will find it will work all right. If any glasses are broken, replace them and repack all nuts.

CORY'S FORCE FEED LUBRICATOR.

At the present time of very fast trains, making long runs between stops, the question of facilities for thorough and positive lubrication of all journals, eccentrics and links of the fast moving engine becomes very important.

The introduction of the device herewith illustrated and described marks a distinct advancement in securing the highly desirable means of oiling all important bearings of the locomotive, while it is running at full speed, and this is fully accomplished direct from the cab, from where it is possible to oil each bearing successively, or any particular bearing repeatedly, that may be giving temporary trouble by heating.

The lubricator is placed, conveniently of access, in the cab, and consists of an oil supply reservoir of one gallon capacity; at the lower part of this reservoir is seated a hollow conical valve A, the cavity in this conical valve will hold about one eightieth of a gallon. This space inside of conical valve is termed the oil discharge reservoir, and connects to oil supply reservoir by small valve B, seated in upper part of hollow valve. The side of hollow valve is perforated by a hole E, one-eighth inch diameter, that can be brought to coincide with any one of the 16 outlet holes, at base of oil supply reservoir, that each connects with a line of pipe to a given bearing. There are 16 notches on the upper rim of lubricator, that when lever is brought to engage with any one of these notches the hole in the side of conical valve then coincides with a given hole, in base, to outlet pipe. When the lever is thus placed for any bearing desired to supply with oil, the valve shown attached to base, and connected to either steam or air pressure, is opened, and pressure enters through small valve C, into oil discharge reservoir, closing valves B and D, and forcing contents of oil discharge reservoir through hole E, and through line of pipe connecting with bearing that is desired to oil. This requires but a moment, when pressure should be shut off, and lever placed midway between any two notches, and in about ten seconds the discharge reservoir will again be filled and ready for discharging to any desired bearing when lever is placed in notch corresponding to bearing to be oiled, and pressure again turned on.

For all main journals, three way tips are furnished for ends of pipe, thus the wedges and jaws are oiled as well as the journal. Thus the engineer has at his command a positive means of oiling all parts of his engine, however fast the engine may be

277

running, and however long distances he is obliged to run without stops, thus preventing any dangerous and destructive heating and cutting of bearings, delays of trains and possible accidents that might occur from rear end collisions by being obliged to stop, cool off and pack hot journals.

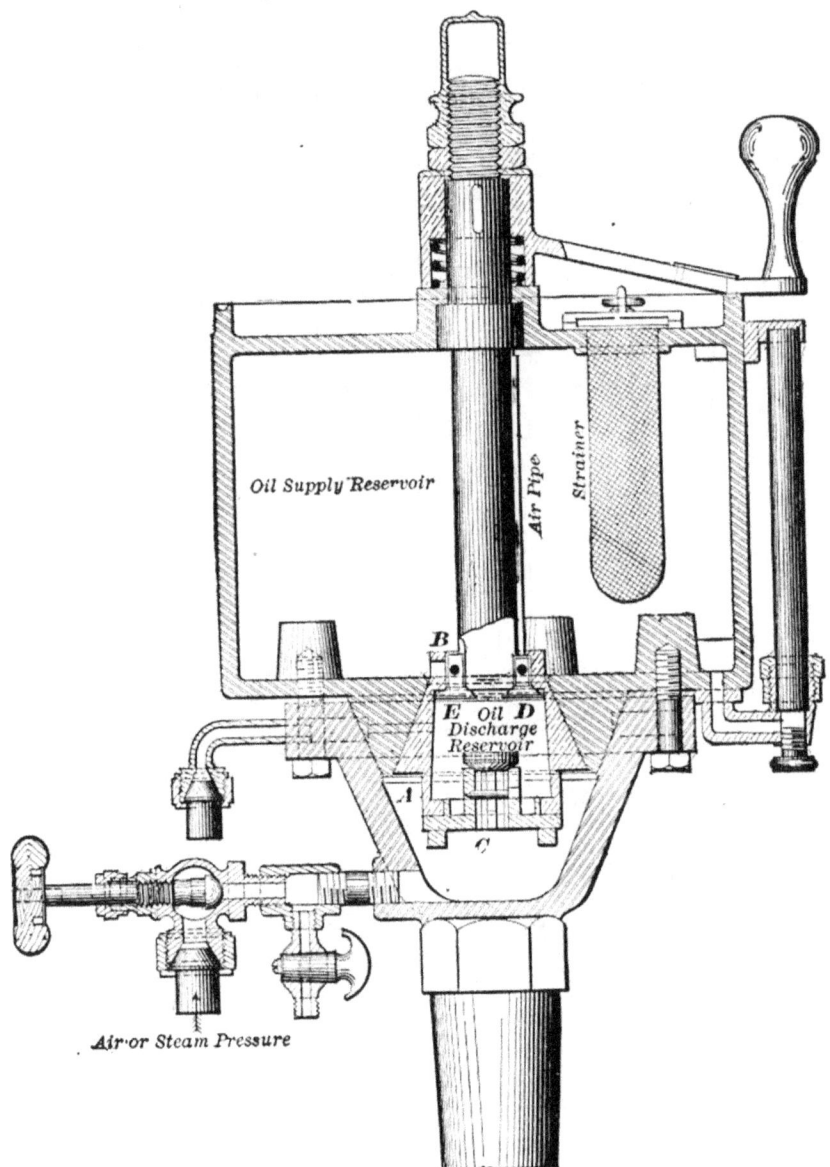

The use of this device is not intended to release the engineer from the responsibility of adjusting his present oil cups, and inspecting and oiling by hand when first taking engine out from terminal station, the same as if engine was not equipped with the force feed lubricator.

There is simply placed at the disposal of the engineer, a gallon of oil that can be forced from the cab to any desired bearing, as occasion requires.

While this lubricator was originally designed for emergencies and long distance runs, it is now being used for oiling at all times all bearings having pipes leading thereto, and is showing saving in oil over hand oiling. It has also been found a great convenience in winter to be able to blow steam to the various parts of the running gear and thereby melt accumulated snow and ice, and thus have cups and bearings in condition for oiling.

There are now a number of locomotives running equipped with these lubricators, and it is claimed some have been in service for four years and have never failed to perfectly perform their work, and none have required repairs of any kind to lubricator or any piping.

The piping may be done with either one-eighth inch wrought iron or copper pipe.

This device was placed upon the market by M. C. Hammett, of Troy, N. Y.

STEAM AND AIR GAUGES.

STEAM GAUGES.

Most all engineers and firemen depend daily on the steam gauge on the boiler head to know the pressure in the boiler, and they are aware that the gauge gets out of order. The engineer, knowing that the engine is not pulling what she should, and thinking that the gauge shows more pressure than is actually on the boiler, reports the gauge. The reported gauge is taken off, taken to the test gauge, and found to be out a little or none. Yet the engineer was correct in thinking the gauge to be incorrect, even when the test gauge did not locate the trouble.

It is possible, owing to the location of the gauge, to have it in error from one to twenty pounds, and even more in extreme cases, and yet have the gauge show correct every time it is tested with the test-gauge. And even if a new gauge, correct with the test-gauge, be applied, under certain conditions, it will be in error in indicating the pressure.

Most steam gauges have something in the form of a horse shoe inside, and this horse shoe is a hollow tube. The pressure in the tube tends to make the horse shoe open further between its two ends. Light connections are fastened to these two ends and a steam gauge hand is attached whereby the opening and closing of the ends of the horse shoe tube cause the hand to turn around. The hand shows on the gauge dial the number of pounds pressure that causes it to move. There are also forms of gauges other than those having the hollow horse shoe tube. Some have a diaphragm which is caused to move by the different pressures that are applied, and a suitable mechanism is applied to transmit the movement of the diaphragm to the steam gauge hand.

Steam gauges are adjusted cold. The fine mechanism is adjusted to resist certain pressures and correctly indicate the same when all parts are at a certain temperature. Almost every one's general experience has taught that metal becomes weaker as it gets warmer—say from fifty degrees upward. Accordingly if a gauge be adjusted and tested with the cold test-gauge at sixty degrees temperature, and then put onto a hot boiler where the gauge is at one hundred degrees or more temperature, the fine hollow tube will open more with the same pressure, than it would when cold. The greater the pressure the more the horse shoe tube will open, and if the tube be weakened by heat the same pressure will make the hot gauge show more pounds on the dial than would a cold gauge.

In order to keep the steam from the working parts of the gauge and thereby keep the gauge cool, a water column is used, called a syphon, wherein the steam condenses. This syphon pipe extends above the gauge, around it, or in a coil, and retains the cool water for the working parts of the gauge.

For an experiment, shut off the gauge from the boiler (the boiler having a known pressure), let the water out of the water column, make all connections tight, and then turn on the steam. If there is eighty pounds pressure in the boiler, the steam entering the gauge and heating all parts will cause the gauge to indicate much in excess of the actual pressure of eighty pounds. As the water fills the column and the inside parts of the gauge cool, the hand will gradually move back to the correct indication. If a connection in the gauge leaks and lets the water out of the hollow tube, the steam getting into the tube will cause the hand to show a greater pressure than is in the boiler.

A very deceptive thing about a hot gauge is this: If the gauge is too near the boiler, the great heat from the boiler will heat all parts of the gauge, including the water column, and cause the gauge to indicate much greater pressure than is in the boiler. Take this very gauge to the cool test-gauge and it will show correct indication. Remember that the gauge gets cool before it is tested with the test-gauge; therefore it will show correctly. If the test-gauge is hot it will also be incorrect.

Unless a second gauge be applied to the boiler, and this second gauge be kept cool, it is not likely that the engineer or fireman will ever know how much the hot gauge is in error. If a safety valve that is set at a known pressure is applied, the error in the gauge may be detected by the popping of the valve. If the steam gauge is too hot to bear the bare hand on any part with ease, it should be moved to a cooler place. Sometimes by placing small pieces of wood between the gauge and the stand, thereby allowing the air to circulate and keep the gauge cool, matters may be greatly helped.

The freezing of water in the gauge will strain the fine parts, thereby causing the gauge to be incorrect when water under pressure is applied.

Short water columns are not as sure to give correct results as long ones. The column should always extend above the top of the gauge. This will ensure the application of water to the gauge, if the connections are tight.

The steam gauge pipe should always be connected direct to the boiler. If it be connected to the boiler head fountain, the steam turret, or any device that has a number of boiler head fittings attached, it will be governed by the pressure in this fountain. When both injectors are working, they taking steam from the fountain, there will be several pounds pressure less in the fountain than in the boiler. This may be detected by noting the boiler pressure when both injectors are working, and then shutting off both injectors at once and noting the instantaneous rise of pressure by the gauge indication. While the boiler pressure may rise after shutting off both injectors, it rises gradually and not instantaneously.

DIFFERENT FORMS OF GAUGES.

Two forms of steam gauges are in general use upon locomotives, one using a diaphragm and the other a hollow seamless brass tube in the shape of the letter C, having an elliptical

UTICA STEAM GAUGE.

section. We have shown an interior view of the three different styles of gauges, of the most improved type.

In the Utica gauge the pressure is behind the diaphragm; the face of the diaphragm is corrugated and a bell crank lever bears against its center. The interior mechanism is so arranged that the slightest pressure inside of the diaphragm will cause it to bulge out, which will affect the position of the hand or pointer.

In the other form of gauge like the Crosby and Ashcroft, the pressure acts on the inside of the brass tubes, which has a tendency to straighten the tubes, thus spreading the ends to which is attached a bent lever which imparts a movement to the hand or pointer by the mechanism shown. The old style gauges of this type had but one tube, but by having the lever attached to the end of two tubes as shown by these improved gauges, the distance the lower lever moves is double what it would be if attached to only one tube.

The steam pipe which connects the gauge with the boiler

ASHCROFT STEAM GAUGE.   CROSBY STEAM GAUGE.

should be bent in the shape of the letter S for all gauges to permit the steam to be condensed in the pipe, as the steam pressure affects the elasticity of both diaphragms and tubes.

THE CROSBY THERMOSTATIC GAUGE.

It is a well known fact that owing to the location of the steam gauge upon a locomotive, its parts become so heated and expand to such an extent that it will make wrong records of pressures. It has been customary to place three small blocks of wood back of the steam gauge, but it was found that even this precaution did not entirely annul the evil effects of expansion. To overcome this defect the Crosby Steam Gauge and Valve Co., of Boston, have placed upon the market a thermostatic water back gauge. The parts which materially effect the correct operation of the gauge are tube springs. It occurs thus: The tube springs having been tested and adjusted to a certain movement under pressure in the ordinary temperature of the factory or where it takes place, will, when the same are heated in use to a high temperature, lengthen by expansion to such an extent that, when they are subjected to the same pressure, their free ends will move through a larger arc than when they were tested. This movement, multiplied by the ordinary mechanism of a steam gauge for transmitting it, causes this increased pressure to appear upon the dial. In such a heated condition of the tube springs, the air produced is sometimes considerable, being several per cent greater than the true pressure, thus deceiving the user of steam into the belief that he is getting a less result, in work, from the indicated pressure than he ought.

This error can be corrected by suitable mechanism in the steam gauge; and the Crosby Company has produced such an one, which, by removal of the dial, is internally shown by the accompanying cut. In the ordinary steam gauge, the bar which transmits the movement of the free ends of its tube springs is made of a homogeneous metal, and when the tube springs are affected under heat, as above stated, it transmits the increased movement just in the same way that it would transmit the intended or designed movement when the tube springs are cold. Thus the error arises. In the improved gauge here shown, this bar is made of brass and steel brazed together, forming a thermal bar C, D, so that, under the influence of high temperatures, it will compensate for the expansion or lengthening of the tube springs and their greater movement thereby under pressure, by retarding simultaneously the motion of the index which records such movement on the dial. The action of this thermal bar is, that its end remote from that where it is attached to the tube springs will drop, or deflect, or move oppositely to the tube springs on account of the action of the temperature upon the two metals composing it, as is commonly understood. This opposite movement retards the index proportionately to the lengthening of the tube springs, as they are both influenced by the same temperature, and thus compels it to keep back to the notation of pressure on the dial where it correctly should be.

In addition to this thermal bar, this gauge has a chamber, A, so constructed that when filled with water or other liquid it not only supplies the Bourdon tube springs B, B, connected to it with all that is required, but serves to equalize the temperature about them. This is important. For unless the tube springs are subjected to a heat greater than 212° Fahr., they do not set when in use; and as it is impossible as made, under ordinary conditions of use, for heat to be transmitted by conduction to such an extent they are secure from this danger. This chamber is located in the gauge case so that it has its connection to it and with the boiler at the bottom. Attached to it are the tube springs, B, B, the index mechanism, E, and the dial, the latter upon the bosses F; and all are independent of the case and are free from any influence of it under heat, excepting at its immediate point of attachment, which is unimportant.

AIR BRAKE GAUGES.

We present herewith two views of the Utica air brake gauge, which is a thoroughly modern gauge. It shows the pressure on the train line pipe and at the same time the pressure in the reservoir. The improvement of this gauge over others is the double spindle. The hands are easily distinguished even at night and no large center bearing is necessary as is the case where one spindle passes through the other. The center spindle of any gauge should be very delicate, as the slightest friction at this point would impair its accuracy.

UTICA AIR BRAKE GAUGE.

TESTING GAUGES.

Most large shops have a hydraulic apparatus for testing gauges. In order to test a gauge intelligently the operator should understand the construction of the gauge to be tested, and the action of the pressure upon the gauge. A gauge is said to be *heavy* when it indicates less than the test gauge, and it is called *light* when it indicates more than the test gauge. A light indicating gauge is often caused by the diaphragm or tubes becoming set or extended, in which case the hand will not come back to the pin, but will indicate 10 or 12 pounds pressure when there is no pressure. By slightly springing the tube or diaphragm will often remedy this evil. A heavy indicating gauge may be caused by the tubes or diaphragm not moving the proper distance at which the dial was marked off at the time it was made and tested, or it may be caused by the dial moving around. In a diaphragm gauge any dirt that would get between the bell crank and diaphragm would cause a heavy gauge. Another cause of this defect is a choking up of the proper pressure opening; lost motion in the mechanism will also cause a heavy indicating gauge. See that the tubes or diaphragm do not leak, and remember the small spring is capable of adjustment. The practice of resetting the hand or pointer is a bad one, and should only be resorted to when all other remedies have failed.

CROSBY PRESSURE GAUGE TESTER.

Mercury columns have long been accepted as the standard for measuring pressures, but are so expensive and difficult to

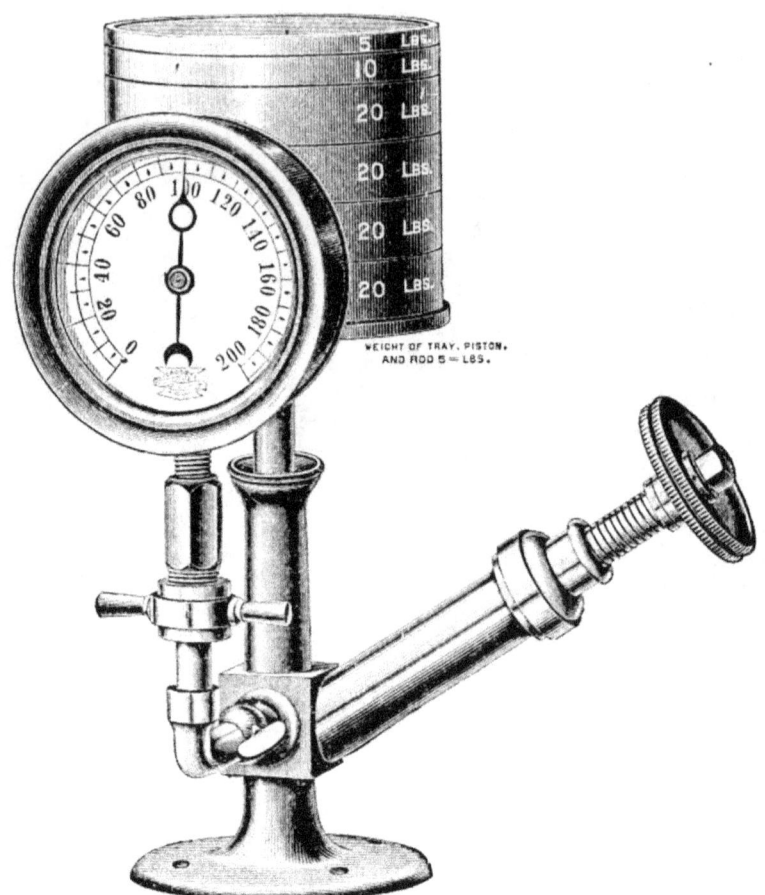

keep in order that a more simple, inexpensive, yet accurate machine is required.

The cut herewith shown represents an improved patent gauge tester. It consists of a stand from which rises a cylinder, having accurately fitted into it a piston with an area of exactly one-fifth of a square inch, which moves freely up and down. Attached to the top of the piston rod is a disc for supporting the weights; each weight is marked with the number of pounds pressure per square inch it will exert on the gauge. From the bottom of the cylinder two tubes project; one forms a standard for holding the gauge to be tested and is furnished with a coupling to connect it and with a three-way cock; the other rises at an inclination and forms the reservoir for oil, having within it a screw plunger for forcing the oil inward or outward.

The machine is set in a neat, strong case, and may be easily carried from place to place. The weights for testing up to 300 pounds are brass cases filled with lead, and are securely packed in a case with strong handles; the additional weights for testing up to 1,000 pounds are of iron.

It is designed and constructed on scientific principles and is a standard of mathematical accuracy.

DIRECTIONS FOR USING.

First, couple the gauge to the arm, using one of the connections furnished. Set the handle of the three-way cock horizontally with "open" on top. See that the screw plunger is in as far as it will go, then remove the cap and pour oil from the can into the cylinder until it is full, then gradually withdraw the plunger and continue pouring in oil until it is out nearly to its limit and the bore of the cylinder is nearly full; then replace the can in the box, with its open nozzle under the cock.

Now insert the piston, which, with its disc, will exert a pressure on the gauge of exactly five pounds. The weights,—one at a time,—may now be placed on the disc, which should be gently rotated to ensure perfect freedom of motion to the piston. Each weight added will exert a pressure on the gauge equal to the number of pounds marked on it.

If, in testing a large gauge, the piston descends to its full length, screw in the plunger and the piston will be forced upward and more weights can be added, as may be required by the limit of pressure marked on the gauge dial.

When the test is completed, remove the weights,—one at a time,—and as the piston rises, withdraw the plunger to make room for the returning oil.

When all the weights have been removed, turn the cock handle to a vertical position, with the end marked × upwards, which will allow the oil in the gauge (but not in the cylinder and pump) to drain into the can.

The oil may be left in the machine, but the piston should be carefully wiped and replaced in the box.

When it is desired to drain the whole machine of oil, set the handle of the cock with the "drain" side upwards and it will all run out. Then turn the cock with the "open" side upwards and so leave it. When setting the cock for another test, see that the open nozzle of the can is under it to catch the drip. Keep the cap on the cylinder when not in use to exclude dust. Use nothing but good light mineral oil, and keep it entirely free from grit.

LOCOMOTIVE POP SAFETY VALVES.

The three forms of locomotive pop safety valves herewith illustrated are the pop valves in most general use upon American locomotives. We have given a brief description of each form. They are all three good reliable pop valves, but, of course, the manufacturer of each valve claims their valves are superior to all others.

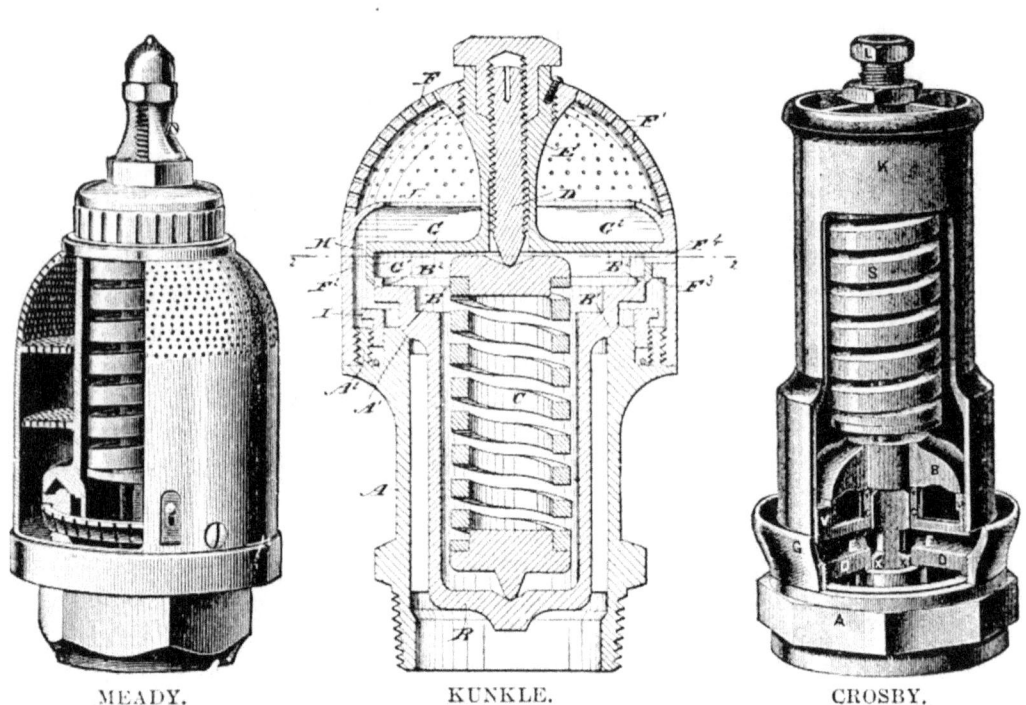

MEADY.   KUNKLE.   CROSBY.

DESCRIPTION OF THE CROSBY VALVE.

The valve proper B B rests upon two flat annular seats V V and W W on the same plane, and is held down against the pressure of steam by the spiral spring S. The tension of this spring is obtained by screwing down the threaded bolt L at the top of the cylinder K. The area contained between the seats W and V is what the steam pressure acts upon ordinarily to overcome the resistance of the spring. The area contained within the smaller seat W W is not acted upon until the valve opens.

The larger seat V V is formed on the upper edge of the shell or body of the valve A. The smaller seat W W is formed on the upper edge of a cylindrical chamber or well C C, which is situated in the center of the shell or body of the valve, and is held in its place by arms, D D, radiating horizontally, and connecting it with the body or shell of the valve. These arms have passages E E for the escape of the steam or other fluid from the well into the air when the valve is open. This well is deepened so as to allow the wings X X of the valve proper to project down into it far enough to act as guides, and the flange G is for the purpose of modifying the size of the passages E E and for turning upward the steam issuing therefrom.

ACTION OF THE VALVE WHEN WORKING UNDER STEAM.

When the pressure under the valve is within about one pound of the maximum pressure required, the valve opens slightly, and the steam escapes through the outer seat into the cylinder and thence into the air; the steam also enters through the inner seat into the well, and thence through the passages in the arms to the air. When the pressure in the boiler attains the maximum point, the valve rises higher and steam is admitted into the well faster than it can escape through the passages in the arms, and its pressure rapidly accumulates under the inner seat; this pressure thus acting upon an additional area, overcomes the increasing resistance of the spring and forces the valve wide open, thereby quickly relieving the boiler. When the pressure within the boiler is lessened the flow of steam into the well also is lessened, and the pressure therein diminishing, the valve gradually settles down; this action continues until the area of the opening into the well is less than the area of the apertures in the arms, and the valve promptly closes.

The point of opening can be readily changed while under steam by screwing the threaded bolt at the top of the cylinder up for diminishing or down for increasing the pressure.

The seats of this valve are flat, and do not cut or wear out and leak so readily as bevelled seats. The valve is made of the best gun metal.

DIRECTIONS.

Setting.—Screw the head-bolt which compresses the spring up for diminishing or down for increasing the pressure, until the valve opens at the pressure desired, as indicated by the steam gauge; secure the head-bolt in this position by means of the lock-nut; for regulating the loss of escaping steam, turn the screw ring G up for increasing or down for decreasing it.

Caution.—Care should be taken that no red lead, chips, or any hard substance be left in the pipes or couplings when connecting the valve with the boiler. Never make a direct connection by screwing a taper thread into the valve, but make the joint with the valve by the shoulder.

Repairing.—This valve having flat seats on the same plane is very easily made tight if it leaks by following these directions, viz.: With an ordinary lathe slightly turn off the two concentric seats of the valve and valve shell or base respectively, being careful that this is done in the same plane and perpendicular to the axis of the valve. The valve will then fit tightly on the valve shell. If no lathe is at hand then grind the valve proper on a perfectly flat surface of iron or steel, until its two bearings are exactly on a plane and with good smooth surfaces; then take the shell and grind its seats in precisely the same manner; rinse both parts in water and put together, and the valve will be found to be tight; to ascertain when the bearings are on the same plane, use a good steel straight edge. Do not grind the valve to its seats on the shell by grinding them together, but grind each part separately as above stated.

MEADY MUFFLED LOCOMOTIVE POP SAFETY VALVE.

Our illustration of the Meady muffled locomotive pop safety valve shows its internal construction. It will be observed that the valve proper projects upward through the perforated casing of the valve, enclosing within it the spring which holds it to its seat; and the upper or outward side of the valve is open to the air at all times, so that when the valve is discharging it is free from any pressure of the out-going steam, which escapes through the perforated casing into the open air without a disturbing noise.

By this design there is no back pressure on the valve, and its component parts so co-operate that the valve rises when it opens to a greater height than is usual in valves of this character.

For tension of the spring and the adjustment of the parts, means are conveniently arranged and provided. In size and utility it is believed to afford all the advantages which are demanded, and to meet all the requirements of an exacting railroad service.

DIRECTIONS.

It should never be meddled with unless it becomes necessary to reset it. In such case, first loosen or remove the acorn check nut above the spring bolt; then holding with a wrench the hexagonal top of the valve, with another wrench turn the spring bolt downward to increase and upward to reduce the pressure, until the valve opens at the desired point as indicated by the steam gauge. To modify the loss of pressure in blowing, slightly withdraw the screw bolt in the base of the valve until it ceases to engage with the ring encircling the valve, then with any pointed instrument inserted into the small opening near the screw bolt turn the ring downward for diminishing and upward for increasing the loss.

DESCRIPTION OF THE KUNKLE POP VALVE.

Our illustration shows the interior construction of this form of safety pop valve. It will be seen that the formation of main valve is entirely different from the other two forms of pop valves herein described. It is a cup valve and extends away below its seat, its seat being formed on its top flange as shown. The cup valve is held to its seat from its lowermost position by a helical spring. The spring, in turn, is held in a central position and adjusted by a stem extending down through the center of the pop. It will be noticed this stem has no head and can be adjusted only by use of a hollow pin wrench; this prevents tampering with the pop after it has once been adjusted to the required pressure. This pop valve is also provided with a disc which can be adjusted from the outside, a slot being cut in the outside shell for this purpose.

It will be seen the semispherical cap is perforated and has a curved lip formed on the inside of the cap above the inlet ports to direct the steam to the middle of the pop.

HOW TO SET LOCOMOTIVE POP VALVES.

Every locomotive boiler is, or should be, provided with two safety pop valves (not necessarily the same make). This is a safeguard in the event either pop should fail to operate. One pop is usually set to carry a few pounds more pressure than the other; however, one good pop is considered sufficient to relieve any ordinary boiler. Test each pop valve separately by tightening down the spring on the other pop and ascertain which is the most sensitive. Whichever one loses the least steam during a discharge should be used; by this we mean, when the pop springs are properly adjusted it should be the first to relieve the boiler pressure, but it is necessary to set the other one first, which can be done by screwing down the spring on the best pop and then set the other one at a pressure of two or three pounds more than working pressure; then gradually adjust the spring of the best one to working pressure. If either pop loses too much steam during a discharge it may be regulated by adjusting the disc until the discharge is normal. Pops are sometimes set by steam gauge testers, but more frequently under steam pressure. Some safety valves are so constructed that the spring can only be adjusted with a hollow wrench. These wrenches are not furnished to the engine, so it is impossible for a roadman to tamper with the pop, grade or no grade.

METALLIC PACKINGS.

The use of metallic packing in locomotive service has become so thoroughly established that it is scarcely necessary to state that since the introduction of this kind of packing the use of hemp and asbestos has been almost entirely discontinued. Various forms of packing are in use, but their success depends more upon the alloy of the metal than upon the method of application. This packing is used upon piston rods, valve stems, air pumps, and brake cylinder pistons. We have illustrated a few of the principal forms of this packing now in use.

UNITED STATES PISTON ROD PACKING.

It will be observed that this packing consists of three babbitt rings (E & F) placed in a vibrating cup (D) whose interior form

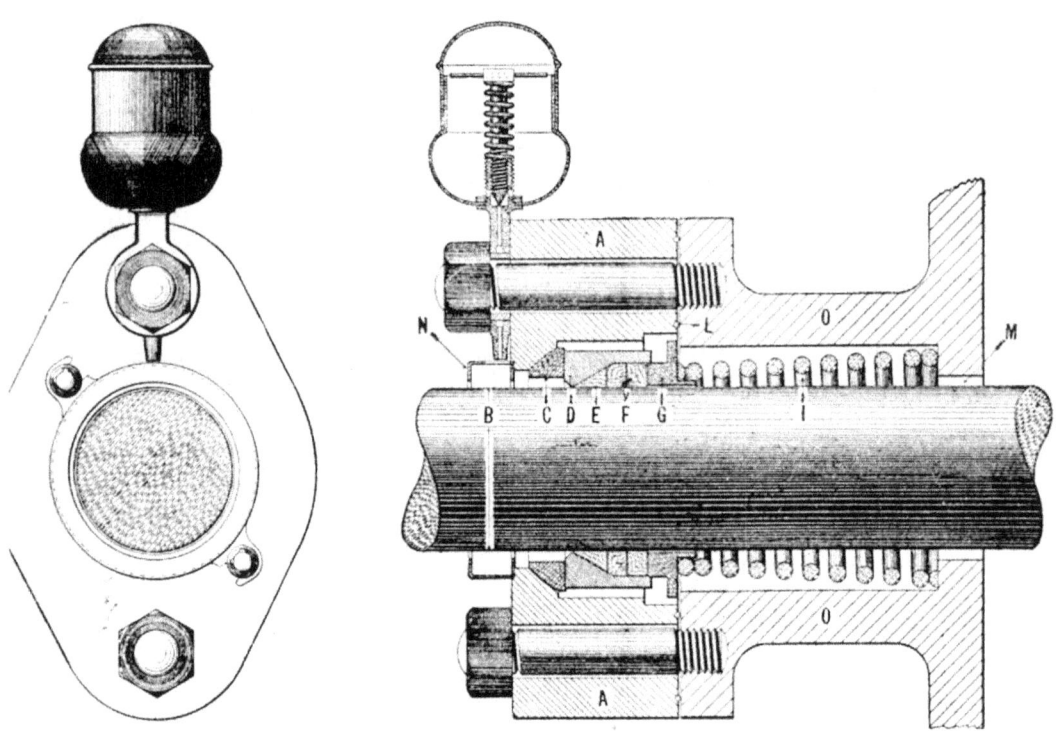

is conical. This cup rests against the flat face of the ball and socket ring (C), which in turn has its bearing against the gland or outer casing (A), the whole being held in place by the coil spring (I) and the flanged follower (G), which prevents any possibility of the parts drawing back with the rod. It will be readily seen from this construction that the surface of the rod must be in good condition and must be round and parallel. It will be noticed that care has been taken to so construct the packing that the soft babbitt rings (E and F) are the only parts which can touch the rod, the idea being that when the rod runs in contact with any hard metal, its surface naturally becomes abraded, and acts as a file upon the packing.

It will also be noticed that, by its construction, the packing may have a direct sliding movement upon the face between the vibrating cup and the ball joint, or may have a rocking motion between ball joint and case, or may combine both at once. By this device the packing never binds the rod, but follows it in all its vibratory movements. Thus the friction is reduced to a minimum.

The lubricating device consists of an oil cup feeding into swab cup (N).

This packing is known as the locomotive or cone packing.

A—Packing Case, or Gland.
B—Piston Rod.
C—Ball Joint.
D—Vibrating Cup.
E and F—Babbitt Rings.
G—Follower.
I—Coil Spring.
L—Copper Wire Joint.
M—Clearance Room through cylinder head.
N—Swab Cup.

UNITED STATES VALVE STEM PACKING.

By reference to the cut of the piston rod packing it will be noticed that the locomotive valve stem packing does not differ apparently in any essential particular from that described and shown on preceding pages. A bush is placed in the stuffing box to carry the weight of the stem and yoke. It is quite natural that in time this support (K) should wear a flat place on the under side of the stem. This flat place would be the length of travel of the stem, and if our packing was placed close to the support, this flat place would travel into the packing and thus cause a leak. Therefore, the distance between the bush and the babbitt rings is equal to the travel of the stem, thus obviating the difficulty; the preventer (8) accomplishing the same results as the flanged follower of the piston rod.

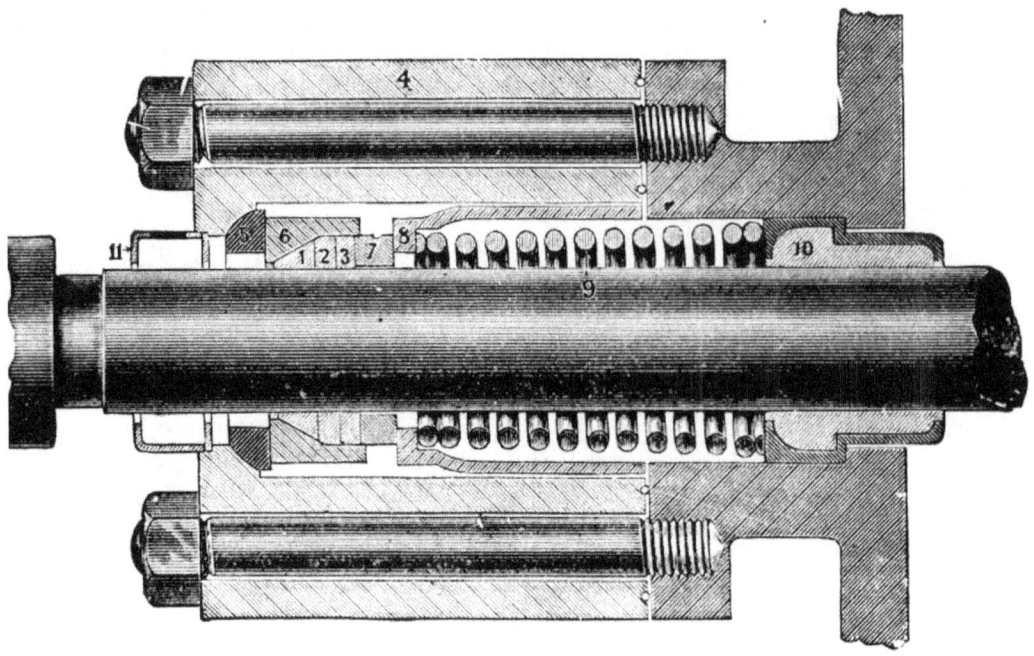

UNITED STATES AIR PUMP PACKING.

It will be noticed the construction of this form of packing is similar to that of the piston and valve stem, the form of the

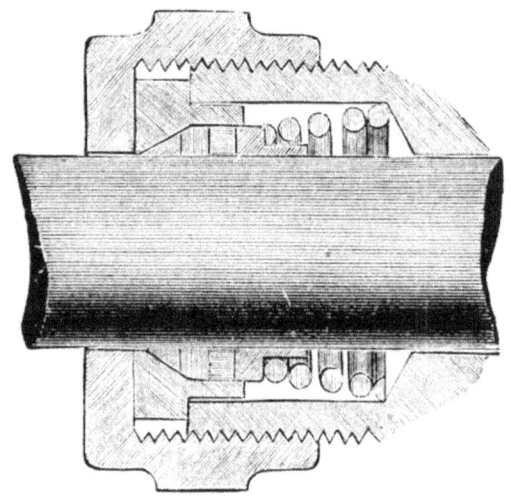

gland only being different; it will be seen that the cone is made small enough to vibrate in the stuffing box, and has a collar which prevents the gland being screwed up tight enough to compress the springs.

THE JEROME METALLIC PACKING.

As may be seen by the illustration this form of packing ring is easily applied; the ring being partly severed, it can be replaced without disconnecting the piston. The shape of the rings also

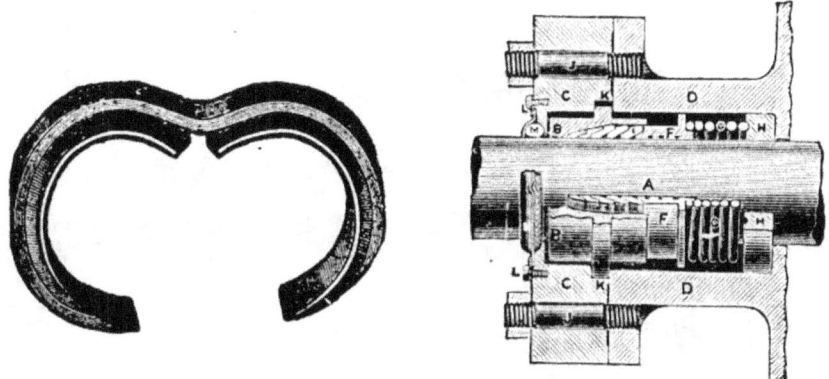

differ from the United States Packing, but otherwise the construction is similar to that already described. This packing has been adopted by many roads.

THE HARRIS PACKING.

Our illustration presents an entirely new form of cylinder and valve stem packing, the principle of its action being quite different from any other form of locomotive packing now in use. Unlike most forms of packing, no spring is used to secure a steam tight fit between the packing rings and the piston, the steam pressure alone keeps the packing tight on the piston, and when no steam is being used in the cylinder the pressure is removed from the packing. Engineers and mechanics will readily perceive the advantage of such an arrangement at early points of cut-off or when drifting down grade. With all other forms of packing with springs the pressure is constant, and therefore the wear on both piston and packing rings.

ITS CONSTRUCTION.

This packing is virtually the Dunbar Cylinder Packing inverted, the rings being L shaped and cut in several pieces. No cone is used, but a wide flat steel band is used to hold the packing in position; the diameter of this band equals the outside diameter of the packing rings, and is made of very light spring steel; it is cut diagonally and can be opened up and slipped over the piston rod without disconnecting, thereby permitting a new set of packing to be applied in a few minutes. A bald joint ring is used with this packing, the same as that used by the United States Metallic Packing, which permits the piston to vibrate; a ring is used in the stuffing box to prevent the packing from entering the cylinder, but the steam is permitted to pass through the stuffing box and circulate around the packing. No lathe work is necessary, the rings being cast in moulds ready for use. It is claimed the piston will not wear out of round with this packing and that it is giving perfect satisfaction, one engine having

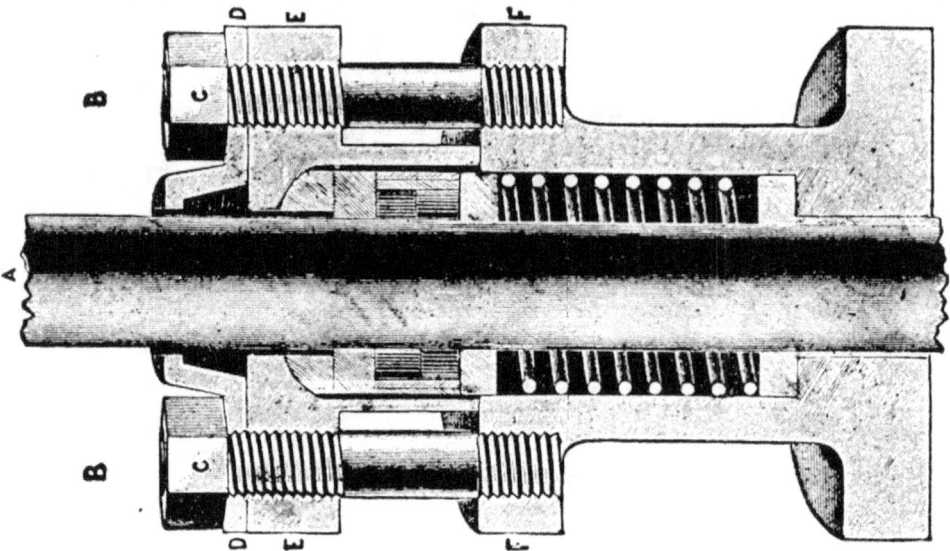

made 75,000 miles up to the present time without the least sign of trouble with packing. It was invented by Mr. Harris, of Louisville, Ky., and is in use upon several engines on the L. & N. R. R.

THE DETROIT VALVE STEM PACKING.

The illustration of this form of packing is self-explanatory.

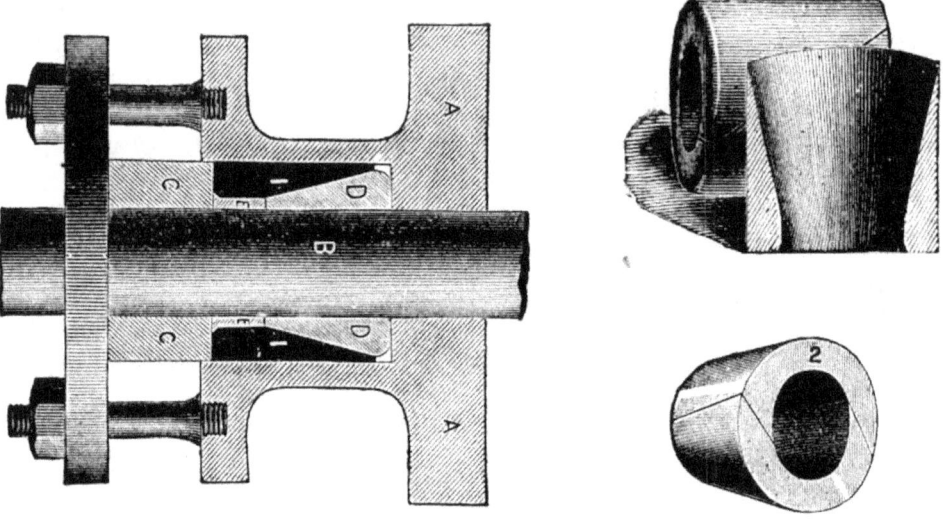

Two rings are used, a male and a female, and the gland can be tightened as the packing wears. This form of packing is not extensively used upon locomotives.

DOUBLE PACKING.

The double packing is practically two sets of United States packing of the locomotive form, arranged in tandem. Instead

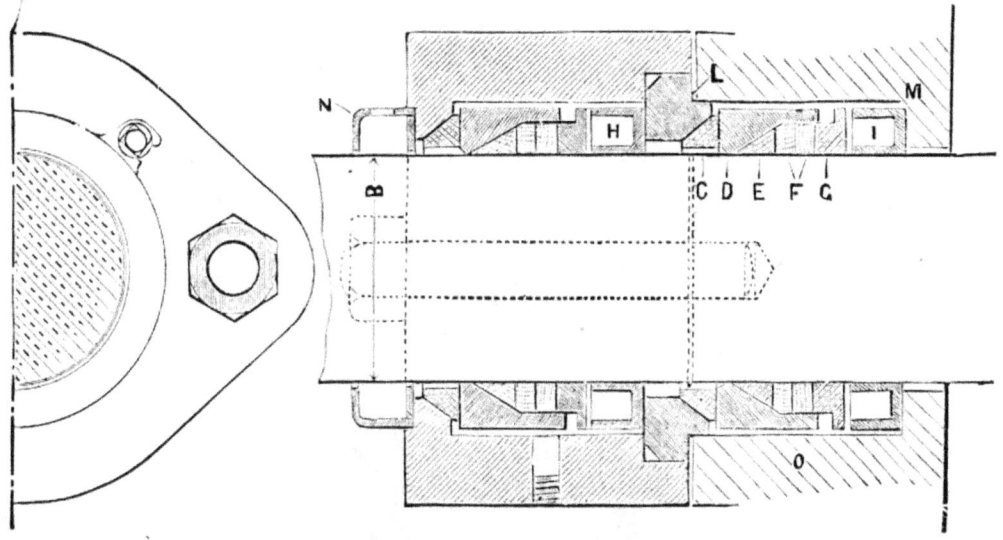

of using the long heavy coil spring for holding packing in place a series of small coil springs are used. This style of packing is used upon very large piston rods.

GENERAL INFORMATION.

TINKER'S LOCOMOTIVE CAB WINDOW.

This device is the invention of Mr. H. W. Tinker, of Boston, Mass. The illustration shows the storm window complete. The improvement consists in attaching to the inside of a regular cab door a specially designed window so constructed as to form a water tight space about 5-8 inch in width between the two panes of glass, which space is filled with water, or, if preferred, any other suitable transparent liquid. The water is heated to a sufficient

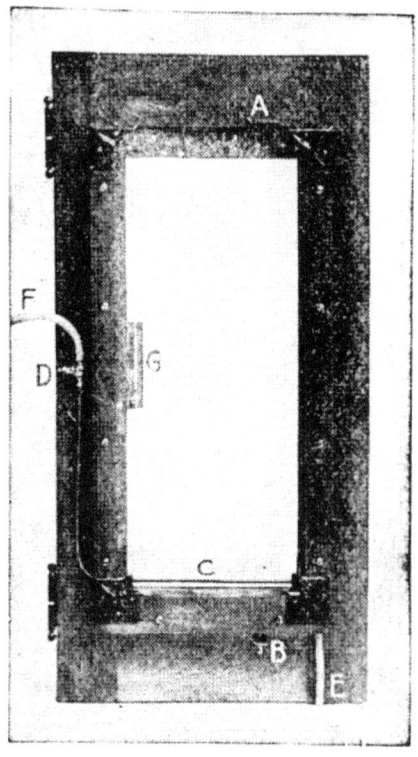

temperature to warm the glass sufficiently so that the snow, ice, frost, etc., will not adhere to its surface, thus providing a clear glass in front of the engineer during the worst storm or coldest weather. The necessary warmth is imparted to the water by means of a heated tube between the glasses, this tube being heated by a small jet of steam passing through it.

The steam is admitted from the boiler through pipe F, passing through brass tube C, and running to waste through rubber tube E, which extends below the running board of the cab. The amount of steam passing through tube C is regulated by cock

D, only a very small jet of steam being required to heat the tube sufficiently to impart the necessary warmth to the water enclosed between the glasses, and this amount can readily be determined by a very little experimenting on the part of the engineer. A thermometer scale is placed between the glasses and immersed in the liquid, by which the temperature of the water can be determined at a glance. The usual temperature required is from 100 to 120 degrees; that is to say, the water needs to be only fairly warm to accomplish the desired result. The space between the glass is filled by introducing the water at cup located at the top of the frame and marked A in the illustration. To remove the water open cock B, at the bottom of window.

The space between the two panes of glass is made water tight by the use of a strip of moulded rubber. Especial attention is called to the fact that there is no pressure between the glasses; that the steam does not come in contact with the water; that the water is only fairly warm; that only a small jet of steam is required; that the appearance of the window when filled with water is like that of one pane of plate glass and not one person in ten would discover the fact that there are two plates of glass with water between. It requires scarcely any attention on the part of the engineer to keep the water at the required temperature, one engineer stating that he hardly ever had occasion to touch the steam regulating cock but once during his run of 128 miles, this time being at a covered station where the train made a stop of ten minutes.

This device has been tried upon the Concord & Montreal, Maine Central, Boston & Maine, and Lake Shore & Michigan Southern Railroads. Judging from the endorsements from engineers who have used this window it should prove a valuable device where the winters are severe.

A NEW FEED WATER HEATER AND CYLINDER LUBRICATOR.

This device is the invention of Messrs. Wallace & Kellogg, of Altoona, Wis., who have kindly furnished us with the following description and claims for their device:

The exhaust-pipe from air-pump branches off near pump, one pipe leading to front end near smoke-box, then branching off, each pipe leading to live-steam ports or steam-chests; said pipes are supplied with check-valves located near steam-chests. The other pipe leading over cab into condenser, coils in tank. When engine is working steam in cylinders, check-valves in said branch pipes close automatically, throwing exhaust steam from air-pump through exhaust-pipe leading over cab, and into condensing coils in tank. The extreme end of said coils is provided with a drain-cock, leading through bottom of tank, thereby providing a means for the condensation to escape. The exhaust-pipe leading over cab is provided with a three-way valve, thus providing a means of throwing exhaust steam to atmosphere at the will of the engineer. When steam is shut from cylinders and engine is at rest, or drifting down grade, check-valves in exhaust-pipe from air-pump open automatically, letting exhaust steam from air-pump into steam-chests and cylinders. Live steam ports are provided with automatic drip-valves situated at lowest point under cylinder saddle, which drain condensation from air-pump exhaust when engine is at rest.

The claims we make are as follows:

First. It is perfectly noiseless. This avoids annoyance to passengers, and the frightening of teams around stations. Consider this advantage alone.

Second. It does not create a draft on fire when air-pump is working, as does the old method of putting exhaust in stack. This saves fuel.

Third. It acts as a lubricator to cylinders and valves when engine is not working steam. The exhaust steam and oil from air-pump circulate in steam-chests and cylinders, keeping them at a uniform temperature, not chilling off in cold weather when engine is at rest, nor overheating, burning off oil on cylinder walls, nor cutting cylinders when engine is drifting down grades, by friction of piston traveling to and fro.

Fourth. It is beneficial to the working of air-pump as there is a partial vacuum formed in steam-chests and cylinders when engine is drifting down grade shut off. It also causes a partial vacuum in exhaust-pipe from pump to cylinders when engine is working steam in cylinders and air-pump is exhausting into condenser in tank. There is also a partial vacuum formed in exhaust-pipe leading to same; this does away with back pressure entirely and gives a free working pump.

Fifth. The relief-valves or suction-valves on steam-chests can be dispensed with, as the air-pump exhausting into same performs their functions.

Sixth. When engine is working steam the air-pump exhausting into condenser acts as a feed water heater, taking waste steam to heat feed water, thereby making a freer steaming engine and affecting a saving of fuel.

Seventh. This appliance reduces the wear on valves, valve-seats, cylinders, etc., to a minimum, by perfect lubrication.

Eighth. Water in tank being warmer than atmosphere, tank never sweats, thus preserving life of paint on tank and keeping it bright and fresh.

Ninth. It also prevents the rusting of the iron in the coal space of tank, caused by the combined action of moisture and coal.

Tenth. This device is automatic in its operation, simple and cheap in its construction, and we guarantee it to make a saving of 50 per cent in cylinder oil and 2 per cent in fuel.

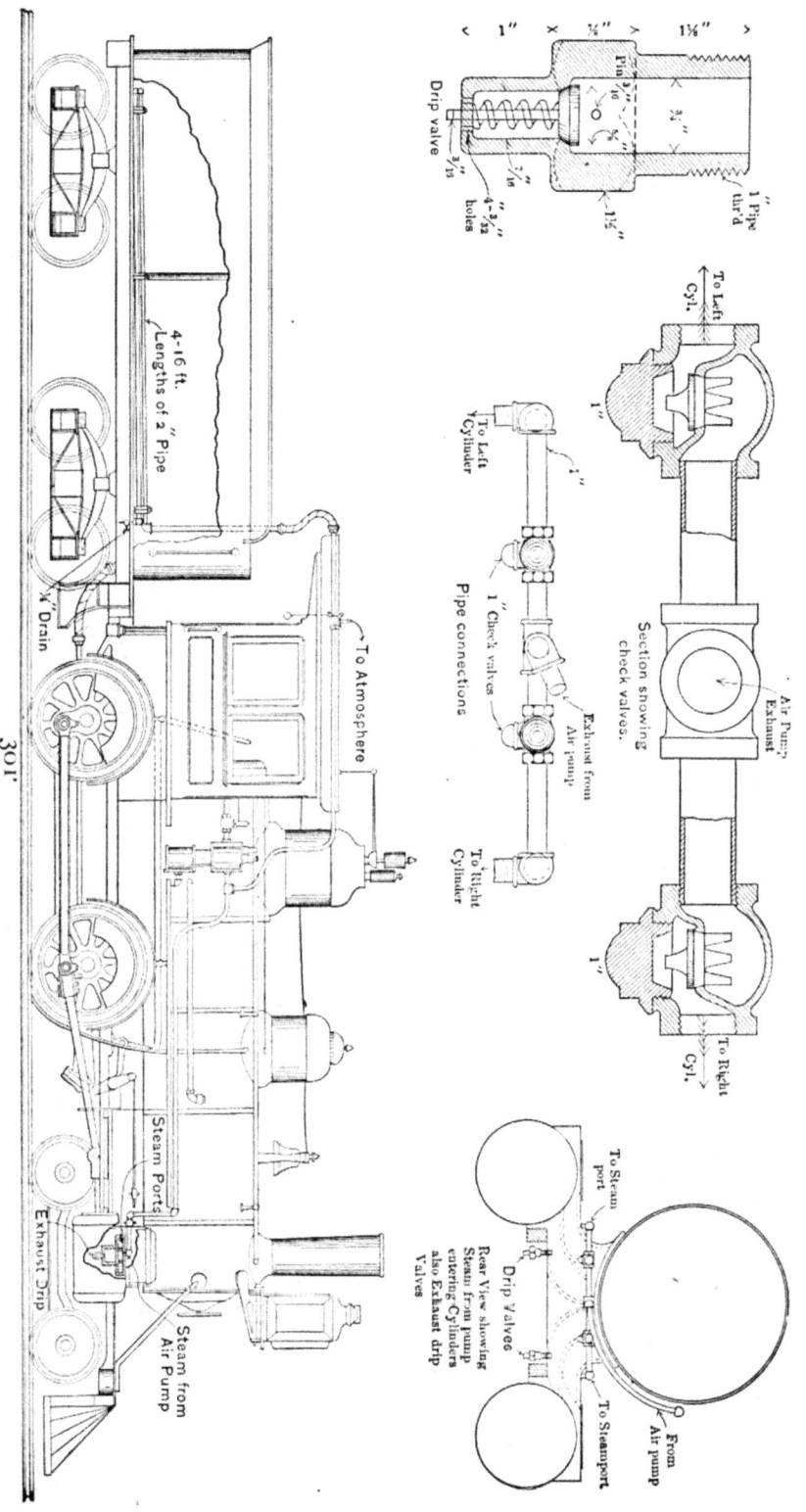

THE PITKIN DOUBLE THROTTLE VALVE.

In locomotives usually a single valve is employed in the steam dome to control the passage of steam from the dome through connecting pipes to the cylinders. On large engines in order to supply sufficient steam to the cylinders, a very large valve has been employed, and much difficulty has been encountered in opening the valve, on account of the outside steam pressure, and it is difficult to keep the valve tight, owing to the expansion of metal in the pipe, valve seat and valve, and by reason of the high pressure on the large surface of the valve.

It was to overcome this difficulty that Mr. Albert J. Pitkin, of the Schenectady Locomotive Works, designed the valve here shown. What follows is his own description of the device: I employ two valves instead of one to control the passage of steam from the dome to both cylinders of the engine, and I operate these valves successively or dissimultaneously by a single set of operating rods and levers, in order that sufficient steam may

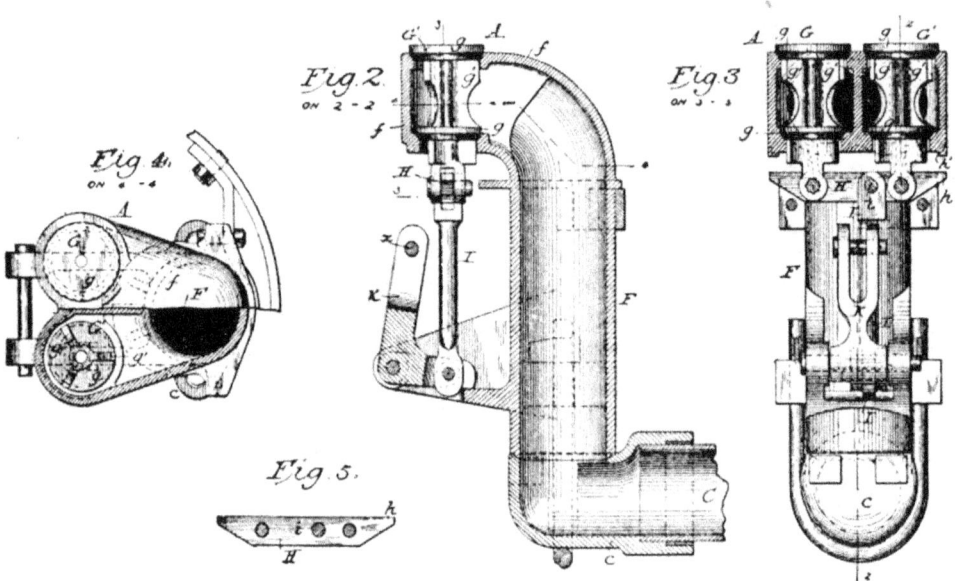

be first admitted to start the engine, and then a larger amount of steam admitted to continue the operation of the engine at high speed. By thus employing two valves, each of them may be smaller than the single large valve usually employed, but their total area is greater, so that steam is supplied to the connecting pipe to its full capacity, and the reduced size of each valve insures less tendency to leakage, or expansion of metals.

In the accompanying drawings Figure 2 is a section of the throttle valve on the line 2—2 of Fig. 3. Fig. 3 is a section on the line 3—3 of Fig. 2. Fig. 4 is a view, partly in plan and partly in section on the line 4—4 of Fig. 2; and Fig. 5 is a detail view of the valve-operating cross head.

Throttle valve is located within the dome of the engine, as usual, and pipes lead from the valve to the cylinders. A rod connects the bell crank lever K, of the valve-operating mechanism with the hand lever within the cab. The pipe C is connected by a coupling c, with an upright pipe or chamber F, which has an enlarged top portion f, extending backwardly and containing valve chambers, and seats for the valves G, G'. These may be of usual well known construction, having upper and lower heads g, and guide ribs or webs g'. Both valve chambers communicate with the pipe or chamber F, and the valves are adapted to admit steam from the dome into the valve chambers at both top and bottom. Each valve is pivotally connected with a cross head H, which is, in turn, connected by a link I to the inner, shorter arm of a bell crank lever K. The upper, longer arm of the lever K is connected to the rod at x, as indicated in Fig. 2.

It will be observed that the pin i, which connects the cross head H to the link I, is located closer to the pivotal connection of the cross head with one valve than to the pivotal connection of the cross head with the other valve. As shown in Fig. 3, the pin i is closer to the valve G' than to the valve G, so that when the bell-crank lever is first operated, the valve G' will be first opened without opening the valve G. A sufficient amount of steam is thereby admitted to start the engine, and steam is admitted to the inside of the valve G, and balances it. As the bell-crank lever continues to turn the end h of the cross head H abuts against the lug h', which acts as a fulcrum on which the cross head turns, and the valve G is thereby also opened, admitting the full amount of steam through both valve chambers to the chamber F, and thence through the connecting pipe C to the cylinders.

ANOTHER BALANCED THROTTLE VALVE.

Balanced throttle valves have been in use on engines of the Chicago & Grand Trunk Ry. for the past four years. The valve has but one seat, which is at the top of the throttle box, and its diameter is equal to the inside diameter of the dry pipe; its connections are the same as for other throttle valves. It is claimed less power is required to operate these valves and the liability to wet steam is diminished. They are known as the Pendry valves.

THE DE WALLACE TRAIN ORDER SIGNAL.

In the De Wallace train order signal, invented by Mr. Harry De Wallace, of St. Paul, Minn., the object sought for (the attainment of which has been fully demonstrated by experiments and trials on railways of the Northwest) has been to provide an accurate and reliable distance indicator and reminder signal, for use by locomotive engineers, for the purpose of preventing forgetfulness, and the miscarriage of train orders and other obligations, and further, to automatically apply the air and stop the train a safe distance short of the order point, in case the engineer fails or neglects to release the same within a given time.

To solve this problem, Mr. De Wallace has devised his purely mechanical train order signal and stopping attachment, which works automatically, is positive in its connections and action, reliable and effective in all its work, and can be supplied and maintained at a moderate cost and expense.

In operation, the device, shown in the accompanying illustration, is placed in the cab, within easy reach and in plain view of the engineer, so that he can observe, set and release it conveniently. It is positively connected with the running parts of the locomotive, whence it derives its power and motion. The preferred connection is made with one of the forward truck wheels or axle, by means of a spiral belt and "V" groove pulleys, although connections may be made at other points and by other positive means.

The revolutions of the wheels give the distance indicated by the pointer on the dials, which shows and measures the travel of the engine or train in miles, with absolute accuracy, whether running forward or backward, and without change or attention. So that by looking at the face of the machine the engineer can tell at any time the exact distance he has run from starting point, and incidentally he is informed at all times as to just where he is at, no matter whether it is light or dark, stormy or foggy.

When orders are received for meeting or passing of other trains, or orders for any other purpose, which require attention or execution at a distance ahead, the engineer, or the conductor and engineer together, set the signal by pulling out one of the triggers or dogs shown on the upper dial and hooking it over the notched edge at the graduation or notch corresponding to the mileage number which appears opposite the name of the station on the schedule card attached to the machine at the right. If he has more than one order, then several of the triggers are used in the same manner, at the same time. The illustration shows settings for four supposed orders: One at twenty-seven miles for Anoka; a second at fifty-six miles for Becker; a third at seventy-five miles for St. Cloud, and a fourth at ninety-one miles for Avon. There are still eleven idle triggers left that may be used for intermediate settings, in case they should be desired or required. In this way, under the present construction of the device, as shown, it will take care of fifteen distinct orders or settings at as many different points. These settings then remain in their respective positions until the hand, which rotates between the two dials, reaches each of the triggers in the order of their setting, and in passing beneath them trips them off, when, the next instant the signal, which is a small whistle, blown by air pressure, begins to sound. The whistle continues to blow during the time it takes the train to run over the first quarter of a mile from the point where the setting is released. If the whistle is not silenced by the engineer within that distance, then the device automatically releases the air from the train pipe and stops the train, independently of the engineer's action or assistance. In other words, it takes the control of the engine and train away from the engineer for the purpose of stopping the same, in case he fails, for any reason, to release the signal before the brakes are set.

The engineer, to release it, should pull out the button "A,"

which projects beyond the center of the dials, and thus retain control of his train. By pulling this button, the further action and opening of the valve will cease. To quiet the whistle, he should thrust in with the finger the button "B" at the lower right hand corner of the case. This slight act will cause the valve to instantly close and prevent the setting of the brakes. At each setting the signal and other automatic acts of the device will be repeated, without further attention, except the releasing by the engineer just referred to.

An engine may engage in switching for indefinite periods of time and distances and no attention or change is required. But if the shift button "D" is pulled out, the engine or train may back up for long distances, and the hand or pointer will be kept going ahead, indicating the miles to correspond with the numbers on the card, and thus enable the engineer to set the same for orders with the same facility when backing up as when going ahead. For ordinary switching this button is not used and the machine will take care of itself. The shift is used merely to save the engineer the necessity of figuring out the backward mileage for setting purposes, as then he can use the card mileage instead.

In starting out on a run the hand should be set as shown in the cut. It should also be set at the same point when the return run is begun. It may be set at any point shown on the dial if desired.

The schedule card bears the names of the stations, sidings, railroad crossings and any other points that may be required for signal purposes. The device may be set for signal on any mile not shown on the cards, there being no disputed or blank mile on the round of the dial. The mileage on the card begins at starting point with zero, and increases in even miles to the end of the run, the same as the hand shows the increased travel of the train. The other side of the card contains the stations and mileage in the reverse order. Each machine is to be equipped with separate cards for each of the different divisions or runs of a road, then when engines are changed from one run to another, only the cards require to be shifted in the holder to make ready for the new run. The card mileage to each station is from one to two miles less than the exact distance, owing to whether it is level or down grade approaching the points named. The figures of the card are used in setting for signal, and are intended to be at least one mile short of the near switch. This is done so that the signal will sound a mile from the point where the train may have to take the siding, and then the automatic stopping of the train, if that should be permitted, will occur before that switch is passed. The letters on the card are intended to be used by the dispatcher, and inserted after the station names in the orders, as a check, to prevent his giving the wrong station for a meeting point. If the check letter and the station name do not agree, then he has made a mistake and those receiving the order will detect it.

By using this device, two trains approaching a station in opposite directions will each be signalled from one to two miles short of the station, and if the signal is not released, the brakes will be set and the trains stopped before they can possibly come together. If they are thus reminded of their orders and duty, then they will not forget them again in so short a distance. Settings may be made for crossings, slow bridges and the like, and the effect and results be the same. If an engineer has difficulty in finding a place in the night, or storm, or fog, he can set the device, and it will locate it exactly and let him know about it. This instrument will stand a speed of 150 or more miles per hour and do good work.

The shaft marked "C," at the lower left corner of the case, is extended and connected with the belt and pulleys attached to the truck axle, and is driven by them. The movement of all the parts of the signal case is very slow, thus ensuring a long working and wearing life. The working parts are all automatically lubricated.

The signal is a small shrill whistle placed inside the case at the point marked "E" in the cut. It is blown by the air supplied by the brake valve. It can be adjusted to any pressure and its pitch varied as desired.

The pipe marked "F" connects the train pipe and the brake and signal valve of the device. The pipe "G" is where the air exhausts when released to set the brakes. When applying the air the machine makes a heavy service application, which may be varied as desired.

The inventor has added to the signal combination what he calls a "disorder signal." The purpose of this is to notify the engineer, in case the train order signal or its connections get out of order, or for any reason fail or cease to do their work. It will also give a warning when any of its own parts become disordered. This makes the invention entirely reliable. If the engineer should learn to rely implicitly upon this signal, and it should fail without his notice, this latter attachment will let him know it at once. If it signals "out of order," then for the remainder of his run he will not rely upon or use it again until repaired or replaced. Many prominent railway officials say that with this feature added, the device is absolutely reliable and free from objection.

Some time ago this invention was subjected to a very severe test upon the Great Northern railway for a period of six or seven weeks, covering a distance of over seven thousand miles. In this test it was attached to one of their large Brooks passenger locomotives in service pulling their coast train between St. Paul and Barnesville, Minn., the run being 220 miles each way and on fast time. Here it operated for one full month, and after that it operated on the same engine for a round trip from St. Paul to Portland, Ore., a distance of nearly five thousand miles. This was done with the inventor's first complete machine, and the tests were said to be a clear demonstration of all his claims for the invention and its principles.

Very recently Mr. De Wallace made a further trial of his second machine (from which the accompanying cut was made) on the Sault Ste. Marie railway between Minneapolis, Minn., and Gladstone, Mich. The distance between these two points, 356 miles, was traversed several times. The locomotive (No. 500) to which it was connected was one of their fine large Schenectadys, used in drawing their Atlantic and Pacific Limited.

In this, as in the former tests, the device measured all distances with absolute accuracy at each degree of speed. It signaled at the exact point set for upon every occasion, and when permitted to do so, it set the brakes and stopped the engine or train immediately after the signal had sounded its measured time.—*Railway Master Mechanic.*

A NEW SPRING SEAT GLOBE VALVE.

The cut herewith shown represents the Crosby spring-seat valve, with a part of the body removed, showing its internal design in section.

A, is the upper disc in which a, a, represents the seat with its conical depression terminating in an annular groove of consider-

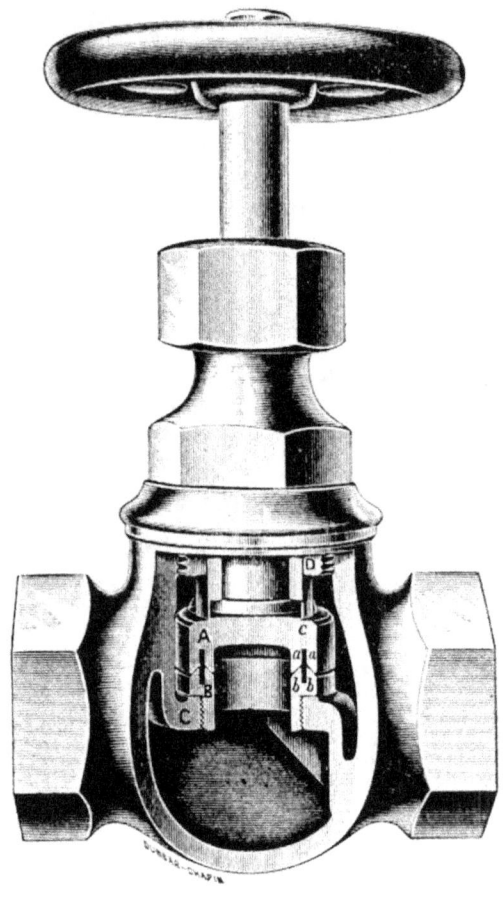

able depth. C, shows the valve body whereon at b, b, in the valve seating B, is constructed a conical seat having in the center thereof a similar annular groove.

These seats respectively of the valve disc and of the valve body so made present a greater contact surface, and by means of the central grooves have a resilient action which permits them to shut tight and remain so. This novel characteristic prevents them from jamming when the valve is closed and leaves them free to accommodate themselves to any variation of temperature. Moreover, when partially open, the outrushing steam or fluid does not abrade their surfaces as ordinarily happens with those of common valves.

All the operative or working parts of this valve are renewable, especially the upper disc A, and the valve seat B of the body C. In case either or both are injured they can be easily and quickly removed and new ones inserted.

The disc is attached to the stem of the valve in the ordinary manner; and is thus quickly detached or another attached upon removing the bonnet.

The seat of the valve body is removed by inserting any flat piece of iron in its opening so that its edges shall impinge on the small lugs which are in this opening. Unscrewing this seat and screwing in a new one is the work only of a moment and requires no special tools nor skill.

The claim is broadly made that by the peculiar design of the seating parts above described this valve will outlast any other valve; yet whenever wear or accident injures these seating parts for a few cents it can be made as good and as serviceable as a new one, without removing it from the piping or disturbing it in its use further than introducing the parts so renewed, so that when the valve is once put in use it is practically imperishable.

These statements are based on an experience covering several years, with the valves working in the most trying places and under the most exacting conditions which could be found. In every instance, reports of their use have been of the most assuring and satisfactory kind, and it needs only a trial of one or more of them in a difficult service to convince any one of their great superiority over all other valves of this kind.

The valve is made either of brass or of iron.

It is manufactured by the Crosby Steam Gauge and Valve Co., of Boston, Mass.

BOILER LAGGING.

The use of wood for boiler lagging has been discontinued upon many leading roads. It was found to be a poor non-conductor. Many master mechanics use a preparation of their own for this purpose. It resembles plaster paris, the ingredients of which differ according to the notion of the official.

Magnesia is considered one of the best boiler laggings in use. It is much cleaner than the plaster preparations. It can be destroyed only by abrasion. It is claimed it never chars like wood, or gets hot like asbestos, but when properly fitted that it will last as long as the boiler. The sectional lagging can easily be removed for inspection purposes.

CHIME WHISTLES.

The peculiar merit of this whistle consists in producing three distinct tones pitched to the first, third and fifth of the common musical scale, which harmonize and give an agreeable musical chord. It is more penetrating than the common whistle, and can be heard at a greater distance. It effectually obviates the harsh, disagreeable noise which has been a source of common complaint in other whistles. This whistle is used extensively on locomo-

tives, and is warmly indorsed by railroad men and by the traveling public. They overcome one of the chief annoyances of railway travel, and serve to distinguish passenger from freight trains. The difference between a chime and plain whistle is in the construction of the bell. The bell of a chime whistle has three different partitions, each of different depth.

ELECTRIC HEADLIGHTS.

While electric headlights are in use upon many roads they have not been universally adopted. They produce a very brilliant and powerful light, yet most roadmen oppose their use, claiming they are very injurious to the eyesight.

The B. R. & P. engine shown on page 481 is equipped with one of these headlights.

THE LOCOMOTIVE TESTING PLANT.

The locomotive testing plant is a machine upon which locomotives may be tested in the shop or round-house before they enter road service. Some few years ago the testing plant at the Purdue University attracted the attention of all progressive railway officials. A complete description of the plant, together with the results of tests, was furnished by Prof. Goss and published in my book "One Thousand Pointers for Machinists and Engineers."

The results which could be obtained by the use of a testing plant were so numerous, and of such importance, that it was then considered probable that most large railway companies

would construct plants for their own use, and a few of them did at an enormous expense, but owing to the various conditions of road service such as wind, damp rail, etc., it was found that the result of tests made upon the testing plant could not be relied upon, and the plants constructed gradually fell into disuse. However, some of the results obtained by Prof. Goss at the Purdue University plant cannot be obtained in actual service, and his papers concerning the various results obtained are still read with much interest by all progressive railroad men. The following are a few of the results obtained by the use of the plant at Purdue:

The traction dynamometer records the draw bar stress under any load.

A small dynamometer shows the power required to move the valves.

Indicators show the steam distribution in the cylinders and the pressure in the steam chest.

A calorimeter shows the dryness or otherwise of the steam.

The speed recorder denotes the speed attained, while a very sensitive indicator (resembling a steam gauge) indicates the speed at all times.

Gauges indicate the rarification of the smoke box and ash pan.

The heat of the smoke box gases is ascertained with a pyrometer.

The counterbalance is tested and the valves adjusted to secure the best results, the coal consumed being weighed and the water measured. All experiments are based upon economy. The results showing units of work performed per ton of coal per horsepower hour.

We have shown one view of the plant for the benefit of those who have never seen a testing plant.

LOCATING BLOWS AND POUNDS.

HOW TO LOCATE A BLOW IN A LOCOMOTIVE.

There is probably no more embarrassing position in which an engineer may be placed, or nothing which reflects more discredit upon his knowledge of the construction and operation of the locomotive than to be unable to locate a serious blow in his engine, which is a constant source of annoyance to himself, besides being the direct cause of an enormous waste of fuel. However, a locomotive is susceptible to so many different blows, some of which are very deceptive, that it is difficult and sometimes impossible to determine the exact location of a blow, but the engineer is expected to locate the blow and report it correctly if it is possible to do so, as it will save much unnecessary labor in the round-house or shop.

We will first call attention to a few of the parts where blows occur most frequently and describe the various sounds, and the action of blows under different circumstances, which may assist in determining the location of the blow before a test is made, we will then explain the correct way to test the engine and determine the location of a blow. A blow may be in the cylinder packing rings, the valve seat, the gibs, rings or rider of a balanced valve, or it may be in the steam pipes or nigger-head, or it may be a crack or a hole in the steam ports. If it is an intermittent or recurring blow, a round roaring, rumbling sound, like whor-r-r-r, you may depend upon it being in the cylinders, and you can usually locate in which cylinder it is by watching the crank pins on a slow pull, as it will usually be worse when the piston is in the center of its stroke. If it is a continuous sharp, shrill sound like whis-s-s-s, it is usually in the valve seat, but a valve sometimes blows intermittently when the valve cocks at one end. If it is a strong continuous blow and you have balanced valves it is possible one of your valve strips, valve springs, rings or rider is broken. But if your engine has a plain slide valve reverse the engine two or three times real quick, as it may be only a cocked valve. If it will not reseat in this way remove the oil plug from the steam chest cover and drive it down. Remember balanced valves do not cock. It may prove to be a sand hole in the valve or between the ports. A steam chest blow is easily distinguished from a steam pipe blow because it will blow straight up the stack and make a clear, singing sound, while a steam pipe blow expends its force in the front end and makes no noise when going out the stack. A steam pipe blow, if very bad, will affect the draft of the fire and when the fire door is open it sounds like a leaky stay bolt. A good indication of a steam pipe leak is the appearance of water in the front end. If you have lost one exhaust it may be a slipped eccentric; as a slipped eccentric will usually cause the valves to sound out of square. A valve yoke cracked or broken on one side only will cause one exhaust to sound out of square while the other three are perfect. When the valve stem breaks off it will usually cause a tremendous blow which will continue so long as the throttle remains open. But if you have a tremendous blow at one point only, and have lost one exhaust, and the three remaining exhausts are perfect, it may be a broken bridge or crack, or a sand hole in the bridge. Notice the cylinder cocks before you stop and see if steam appears at only one cylinder cock when the piston is at one end of the cylinder, and at both cocks when at the other end. If so it is a very good indication of a broken bridge, but examine your eccentrics as soon as you stop. When an engine has a bad blow when in full gear which disappears when hooked up a few notches, it indicates that the valve travels too far and opens the exhaust port to direct steam chest pressure. This is sometimes caused by the top arm of the tumbling shaft working loose, perhaps the key is lost. When the exhaust nozzle is gummed up it produces a sort of asthmatic wheeze, or whistle, which is sometimes mistaken for a blow. When two exhausts are heavy and two very light you may have blown out a nozzle tip, providing you have double nozzles. When the dry pipe leaks the engine will work water through the cylinders and when standing in the round-house it may be discovered by a constant leak at the cylinder cocks.

A leak at the bottom of the exhaust pipe will not cause a blow, but will affect the exhaust.

METHOD OF TESTING.

From the descriptions we have given of different blows, you can usually determine about where the blow may be found and proceed to test that particular part without giving the engine such a severe test as we have outlined, as this chapter necessarily covers all kinds of blows. We will first test the steam chest, and afterward the cylinders. It is an easy matter to determine which cylinder a blow is in, but it is sometimes very difficult to locate which steam chest, so follow these instructions closely.

Place each rocker arm in a vertical position alternately, block the wheels, open the cylinder cocks and give the engine a little steam. If no steam appears at either cylinder cock you may depend the valve seats are tight. If your engine has balanced valves test the valve strips, rings or riders. A blow of this kind is sometimes very difficult to locate, but it can be done, viz.: [The McDonald valve, which we have illustrated and described on page 84, has a cock on top of the steam chest for this very purpose.] If your engine has drain cocks screwed into the exhaust port, go under the cylinders and open the cocks and have the fireman give a little steam; if steam appears at either cock that is the side your blow is on. Another way is to open the front end if you have a double nozzle, and you can see which side blows; if a single nozzle, climb up on the boiler and by feeling the draft on each side of the stack with a broom, or lighted torch if at night, you can usually notice a difference in the draft. On whichever side the draft is the strongest the blow will be in the opposite chest. Or, you may put a little fresh coal in the fire box and watch its action on the smoke. This kind of a blow can sometimes be located by the increased friction, which will cause the valve stem to jerk when in motion, or it may be discovered by placing the crank pins on center alternately and handling the reverse lever under steam pressure, the blow will be on the side that handles the hardest while the pin is on the quarter (not the center). Hoping these suggestions will assist you, we will return to the valve seat.

Now if steam does appear at both cylinder cocks on one side while the steam ports are covered, it is evident that the valve seat on that side leaks, providing the opposite side is tight; the leak may be in the valve seat or beneath the false seat, or if the valve has inside clearance it may be a flaw in the valve. If steam appears at only one cylinder cock on only one side of the engine, while the ports are covered, it may be a sand hole between the supply port and the steam port, but it is more probable a false seat loose on one side. If steam appears at the forward cylinder cock, the forward end of the false seat is loose, and if at the back cylinder cock, the back end. Now if steam appears at both cylinder cocks on both sides it is evident that the valves on both sides blow.

We will now proceed to test the cylinder packing, placing each main pin on either quarter alternately, and with the reverse lever in the forward notch give the engine a little steam. If steam escapes at only one cylinder cock the cylinder packing on that side is all right, but before leaving it place the reverse lever in the back notch and try it there. Now if steam appears at both cylinder cocks when one port is open, and at only one cock when the other port is open it indicates a broken bridge, (although a broken valve strip or ring might cause this, or a sand hole in the bridge below the valve seat); which particular bridge is broken may be determined by noticing which port is open when it shows steam at both cylinder cocks; if the forward port is open then it is the forward bridge, and vice versa. A broken bridge can usually be determined from a crack, or sand hole by a tremendous blow. If steam appears in great volume at both cylinder cocks when the lever is in both motions, it is then impossible to say whether it is a broken valve seat or broken cylinder packing rings, so have the cylinder head removed first; if it is all right then you know it is the valve seat. A broken cylinder packing ring can usually be distinguished from one that simply leaks by the volume of the blow. A packing ring that leaks will also show steam at both cylinder cocks when in both motions, but it will not be such a heavy blow as a broken ring will produce. Most every engineer has had some experience with packing rings that simply blow and can distinguish it from anything extraordinary, such as a broken packing ring or bridge. When you raise the steam chest cover first examine the rings, or valve strips and springs. See that the valve strips do not fit too tight on the ends, for the long strips expand 1-32" more than the valve and often cramp the short strips. Next examine the valve seat and face, then the bottom joint of a false seat, and the pressure plate, and face off all joints that need trueing up. Now examine the valve carefully for sand holes. If you cannot locate the blow elsewhere then fill the supply ports with water (one at a time); open the cylinder cocks and see if it leaks water; if it does there is certainly a sand hole or crack, if not then fill the cylinder and steam ports and see if it leaks into the exhaust cavity. Open drain cock at bottom of the exhaust cavity.

CAUSE OF VALVES SOUNDING "OUT OF SQUARE."

Sometimes an engine's valves will sound "out of square" when the valve gear is perfectly adjusted. The following are some of the causes which may produce such an effect:

Driving wheels improperly quartered.
A main driving axle bent.
A patch inside the steam ports, or ports of different size.
Cylinders of different size (not compound engines).
Eccentrics of different throw.
Links of different radius.
A hole in the petticoat pipe or stack.
A leak at the exhaust pipe joint.
A valve-yoke cracked on one side.
Cylinders working loose on the frames.

REPAIRING CYLINDER PACKING.

Engineers are not expected to do this kind of work and should only do it in case of emergency out on the road. If your packing blows report it without delay. When it becomes necessary to overhaul the cylinder packing, if steam is up too much attention cannot be given the throttle while the cylinder-head is off, otherwise you may get a severe scalding.

First close the throttle and tighten the thumb screw, then open the cylinder cocks. Now move the engine until the piston-head is within one or two inches of its extreme forward travel.

If you cannot move the engine then the main rod must be disconnected, but before doing this, scribe a line on the guides true with the back end of the crosshead and when recoupled again the crosshead should be true with the same line (the liners sometimes become misplaced). Also place a wooden block between the guides in front of the crosshead; this is merely for safety to prevent knocking out a front cylinder head. Now when you have the rod disconnected, move the crosshead forward within an inch or two of its extreme forward travel. By moving the reverse lever you can then cover both steam ports on that side to keep the steam out of the cylinder, then take off the cylinder-head and follower, make a mark on the cylinder-head; also on the follower before you take them down, so that you can put them up to the same places. (Sometimes the holes only fit one way.) Be careful to lay the cylinder-head nuts and follower bolts in such a manner when you remove them that you will be able to put them back where you found them. Now if solid ring packing, peen each packing ring inside until it fits the cylinder nicely; if it is dunbar packing with springs, open out the springs until they will hold the packing securely against the cylinder. If it is a solid piston head the piston and head must be removed before you can remove the packing rings; when replacing the packing use a pair of inside calipers; hermaphrodites, or a pointed stick the right length will do; set the piston slightly above the center of the cylinder.

Be sure that the packing lays against the cylinder all around; if it does not, the packing should be taken out and the cause ascertained and remedied, if possible. After the packing is set out, clean off the follower and put it on. See that the heads of the follower-bolts press the follower, and that their ends do not touch bottom; then clean the cylinder-head joints, and put the head on. In screwing up follower-bolts and cylinder-head nuts, be sure to get them solid. Use judgment especially in screwing the cylinder-head nuts, or you will break the studs. When the head is put on, see that the joints lay close together all around; then put the top nut on and run it on until it just comes against the head; then put the bottom nut on in the same manner; then a nut on each side, and draw them, watching that the joints are together all around; then put on all the nuts and draw them equally all around the head. You will sometimes find, after removing the follower, that the packing is not slack, although it seemed so before the follower was taken off. This shows that the packing was clamped and held by the follower, and that the packing is too long—it was follower-bound. This can be remedied by placing a piece of wrapping paper between the follower and spider. Packing does not need setting out if it blows only a little at starting—you rob the engine of its power by having it too tight. It should not be snug enough to prevent an engine drifting freely. Packing should not be allowed to run longer than two months without being examined.

LOCATING A POUND IN A LOCOMOTIVE.

For lack of a better word we have classed all the disagreeable and annoying jerks and sounds familiar to the engineer and fireman under the heading "Pounds in an Engine." In actual practice they are called clicks, knocks, jerks, thumps and pounds, and like a blow, they are sometimes very difficult to locate, in addition to the annoying sound. If a serious pound is neglected it may be the cause of disabling the engine. For this reason it should be located and reported without delay, and thereby relieve the engineer of further responsibility.

An experienced engineer who is familiar with the various sounds produced by a locomotive, can very often locate a pound by its particular sound. One of the most deceiving pounds to locate upon a locomotive is a loose piston head (or spider); these knocks have deceived many old runners; they usually come suddenly and sound as if there was an inch lost motion somewhere, when in fact it may not be more than the thickness of tissue paper. Such a knock is generally mistaken for a loose wedge, driving brass, crosshead, or main rod brasses; the noise is similar to that made by a crosshead being loose on the piston rod, and occur when passing both centers.

METHOD OF LOCATING A POUND.

When you have determined which side of the engine the pound is on, the best way to locate it is to place the crank pin on that side on the top quarter, block the driving wheels and have the fireman give the engine a little steam and reverse the engine a few times while you examine all points on that side where there is liable to be a pound.

The reason the crank pin is placed upward when you wish to find a knock or thump, is because the crank pin is freer to move at that point; if the pin was downward the weight of the engine would have to be moved before you could find a thump in a driving box or frame; if the pin was on either of its centers you could get steam in only one end of the cylinder when the reverse lever is moved.

Knocking or pounding may be caused by an insufficient oiling of the cylinders, main shaft, main crank pin, or cross head pin or lost motion in any of the reciprocating parts, such as the main rod brasses, loose cross head pin or loose crank pin, or a bad fit between piston rod and cross head or piston. If a piston strikes the cylinder head it will cause a knock, too loose wedges, or loose knuckle pin or bushing, or loose middle connecting brasses. Wedge down, or when wedges are stuck they will also cause a thumping sound. A broken engine frame or loose cylinders or loose deck will also cause a bad pound. Very loose pedestal braces on light frames will sometimes cause a pound. Imperfect fitting up of the shoes and wedges while undergoing repairs is also a cause of many pounds; when finished the shoes and wedges should be perfectly parallel and not bind the box on either top or bottom, leaving the other end loose, which is sure to cause a pound. Another cause of pounding in the driving boxes is loose oil cellars; cellars should be a good, close fit into the driving box, as the weight of the engine and the constant wear of the brass tends to close the box at the bottom, which, if permitted, would eventually cause a pound. The driving box wears the surface of the shoe and wedge against which it rubs, while the surface extending above the box, not being subject to wear, naturally causes a shoulder to be worn on both shoe and wedge, which is a frequent cause of the box sticking and causing a pound; it is for this reason so many roads have the top ends of the shoes and wedges planed off below the wearing surface, and many roads have discontinued the practice of beveling the top edges of the driving boxes to admit oil, as they catch too many cinders, which causes the shoes and wedges and boxes to cut. Oil holes are drilled through from the oil cellar on top of the box to the wearing surface of the box. Loose driving brasses, either circle or gib brasses, will cause a pound. Square bottom spring bands, or poorly fitted spring saddles or any interference with a free movement of the equalizer will also cause a knocking sound. See that the springs do not rub the boiler, or the saddle strike the engine frame. Pounds are also caused by wet steam or foaming, causing water in the cylinders. Cylinders of unequal size in a single expansion engine will cause unequal stresses and produce a jerking motion. Excessive back pressure will also cause a pound in the cylinders. Imperfectly balanced driving wheels will create a jerking of the engine. Too great or too small a compression will also cause a jerk when passing over the center. For a high speed the compression should be almost, if not quite, equal to the initial boiler pressure. Lost motion in the valve gear will often cause the reverse lever to rattle; badly distorted valves which cause an unequal distribution of steam often create a jerking sensation when hooked up near the center notch. Side rods (except the middle connection) will not pound, but will rattle. A loose follower bolt will usually cause a knocking in the cylinder when steam is shut off. Packing springs sometimes cause a sort of clicking sound.

NEW STYLE DRIVING BOX.

A driving box has been invented which is intended to prevent pounds in the box or brass. It resembles the ordinary driving box with three gib brasses, the improvement being that the two side gibs are adjustable, having lugs in their center which protrude through the box and bear against the shoe and wedge, the top gib being pressed into the box in the usual manner. The side brasses extend considerably below the center of the axle, thereby making the bearing about two-thirds the circumference of the shaft.

SETTING UP WEDGES.

An engineer should have steam up when the wedges are adjusted, because the parts of the frame that lay against the fire box become hot and expand, or get longer than when the boiler is cold. The crank pins should be on the top quarter when the wedges are set up; a block may be placed on the rail back of the wheel and the engine moved back against it, or the wheel may be pinched forward with a pinch bar, the object being to jam the driving box up solid against the shoes so that whatever lost motion there may be between the jaws will be at the back of the box, so the wedges will slip up freely. When you have all the boxes up tight against the shoes set the wedges up moderately tight with a 12-inch monkey wrench and tighten the jam nuts with a large wrench, because the pedestal braces, or binders, are tight before setting up the wedges. The reason the crank pins should be upward on the side that the wedges are being adjusted is because if the pins were on center and the side rods were too long or too short they would necessarily force one of the driving boxes against the wedge instead of the shoe; and if the crank pins were below the centers when the wheel was pinched forward the side rod would have a tendency to draw back on the crank pin and would therefore draw the box back against the wedge. While if the crank pins were on the top quarter the tendency would be to force all the boxes on that side against the shoes. It would be well to slack all the rod keys before setting up the wedges if time will permit, and tram the wheel centers when you have finished. A very good way to set a wedge is to screw it up as far as possible, then at the top of the wedge, scribe a line on the frame, then draw it down about $\frac{1}{8}$ of an inch and jam the nuts on the wedge bolt. If there is lost motion at the head of the wedge bolt you must make allowance for it. If there is another wedge bolt passing through the jaw tighten it and be sure that it does not extend through the wedge and rub on the driving box; measure its length.

KEYING UP RODS.

Side Rods.—Owing to the general adoption of solid end rod brasses this operation will soon be a thing of the past. Solid end rods are much safer, and prevent many accidents with rod straps and broken bolts and save many crank pins that would otherwise be broken on account of improperly keyed side rods, but as many locomotives are still in use which have strap rods with keys, we shall explain the proper method to set up rod keys. We will begin with the side or parallel rods. First select a level piece of straight track without any high or low joints, place the crank pins parallel with the wheel centers on one side, and beginning with the middle connection, if there is one, and key the forward pins last on either eight, ten, or twelve-wheeled engines; drive the keys down moderately tight with a soft hammer, or hand hammer (not the coal pick), then key the other pins on that side the same way, then try the rods on the other centers opposite to those on which they were keyed, and see that they will move latterly and do not clamp the pin; it is well, also, to try them on the quarters, as old pins may be worn out of true; now key the rods on the opposite side of the engine exactly the same way, leave your side rods free; it is no injury to have them rattle a little, especially if the driving boxes are well worn; when finished tighten all set screws.

Main Rod.—When keying up a main rod place the crank pin on the *forward top eight,* which is midway between the forward center and the top quarter. The reason for this is that the greatest wear on the pin takes place under the greatest pressure of steam; in others words, when the valve has reached its greatest opening, a brief study of the valve's motion will show you that steam is first admitted to the cylinder at or near each dead center, and as the average cut-off for all kinds of service is considered half stroke, it follows that steam will be cut off when the crank pin is at or near the quarter in each stroke; therefore when an engine is in the forward motion the wear on the pin will take place while the crank pin moves from the forward center to the lower quarter, and from the back center to the top quarter, must open and close with each stroke while the crank pin passes through this space. It therefore follows that the valve will have reached its greatest opening when the crank pin is at or near a point midway between the forward center and the bottom quarter and also between the back center and the top quarter. These points are called the lower front eight and the top back eight, respectively; the pin must therefore be the smallest at these two points, and as a main rod brass should always be keyed on the largest part of the pin, the pin should occupy a position directly opposite to the two points mentioned; therefore place the pin on the forward top eight. Key the brass moderately tight, then see that you can move it sideways at both front and back ends, then tighten your set screws; some engines have so very little clearance that it is sometimes safer to key the main rod with the pin on dead center. Engines should have steam up when keying up the rods as the frames will sometimes expand enough to bind the brasses, besides you will need steam to move the engine; for the same reason the valves should be adjusted while the engine is hot.

HOT BRASSES OR ECCENTRIC STRAPS.

Hot bearings may be caused by a lack of lubrication, over-pressure, grit in the bearings or too high rotation, speed, etc.; much also depends upon the alloy of the metal and the dimensions of the bearing; the larger the diameter and the longer the bearing the less liability of heating so long as everything is kept in line. Tallow, soap or graphite mixed with lard oil are good cooling lubricants for a hot bearing. When a driving brass gets very hot you should relieve the box, as it may be caused by excessive weight; this may be done by running the wheel up on a wedge and then blocking solid between the spring saddle and the frame. Slacking the wedge slightly is another remedy. Many engines are now supplied with a small water pipe which may be turned onto each bearing. When a rod brass starts to throw the babbitt don't stop until it is all thrown out; then stop, examine the oil cup and the hole in the strap and see that it is not choked up. Slack the rod key slightly, give it a good oiling and proceed. An engineer usually knows the condition of his eccentric straps, so if you smell a hot bearing and think it is an eccentric strap shut off steam, but don't touch the reverse lever, for if you try to "drop her down in the corner" it will invariably break the eccentric strap. Wait until she stops and then slack the strap bolts a little, and put in a little more liner if convenient. Never cool a cast iron eccentric strap with water. If a truck or tender brass runs hot, replace it or repack the cellar and keep it well oiled. Turn the small hose on it if you have pipes to each box.

OILING.

In oiling an engine do not fail to oil the sides of all boxes and bearings; hot boxes and bearings are often caused by neglecting this. Hot boxes are mostly caused, however, by the sponging settling away from the journals. Oiling on top of a box will be of little benefit unless the sponging presses the journal from the oil cellar. All rod and guide oil cups should be taken out and cleaned every week. If an engine is low on one side it may cause the back end of main rod brasses to heat or cross head to work badly on that side.

WHAT AN ENGINEER SHOULD ATTEND TO BEFORE LEAVING AN ENGINE HOUSE.

When an engineer is required to take an engine out, he should try how much water is in the boiler, and see that the fire is in good condition. Examine every part of the engine and tender, to see that nothing is broken. Test the rod-keys; don't depend on set-screws; try to drive every key back with a soft hammer; examine every bolt and nut; test the brakes; see that they apply and release easily, especially if using automatic or air brakes. See that you have a good supply of fuel and a tank full of water; all necessary tools, especially for firing, and that all lamps and signals are in good order and ready for immediate use, and that there is sufficient oil and tallow and a box full of sand. You should also have a pinchbar and a pair of jacks in good order. If you have no jacks get four wedges of oak, about three feet long, four inches square at one end tapered to an edge one way at the other end. You should also have an axe and blocking for cross heads, and a hand saw.

STARTING A TRAIN.

In starting a train, always have the reverse lever in extreme full throw, never cut back; after the train is well under way cut back one notch at a time; cut back as far as circumstances will permit; run with throttle wide open; you save steam better by cutting back with the reverse lever, if the notches in the reverse rack are close together, than by closing the throttle, and if drifting without steam, place the reverse lever at full stroke, allowing the valves to have full travel.

The reason that a train should not be started without having the reverse lever set at full stroke, is because there is a greater pressure on the valves when steam is first admitted to the steam chests than after the engine gets under headway; while the engine is running and using steam a thin film of steam is between the valves and their seats all the time, so the valve is not in actual contact with the seat, and if the valves are allowed to move over a portion of their seats without having steam (which of itself acts as a lubricant), or something to relieve the friction between them, the seats will in time become worn in the centers; for the same reason the lever should be placed at full stroke while the engine is drifting without steam to prevent, as much as possible, the wearing of the valve seats in the center, and from wearing a rounding face on the valve.

The reason why an engine saves steam by running with the throttle wide open and using the reverse lever to control the admission of steam to the cylinders, is that as the engine is cut back the lead on the valve increases and admits more pressure in the cylinder just where it is needed; that is, when the engine is at its weakest point right after passing its centers and nearly all of the full effective pressure of steam in the boiler remains in the cylinder nearly all of the time that the valve remains open, a little before the valve closes, the pressure entering the cylinder becomes what is known as wire drawn or weak steam. The same thing occurs if you run with your throttle partly closed, the steam becomes wire drawn before it reaches the steam chest.

If the throttle is wide open almost all of the full effective pressure of the boiler is always in the steam chest, and that pressure is at the valve ready to give its full impulse up to the piston, and is more effective than if the pressure was less and longer continued by having the throttle partly closed and having the valve travel further. If you use steam in a cylinder until the piston has traveled twelve inches, when the engine will do the work by the piston traveling but eight inches, you are wasting at least one-third of that extra steam, because you do not get the full effective pressure against the piston when it is most needed. If you strike a nail with a force of 120 pounds it is driven farther than if struck with a force of 100 pounds. So is a piston driven with more force and a greater distance with 120 pounds pressure than it is with 100 pounds pressure.

Another point is that after a piston has moved five or six inches from the end of the cylinder not near as much steam is required to move the engine for the next third of the half revolution of the wheels, because the leverage on the driving wheel increases as the crank pin leaves its center until the pin is up or down, the leverage decreasing gradually until the pin reaches the other center, but before you reach that point you have had the full benefit of all of the expansive force of 120 to 140 pounds of pressure, and it is well known that the higher the pressure the greater the expansive force of steam, and by closing the throttle the motion of the piston takes the pressure out of the steam pipes faster than the throttle allows it to flow; immediately after passing the center you want the highest initial pressure in the cylinder that you can get there, and when the steam has done its work in that end of the cylinder you want to get rid of it as soon as possible, because we want to get steam into the other end of the cylinder, and do not want pressure on both sides of the piston at the same time. If the pressure could all be let out of one end of the cylinder as soon as steam enters at the other end, and if there was a free passage for the exhaust after the steam is discharged from the cylinder, there would be no back pressure to overcome; but then we need some steam in that end of the cylinder for a cushion for the piston while passing the center. The best constructed valve motion gives the smallest amount of back pressure, allowing just enough compressed steam to form a cushion and to give the full boiler pressure against the piston as soon as the valve opens, and the outside lead that is gained by the valve being cut back is also gained on the exhaust cavity in the valve, and thus you begin to release the steam sooner than if you used the throttle instead of the reverse lever. Of course, there is more compression of steam between the cylinder head and piston at the end of the stroke, on account of increased lead, and the exhaust is somewhat choked, because the cavity of the valve does not open so wide, when the engine is cut back, but four or five inches of steam saved at each end of a cylinder more than makes up for all these losses. Your steam gauge will show that your engine will steam better with any train it can haul at maximum speed with valves cut back as far as circumstances will admit with throttle wide open, than with reverse lever two or three notches ahead with the throttle partly closed. The exceptions to this rule are very rare.

A WORD TO YOUNG RUNNERS.

One of the hardest things for a young runner to learn is the art of letting good enough alone. He usually monkeys with his wedges, rods or diaphragm or petticoat pipe or something that if left alone would answer a better purpose. He cannot be too careful about oiling, but this does not mean to wash the engine down with a half gallon of oil at the beginning of a trip and then letting her run until the babbitt begins to fly, but at every station when there is time enough to put a few drops on every wearing surface and lay his hand on every such surface expecting to find it hot. Another thing he will find to be a great satisfaction to him is the thorough inspection of every part of the engine and tender before starting on a trip. It will give him more confidence while running that he is going to get through all right, if he satisfies himself that to the best of his knowledge the engine is in good condition before he starts, and if anything should go wrong he will not have himself to censure for want of careful examination. Examine your engine and tender thoroughly before leaving the engine house at the end of each trip and report all needed repairs. If there are any defects that can be remedied, attend to them as soon as possible, no matter how trifling or unimportant they may seem.

COMPOUND LOCOMOTIVES.

The compound steam engine was invented by Mr. Jonathan Hornblower in 1776. The principle of compounding is to use the same steam in more than one cylinder before release, the steam from a high pressure cylinder being exhausted into a low pressure one. Soon after the locomotive era efforts were made to utilize the compound principle upon the locomotive, but Mr. John Nicholson, an English "engine driver," was the first to actually build and run a compound locomotive; this was in the year 1850. The compound principle progressed slowly in locomotive practice until about fifteen years ago, when the subject of fuel economy by compounding on locomotives was taken up actively in France, Russia, Germany and England, and stories of successful compounds started American locomotive builders to designing engines of this class. Some of the inventions made by our designers were, and are, fearful and wonderful, but these have already found, or soon will find their proper place in the scrap heap.

Although compound locomotives have been in use in this country for many years, a great diversity of opinion exists at the present day regarding their merits; while they have given excellent results in some cases, on many roads where they were given a fair trial the additional expense for repairs largely offset their economy of fuel and their use was discontinued, but upon some of these same roads the compound has since been adopted.

While many of the best forms of compounding now in use possess disadvantages under the present mode of construction, yet they embrace the correct principle, that of economy, and will, no doubt, when improved eventually become the established type of American locomotive; in fact they are a thoroughly established class at the present day, and have come to stay. They are gradually replacing single expansion engines, and the man that expects to continue railroading will find it necessary to understand the operation of compound locomotives, regardless of his personal views of this class of engines. For this reason we have devoted considerable space in this book to compound locomotives, describing in detail most every form of compounding now in use in this country. Numerous patents have been taken out for compounds that have not been further developed, which with other forms that have proved failures and a few in the experimental stage we have omitted.

Three forms of compounding are in use in this country. They are: First, the Vauclain system, which is a four cylinder compound with two cylinders on each side of the engine, one high pressure and one low pressure, and one cylinder above the other. Second, the Tandem compound, a four cylinder engine with a high and low pressure cylinder on each side, the axis of each cylinder being in a horizontal line; few of this kind are in use. Third, the cross, or two cylinder compound with a high pressure on one side and a low pressure cylinder on the opposite side. The cylinders of this form of compound are made of different sizes in order to equalize the power on each side of the engine; a small cylinder with a high pressure in it will exert as much force as a large cylinder with a less pressure, and they are proportioned to each other in order to obtain an equal amount of power on each side of the engine.

With the other two forms of compounding the power is equal on each side and the cylinders are proportioned to each other to give the best results; the low pressure usually having about three times the area of the high pressure cylinder.

The cross compound engines have a separate exhaust for each cylinder. An intercepting valve is used and it can be worked simple at will or in case of a break down on either side. When working simple the pressure of live steam in the low pressure cylinder is controlled by an automatic regulating valve and cannot exceed one-half of the boiler pressure.

The same valve gear is used upon compound locomotives as upon single expansion engines. The most successful compound will be found to be the one which is the least departure from ordinary practice and having the smallest number of parts, the lightest reciprocating parts, and least number of steam joints, and be the simplest both in mechanism and in the handling of steam from boiler to exhaust.

It is not claimed for compound locomotives that a heavier train can be hauled at a given speed than with a single-expansion locomotive of similar weight and class. No locomotive can haul more than its adhesion will allow; but the compound will, at very slow speed on heavy grades, keep a train moving where a single-expansion locomotive will slip and stall. This is due to the pressure on the crank-pins of the compound being more uniform throughout the stroke than is the case with the single-expansion locomotive.

The principal objects in compounding is to effect fuel economy, and this economy is obtained:

1. By the consumption of a smaller quantity of steam in the cylinders than is necessary for a single-expansion locomotive doing the same work.

2. The amount of water evaporated in doing the same work being less in the compound, a slower rate of combustion com-

bined with a mild exhaust produces a higher efficiency from the coal burned.

In a stationary engine, which does not produce its own steam supply, it is of course proper to measure its efficiency solely by its economical consumption of steam. In an engine of this description the boilers are fired independently, and the draft is formed from causes entirely separate and beyond the control of the escape of steam from the cylinders; hence, any economy shown by the boilers must of necessity be separate and distinct from that which may be effected by the engine itself. In a locomotive, however, the amount of work depends entirely upon the weight of the driving-wheels, the cylinder dimensions being proportioned to the weight; and whether the locomotive is compound or single-expansion, no larger boiler can be provided, after allowing for the wheels, frames, and other mechanism, than this weight permits. Therefore, the heating surface and grate area are practically the same in both types, and the evaporative efficiency of both locomotives is determined by the action of the exhaust, which must be of sufficient intensity in both cases to generate the amount of steam necessary for utilizing, to the best advantage, the weight on the driving-wheels. This is a feature that does not appear in a stationary engine, so that the compound locomotives cannot be judged by stationary standards and the only true comparison to be made is between locomotives of similar construction and weight, equipped in one case with compound and in the other with single-expansion cylinders.

One of the legitimate advantages of the compound system is that, owing to the better utilization of the steam, less demand is made upon the boiler, which enables sufficient steam-pressure to be maintained with the mild exhaust, due to the low tension of the steam when exhausted from the cylinders. The milder exhaust does not tear the fire, nor carry unconsumed fuel through the flues into the smoke-box and thence out of the smoke-stack, but is sufficient to maintain the necessary rate of combustion in the fire-box with a decreased velocity of the products of combustion through the flues.

The heating surfaces of a boiler absorb heat units from the fire and deliver them to the water at a certain rate. If the rate at which the products of combustion are carried away exceeds the capacity of the heating surfaces to absorb and deliver the heat to the water in the boiler, there is a continual waste that can be overcome only by reducing the velocity of the products of combustion passing through the tubes. This is effected by the compound principle. It gives, therefore, not only the economy due to a smaller consumption of water for the same work, but the additional economy due to slower combustion. It is obvious that these two sources of economy are independent.

The improved action of the boiler can be obtained only by

the use of the compound principle, while the use of the compound principle enables the locomotive to develop its full efficiency under conditions which in a single-expansion locomotive would require a boiler of capacity so large as to be out of the question under the circumstances usually governing locomotive construction. It is therefore evident that where both locomotives are exact duplicates in all their parts, excepting the cylinders, the improved action of the boiler is due entirely to the compound principle, and the percentage of economy should be based upon the total saving in fuel consumption, and not upon the water consumption, as in stationary practice.

We quote the following article regarding compound locomotives from the *Railway Master Mechanic:*

Had anyone predicted in 1890 or 1891 that it would require eight or nine, or more, years to determine what conditions of traffic could be more cheaply handled with compound locomotives than with simple locomotives he would have been considered, at best, a very poor prophet. Nevertheless such a prophecy would have been entirely fulfilled, and even after so many years it is yet a question in the minds of many officers as to just how far the compound locomotive can be used economically when the first cost, cost of fuel and cost of repairs, and cost of lost time due to the latter, are given due consideration. Attention is attracted again to the compound locomotive because quite recently orders have been placed for them by roads which, heretofore, have not been inclined to consider them favorably.

When the compound was first being considered there was much discussion as to whether the cost of repairs to it would be greater than the cost of repairs to the simple locomotive, and in the minds of some the question has not been decided yet, while for others it has been settled, apparently, sometimes favorable to the compound and possibly quite as frequently otherwise.

It is entirely reasonable to expect an increase in cost of repairs to compound locomotives, because every attachment or device added to the equipment must increase the cost of attention, but to just what extent such increase in cost will show in the records will depend upon the basis of such records, and also upon the basis of comparison, and upon the period in the service of each type which is taken for comparison. It is manifestly unfair to compare a locomotive of either type and on either basis with a locomotive built two or three years previous to it, because the former would have been built correspondingly stronger to handle the ever-increasing weight of train; and if the comparison is based on the engine mile record the inequality is even more pronounced. Records which are to be used for comparing compound and simple locomotives should be kept very carefully; the same number of each type, probably not less than ten, should be operated similarly on the same division of road, and the average loading of each should bear the same ratio to capacity of each in both cases. Moreover, those who work the compounds and those who have the care of them should be as familiar with the mechanism and the method of operation of this type as those are who handle the simple locomotives with their type. This last is a consideration which is seldom fulfilled, if, indeed, even considered. This oversight, or neglect, affects unfavorably the record of compound locomotives; but, on the other hand, the general practice of taking for comparison the first year or two of service of the compound operates decidedly to the advantage of the latter. It is generally understood that the cost of repairs after about two years of service increases more rapidly for the compound than for the simple locomotives, and, while designers and builders have put forth strenuous efforts to overcome this difference, it is hardly to be expected that the rates of increase in cost of repairs can be made the same for both types. For this reason records from compound locomotives which are to be used for comparison with similar records from simple locomotives should extend over the period of the first four or five years of service, or should cover a period which shall include the cost of the first general repairs and a year or two of service after such repairs.

The compound locomotive has had much with which to contend and not the least obstacle has been prejudice. The greatest, however, has been ignorance—ignorance as to the methods of handling and caring for the type and ignorance as to the service suited for it. This ignorance has been shared by the designers and builders to no less degree than by the railroad men, and of course is due entirely to lack of experience. Lack of experience in handling and caring for the compound has been, probably, the direct cause of the reputation which this type has of being a "good shop engine," but it is a condition which was to have been expected and which, if entirely surmounted at all, must be overcome gradually.

Reference has been made above to the ignorance of the service to which the compound type is best suited, and it is believed possible to show that such ignorance is quite general. The opinion is all but universal that a simple locomotive, doing its work with the cut-off averaging from one-quarter to one-third the stroke, with uncovered cylinders and other arrangements as at present, cannot be improved upon very much by simply providing it with compound cylinders; in other words, when the simple engine works at a cut-off of one-half or a greater proportion of the stroke, it is working under conditions which favor the compound, and the fineness of the problem lies in determining the exact location of the neutral line where the simple and the compound can be operated equally advantageously.

In practice, no effort has been made to locate this line, contentment being reached in selecting such service in which it was supposed there could be most probability and most possibility that the compound would be most favored, and it is in the selection of such service that it is desired to raise question now. Judging from the class of service for which compounds usually have been selected, it has been the quite general opinion that they should give best results in mountain service; but while the compound is working more advantageously up the heavy grades, the simple locomotive runs down such grades under conditions just as favorable to itself, and works over the level track and lighter grades under conditions, possibly, just as favorable to it. The net advantage to either depends on the ratio of the heavy grades, in both directions, to the total distance both ways. It will follow from this that the conditions ideally favorable to the compound locomotive are those under which the simple locomotive is worked at about one-half cut-off, or greater, over the entire round trip run, and these conditions can be more nearly fulfilled on a level track with full loading in both directions.

The substantiation of much of the foregoing is found in the results obtained on the Lake street elevated railroad, in Chicago, when it was operated by steam locomotives. The track was practically level; the traffic and service did not vary from day to day, but did vary during the day; the several locomotives of the compound and simple types were built at the same time and were identical except as to design of cylinders and the difference in machinery made necessary by the difference in cylinders. The evidence is most tersely presented by saying that after the experience of several months with both types, no simple locomotive was used when there was a compound available for service.

THE INTERCEPTING VALVE.

An intercepting valve is a small valve controlling the passage of steam to the cylinders of compound engines. It received its name of "intercepting valve" from the fact that it cut out connection between the receiver and the low-pressure cylinder while the engine was working live steam in both the high and low-pressure cylinders at starting. After starting, the high-pressure cylinder exhausts into the receiver, and the resulting pressure moves the intercepting valve so as to cut off live steam from the low-pressure cylinder, and at the same time open communication between the latter and the receiver, and allow the engine to work as a compound. This valve on the Vauclain compound is simply a plug cock which admits live steam to the high and low-pressure cylinders, and is called a "starting valve," for the reason that when starting a heavy train, it is opened to admit steam from one end of the high-pressure cylinder to its opposite end, and from there through the exhaust passage to the low-pressure cylinder, and thus work as a simple engine. This valve is also a cylinder cock for the high-pressure cylinder, and performs its function as such, simultaneously with those on the low-pressure cylinders, being opened and closed by the same lever.

BALDWIN COMPOUNDS.*

THE VAUCLAIN SYSTEM.

This system of compounding is the invention of Mr. Samuel M. Vauclain, a member of the firm of Messrs. Burnham, Williams & Co., proprietors of the Baldwin Locomotive Works. The system bears the inventor's name. It has both high and low-pressure cylinders on each side, the steam from the high-pressure cylinders exhausting into the low-pressure cylinders on the same side and thence out through the stack, making each independent of the other, with an equal pressure exerted on each side, one cylinder being above the other, with both pistons fastened to the same crosshead. On engines that stand high above the rail the high-pressure cylinder is above the low-pressure one, while on engines that are very low the high-pressure cylinder is placed below to clear the rail.

DESCRIPTION.

In designing the "Vauclain" system of compound locomotives, the aim has been:

1. To produce a compound locomotive of the greatest efficiency, with the utmost simplicity of parts and the least possible deviation from existing practice. To realize the maximum economy of fuel and water.

2. To develop the same amount of power on each side of the locomotive, and avoid the racking of machinery resulting from unequal distribution of power.

3. To insure at least as great efficiency in every respect as in a single-expansion locomotive of similar weight and type.

4. To insure the least possible difference in cost of repairs.

5. To insure the least possible departure from the method of handling single-expansion locomotives; to apply equally to passenger or freight locomotives for all gauges of track, and to withstand the rough usage incidental to ordinary railroad service.

The principal features of construction are as follows:

---

*These articles were especially prepared for this book by the Baldwin Locomotive Works.

Passenger Engine No. 385, Philadelphia & Reading Ry.
Baldwin "Vauclain" Compound.
Description Given on Page 480.

CYLINDERS.

The cylinders consist of one high-pressure and one low-pressure for each side, the ratio of the volumes being as nearly three to one as the employment of convenient measurement will allow. They are cast in one piece with the valve-chamber and saddle, the cylinders being in the same vertical plane, and as close together as they can be with adequate walls between them.

Where the front rails of the frames are single bars, the high-

Fig. 1.

pressure cylinder is usually put on top, as shown in Fig. 1, but when the front rails of frames are double, the low-pressure cylinder is usually on top, as shown in Fig. 2.

The former (Fig. 1) is used in "eight-wheel" or American type passenger locomotives, and in "ten-wheel" locomotives, while the latter (Fig. 2) is used in Mogul, Consolidation, and Decapod locomotives; for the various other classes of locomotives the most suitable arrangement is determined by the style of frames.

Fig. 3 shows the arrangement of the cylinders in relation to the valve.

The valve employed to distribute the steam to the cylinders is of the piston type, working in a cylindrical steam-chest located in the saddle of the cylinder casting between the cylinders and smoke-box, and as close to the cylinders as convenience will permit.

As the steam-chest must have the necessary steam passages cast in it and dressed accurately to the required sizes, the main passages in the cylinder casting leading thereto are cast wider than the finished ports. The steam-chest is bored out enough

Fig. 2.

larger than the diameter of the valve to permit the use of a hard cast-iron bushing (Fig. 4). This bushing is forced into the

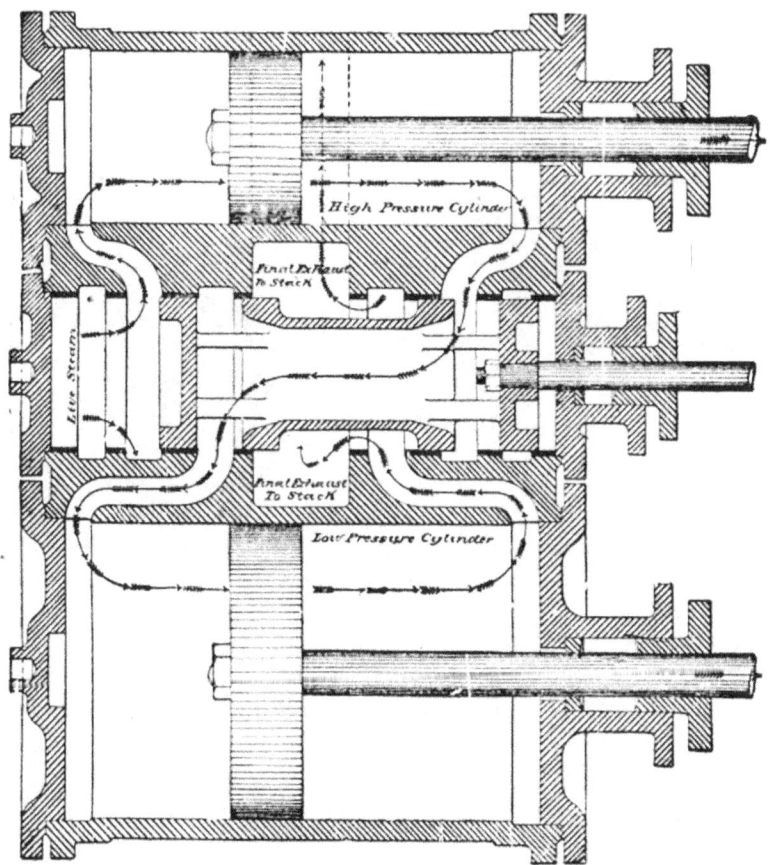

Fig. 3.

steam-chest under such pressure as to prevent the escape of steam from one steam passage to another except by the action of the valve. Thus an opportunity is given to machine accurately all the various ports, so that the admission of steam is uniform under all conditions of service.

The valve, which is of the piston type—double and hollow,—as shown by Fig. 5, controls the steam admission and exhaust

Fig. 4.

of both cylinders. The exhaust steam from the high-pressure cylinder becomes the supply steam for the low-pressure cylinder. As the supply steam for the high-pressure cylinder enters the steam-chest at both ends, the valve is in perfect balance, except the slight variation caused by the area of the valve stem at the back end. This variation is an advantage in case the valve or its connection to the valve rod should be broken, as it holds them together. Cases are reported where compound locomo-

Fig. 5.

tives of this system have hauled passenger trains long distances with broken valve stems and broken valves, the parts being kept in their proper relation while running by the compression due to the variation mentioned. To avoid the possibility of breaking, it is the present practice to pass the valve-stem through the valve and secure it by a nut on the front end.

Cast-iron packing rings are fitted to the valve and constitute the edges of the valve. They are prevented from entering the steam-ports when the valve is in motion by the narrow bridge across the steam-ports of the bushing, as shown in Fig. 4. The operation of the valve is clearly shown by Fig. 3, the direction of the steam being indicated by arrows.

When the low-pressure cylinder is on top, as shown by Fig. 2, the double front rail prevents the use of the ordinary rock-shaft and box, and the valve motion is then what is called "direct acting," changing the location of the eccentrics on the axle in relation to the crank-pin.

When the low-pressure cylinder is underneath, the rock-shaft is employed, and the eccentrics are placed in the usual

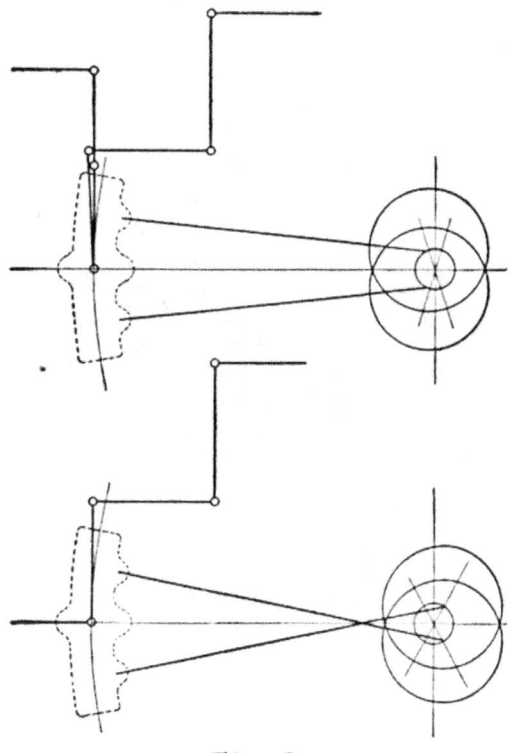

Fig. 6.

position; the valve motion is termed "direct acting." Fig. 6 shows the relation of the eccentrics with and without the rocker-shaft. Great care should be taken by mechanics, when setting the valves on these locomotives, to observe this difference and not get the eccentrics improperly located on the axle. If the crank-pin is placed on the forward center, the eccentric-rods will not be crossed when the rocker-arm or indirect motion is used, but will be crossed when no rocker-arm or direct motion is used. Serious complications have arisen from this being disregarded.

Various methods have been employed to transfer the motion from the link to the valve rod, that most commonly used being a small cross-head sliding between two guide bars, as shown by

Fig. 7. These parts are thoroughly case-hardened, and with reasonable care should wear indefinitely. It is preferable, however, to use a rock-shaft when possible, as there is then less departure from ordinary locomotive practice.

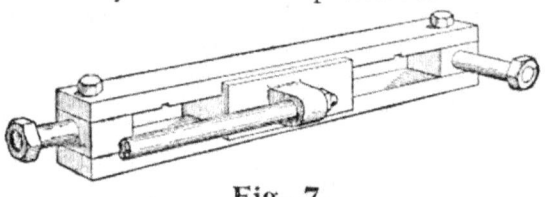

Fig. 7.

The cross-head is shown by Fig. 8. It is made of open-hearth cast steel and is machined accurately to size. The bearings for the guide-bars are covered with a thin coating of block

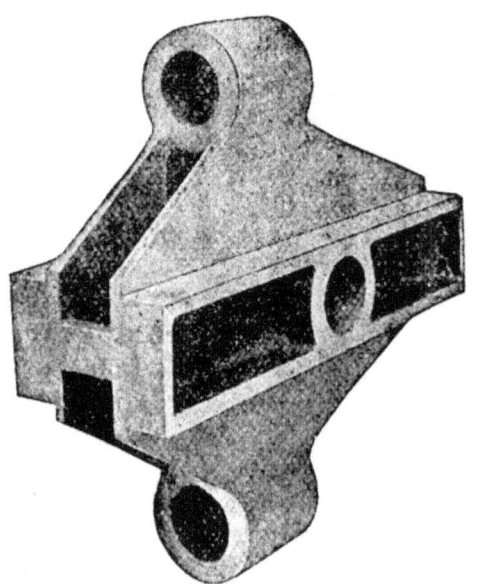

Fig. 8.

tin, about 1-16" thick, which wears well and prevents heating. The holes for the piston rods are bored so that the piston rods will be perfectly parallel, and are tapered to insure a perfect fit.

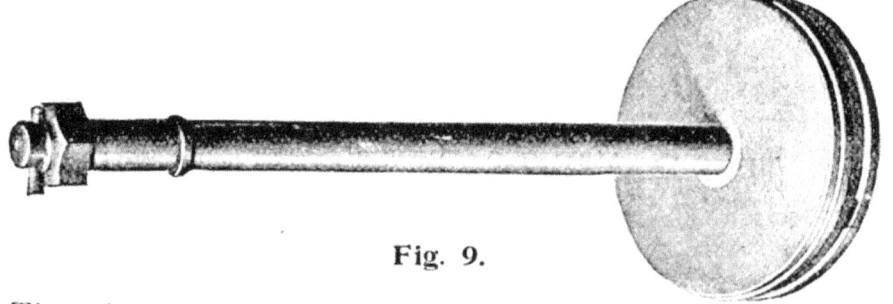

Fig. 9.

The piston shown by Fig. 9 is made with either cast-iron or cast-steel heads, and is as light as possible. The rods, which are of triple-refined iron, are ground perfectly true to insure good service in connecting with metallic packing for the stuffing boxes. The diameter of both piston rods is the same, both having equal work to perform. They are made large enough to

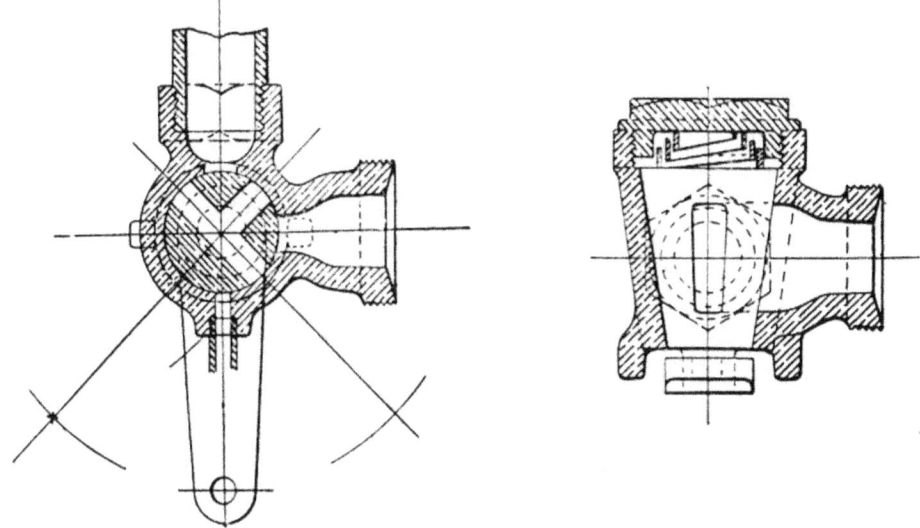

Fig. 10.

resist strains due to any unequal pressure that may come upon them in starting the locomotive from a state of rest. The crosshead has a shoulder which prevents the piston rod being forced

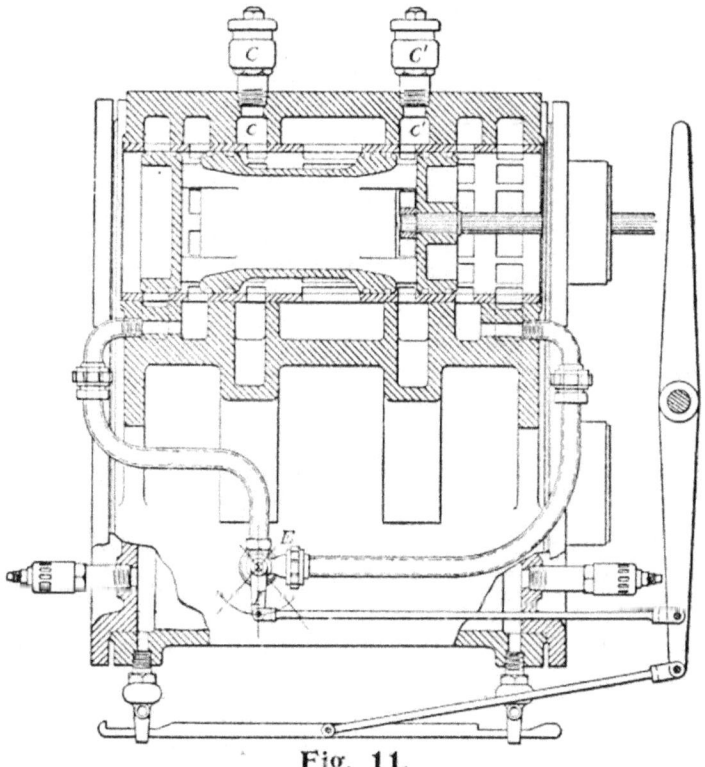

Fig. 11.

into the cross-head, and at the same time permits the cross-head end and the body of the piston rod to be of one diameter, thus permitting vibratory strains to act throughout the entire length of the rod instead of concentrating them at the shoulder next to the cross-head. The piston rods are secured to the cross-head by large nuts, and these in turn are prevented from coming loose by taper keys driven tightly against them.

It is obvious that in starting these locomotives with full trains from a state of rest, it is necessary to admit steam to the low-pressure cylinder as well as to the high-pressure cylinder, which is accomplished by the use of a starting valve (Fig. 10). This is merely a pass-by valve which is opened to admit steam to pass from one end of the high-pressure cylinder to the other end and thence through the exhaust to the low-pressure cylinder. This is more clearly shown at E in Fig. 11. The same

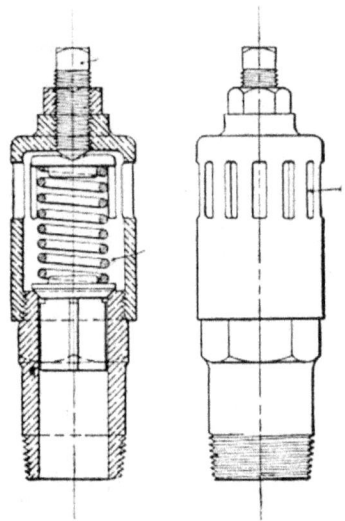

Fig. 12.

cock acts as a cylinder cock for the high-pressure cylinder and is operated by the same lever that operates the ordinary cylinder cocks, thus making a simple and efficient device, and one that need not become disarranged. This valve should be kept shut as much as possible, as its indiscriminate use reduces the economy and makes the locomotive "logy."

As is usual in all engines, air valves are placed in the main steam passage of the high-pressure cylinder. Additional air valves, marked C and C' in Fig. 11, are placed in the low-pressure cylinders to supply them with sufficient air to prevent the formation of a vacuum, which would draw cinders into the steam-chest and cylinders.

Water relief valves (Fig. 12) are applied to the low-pressure cylinders, and attached to the front and back cylinder heads, to prevent the rupture of the cylinder in case a careless engineer should permit the cylinders to be charged with water, or to relieve excessive pressure of any kind.

In all other respects the locomotive is the same as the ordinary single-expansion locomotive.

OPERATION.

It is not surprising, in view of their difference of opinion respecting single-expansion locomotives, but there has been much controversy among engineers and firemen in regard to the operation of compound locomotives of this system. The first thing the engineer must learn is to use the reverse lever for what it is intended, that is, he must not hesitate to move it forward when ascending a grade if the locomotive shows signs of slowing up. The reverse quadrant is always so made that it is impossible to cut off steam in the high-pressure cylinder at less than half stroke, which avoids the damage that might ensue from excessive compression. It is perfectly practicable to operate the engine at any position of the reverse lever between half stroke and full stroke, without serious injury to the fire. When starting the locomotive from a state of rest, the engineer should always open the cylinder cocks to relieve the cylinders of condensation, and as the starting valve is attached to the cylinder cocks, this movement also admits steam to the low-pressure cylinder and enables the locomotive to start quickly and freely. In case the locomotive is attached to a passenger train and standing in a crowded station, or in some position where it is undesirable to open the cylinder cocks, the engineer should move the cylinder cock lever in position to permit live steam to pass by into the low-pressure cylinder, thus enabling the locomotive to start quickly and uniformly, without any of the jerking motion so common on two-cylinder or cross-compound locomotives. After a few revolutions have been made and the cylinders are free from water caused by condensation or priming, the engineer should move the cylinder cock lever into the central position, causing the engine to work compound entirely. This should be done before the reverse lever is disturbed from its full gear position. The reverse lever should never be "hooked up," thereby shortening the travel of the valve, until after the cylinder cock lever has been placed in the central position. It is often necessary to open the cylinder cocks when at full speed, to allow water to escape from the cylinders, especially when the engineer is what is commonly called a "high-water" man, and in such cases no disadvantage is experienced and the reverse lever need not be disturbed. The starting device should not be used for any purpose other than the "starting" of the train. After the train is in motion it should not be used. Cases have been observed where the engineers use it all the time and have the reverse lever "hooked up" in the top notch (half stroke), in consequence of which the locomotive will slow down to a low speed whilst burning an excessive amount of coal. Such running must result in general dissatisfaction.

The starting device is useful in emergencies, as, for instance, when starting with a heavy train on a grade, if live steam is admitted to the low-pressure cylinder sufficient additional power is obtained to start the train and take it over the grade. This should be resorted to only in emergencies, and allowance should be made for the extra repairs caused by frequent cases of this kind.

On account of the very mild exhaust, the fireman should carry the fire as light as possible. A little practice will enable him to judge how to get along with the least amount of fuel.

The following diagram (Fig. 13) shows the difference in the amount of water required to do the work at various points of cut-off in compound and single-expansion locomotives. The upper line shows the rate of water consumption per horse-power developed for several points of cut-off in single-expansion loco-

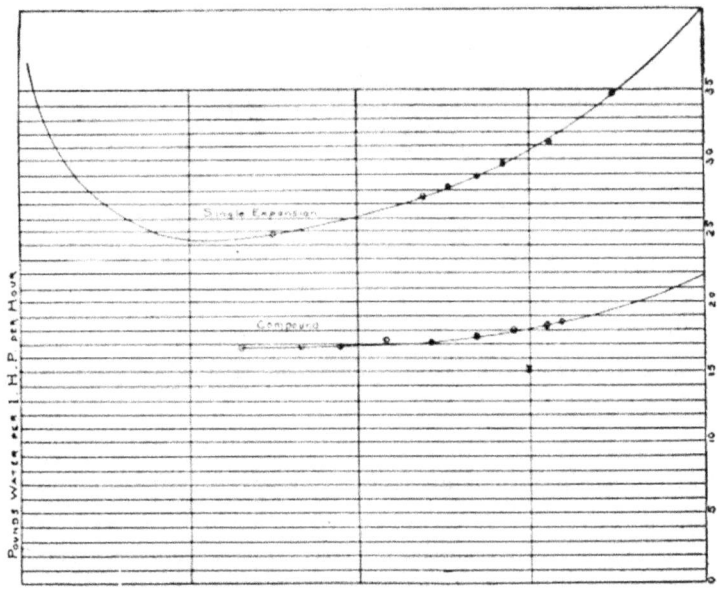

Fig. 13.

motives, whilst the lower line shows the same for compound locomotives. It will be observed that the most economical point of cut-off is about one-quarter stroke on the single-expansion locomotive, and about five-eighths stroke on the compound locomotive.

It is also noticeable that the water rate per horse-power varies very little on the compound locomotive when the reverse lever is moved toward full gear or longer cut-off, but in the single-expansion engine it increases rapidly, causing engineers to remark that they cannot "drop her a notch" on account of "getting away with the water." This does not occur with the compound locomotive when the reverse lever is moved forward towards full gear, and no engineer should open the pass-by valve, admitting live steam to the low-pressure cylinder, until the last

THE LOCOMOTIVE UP TO DATE. 343

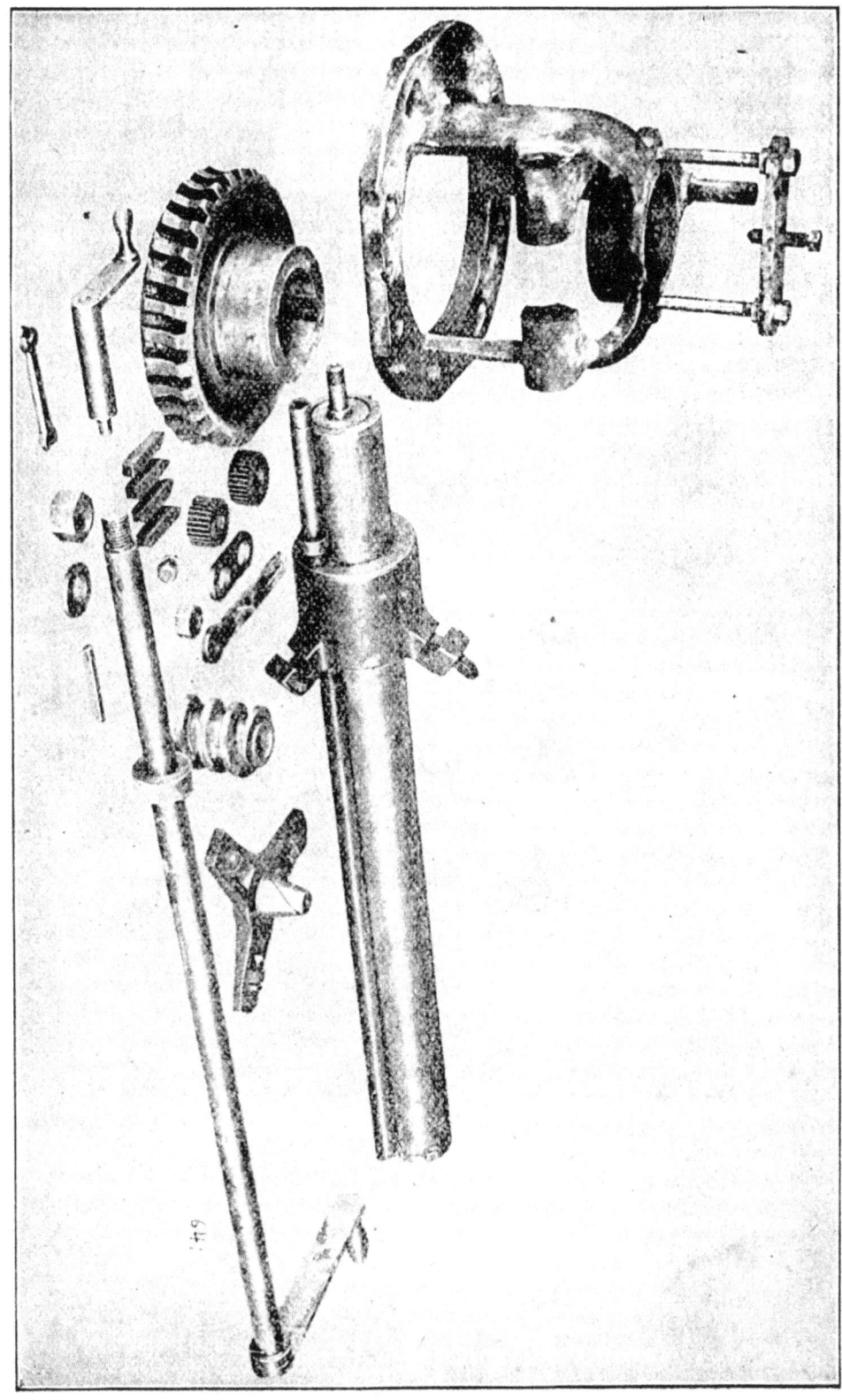

Fig. 14

notch has been used on the quadrant and the engine is about to stall.

It is also desirable to move the reverse lever forward a notch before the locomotive slows down too much, as it is better to preserve the momentum of the train than to slow down and again have the trouble of accelerating. In this way both coal and water are wasted.

If these instructions are observed the locomotives will work satisfactorily.

REPAIRS.

On account of their great simplicity to single-expansion locomotives, mechanics familiar with the latter have no difficulty

Fig. 15.

in understanding these compound locomotives. There is no new element of repairs introduced,—no complicated starting or reducing valves, such as are common to other systems of compound locomotives.

The cross-heads, when badly worn, may, in a short time, be retinned by any coppersmith; in fact, an ordinary laborer can be taught this in a few days. The cross-head is heated warm enough to melt solder, and is then cleaned and wiped with solder, using dilute muriatic, such as tinsmiths use in soldering. Block tin is then poured against the surfaces so prepared, to which it adheres. A piece of iron placed alongside the cross-head can be used to regulate the thickness.

The cross-head is then put on a planer to true it up, care being used not to let the tool "dig in" and tear off the tin.

The pistons are treated the same as in ordinary single-expansion engines. The packing rings in the low-pressure cylinder require renewal more frequently than those in the high-pressure cylinders. It is also more difficult in compound cylinders to detect faulty packing rings, and they are sometimes noticed only by the locomotive failing in steam and in not making time on the road.

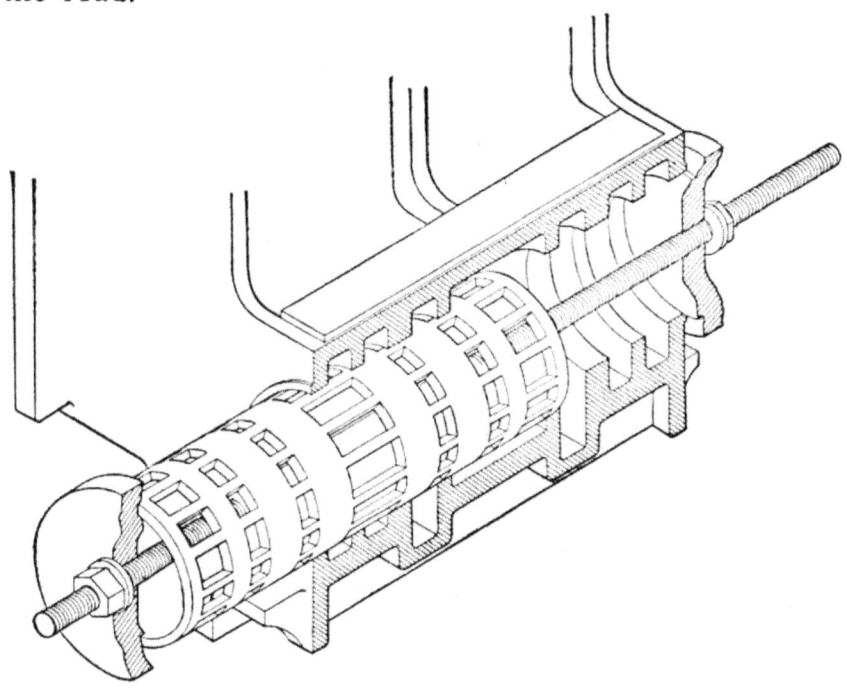

Fig. 16.

The piston valves should last a long time if properly lubricated, but when the bushing (Fig. 4) and valve (Fig. 5) are worn enough to require attention, the bushing should be bored out and new rings put in the valve; very often it is not necessary to bore the bushing, merely to put new packing rings in the valve.

After the bushings (Fig. 4) have been bored several times, larger valves may be fitted to them so as to have as little play as possible. Figs. 14 and 15 show a very convenient type or boring bar for boring out the bushings; it will be noticed that the work can be done without taking down the back head of the steam-chest. It is possible with this tool to bore out the bushings in less time than required to face a valve seat on a single-expansion locomotive.

When putting new bushings in the steam-chest, the device

shown in Fig. 16 may be used, which gives the required power and is slow enough to permit the bushing to accommodate itself to the cylinder casting.

When extracting old bushings, it is best to split them with a narrow cape chisel—they are only fit for scrap when removed, and can be much more quickly removed this way than to attempt to draw them out with draw screws.

Enough attention should be given the starting valves to insure their moving in harmony with each other. Engineers sometimes strain the cylinder cock shaft, which causes one starting valve to open and the other to remain shut; this causes the exhaust to beat unevenly, and the engineer is apt to complain that the valves are out of square. Before altering the valve motion on these engines, make sure that the starting valves open and close simultaneously, and examine low-pressure pistons and piston valve for broken packing rings. In one case an engineer ran his locomotive two days without any piston head on one of the low-pressure pistons, and even then could not tell what was the matter, only that the locomotive sounded "lame" and did not make good time with the train. Men were put to work to locate the trouble, and found it, to the great surprise of the engineer.

ACCIDENTS WITH THE VAUCLAIN COMPOUND.

VALVE STEM.

Clamp valve in center of seat to cover all steam ports, remove main rod on that side and clamp cross-head secure.

HIGH PRESSURE PISTON OR CYLINDER HEAD.

Remove the broken piston and clamp heavy board on end of cylinder, making steam-tight joint. The steam will then pass direct to the low pressure cylinder, making it high pressure.

LOW PRESSURE PISTON OR CYLINDER HEAD.

If piston or head is broken you may still proceed, using high-pressure cylinder; the steam will then exhaust through broken cylinder head instead of through stack, so if the escaping steam obstructs your view you had better disconnect that side.

MAIN ROD.

Clamp valve in center of seat to cover all ports, remove broken rod and clamp cross-head secure—run in with one side. If your engine has direct motion, and valve stem is fastened to the cross-head, clamp cross-head secure in the center of the guides.

EQUALIZING VALVE.

If the small equalizing valve in the end of the main steam valve is broken or removed it will convert that side into a high-pressure engine, the work being done by the low-pressure cylinder; the high-pressure piston then would be approximately balanced. A head broken out of the main steam valve would have the same effect.

BALDWIN TWO-CYLINDER COMPOUND LOCOMOTIVE.

The Baldwin two-cylinder compound locomotive is illustrated and described in the following pages. In designing this type of compound locomotive, the essential feature sought was to pro-

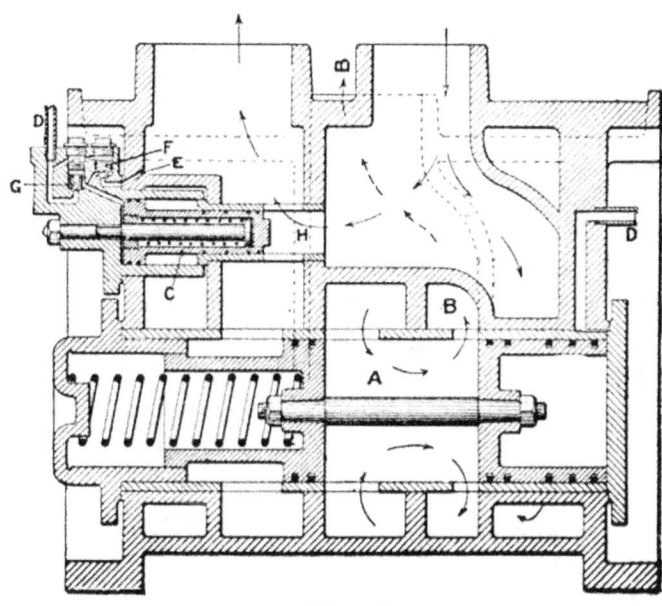

Fig. 1.

vide an intercepting and reducing mechanism which would permit the engine to work as a single-expansion engine until such time as the engineer desired to change it to a compound. To obtain this result, the normal condition of the parts employed is such that the engine will start at any point of the stroke and any position of the crank.

To accomplish this result an intercepting valve A is employed, consisting of two pistons, connected by a distance-bar or rod, also a reducing valve C, both of which are placed in the saddle casting of the high pressure cylinder.

The function of the intercepting valve A is to divert the exhaust steam from the high-pressure cylinder either into the atmosphere, when working single-expansion, or into the receiver, when working compound, and is operated at will by the engineer.

The function of the reducing valve is to admit live steam at reduced pressure into the receiver when engine is working single expansion, and also to close itself instantly when the intercepting valve is changed to the position which causes the engine to work compound; it being evident that the receiver

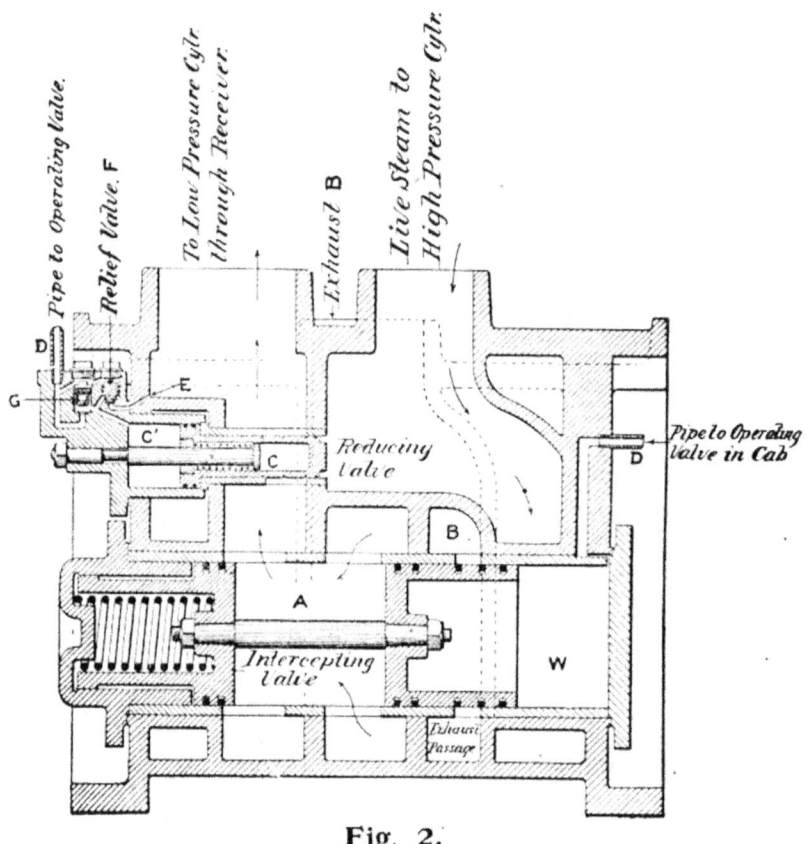

Fig. 2.

needs no live steam from the boiler when receiving its supply from the exhaust of the high-pressure cylinder. A further function of the reducing valve is to regulate the pressure in the receiver so that the total pressure upon the piston of the two cylinders will be equalized. The intercepting and reducing valves are both cylindrical in form, placed in bushings in which suitable ports are cut. The movement of the valves A and C in one direction is caused by steam-pressure against the action of suitable coil-springs. These springs cause a return movement when the steam pressure is withdrawn. These valves are controlled by steam supplied through pipes D from the operating valve in the cab.

Consolidated Engine No. 62, Norfolk & Western Ry.
Baldwin Two Cylinder Compound.
General Dimensions Given on Page 493.

The reducing valve C is operated automatically by the pressure in the receiver when not closed permanently by the steam in the operating pipe D. For this purpose the port E is provided, connecting the receiver with the large end of the reducing-valve piston, under the poppet-valve F, which remains open as long as the engine is working compound, the poppet-valve G being held closed by the same pressure, thus preventing its escape to the atmosphere. This pressure, acting on the large end

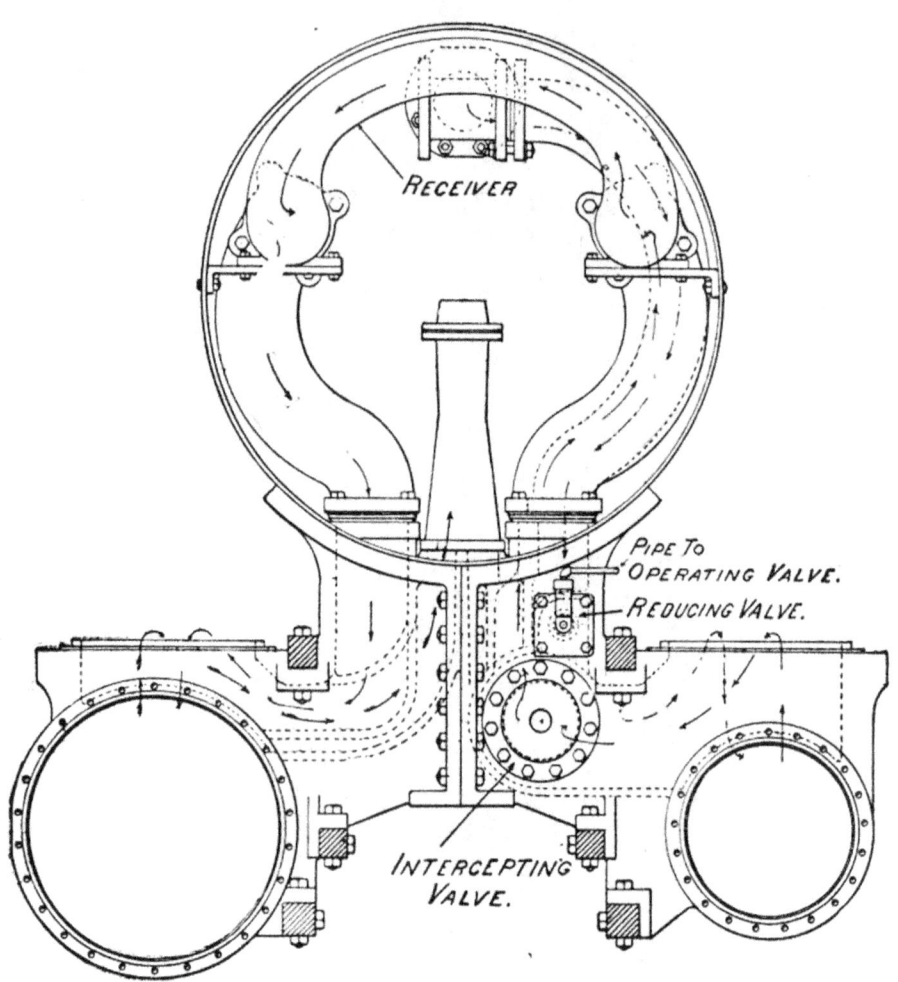

Fig. 3.

of the reducing valve, causes it to close the steam passage H between the live-steam passage of the high-pressure cylinder and the receiver when the receiver-pressure becomes excessive, and *vice versa* when deficient, the two ends of the reducing valve being so proportioned that equal cylinder power will be given to both sides of the engine. When the engine is in its normal condition, the lever of the small operating valve in the cab is placed at position marked "simple," and the engine is then in position to work as a single-expansion engine, the steam pressure against the intercepting valve and the large end of the reducing valve being relieved, permitting the valve to assume (by the action of the springs) the positions shown in Fig. 1. The ports of the intercepting valve A stand open to receive the exhaust steam from the high-pressure cylinder and deliver it to the auxiliary exhaust passage B, and through it to the atmosphere. At the same time the reducing valve is wide open, connecting the live-steam passage H with the receiver, through which live steam enters and charges the receiver, from which the low-pressure cylinder derives its supply. The receiver pressure is governed by the automatic action of the reducing valve, previously explained.

Thus the engine can be used as a single-expansion engine to make up trains and start them, and at the will of the engineer the operating valve in cab can be moved to position marked "compound," which will admit steam to the supply pipes D, thence to the cylinder W and cylinder C, changing the intercepting and reducing valves instantly and noiselessly to position, as shown in Fig. 2, diverting the exhaust from the high-pressure cylinder to the receiver, instead of to the atmosphere, and closing the passage H between the live-steam passage and the receiver at the reducing valve. All the parts liable to wear are made of the piston type, depending on packing rings instead of general joints for prevention of leakage, and at the same time prevent the hammering action so common with valves of the poppet type automatically arranged.

At any time the engineer may desire to increase the traction power of the engine, as, for instance, when the danger of stalling, by moving the lever of the operating valve to position marked "simple," the engine is again changed instantly to a single-expansion engine.

ACCIDENTS WITH THE BALDWIN TWO-CYLINDER COMPOUND.

FOR MAIN ROD, HIGH PRESSURE.

Remove broken rod, block crosshead at back end of guide, clamp high-pressure valve in center to cover both ports. Place engineer's operating valve at point marked "simple." The steam will then pass through the reducing valve to the receiver and thence to the low-pressure cylinder.

FOR MAIN ROD, LOW PRESSURE.

Remove broken rod, block crosshead at back end of guide, disconnect valve rod and close plug cock in pipe leading to intercepting valve. Place engineer's operating valve at point marked "compound." The steam will then act in the high-pressure cylinder and pass to the exhaust without entering the low-pressure cylinder.

FOR VALVE STEM ON EITHER SIDE.

Same remedy as for broken rod.

FOR INTERCEPTING AND REDUCING VALVE.

The intercepting and reducing valves are automatically operated by steam pressure. Access can be easily had to either of these valves by removing the head of their respective cylinders, and any failure of the valve to act can be readily ascertained and remedied.

RICHMOND SYSTEM, COMPOUND LOCOMOTIVES.*

This type belongs to the two-cylinder class, having a high-pressure cylinder on one side and a low-pressure cylinder on the other, and when working as a compound engine steam enters the high-pressure cylinder direct, as in simple engines, and is exhausted through the receiver into the low-pressure steam chest and cylinder and thence exhausted up the stack in the usual way. In the passage from the high-pressure to the low-pressure cylinder are three valves, of which a detail description follows, and have the following functions:

(1) They enable the low-pressure cylinder to be supplied with sufficient live steam for starting before the high-pressure cylinder exhausts.

(2) When desired they open a separate exhaust from the high-pressure cylinder up the stack, at the same time close communication between the high and low-pressure cylinder, and admit live steam to the low-pressure cylinder, thus virtually converting the engine into a simple engine, as both cylinders receive live steam supplied direct from the boiler and each has its separate exhaust up the stack. This enables the engine to develop from 20 to 25 per cent more power for starting heavy trains and prevents stalling on heavy grades.

DESCRIPTION OF INTERCEPTING VALVE.

The various drawings show sections through the low-pressure cylinder saddle, with the valves in their four relative positions. Figure 1 shows the position of valves in starting automatically just after the throttle is opened. Fig. 2 shows the same at maximum pressure in low pressure steam chest.

Figure 3 shows the position when working compound, and Fig. 4 shows position when working simple. The high-pressure cylinder exhausts into a receiver, which is placed inside the smoke box and opens into the chamber "F." The intercepting

---

* This article was especially prepared for this book by the Richmond Locomotive and Machine Co.

Engine No. 673, Canadian Pacific Ry.
Richmond Compound.
General Dimensions Given on Page 502.

valve, as shown at "V" in the several views, has a piston on its outer end, which acts as an air dashpot, preventing any slamming of the valve. Around the steam of this valve is a sleeve "L," which slides on the stem, and acts as an admission and reducing

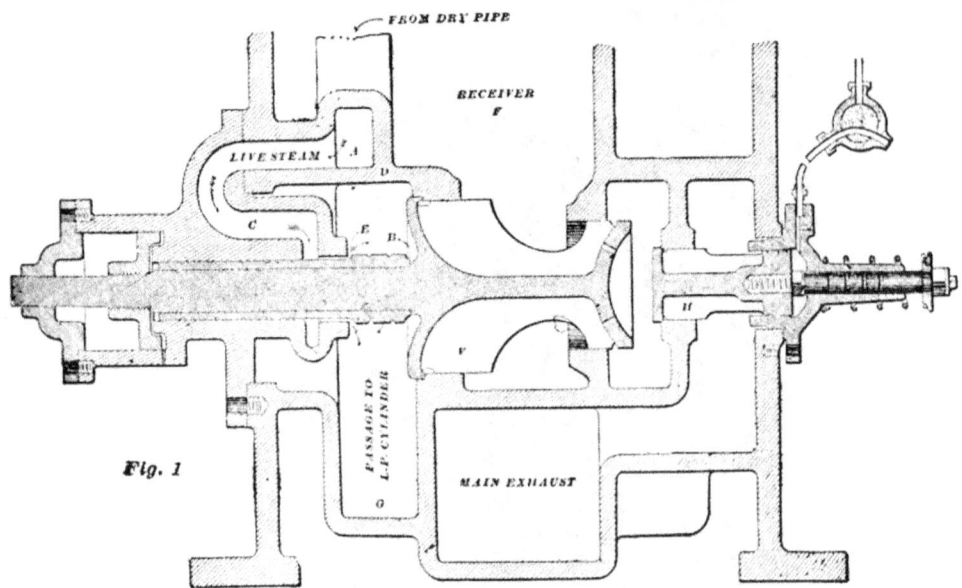

Fig. 1

valve to the low-pressure steam chest when starting and when working simple. Valve "H" is a plain bevel-seated winged valve, and is called the emergency valve, as by its use the engineer can, at will, operate as a simple engine. When starting,

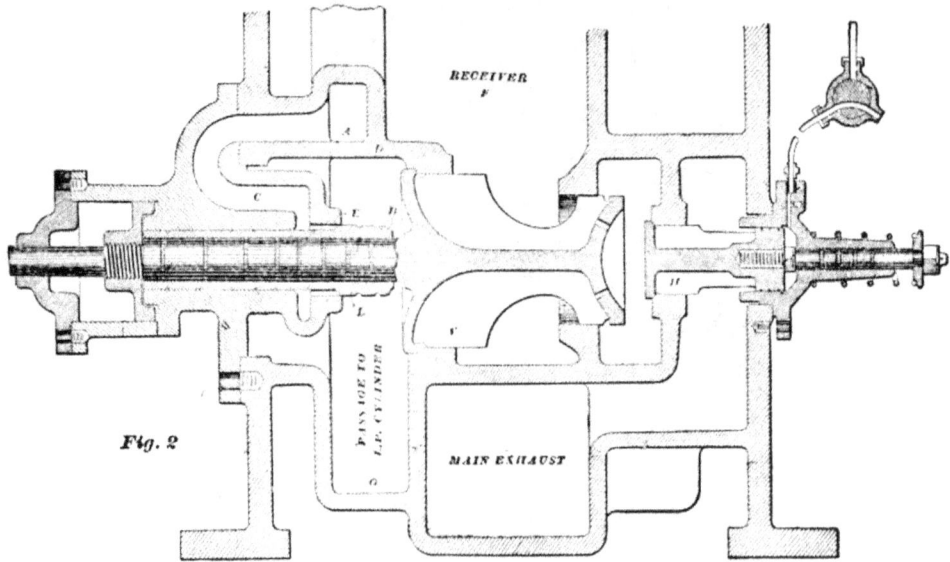

Fig. 2

steam from the boiler goes to the high pressure cylinder in the ordinary way, and also to the port "C" through a two-inch steam pipe connected to the dry pipe. When the throttle is opened, no matter in what position the valves stand, there is no pressure in the receiver "F," and the pressure on the shoulder "E" of the the receiver and letting steam past the shoulder "E" into the low-sleeve "L" moves the sleeve and valve "V" to the right, closing pressure steam chest "G," as shown in Fig. 1.

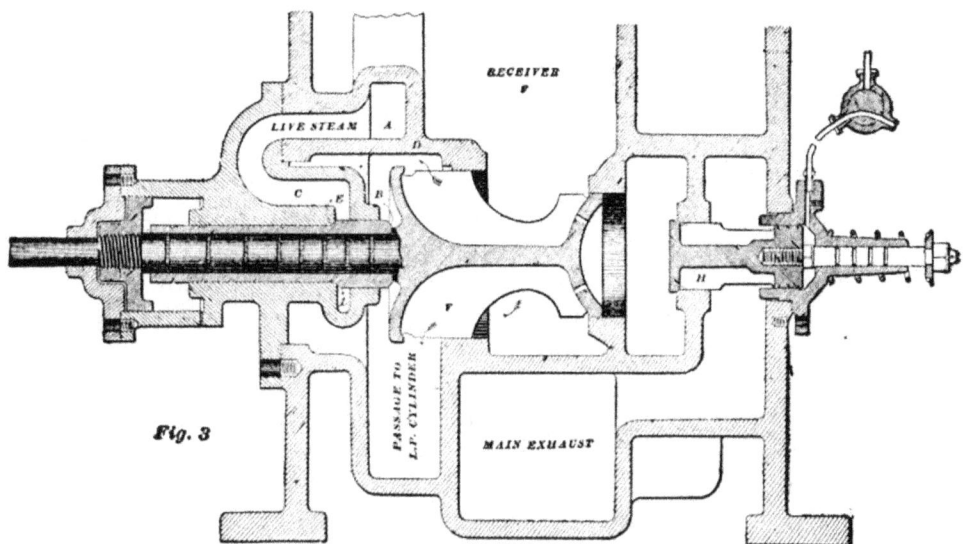

Now, since the area of the end "B" of the sleeve "L," shown in Fig. 2, is say twice that of the shoulder "E," half of the boiler pressure will move the sleeve "L" to the left, cutting off steam through port "C," and thus equalizing the work in both cylinders, since the reduced pressure is thus maintained in the

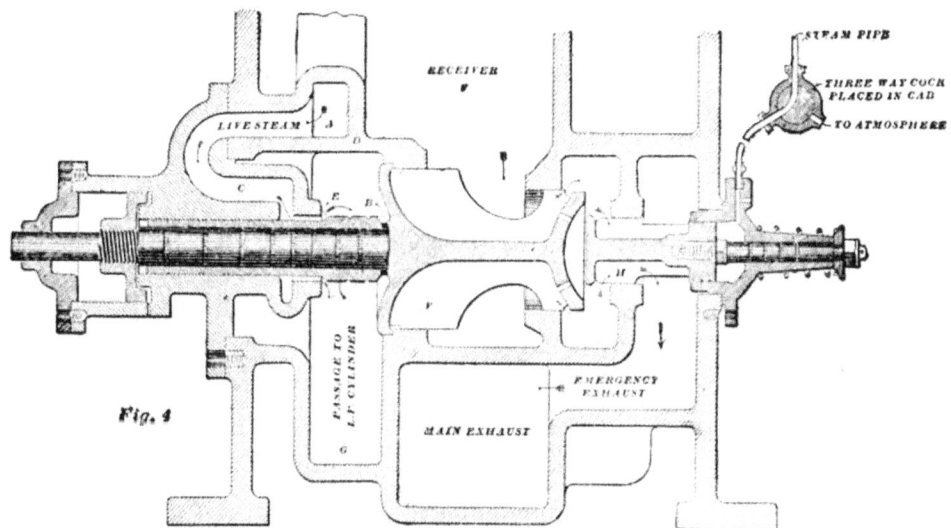

low-pressure steam chest by the reciprocating action of the sleeve. After, say one-half revolutions, the pressure accumulates in the receiver "F," due to the exhaust from the high-pressure cylinder, and acting against the large face of valve "V," moves this valve to the left, carrying the sleeve with it, thus opening a straight connection between the high-pressure exhaust and the low-pressure steam chest, and at the same time permanently cutting off live steam from port "C," as shown in Fig. 3.

In starting on grades, or when exercising maximum power, the engineer can move the three-way cock in the cab, letting boiler steam behind the piston on the emergency valve "H," and holding it open against its spring. This exhausts the small cavity "J," in which the pressure is equalized with the receiver through holes in the rear end of valve "V," and then the valve, being unbalanced, moves with the sleeve "L" instantly to the right, assisted by steam pressure on the shoulder "E" of the sleeve. The high-pressure cylinder has now a separate exhaust around the end of valve "V," through valve "H," into the main exhaust, since the intercepting valve remains closed, due to no accumulated pressure in the receiver "F." The low-pressure steam chest then gets reduced pressure steam direct from the boiler through port "C," and reducing valve "L," as shown in Fig. 4.

Except when working simple, the valves act entirely automatically. The lubricator to the low-pressure cylinder enters port "A," and thus ensures constant lubrication to the intercepting and reducing valve. Owing to the small area of port "C," and the contracted exhaust through valve "H," the engine develops less power as a simple engine than as a compound, at a speed of over say eight or ten miles an hour, and thus the runner is compelled to work compound.

On compound locomotives difficulty has been experienced by rough riding when the engine is shut off and running at moderate or high speeds, due to air compression in the large low-pressure cylinder. This has been overcome in the Richmond type by the introduction of "over-pass valves" which automatically open a passage from one end of the cylinder to the other of sufficient size to allow the air to circulate freely from one end to the other when the engine is shut off, thus effectually preventing any compression and consequent rough riding. These valves also reduce the effect of fanning the fire which is caused in some types by the air that is pumped through the cylinder and exhausted up the nozzle, and consequently effects quite a saving in coal when rolling down long grades.

DIRECTIONS FOR OPERATING RICHMOND SYSTEM.

With ordinary trains the engine will start smoothly with the emergency exhaust valve in the compound position—that is, with the handle of the operating valve in the bridge pipe pointing ahead; but with heavy trains the start should be made with the valve in the simple position, or with the handle pointing to the rear; it should be left there, however, as short a time as possible, usually six or seven car lengths being sufficient distance, as when running in this position the exhaust has a very severe action on the fire, besides increasing the steam consumption.

The cylinder cocks should always be opened when starting, as condensation is very rapid for the first few revolutions, especially in the high-pressure cylinder. Water resulting from this condensation is carried over to the low-pressure cylinder, and increases the danger of blowing out a cylinder head should slipping occur.

USE OF EMERGENCY VALVE WHILE RUNNING.

Only use the simple feature (the emergency) with the reverse lever in the corner, and cut it out before pulling the lever back. It should only be used to avoid stalling, and at speed not over six to eight miles per hour.

USE OF REVERSE LEVER.

In cutting the reverse lever back, do not expect to find the best running position as near the center of the quadrant as in a simple engine. You will find that the engine does better work with the lever nearer the corner, just how far depending, of course, on the varying conditions of load, grade and speed; but in general do not cut the lever back too far.

It is also to be noted that you can drop your lever lower without tearing the fire than with a simple engine, and this should be done readily when making a run for a hill.

THROTTLING.

The throttle should in general be run as wide open as possible, but much judgment should be exercised in this respect, as it is frequently better policy to drop the lever a notch and throttle slightly. No hard and fast rule can be given for this, as every engineer knows from his experience.

LUBRICATION.

Lubrication of the cylinders is accomplished by means of one side of the lubricator only, viz.: the one connected to the high-pressure steam chest. It will be noted that the oil pipe on the low-pressure side does not lead to the steam chest at all, but is connected to the live steam passage of the intercepting valve, so that when the engine is running compound no oil can get past the seat on the inner end of the reducing valve, resulting in a positive waste of oil if fed with the valve in that position.

When running feed from six to ten drops per minute through the left side of the lubricator only, varying the feed as quality of the steam or duty of the engine offers. If the water is raising badly more oil is, of course, required.

Before starting allow a few drops of oil to pass through the right side of the lubricator to lubricate the intercepting valve, but shut it off immediately when the engine is thrown into compound. When starting compound a small quantity (about 6 drops) is sufficient. When it is necessary to work the engine simple, reduce the feed on the left side and start the feed on the right side. The oil then passes through the reducing valve with the live steam into the low-pressure cylinder.

Avoid feeding oil into the intercepting valve as much as possible, as it has a tendency to gum up the small packing rings.

On the emergency exhaust operating valve in the bridge pipe will be noticed a small oil cup. Fill this cup with cylinder oil once every two days for the purpose of lubricating the emergency exhaust valve. *Do not flood this valve with oil.* The steam merely forces this valve off its seat, and is exhausted through the same pipe that admitted it, naturally entraining most of the oil and preventing its loss.

The over-pass valves are located on the outside of the low-pressure cylinder along the side of the steam chest, and operate automatically to open the connection between the two steam ports in the cylinder when drifting, allowing air to pass from one end of the cylinder to the other. The valves shut again when the throttle is opened. The valves should be cleaned at intervals, especially when the engine is new, otherwise the sand and grit will cause them to stick.

ACCIDENTS WITH RICHMOND COMPOUND LOCOMOTIVES.

The simplicity of this system and general arrangement of its parts is such that in the case of a break down it is only necessary to convert the engine into simple by operating the three-way cock in the cab and proceeding exactly as in the case of an ordinary simple engine when disabled.

SCHENECTADY COMPOUND.*

COMPOUND TWELVE-WHEEL LOCOMOTIVE—NORTHERN PACIFIC RAILWAY.

A noble engine, indeed, is that which is the subject of our illustration. It is a twelve-wheeled two-cylinder compound built by the Schenectady Locomotive Works for especial service as helper on a 17-mile, 116-ft. grade on the Northern Pacific Railway between Helena and Missoula. It is undoubtedly one of the most powerful locomotives of its type in the country, but, with all its enormous weight and size, it presents a neat and symmetrical appearance, as may be seen from our perspective view. In addition to this perspective we give a bit of detail showing the peculiar method of attaching the low-pressure cylinder to the

---

* This article was especially prepared for this book by the Schenectady Locomotive Works.

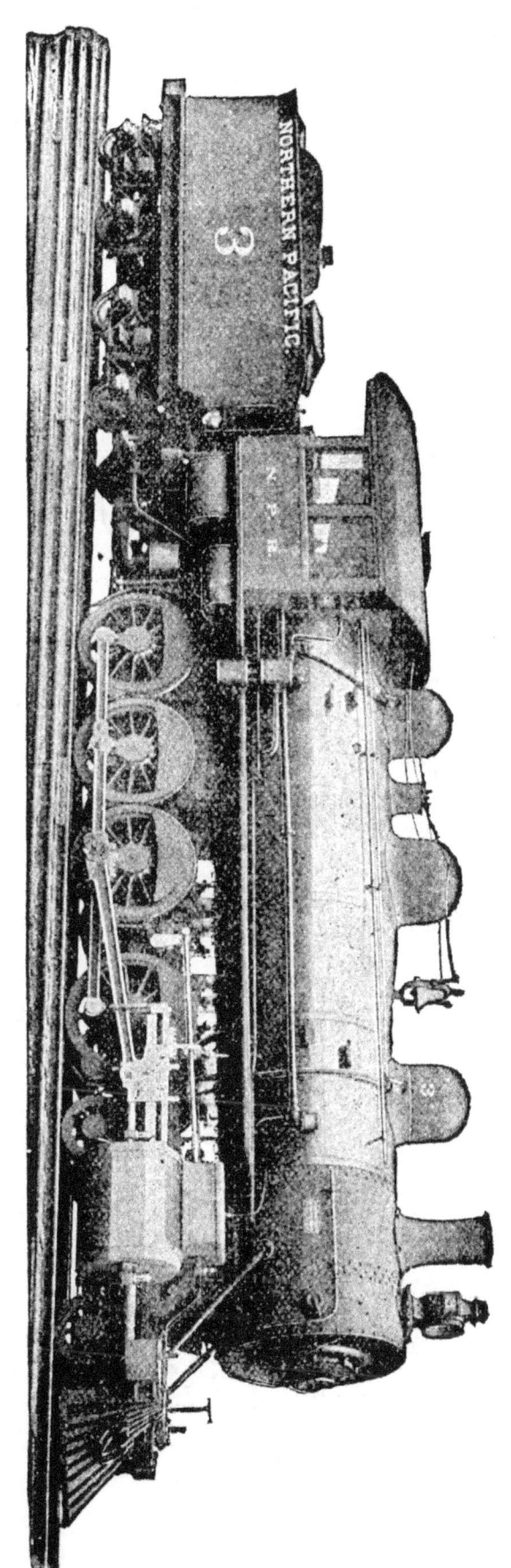

Engine No. 3, Northern Pacific Ry.—Compound Freight.

Schenectady.

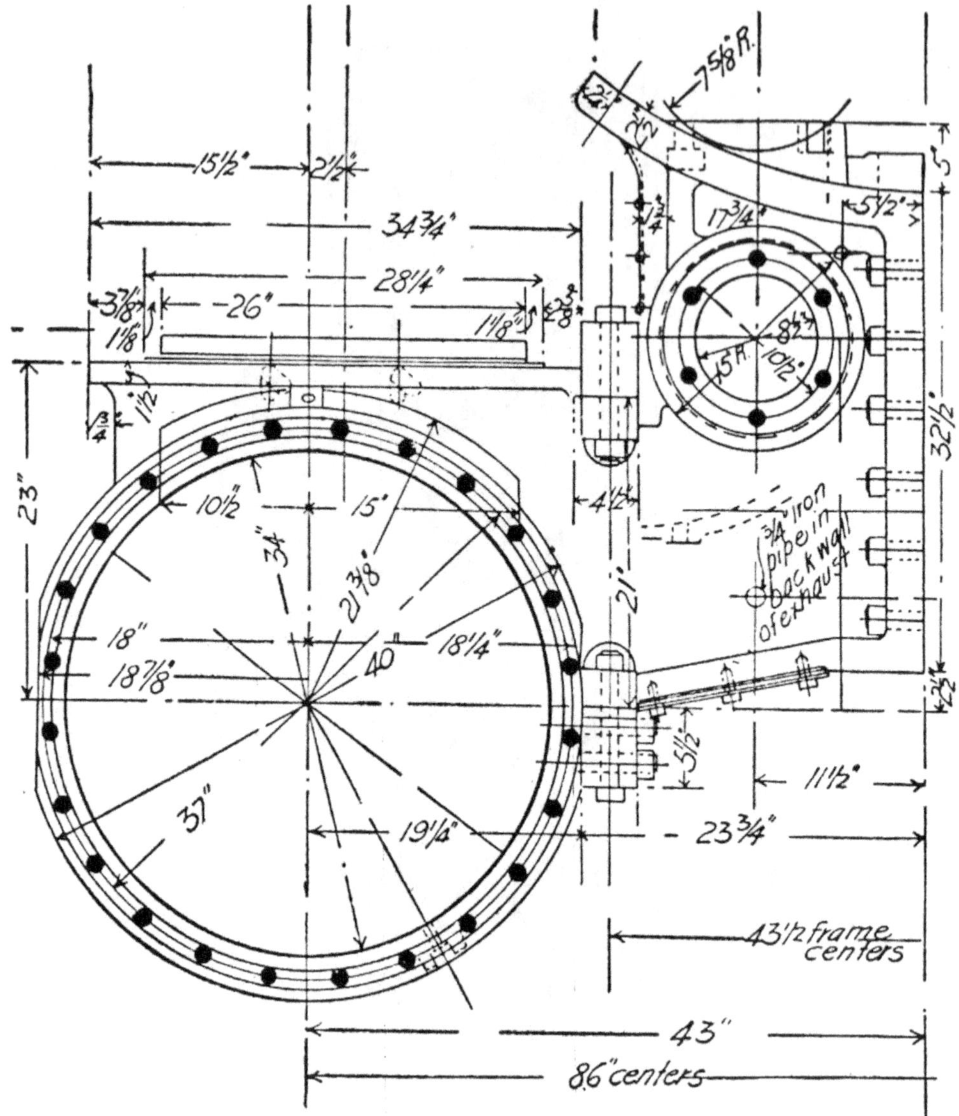

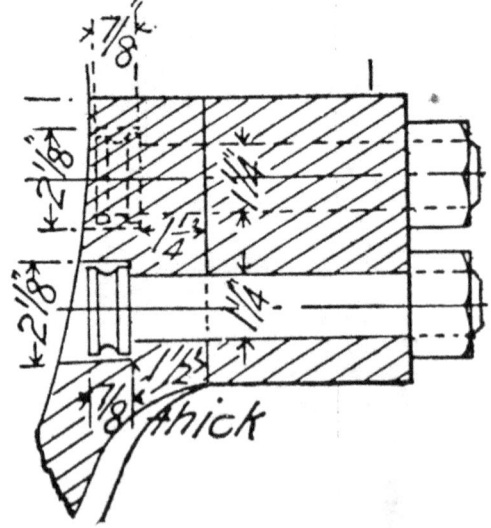

Fig. 1.

frame, and a series of views of the new intercepting valve (the design of Mr. A. J. Pitkin and Mr. J. E. Sague, of the Schenectady works), with which this locomotive is fitted.

The cylinders of this engine are 23 and 34x30 in.; the drivers are 55 in. in diameter; the boiler, of the wagon top type, being 72 in. in diameter at first ring and 30 ft. 7 in. long and having a total heating surface of nearly 3,000 sq. ft. and a grate area of 35 sq. ft.; the engine weighs 186,000 lbs., of which 150,000 are on the drivers, and it has a draw-bar pull of from 35,000 to

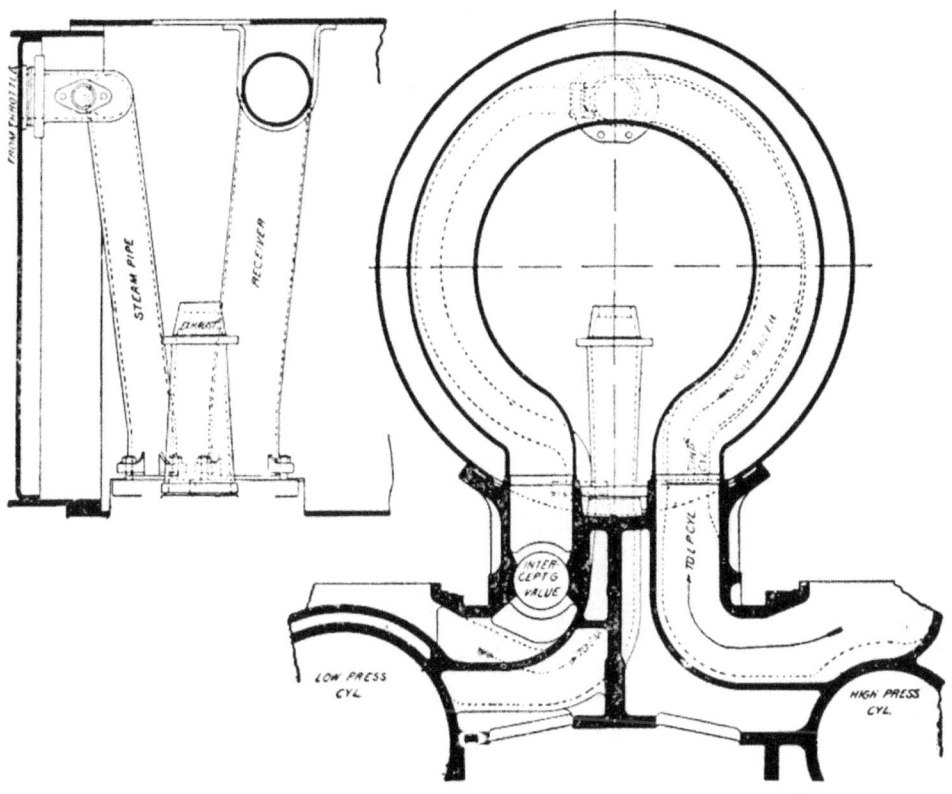

Fig. 3.

40,000 lbs., and develops 1,200 horse power for three consecutive hours at 16 miles per hour.

The extended wagon top boiler has radial stays and is designed to carry 200 lbs. pressure. Its fire-box is 120 in. long and 42 in. wide. It presents 206.51 sq. ft. of heating surface, which, with 2721.6 sq. ft. that the water tubes supporting the brick arc give, makes a total of 2943.4 sq. ft. of heating surface.

The method of attaching the low-pressure cylinder to the frames is shown in Fig. 1, the interesting feature being the connection made with the lower bar of the frame. Seven horizontal 1¼ in. bolts are used here and they extend through the bar and into and through the cylinder walls. From the detail given in

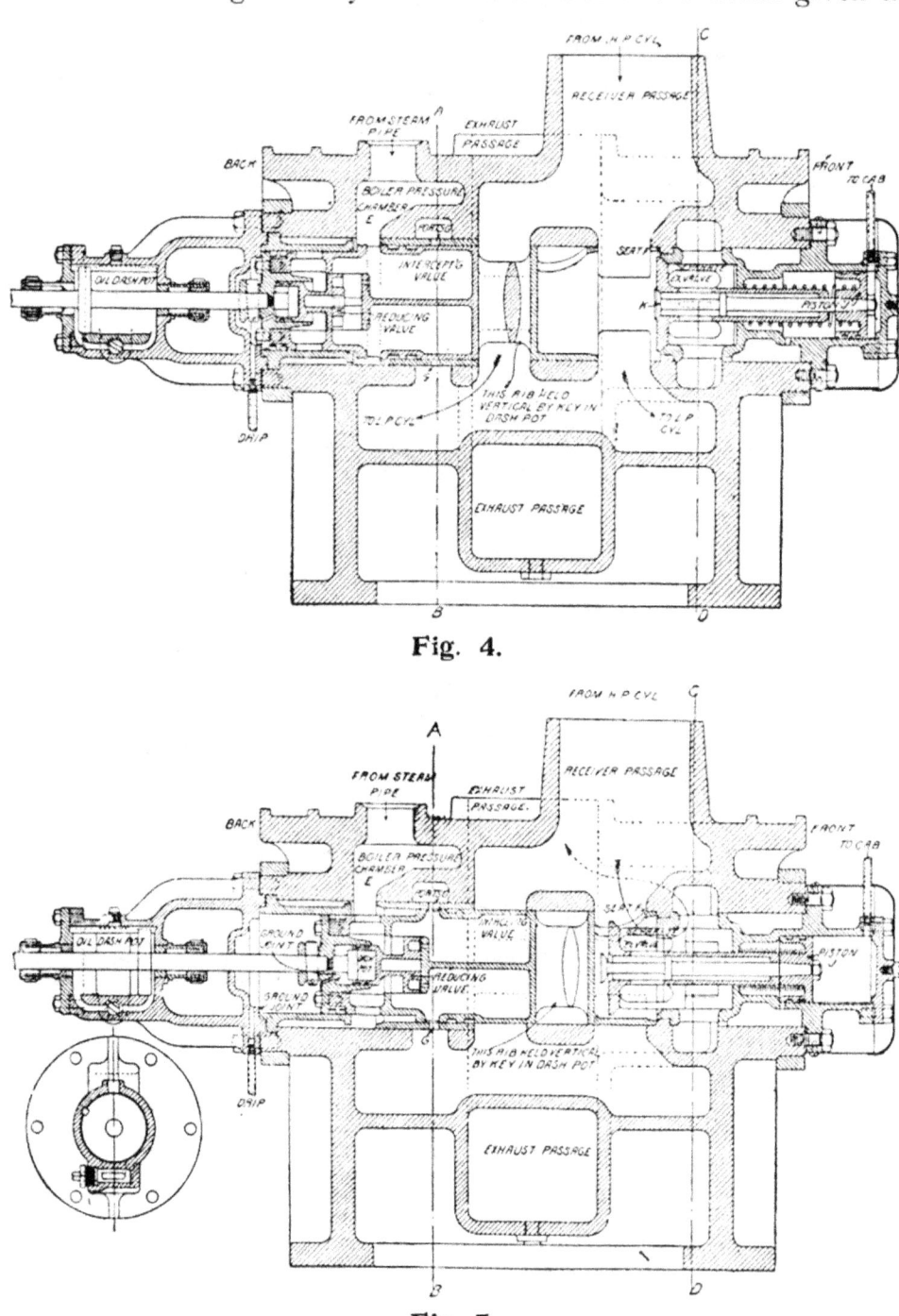

Fig. 4.

Fig. 5.

Fig. 2 it will be seen that these bolts have 2 in. grooved heads which are counter sunk in the inner face of the cylinder walls. The counter bore about these heads is filled with babbitt when the bolts are in position and the cylinder boring leaves, of course, a perfect finish.

To the intercepting valve used on this engine we now wish to give extended attention; we will say, however, before opening our account of this intercepting device that in trials of this engine on the New York Central Railroad before it was sent west, some tests in the way of changing from compound to simple were made while working under 200 lbs. pressure, full stroke at 15 miles per hour. This rather hard imposition was borne without the slightest evidence of complaint from the engine for five successive times.

THE INTERCEPTING AND SEPARATE EXHAUST VALVES.

Fig. 3 gives sections through the smoke arch and cylinder saddle and intercepting valve, and shows the intercepting and

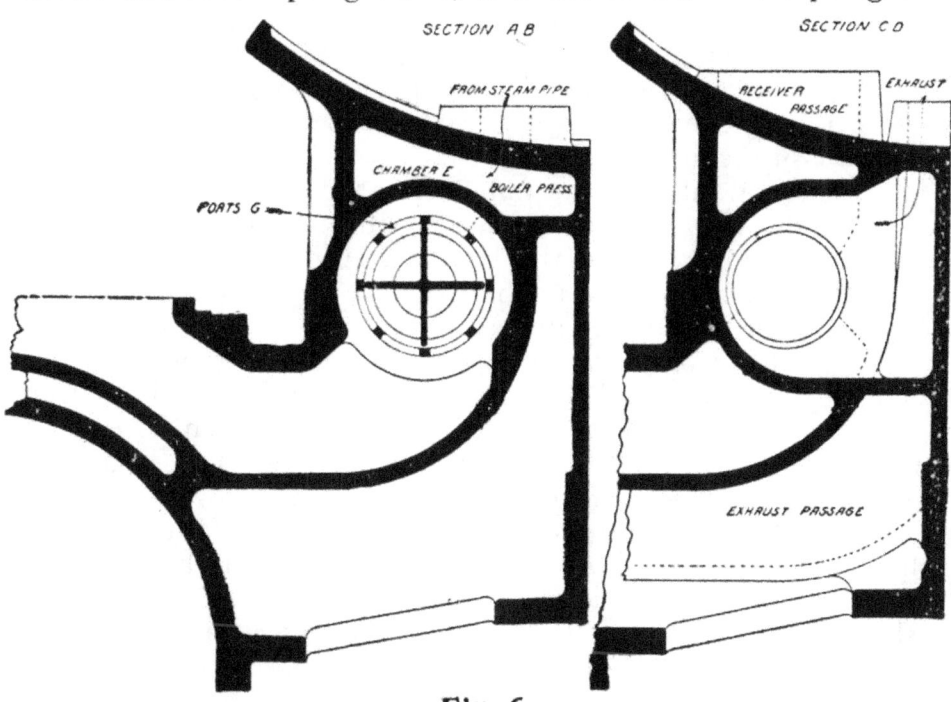

Fig. 6.

separate exhaust valves in the position taken when the engine is working simple.

Fig. 4 shows the same section through the low-pressure cylinder as in Fig. 5, but shows the intercepting and separate exhaust valves in the position assumed when the engine is working compound.

Fig. 6 shows two transverse sections through the low-pressure cylinder saddle at AB and CD, for location of which see Fig. 4. Section AB shows the passages for admitting live steam into the low-pressure cylinder, and section CD shows the outlet passage from the separate exhaust valve to the main exhaust pipe.

With the arrangement of valves shown in these figures the engine can be started and run either compound or simple, and can be changed from compound to simple or from simple to compound at the will of the engineer, with any position of throttle and at any point of cut-off. The part which each valve does in accomplishing this is as follows:

The separate exhaust valve when open allows the steam to exhaust direct from the high-pressure cylinder to the atmosphere without going through the low-pressure cylinder, thus working the engine simple and when closed causes the steam from the high-pressure cylinder to go through to the low-pressure cylinder, thus working the engine compound. The intercepting valve closes the passage between the cylinders when the separate exhaust valve is open, so that steam cannot go from the high-pressure cylinder to the low-pressure cylinder, and it also admits steam to the low pressure cylinder direct from the dry pipe through the reducing valve. When the separate exhaust valve closes, the intercepting valve opens the passage between the cylinders and cuts off the supply of steam from the dry pipe to the low-pressure cylinder.

The reducing valve works only when the engine is running simple and throttles the steam passing through it, so that the pressure of steam going to the low pressure cylinder is about one-half the steam pressure in the dry pipe.

The intercepting and reducing valves are worked automatically by the steam pressures acting on the difference of areas of the ends of the valves, and their movement is cushioned by dash pots. The separate exhaust valve is operated by the engineer by means of a three-way cock in the cab. To open the separate exhaust valve the handle of the three-way cock is thrown so as to admit air or steam pressure against the piston "J." Pulling the handle back relieves the pressure against "J" and the spring, which is shown in the figures, shuts the valve.

All the engineer ever has to do in connection with the operation of the valves is to pull the handle of a three-way cock in the cab, one way or the other, according as (he) wishes the engine to run simple or compound. The engineer uses this handle under the following conditions:

First, to start simple. Under ordinary conditions this is not necessary, but if the maximum tractive power of the engine is needed to start a heavy train, the engineer pulls the handle of the three-way cock so as to admit pressure on the piston "J," which is then in the position shown in Fig. 4. This will force the piston "J" into the position shown in Fig. 5, which opens the separate exhaust valve and holds it open. As soon as the throttle is opened, steam at boiler pressure enters the chamber "E" and forces the intercepting valve against the seat "F," as shown in Fig. 5. Steam enters the high-pressure cylinder and is exhausted through the reservoir through the receiver pipe and separate exhaust valve to the atmosphere, as shown in Figs. 3 and 5. Steam also enters the low-pressure cylinder from the chamber "E" through the reducing valve and ports "G," shown in Figs. 5 and 6, and is exhausted in the usual way. The steam is prevented from reaching the low-pressure cylinder at boiler pressure by going through the reducing valve. As will be seen from Fig. 5, the valve is partly balanced by the cylinder open to the atmosphere, and boiler pressure acting on the unbalanced area throws the valve to the right. When the pressure on the right of the valve becomes high enough it will throw the valve to the left, because it acts on the whole area of the valve, and in so doing throttles the steam to the proper pressure for the low-pressure cylinder.

Having started the train in this way, when the engineer wishes to change the engine from running simple to running compound, he pushes the handle of the three-way cock to its first position, which relieves the pressure on the right of the piston "J," and the spring throws that piston to the right, into the position shown in Fig. 4, closing the separate exhaust valve. As soon as this valve is closed the pressure in the receiver rises and presses the intercepting valve to the left, against the pressure in the chamber "E," which only acts as an unbalanced area of the valve. The receiver pressure holds the intercepting valve to the left, as shown in Fig. 4, closing the ports "G" and opening free passage from the high pressure cylinder to the low pressure cylinder, and the engine works compound.

It will be noticed that while working compound, which is the usual way of working the engine, the intercepting and reducing valves are both held against ground joint seats which prevent the leakage of steam that may have leaked past the packing rings.

Now with the engine running compound, if the engineer wishes to run the engine simple, because of a heavy grade, he pulls the handle of the three-way cock the same as for starting simple. This will open first the by-pass valve "K," and then the separate exhaust valve, the by-pass valve relieving the pressure more gradually than if the large valve were opened at once. As soon as the separate exhaust valve is open the pressure in the receiver drops and the intercepting valve is forced against the seat "F," by the pressure in the chamber "E," and the engine runs simple as before. When the grade is passed, the engineer pushes the handle of the three-way cock over and the engine begins to work compound.

To start the engine compound the separate exhaust valve is left closed, as in Fig. 4, and when the throttle is opened the intercepting valve will be forced against the seat "F" by the pressure in the chamber "E," as shown in Fig. 5. The low-pressure cylinder will then take steam through the ports "G" and the high-pressure cylinder will exhaust into the receiver for a few strokes of the engine. This will raise the pressure in the receiver and force the intercepting valve into the position shown in Fig. 4, closing the ports "G" and the engine will run compound.

The combination of the automatic intercepting valve with the separate exhaust valve permits the engine to be changed from simple to compound, and the reverse, very smoothly and without danger of jerking the train, and in recent tests the engine was changed from compound to simple and the reverse repeatedly, when operating at a maximum power, with the throttle remaining wide open.

ACCIDENTS WITH THE SCHENECTADY COMPOUND.

MAIN ROD HIGH PRESSURE.

Remove broken rod, blocking cross-head at front end of the guides. Clamp the high-pressure valve forward to clear the exhaust port; the steam will then pass through the exhaust port on the high-pressure side into the receiver, thence to the low-pressure steam chest. The low-pressure cylinder then acts as a high-pressure engine. Open the throttle valve moderately on account of the large area of low-pressure cylinder.

MAIN ROD LOW PRESSURE.

Remove broken rod, secure cross-head at back end of guides; clamp the low-pressure valve back far enough to clear exhaust port. Exhaust steam can then pass through low-pressure cylinder and out through the stack.

VALVE STEM ON EITHER SIDE.

The same remedy as for a main rod on that side; also remove the main rod and secure it as previously instructed.

INTERCEPTING VALVE.

Should the intercepting valve become disabled, and you can clamp the poppet valve open, do so; if not, then remove the back head from the intercepting valve steam cylinder, and push the piston forward, putting in a block to hold it in that position; then put on the head, which will prevent the steam in the receiver closing the poppet valve. The same may be done with the small piston which moves the valve, admitting steam to the intercepting valve steam cylinder; this will also prevent closing the poppet.* Live steam would then be admitted to both cylinders for starting.

Mr. J. E. Sague, mechanical engineer for the Schenectady Locomotive Works, says of break downs, that all the compound locomotives now being constructed by their company are provided with a separate exhaust valve which enables the engine to be run as a simple engine when desired. Therefore the same rules that apply for break downs on simple engines will apply to their compound engines.

PITTSBURG COMPOUND.*

This system is of the two-cylinder type, with the Colvin starting valve. It can be worked as a compound, or steam can be admitted directly into the low pressure cylinder, as desired. When both cylinders are using steam direct from the boiler a reducing valve is used on the low-pressure side to equalize the force exerted in the two cylinders.

DESCRIPTION AND METHOD OF OPERATING.

The following description of the mechanism employed in operating the two-cylinder double-expansion locomotives, built by the Pittsburgh Locomotive Works, is to illustrate briefly the simplicity of construction and the principal advantages peculiar to this system.

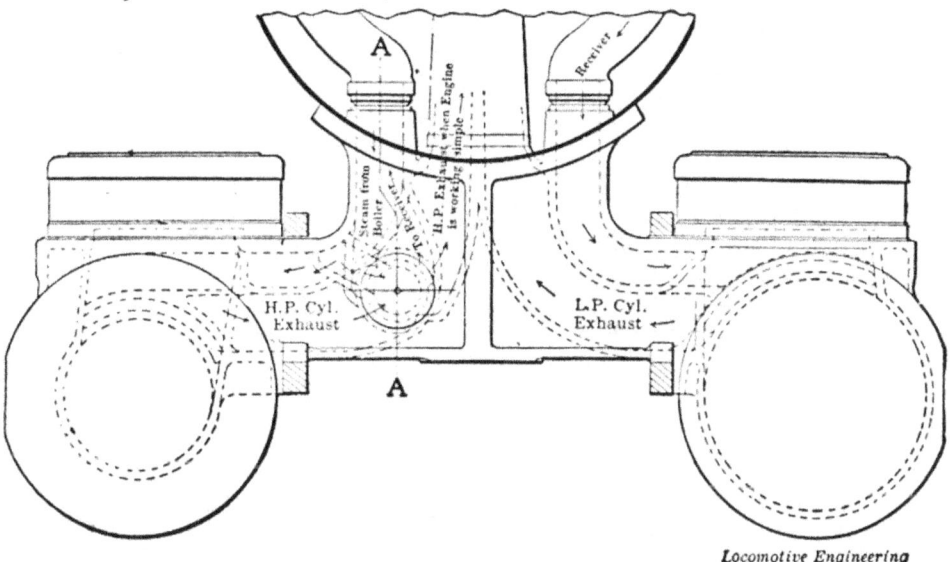

Locomotive Engineering

Fig. 1.

Figures 2 and 3 show the general arrangement of intercepting valve and reducing valve as placed in saddle on high-pressure side. Figure 2 shows position of intercepting valve when working single expansion, the passage to receiver being open to reducing valve, which is free to act and admit live steam to receiver. The high-pressure exhaust is also open to

---

*This article was especially prepared for this book by the Pittsburg Locomotive & Car Works.

the atmosphere as shown. In Figure 3 the intercepting valve is shown in forward, or proper position for working double expansion, with passage between reducing valve and receiver closed, while the exhaust from high-pressure cylinder is diverted from the open air, as indicated. In or near the cab is placed an air or steam reversing cylinder, actuated by movement of the reverse lever or reach rod, as follows: When lever is down, or at full stroke, at either end, the intercepting valve is in position indicated by Figure 2, permitting admission of live steam to receiver, but the moving of reverse lever one or more notches

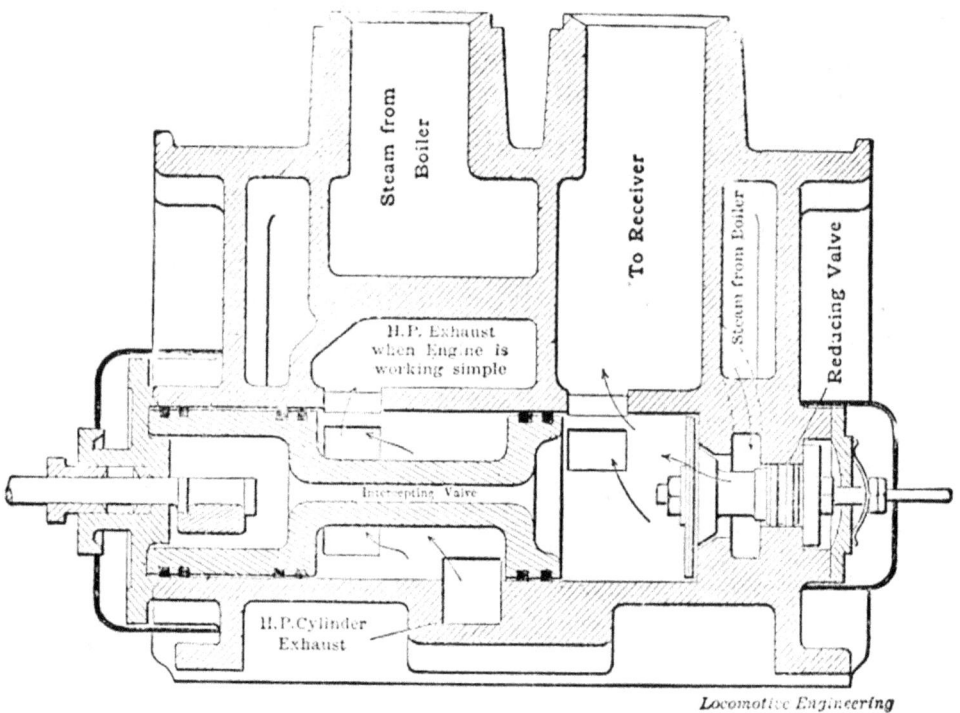

Fig. 2.

opens valve which admits pressure to reversing appliance, and intercepting valve is moved to position shown by Figure 3. The dropping back of lever to full stroke again changes the valve and engine is thrown into single expansion as before. It will thus be seen that the intercepting valve never moves automatically,— it being entirely under the control of the engineman. This is particularly desirable in switching, or shunting service, as by placing reverse lever in first notch, or full stroke, and closing valve in pipe connecting boiler with reversing cylinder, the locomotive can be worked single expansion as long as desired and at any position of cut-off.

The intercepting valve is of the plain piston type with cast iron packing rings designed to separate the high and low-pressure passages. The reducing valve is also exceedingly simple; it has the only movable ground joint in the system, and its loca-

Engine No. 4, Lake Superior & Ishpeming Ry.
Pittsburgh Compound.

tion between packing rings at either end does not impair its usefulness or action if the joint itself leaks. Both are in one chamber where they can be easily reached by hand.

The object of the reversing mechanism in the cab is to relieve the engineman of the necessity of converting the engine from single to double expansion by hand. The hand arrangement is always present, however, ready to be used in the event of any accident occurring to reversing cylinder. Failure of operation is therefore practically impossible.

In setting up, the reducing valve should first be placed in position. That this may be accomplished without it being necessary to put piston end of valve with packing in through the

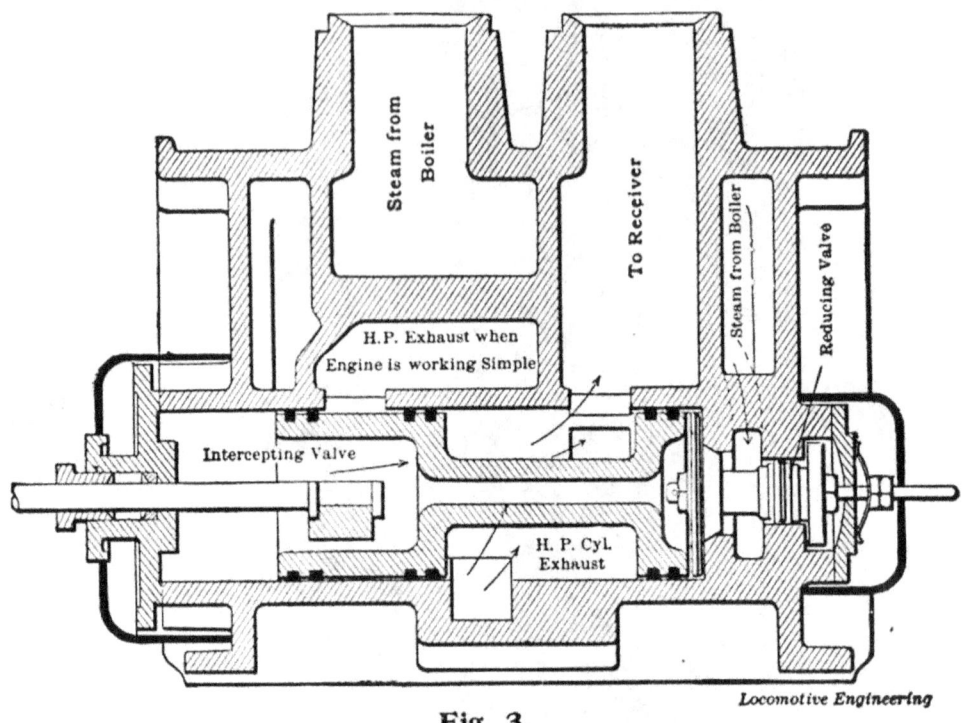

Fig. 3.

intercepting valve chamber, in which case the packing would drop into the ports, the reducing valve has been made in two sections and bolted together with a through bolt. After taking valve apart by unscrewing the nut on the small or piston end, place the valve, with bolt in position, through the intercepting valve chamber and against its seat; take the small or piston end, with packing rings in position, and place it in the small sleeve furnished for the purpose, place sleeve against valve chamber to prevent packing rings from dropping out and push piston in against its seat on the valve; remove sleeve and put follower in place and secure by nut. After securing to head, place the small coil spring in position on bolt and adjust with two nuts. This spring should only be tight enough to keep valve from rattling when engine is running with steam shut off and so that valve can be easily moved by hand.

In placing intercepting valve in position care must be taken that the cut in packing rings pass directly over bridge in center opening, otherwise rings will open out and catch in this port when being pushed into place. When putting valve in use the sleeve furnished with engines for holding packing rings in place so that they can be entered without rings opening out. After engine has run about one week the intercepting valve should be taken out and the rings cleaned with head-lamp oil. This should be done without removing rings from the valve.

Obviously, the object of a double expansion locomotive is to economize in the use of fuel and water, in connection with which the following instructions for the use of the engineman are recommended:

1. Convert the locomotive to double expansion as soon as

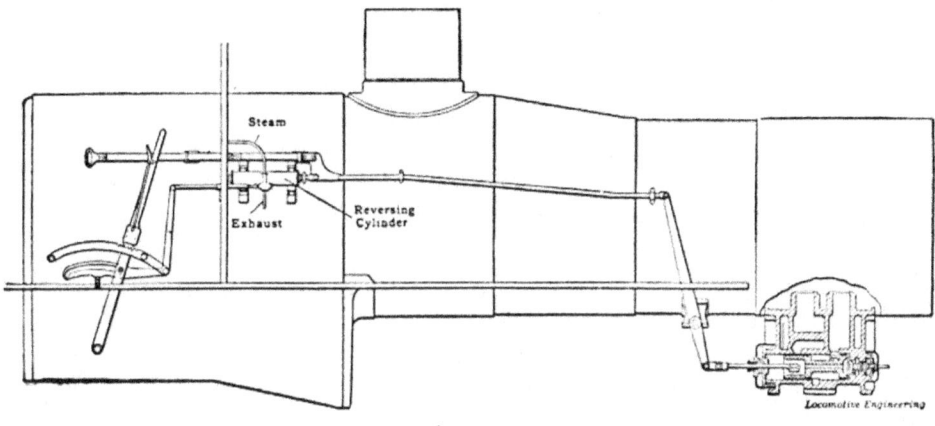

Fig. 4.

possible after starting train, three or four revolutions being sufficient in single expansion.

2. When climbing grades do not change engine from double to single expansion unless speed is reduced below four miles per hour, and convert again to double expansion at the earliest possible moment.

3. Care should be taken to close throttle to some extent before changing from double to single expansion, or the engine will slip badly on account of increased power in cylinders.

4. Engine should always be run in single expansion when not using steam. If, at any time, this is not desirable, a little steam should be worked in cylinders.

5. In oiling cylinders the greater part of the valve oil should be put in the right-hand or high-pressure cylinder,—one drop in the low-pressure to five or six in the high-pressure cylinder. The back end of the intercepting valve should be oiled at least once a week and cup at front end for lubricating reducing valve each trip.

6. It is necessary that an engine constructed on this system should move into double expansion immediately on reverse lever being hooked up one or two notches, and into single expansion on lever being put into corner notch. This applies to both forward and back gear. If this does not work exactly it should be reported and attended to at once. The rod connecting the reversing cylinder and intercepting valve is forward in single expansion and back in double expansion, both going ahead and backing up.

7. To insure engine steaming freely care should be taken to keep a thin fire and without holes.

ACCIDENTS WITH THE PITTSBURG COMPOUND.

In the event of either side being disabled the intercepting valve should be placed in position for single expansion, or at the back end of intercepting valve chamber. If it is desired to use the right-hand or high-pressure side only it will be necessary to prevent high-pressure steam from entering receiver and, consequently, steam chest of low-pressure cylinder. To do this the reducing valve, in front of cylinder saddle, must be prevented from operating. This can be accomplished easily by removing spring, putting in its place a clamp made of wood or iron, against which the lock nuts may be screwed, thus holding valve securely to its seat.

If the high pressure side be disabled, and it is desired to use the left-hand or low-pressure cylinder only, it will be necessary to proceed in precisely the same manner as would be in the case with a single expansion locomotive, viz., by placing the main valve in central position, covering both steam and exhaust ports. The low-pressure cylinder would then be operated by live steam admitted through reducing valve and receiver.

BROOKS COMPOUNDS.*

DESCRIPTION OF BROOKS TWO-CYLINDER COMPOUND LOCOMOTIVE.

JOHN PLAYER, INVENTOR.

This is of the cross compound type, having the high-pressure cylinder upon one side and the low pressure upon the other, and between them a combined admission, pressure regulating and intercepting valve, located either in the smoke box or the cylinder saddle, which immediately upon opening the throttle valve, admits live steam at reduced pressure to the low-pressure cylinder, this pressure being regulated in such ratio as desired, the intercepting valve at the same time automatically closing and

---

* These articles were especially prepared for this book by the Brooks Locomotive Works.

preventing the live steam pressure from working against the high-pressure piston, the reducing valve remaining open until such time as the pressure in the receiver pipe on the high-pressure side of the intercepting valve becomes equal to or

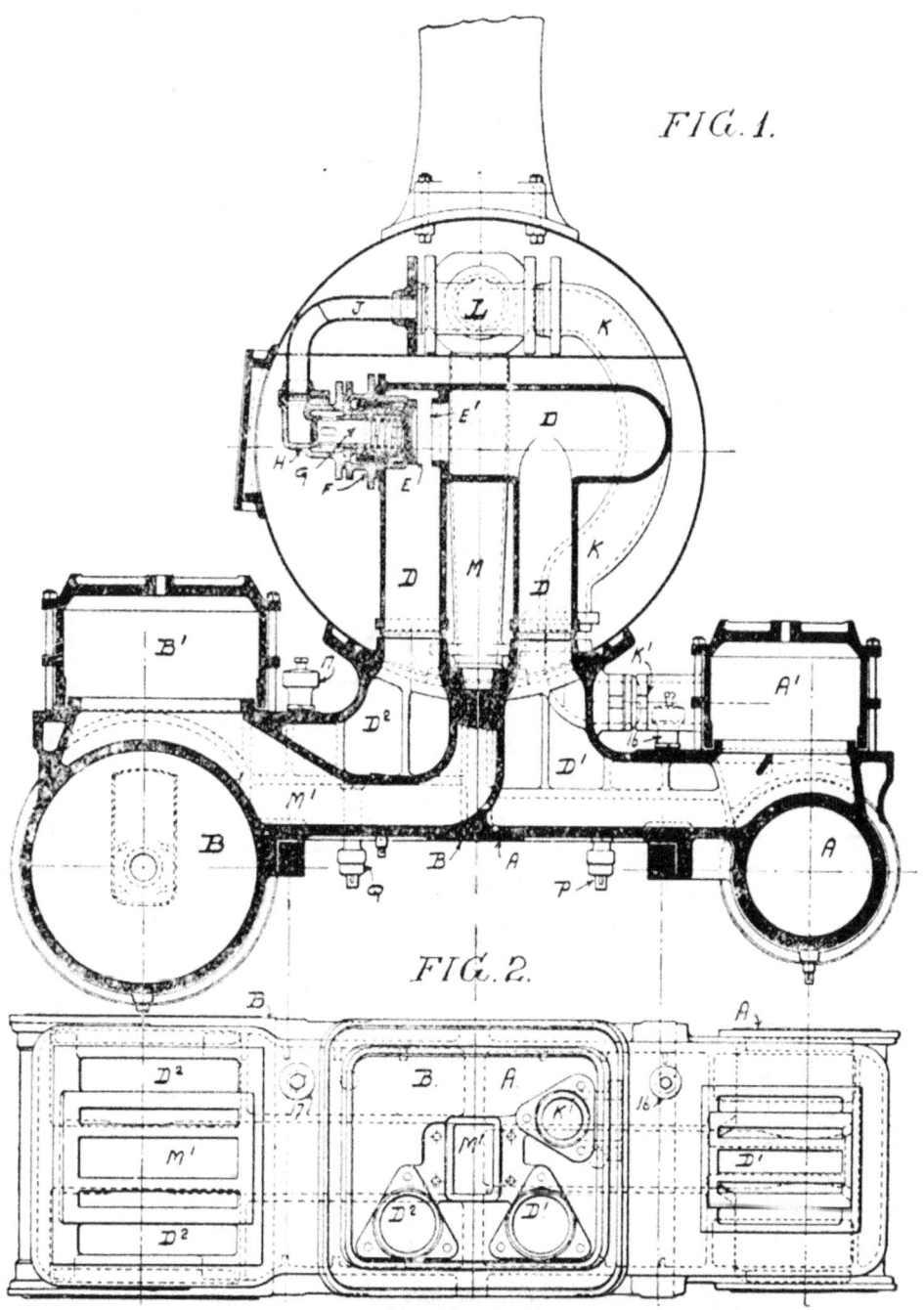

FIG. 1.

FIG. 2.

slightly in excess of that on the low-pressure side, when the pressure regulating valve automatically closes and the intercepting valve opens simultaneously, the first cutting off the supply of live steam to the low-pressure cylinder, the second open-

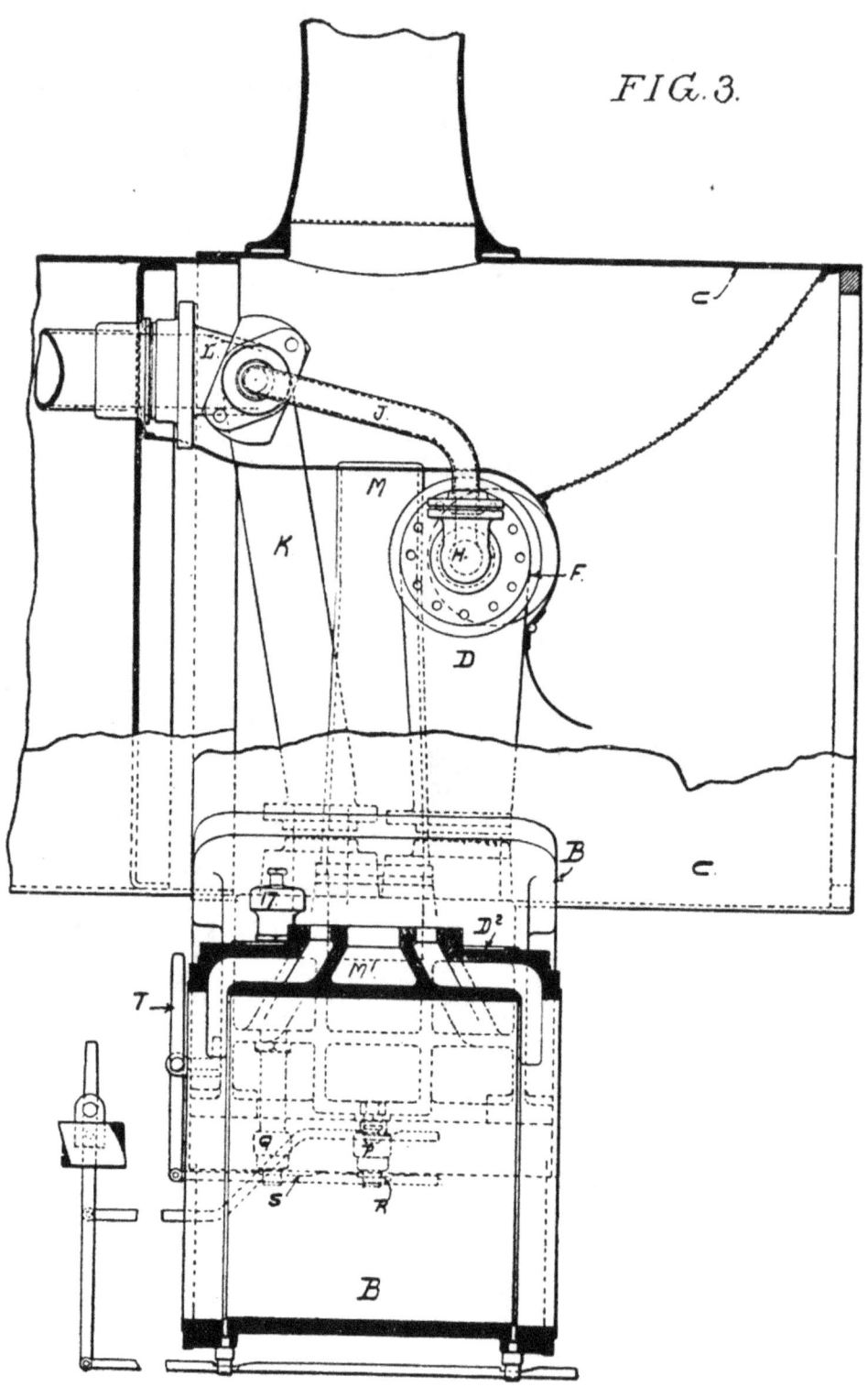

FIG. 3.

ing connection between the two ends of the receiver and allowing the high-pressure exhaust steam to act directly on the low-pressure piston and at the same time locking the pressure regulating valve upon its seat and preventing the further admission of live steam to the low-pressure cylinder, these valves remaining in this position during the time the throttle valve is open. In order to give the engineman full command of the locomotive at all times, controlling valves are provided in the receiver. These in the illustration are shown upon the bottom of the receiver and are connected to the cab by suitable levers. These, however, may, if desired, be of larger area and arranged in the upper portion of the receiver, connected with the exhaust pipe and arranged to work automatically in combination with the intercepting valve, so that the locomotive can be worked as a simple engine when required. However, on account of the arrangement of combined admission and pressure regulating valve which at all times when necessary admits sufficient steam to the low-pressure cylinder to give the locomotive its maximum power,

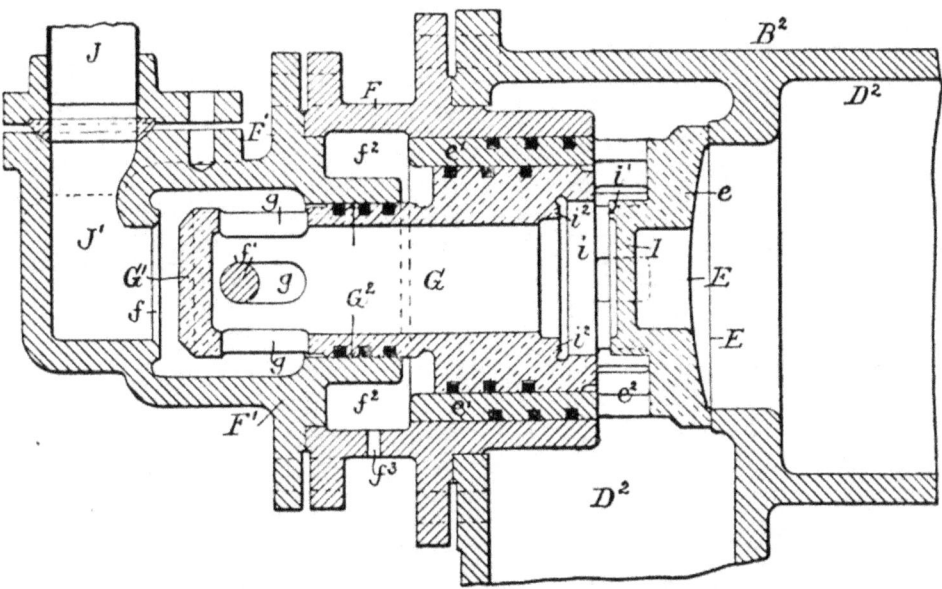

FIG. 4.

the use of such a separate exhaust valve, whereby the engine can be worked simple for long periods, has been found in practice absolutely unnecessary, the arrangement of admission and pressure regulating valve previously referred to performing all the requirements of a simple locomotive.

This locomotive operates as follows: The live steam is admitted to the high-pressure steam chest through the steam pipes and passages and operates upon the high-pressure piston in the usual manner. At the same time steam is admitted to the high-pressure end of the pressure regulating valve through the connecting pipe, causing the valve to open, passing thence through the hollow portion of the valve, causing the intercepting valve to automatically close against its seat. This steam flows through the passages of the intercepting valve into the low pressure end of the receiver, and, acting upon the large end of the pressure regulating valve, causes it to partially close as soon as the requisite pressure is obtained and thereafter regulates the amount of steam admitted by the pressure regulating valve, maintaining an even pressure in the receiver. The reduced pressure steam thus admitted to the receiver acts upon the low-pressure piston in the usual manner. As soon, however, as the high-pressure cylinder has exhausted sufficient steam in the high-pressure end of the receiver to overbalance the intercepting valve, this valve opens automatically, at the same time lock-

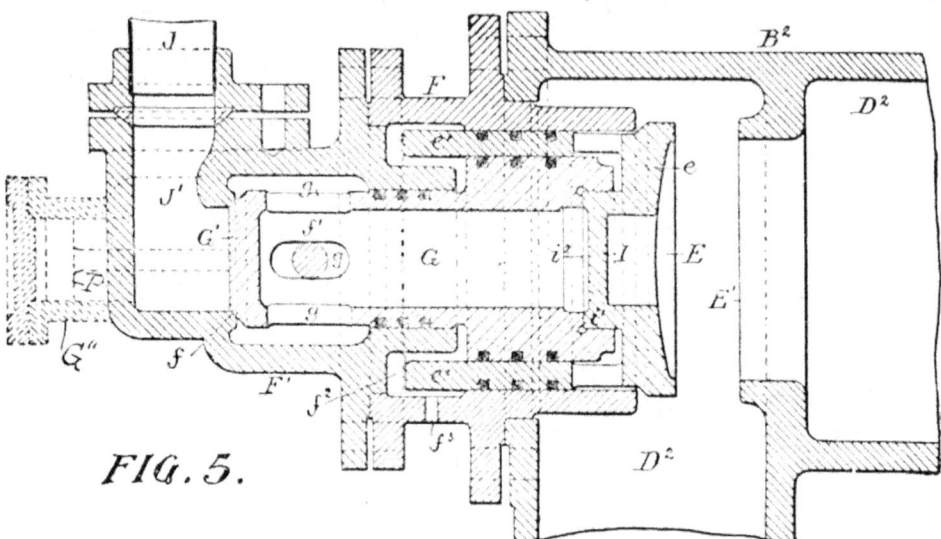

FIG. 5.

ing the pressure regulating valve against its seat. The exhaust steam from the high-pressure cylinder then flows through the receiver and acts directly upon the low-pressure piston, the pressure of this exhaust steam, even when considerably reduced, being sufficient to keep the pressure regulating valve closed through the action of the combined valves at all times.

ACCIDENTS TO BROOKS TWO-CYLINDER COMPOUND.

MAIN ROD, LOW PRESSURE.

Remove broken rod and block the crosshead securely at back end of guide. Clamp low-pressure valve over ports. Open controlling valves underneath saddle wide and run in on one side.

MAIN ROD, HIGH PRESSURE.

Disconnect this side same as for simple engine and run in on one side. Steam will pass to low-pressure cylinder through reducing and intercepting valves.

VALVE ROD, EITHER SIDE.

Same as for main rod on same side.

INTERCEPTING VALVE.

Should the intercepting valve become broken, so as to leave an opening between both ends of the receiver, disconnect the steam pipe to high-pressure end of reducing valve and insert blind gasket to shut off flow of steam.

REDUCING VALVE.

Same remedy as for intercepting valve.

DESCRIPTION OF BROOKS TANDEM COMPOUND LOCOMOTIVE.

JOHN PLAYER, INVENTOR.

This is of the four-cylinder type, and consists in the combination of a structure containing the low-pressure cylinder and saddle and low-pressure steam chest and another structure containing the high-pressure cylinder and steam chest, attached preferably to the forward end of the low-pressure cylinder and having a steam chest communicating with the steam chest of the low pressure cylinder. The steam is supplied to the high-pressure valve chest through suitable connecting pipes and the low pressure cylinder exhausts through the saddle in the usual way. The high pressure steam chest is fitted with a hollow piston valve, having an internal admission, the low pressure steam chest being fitted with either a balanced slide valve or an external admission piston valve. On account of the high pressure valve having an internal admission and the low pressure valve an external admission, it is necessary for these valves to travel in opposite directions. This is accomplished by the insertion of a rocker arm in the receiver, having arms of the desired ratio to give the different relative travel to the high and low-pressure valves, the low-pressure valve being connected to the rocker arm and driven in the ordinary manner, the intermediate rocker being connected by suitable rods to the low-pressure valve yoke and high-pressure valve, the arms of this rocker being arranged so that the travel of the high-pressure valve is different from that of the low-pressure valve, and consequently the different points of cut-off are caused to vary from those of the low-pressure valve, thereby causing a better relative distribution of steam in the high and low-pressure cylinders.

One of the chief advantages of this type of compound is that the castings for the tandem compound cylinders are and can always be made absolutely interchangeable with those of a simple engine, this arrangement necessitating no change whatsoever forward of the crosshead and valve rod keys or in the

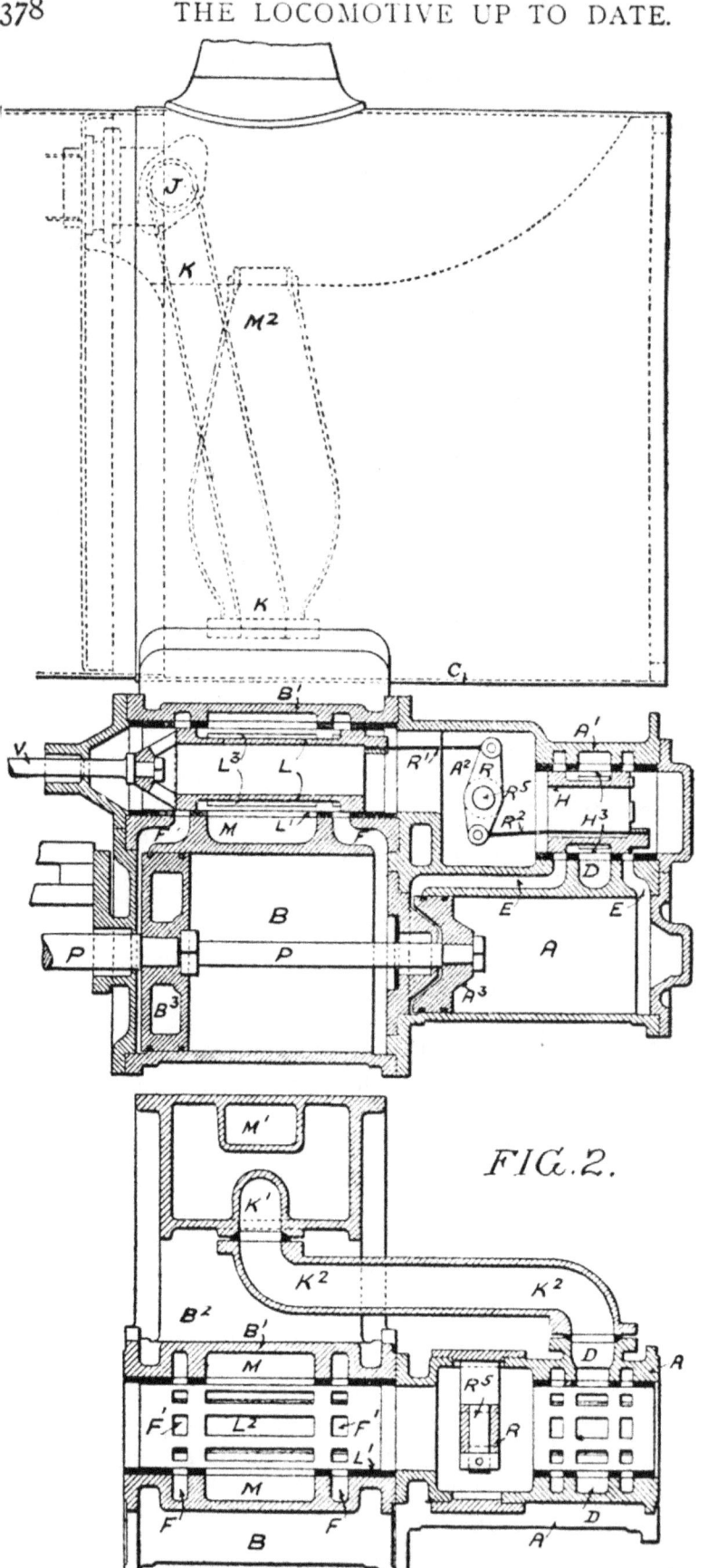

FIG. 1.

FIG. 2.

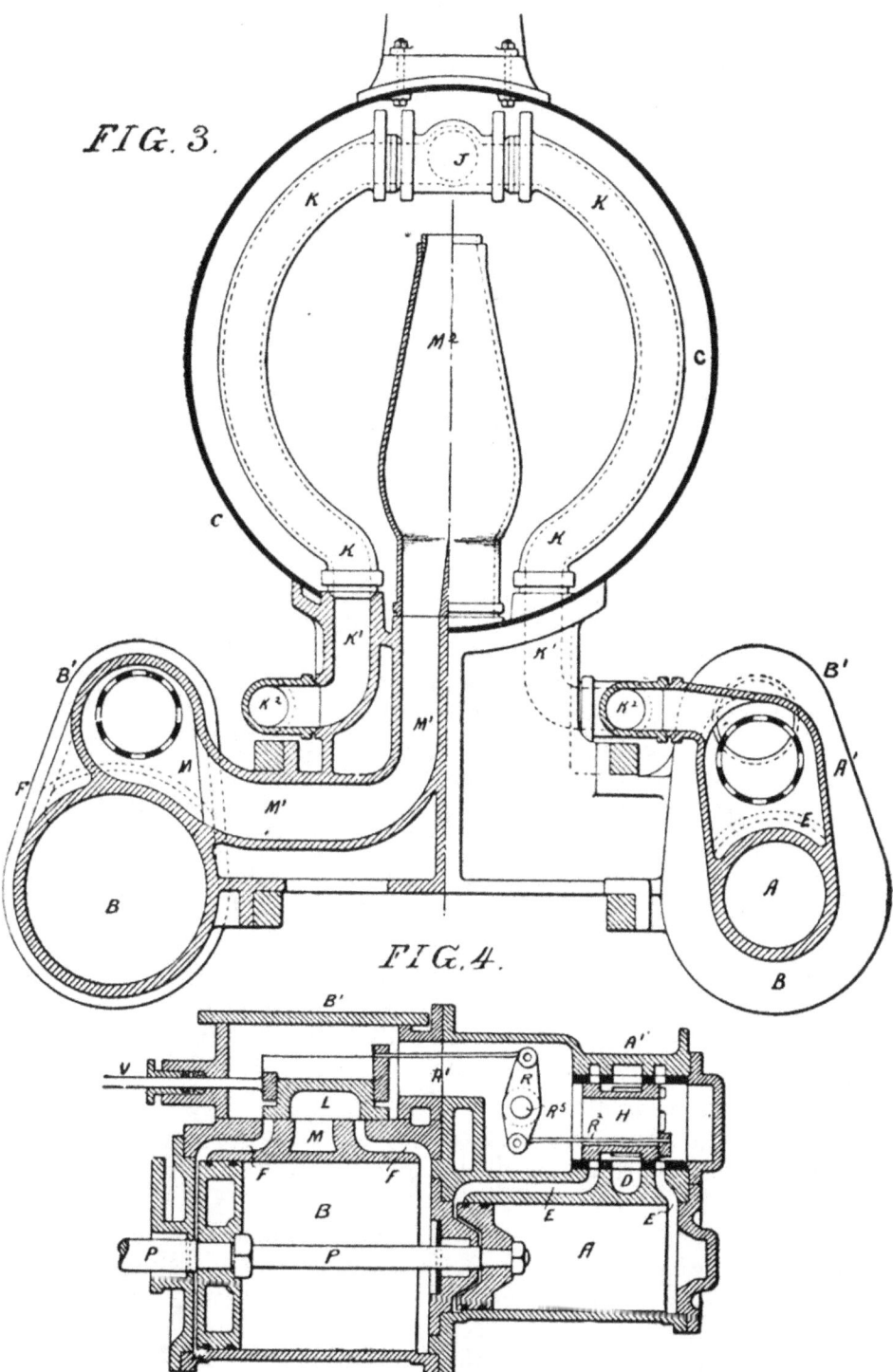

FIG. 3.

FIG. 4.

steam and exhaust pipes in the smoke box, the compound cylinders thus giving no more cost for application than would be the case in applying a new pair of simple cylinders to an engine.

The high-pressure cylinder is generally located ahead of the low pressure. This, however, is not necessary and in some types of locomotives having four-wheel trucks it is preferable to place the high-pressure cylinder back of the low pressure, thus materially reducing the weight and rendering the parts more accessible. The pistons of the high and low pressure cylinders are arranged upon the same rod and the intermediate head between the high and low pressure cylinders is fitted with suitable metallic packing, as is clearly shown in the illustrations. The low-pressure steam chest is provided with a reducing and starting valve connecting with the high-pressure steam pipe. This valve is permitted to operate automatically when the reverse lever is in full forward or full back gear. In the intermediate positions of the lever, this reducing valve is locked to its seat by a suitable spring, so that it is rendered inoperative and the engine must necessarily work compound at all times under all conditions of steam pressure when the reverse lever is in any other position except full gear. The use of this combined starting and reducing valve permits the introduction of steam into the low-pressure cylinder at an equivalent to the maximum pressure obtained in this cylinder when the engine is working compound. Of course, as soon as the engine has made one complete revolution and the receiver is charged with exhaust steam from the high pressure cylinder, the starting valve becomes inoperative, thereby necessitating the engine to work compound.

This engine operates in the following manner: The steam is admitted to the high-pressure steam chest through the connecting pipes into the admission cavity surrounding the high-pressure piston valve; thence is admitted to the high-pressure cylinder through the induction ports, the exhaust from the high-pressure cylinder passing into the receiver; that from the front end of the cylinder passing through the hollow piston valve. The low-pressure steam chest being also of large area, somewhat increases the receiver capacity, so that a uniform pressure is maintained therein, the steam being admitted to the low pressure cylinder by the external edges of the valve and exhausted therefrom in the usual manner through the central exhaust passage.

The reducing and starting valve being automatic, responds absolutely to all variations of pressure and allows the engine to start without jerking or slipping, as is the case when high-pressure steam is wire drawn into the low-pressure cylinder.

This type of compound has been in successful operation for the past six years, showing excellent economy in fuel, water and repairs.

ACCIDENTS TO BROOKS TANDEM COMPOUND.

MAIN ROD.

Disconnect and block valves, same as for simple engine, and run in with one side.

VALVE ROD.

Same remedy as main rod.

FRONT HEAD OF HIGH PRESSURE CYLINDER.

Remove front steam chest head; disconnect high-pressure valve and block it to cover ports. Throw reverse lever in full forward gear and disconnect and block starting valve rod on same side. This will allow the reduced pressure steam to operate in low-pressure cylinder.

ROGERS SYSTEM OF COMPOUNDING.*

THIS SYSTEM IS OF THE TWO-CYLINDER TYPE.

Fig. 1 is a view from the front showing the H. P. cylinder on the right hand side of the engine. The intercepting valve and reducing valve are located in the saddle of that cylinder at A.

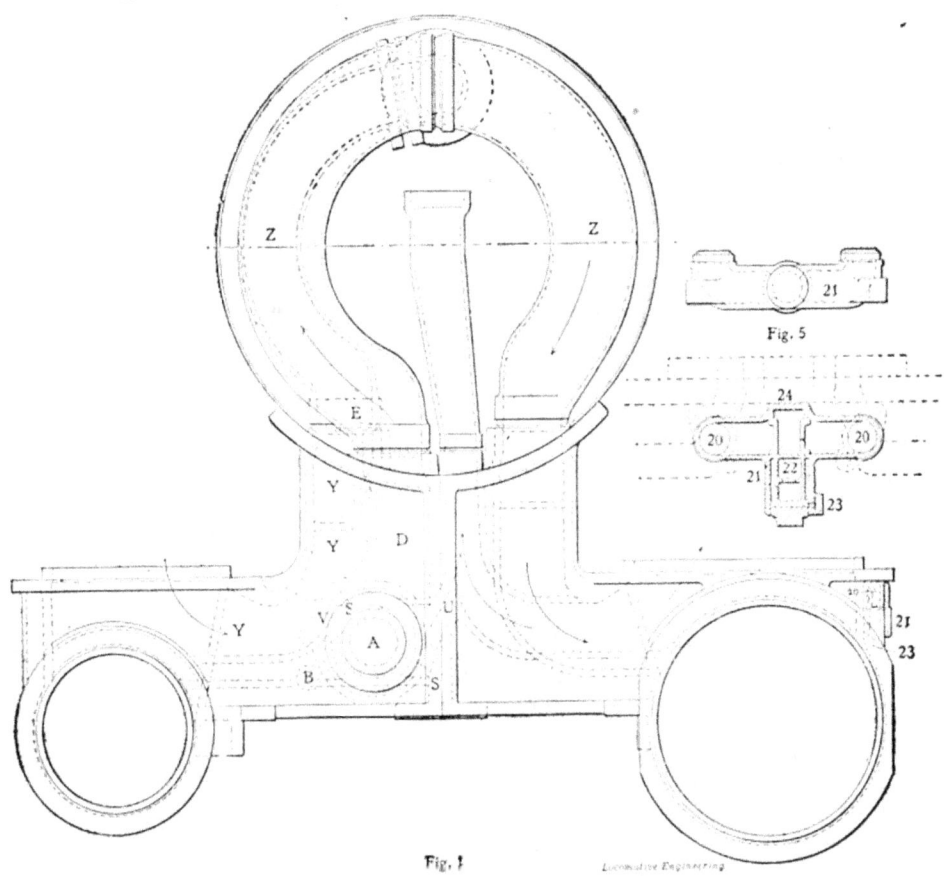

Fig. 1

The regulating valve (Fig. 4) for changing from simple to compound, and *vice versa*, is located at B (Fig. 1) and consists of a small cylinder 1¾" diameter, in which is a slide valve of the common D type operated by a lever, and a rod from it to the cab.

Fig. 2 is a sectional view of the intercepting valve referred

---

*This article was especially prepared for this book by the Rogers Locomotive Works.

to, as located in the saddle of the H. P. cylinder. The regulating valve, for simplicity in explaining its operation, is shown in both Fig. 2 and Fig. 3 as being located directly under the intercepting valve arrangement, but its position in the cylinder saddle is to the left of A, or on the quarter, as shown at B (Fig. 1). C is the exhaust passage of the H. P. cylinder. D is a steam chamber above it through which the exhaust steam passes on its way to the receiver, when the engine is working compound. E is where the receiver is connected. F is the "regulator" referred to. G is a circular chamber around the sleeve H, in which the cylindrical part of the intercepting valve is located. This chamber G is connected with the H. P. steam passage Y Y in the

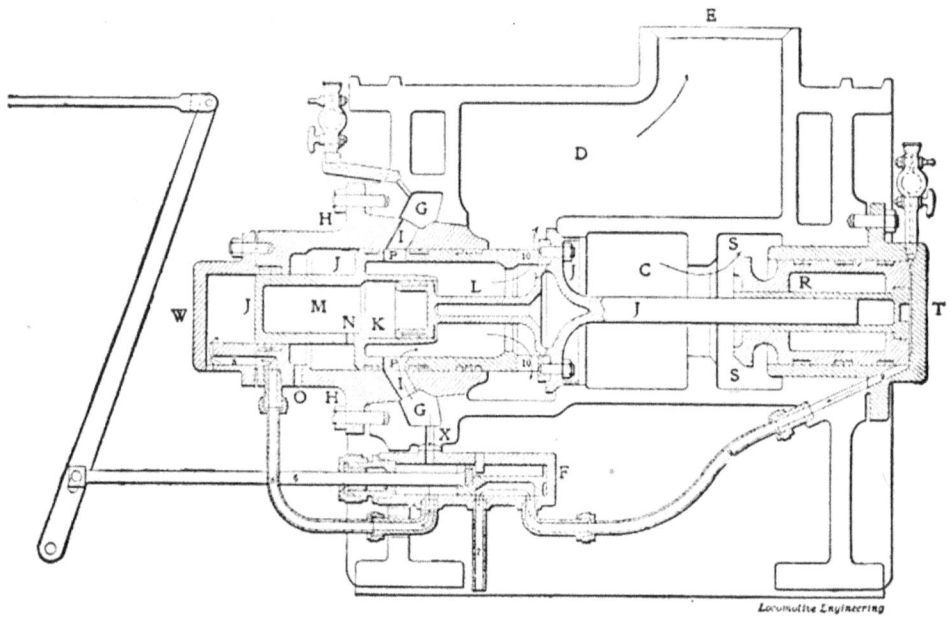

Fig. 2
INTERCEPTING VALVE IN SIMPLE POSITION.

saddle by a port shown at V (Fig. 1), and is practically a part of it. I I (Fig. 2) is a series of holes in the sleeve H, 1" diameter, connecting G with the space around the reducing valve L by a series of corresponding holes P P in the cylindrical body of the intercepting valve J J J J. The intercepting valve J has another series of holes (near the valve face) shown at 10, to the right of and outside of the seat of the reducing valve. Fig. 2 shows the intercepting valve J on its seat (closed), and the reducing valve L off its seat (open), as it is when working simple, except that L is then brought near enough to its seat to "wire draw" the steam and thus reduce the pressure behind L to practically one-half of that in front of it, so that the pressure in D, by the operation of the reducing valve, will never be more than half that in G. K is a cylinder in which the piston part of L (the reducing valve) moves. It is practically one-half the area of the valve L. M is a chamber, or rather an extension of K, and is always open to the atmosphere through the two holes N and O. R is the exhaust valve in the chamber S S through which the exhaust from the H. P. cylinder passes to the outlet at U (Fig. 1), where it enters the exhaust passage of the L. P. cylinder, and passes from there on to the stack through that exhaust pipe whenever the engine is working simple, both cylinders then exhausting through the one pipe. T is a sleeve or cylinder in which R has its movement (about two inches). The "guide" or stem of the intercepting valve extends over into the exhaust valve R, as shown, and serves to keep both valves central, but allows each to act independent of the other.

W is a short cylinder on the end of the sleeve H, into which the small end (about 4" diameter) of the intercepting valve J

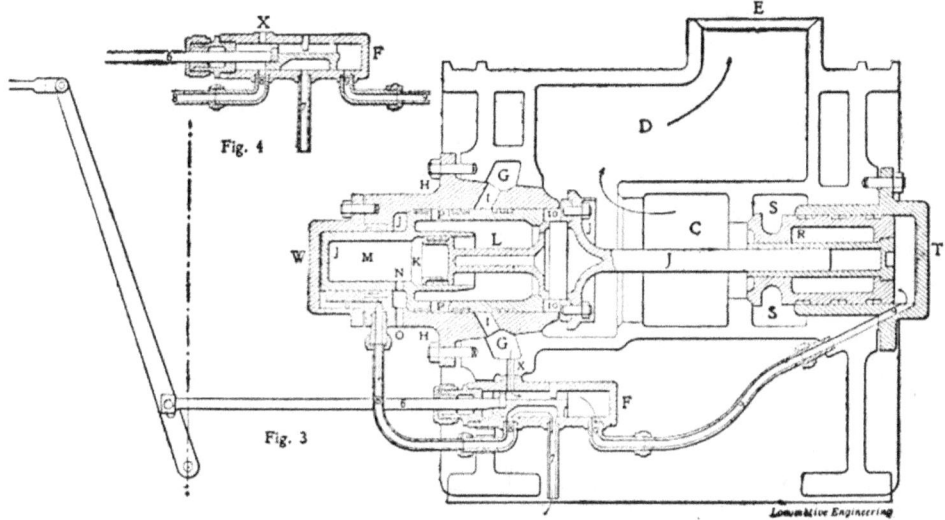

INTERCEPTING VALVE IN COMPOUND POSITION.

moves when opening as at Fig. 3 (working compound). It is fitted with packing rings as shown. Y Y, Fig. 1, is the steam passage in the H. P. cylinder saddle, which is double walled next to the exhaust passage. Z, Fig. 1, is the receiver pipe, 10" diameter, and in two parts. X, Fig. 3, is a $\frac{1}{2}$" port connecting the interior of the regulator (F) with the chamber G, therefore, whenever there is steam in G, the regulator also receives it through the port X. When the valve of the regulator F is in the position shown in Fig. 2, and the throttle is open, steam from Y and G passes into F and through the $\frac{1}{2}$" pipe and port 8 into W and pushes the intercepting valve (to the right) onto its seat, as shown in Fig. 2, at the same time causing the holes I and P to correspond, and steam from G to flow into the chamber around L and on out through the holes 10 into D, but as the cylinder K of the reducing valves piston and chamber M are open to the atmosphere through the holes N and O, the piston in K will be moved with the pressure behind it to the left, and L be drawn towards its seat, but as the valve is of larger area than its piston it will not entirely close until the pressure in D (a back pressure on L) combined with the pressure on the piston of L towards M, equals or exceeds the pressure on the piston side of L so that the reducing valve automatically prevents the pressure behind L (in D) from being more than about one-half that in front of it (in G) by "wire drawing" between L and its seat, giving to the L. P. cylinder steam at a pressure no more than about one-half that going to the H. P. cylinder at the time, whenever the engine is working simple. When the regulator F is in the position shown in Fig. 2 the chamber in the end of T (back of R) is in exhaust through the $\frac{1}{2}$ inch pipe and ports 9 and 7 to the atmosphere, and R will remain as shown (open), consequently the engine in that case will work simple; the exhaust going out through S S to the L. P. cylinder exhaust passage, Fig. 1.

If the regulator valve is moved to the position shown in Fig. 3, steam will then flow through the pipe 9 9 into T, push and hold R to its seat and at the same time W will be in exhaust, and as R is then closed, a very slight pressure in C will open the intercepting valve, moving it to the left, to the position shown in Fig. 3. The exhaust from the H. P. cylinder then passes up through D to the receiver, and the engine works compound. When J is open (Fig. 3) the holes P do not then correspond with the holes I in the sleeve H and no steam from G can pass to the receiver, the reducing valve being then rendered inoperative, and as J at its shoulder seats itself on the offset at O no leakage of steam to the atmosphere can occur in case any should get by the packing rings; J has a movement of about three inches. The regulating valve F is operated by a $\frac{3}{4}$ inch stem (6) extending out and connected to a lever as shown.

If the regulating valve is moved to its central position as shown in Fig. 4 when the throttle is closed, and the latter is then opened, steam will flow in at X, and as both ports 8 and 9 are uncovered the chambers T and W will be filled and the valves R and J instantly closed on their seats. D (the receiver) would then be filled with steam from G and L and holes 10 (at the reduced pressure), but as soon as the pressure in C, from the exhaust of the H. P. cylinder, equals about 20 per cent of the pressure going to the H. P. steam chest at the time, the influence of the receiver pressure on the larger diameter towards the atmospheric chamber at the outlet O will move the intercepting valve to the left (open), to the position shown in Fig. 3, against the pressure on the smaller diameter in W and it will remain open as the total of the receiver pressure on the larger diameter of the intercepting valve towards the atmospheric chamber is greater than that of the H. P. cylinder steam on the smaller diameter of W.

The intercepting valve on its face side has a thin flange $\frac{1}{8}''$ thick projecting $\frac{1}{8}''$ beyond the valve seat into the $8\frac{1}{2}''$ diameter opening, as shown in Figs. 2 and 3. This is to insure the seating of the intercepting valve and to prevent any flow of steam from G to C and out through S, after the holes P and I begin to coincide (holes begin to open) and before J completes its movement in closing, as might otherwise occur. The outer edge of this flange practically closes the hole at C as soon as its edge comes in line with the valve seat, and just as the holes P and I begin to correspond.

If the engine is started "simple," the regulator being as per Fig. 2, it will continue to work "simple," but by changing the regulator to either the position of Fig. 3 or Fig. 4, it will automatically go to "compound" on account of R closing and being held on its seat, and the pressure in C accumulating to about 20 per cent of that going to the H. P. cylinder at the time as explained in the case of the regulator being in the position shown in Fig. 4. If, however, the regulator is moved to the position shown in Fig. 3 the intercepting valve will open with almost no pressure in C as W is then in exhaust. After starting simple, and when it is desired to work compound, it is best to move the regulator, first, to its middle position as per Fig. 4 (without closing the throttle), as the intercepting valve will then, as explained above, move back (open) against the pressure in W without shock or jar, and after that, the regulator should be moved as per Fig. 3 and remain there until it is desired to start "simple" again or to go from compound to simple. In the position of Fig. 3 there is little or no useless movement of the intercepting valve, caused by closing and opening the throttle while running, or from sudden and wide variations in the steam pipe pressure, and it is therefore best to keep the regulator in that position (Fig. 3) while running, under all ordinary circumstances. To prevent as far as possible the effects of compression in the L. P. cylinders when running "shut off" at high speeds, as for instance down grades, when no steam is needed in the cylinders, a "by-pass" arrangement similar, in some respects, to that called the "LaChatalier" arrangement. Fig. 5 is used on the outside of the L. P. cylinder below the steam chest seat. The dotted lines show the two port openings of the cylinder below the valve seat. A hole 2" diameter is made into each as shown at 20, connecting these by a short pipe 21. In this pipe is a cylindrical valve 22. When the valve is down in its chamber, as shown, the port through pipe 21 is open, leaving a free passage from the port to one end of the cylinder to that of the other.

To the lower end of this valve chamber at 23 is connected a $\frac{1}{2}''$ pipe from the steam chamber in the saddle of the H. P. cylinder, so that when that cylinder receives steam this chamber receives it also at 23, and the valve 22 is raised up to the top of its chamber 24, closing the passage through 21, and remaining there as long as the H. P. cylinder is receiving steam, but when the throttle is closed and the chamber Y of H. P. cylinder is empty, the valve 22 falls by gravity to the position shown, leaving the passage open, that is, it is closed when steam is being used and open when the throttle is closed.

The sleeve H, Figs. 2 and 3, is tapered on the outside and makes a ground joint with the saddle casting on each side of the chamber G as shown. To remove H and the intercepting valve arrangement, it is only necessary to remove the nuts from the studs holding H in place, unconnect the pipe 8 and the exhaust pipe attached to O.

The small cylinder W can be taken off by removing the nuts from the studs holding it. The cylinder T and its valve R can be removed by simply removing the nuts from the studs in its flange. The "regulator" can be removed by unconnecting its pipes 8 and 9 and exhaust 7, taking off the clamp that secures the joint at X and removing two bolts through its flange at outside of cylinder saddle.

The rod from the top of the regulating lever to cab has at that end three notches, which hold the valves in either of the three positions as shown in Figs. 2, 3 and 4, that may be desired.

The packing rings in J R and at the end of W are all as shown, and their position will indicate what purpose they serve.

To operate this arrangement it is only necessary to move the rod extending to the cab to the notch corresponding with the desired position of the regulating valve as shown in Figs. 2 and 3 and 4.

The movement of all the other various parts of the compound arrangement then occurs automatically as explained above by the pressure after the steam has passed the throttle valve.

The sleeve H, cylinders T and W, valve R, regulator F and its valve are of cast iron. The intercepting valve J both cylindrical part and valve proper (the two parts lipped and seamed together by studs and lock nuts as shown) and the reducing valve L are all of forged steel.

Packing rings in R at W in the piston of L and in J are all cast iron except the $1\frac{1}{4}$" wide ring in J, which passes over the holes I by the movement of J. That ring is of steel with a "spring ring" under it. The sleeve in R and oil cups at G and T, Fig. 2, are of brass.

ACCIDENT WITH THE ROGERS COMPOUND.

In case of accident rendering either of the cylinders useless, the engine can be run by the other cylinder as follows:

If, for instance, the H. P. cylinder cannot be used, unconnect its valve stem, place and secure the valve so as to cover the cylinder ports, place the regulating valve (Fig. 4) in the position it is shown in Fig. 2 (the same as when the engine is working simple). The engine will then run by its L. P. cylinder, which will receive its steam in the same way as when working simple, that is, from the steam passage Y through V and the annular chamber G, holes I P through L (the reducing valve) holes 10 and chamber D and the receiver. The position of the regulating valve (Fig. 2) keeps the intercepting valve closed as shown by the steam pressure in W.

In case the L. P. cylinder cannot be used the engine can be run by the H. P. cylinder by unconnecting the valve stem of the L. P. cylinder, placing and securing its valve so as to cover the ports and placing the regulating valve in the position shown in Fig. 2, that is, the same as when the engine is working simple. The H. P. cylinder will in that case receive its steam in the usual way. The intercepting valve being closed H. P. steam would pass through the reducing valve to D and fill the receiver Z, but as the valve in the L. P. steam chest would cover the ports it could not go beyond filling the receiver and steam chest. The exhaust valve R being open the H. P. cylinder will exhaust through S and U into the exhaust pipe, the same as when the engine works simple.

It should be remembered, therefore, that whether the engine is run by its H. P. or its L. P. cylinder the regulating valve (Fig. 5) must in such case always be in its extreme forward position as it is shown in Fig. 2, when the engine is ordinarily working simple.

As the chamber of valve 22 of the by-pass arrangement (Fig. 5) is connected with the H. P. steam passages, that valve will always keep the by-pass closed whenever the engine is using steam in either of the two cylinders.

COOKE COMPOUND.*

THIS SYSTEM IS OF THE TWO-CYLINDER TYPE.

The intercepting device consists of two cylindrical valves E, each having an open end, and a passage in one side through which the exhaust steam passes when engine is running simple.

The intercepting valves are operated by pistons F, to which they are attached, as shown in Fig. 2. Fig. 1 shows valve chambers with valves removed, and also the connection between one of the intercepting valves and the reducing valve D. The reducing valve D is connected to live steam passage in high-pressure cylinder and to steam passage in low-pressure cylinder.

---

*This article was especially prepared for this book by the Cooke Locomotive Works.

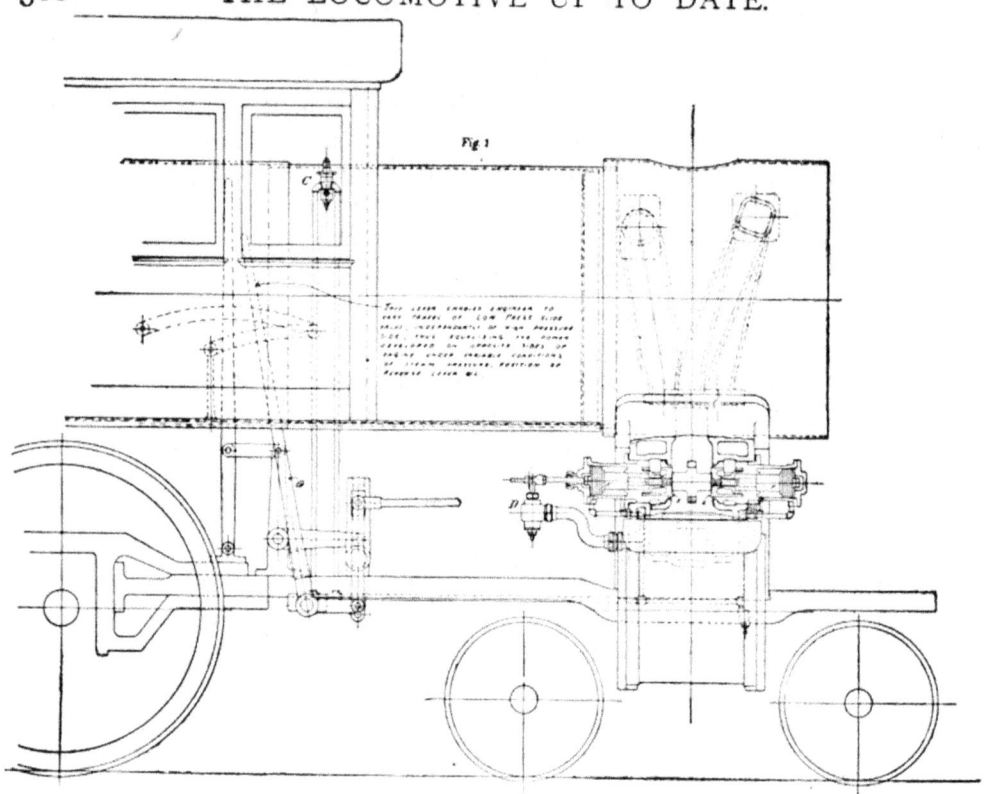

Fig. 1

TWO WAY COCK C.—To run engine Simple, in starting, or whenever necessary to develop maximum power, pull handle to back position, admitting steam to outer ends of intercepting valve cylinders; this closes communication between receiver and low pressure cylinder, diverts high pressure exhaust to stack, and opens reducing valve D which admits live steam to low pressure steam chest.

To run engine Compound, place handle in forward position, admitting steam to inner ends of intercepting valve cylinders; this opens communication between receiver and low pressure cylinder, closes passage between receiver and exhaust to stack, and shuts reducing valve D.

After changing from simple to compound, or vice versa, handle should be placed in mid position, cutting off steam from intercepting valve cylinders.

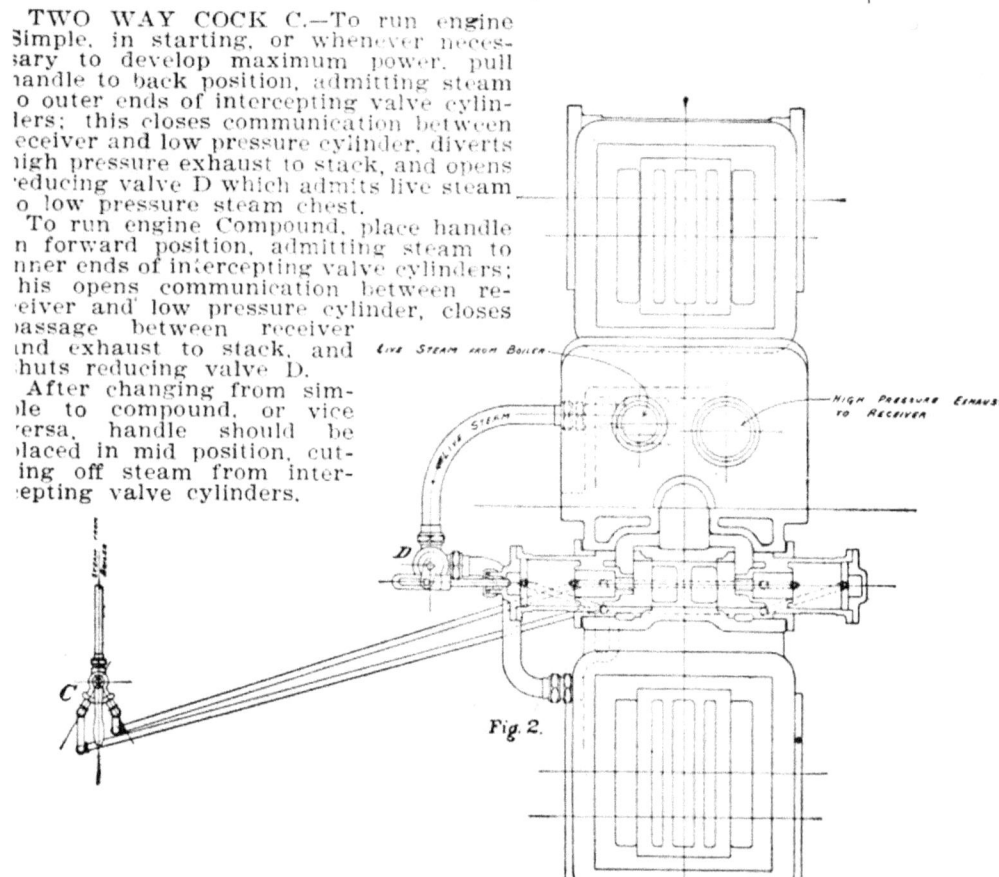

Fig. 2.

Engine No. 1, Delaware, Lackawanna & Western Ry.
Cooke Compound.
General Dimensions Given on Page 50.

390   THE LOCOMOTIVE UP TO DATE.

The intercepting valve pistons are operated by steam, which is admitted through pipes at either side of pistons.

The movements of intercepting valves are controlled by the engineer's valve C in cab, which is connected as shown on Fig. 1.

Fig. 4 shows intercepting valves drawn back into compound position, giving a free passage for exhaust steam from high-pressure cylinder to low pressure steam chest by way of receiver, as indicated by arrows.

Fig. 5 shows intercepting valves closed or in simple position,

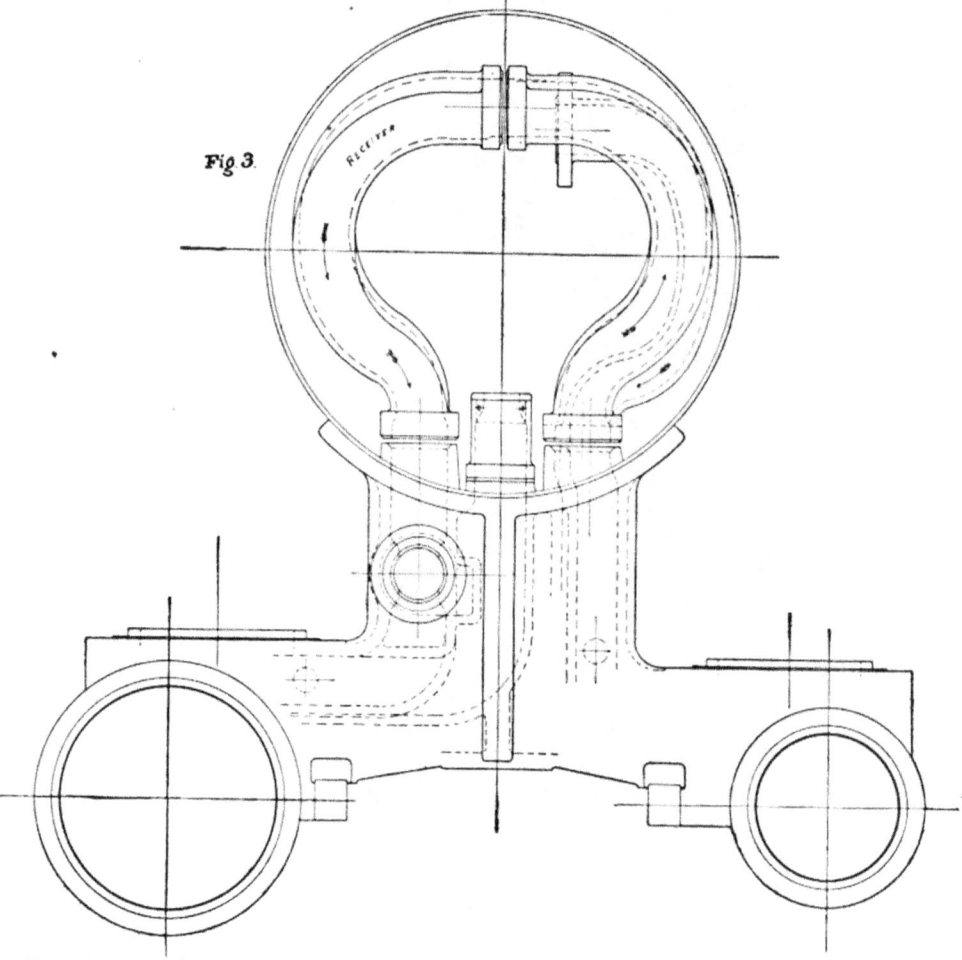

Fig. 3.

allowing exhaust steam from high-pressure cylinder to pass directly to exhaust pipe. When valves are in this position steam is being admitted to steam chest of low-pressure cylinder through reducing valve, which has been opened by the movement of the intercepting valve, and which remains open as long as engine is running simple.

Fig. 6 shows valves during movement from simple to compound position with exhaust passage A closed before the openings H in receiver are uncovered.

In starting, the handle of engineer's valve C is pulled to back notch, thereby admitting steam to outer ends of intercepting valve cylinders, and moving valves to simple position, as shown in Fig. 5.

The engine may be run simple as long as desired.

To change from simple to compound the handle is pushed to forward notch, admitting steam to inner side of intercepting valve cylinders, and moving valve to compound position, as shown in Fig. 1.

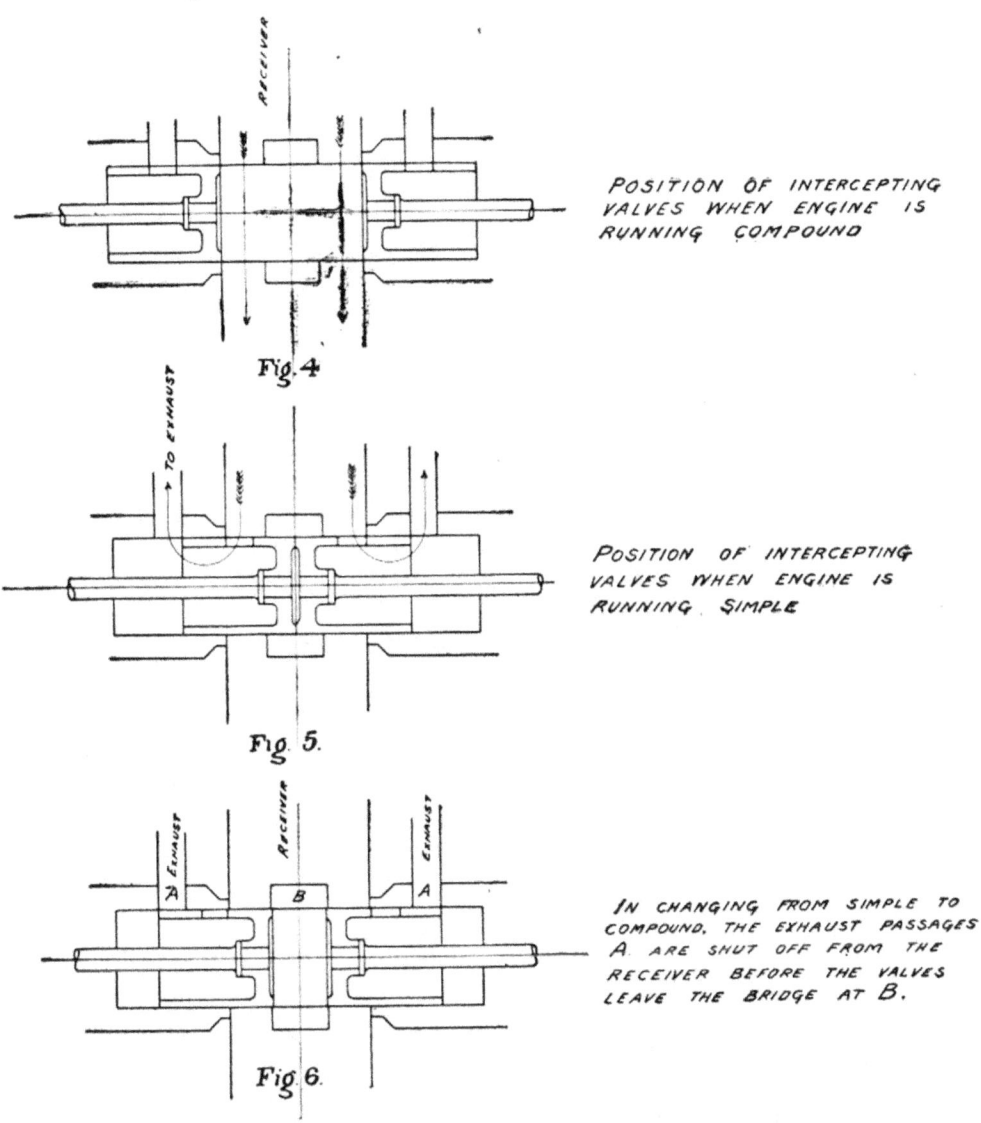

POSITION OF INTERCEPTING VALVES WHEN ENGINE IS RUNNING COMPOUND

Fig. 4.

POSITION OF INTERCEPTING VALVES WHEN ENGINE IS RUNNING SIMPLE

Fig. 5.

IN CHANGING FROM SIMPLE TO COMPOUND, THE EXHAUST PASSAGES A ARE SHUT OFF FROM THE RECEIVER BEFORE THE VALVES LEAVE THE BRIDGE AT B.

Fig. 6.

After changing from simple to compound, or *vice versa* the handle should be placed in middle notch, as shown in Fig. 1, in which position all steam is cut off from intercepting valves.

In Fig. 2 a cut-off lever G is provided by which the travel of slide valve on low-pressure side may be increased or decreased independently of the valve on the opposite side.

ACCIDENT WITH THE COOKE COMPOUND.

In case of accident to either side of engine the opposite side may be run for any length of time as a simple engine by disconnecting and blocking slide valve on injured side in the same manner as done with ordinary engines.

THE DICKSON COMPOUND LOCOMOTIVE.*

The compound locomotive furnishes the only fundamental means of saving fuel and water compared with simple locomotives, all other means being secondary and as applicable to compound as to simple engines. Compounding locomotives is the application to locomotives of the principle now almost exclusively used in pumping, mill and marine engines, by means of which enormous savings in fuel have been made. In marine service, the extension of the compound engine, known as the triple expansion engine, has rendered it possible for steamships to make great speed without carrying sufficient coal to seriously encroach upon their cargo-carrying space.

Briefly considered, the difference between the simple and compound locomotive is as follows: The simple or ordinary locomotive consists of two independent steam engines turning the same axle. All compound locomotives consist of two engines, one of which receives steam directly from the boiler, and the other receives the steam exhausted from the first cylinder. This is a means of saving steam long since known and used in all engines except locomotives. The saving of coal in such locomotives, therefore, results primarily from the saving in steam. There is, however, a secondary saving, produced by the less violent effect of the exhaust steam upon the fire. In the simple locomotive the exhaust produces such a violent draft that great quantities of unconsumed coal are drawn from the fire-box and thrown from the smoke-stack, and is often heard falling upon the roofs of the cars and felt entering the open car windows. With the Dickson compound locomotive the exhaust steam escapes at so low a pressure that very little is ejected from the stack, thus leaving nearly all the coal to be consumed, and rendering travel much more comfortable.

This secondary saving is of as much importance, as far as the fuel is concerned, as the other. The saving of water is of value wherever water is purchased, or is scarce. Wherever water

---

*This article was especially prepared for this book by the Dickson Locomotive Works.

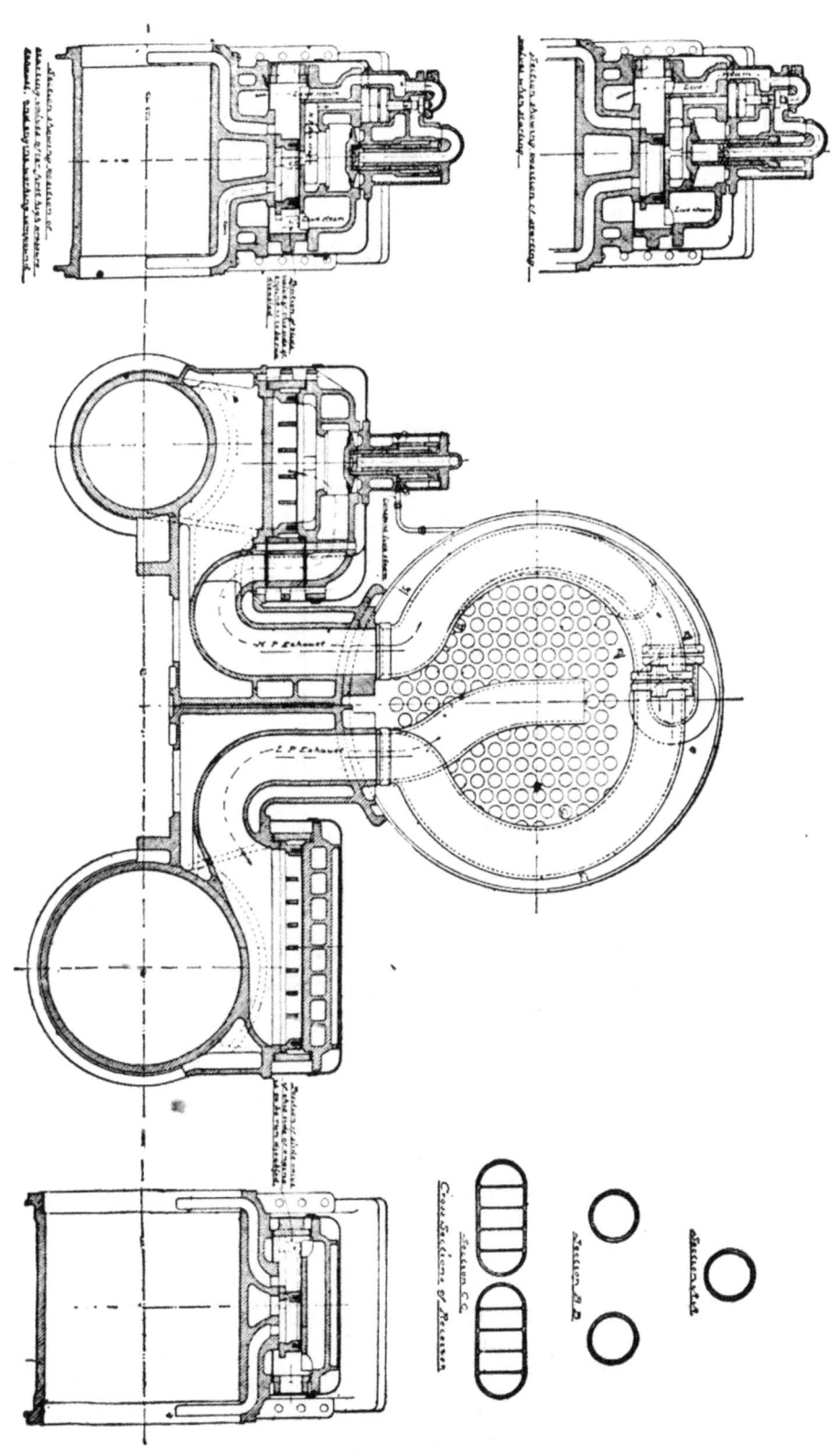

394   THE LOCOMOTIVE UP TO DATE.

is of bad quality it is of the first importance that as little as possible should be used in the boiler.

Compound locomotives do not cause fires, and the reduction of the fire risk in forests, lumber districts and town is an important advantage.

In bad water districts the reduction of water use by 15 to 20 per cent must result in greater life and diminished leakage of boiler tubes.

The greatest savings to be realized by compound locomotives

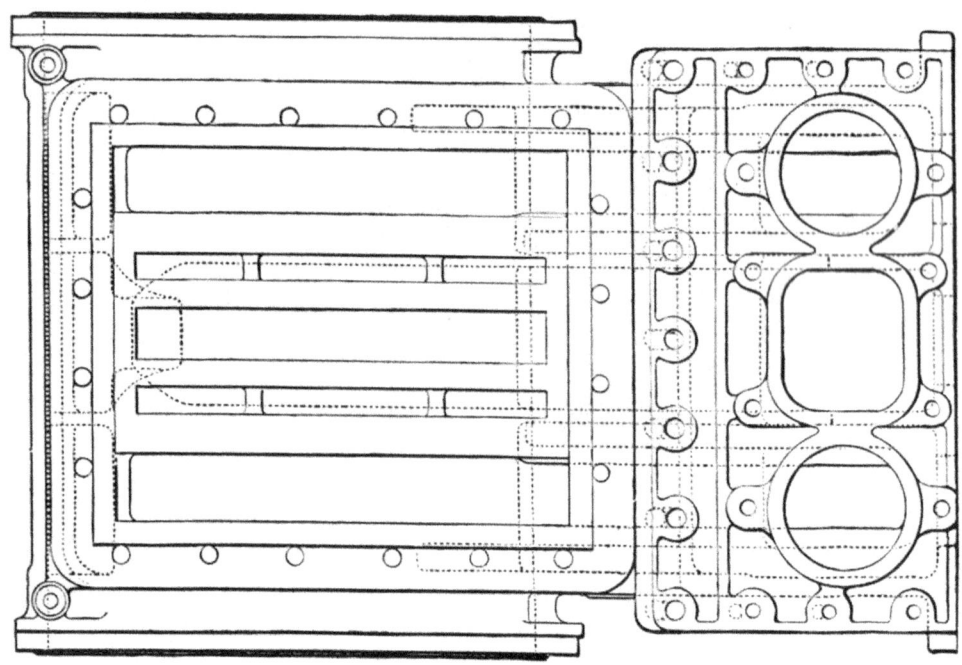

are to be found in freight and local passenger trains, but the saving in express passenger work is but little less if the engine is designed with care for fast work.

The Dickson compound locomotives are of the two-cylinder type, and are built under the Dean patents. They are made to start automatically or they can be worked at will as simple engines, and will be made in this respect as the purchaser prefers.

The starting valve is placed on top of the high pressure steam chest, can be removed in a few minutes and can be replaced by a spare valve in equally quick time. Its moving parts have only vertical motions and have not even their weights to wear them out. They are perfectly cushioned and yet have neither stuffing boxes nor dash pots as others have. It can confidently be asserted that the starting valve will last as long as any part of the engine.

An advantage of the position of the starting valve of our engine compared with those compound locomotives that have the valve in the cylinder casting, is that in case of a broken cylinder a new cylinder can be easily made as that of a simple engine.

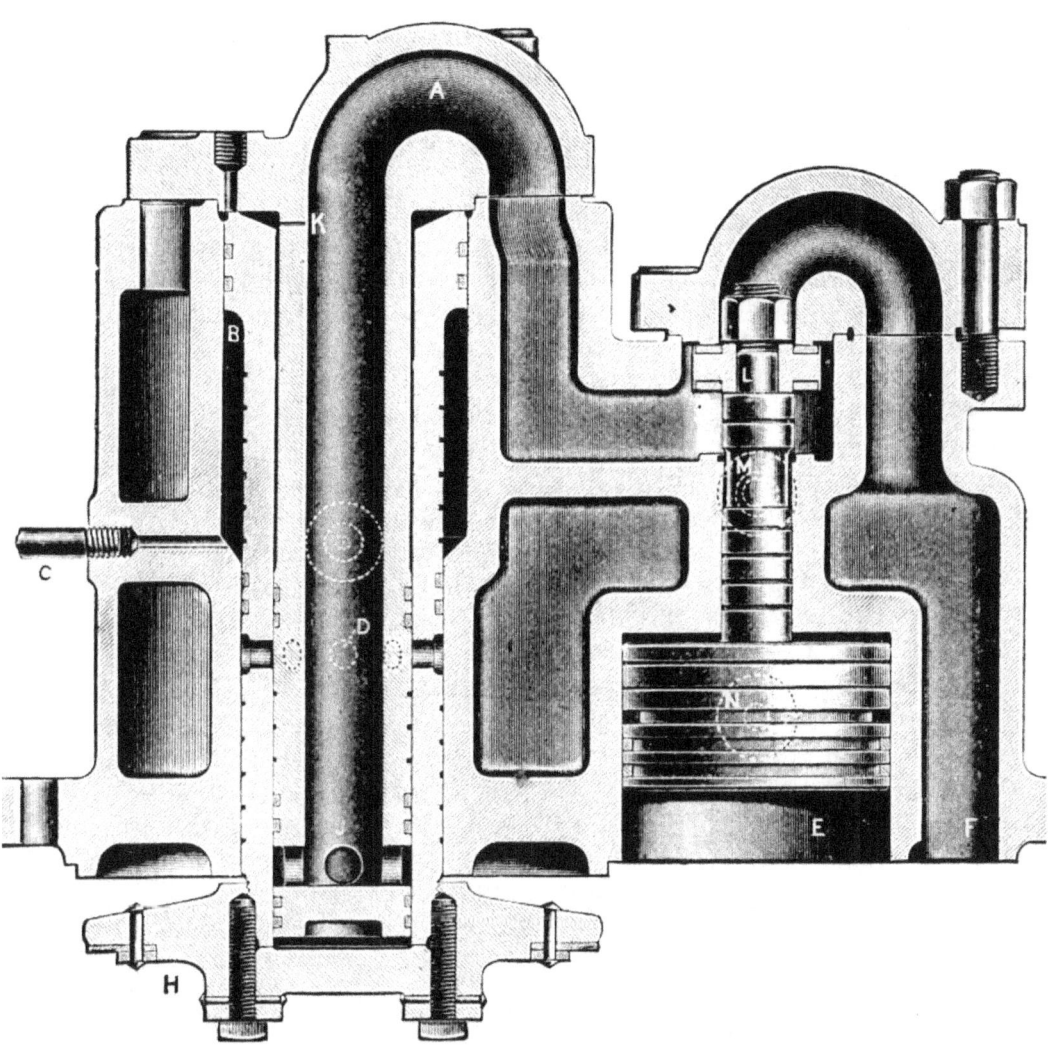

Incidentally the weight of the two sides of the engine are equal in consequence of this feature.

Great attention is paid to utilizing the superheating capacity of the hot smoke box gases by making the high-pressure exhaust pipe branch into two receivers of oval cross section, thus exposing a large surface for reheating the exhaust steam of the high-pressure cylinder before it is used in the low-pressure cylinder.

Each receiver has its own opening into the low-pressure

steam chest, and thus large passages are provided where there is the greatest demand for steam, this demand being caused by the large size of the low-pressure cylinder. In this connection it is important to note the fact that the starting valve, which in all other designs presents somewhat indirect passages, is in our design placed where the velocity of steam between the two cylinders is least.

The two features of the design mentioned in the last paragraph reduce the loss of steam effect between the cylinders to a minimum and make this engine very effective for fast work.

DESCRIPTION OF THE AUTOMATIC STARTING VALVE.

The starting valves consist of two valves in one casting, which is bolted to the top of the high-pressure steam chest cover and makes therewith a scraped or ground joint.

One valve, marked L, is called the converting valve, and the other, marked H, is the intercepting valve.

The intercepting valve is screwed to an annular stem which is enlarged at the top above B. The space B is subjected to constant boiler pressure or air pressure from the air brake reservoir and its normal position is therefore at the top of its travel after the engine has started, in which position it is locked by the pressure referred to. At such times, also, the converting valve is at its highest position, being there held by receiver pressure acting on the piston N at E.

In order to explain the action of the valves, it will be assumed that the engine is running with the throttle open and the valves in the positions described above. The throttle is closed on approaching a station, or for other reasons, and the steam is all worked out of the receiver. The converting valve then falls by the weight, the jar of the engine and the suction of the low-pressure cylinder, until the valve L seats on its under side. Upon opening the throttle, live steam from the high-pressure steam-chest flows through a hole in the cover coincident with the passage F, thence past the valve L, into the central tube AJ. It also passes through the small hole K and acts upon the top of the intercepting valve stem, which top being larger in area than the annular area of the under side B, overcomes the pressure at the latter point and forces the steam or air back into the boiler or air brake reservoir, as the case may be, and thus causes the intercepting valve to slowly descend until it reaches its seat.

Just before the intercepting valve reaches its seat, and after it enters a raised edge around the seat, the holes D approximately coincide with and are fed with steam by a groove around the outside of the central tube, joining the holes J through this tube. The lower edge of the groove on the outside of the intercepting valve stem, joining the holes D, is at this time below the

THE LOCOMOTIVE UP TO DATE.

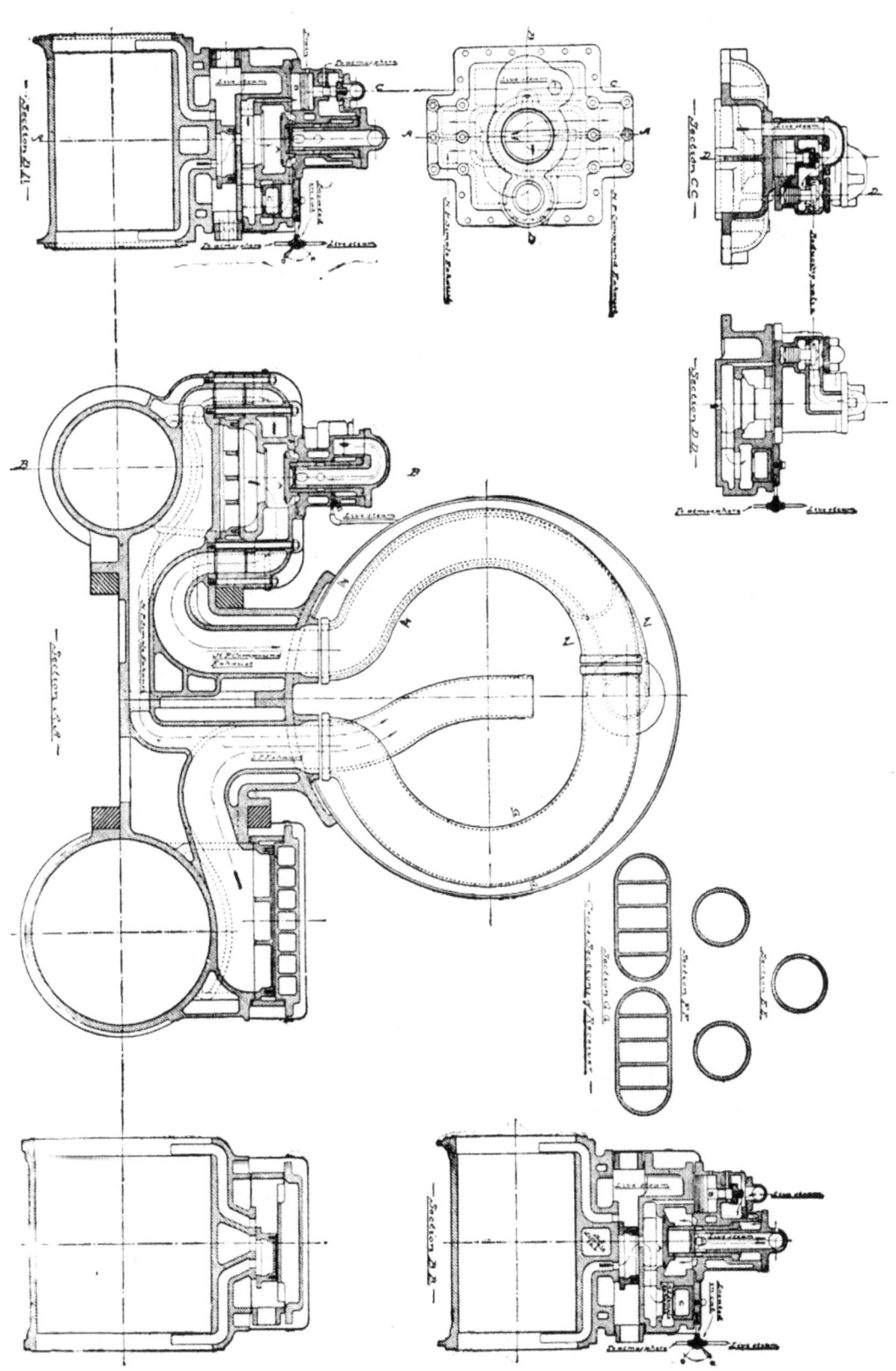

lower edge of the intercepting valve casing, and thus the steam flows into the receiver and over to the low-pressure steam chest and starts the low-pressure piston.

After the engine starts, the first high-pressure exhaust soon occurs and moves the starting valves into their running position. The engine is brought into its permanent compound condition, as follows:

The high-pressure exhaust port is in the balance shield of a Richardson balanced valve having its top removed (as is not infrequently done in simple engines), instead of being between the steam ports, and thus the exhaust steam is led through a passage above the exhaust port to the cavity E, and closes the converting valve L. This shuts off the supply of live steam and at the same time allows the steam in the central tube A to escape to the atmosphere around the reduced diameter of the converting valve stem, and hole M in the casting.

The escape of the steam from A permits the steam above the intercepting valve stem to escape through the small valve hole K, and therefore the constant pressure at B from the pipe C is free to lift the intercepting valve to its highest position. The engine is then permanently compound, and as above mentioned, the intercepting and converting valves are locked in position by pressure. This prevents vibration of the valves while the engine is running either with or without steam.

The converting valve piston has two packing rings near the bottom, and above them a groove. The function of the groove is to collect any steam that leaks by the packing rings and to allow it to escape to the atmosphere through the hole at N, so that it will not collect above the piston and tend to depress the converting valve by an accumulation of pressure.

By disconnecting the pipe C and unbolting, the whole valve casing can be removed in a few minutes. By taking off the cover above the converting valve the valve disc can be examined, and removed from the stem. The central tube of the intercepting valve can be taken out and a portion of the intercepting valve examined.

The intercepting valve cannot slam when moving in either direction, as steam or air is slowly squeezed out in both motions, nor can the converting valve when moving upward, as the space above the piston N acts as a dash-pot. The descent of the converting valve is not violent, and is held down by the suction of the low-pressure piston when the engine is running without steam.

Both the converting and intercepting valves are faced with Pratt & Cady's vulcanized asbestos, which makes a soft tight fit, is easily replaced, and lasts about four years.

CAN BE WORKED AS A SIMPLE ENGINE AT WILL.

The description of the automatic starting valves applies to this as far as it goes, but there is added in this engine a reducing valve between the converting and intercepting valves. The reducing valve is a cast iron vertical working plug in the valve casing, so made that it will reduce the pressure of the steam that goes to the low-pressure cylinder inversely proportional to the area of the piston of that cylinder.

In the steam chest cover, in front of the starting valve, there is a valve C which is held to its seat when the engine is working compound, by live steam, but which, when the atmosphere is in communication with the space above, will move with two or three pounds per square inch and thus permit the high-pressure cylinder to exhaust direct to the atmosphere. The passages through which this exhaust reaches the atmosphere are in the steam chest cover, the steam chest, the cylinder between the steam ports, and a horizontal passage to the low-pressure cylinder exhaust passage.

If the valve C is held on its seat the engine starts as an automatic compound. If the valve C can lift in virtue of there being no pressure above it, insufficient pressure will accumulate to raise the converting valve. Live steam will therefore pass by the converting valve and hold down and flow through the intercepting valve to the receiver and low-pressure steam chest and the engine works simple. If the steam is turned on to press down the valve C the converting and intercepting valves rise and the engine works compound. These changes will not be quick and harmful to the valves because the valve C will descend slowly as the steam passes through a very small admission hole, and in rising it cannot slam as the steam above cushions it. For these reasons the change from simple to compound and *vice versa* can be safely made when the engine is running at any speed.

THE RHODE ISLAND COMPOUND.

This form of compound is of the two-cylinder type. The engine starts with steam in both cylinders and automatically changes to a compound at any desired receiver pressure. The engine may be worked simple or compound, as desired, by simply moving a lever in the cab.

The intercepting valve used is a patent of the Rhode Island Locomotive Works, but it did not work satisfactorily and the Company is not at present building compound locomotives, but as there are a few of these engines still in service we shall state what should be done in case of accident to this system of compounding.

ACCIDENTS WITH THE RHODE ISLAND COMPOUND.

MAIN ROD LOW PRESSURE.

Remove the broken rod and clamp the cross-head at the back

end of the guides. Clamp the low-pressure valve central on its seat to cover all ports; open the exhaust valve in the receiver to the exhaust nozzle and proceed with one side, as with simple engine.

MAIN ROD HIGH PRESSURE.

Remove all broken rod and secure the cross-head at the back end of the guides; clamp valve in center of ports, using low pressure as a high-pressure cylinder. Open throttle gradually on account of the large area of the low-pressure cylinder.

VALVE STEMS, EITHER SIDE.

The same as for broken main rods.

INTERCEPTING VALVE.

If the piston head of the intercepting valve which closes the receiver to the low-pressure steam chest gets broken or cracked, the engine may be run without using live steam in the low-pressure cylinder by removing the back head of the intercepting valve cylinder, and moving the valve to the compounding position and blocking it secure. The hole in the intercepting valve which would allow live steam to escape from the low-pressure steam chest to the atmosphere prevents the exhaust valve in the receiver being opened and engine run as a high-pressure engine.

RECEIVER.

In case of a broken or cracked receiver, open the exhaust in the receiver and run as a high-pressure engine.

THREE-CYLINDER LOCOMOTIVES.

There are a good many three-cylinder locomotives in use to-day in different parts of the world, but they are a disappointment to their friends. The extra cylinder puts nearly a third of extra mechanism upon the engine without bringing any advantages to speak of.

American mechanics were early in the field with three-cylinder locomotives. The "Washington" and "Ohio," three-cylinder engines, were rebuilt at Wilmington shops by Louis Fleicker some time about 1850. The "Washington" was a 30-ton engine, carried 90 pounds steam pressure, had two outside cylinders 13½x20 inches and one middle cylinder 16x20 inches, 20-inch stroke and a 5-foot wheel without the tire.

The "Ohio" was a 35-ton engine, carried 95 pounds steam pressure, had two outside cylinders 13x20 inches and one middle cylinder 16x20 inches, 20-inch stroke and a 5-foot wheel without tire.

They were both wood-burning engines. Mr. Fleicker was master mechanic of the Wilmington shop and the designer of this style of engine. Previous to this time they were "Norris" engines.

BREAK-DOWNS,

OR

"ACCIDENTS TO LOCOMOTIVES."

This chapter is intended for the purpose of aiding, or reminding, locomotive engineers of their duty in case of an accident to their engine. If your engine is a compound and any accident happens to the cylinders, or valve gear, see our chapter on "Compound Locomotives" and accidents with compound locomotives.

The prudent engineer who inspects his engine regularly and replaces the loose bolts, tightens nuts and keys, looks for defects and carefully examines any cracks, flaws or other defects upon his engine is seldom troubled with those annoying and sometimes dangerous break-downs while on the road. An observant engineer who becomes familiar with the regular and equal exhausts of his engine often detects a lame exhaust in time to stop and prevent a serious accident to his engine, as any defect in the valve motion, such as a loose eccentric, strap bolt, loose blade bolts, a loose valve stem key, a broken valve yoke or slipped eccentric, will cause an imperfect exhaust. But breakdowns sometimes occur after a man has done everything human foresight can aid in doing to prevent an accident. These pages were written to assist those who are so unfortunate as to meet with a mishap on the road, who may through excitement or lack of knowledge omit, or forget, to do something of great importance, which may cause a worse breakage than the first. Numerous cases may be cited where engines have been brought in without performing the amount of work which we say becomes necessary, but those engineers did not take the safe side, and had other breakages occurred they would have been held strictly accountable for their neglect. In case of an accident we presume the engineer will first comply with his book of rules regarding signals, flagman, etc., and will not neglect his boiler while working on a disabled engine. If the engine is in the ditch, or the crown sheet not properly protected, kill the fire immediately. If you cannot secure water gravel will extinguish the fire. These pages only treat on the mechanism of the locomotive. All well equipped engines are supplied with a sufficient number of jacks, hand tools, wrenches, clamps, blocks, etc., to be used in case of an accident, and every thoughtful engineer will see that his engine is supplied with such things before starting out on the road.

DISCONNECTING BOTH SIDES.

This implies that the engine is dead and must be towed in. Remove both main rods and both valve rods, but it will not be necessary to block either, if the crank pins clear the cross-heads. Do not remove the side rods or eccentric straps unless it is necessary; and when it is considered necessary be sure to take the precautions previously explained.

In freezing weather, if the fire is down, all water should be drained out of the injectors, pumps, feed and branch pipes. If there are not frost plugs slack the joints and let the water out. If there is danger of the water freezing in the boiler run it out of both boiler and tank. See that all oil cups are well filled before starting. Most all roads are very strict regarding the speed of dead or disconnected engines, as the engine is not then counterbalanced perfectly, and is therefore very injurious to the track. Some of the best roads limit the speed of all heavy engines which are disconnected on one or both sides, or which have the side rods removed, or dead engines hauled in a train, to twenty miles per hour.

ENGINE OFF TRACK.

If engine is in such a position as to leave crown sheet or flues unprotected by water, draw the fire; or if you cannot draw it, smother it with earth or sod, snow or fine coal. Ask headquarters at nearest telegraph office for assistance.

The first thing to do after protecting yourself from approaching trains, is to see if the boiler is high enough at either end to leave either the crown sheet or the front ends of the flues unprotected by water, for with a hot fire if these parts are not well covered with water, either of them may become red hot and may burn or even melt the metal in a short time.

Most engines, if not off very badly or too far away from the track, will help themselves on without the aid of another engine by using blocking under the wheels. If you have jacks they will aid materially by setting them to push the engine. Engines can usually be put on the track easier by moving them in the direction opposite to that in which they ran off; that is, if they get off by running ahead, they should be moved backward to put them on the track.

Do not ask head-quarters for help if you can possibly avoid it.

WATER FOAMING IN THE BOILER.

The surface skimmer, which is a part of the Hornish mechanical boiler cleaner, is admitted by leading railway men to be the best device yet invented to prevent foaming, or to settle the water in a foaming boiler. Other surface blow-offs are in use and giving good results, but the Hornish skimmer is the full width of the boiler and will remove all impurities from the surface of the water in a few seconds, and as it is oil and impurities in the water that causes a boiler to foam, and as they always come to the surface of the water, they can be easily blown out, but if your engine is not equipped with any kind of a surface blow-off. Then open the cylinder cocks, close the throttle valve gently until the water settles solid, then try how much water there is in the boiler. Put pumps and injector to work, if necessary. Open the throttle gently and work the foul water through the cylinder; close the throttle often and try the height of the water. Be very careful in admitting steam into the cylinder, or you will knock the packing down or the cylinder-head out. If the cause of foaming is found to be grease in the tank, flow the tank over when you take water, and if you can get about one-fourth of a peck of unslacked lime put it in the tank. A piece of blue-stone about the size of a hickory nut put in the hose back of the screen will prevent foaming. This can be had in any telegraph office.

BURST FLUE.

Reduce the steam pressure and plug the flue. A wooden plug will answer if you have no iron plug. If you can get at the flue from the fire door (a long iron rod is usually furnished each engine for this purpose) you can plug it without drawing the fire, or a sharpened pole or stick of wood long enough to reach the flue may be driven into it. The wood will not burn inside the flue sheet. If this cannot be done then cover the fire dead, open the blower enough to carry off the smoke in the fire box, then lay a board on top of the coal and go into the fire box and calk or plug the flue. Of course this cannot be done if there is a brick arch in the fire box.

BROKEN VALVE YOKE.

A valve yoke usually breaks off at the neck of the valve stem. It can be readily discovered in the exhaust by a tremendous blow. If the valve is pushed far enough ahead it will blow; if not, it is often mistaken for a slipped eccentric (examine the eccentrics first). It may be discovered in this way: Place the crank pin on top or bottom quarter and reverse the engine; if the steam still continues to come out of the back cylinder cock you may depend it is usually the yoke. (See how to locate blows.) A great diversity of opinion exists regarding the best remedy for this kind of a break. The old and safest way is to raise the chest cover and block the ports central, replace the cover, remove the valve rod and main rod and block the cross head at the back end. But this remedy requires much time and labor and as *time* is a very important consideration on the road, and as there appears to be no mechanical objections to the other

DISCONNECTING.

In order to avoid an unnecessary repetition of words, and to assist in condensing this chapter, we will first explain what should be done to disconnect one side of an engine; and next what should be done and what precautions should be taken when it is necessary to disconnect both sides of a locomotive. Afterwards throughout this chapter we will simply say "disconnect one side" or "disconnect both sides," as may be necessary. Some writers on this subject repeat the method of disconnecting every time it is referred to, but we have no fear that this book will be too small and are not trying to "fill it up."

DISCONNECTING ONE SIDE.

This necessarily implies that the engine is to continue its trip. Remove the main rod on one side and place the liners and brasses back in the straps just as you found them. Secure the cross-head near the back end of the guides with a cross-head clamp, if you have one, if not, then with hard wooden blocks, securing the blocks with a rope so they cannot work out. Don't move the cross-head clear back to the striking point as the cylinder packing rings may get down into the port or counterbore. Remove the valve rod and secure the valve stem with a valve stem clamp, if you have one, if not, set the valve central upon its seat and cramp the valve stem by tightening the gland on one side. Most engines that use metallic packing are supplied with a valve-stem clamp made to hold the valve central upon its seat; but if you have none, the valve can easily be set to cover the ports by opening the cylinder cocks and giving the engine a little steam. Then adjust the valve stem until steam is entirely shut off from both cylinder cocks. Do not remove the eccentric straps or side rods unless it is necessary. Whenever the eccentric straps are removed on one side the top of the link should be tied to the short arm of the tumbling-shaft to keep it from tipping over, which would prevent reversing the engine. If it is necessary to take one side rod down, remove the one directly opposite to it; if this cannot be done, then remove all the side rods. Do not remove the eccentric blades, leaving the straps on the eccentrics, unless they will whirl and clear everything in all positions; otherwise they might punch holes in the fire box.

If the side rods have been removed from a ten-wheeled engine, or pony engine, see that the forward crank pins will clear the cross-head in all positions; if not, *take no chances*, but disconnect both sides, blocking both cross-heads clear forward or wherever they will clear the crank pins and have the engine towed in.

methods, providing the cross head is securely fastened, we will state the other remedies. Disconnect the valve rod and push the valve clear ahead, remove the stem if it would blow out, and use a gasket back of the gland, or hold the valve stem intact with valve stem clamp. Block the cross head at the front end, and proceed; the pressure will hold the valve forward and if it should move it can do no harm, providing the cross head is securely blocked. Another way is to remove the release valve, push the valve clear back, fit a block into the release valve long enough to hold the valve back, then block cross-head at back end. Still another way is to push the valve stem forward and clamp it by cocking the gland, then block cross-head at the front end. If the yoke is only broken at one side of the valve it will only affect one exhaust. When the yoke pushes the valve forward the valves will sound all right, but when it pulls the valve back the engine will be lame. With careful handling you may finish your trip without breaking the other side. Work the engine in full gear with a light throttle.

BROKEN STEAM CHEST OR COVER.

This is a very troublesome mishap. If you think that the chest is only cracked, remove the casing, and if it is only cracked on one side by wedging between the chest and studs you may be able to close the crack enough to get in. But if it is a bad crack, or a cover, and you have no way of clamping the broken parts, disconnect that side and then the quickest remedy would be to use a blind gasket at one end of the steam pipe. But that is considered impracticable, owing to the corrosion of the bolts and nuts, the netting and a very hot front end. So remove the chest cover and plug up the supply ports, not the steam ports, with wood and clamp the plugs with steam chest studs. If steam enters from side of chest use a gasket there. If the chest is completely knocked off, clamp your wooden plugs with old bolts and fish plates, or whatever you can find.

If you cannot make the ports tight, you can leave the train and run under light pressure to the nearest telegraph office and report to headquarters condition of the engine. Remember many steam chests may be saved, when you have no relief valves, by opening the throttle slightly as soon as you reverse the engine, for a reversed valve gear is virtually an air pump, and if the air cannot enter the boiler it must escape somewhere when compressed.

BROKEN PISTON ROD.

When a piston rod breaks it invariably knocks out the front cylinder head. When the piston is entirely out of the cylinder or if it can be easily removed, then it will not be necessary to remove the main rod. Simply clamp the valve stem central as previously explained. Remove all loose broken pieces and proceed.

BROKEN PISTON OR VALVE STEM GLANDS.

If the gland breaks in two try and wrap it with bell cord or wire. If a lug breaks off make a wooden clamp; you have two nuts on each stud, so remove one of the nuts and tighten the other up against your temporary clamp. If you loosen one stud wrap wire or rope around the steam chest and try to hold it secure, or remove part of the packing and shove gland in further and try to hold it with one stud. If all other remedies fail disconnect one side.

BROKEN VALVE ROD.

Disconnect on the broken side.

BROKEN CRANK PIN.

If it is a main rod disconnect one side and remove all side rods. If a back pin on an eight-wheel engine simply remove both side rods; if a back pin on a ten-wheeled engine remove the back pin of side rods only. If a front pin on a ten-wheel engine remove the forward pair of side rods only.

BROKEN CROSS-HEAD.

A slight break, such as a gib or plate, may sometimes be clamped so you can proceed, but be careful that the clamp does not strike the guide block at extreme travel of the cross-head. If it is a bad break disconnect the broken side. If the piston is not broken push it against the forward cylinder head and then block the cross-head in that position. If the cross-head is broken so that the cross-head cannot be blocked, the safest way is to remove the piston. If it cannot be taken out set the valve so as to admit steam to the back end of cylinder only, and clamp valve stem securely in this position.

BROKEN MAIN ROD OR STRAP.

Disconnect on the broken side.

BROKEN SIDE ROD OR STRAP.

Remove the broken rod and the parallel rod directly opposite to it. If it is a ten-wheeled engine and this cannot be done remove all the side rods. If a front or back rod or strap on a twelve-wheeled engine, remove the broken rod and the one directly opposite to it, if this can be done, and leave the others up.

BROKEN ROD SET SCREWS.

When it is required to remove a key from a rod, if the set screw is broken and cannot be backed with a chisel, if in the back end of a main rod, take the strap bolts out of that end of the rod, and block the cross-head; then with a pinch bar move the engine until the key is loose. If the set screw is broken in a parallel rod, take the bolts out of the strap where the screw is broken, block the other drivers and with pinch bars slip the wheels until the key is loose.

BROKEN CYLINDER HEADS.

BACK HEAD.

Disconnect the engine on broken side. If it is necessary to remove the guides and broken head then remove the piston also.

FORWARD HEAD.

Disconnect one side of the engine. Another method advocated by many, but practiced by few, by which three-fourths of the power of the engine could be retained, is to remove the steam chest cover and plug up the forward steam port with wood and proceed working both sides. This method is impracticable, owing to the shape of the steam port cavity on most engines, and the time it would require, as *time* is usually the most important factor, besides the improbability of the block remaining intact.

BOTH FORWARD HEADS.

This is a subject which has received considerable attention from railroad men. It is a sort of a "catch question" propounded to test a man's knowledge of valve motion, for it is very seldom such an accident occurs, and when it does happen ninety-nine engineers out of every hundred either telegraph for new cylinder heads or have their engine towed in. However, if both pistons were all right and both forward steam ports were properly blocked the engine could handle itself in this condition. But such a method is impracticable on account of the shape of the ports and the improbability of getting the port securely blocked without interfering with the movement of the valve. Right here it may be well to state that it is the steam port which should be blocked and not the supply port.

BROKEN GUIDES, BLOCKS OR BOLTS.

If any of the bolts break try and replace them. See that all nuts are tight, or they may be the cause of springing the piston. If a guide bar is broken badly disconnect one side.

BROKEN GUIDE YOKE.

If a yoke is bent or broken and will not hold the guides secure disconnect one side.

BROKEN ROCKER BOX.

If it cannot be clamped or blocked secure then disconnect that side. If you can remove the rocker arm it will save taking down both eccentric straps, but if you cannot remove the rocker without difficulty then remove both eccentric straps and tie the top of the link to the short arm of the tumbling shaft, to prevent the link from tipping over.

SLIPPED ECCENTRICS.

Since the practice of keying eccentrics to the shaft has become so general these mishaps occur less frequently than formerly. Yet every engineer should know how to reset a slipped eccentric. This accident is easily detected, for without any warning the exhaust will become very irregular and you may lose one exhaust entirely. You should immediately stop and go down under the engine, when the cause will be quickly discovered. If you have previously studied the positions of the different eccentrics it will be an easy matter to reset it, but if you are not familiar with the relative positions of the eccentrics you may experience considerable trouble. Every engineer and fireman should familiarize himself with the correct position of each eccentric, and the practice of marking each eccentric and shaft is a very wise precaution, although many roads prohibit the practice of making a chisel mark on the shaft. On all standard engines with indirect motion when the crank pin is on the forward center the go-ahead eccentric will be above the pin, and the back-up below the pin, and when the pin is on the back center their positions are reversed. The rib of each eccentric is set about the third spoke away from the pin. The spokes in different wheels may vary, but so does the lead and lap. Remember they should never be at right angle with the pin, but each should incline slightly toward the pin. Perhaps the quickest way to set a slipped eccentric approximately correct is that old way of marking the valve stem, which is as follows: Place the crank pin on either dead center on the side that has slipped; the forward center is the most convenient; if the forward eccentric has slipped (the forward motion is usually attached to the top of the link), place the reverse lever in extreme back notch of quadrant, and with a knife or some other sharp instrument scratch a line on the valve stem as close as possible to the gland. Now place the reverse lever in extreme forward notch and move the eccentric around until the same line on the valve stem appears, then set the eccentric in that position. If the crank pin is on the forward center the center of the eccentric should be above the pin, and if on back center below the pin. If the back-up eccentric slips go through the same performance in exactly the reversed manner, by placing lever in forward notch and marking the stem, then put in back notch and set the eccentric same as before. If both eccentrics on same side slip set each eccentric as near as possible to the positions we have previously explained. Now place crank pin on forward center, block the

THE LOCOMOTIVE UP TO DATE. 409

wheels, open the cylinder cocks, place reverse lever in back notch and give engine a little steam. Now move back eccentric until steam appears at front cylinder cock, then fasten the eccentric. Now place lever in forward notch and move eccentric until steam appears at same place (forward cylinder cock); then fasten the eccentric and you are done.

Another method is to get the engine on its dead center as near as you can by the eye, and if the forward motion is slipped, hook the reverse lever clear back, then clamp the valve stem so that the valve cannot move, then take out the bolt that connects the forward motion eccentric rod to the link, then throw the reverse lever all the way ahead, being careful that the valves do not move, then by moving the slipped eccentric until you can put the jaw bolt in, the eccentric will be near enough right to run in; only be careful that the eccentrics are not in the same position on the axle, or you will have both set run backward—one eccentric must point up, the other down. A back motion eccentric can be set in the same manner, only the reverse lever must be thrown ahead, then clamp the valve stem, then take the jaw bolt out of the back motion rod and move the back motion eccentric until the bolt will go in without moving the valve or rocker arm. When an engine is on its dead center, the valve should be in exactly the same position when the reverse lever is in extreme back notch as when in extreme forward notch; so if the valve rod does not move while the pin is out and the reverse lever is being moved, the eccentric will be nearly right; after the pin is put in the valve rod will move while the lever is being moved, but in the extreme notches will show that the rod is in exactly the same place.

Many experienced men who are familiar with the position of each eccentric ignore these old rules and set each eccentric by eyesight, when their correct positions are not marked.

SLIPPED ECCENTRIC BLADE.

This defect is easily detected by the irregular exhausts. It can be reset by placing the crank pin on either center; place reverse lever in forward notch, open the cylinder cocks, then adjust the blade until steam appears at the front or back end of the cylinder according to the position of the crank pin. If pin is on forward center steam should appear at front end of cylinder, and *vice versa;* or it can be set by marking the valve stem same as for slipped eccentric.

BROKEN ECCENTRIC STRAP, ROD OR ECCENTRIC.

In the event of a break-down of this nature the safest remedy would be to remove both eccentric straps and tie the top of the link to the tumbling shaft arm to prevent it from tipping over, then disconnect the engine on the broken side. If only the jaw of a back motion strap breaks off or holes break out, it is sometimes possible to fasten the blade to the other strap with a long bolt. Remove the broken strap and if the temporary bolt used, and also the link, will clear everything during one complete revolution of the wheel, then you may proceed slowly without disconnecting, but do not try to run backwards. If the forward motion eccentric or strap is disabled you must disconnect the engine on that side, but it will not be necessary to remove the back motion eccentric strap, but if it is left on, the link lifter on that side must be removed.

BREAKING OF ECCENTRIC STRAP BOLTS.

Some trouble has been caused on the Chicago, Burlington & Quincy Railroad by the breaking of eccentric strap bolts on the heavy engines in fast service. The straps in which the breakage occurred were like Fig. 1, and the bolts were increased from $\frac{7}{8}$ inch to $1\frac{1}{8}$ inches diameter, with the idea that more area of metal would meet the difficulty. These bolts are seen to be in both tension and shear. A change of strap to the shape shown

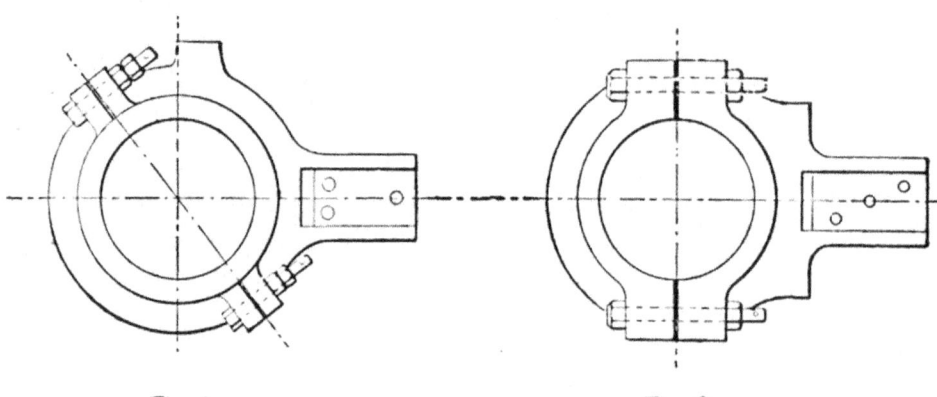

Fig. 1   Fig. 2

in Fig. 2, where the bolts were subjected to tension only, effected a complete cure, with $\frac{7}{8}$-inch bolts. The design of strap, Fig. 1, was one of the common forms of strap construction several years ago, and quite generally used when boiler pressures were lower than they are now. They furnish a good illustration of the effect of position on resistance of parts to stresses. Cast steel eccentric straps are used upon the Erie, a brass band being fit between the eccentric and the strap.

BROKEN TUMBLING SHAFT, ITS ARMS OR STANDS.

Should either tumbling shaft stand get broken so badly that it cannot be clamped some way and used, it is sometimes possible to use a wooden block in its place. Should the top arm of tumbling shaft get bent or broken so it cannot be used, it is sometimes possible to use a pinch bar or other iron rod across the frame to hold the short arms in working position. If a pinch bar or rod be used, see that it will not interfere with the eccentric rods and fasten it securely. If a rod cannot be used, then the links must be blocked. Fit a piece of wood in the slot between the top of the link and link-block for the link to rest on; make the block long enough to hold the link as high as you desire to have the engine cut off, and tie the block securely in its place. It is not necessary to block both links. Run carefully. If you wish to reverse the engine you must reverse the blocks to the lower end of the link. If one short arm of the tumbling shaft gets bent or broken, remove the link lifter on that side and block one link as previously explained. Should both the short arms get broken, remove both link lifters and block both links.

BROKEN LINK HANGER, EITHER PIN, OR LINK SADDLE.

Same remedy as for short arm of tumbling shaft.

BROKEN LINK.

Disconnect one side, remove both eccentric straps and remove broken link or fasten top of it to tumbling shaft arm with bell cord. If both links should break prepare to be towed in.

BROKEN EXTENSION ROD.

The engine truck sometimes bends or breaks the long extension rod which connects the link block and rocker arm on a ten-wheeled engine. Should this happen disconnect one side and remove the broken rod.

BROKEN REACH ROD.

Same remedy as for top arm of tumbling shaft.

BROKEN REVERSE LEVER.

Should the break occur in the reach rod hole or below it, then you must apply the same remedy as for top arm of tumbling shaft, or reach rod. Should the break occur above the reach rod connection you can usually hold it intact by fitting blocks inside the quadrant, if a solid quadrant, then any place you can secure a brace.

BROKEN ROCKER ARM.

Should a top arm break disconnect that side. Should the bottom arm break, disconnect one side, remove the broken part of the rocker; if the link will then clear everything you can leave the eccentric straps up. But you must be certain that it will clear everything in both gears. If in doubt, remove the eccentric straps, then tie the top of the link to the short arm of the tumbling shaft to keep it from tipping over, which might prevent reversing the engine.

BROKEN ROCKER PINS.

If the top one should break replace it if you have an old one; if not, then disconnect that side. If the bottom pin breaks it is sometimes possible to remove the top pin and turn the bottom arm up high enough to clear the link, and then tie it up to the guide yoke. But you must be certain that it will clear the link when in full gear in each motion. If you are in doubt remove both eccentric straps, then tie the rocker forward or back to clear the link, and tie the top of the link to the short arm of the tumbling shaft to keep it from tipping over. You can use bell cord or wire for this purpose, then disconnect on broken side.

BROKEN VALVE OR BRIDGES.

Remove the steam chest cover and ascertain the extent of the breakage. If serious, remove the broken parts and disconnect engine on the broken side. Now if the valve can be used to cover both steam ports, set it central and hold it in this position with a valve stem clamp, if you have one; if not, use blocks. If the valve cannot be used secure a large wooden block for this purpose. If you can block the supply ports, do so, then screw down the steam chest cover and proceed. The practice of using a board under the valve cannot be done with balanced valves. The steam chest cover may be used to secure blocks over the receiving ports.

BROKEN VALVE STEM OR ROD.

Remove broken parts and disconnect engine on broken side. If the valve stem or yoke is broken inside the steam chest, remove the cover and block the valve central.

BROKEN THROTTLE ROD.

This is a serious mishap. If you cannot shut off steam be positive it is the throttle and not the tallow cups leaking, if tallow cups are used. If the throttle is open, do not attempt to proceed with your train, unless your air brake is in good condition. Keep the boiler pressure low and the engine can be controlled by using the air brake and reverse lever for stopping.

As soon as you find that you cannot close the throttle or prevent steam from entering the cylinders, if you have a high pressure of steam, the engine will be apt to slip its drivers. In this case do not use sand, but control the engine with the reverse lever until the pressure is reduced so the engine will not slip. If much sand is used there is danger of damaging the machinery. If the throttle valve will not open it is probable the rod has become disconnected inside the boiler. If the engine has steam chest tallow pipes leading from the cab, you may be able to admit enough steam to the cylinders through these pipes to run the engine light. If not, the only remedy would be to kill the fire and raise the dome cap, but this is not expected of roadmen So prepare to be towed in.

BROKEN SMOKE BOX FRONT.

Board it up close, using the front end bolts to hold the boards.

PUMPS WILL NOT WORK.

See if there is plenty of water in the tank and that the tank valve is connected; open the heater cock a few seconds, then open the pet cock, then close the heater and try the pump. If it will not work then slack down the lower pump joint. If the water flows freely move the engine about a dozen revolutions, then tighten up the joints. If the water does not flow freely, the feed-pipe, strainer or hose is choked (inside lining of hose may be torn loose) and must be cleaned out. If the pump will not work when the water flows through the joints, take the lower valves out and see that they are free. If you still find nothing wrong the pump must be repaired at the shop.

But little more can be said about failures of pumps. If the remedies given do not cause them to work, it is not probable that you will have tools with you or time to do necessary repairs such as changing lift of and facing and grinding valves.

BOTH INJECTORS FAIL TO WORK.

Kill the fire and prepare to be towed in, but you should use every effort to keep them working. See chapter on "Injectors."

LUBRICATOR WON'T WORK.

Shut off steam, disconnect oil pipes and oil through the pipe frequently.

POP OR WHISTLE BLOWN OUT.

Start both injectors immediately in order to retain as much water as possible. Kill or smother the fire, and as soon as sufficiently cool drive a soft wooden plug into the hole. See that you have sufficient water in the boiler, then fire up again, but keep the steam pressure low and proceed.

HAND HOLE PLATE OR PLUG BLOWN OUT OR HOLE IN BOILER.

Kill the fire immediately. Repack hand hole plate and replace it; if bolt is broken use some other bolt off engine, but see that it clears the eccentric strap. If a plug, make a soft wooden plug and drive it in tight. Refill the boiler, if near a water tank fire up, keep steam pressure low and proceed. If you have no way to refill the boiler prepare to be towed in.

BROKEN WATER GLASS.

Shut off both cocks and use gauge cocks, which should be frequently used even if you have a water glass.

BROKEN DRIVING SPRINGS OR HANGERS.

With the heavy engines now in use road men are not expected to jack up the engine, and even if you have a small engine the quickest way is to use wedges on the rail, when possible to do so, as time is usually an important consideration; but it should be done carefully or you may break other springs or hangers or engine may leave the rail. If an eight-wheeled engine and a forward spring or hanger should break, place a fish plate or other piece of iron between the top of the back box and frame on the broken side, which will save raising the wheel that much higher and permit of a thinner wedge being used; now place a wedge on the rail and run the back wheel up on it, which will take the weight off the forward box. Now block solid with wood between the top of the forward box and the frame, remove spring saddle if necessary; now let engine down, remove the fish plate from the back box and run the main wheel up on the wedge, which will take the weight off the back box and relieve the equalizer. Now pry up the front end of the equalizer and block it solid. Then let engine down, remove all loose parts and you are done. If the back spring or hanger is broken go through the same performance in the reversed manner by running the main wheel up on wedge first. If it is a mogul, or ten-wheeler, to raise the weight off the main wheel run the forward wheel up on the wedge, and to raise the weight off the forward pair, run the main wheel up on the wedge. If a mogul, and a forward spring or hanger is broken, you may have to remove both forward springs and block on top of both forward boxes, but if it is only a hanger, remember a chain may sometimes be used to replace it. When you block both forward boxes also block intermediate equalizer to truck. When the springs and equalizer are below the frame proceed in the same manner, then block or chain up the equalizer until level and remove or secure broken springs and hangers. When the spring hangers straddle the frame it is sometimes possible to block between the hanger and frame. Should the large spring below the frame and between the drivers break, block the top of both boxes, or block between both long hangers and the bottom of frame, and remove or secure broken spring and equalizers.

Should the small coil spring hanger back of the rear drivers break it may be possible to remove one of the small equalizers which ride the back box. If so, block on top of box; if not, and you cannot hold the spring hanger any other way, you may be able to chain the back end of small equalizers to frame; if not, you must let the frame ride the box, but run very slow.

BROKEN EQUALIZERS.

Raise the engine the same as for a broken spring or hanger when possible to do so. If an equalizer on a standard eight-wheeled engine, block on top of one box and block up the loose end of the equalizer, when possible, the same as for a broken spring or hanger; if it cannot be used remove the equalizer and block on top of both boxes. If an equalizer below the frame do likewise, or chain it up. If forward equalizer on a ten-wheeled engine, block on top of the forward and main boxes, and block up forward end of back equalizer. If it is the cross equalizer on a mogul, block on top of both forward boxes and block on top of the back end of the long intermediate equalizer that goes to truck. If the intermediate equalizer breaks, block between the boiler and the cross equalizer. If it is the cross equalizer on a four-wheeled pony, block on top of both forward boxes. When this equalizer is below or between the frames it is sometimes possible to block between the hangers and the frame. If a small equalizer that rides the back box, block on top of the back box and chain up the back end of the bottom equalizer. If it is a truck equalizer, block on top of truck boxes between the box and truck frame. Always remove or secure all loose parts.

BROKEN EQUALIZER STANDS.

If the stand breaks then you must use the same remedy as for a broken equalizer, but if only the bolts break you may be able to find some old bolts to replace them, or take bolts off some other part of the engine that will fit, and the loss of which will not impair the working of the other parts.

BROKEN ELIC BOLT ON MOGUL.

If an elic bolt should break you must block up between truck axle and the forward end of the long intermediate equalizer. A truck brass is very handy for this purpose.

BROKEN ENGINE TRUCK SPRING OR HANGERS.

First raise the front end of the engine with jacks. If it is a four-wheeled truck pry up the frame on the broken side and block between the equalizers and the truck frame, close to the spring band, keeping it up level with the other side. If a mogul truck, then block between the top of the truck box and the truck frame.

BROKEN CENTER CASTING.

If a truck center casting should break on a standard four-wheeled truck, jack up the front end of the engine; if you can find two short rails run these across the top of the truck equalizers and under the center casting; if you cannot find any block up both sides between the truck frame and cylinder saddle, but in this case run very slow when rounding curves, as engine won't track very well. If the top or male casting breaks you must block up the same way. If the pony truck center castings break, block between the truck frame and the engine frame on each side, but round curves slowly.

BROKEN DRIVING AXLE, WHEEL OR TIRE.

An accident of this kind is very serious, as it may strip one side of the engine or perhaps disable it. But in most cases of this kind the engine can be blocked up so as to reach the nearest siding, if not the terminal. If the injury is slight, by running slow and being careful the nearest siding may be reached where you can notify headquarters and where the engine should be properly blocked to run to the shop without assistance; but it implies considerable work and good judgment of the engineer. We shall endeavor to treat these various mishaps separately.

MAIN WHEEL BROKEN OFF OF THE AXLE.

Disconnect on the broken side and remove all side rods and broken wheel. The eccentrics will prevent the other main wheel from leaving the rail even though it be a blind-tire. Use a jack and raise the axle on the broken side, remove the oil cellar and fit a hard wood block between the driving axle and pedestal jaw or binder, then use old rod keys or any kind of iron and block between the spring saddle and the top of the frame to keep part of the weight off the axle at this point, then remove the jack. Now raise the engine slightly on the broken side and block between the top of the driving box or boxes nearest the main wheel on the broken side, using hard wood or iron. If it is an eight-wheeled engine, also block between the engine truck equalizers and the truck frame on both sides, as additional weight will be imposed upon the truck. If it is a consolidated engine it will only be necessary to block on top of the two boxes nearest to the main box. Then let the engine down and proceed slowly.

TIRE BROKEN ON MAIN WHEEL.

If it is only a bad crack or even if broken through, if the tire is still on the wheel, let the fireman stand on the steam chest and watch it while you run slow and try to reach the nearest siding. Now if the tire is very loose, take off the rods and remove the tire. Disconnect on the broken side and remove all side rods. Then place a jack under the axle and raise the broken side, remove the oil cellar and fit a hard wood block between the axle and the pedestal jaw or binder, and also block under the spring saddle in order to keep the weight off the box. Now remove the jack from beneath the axle and raise the engine on the broken side and block between the top of the driving box and the frame on the box or boxes next to the main wheel on that side. If it is an eight-wheeler, or consolidated, engine block the boxes the same as for main wheel broken off axle. Now let the engine down and if the wheel clears the rail proceed slowly. If it will not clear the rail block the engine a little higher on the broken side. If you cannot remove the broken tire, and it will not permit the wheel to revolve, disconnect on both sides and prepare to be towed in. Block up the main pair of wheels high enough to clear the rail on both sides, then block on top of all other driving boxes, and if an eight-wheel engine block on top of the truck equalizers.

MAIN WHEEL CRACKED.

If the wheel is not too badly broken, watch it, and run slow to the first siding. Disconnect on broken side and remove all side rods, which will take the strain off the crank pin on the broken side; you may then be able to proceed. If the break is of such a nature that you consider it unsafe to proceed with the engine alone (or light), then block the wheel up the same as for a broken main tire, and see that the tire clears the rail, then proceed slowly.

FORWARD WHEEL BROKEN OFF THE AXLE.

Remove both side rods between the forward and the main pair of wheels; if this cannot be done, then remove all side rods and remove the broken wheel or chain it up to the frame. Now use a jack and raise the axle on the broken side; remove the oil cellar and fit a hard wood block between the axle and the pedestal jaw or binder. Block under the spring saddle on the broken side. Now if the forward pair of wheels have blind tires, then unless there are collars on the axle, the opposite wheel must also be raised and blocked to clear the rail; in this case raise the front end of the engine and block on top of each main box and on both truck equalizers, and if a pony truck then block between the truck frame and the engine frame on both sides. If it is a six-wheel connected engine without an engine truck, block on top of the main boxes and under the back boxes, which will tend to counter-balance the weight. If necessary to disconnect the forward pair of springs on a mogul an old truck brass may be used between the truck axle and the long equalizer. But if the forward pair of wheels have flanges or collars on the axle, let the good wheel run on the rail and block on top of its driving box and on top of the main box on the broken side and on the truck equalizer as explained. See that the forward crank pin on the good wheel will clear the cross-head in all positions; if it will not, then both sides of the engine must be disconnected and the engine towed in; but assuming that it will clear, then proceed slowly, especially rounding curves.

TIRE BROKEN ON FORWARD WHEEL.

If the tire is on the wheel try to keep it on until the nearest siding is reached; then remove both side rods between the forward and the main wheels; if this cannot be done, remove all side rods, then, if the tire can be removed, take it off, and block up the wheel. If you cannot remove the tire, block up the wheel on that side so the tire will clear the rail, but if the tire is bent, or twisted, in such a manner as to prevent the wheel turning, then both wheels must be blocked up to clear the rail; block them the same as for the forward wheel broken off of the axle.

Some engineers claim that the side rods may be left up by simply slacking both rod keys, but it is not considered good practice. It will be much safer to remove the two forward parallel rods, if it can be done conveniently; proceed carefully.

FORWARD WHEEL CRACKED.

If the break occurs at, or near, the crank pin hub remove both side rods between the main and forward pair of wheels and all side rods, if necessary. Now if the wheel is not too badly broken, it is probable you can go ahead, but maintain a close watch on the broken wheel—run slow. If it is not safe to run this way you must block up one or both wheels the same as for a broken tire.

BACK WHEEL BROKEN OFF OF THE AXLE.

Remove both side rods between the main and the back pair of wheels, and all of the side rods, if necessary. Place a jack under the axle and raise the broken end; remove the oil cellar and fit a hard wood block between the axle and the pedestal brace, or binder, on that side, also block under the spring saddle and let the good wheel remain on the rail, and block on top of its box, and drive wedges between the drawbar and the chafing iron and block on top of both main boxes. Leave your train and proceed cautiously.

If the axle should be broken between the two driving boxes, try to remove both wheels. If this cannot be done, they must both be blocked up high enough to clear the rail, and if it is a heavy engine, part of the weight of the engine must be transferred to the tender. Block between the equalizers and the engine truck frame on both sides, then raise the back end of the engine and block on top of both main boxes and wedge between the draw-bar and the chafing iron. If this can be done, you may still run the engine light, but if it cannot be done, then run a short rail into the fire door and chain it to the draw-bar and block up under the back end of the rail on the tender. If the latter method is followed the engine must be towed in, so disconnect on both sides.

TIRE BROKEN ON BACK WHEEL.

Try to keep the tire on the wheel until you reach the nearest siding; run very slow. Then remove both side rods between the main and rear pair of wheels and all side rods; if necessary, also remove the tire, if you can, and block the wheel up to clear the rail, leaving the opposite wheel on the rail. If you cannot remove the broken tire, and it will not allow the wheel to revolve, both wheels must be blocked up to clear the rail. Block them up in the same manner as for the "Back wheel broken off of the axle."

BACK WHEEL CRACKED.

Run slow to the nearest siding; then remove both parallel rods between the main and rear driving wheels; if this cannot be done, then remove all the side rods. By removing the rod you take the strain off of the crank-pin and unless the wheel is broken very badly you may proceed slowly. If you consider it unsafe to run in this way, then block the wheel up to clear the rail the same as for a broken tire on back wheel.

CAUSE OF TIRES CUTTING.

One side of an engine being higher than the other will cause tires to cut on the low side. If the driving axles are not square, or at right angles with the cylinders and center casting, or if they are not an equal distance apart, the wheels that are too far back will cut their flanges. When engine truck wheel flanges are cut it is an indication that the engine is not in the center of the truck. The front of the engine should be moved toward the cutting side. An engine not being central upon its truck may also cause the forward driving wheel flange to cut on the side opposite to which the truck flanges are cutting.

BROKEN ENGINE TRUCK WHEEL OR AXLE.

If a piece of the flange is broken off a truck wheel run very slow, especially over frogs, switches and crossings. If a piece be broken out of a truck wheel, it is sometimes possible to chain the wheel, or place a timber across the rail in front of it, so it will slide to the nearest siding, where that pair of wheels must be removed or blocked up to clear the rail. If it is a four-wheeled truck, chain the broken end of the truck frame to the engine frame, then raise the engine and block solid between the truck box and truck frame and between the truck frame and engine frame above the good pair of wheels. Also block on top of the forward driving boxes, leave your train and run slow, especially rounding curves. If it is a two-wheeled truck, the truck frame should be chained to the engine frame and then the front end of the engine should be raised high enough for the truck wheels to clear the rail and then block solid between on top of the two forward driving boxes, and under the two back driving boxes (don't break the back driving box cellars). This will help to hold up the front end of the engine. Place fish plates on top of the back driving boxes before you raise the front end of the engine, which will prevent breaking the driving springs or hangers. If a truck wheel is broken off or the axle badly bent block up the same way. It is sometimes easier to remove the truck entirely.

ENGINE TRUCK FRAME.

Raise weight of engine and place pieces of heavy iron between the equalizers and the truck frame, or a piece of rail may be chained to it for a splice.

BROKEN TENDER WHEEL OR AXLE.

If you can find a piece of a rail the proper length, or a cross tie will answer, place it across the top of the tank directly over the broken pair of wheels, block under the rail or tie to protect the flange on the top of the tender, jack up the broken pair of wheels to clear the rail and while in this position chain the truck to the rail above the tank on both sides.

BROKEN ENGINE FRAME.

If a bad break occurs between the main driving axle and the cylinders, and the break opens up very bad, you should use a little discretion—give up your train or disconnect that side as your judgment dictates—and under no circumstances let another engine pull on you. Report a broken frame as soon as possible.

BROKEN DRAW BAR.

If the engine has safety chains they will hold the tank, but not always a heavy train. If the engine is not equipped with safety chains then secure a chain from the tank box or caboose and chain the tank to the deck. Safety chains should not have more than 4" of slack.

BROKEN DRIVING BRASS.

If a driving brass breaks and is cutting badly, run that wheel up on a thin wedge; then use an iron block between the top of frame and spring saddle, which will take the weight off that box.

BROKEN WEDGE BOLT.

It is sometimes possible to screw the nut half way onto each part of the broken bolt and thereby hold it up in place. If this cannot be done, then with a wire try and fasten a nut under the wedge to hold it up.

COMBUSTION.

It has been estimated by competent authority that the amount of fuel and combustible gases which are drawn through the flues unignited, *absolutely wasted* by the excessive draught of the exhaust as at present arranged in our locomotives, constitute from thirty-five to fifty per cent. This immense waste of fuel is the basis for a strong anticipation of a more economical motive power to supersede that of steam on our railroads, and has been the subject of many ideas to modify the effects of the exhaust upon the fire in starting and attaining headway with trains, and from the overworking of engines after this has been obtained. For a number of years practical railroad men have been experimenting in the effort to overcome this enormous waste of fuel and the various devices invented for this purpose are too numerous to mention. We will not attempt to enter upon a lengthy discussion of this subject in this book, but simply present a few of the latest devices produced in this line. The principal parts of the locomotive upon which experiments have been made are the exhaust nozzles, the smoke box and smoke stack, the ash-pan, grates, fire-box and flues.

CORRECT SIZES FOR EXHAUST NOZZLES.

Owing to the changing conditions under which a locomotive must perform its work, no definite proportion can be given for the diameter of a nozzle that would give the best results under all circumstances. However, the nozzle used should be just small enough to furnish the required draft, and should be left as large as possible to reduce the back pressure in the cylinders.

It is not necessary that a single nozzle should contain twice the area of a double nozzle, as the exhaust from each cylinder is not simultaneous; yet they occur in such rapid succession that if a single nozzle was made no larger than a double nozzle, it would produce an enormous back pressure in the cylinders, and choke the engine.

Each engine is usually supplied with three nozzle tips. The following sizes have been found to give good results with cylinders 18x24: Single nozzle, $4\frac{1}{2}''$, $4\frac{5}{8}''$, $4\frac{3}{4}''$; double nozzle, $3\frac{1}{4}''$, $3\frac{3}{8}''$, $3\frac{1}{2}''$.

With 19x24 cylinders: Single nozzle, $4\frac{1}{2}''$, $4\frac{3}{4}''$x5". Double nozzle: $3\frac{3}{8}''$x$3\frac{1}{2}''$x$3\frac{5}{8}''$; and other sizes in proportion to the cylinders.

NOZZLE REAMERS.

Quite a number of different kinds of reamers are in use for cleaning out exhaust nozzles while in position in the smoke box. They are long and are used through the stack.

ADJUSTING THE PETTICOAT, CONVEY OR DRAFT PIPE.

When an engine burns the fire more at the back than at the front of the fire box the draft pipe is too low. If the pipe is adjustable, raise it; if the fire burns most at the front of the fire box the pipe should be lowered. If an engine tears the fire, the exhaust nozzles are either too small or need cleaning out.

The conditions of the petticoat, convey or draft pipe are different to that of a diaphragm in an extended smoke box, although each produce the same effect upon the draft, but while the diaphragm has its vacuum at its lower edge only, the draft pipe has a vacuum above as well as below after each exhaust. So if the fire burns most at the front end of the fire box it shows that the pipe is too high and the greatest vacuum is at the bottom of the pipe; the top flues not having sufficient draft through them will fill up with ashes and coal and become choked, and a large amount of sparks will be apt to accumulate in the smoke box; so if the top flues become choked, and you find many sparks in the smoke box, lower the draft pipe, but if the fire burns too fast at the back of the fire box and the lower flues become choked, raise the draft pipe. Engines having draft or convey pipes and spark arresters in the smoke stacks pulverize the sparks against the nettings or cones in the top of the stacks, and after they become small enough to pass through the netting they are thrown out. If an engine does not steam well that keeps its flues and front clean with a light fire the fault is not with the draft pipe, but most likely the exhaust nozzles are too small; the exhaust does not fill the stack, but goes through it tearing and chopping the fire, taking with it green coal and unconsumed gases. It is very important that the exhaust nozzles be properly located as to height and size, so that the escaping steam will just fill the smoke stack at its base. If an engine does not steam well the exhaust may be loose in the smoke box. If you find the sparks blown away from the joint it is positive evidence of a leak.

ADJUSTING THE DIAPHRAGM.

If an engine having an extended smoke box and diaphragm burns the fire too hard at the front of the fire box, the apron on the diaphragm should be raised; or if the fire does not burn well and the inside of the fire door becomes black, the draught is choked either by stopped up flues, or the apron is too low. When an engine has proper draught the inside of the fire-door becomes white while running.

The object of raising the apron on the diaphragm in an extended smoke box, if the fire burns too fast at the front of the fire box, is to give more draught to the top flues. The lower the apron is the harder the draught acts on the lower flues. The exhaust fills the stack as it passes out of it, leaving a void or partial vacuum in the stack and part of the smoke box. The gases from the fire box rush at once through the flues toward the point where the greatest vacuum exists in the smoke box— this point is at the lower edge of the apron—so the height of the apron as it regulates the height of the vacuum, influences the draught in the upper or lower flues; for a trial, one-fourth of an inch is enough to raise or lower the apron at a time. This amount of change has a great effect on the fire. The sparks from the fire box of an engine having an extended smoke box are thrown by the draught into the extension out of reach of the vacuum and cannot rise unless the extension is nearly full. If the netting is in good condition and properly fitted fine dust will be thrown, but no large sparks.

HINTS TO FIREMEN.

If using soft coal do not carry over ten or twelve inches of fire in the center of the fire box; keep the sides and corners a little higher; aim to fire in the corners and sides more than in the center. If the boiler will not steam well with a light fire, more air is probably needed at the front of the box. Leave the fire door open a little way for a few seconds after putting in coal, it helps to consume the smoke. Two shovel fulls is enough at one time if put on the bright spots. No boiler will steam well with the fire box and flues full of smoke. If you have occasion to use the hook, be careful not to mix the green coal with that partly consumed. Do not use a slash bar if it can be avoided, and be careful not to get green coal on the grates. If the box has an arch, keep a good space open between the arch and the fire. If the engine has a heavy train, it will need a heavier fire than with a light train and a fast run; always make calculations to fire according to train and speed. Hook out all clinkers from the fire as soon as you find them. Do not fire much while pumps or injectors are on full. If the engine has ash pan dampers use them when necessary. If there is more steam than is needed the dampers should be closed; a certain amount of air is necessary to make a fire burn as it should; if too much air is admitted the gases will be chilled; if too little, they will not ignite; no rule can be made for the exact amount of air required, because different kinds of coal require varying quantities of air; only keep a bright fire low in the center of the box where the most air is needed and watch when the greatest flame appears in the fire box with the least smoke going out of the stack; attend to the fire often, and do not use lumps of coal larger than an egg.

Keep the ash pan clean or the grates will burn out. If firing an engine hauling a passenger train, on approaching a station, as soon as the throttle is closed, put the blower on lightly and open the fire door about half an inch; when nearing the end of the trip let the fire run low. Do all you can to help the engineer, but do nothing without first knowing that he wishes it done. Keep all tools and cans clean and be ready and willing to aid him. Try to learn what he does and how he does it, trying to anticipate his wishes.

MUFFLING THE EXHAUST IN STATIONS.

The unearthly, ear-piercing noise of the exhaust of a locomotive as it backs out of the station after having brought its train in on time, perhaps, is one of the annoying features of travel which can and should be remedied. This nuisance is particularly true of the suburban service, where the trains have to be got out of the way about as soon as the passengers are out of them. Then the engine backs out, with its sharp, ear-splitting noise, past the passengers who are hurrying out of the station, anxious to get out of range.

A few roads have taken steps to prevent this annoyance, and have accomplished it very easily and cheaply, simply having an exhaust damper which is turned down over the tips while in the station, and thrown up out of the way as soon as the open air is reached.

The device was used on the Boston & Maine Railroad, and a similar device has been in use on the Pennsylvania lines for some time. They cost very little to apply, and they save a heap of cuss words on the part of the male passengers, as well as tall thinking by those who feel duty bound to say nothing, no matter what they think.

CARRYING WATER IN THE BOILER.

The water in the boiler should be kept as near a uniform height as possible. Engines rarely steam well when boilers are pumped full then allowed to run low in water, besides such pumping is wasteful of fuel and destructive to flues; a good runner will feed the boiler so that the height of the water will not vary. Never use pumps or injector if it can be avoided unless you have a bright fire. On approaching a grade, always try to have a good supply of water in the boiler, and if you have to apply pump or injector on a descending grade, be sure to have a bright fire. If you have no occasion to supply the boiler on a descending grade, level the fire and cover it, or arrange it so that the pressure of steam will not increase.

After standing at a station or after running awhile without using steam and having the injectors at work, unless the safety valves are blowing, you cannot depend on having the full pres-

THE LOCOMOTIVE UP TO DATE. 425

sure of steam as shown on the gauge. The pressure is really there, but a great deal of it is only dead steam and as soon as you open the throttle the gauge will fall back and show less pressure.

A VARIABLE EXHAUST NOZZLE.

The device herewith illustrated is the invention of Messrs. Wallace & Kellogg, of Altoona, Wis. (both of whom are engineers on the St. P., M. & O. R. R.). The principle of the variable nozzle has been recognized as being correct for twenty years or more, but the great trouble experienced with such nozzles was

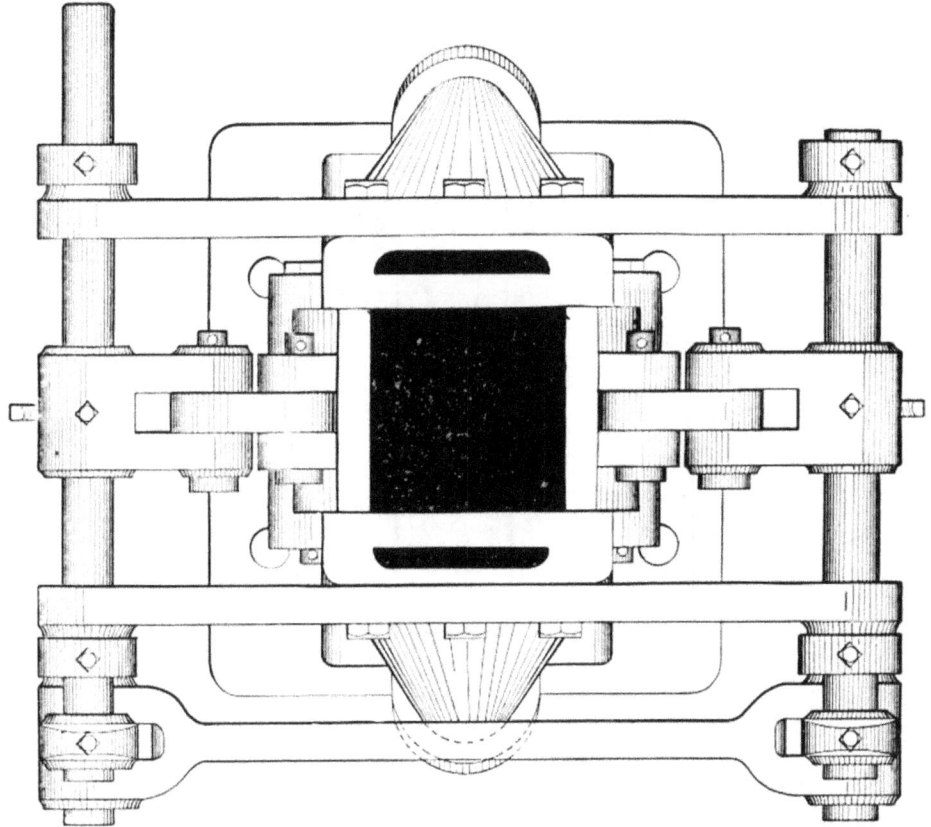

in their gumming up, but this objection seems to be eliminated in this form because it works automatically from the reverse lever connection or connected to the forward end of the reach rod. The illustrations of this device are self-explanatory. The device has been tried upon the Duluth & Iron Range R. R. Co. and Mr. L. H. Bryan, general foreman of the road, writes us as follows concerning same: "We have been experimenting with the Wallace & Kellogg form of variable nozzle for the past two years. We have put on two this winter and they give as good satisfaction as the one already on—we will probably adopt them. This is a square nozzle and the variation is made by two ears or hinges which open and close as the reverse lever is moved from center to corner, so that the more steam that is being used, the larger the nozzle. We connect to forward end of reach rod and with a fine notched quadrant set them to open one-half a square inch to each notch. There being fourteen notches each side of the center we get a difference of seven square inches between center notch and either corner; the engines are the best steamers on the road and handle their trains better on the hills because of less back pressure and burn less coal. The patentees offered to apply them for one-half the saving shown in three years. I understand there is one in use on the C. St. P. M. & O. Ry., which is also giving satisfaction. On our 22 and 26 engines the nozzle area is as follows: In center, 22"; second

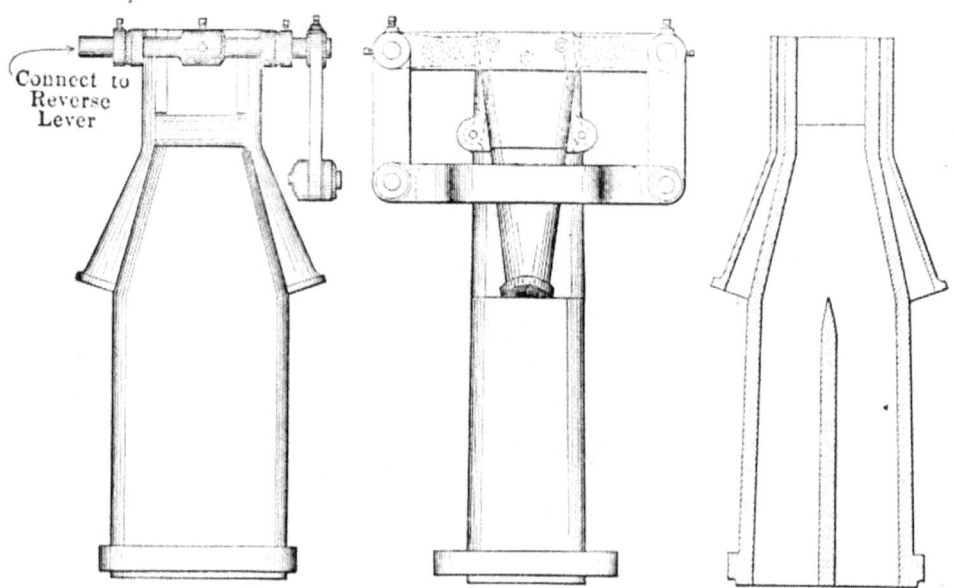

notch, 23"; fourth notch, 24¾"; sixth notch, 25½"; eighth notch, 26 9-16"; tenth notch, 28¼"; in the corner, 29 11-16".

"On the engine that the experiments have been made with, it has shown a saving in fuel from $59.00 to $97.00 per month for eight months; coal rated at $2.58 per ton; and making from 5½ to 10 miles per ton above the general average."

THE SMITH TRIPLE EXPANSION EXHAUST PIPE.

This device is the invention of Mr. John Y. Smith, the originator of the Smith vacuum brake. In the cuts of the front and side views shown *A A* represent air passages, *S S* exhaust steam passages, and *B* an annular blower, forming part of the nozzle. This is an entirely new departure in the construction of exhaust pipes for locomotives. Its distinguishing features are that the exhaust steam is not restricted after it leaves the cylinders, and the gases and heated air in the smoke arch are mingled with the exhaust in the exhaust pipe. The exhaust steam is thus super-heated and expanded, and a powerful, prolonged, pulsating blast is created, which keeps the fuel in a constant state of agitation, and produces more perfect combustion. Some of the beneficial results claimed are: Reduction of back pressure to a minimum (area of nozzle being greater than the steam ports); prevention of ejection of sparks from smoke stack; almost complete absence of noise from exhaust; prevention of formation of cinders in fire box, and a large saving of fuel. A reduction of back pressure in the cylinders without impairing the draft of the fire has long been an unsurmountable obstacle to designers of locomotives, but it is claimed that an

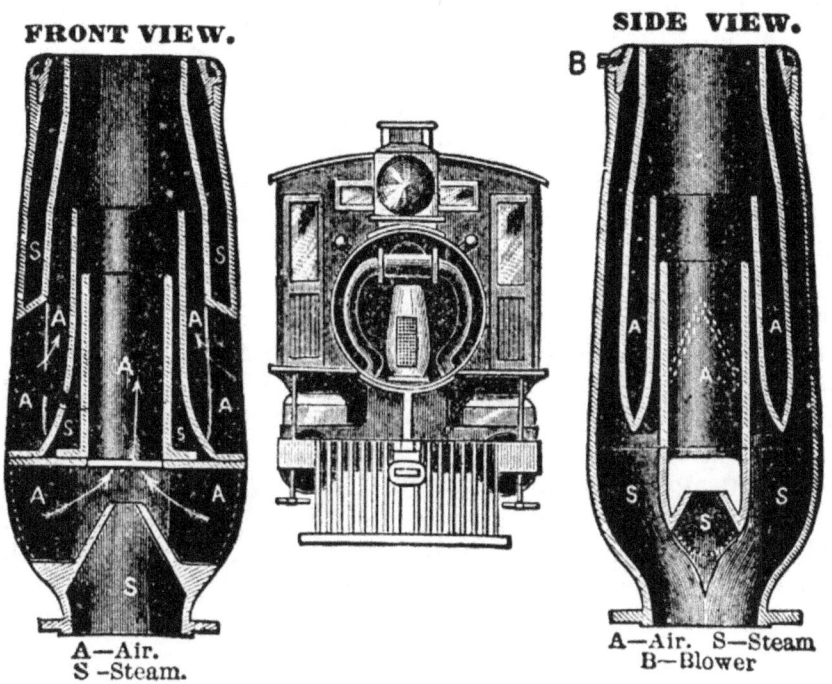

THE SMITH TRIPLE EXPANSION EXHAUST PIPE.

engine equipped with this pipe will pull from thirty to sixty tons more than with the ordinary exhaust pipe. The pipe can be used with either straight or diamond stacks, in long or short front ends, and on locomotives burning hard or soft coal, wood or coke.

ENGINE NO. 1. N. Y. O. & W. RY.

REBUILT BY

THE COOKE LOCOMOTIVE WORKS.

We give herewith an illustration of one of a number of eight-wheel locomotives built by the Cooke Locomotive & Machine Co. for the New York, Ontario & Western Railroad, the distinctive feature of which is the wide fire box set up over the drivers.

How to get rid of the enormous accumulation of "culm" or

refuse from their anthracite mines was a problem with which Mr. J. E. Childs, General Manager of the Ontario & Western, wrestled for some time, and at length decided to kill two birds with one stone by burning this refuse in the boilers of their locomotives. But this necessitated a new type of boiler and this part of the problem was turned over to Mr. Geo. W. West, Superintendent of Motive Power, who contracted with the Cooke Locomotive & Machine Co. for the rebuilding of one of their engines, which has since been known as No. 1. It has 17x24 in. cylinders, weighs complete 110,000 lbs., of which 76,000 lbs. are on the drivers and 34,000 lbs. on the truck; the boiler is 56 in. diameter and contains 197 ft. 2 in. x 11 ft. 6 in. tubes; the fire box is 84 in. wide by 108 in. long.

This engine was designed and built in order to demonstrate whether an engine of these dimensions and weight would give better results than a compound engine, also whether an engine of this weight in passenger service with cylinders 17x24" and a constant boiler pressure of 180 pounds would not be better than one with 18x24" cylinders, and lagging for steam on heavy grades. It was intended, also, to demonstrate whether such an engine could not be run at a much reduced cost of fuel. This engine has been in constant service since November 23, 1895, and has met the highest expectations of its designer. It has made a remarkable record, cutting down the fuel cost one-third, steaming freely, and has run 79 miles per hour hauling its regular train. Encouraged by these results the road proceeded to change over a number of other engines. Up to the present time the Cooke works have rebuilt eight and are now at work on the ninth eight-wheeler, two moguls and one consolidation, in addition to which they have built two new locomotives—all on lines similar to No. 1.

Mr. West has kindly supplied some data giving the performance of No. 13, the results being compared with No. 77, one of the old style. It will be noted that the comparison covers a period of two months or long enough to equalize weather and traffic conditions. This data shows a saving in fuel cost of more than fifty per cent. Commenting on this showing the builders say: "The results here described may not be unworthy the consideration of the management of some of our western roads who in some cases are adopting locomotives with considerable complication of mechanism in the hope of effecting a saving of ten or possibly twenty per cent in fuel. Here we have a saving of over fifty per cent, with absolutely no added complication. While anthracite refuse is not one of the troublesome problems in the west, may it not be practicable with this pattern of fire box to secure much more economical results from some of the low grade coals used in the west or to profitably utilize coal which is now considered too poor for locomotive consumption? In

Passenger Engine No. 1, New York, Ontario & Western Ry.
Cooke.
General Dimensions Given on Page 430.

fact, we know that bituminous coal mixed with anthracite slack, and bituminous coal alone, of the very poorest grade, have been burned under some boilers of this type with most economical and satisfactory results. The wide fire box was tried on some of the western roads some years ago, and the results not being entirely satisfactory, it was abandoned. The form used in the Ontario & Western locomotives is an improvement on the original pattern, the principal change being the omission of the combustion chamber which was a prominent feature of the original wide fire box. As constructed for the Ontario & Western it is just as safe and does not require any more attention or repair than any other form of radial stayed fire box."

Appended is the statement of performance above referred to:

Statement of performance of Engine No. 13 (same class engine as No. 1), with fine anthracite coal and Engine No. 77, with bituminous coal, during test made in months of November and December, 1897, between Delhi and Utica on trains No. 13 and 14.

	Mileage.		Fuel.		Lbs. per mile.		Cost per mile.	
No.	Engine.	Car.	Lbs. Coal.	Cost.	Engine.	Car.	Engine.	Car.
13	1372	4258	98,596	26.88	71.8	23.1	1.959c	0.631c
77	1346	4193	67,007	53.60	49.8	16.0	3.982c	1.276c

These engines are considered as near perfect as an engine can be built for burning cheap fuel. As may be seen by the illustration, it is of the eight-wheeled type, the rear driving axles being under the fire box. The fire box is of the Wooten type, and the cab is over the center of the boiler and entered from the front, as is usual in this arrangement. The boiler is supplied by two No. 8 Monitor injectors. The Smith triple expansion exhaust pipe is used, also the Leach track sanding apparatus, and Nathan triple slight feed lubricator.

The following are the dimensions, weight, etc., of engine No. 1, N. Y., O. & W. R. R.:

Total weight of engine in working order.................... 110,000 lbs.
Total weight on drivers.. 76,000 lbs.
Total weight of tender with fuel and water................. 80,000 lbs.
Total weight of engine and tender in working order........ 190,000 lbs.
Driving wheel base.. 8 ft. 6 in.
Total wheel base of engine................................... 23 ft. 1 in.
Total wheel base of engine and tender........................ 47 ft. 2 in.
Height from rail to top of stack............................. 14 ft. 11 in.
Diameter of driving wheels outside of tires.................. 68 in.
Cylinders.. 17x24 in.
Steam ports.. 1½x15½ in.
Exhaust ports.. 3x15½ in.
Kind of valves... Richardson Balance
Outside lap.. 13-16 in.

Lead	1-32 in.
Maximum travel of valve	5¾ in.
Number of boiler tubes	197
Heating surface fire box	130 sq. ft.
Total heating surface	1,166 sq. ft.
Grate area	63 sq. ft.
Fire box, width 7 ft.; length 9 ft.	Wootten Type
Boiler, inside diameter, smallest ring	55 in.
Maximum boiler pressure	180 lbs.
Exhaust pipe	Smith triple expansion

THE DE-LANCEY EXHAUST NOZZLE.

This is another form of variable exhaust nozzle, as may be seen by the illustrations. It is the invention of Mr. John J. De-Lancey, of Binghamton, N. Y., who describes his device in the following words:

The object of my invention is to provide a new and improved exhaust-nozzle for locomotives, serving to regulate the exhaust of the engines, and thereby regulating the draft in the boiler.

Figure 1 is a side elevation of the improvement as applied to a locomotive, parts being broken out. Fig. 2 is an enlarged plan view of the improvement. Fig. 3 is a transverse vertical section of the same. Fig. 4 is a sectional side elevation of the same on the line $x\,x$ of Fig. 2, and Fig. 5 is a plan view of a modified form of the plate.

The improved exhaust-nozzle A is provided with a plate B, fitted onto the upper end of the exhaust-pipe C, which may be double, as is illustrated in Fig. 3, or single—that is, the two exhausts of the engines of the locomotive running into a single exhaust-pipe.

The plate B is provided with apertures D of the same size as the apertures at the upper ends of the exhaust-pipe C, so that when the plate B is in a central or normal position the apertures D of the plate B fully register with the openings in the end of the exhaust-nozzle. The plate B is fulcrumed in its middle on a pin E, projecting from a bar F, supported on brackets G, secured to the sides of the exhaust-pipe C, the said plate being held in place on the brackets by nuts H, screwing on the threaded ends of the said brackets G, as is plainly illustrated in Fig. 4. The pin E, after passing through the plate B, also passes a short distance into the top of the exhaust-pipe C, so as to form a secure bearing for the plate B. On the top of the latter, at its sides in the middle, are arranged offsets I, onto which fits the under side of part of the bar F in such a manner that the plate B is free to turn on its pivot E, and at the same time is held securely against the upper end of the exhaust-pipe C to prevent the plate from being lifted upward by the force of the exhaust-steam.

From the plate B projects to one side an arm J, pivotally

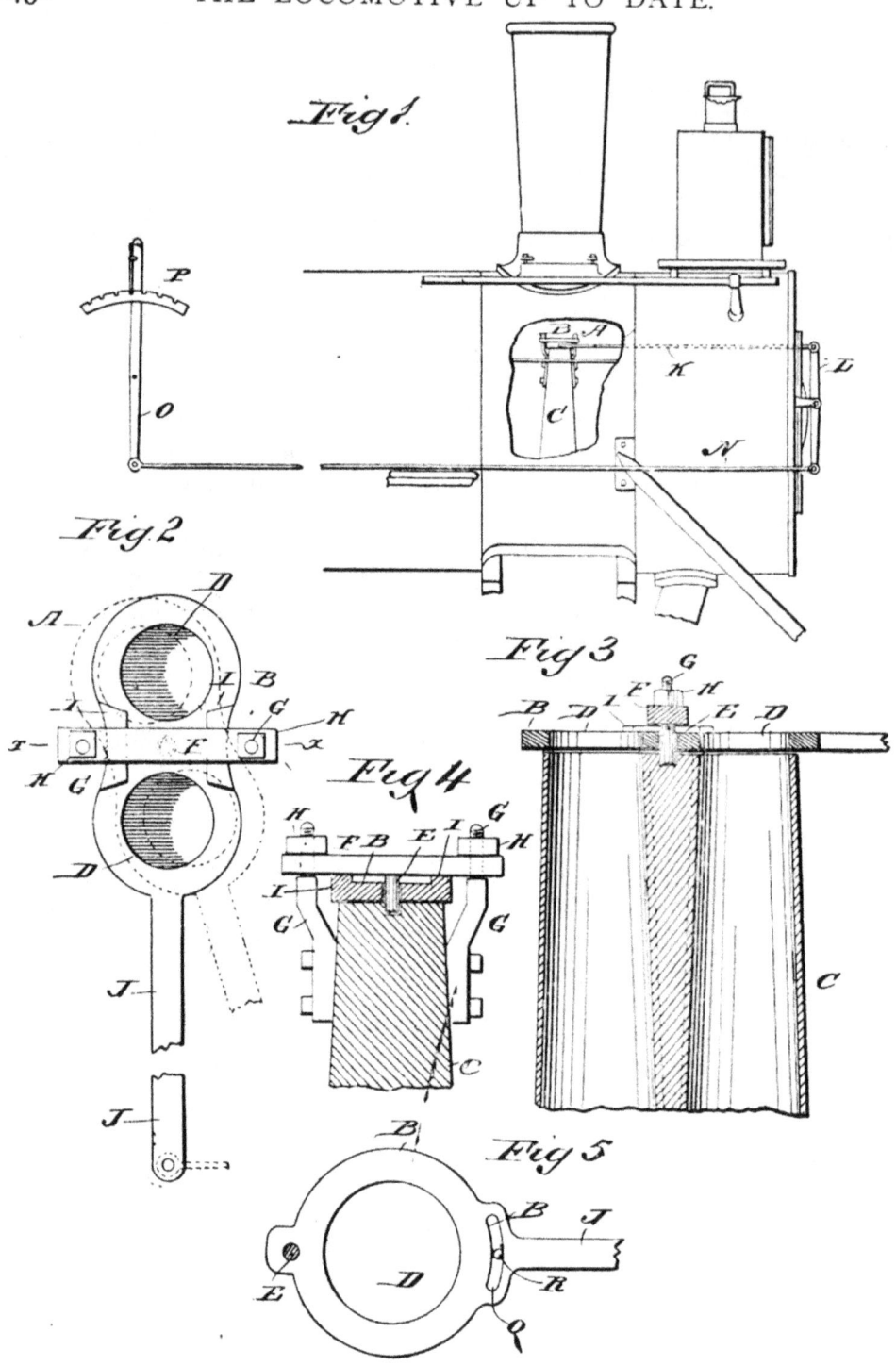

connected by a link K with a lever L, fulcrumed on the outside at the front end of the locomotive-boiler, the link K passing through the said front end. The lever L is also pivotally connected by a link N, extending along the outside of the locomotive, with a lever O, pivoted on the cab of the locomotive and extending into the same so as to be within convenient reach of the engineer in charge of the locomotive. The lever O is adapted to be locked in place in any desired position by the usual arrangement connected with a notched segment P, as shown in Fig. 1.

When the lever O stands in a vertical position, as illustrated in the said figure, the openings D in the plate B fully register with the openings in the exhaust-pipe C. In this position the exhaust-steam can pass freely out of the exhaust-pipe C through the smoke-box and smoke-stack of the locomotive, so as to cause considerable draft in the fire-box of the boiler. When it is desirable to increase the amount of draft in the fire-box of the locomotive, the engineer in charge of the locomotive operates the lever O either forward or backward, so that the lever L swings and imparts a swinging motion by the link K and the arm J to the plate B, which latter moves across the top of the exhaust-pipe C, and part of the openings of the latter are cut off or diminished in size, so that the exhaust of the engine is retarded, and consequently the draft in the smoke-box and smoke-stack is increased, so that a consequent increase of the draft takes place in the fire-box of the locomotive.

It will be seen that the two openings in the exhaust-pipe are diminished in size alike by moving the plate B, and it is immaterial in which direction the engineer moves the lever O, as the cut-off takes place either way.

NEW SOUTH WALES EXHAUST PIPE.

The curious form of exhaust pipe here shown was invented by Messrs. Stewart and Buckley, two railroad men residing at Penrith, N. S. W.

Regarding this device Mr. Stewart says the following: "It will create about 5 inches of vacuum on the exhaust side of the piston, and thus relieve the engine considerably; and if used in connection with compound engines, it will go far to making them a success. This pipe, with its cones and ports marked F, acts just the same as an injector. It will pick up water and throw it a considerable distance. It does away with the necessity of two exhaust pipes, and still gives a good vacuum in the smoke box. But the principal thing it was invented for is to relieve the engine of back pressure—a thing American engineers, above any, recognize the necessity of. The two cones with the ports F being enclosed in a vacuum chamber, enables the exhaust steam from one engine to create a vacuum in the exhaust pipe of the other, thereby enabling the engine to work as with a condenser. The vacuum chamber is made in two parts and fastened together with three clips, so that the cones which are held in place by the chamber can readily be removed.

"I was under the impression that this invention would have

speedily been given a trial in America. But perhaps I have not gone the right way about it. If so, will some of your interested

readers write me? I shall be glad to further explain, if necessary, or send drawings."

BELL FRONT END ON THE GREAT NORTHERN MASTODON.

This design of interior arrangement for smoke boxes is the invention of Mr. J. Snowden Bell, of Pittsburg, Pa. Mr. R. J.

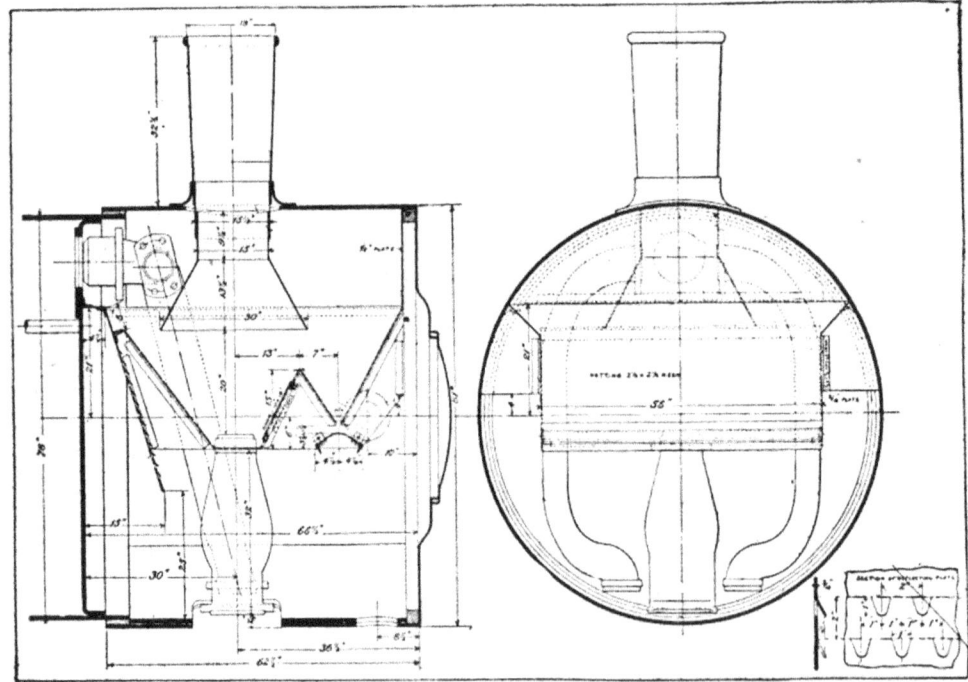

Gross, Vice-president of the Brooks Locomotive Works, the builders of the engine, say: "The reports from the Great Northern regarding the steaming qualities of these engines are very encouraging as the engine went into service without a single change in the exhaust nozzles, netting or position of diaphragm."

The Bell design dispenses with an extension and is in line with the statement of the committee as exhaust and steam passages in its report at the Master Mechanics' Association convention of 1894, which reported in this convention as follows: "An increase in the length of the smoke box over and above that necessary to get in a cinder pocket in front of the cylinder is unnecessary and undesirable as the long smoke box greatly decreases vacuum. Sufficient area of netting can be put into a smoke box which is long enough to give room for cinder pocket in front of the cylinder saddle."

This position of the committee appears to be sustained in the use of the Bell front, and its later design dispenses with the cinder pocket. It also retains in use the present standard of straight stack. Some other roads using this front are the Baltimore & Ohio on mountain and other engines; the Wisconsin Central, on old and new engines now building, and on the Illinois Central on its fast passenger trains.

WARREN'S DRAFT EQUALIZER.

We herewith show three views of Warren's draft equalizer which has been applied to more than twenty locomotives on the

T. P. & W. Ry. Co. It has been in use since 1895, giving very good satisfaction. It was invented by Mr. W. B. Warren, general foreman of the T. P. & W. Ry., of Peoria, Ill.

DESCRIPTION.

The object of this invention is to provide a construction of exhaust passages and smokestacks which will create (with larger exhaust nozzles) a greater and more equal draft through the flues of the boiler, thereby saving fuel, a better steaming engine, and making greater speed with less weight of motive power.

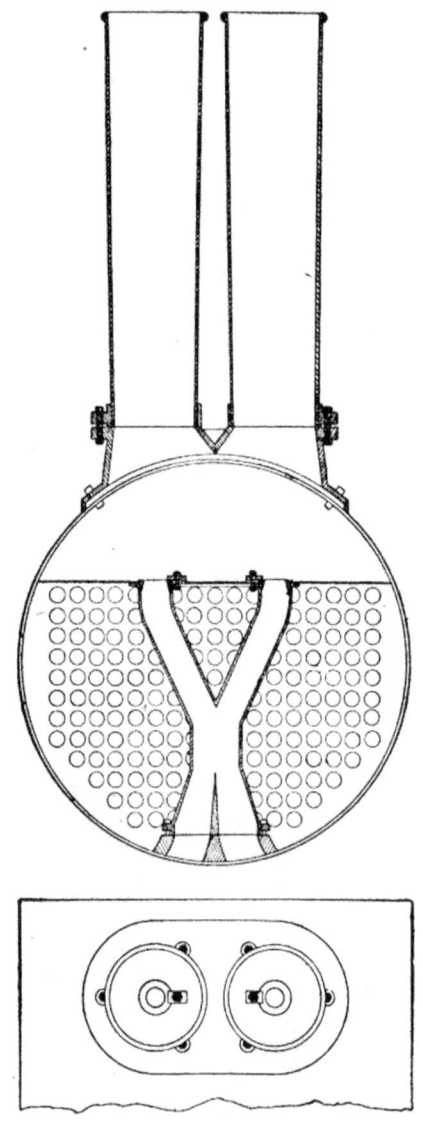

Referring to the cuts, the following expiains its workings: Each exhaust of steam from each cylinder of engine exhausts through both smokestacks, thus accomplishing the purpose of creating a greater and more equal draft or vacuum through all the flues that can be produced when exhausting steam into one smokestack on center of boiler, the single stack tending to produce a draft through the center flues and not through the side and center flues, as in case of my invention, thereby losing

the greater advantage of the heating surface in the side flues and sides of boiler.

THE SPECIAL ADVANTAGES CLAIMED OVER SINGLE STACK, ARE

First. Greater draft on sides of boiler.
Second. Produce greater amount of steam with larger exhaust nozzles, thereby reducing back pressure.
Third. Greater speed with less weight of motive power.
Fourth. Forty per cent larger area of smokestack opening.
Fifth. Twenty per cent larger area of exhaust nozzle opening.
Sixth. Saving in fuel.
Seventh. Saving on wear of boiler flues.
Eighth. Greater steaming capacities with smaller boilers.

LORD'S IMPROVED DRAFT REGULATOR.

The device herewith illustrated is the invention of Mr. C. B. Lord, Astoria, L. I., New York. In this device the use of exhaust steam for a draft is entirely dispensed with, the exhaust steam escaping through a partition in the smoke

PRESENT ARRANGEMENT OF EXHAUST—HEAVY FIRE.

stack or through a separate stack. A small perforated blower head taking live steam from the boiler being used to create the draft. This device has been tried upon the King's County Road of Brooklyn, N. Y., and is said to have given

NEW ARRANGEMENT OF EXHAUST—LIGHT FIRE.

excellent results, showing a considerable saving of fuel since the test. The inventor has designed a new tube blower, having 16 small nozzles with 3-32" openings. The following is Mr. Lord's description of his improvement:

"The simple and feasible plan here represented insulates the direct effect of the exhaust upon the fire and reliance is placed upon the continuous and moderate draught of the blower blast (being equivalent to the draught produced by a modified exhaust) for generation of steam, thereby retaining and utilizing the present waste of fuel. Its requisite aid would be an adequate under-draught, which, combined with the large increase of nat-

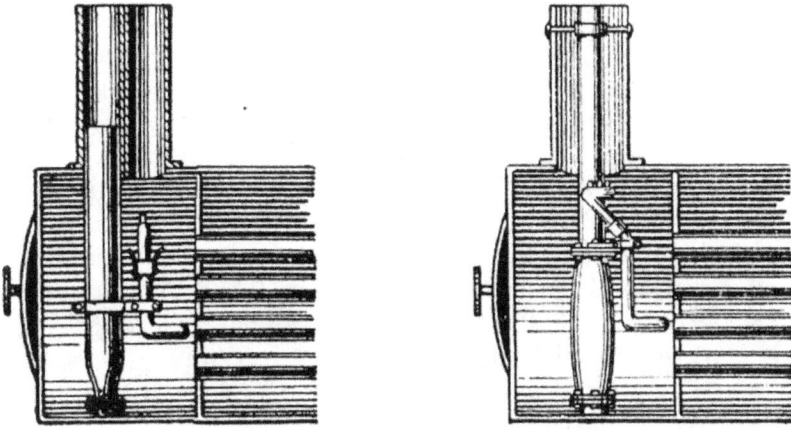

DOUBLE STACK ARRANGEMENT. CENTRAL ARRANGEMENT.

ural draught created by the non-use of arches, netting, petticoat pipe, etc., and the compulsion of carrying light fires, would vastly improve the conditions of combustion. The insulation of the exhaust would secure especial economy in the handling of local trains of both classes, also in shifting and yard service, and the engineer would have no option in the manner of using this device (which is allowed in no other devices of this nature),

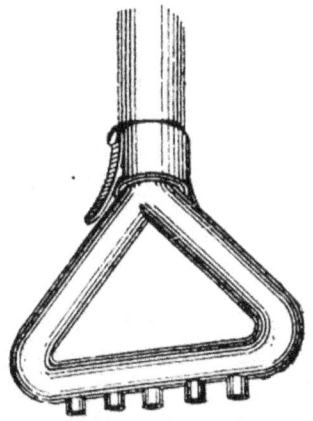

DETACHABLE BLOWER HEAD.

except an intelligent manipulation of the cut-off and feed water, would produce a compulsory uniform result as regards the consumption of fuel by the exhaust blast.

"The manifest benefits of this re-arrangement of draught would be: Its great economy of fuel through its maintenance of a steady, reliable and graduated draught upon the fire under any circumstance of exhaust.

"Additional security and longevity of the boiler and avoidance of expensive repairs to the same. The proper adjustment of reverse compression in the cylinder by means of an exhaust tip (detachable and optional to use) inside of expansion joint of the exhaust pipe. The consequent development of power with corresponding decrease in fuel consumption.

"The saving of the large aggregate amount of fuel wasted by an engine's slipping, and the impossibility of throwing fire or cinders to cause damage or annoyance. A perceptible reduction in the noise of the exhaust, desirable under many conditions. The design of tube blower shown in this plan claims greater effectiveness with the use of less steam than the style now in use, and is calculated to be of a shape and dimension to agitate the entire width of air space at the base of any sized stack, and being detachable renders it convenient to quickly substitute for a duplicate in case of corrosion or disablement without resort to the repair shop. The simplicity and small cost of this device, together with the fact that it has never been tested, should strongly commend it for a thorough and impartial trial to those officials whose duty requires especial interest in fuel economy on our railroads, the results of which would doubtless recommend its adoption and prove largely remunerative, not only to the stockholders, but all those connected with the mechanical department of such roads, inasmuch as through its economies the use of the steam locomotive would be prolonged indefinitely."

A NOVEL SPARK-ARRESTER—NORTHERN PACIFIC R. R.

The spark-arrester shown herewith at first sight appears like a useless monstrosity. But it is not. It is a thoroughly satisfactory construction that is proving its worth in daily service on the Northern Pacific. It was designed by Mr. H. H. Warner, Master Mechanic of that road at Tacoma, Wash. It is used upon a district of that road—about 300 miles long—upon which is found a local deposit of fuel strongly lignite in character. This fuel had been used, but always with much attendant danger from fire, all devices that had been tried failing to prevent the emission of sparks. Finally this device was made and has fully met the difficulties encountered in attempting to use the fuel, which was, aside from its spark-making, otherwise desirable.

Despite its ugliness "it does the business" very completely. It constitutes a complete arrester of all fire sparks from the lightest of coal, requires no netting, is economical in fuel, necessitates no dumping or clearing of cinders and is simple in construction and durable in service.

As may be noticed in the diagram of the device and boiler that we give, the exhaust pipe is shaped as a reverse curve. This shoots the products of combustion from the front of the smoke box through a petticoat pipe which is so placed as to

THE LOCOMOTIVE UP TO DATE.

A NOVEL SPARK ARRESTER.—NORTHERN PACIFIC RY.

gather all the contents of the box freely and to pass them on through the convex pipe attachment to the spark chamber on the top of the boiler. In this chamber is a dash plate, as shown, scattering the sparks, there being ample space above and beneath it for the passage of the sparks. The heaviest of the sparks fall to the bottom of the chamber, all being extinguished by contact with the exhaust. At the lower corners of the chamber are exhaust pipes intended to convey the sparks, which are deposited at the base of the chamber, back to the fire box. This is effected by a strong draft through the pipes created by steam jets placed at the connection of the pipes to the fire box. The light extinguished sparks are conveyed to atmosphere by the exhaust through the stack.

One can become accustomed to ungainly-looking devices, as this certainly is, especially when, as in this instance, it has be-

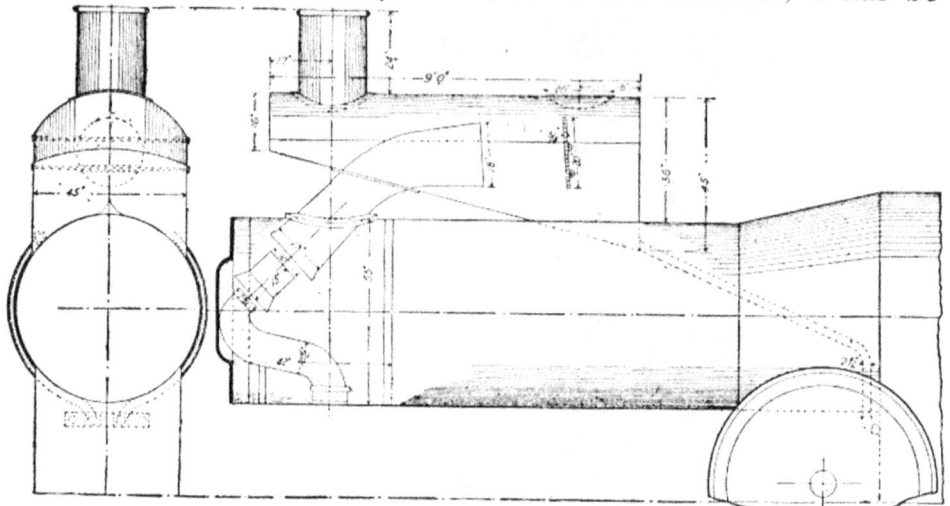

come an acknowledged success in its operation and all danger from fire is eliminated by its performance. This arrangement is fitted to a number of engines running through the lignite district. It is stated that they are free steamers and that in the use of lignite they are economical.

CANBY'S DRAFT REGULATING APPARATUS.

The inventor, Mr. Joseph C. Canby, of Orange, Luzerne County, Pa., has furnished us with the following description of this device:

My invention relates to draft-regulating apparatus for locomotive and that class of boilers; and it consists of a smoke-stack with an adjustable petticoat or mouthpiece to equalize the draft through all the flues, also an arrangement of pipes and valves to introduce fresh air into the smoke-stack to check the draft without opening the fire-door and letting the cold air in onto the boiler and tubes, thereby making a great saving in the fuel and being better for the boiler and flues.

Figure 1 represents the front view of the boiler with the automatic draft-regulator attached. Fig. 2 is a horizontal section of front of boiler, showing smoke-stack and rock-shaft. Fig. 3 is a longitudinal section of the smoke-box and boiler, showing the connection of the valve *N* and regulator *O* and the connection of arm *J* to the cab *R* by the rod *K*.

A A' A'' represent the sections of the smoke-stack, or, as familiarly called, "petticoats," arranged with lugs *B* on the sides with slides *D D'*, having slots and set-screws *E E'*, by which they are adjusted to the space required between them, thereby

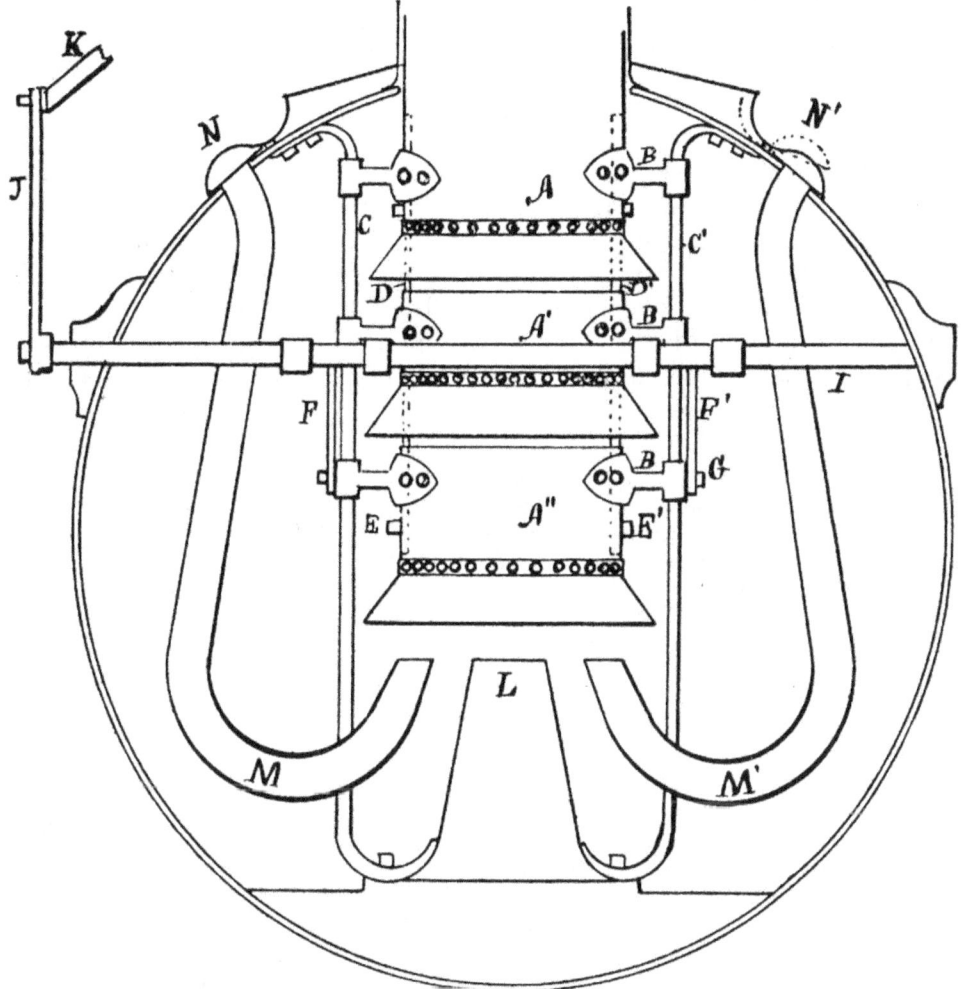

Fig. 1.

enabling the engineer to equalize the draft in the fire-box, as experience shows that when the draft is nearest to the bottom of the smoke-jacket the draft is strongest on the back end of the fire next the flue, and by decreasing there and increasing it in the top flues the draft is made stronger in the front part of the fire-box. This more nearly equalizes the combustion of the fuel. The connecting-rods *F F'* are attached to the lugs *G* and the arms $H H'$ project from the rocking shaft I, which is operated by the arm J and rod K, which runs to the cab R. By pulling or pushing the rod K the petticoats are raised and lowered, thus increasing and decreasing the distance from the

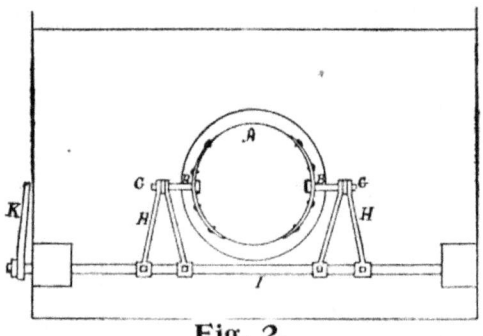

Fig. 2.

exhaust-nozzle L, thereby increasing or diminishing the draft. The air-tubes $M M'$ turn up alongside the exhaust-nozzle L, and are opened and closed by valves $N N'$ on the outside of the boiler. The valves are operated by a pressure-regulator O, so adjusted that they are opened by the steam when it passes a given pressure. This operates on the crank P and connecting-rod Q to open the valve, thus admitting air to the smoke-box

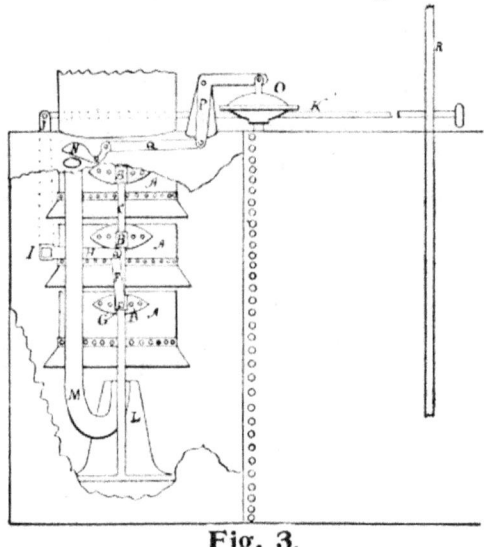

Fig. 3.

and decreasing the amount drawn through the tubes and decreasing the consumption of the coal and obtaining the full benefit for all fuel consumed without letting the cold air in onto the hot iron. By this means we have the combustion automatically regulated, also obtaining the greatest amount of heat from the fuel consumed.

INCRUSTATION.

MODERN BOILER CLEANING.

As this is a subject of vital importance to every master mechanic and engineer we have endeavored to secure all the information possible concerning the subject.

The following is from *Power*, October, 1898.

"The savings to be effected in power generation to-day consists more in the overcoming of simple practical difficulties in the use of that which we already have than in any revolutionary invention. The man who could supply a simple, inexpensive means of furnishing steam boilers with pure water which, when evaporated, would leave nothing behind it, would do more to decrease the average cost of power production than the man who develops the compound engine. Such a process would have to be so cheap in first cost as to warrant its use in comparatively small plants, and so simple as to require attendance of no higher order than that found about the ordinary boiler plant."

While chemicals and boiler compounds are largely used at the present time there are many who believe that, owing to the circulation, boilers can be kept clean by mechanical means, as it is a well known fact that heat will separate the impurities from the water. Many mechanical devices have been used for this purpose, but we believe the correct principle to be more closely followed by the inventors of the boiler cleaners herewith shown than any we have yet seen.

Each article describing these devices were especially prepared for this book by the inventors themselves, and they each appear to have a pretty thorough knowledge of the subject. We have reproduced the article prepared by Mr. Hornish in full, owing to the vast amount of information it contains, which will be very interesting reading to anyone interested in this subject. Mr. McIntosh's "Blow-off Cock" is doing good work and giving entire satisfaction upon several different roads, while his "Surface Blow Off" has been applied to several engines on the C. & N. W. Ry. The writer has a personal knowledge of the success of Mr. Hornish's former cleaner, which has been in use for more than ten years. Perhaps no better recommendation could be given his new cleaner than the fact that it was accepted on sight by Mr. Robert Quayle, Superintendent of M. P. & M., of the C. & N. W. Ry., who is recognized as being one of the most progressive railway men in the country.

We have also shown the McLeod revolving check valve referred to in Mr. Hornish's article, and also a brief article on "Stay-Bolts" prepared by the Hartford Steam Boiler Inspection and Insurance Co., which substantiates the theory advanced by Mr. Hornish.

THE McINTOSH SURFACE BLOW OFF.

The surface cock is especially useful when a locomotive is working hard and the water in boiler is violently agitated, carrying matter in suspension upward and forward in the natural course of circulation within reach of the funnels, as shown in the

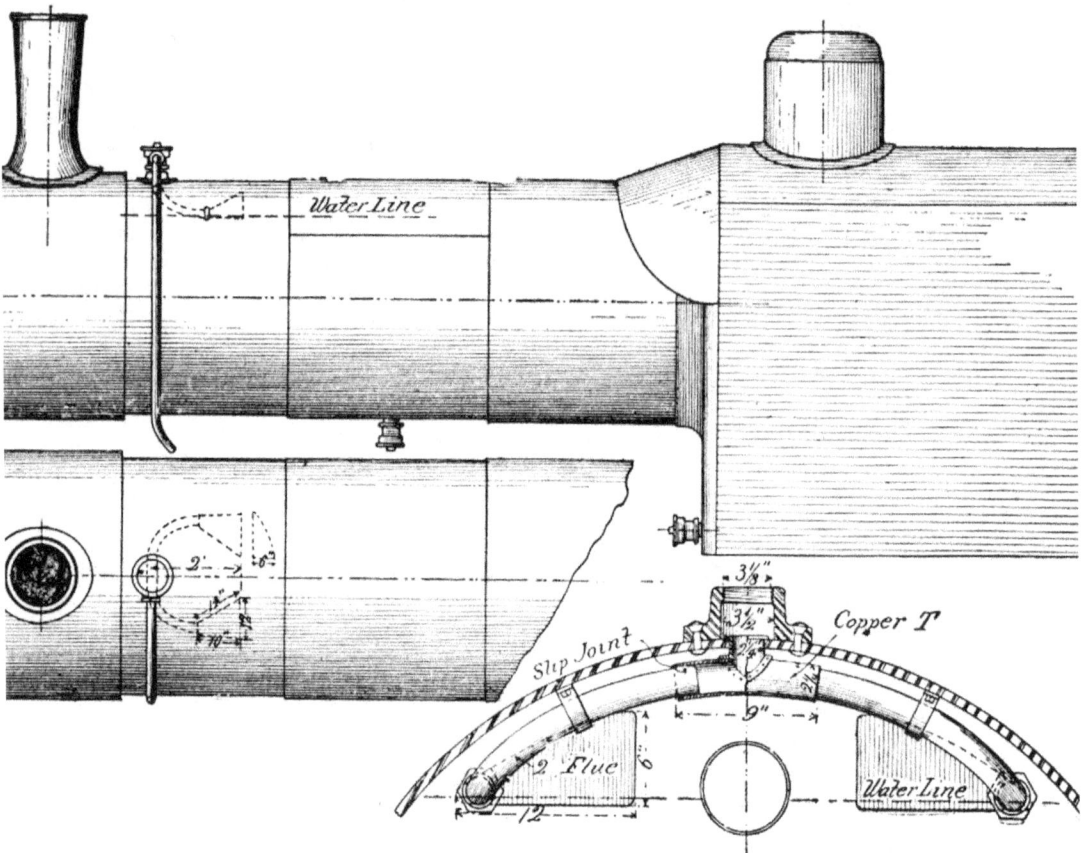

surface blowing off device illustrated herewith. It is obvious that under such conditions frequent opening of the surface cock for a few seconds at a time will blow out a great deal of floating matter that would otherwise settle down when the agitation ceased, a considerable part adhering to the flues; but a few pails full of water and a corresponding small amount of heat is lost in this manner which is far more than compensated for in the improved condition of water remaining in the boiler. The surface cock is particularly useful in regions where alkali water must be used for boiler feeding and which frequently causes foaming or priming to the extent of rendering a reduction of train imperative, and in some cases practically disabling the engine; with such water a judicious use of the surface blow-off cock controls the worst case of foaming and enables an engine to proceed with full train at all times. Engines equipped with four cocks, as shown in cut, can be run for months without washing out, a valuable consideration with operating officials in busy times, when laying engines in daily for washing out means a direct loss of revenue.

The surface cock can either be located on top or bottom of boiler, the only difference being in the length of pipes leading to the funnels. The apparatus is very simple, consisting of a two-inch copper tee expanded into a hole in boiler, surrounding which a flange is riveted and into this the blow-off cock screws. Into each branch of the tee is slipped an old flue, which terminates in a funnel at the water line.

THE McINTOSH BLOW-OFF COCK.

The accompanying cross sectional views of the duplex and single valve fluid pressure blow-off cocks require but little explanation. Air or steam is admitted through the elbow on cylinder cap, forcing down the piston, the stem of which opens the valves, the combined influence of the spring and boiler pressure returning them to place when pressure is removed from piston; this is done through a three-way cock convenient to the engineer in cab.

The duplex is the preferable form of cock, possessing the additional safety feature of the independent inside valve, which in the event of an accident and outer casing being broken off would retain boiler pressure. The disc nut or key is for seating or removing the inner casing of the duplex cock, a partial end view of which is shown; into the oppositely disposed internal recesses of this casing enter the lugs of the disc nut that affords, when secured by the clamping ring, a positive means of seating or removing the casing with a common wrench, engaging the square or otherwise angular head of the key. Both valve seats of the duplex cock are removable and can be quickly replaced when desired. The single valve cock locks the absolute safety feature of the duplex, but is itself safer than any other single valve cock made. Owing to its valve being internally seated, it also furnishes a direct opening to the atmosphere. This was the original fluid pressure cock, but was not put on the market on account of the more desirable features of the duplex design. There is a demand, however, in some quarters for a cheaper cock, and the single valve is furnished when specified. Both forms are now provided with means of opening or closing by hand through the medium of the hollow handle shown on cap; it is obvious that the piston can either be forced in or pulled out by screwing the handle in the desired direction. The position

448   THE LOCOMOTIVE UP TO DATE.

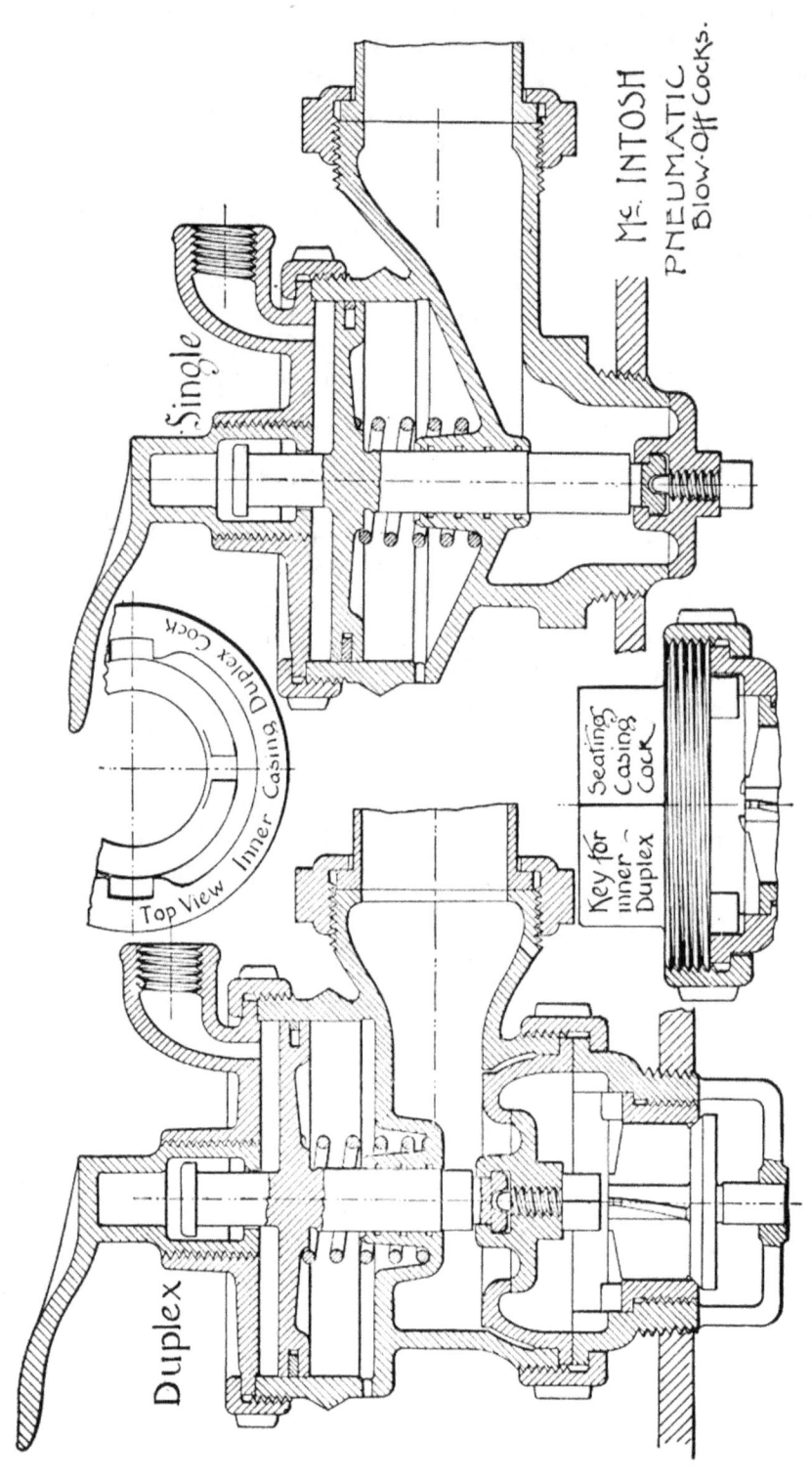

it occupies in cut is normal—where it should be placed when not in action.

It will be noted that there are no packing glands of any kind about these cocks and that all parts have the greatest freedom of movement.

Until recently the only means employed to remove incrusting material carried into locomotive boilers with the feed water was to remove flues frequently either in full sets or partially and scrape and wash out the accumulation and incidentally renewing the fire box plates that had burned out. During the past five years soda ash and other chemical compounds have been used to separate the incrusting matter from the feed water when it settles to the bottom of the boiler to be blown or washed out frequently, otherwise the remedy would aggregate the unfavorable conditions by increasing the density of the water about the lower flues and fire box sheets to such an extent that it would fail to carry away the heat and allow the plates or flues to become overheated.

The old form of plug cock was found entirely inadequate for the extra work imposed upon it, in fact, never was used to any extent, usually requiring the combined efforts of engineer and fireman to open, with the assurance that it would afterwards leak until refitted. These conditions led to the development of the pneumatic cock, which can be opened without effort and as often as desired; twenty times daily is not infrequent in bad water regions. Owing to their ease of manipulation and reliability there is no trouble in having engineers use them as required, and the functions they are designed for are performed.

Of course it is well understood among mechanical men that the proper place to purify water is before it goes into the boiler, but to do so involves a large initial expense in settling tanks and a constant outlay for chemicals, and but little has been done as yet in that direction. It therefore remains for railway officials to take advantage of the means at hand that will bring about immediate results, several of the leading roads in the country having their boiler repairs reduced fifty per cent by the use of soda ash and the adoption of better means of blowing and washing out.

THE REED BLOW-OFF COCK.

The construction of this device is very simple and not liable to get out of order, and is of convenient form for regrinding, and can be opened by steam pressure from the boiler to which it is attached. The illustration is sufficiently clear for a correct understanding of its mode of action. Steam passing through the pipe O enters the cylinder F, forcing the piston H downward and also the valve I permits water to escape from the boiler at E. When steam pressure is shut off in the pipe O, the spring K, together with the boiler pressure, forces the valve I onto its seat *d*, thereby closing the discharge. It can be seen

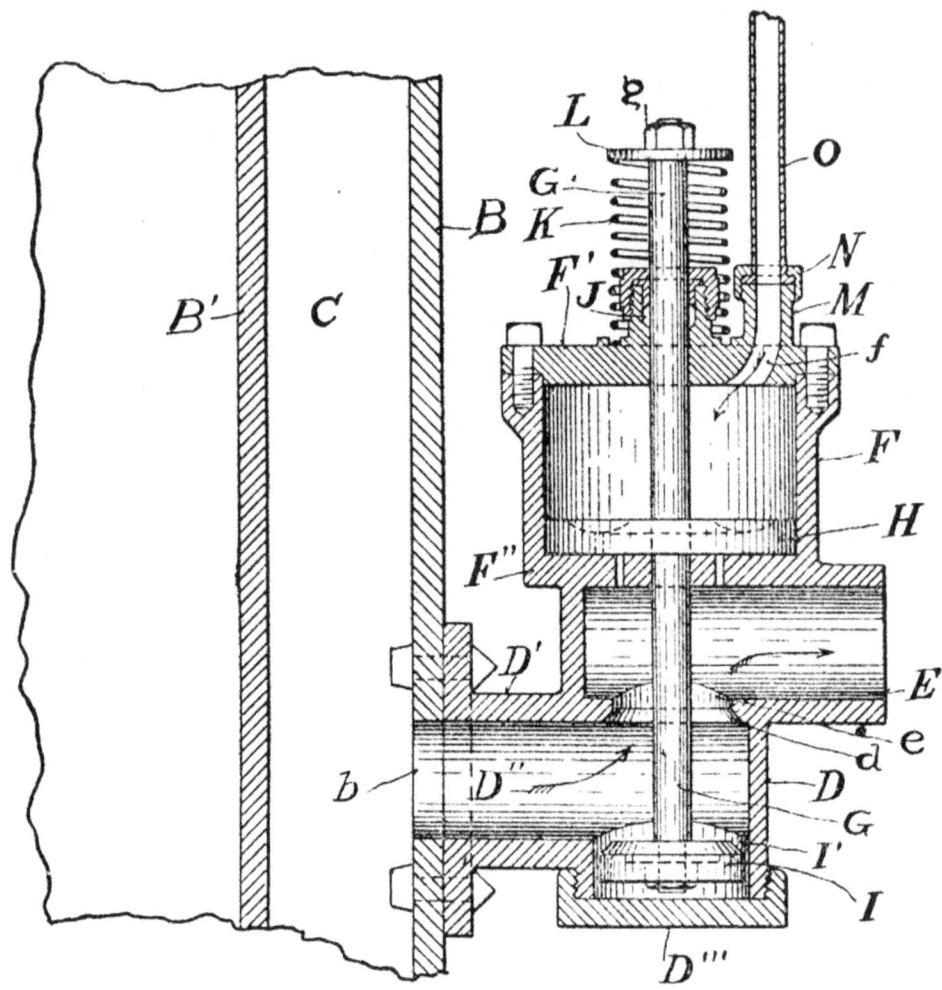

that the piston and valve can be easily removed by simply removing the caps F' and D'''. This device is the invention of Mr. Geo. Reed, of Forest City, Mo.

THE NEW HORNISH MECHANICAL BOILER CLEANER. *

The inventor's first experience with a locomotive was one year's work as boiler washer, washing out several engines each day, and had to do them all more or less hurriedly. He had forced upon his attention the great loss sustained by a railway company in the course of a year, due not only to the cost of

---

*This article was especially prepared for this book by the inventor, Mr. Frank W. Hornish, of Chicago, Ill.

washing out the boilers, but also to the gradual decrease in the efficiency of an engine from one cleaning to another. To invent a mechanical cleaner that would do away with the old expensive way of washing out, and at the same time keep the engine always at its maximum efficiency, as far as the boiler is concerned, was the task he set for himself. He succeeded so well that every engine on the road he was connected with has his old cleaning attachment on it and has had for several years.

This cleaner gives good results; but the inventor was not satisfied with it and has gone on experimenting until he has now a much superior one. The following is the inventor's own description of his old and new cleaners:

The use of my old cleaner for 10 or more years upon several thousand stationary and steamboat boilers and on more than 100 locomotive boilers (P. D. & E. Ry. and E. & T. H. Ry.) and several other roads, had given me the opportunity to discover its deficiencies in actual use, especially when used on locomotives. My old cleaner was not entirely suitable for a locomotive, owing to its outside attachments, yet it proves that most of the impurities in the water come to the surface when steam is up, and there is a circulation in the boiler; and that the solid matter held in solution is separated by heat and can, while the engine is steaming, be removed from the surface of the water by mechanical means.

DESCRIPTION OF THE OLD CLEANER.

This device consists of a skimmer which is placed as far forward in the boiler as the braces will permit. It faces the cab and reaches from side to side of boiler, and from top of boiler to top of flume, with space enough left for the dry pipe to pass through. The skimmer is set in line with the natural circulation of the boiler, and it makes an artificial circulation of its own, and carries the impurities it collects to the drum to settle, and as sediment seeks the coolest and quietest place, this drum makes a perfect catch-basin for the solid matter to settle in. As the water goes in at one side of the drum at the top, and out at the top of the other, on its way to the leg of the boiler, there is no circulation in the lower part of the drum, and the impurities are caught there. The pipe leading to the leg of the boiler from the drum has a check which opens towards the boiler, and when the cleaner is blown off this check closes from back pressure, thus making a perfect surface blow-off the full width of the boiler when skimmer is blown off, and an automatic continuous draw-off with a large capacity for storing the sediment between the times of blowing off. In the bottom of the drum is the draw-off head, extending the full length of the drum. This head empties the impurities which are over and around it before it discharges any of the water or steam, so that nothing but the solid matter

452 THE LOCOMOTIVE UP TO DATE.

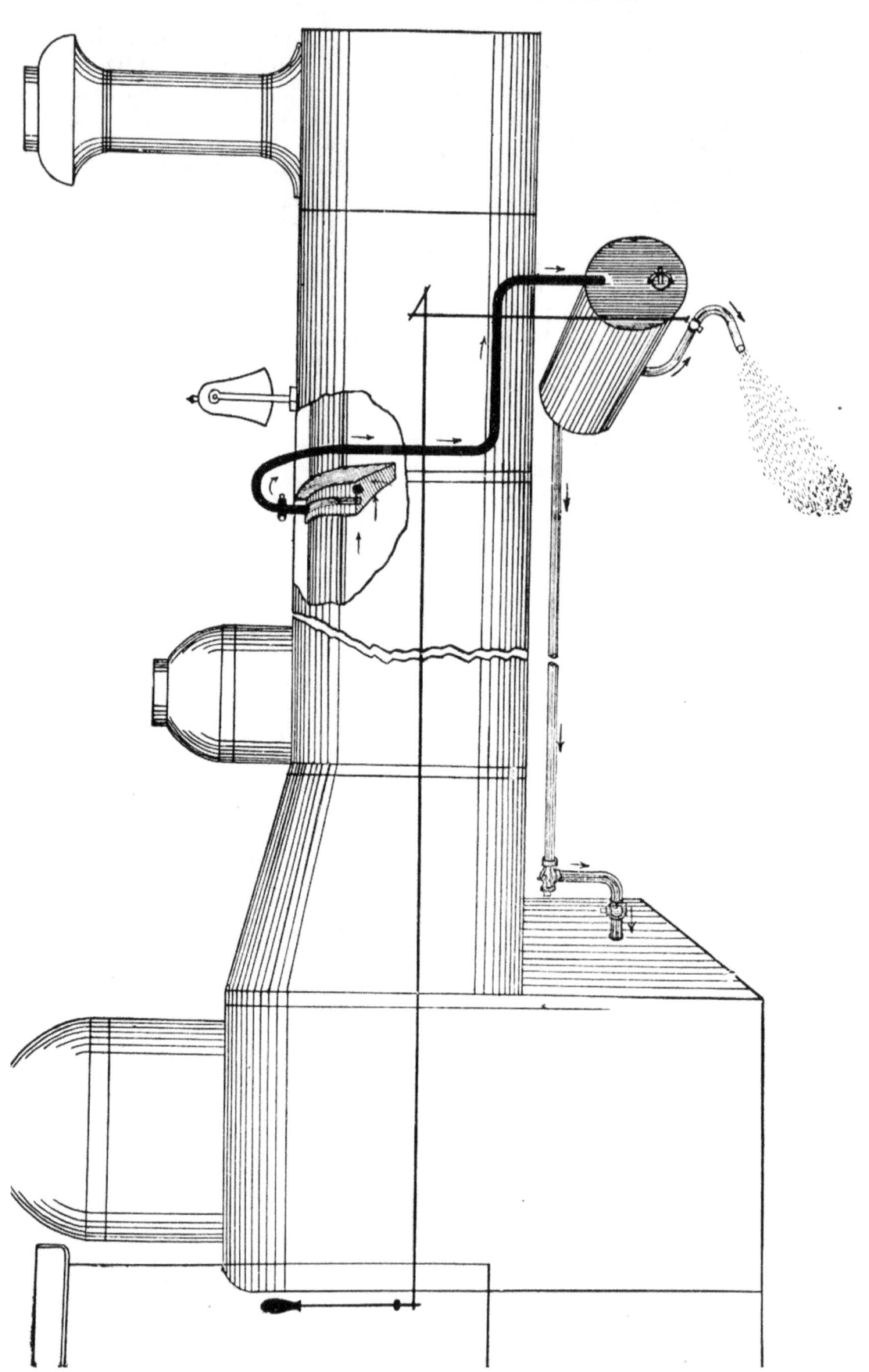

in the drum can be discharged until it has been emptied of its settlings. This draw-off head reduces the weight of the water to a minimum in blowing off, in fact no water is wasted at all that is fit to use, for the head is shut off as soon as the water becomes clear. One of the disadvantages of this style of cleaner is that the circulating pipes and the drum are on the outside and take up too much space, and a locomotive is already incumbered with too many outside appliances that cannot be dispensed with. Another disadvantage is that as a surface skimmer cannot catch all the impurities in the water, there is a slow accumulation in the leg of the boiler, and if not looked after very closely will cause trouble, and herein lies the danger of this device. Too much confidence is likely to be placed in its merits and it may be neglected and the engine run so long between washouts that the leg of the boiler fills up too far and allows the boiler to be burnt out.

DESCRIPTION OF THE NEW CLEANER.

My new cleaner obviates all this, as there is a device in the leg of the boiler which removes all the sediment deposited there. The new cleaner will permit an engine to be run the entire year without being washed out, for the cleaner prevents scaling and makes foaming impossible. It is, however, neither necessary nor advisable to let the engine go so long a time without washing it out in the ordinary manner. This cleaner is the product of 20 years of study and experiment carried on by a practical locomotive machinist and engineer. Neither money, time nor pains have been spared in perfecting it.

Owing to the large amount of water evaporated in a locomotive boiler, the water soon becomes foul and thick, because the solid matter is left behind and these impurities are kept at the surface when the engine is working hard, thus forming a sort of blanket on the surface. Foaming is caused by the steam forcing its way through this blanket. My device keeps the surface clean, and hence makes foaming impossible. A practical test on Chicago and Northwestern Ry. has proven this.

Chicago, Ill., Dec. 28, 1897.

Mr. Frank W. Hornish.

Dear Sir:—In regard to your cleaner placed on engine No. 806 will say the engine has made 1,304 miles without washing out. I reported the fact to traveling engineer, and he told me to report her washed out, to find out how much mud there was in the leg of the boiler, but have not heard.

I think the mileage I have made is a good showing for the cleaner. Five hundred and fifty miles is about all we could make without cleaner and do good work. I do not know how long I could run her without washing out, as she has not given me any trouble in making the mileage stated on account of foaming. Respectfully yours,

CHAS. PRIESTER, Engineer.
104 W. Avers Ave., Chicago.

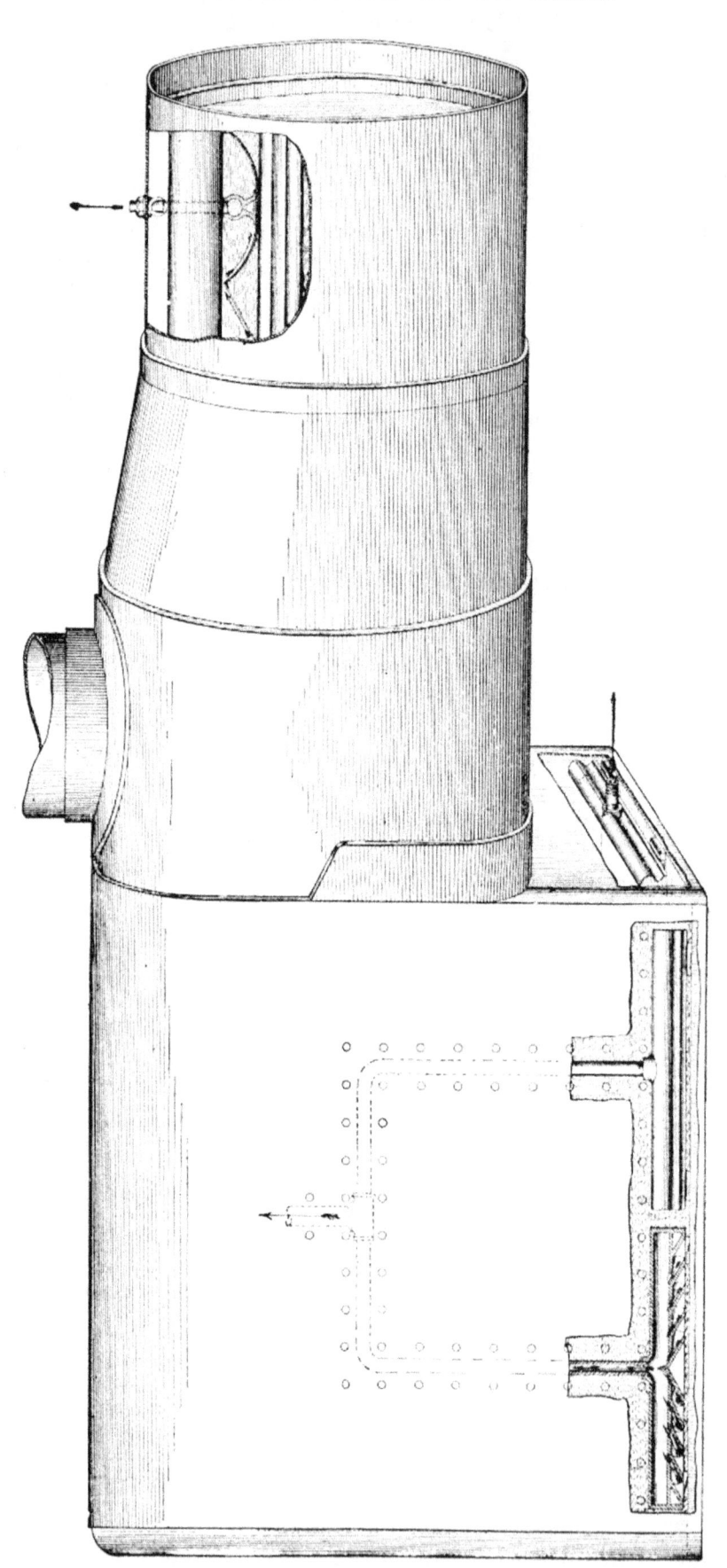

THE LOCOMOTIVE UP TO DATE.   455

Chicago, Ill., Feb. 26, 1898.

Mr. F. W. Hornish,
    Chicago, Ills.

Dear Sir: I am pleased to state that the boiler cleaner you have fitted to our engine No. 806 has given entire satisfaction. It has been of material advantage in preventing foaming, and because of its excellency in this respect, and the fact that it removes considerable sediment from the water in the boiler, we have been able to run twice as many miles between the times of washing out as formerly. I am pleased to give you this testimony.

I believe that the mud ring device will be just as successful in its way as the cleaner or skimmer has been. The two devices are so much alike in principle that the success of one practically guarantees the success of the other. I have written to you at considerable length in this matter, because I think that you feel that we have not followed this matter closely, or given your device the attention it deserves. If you have this feeling, I assure you it is a mistake. We have followed the workings of this device closely, and what I have stated concerning its workings in the first part of this letter I state to you most cheerfully.

Yours truly,
W. H. MARSHALL,
Asst. Supt. M. P. & M., C. & N. W. Ry.

The above is part of a letter which shows the good work the cleaner is doing under adverse circumstances. Only one-half is

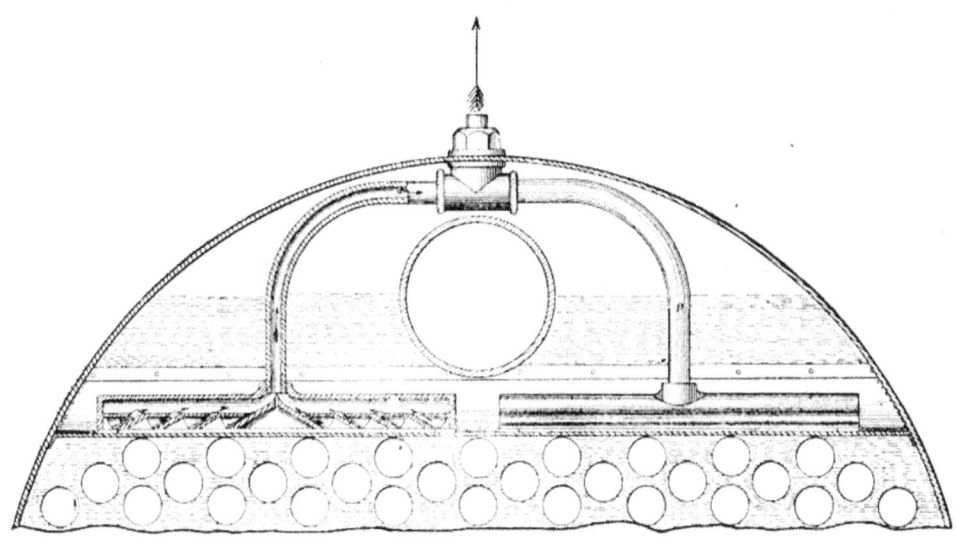

in use (the surface skimmer) and that was put on at a disadvantage. The skimmer is too far forward, and in two parts, leaving 10 inches space for dry pipe to pass (this open space reduces its efficiency), and is not fitted tight to either the top or the sides of the boiler. What it has done in that case is no more than an apology for what it can do alone if set properly.

My new cleaner is in two parts, all inside the boiler, except the air valve and the blow-off pipe. One part is in the forward end of the boiler, and reaches from side to side, using the front head as a back to which it is riveted. It extends under the dry pipe and is also riveted to the sides. The space between the flues and the dry pipe is used for the skimmer. It makes a perfect surface skimmer the full width of the boiler, and at the same time forms a basin holding from 20 to 25 gallons.

This makes a large storage capacity in which to collect and settle all the impurities that the skimmer takes from the surface. The basin holds the skimmings and settlings which can be blown off at the will of the engineer. The impurities are carried to the skimmer by the natural circulation in the boiler.

The fluctuation of the water line does not affect the working of this device.

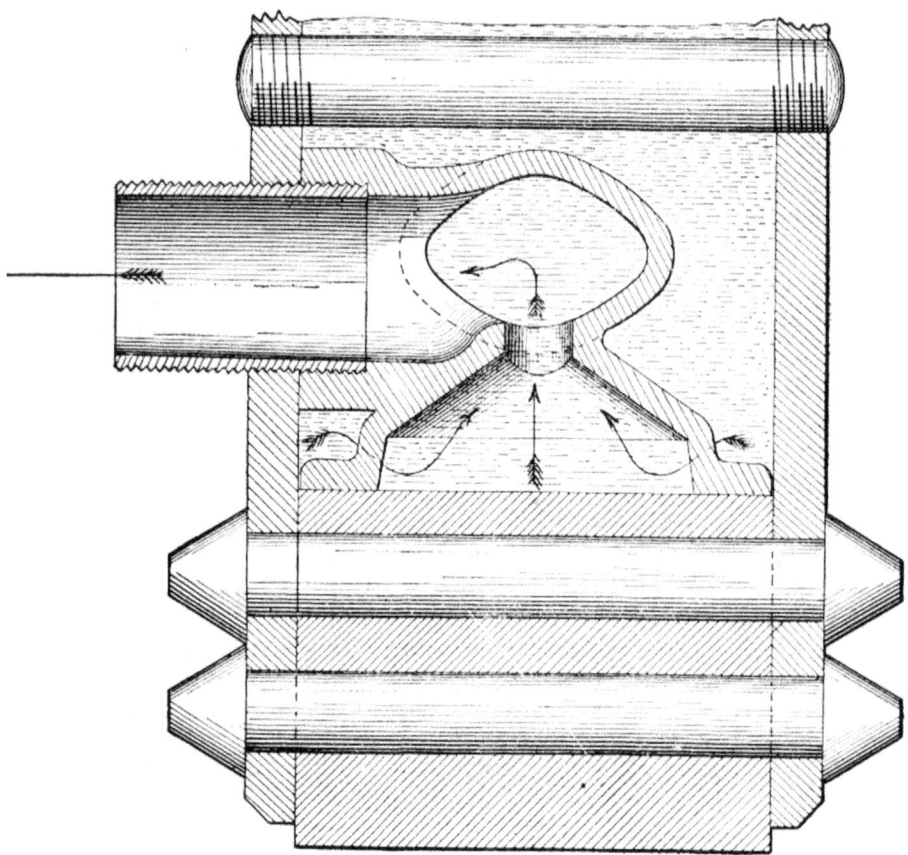

This skimmer takes the place of the skimmer drum and pipes in the old device, and is all inside the boiler and much more efficient as well as more simple.

In this skimmer is a device for blowing off what solid matter the skimmer catches. This practically prevents any waste of water at all, when the skimmer is blown off. It makes a perfect surface blow-off, extending the full width of the boiler, and between the times of blowing off there is a continuous automatic drawing off.

The other part is in the leg of the boiler and cannot be put in except when the boiler is first built, or when a new fire box is put in. It is the same kind of a device that empties the skimmer so nicely. It sits on top of the mud ring, the suckers facing down, which are raised one-half inch above the mud ring by legs on the draw-off head. The opening of all the suckers are the same size, but the small end of the suckers vary in size. If they were all the same size, the openings nearest the center would pass all the sediment, and those further along the head would not pass their shares. This is due to the fact that all liquids under pressure will seek the nearest outlet first. To overcome this, the small end of the sucker nearest the opening in the center of the draw-off head is made the smallest, and the size of the others increases with the distance from the center. The matter surrounding this is always watersoaked, and is a soft slush, which is easily removed by the pressure in the boiler.

As the impurities are over and around the draw-off head, in front of the water and steam, the pressure in the boiler pushes them out through the suckers into the blow-off pipe before any water or steam can pass through. There is no waste of water, as the blowing off is stopped as soon as water shows clear. This is what reduces the waste of water to a minimum. This draw-off head is a great improvement over the old head used in my old cleaner, and yet 10 years' use has never known one of those to stop up, and even if they should stop up from neglect, they can be cleaned from the outside by forcing water through them. They cannot scale up since there is no heat next to them from the fire box. These heads are a certainty and no experiment.

Each part of the cleaner is a companion to the other, and what one leaves undone the other does. The two must be used together to obtain perfect results.

When the water is boiling the impurities come to the surface in a boiler the same as in an open vessel. The circulation in a locomotive boiler is down to the belly of the boiler after water leaves the check, then down the forward leg along the side and up the back leg over the crown sheet, and forward over the top of the flues, striking the front head of the boiler, from which point it starts to repeat the same course over again. It is here at the front end of the boiler and at the surface that the skimmer intercepts the impurities, settling and removing them before they have time to touch the hot flues or hot sheets. This not only prevents foaming, but leaves nothing in the boiler to make scale and lessens the accumulation in the leg of the boiler.

Nothing but water will make steam, so keep grease and compounds out of the boiler. Heat is the best agency known for separating solid matter from water. A boiler is the best contrivance yet devised by man for heating water; as most of the solid matter contained in the feed water comes to the surface when steam is up, that is the best place to remove it. The surface skimmer does this before it starts to make the second round with the circulation. What little is left goes to the leg of the boiler and is removed by the mud ring device. This absolutely prevents foaming, reduces the number of washing outs and prevents incrustation and corrosion, and is far in advance, and much more practical than purifying the water before entering the boiler, and after the first cost costs nothing.

It is the soft matter in the boiler that is dangerous, and not the hard scales; so the removing of that part removes the danger. Hard scale is a good conductor rather than a poor one, as is generally supposed. A thin scale is an advantage rather than a disadvantage, as it prevents pitting.

This device does not interfere in the least in cleaning out the old way, or in the use of compounds. If scale should form from neglecting to use the cleaner or any other cause, compounds can be used, and what they dissolve and throw down the cleaner will remove at the surface and at the leg. It will not be necessary to remove the plugs or the hand hole plates.

As long as there is a circulation in the boiler this cleaner is at work. While the engine is either idle in the round-house or on the road, it is skimming the surface and storing up the impurities in the skimmer to be blown off at will. This reduces both the number of blow-offs and the waste of water. As both cleaners are blown off before entering the roundhouse, there is little or nothing left in the water, and what there is it catches while the engine is idle to be blown out again as soon as it leaves the round-house. This makes a clean boiler at all times.

A boiler scales most when the water is let out, while it is still hot, as is the case when washing out in a hurry. The flues and sheets are hot and the water is loaded with impurities, which settle and attach themselves to the hot places in the boiler. The amount that attaches at any one time is not great, and the oftener boilers are washed out in a hurry the faster they scale.

If boilers were allowed to become cold before removing the water, very little scale would form. It is washing out hot boilers in a hurry that causes them to scale so fast and leak.

As water in the boiler is practically pure when using this device, it is almost equal to distilled water, and experience has proven that such is a good scale remover. As soon as all loose matter is removed from the water in the boiler the distilled water begins to act on the scale and eats into and dissolves it as a compound would (pure water is the best of compounds). The cleaner then takes and removes it the same as it does the solid matter in the feed water. The distilled water acts like a sponge and re-absorbs the solid matter, removes and carries it to the skimmer and mud ring by the circulation.

The fact that this cleaner removes all solid matter from the boiler as fast as it enters, and allows no accumulation, settles the scale question. Since there can be no scale where there is no scale forming matter, hot places cut no figure. Some say a boiler will scale because the injector, injector pipe and the check scale from the water passing through them to the boiler when the water is comparatively warm. It is the water coming from the boiler through these parts caused by a leaky check and injector that causes them to stop up, and not the water going through them to the boiler.

The water going through an injector is not hot enough to remove the solid matter that is held in it by solution. Water that stops up and eats out pipes before it reaches the injector or pump is not fit for use in making steam. If there was a continuous circulation through the injector, the pipe and the check, there could be no corrosion; but on the other hand, it would keep those parts scoured out.

It is a leaky check and injector that stops them up, and this has been proved beyond all doubt by the use of *McLeod's non-fouling turbine check*, which revolves with great rapidity while in use, and never seating itself twice in the same place, makes a

tight seat and clean cage, and foreign matter in these parts is almost unknown; as it always seats itself with a twist it has a tendency to regrind itself every time the injector or pump is shut off.

When a check leaks it causes the impurities to leave the boiler when check is not in use by this route, and as the injector is apt to leak also from the impurities which the leak passes all the parts will have become foul. The water in the forward end of the boiler contains most of the impurities, and it is there that the check enters.

When a boiler is steaming hard the crown sheet is the hottest place, and it is from there that most of the steam is taken for the throttle. The water is higher there, for the steam in its rapid passage to the cylinders raises the water and the incline on the surface causes the impurities to gravitate toward the forward end of the boiler. This is one reason why the skimmer is placed there. A leaky check will draw these impurities through itself and deposit them in the injector and injector pipe. The circulation from a leak is so slow that there can be no scouring out and the sediment has time to harden.

My cleaner will obviate this trouble, and since the check comes under the skimmer it catches all the impurities that are on the surface of the water and discharges them before they get a chance at the check; there can be no trouble from the parts mentioned. The use of McLeod's check with the cleaner is a good combination, as it makes relief from these parts doubly sure.

This insures an easy working injector, as the check passes no hot water from the boiler to the injector when not in use. It allows the injector to become cold when not working, making the injector start easier, and avoiding any scaling in these parts, since cold water makes no scale except in very rare cases, and such water should never be used. The rapid revolution of this check makes a partial vacuum and relieves some back pressure. It never sticks; two years of continuous use with the worst of water has settled that point.

It is further stated that solid matter in the water will touch the hot sheets and flues in its circulation, and burn itself there. Such is not the case. The solid matter will not attach itself to the hot sheets while there is a circulation until the water has become saturated with impurities beyond its limit to carry them along with the circulation; until the limit of saturation is reached the circulation will prevent scaling. The cleaner keeps the water so pure that it readily absorbs all loose matter and carries it to the surface skimmer and to the leg of boiler before it has a chance to stick to the hot places.

Incrustation and corrosion cause most boiler explosions, due to the impurities contained in the feed water; and when the impurities are removed the cause goes with them.

The advantages to be derived from a clean boiler are too numerous to mention in the space to which I am limited in this article. My device always insures a clear water glass and clean gauge cocks and the blowing off of these never soils the boiler head.

As foaming is eliminated, no solid matter goes over with the steam to cut the valves, valve seats, valve stems or packing. As no solid matter ever enters the cylinders they too are kept smooth, and when free from grit their wearing surface soon acquires a gloss that insures easy working and long wear, together with a minimum use of oil. This is of great value for balanced valves, which are now coming into use with high pressure.

When this cleaner is used flues seldom leak, they wear longer, boiler repairs are reduced to the lowest cost, the engine is kept from the back shops a greater length of time, a vast amount of water is saved now used in washing out and refilling; also a saving of fuel by saving a boiler of hot water now lost each time in washing out. The time now lost in washing out could be saved and used to move cars, and such mileage made is a clear gain.

Work that now requires one thousand locomotives can be done with nine hundred or less, with greater ease and less expense. Better time is made, and there can be no foaming. Time lost in slowing up from this cause and reduction of load is done away with. The extra number of useful miles that can be made in a year is considerable for one locomotive alone.

The advantages to the mechanical department can be figured to a nicety. But the benefits to the transportation department when in a rush are not so easily stated, but amount to many times more in dollars and cents. The expense of compounds is avoided and any water that comes handy can be used, and the worse it is the better the cleaner will show up, for it can take out the solid matter as fast as it can be fed to the boiler and separated by heat. Heat removes all solid matter in the water by solution, and this matter being very light is kept at the surface when the engine is steaming, and is blown off by the surface skimmer. The solid matter that is held in the water by suspension is heavier and goes more readily to the bottom. It is there easily removed by the mud ring device.

It is the effective heating surface that counts and not the large amount. The more rapid the circulation is the more times it will pass over the heated surface in a given time, and a smaller heating surface with a rapid circulation is better than a larger heating surface with a more sluggish movement of the water.

Fewer tubes in a locomotive boiler would be an advantage. The excessive number now in use impedes the circulation and reduces the water space. The thinner the liquid is the more rapidly it will circulate, and as it becomes thicker its movements will become slower. In a boiler the water starts to thicken from the time the water begins to evaporate and continues to thicken until it is too thick to use.

The cleaner keeps the water as thin as it is possible to make it, and the circulation is always up to its maximum rapidity. The rough scale also retards the movement of the water and reduces the circulation, and the space it occupies lessens the amount of water contained in the boiler. The larger the amount of water in a boiler the more evenly it will steam.

The water in a boiler should be of the same temperature throughout to prevent unequal strains. A good circulation will bring this about.

The cleaners are so placed that they do not interfere in the least in overhauling the engine and need not be removed. The skimmer never interferes in the removal of dry pipe at any time.

Their first cost is almost their last cost, and they will lengthen the life of the boiler, for a clean boiler is a long liver.

The objection to the ordinary surface blow-off is the extravagant waste of water required to produce good results. It requires from three to five blowings per trip, with a waste of two or three gauges of water each time used. This water has been heated and its waste is a waste of fuel, and on a road with a large number of engines the quantity of water and fuel per year required for such a device will cost a large sum of money, and in case of a scarcity of water it could not be used at all.

My first cleaner possessed this defect and I was some time in overcoming it. Two gauges of water were required every time it was used, but since the adoption of the new draw-off head there is no waste of water that can be noticed, so far as can be seen in looking at the water glass while the cleaner is being blown off. The use of the blow-off cock in the leg of the fire box while on the road is dispensed with, and any trouble caused by its sticking is avoided, for it is used only upon leaving and just before entering roundhouse.

The water entering check being cooler than the water in the boiler has a greater specific gravity and travels downward as soon as it enters the boiler, and it is a fact that cold and warm water are poor mixtures, the same as cold and warm air. The gulf stream is a perfect illustration of this. It being warmer than the water of the ocean, it does not mix with the salt water and its banks are as clearly defined in the ocean as the banks of any river on land. The ocean is full of cold currents, and before they will mix with the warmer water of the gulf stream will dive down and pass under.

There are boilers set whose ash-pit doors are closed and the air supply is drawn down the chimney at the same time that the hot gases are escaping and there is no mixing, owing to the difference in their temperature. The fact that water of different degrees does not mix is what causes so much trouble to stay bolts. The water leaving the check travels down to the belly of the boiler and along its bottom to the forward leg and down the forward leg and along the side sheets and up the back head of the boiler over crown sheet and top of flues to the front head, and then down as before, the cooler feed water always following the cooler outside sheets.

It is this fact that makes such a difference in the two fire box sheets by causing a difference in their expansion that cracks stay bolts. If water was delivered to the boiler as hot as that contained in the boiler, which can be done, a great saving would result in boiler repairs. A practical test has demonstrated that live steam taken from the boiler to heat the feed water resulted in a saving of fuel. A heater in the boiler will not fill up when the cleaners are in use and there is a tight check, for it cannot become filled with scale. If used without the cleaner it soon becomes filled up and cannot be fixed without much trouble and expense, and it would always be a source of danger.

The importance of fresh water for steaming purposes cannot be over estimated. The wonderful performance of the *Oregon* was due to fresh water alone. (See the *American Machinist* of August 25th, 1898, on the *Oregon's* wonderful trip.)

The absolute power of an engine depends upon the boiler. Men with only a theoretical knowledge of this subject say it cannot be done. But practical railway men who have used my old cleaner for several years and have investigated the workings of the new one are convinced that it will do all that is claimed for it.

The new Hornish mechanical boiler cleaner is adapted to all classes of boilers, locomotive, stationary, upright, water-tube, marine and traction engine boilers.

DIRECTIONS FOR ITS USE UPON LOCOMOTIVES.

Use surface blow-off after leaving and before entering roundhouse at each end of the division, also once or twice between terminals on each trip, blowing from one to one and a half minutes each time. The best time to blow off the surface cleaner is when the engine is doing the hardest work. Should boiler foam from any cause, use surface blow-off for instant relief. The mud ring device is used only at each end of terminals just before and after leaving the roundhouse.

CONCERNING STAY-BOLTS.*

It is a notorious fact that stay-bolts in water-legs and other similar places are liable to break off without any apparent reason. Sometimes they will last for years, and again they may give way entirely within a month or two. This being the case, it is highly desirable that we should discover the cause of the breakage, if possible, so that we may devise some means of preventing or, at least, of diminishing it as much as possible.

The usual way of detecting broken stay-bolts is by sounding them with a hammer. A stay-bolt that is entirely broken off gives a sound quite different from that given by one that is unbroken, and an experienced man will rarely make a mistake in marking the bolts that are broken off. (It is better to make this examination with the boiler under a slight hydrostatic pressure, so that the broken ends of the bolt may be separated somewhat.) When two inspectors can work upon the same boiler one of them often goes inside of the furnace and holds a sledge against the inner end of each stay-bolt, while the other one taps the bolt on

---

*By permission of The Hartford Steam Boiler Inspection and Insurance Company, of Hartford, Conn.

the outer end. In this way the man inside can tell with greater certainty whether the bolt is broken off or not by the way in which the sledge that he holds responds to the blows. If the bolt is sound the shock is transmitted to the sledge very plainly, but if it is broken off the sledge remains passive and gives little or no response. Still another method, which is often adopted in doubtful cases, consists in the use of the teeth of one of the men in place of the sledge. In applying this method the inspector rests one end of a prick-punch or other similar object against the bolt and brings his front teeth to bear against the free end of the punch. When the stay-bolt is tapped at the other end this affords a very delicate and certain means of detecting a fracture if it exists.

These various methods enable skilled men to detect with al-

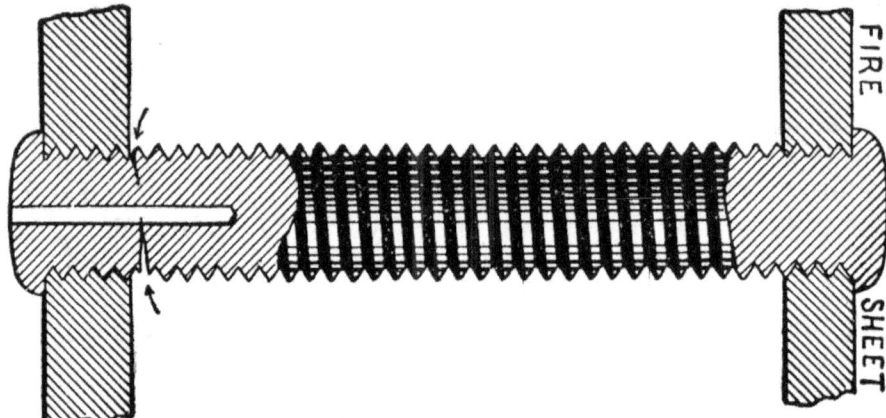

Fig. 1.

most absolute certainty all stay-bolts that are broken entirely off, but they are all indefinite and unsatisfactory when the bolt is only partly broken.

A good man will often detect such a partially broken bolt if the fracture is extensive, but it is evident that there can be no certainty about the test in such cases, and experiment shows, in fact, that the best inspectors cannot detect such partially broken bolts with any precision. This fact is well illustrated in a communication from Mr. T. A. Lawes, presented at the recent meeting of the Railway Master Mechanics' Association: "So far as my investigation goes," said Mr. Lawes, "partially broken stay-bolts are never detected by the old methods, and must be regarded with suspicion. For some years hollow stay-bolts and drilled stay-bolts have been used to a limited extent, but for some reason—or no reason—neither one has been put into general use, although, as I believe, the protection afforded is invaluable. To satisfy myself as to an inspector's ability to detect broken and partially broken stay-bolts, I have had the stay-bolts in 13 locomotives drilled during the past year. The plan adopted was to have the inspector locate all the broken stay-bolts he could find, after which the stay-bolts in the fire-box were drilled, including those marked by the inspector as broken. In the first fire-box tested in this manner the inspector found 39 broken stay-bolts. After drilling and testing under water pressure these were all found to be broken—and in addition to these we found 50 other broken which the inspector was unable to detect by the hammer test. This surprising result led me to examine the broken stay-bolts critically, and I found that those detected by the hammer test were broken entirely off, while the additional ones that were found by drilling holes in the ends were only partially broken off. After testing the stay-bolts in 12 fire-boxes and finding that the ratio of wholly broken stay-bolts to those that were only partially broken ran about the same as in the

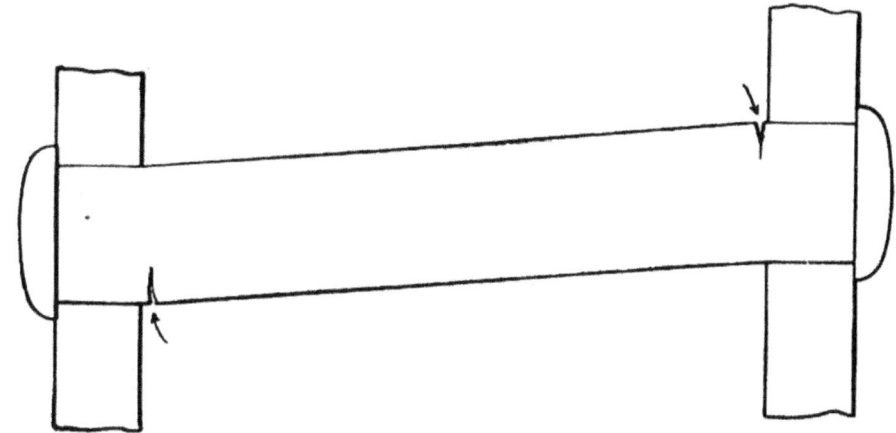

Fig. 2.

first fire-box tested, I concluded to try the method of testing under boiler pressure, and having a helper hold on while the inspector gave the hammer test, but with no better results. I desire, now, to direct your attention particularly to the 13th and last fire-box tested for broken and partially broken stay-bolts by the two methods. I consider it the most severe comparative test of all, from the fact that three inspectors in turn did their level best to locate broken and partially broken stay-bolts by the hammer test, after being informed that the stay-bolts were to be drilled when they had completed their examination. They were given all the time they needed for a careful and accurate inspection. The result was that the hammer test located four broken stay-bolts and the drilling test discovered 46 others that were only partially broken. A careful record of the broken and partially broken bolts detected in 13 engines shows that 440 were discovered by the hammer test and 619 additional ones by the drilling test. To me these facts constitute conclusive evidence that partially broken stay-bolts cannot be detected by the hammer test, and I believe that great risk is run by not drilling th[em] or using hollow stay-bolts, and I am satisfied that either o[f these] precautions will prevent many boiler explosions where we n[ow] hear the verdict, 'Cause of explosion unknown.'"

Fig. 3.

When we take Mr. Lawes' experiments in connection with the probability that partially broken stay-bolts will not be detected by any form of hammer test, we have an amount of evidence which really amounts to a demonstration of the inadequacy of the usual methods for detecting stay-bolts that are fractured, but not wholly parted. One of the most striking things about stay-bolt failures is that the break almost invariably occurs at the outside end of the bolt—that is, at the end which is away from the fire-sheet. Occasionally a bolt may give out elsewhere, but the exceptions are so rare that we can hardly doubt that when they do occur they are determined by some accidental flaw in the metal. We mention this peculiarity of stay-bolt failures because it suggests a cause for such failures, as well as an excellent method of detecting partial fractures when they do occur. We shall discuss the cause of the trouble presently, but before doing so we wish to refer to the way in which partial breakage may be detected by drilling, as Mr. Lawes has suggested in the passage quoted above. This article has been in use a number of years, though it is by no means generally employed, even at the present day. So far as we remember it was first tried about twenty-two years ago by Mr. Wilson Eddy, who was then Master Mechanic for the Boston & Albany Railroad. It consists in drilling a hole, say one-eighth of an inch in diameter, centrally into the end of the stay-bolt for a distance of about one inch or an inch and a quarter, as indicated in Fig. 1. The stay-bolt is of course weakened to some extent by this proceeding, but the loss of strength is slight. Thus it is easy to show that an eighth-inch hole drilled in this way into a stay-bolt that is $\frac{3}{4}''$ in diameter at the base of the thread diminishes the strength of the bolt by less than 3 per cent. Similarly a 3-16″ hole in a stay-bolt that is 1″ in diameter at the base of the thread weakens it only about $3\frac{1}{2}$ per cent. As the stay-bolt invariably breaks just inside of the outer sheet, the hole must be drilled into the outer or visible end of the stay-bolt. It will be seen, from Fig. 1, that as soon as a fracture extends inward so as to reach the hole at any point water will blow out at the end of the stay-bolt, so that attention will be called to the defect at once.

Coming now to the reasons why stay-bolts break so universally at the outer ends, close up to the sheet, we can only offer a tentative explanation, which may have to be revised in the future, though it appears to agree very well with the facts now known. In the first place, we must admit that there is more or less relative movement between the inner and outer sheets of the furnace or water-leg, due to the variations of temperature to which the two are subjected. That this relative motion does take place is not only evident from the nature of the structure, but it has also been proved, experimentally, that it does occur, and a relative motion of more than a quarter of an inch has sometimes been observed. This relative motion would cause the stay-bolts to "rock" back and forth somewhat, as indicated in Fig. 2. If the sheets are both too thick to "buckle" to any sensible degree, a bending strain would be thrown upon the stay-bolt, which would be most severe at its two ends, as indicated by the arrows. The bolts would therefore be bent backward and forward at these points every time the furnace sheets moved relatively to each other, and in the course of time, after these bendings had been repeated often enough, we should probably find cracks started at both ends of the bolt. Even if the bending were not severe enough in itself to cause these cracks it would in all probability be sufficient to keep the fibers of the iron constantly opened up to the action of the water, and the iron would be continuously oxidized at these places, so that the final result would be much the same as though the failure were due directly to the bending strains.

Thus far we have assumed that the two sheets are of equal thickness, and that neither of them is thin enough to "buckle"

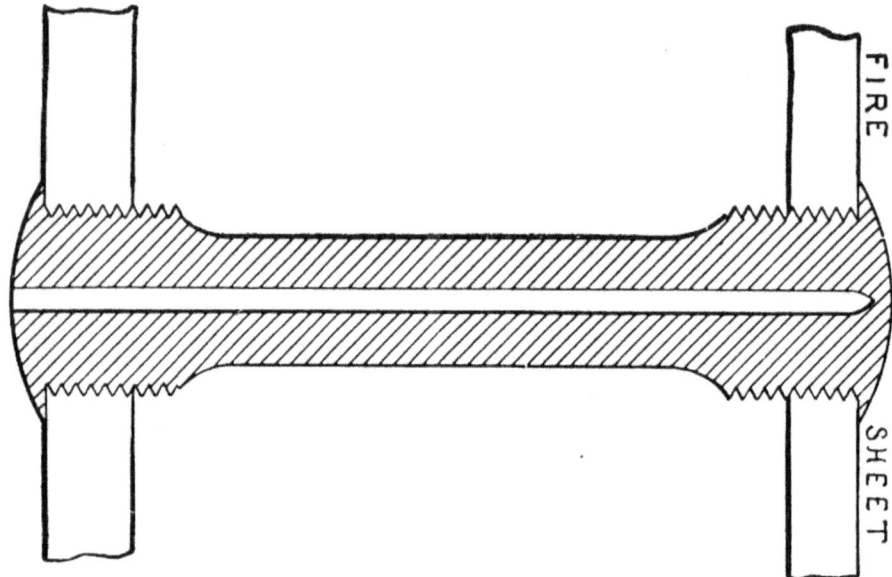

Fig. 4.

to any sensible degree. But in actual boiler construction the fire-sheet is almost invariably considerably thinner than the outside sheet, and we may suppose that the thin inside sheet does "buckle" to some extent under the varying motions of the furnace, as is suggested on a greatly exaggerated scale, in Fig. 3. If such a buckling action takes place the inner ends of the stay-bolts would be thereby relieved of a considerable part of the bending strain that would otherwise come upon them, while the outer ends, being secured to a considerably thicker and stiffer sheet, would be tried just as severely as before. Hence we should expect the failures to occur at the outside ends of the stay-bolts, which is the observed fact.

An objection which is sometimes urged against the form of stay-bolt shown in Fig. 1 is, that although the drilled hole is of no special importance so far as the tensile strength of the bolt is concerned, it does weaken it to some extent, and tends to localize the bending strains at the very point where experience has shown the bolt to be weakest. We do not consider this argument to be of sufficient force to condemn the practice of drilling, for the slightly increased liability of breakage, due to the hole, is far more than counterbalanced by the increased security that the hole affords against bolts which might otherwise be partially fractured without our knowing it.

Various plans have been proposed for preventing the stay-bolts from bending locally near the outside sheet, but most of these involve an initial outlay in building the boiler, which will probably prevent their general adoption. The form of stay-bolt shown in Fig. 4 meets all the requirements of the case without being too expensive for practical use. It was recommended by a committee of the Southern and Southwestern Railway Club in these words: "The committee believes that hollow stay-bolt iron should be used; that the thread should be the U. S. standard, 12 threads per inch; that the inner ends of the holes should not be opened after the stays are headed up, but that the outer ends of the holes should be kept open; that $\frac{7}{8}$" stays should be turned down between the sheets to a scant $\frac{3}{4}$" inch." It will be seen that in this construction there is no local weakening of the bolt, due to the hole. Such weakening as there is extends uniformly along the entire length of the bolt. The object of turning down the central portion of the bolt is to make this portion more flexible than the end of the bolt, so that when the bolt is bent by the motion of the sheets the flexure will be distributed along the entire length of the bolt instead of being concentrated at the outside end.

We are inclined to believe that better results will be obtained by using larger stock and turning down the bolt more in proportion to its diameter than the committee has recommended in the passage cited above. Thus we know of one case in which the stay-bolts of a considerable number of locomotives has given much trouble until they were replaced by others that measured $1\frac{1}{4}$" over the thread, and were turned down to $\frac{3}{4}$" in the central part. None of these new bolts have broken, although they have been in use for a considerable time. (In this case neither set of stay-bolts was drilled.)

MODERN LOCOMOTIVES.

THE LARGEST LOCOMOTIVE IN THE WORLD.

ENGINE 95. UNION RAILROAD.

The magnificent engine shown on the first page of this book is one of two locomotives recently built by the Pittsburgh Locomotive and Car Works for the Union Railroad of Pittsburgh and completed during October, 1898. They are the heaviest locomotives ever built up to the present time, and are notable not only for their size and weight, but for the fine designing, which did not put in one pound of weight that was not necessary for the strength of the parts to which it belonged.

These engines are of the consolidation type and each weigh 230,000 lbs., of which 208,000 lbs. are on the drivers. They have a total heating surface of 3,322 sq. ft., and a grate area of 33.5 sq. ft. The cylinders are 23x32 in. and the drivers are 54 in. in diameter. The engines have a traction power of 53,292 lbs.

These engines were designed to haul ore.

The Union Railway, for which they were built, is a part of the Carnegie system, connecting the Duquesne furnaces, Homestead Steel Works, and Edgar Thomson Steel Works, and extends, nominally, from Munhall to North Bessemer, Pa., a distance of about 12 miles. Some four miles of the line has a grade of 70 ft. to the mile, while for about 2,000 feet (commencing at the yards near the Edgar Thomson Steel Works, and passing up over the line of the Pennsylvania Railroad, and ending at the foot of the 70 ft. incline), there is a grade of 2 4-10 per cent.

The locomotives are being operated daily upon this line, and steam freely, and, so far, appear not to be extravagant in the use of fuel and water.

The cylinders of these engines are of the half saddle type, made heavy, and have great depth, longitudinally. A steel plate 1½ in. thick, and of the same width as the bottom of the saddle, extends across, and is bolted to the lower frames, and to this plate, as well as to the frames, the cylinders are securely fastened. Heavy bolts passing through the top frame bars, front and back of saddle, form additional transverse ties, and relieve the saddle casting from all tensile strains. The longitudinal strains usually transmitted to the cylinders through the frames are largely absorbed by the use of a casting extended from the bumper beam well up to the saddle, securely bolted to the top and the bottom front frame. This casting also acts as a guide for the bolster pin of the truck. This method of relieving cylinders of longitudinal stress was introduced by the Pittsburgh Locomotive Works nearly two years ago, and has proven in practical use, on a large number of locomotives, to be of great value in reducing breakage of saddle castings. The frames are 4½ in. wide and have been cut from rolled steel slabs made by the Carnegie Steel Company and weigh 17,160 lbs. per pair, finished. The main driving wheels are cast steel centers; the centers of the other wheels are cast iron. The journals of the driving wheels are 9x12 inches, and of the truck 6x10 inches. The main crank pin is 7x7 inches. As already mentioned, the size of the cylinder is 23x32 inches, and the piston rod is 4½ inches diameter. The main rod is 9 feet 10½ inches from center to center. The steam ports are 20x1¾ inches and the exhaust ports 3¼ inches wide; the bridge is 1¼ inches wide.

The valves have 1-inch outside lap and a set of 1-16 inch lead in full gear. The travel is 6 inches. The boiler was tested with water of 300 pounds, and it is designed to carry a working pressure of 200 pounds.

Appended are the leading details of these engines:

GENERAL DESCRIPTION.

Type	Consolidation
Name of builder	Pittsburgh Locomotive Works
Name of operating road	Union Railroad
Gage	4 ft. 8½ in.
Kind of fuel to be used	Bituminous coal
Weight on drivers	208,000 lbs.
Weight on truck wheels	22,000 lbs.
Weight, total	230,000 lbs.
Weight of tender, loaded	104,000 lbs.
Weight, total of engine and tender	334,000 lbs.
Tractive power	53,292 lbs.

DIMENSIONS.

Wheel base, total, of engine	24 ft. 0 in.
Wheel base, driving	15 ft. 7 in.
Wheel base, total (engine and tender)	54 ft. 9½ in.
Length, over all, engine	29 ft. 8¾ in.
Length over all, total, engine and tender	65 ft. 3½ in.
Height, center of boiler above rails	9 ft. 3⅜ in.
Height of stack above rails	15 ft. 6 in.
Heating surface, fire-box	205.5 sq. ft.
Heating surface, tubes	3116.5 sq. ft.
Heating surface, total	3322 sq. ft.
Grate area	33.5 sq. ft.

WHEELS AND JOURNALS.

Drivers, number	Eight
Drivers, diameter	54 in.
Drivers, material of centers	Steeled cast iron
Drivers, material main centers	Cast steel
Truck wheels, diameter	30 in.

Journals, driving axle, size.................................... 9x12 in.
Journals, truck axle, size...................................... 6x10 in.
Main crank pin, size... 7x7 in.

CYLINDERS.

Cylinders, diameter.. 23 in.
Piston, stroke .... 32 in.
Piston rod, diameter... 4½ in.
Piston and valve stem packing.................................Metallic
Main rod, length center to center.........................9 ft. 10½ in.
Steam ports, length.. 20 in.
Steam ports, width... 1⅝ in.
Exhaust ports, length.. 20 in.
Exhaust ports, width... 3¼ in.
Bridge, width .... 1½ in.
  Cylinders and valves oiled by sight feed lubricator.

VALVES.

Valves .... Balanced
Valves, greatest travel.. 6 in.
Valves, outside lap.. 1 in.
Valves, inside lap or clearance................................ 0 in.
Valves, lead in full gear...................................... 1-16 in.

BOILER.

Boiler, type of.....................Straight with sloping back end
Boiler, water test... 300 lbs.
Boiler, steam test... 220 lbs.
Boiler, working pressure....................................... 200 lbs.
Boiler, material in barrel.................................Carnegie steel
Boiler, thickness of material in barrel........................ ⅞ in.
Boiler, diameter of barrel at front sheet...................... 80 in.
Boiler, diameter of barrel at throat sheet..................... 83½ in.
Boiler, diameter of barrel at back head........................ 74⅝ in.
Seams, kind of..Horizontal, butt joint, double welted and sextuple riveted
Seams, kind of, circumferential..........................Double riveted
Thickness of tube sheets....................................... ⅝ in.
Crown sheet supported by stays.........................1⅛ in. diameter
Dome, diameter... 32 in.
Safety valves................two 3 in. open pops and one 3 in. muffler
Water supplied through.........................Two No. 11 injectors

TUBES.

Tubes, number 355
Tubes, material ....Knobbled charcoal iron
Tubes, outside diameter.. 2¼ in.
Tubes, length over sheets...................................... 15 ft. 0 in.

FIRE-BOX.

Fire-box, length .... 10 ft. 0 in.
Fire-box, width .... 3 ft. 4¼ in.
Fire-box, depth, front... 76¾ in.
Fire-box, depth, back.. 69 7-16 in.
Fire-box material ....Carnegie fire-box steel
Fire-box, thickness of sheets, crown........................... 7-16 in.

Fire-box, thickness of sheets, sides and back........................ ⅜ in.
Fire-box, thickness of sheets, tube................................ ½ in.
Fire-box, brick arch..............................Supported on studs
Fire-box water space, width...........Front 4 in., sides 4 in., back 4 in.
Grates.................................Cast iron, rocking pattern

SMOKEBOX.

Smokebox, diameter... 83¼ in.
Smokebox, length from tube sheet to end....................... 68⅝ in.

OTHER PARTS.

Exhaust nozzle .... Single
Exhaust nozzle, diameter..................................... 5¾ in.
Exhaust nozzle, distance of tip below center of boiler.......... 5⅝ in.
Netting, size of mesh... 2x2 in.
Stack .... Taper
Stack, least diameter.. 17 in.
Stack, greatest diameter...................................... 18 in.
Stack, height above smokebox............................... 2 ft. 9 in.
Track sander... Pneumatic
Power brake .... Westinghouse-American

TENDER.

Type ....With swivel trucks
Tank capacity, water.................................... 5,000 gallons
Tank capacity, coal.. 10 tons
Kind of material in tank............................... Carnegie steel
Thickness of tank sheets............................. ¼ in. and 5-16 in.
Type of under-frame................................... Steel channels
Type of truck.. Diamond
Truck with rigid bolster..
Type of truck springs............................... Double Elliptic
Diameter of truck wheels..................................... 33 in.
Diameter and length of axle journals.......................... 5x9 in.
Distance between centers of journals........................... 76 in.
Diameter of wheel fit on axle................................. 6⅜ in.
Diameter of center of axle..................................... 5⅜ in.
Length of tender frame over bumpers..................... 22 ft. 11¼ in.
Length of tank... 20 ft. 6 in.
Width of tank.. 9 ft. 8 in.
Height of tank, not including collar........................... 56 in.
Height of tank over collar.................................... 68 in.
Type of back drawhead............................... M. C. B. Coupler

ENGINE 160. GREAT NORTHERN RY.

The handsome engine shown on the second page of this book is one of two Mastodon twelve-wheeled freight locomotives recently built by the Brooks Locomotive Works for the Great Northern Ry.

One of the first things that strikes the beholder is the symmetry in the design of these huge engines. They were designed by Mr. J. O. Pattee, Supt. M. P. of the Great Northern Ry., in collaboration with the staff of the Brooks Locomotive Works;

they were designed for heavy mountain service on 2.2 per cent grades. The total weight of these engines in working order is 212,750 pounds, 172,000 pounds being on the drivers, and previous to the construction of the large Pittsburg engines they were the largest locomotives in the world.

The essential proportions from which we find out the power of the engine are cylinders, 21x34 inches, driving wheels 55 inches diameter, and working boiler pressure 210 pounds per square inch by gauge. Taking 90 per cent of the boiler pressure as reaching the cylinders in starting or on a slow pull, the tractive force is 46,300 pounds. That means that the engine is capable of putting a pull of over 23 tons on the drawbar. It also means that the engine will pull over 7,700 tons on a level, straight track, assuming the resistance to be overcome as not more than 6 pounds per ton of load, included in engine and train.

Men not accustomed to the measuring of power can scarcely grasp the full significance of the figures quoted. It will be more of an object lesson to say that this engine is about as powerful as six of the 15x22-inch locomotives which were so popular only a few years ago.

The boiler of the Player-Belpaire type is $87\frac{1}{4}$ inches diameter at largest part, and 78 inches at smallest ring. The fire-box is 124 inches long and $40\frac{1}{2}$ inches wide, 86 inches deep in front and 79 inches at back end. The boiler has 376 tubes $2\frac{1}{4}$ inches diameter, 13 feet $10\frac{3}{4}$ inches long, these with fire-box giving a total heating surface of 3,280 square feet.

The boiler, dome, steam chests, cylinders and cylinder saddles are lagged with Sal Mountain asbestos in order to reduce radiation and condensation to a minimum, as the engines are to be operated in the mountains of Montana, where the temperature frequently descends in inverse ratio of the altitude. The nozzles are set 30 inches in front of flue sheet and $36\frac{1}{4}$ inches from the door. The improved Bell spark arrester is used.

The valves are of improved piston type and absolutely balanced. They are 16 inches in diameter, as large as the pistons of many locomotives still in service. The piston rods are hollow, $4\frac{1}{2}$ inches diameter, and are extended through the front head. The extension is $3\frac{1}{2}$ inches diameter and is covered with extra heavy wrought iron tube.

The link radius is 40 inches, eliminating the bad work of a bent eccentric rod. The center of the valve rod is $3\frac{1}{2}$ inches above upper rocker pin, for the purpose of giving lateral clearance between the rod and the front driving tire. The rod has a knuckle joint and is dropped again at its junction with the stem. The steam ports are $18 \times 1\frac{3}{4}$ inches, and the exhaust port is 50x9 inches. The bridge is $6\frac{3}{4}$ inches wide. The valves have $1\frac{1}{8}$ inches outside lap and $\frac{1}{8}$ inch inside clearance, with $6\frac{1}{2}$ inches maximum travel.

To control the tremendous force transmitted from the cylinders of these engines the main frame is forged solid 5 inches wide and 5 inches deep at the jaws and 4 inches deep at intermediate points of the top. The lower part is 3¼ inches thick at the jaws and 2½ inches thick between jaws. The journals of the driving axles are 9x11 inches, those of the engine truck 5½x12 inches. The main rod bearing is 6½x6½ inches; side rod bearings, 7⅜x5 inches; wheel fit, 7⅞x7 13-16 inches. The driving-wheel centers are of cast steel, made by Pratt & Letchworth, and Krupp tires are used for engine and tender.

The engine is equipped with Crosby safety valves, Monitor injectors, Nathan lubricators, Jerome gland packing, Leach sanders, New York air brake, French springs and Curran chime whistle.

Every care has been taken to make the engine as light as possible consistent with the required strength. To this end wheel centers, engine truck wheel centers, driving boxes, driving box saddles, spring fulcrums, pistons, cylinder heads (front and back), crossheads and guide yoke ends are made of cast steel. Cylinder head casings, smoke box front and door, smoke-stack base, dome casing and sand-box casing are of pressed steel. Steam pipes of malleable iron, piston rods of O. H. steel, hollow and extended in front, crank pins of O. H. steel, crosshead pins of O. H. steel and hollow, rockers of hammered iron and hollow, rods of hammered iron, reverse lever, Player patent.

The heavy weight, the long cylinder stroke (34 in., the longest that we have ever known to be used in locomotive practice), the enormous heating surface, and the combination of these and other brobdignagian dimensions in a finished machine which still possesses the fine symmetry as revealed in our half-tone perspective of the completed engine—all these combine to make this a truly remarkable engine. The results of its performance in service will be noted with no inconsiderable interest. The fact that it will go into a service which will, as we understand it, be to a considerable extent comparable with that which the heavy compound mastodons are now meeting on the Northern Pacific, will tend to even more closely center attention upon the performance of these later engines.

GENERAL DIMENSIONS, ETC.

How many and dates of delivery	Two. Dec., 1897
Gauge	4 ft. 8½ in.
Wheel base, total, of engine	26 ft. 8 in.
Wheel base, driving	15 ft. 10 in.
Wheel base, total, engine and tender	54 ft. 3¼ in.
Light over all, engine	41 ft. 4 in.
Light over all, total, engine and tender	64 ft. 1½ in.
Height, center of boiler above rails	9 ft. 5 in.
Height of stack above rails	15 ft. 6 in.

Heating surface, fire-box... 235 sq. ft.
Heating surface, tubes... 3,045 sq. ft.
Heating surface, total... 3,280 sq. ft.
Grate area... 34 sq. ft.

WHEELS AND JOURNALS.

Drivers, number .... Eight
Drivers, diameter... 55 in.
Drivers, material of centers... Cast steel
Truck wheels, diameter... 30 in. centers cast steel spoke
Journals, driving axle, size... 9 x11 in.
Journals, truck axle, size... 5½x12 in.
Main crank pin, size { Main rod bearing, 6½x6½ in.; coupling rod bearing 7⅜x5 in.; wheel fit 7⅞ in. diam. 7 13-16 in.

CYLINDERS.

Cylinders, diameter .... 21 in.
Piston, stroke... 34 in.
Piston rod, diameter... 4¼ in.
Kind of piston-rod packing... Jerome
Main rod, length center to center... 8 ft. 10 in.
Steam ports, length... 18 in.
Steam ports, width... 1¾ in.
Exhaust ports, length... 50 in.
Exhaust ports, width... 9 in.
Bridge, width .... 6¾ in.

VALVES.

Valves, kind of... Piston
Valves, greatest travel... 6½ in.
Valves, outside lap... 1⅛ in.
Valves, inside lap or clearance... ⅛ in. clearance
Valves, lead in full gear... 0
Valves, lead .... Variable

BOILER.

Boiler, type of..... Player Patent Improved Conical Connection Belpaire
Boiler, steam working pressure... 210 lbs.
Material in barrel... steel
Thickness of material in barrel... ⅞ and 15-16 in.
Diameter outside, smallest... 78 in.
Diameter outside, largest... 87⅛ in.
Seams, kind of, horizontal... sextuple lap
Seams, kind of, circumferential... triple lap
Thickness of tube sheet... ¾ in.
Crown sheet stayed with... improved system direct stays
Dome, diameter... 30 in.

FIREBOX.

Type... Horizontal over frames
Length... 10 ft. 4 in.
Width... 3 ft. 4½ in.
Depth front... 86½ in.
Depth back... 79 in.

Material .... Steel
Thickness of inside sheets.. ⅜ in.
Brick arch .... none
Mud ring............................... Riveting double, thickness 4 in.
Water space at top.................. Front 4 in., sides 7 in., back 5 in.
Grate, kind of.. Cast iron rocking

SMOKE BOX.

Diameter outside .... 81 in.
Length from flue sheet... 67 in.

OTHER PARTS.

Exhaust nozzle.................... Single, permanent, 6 in. below center
Exhaust nozzle, diameter........................ 5⅛ in., 5⅜ in., 5⅝ in.
Netting, type of........................... Improved Bell spark arrester
Stack .... taper
Stack, least diameter.. 15⅞ in.
Stack, greatest diameter... 18 in.
Stack, height above smoke box................................. 2 ft. 8½ in.

TENDER.

Type .... Swivel trucks
Tank capacity for water.. 4,670 gal.
Coal capacity .... 10 tons
Kind of material in tank... Steel
Type of under-frame............................... 10 in. channel steel
Type of truck..................... 4 wheeled Great Northern Standard
Truck .... Rigid
Type of truck spring.. ½ elliptical
Diameter of truck wheels.. 33 in.
Diameter and length of axle journals............................ 4¼x8 in.
Distance between centers of journals......................... 6 ft. 3 in.
Diameter of wheel fit on axle..................................... 5⅜ in.
Diameter of center of axle... 4¾ in.

NAMES OF MAKERS OF SPECIAL EQUIPMENT.

Wheel centers.. Pratt & Letchworth
Tires .... Krupp
Axles .... Pennsylvania Steel Co. billets
Truck wheels, tender............... Krupp No. 4, engine, B. L. W.
Sight-feed lubricators .... Nathan
Couplers .... B. L. W.
Safety valves .... Crosby
Boiler covering .... Sal Mountain Asbestos
Sanding device........................... 1 pair Leach double sanders

PASSENGER ENGINE 1101. CHICAGO, ROCK ISLAND & PACIFIC RY.

The illustration on page 479 shows one of three engines recently built by the Chicago, Rock Island & Pacific Ry. at their Chicago shops. These engines were designed by Mr. George F. Wilson, Superintendent of Motive Power and Equipment of that road, especially to meet the requirements of fast passenger service. They are very large engines and are generously treated as to steam space, steam passages, heating surface, cylinder dimensions, etc., the cylinders being 19½"x26". The dry pipe is 8" in diameter and the steam ports are 1½x20", and the exhaust ports 3½"x20", and a 5" nozzle is used. The steam passages to the cylinders are insulated with air spaces, and magnesia covering is provided for chests and heads. Large Richardson balance valves are used which weigh 161 lbs. each. As may be seen, the engine is of clean, symmetrical design. The 61" boiler, wagon top, radially stayed, though designed by Mr. Wilson, was built for him by the Brooks Locomotive Works. The 9 ft. fire-box is carried between the frames.

A feature of the engine is its drivers, especially designed by Mr. Wilson. They are 72 in. over centers and 78 in. over tires. They are of cast steel. Each main center weighs 2,461 lbs.; each back center 2,346 lbs. There is a cast iron disc let into the hub for a bearing surface against the boxes. The bearing surfaces of the latter are babbitted.

The lifter shaft, rocker arms, and cross-heads are of cast steel, and the guides of cast iron. The eccentric straps are of brass, weighing 90 lbs. per pair. The tank capacity is 4,000 gallons, and carries 8 tons of coal.

An idea of the size of these engines can be seen by the mail car attached, which is of standard size.

The main dimensions of these engines are as follows:

Cylinders	19½x26 in.
Driving wheel centers	72 in.
Diameter of boiler	61 in.
Diameter of back head of boiler	71¼ in.
Fire-box, length	9 ft.
Fire-box, width	2 ft. 8⅞ in.
Fire-box height, front	6 ft. 7¾ in.
Flues, outside diameter	2 in.
Flues, number	296
Flues, length	11 ft. 7 in.
Boiler pressure	190 lbs.
Grate surface	24.5 sq. ft.
Heating surface, in tubes	1795 sq. ft.
Heating surface, in fire-box	193.3 sq. ft.
Heating surface, total	1988.3 sq. ft.
Steam ports	1½x20 in.
Exhaust ports	3½x20 in.
Travel of valve	6 in.
Eccentric throw	6 in.
Outside lap	1⅛ in.
Inside lap	0
Height top of rail to top of stack	15 ft. 6 in.
Height top of rail to center of boiler	8 ft. 5¼ in.
Total weight of engine	125,000 lbs.
Weight on driving wheels	83,000 lbs.
Weight on truck	42,000 lbs.
Weight of tender, loaded	78,000 lbs.
Driving wheel base	8 ft. 6 in.
Length of frames	30 ft. 9¾ in.

Passenger Engine No. 1101, Chicago, Rock Island & Pacific Ry.
Built by the Railroad Company.
General Dimensions Given on Page 478.

The boiler is of carbon steel, the coal gate in tender is of A. T. & S. F. design; the driver centers are from the American Steel Company; the standard C. R. I. & P. shaking grate is used; Page wheels are used on the trucks and tender; the McIntosh blow-off cock (illustrated elsewhere in this book) is used; the Westinghouse air brake is used throughout, the American going on the drivers.

PASSENGER ENGINE 385. PHILADELPHIA & READING RAILROAD.

The single driver, Baldwin "Vauclain" compound, illustrated on page 333 represents another radical departure from well established American practice of locomotive construction. Single drivers have been used in this country before, but they were never so popular as in England, and this is the first compound single driver built in America. This engine was built by the Baldwin Locomotive Works in June, 1895, from designs by Mr. L. B. Paxon, Supt. M. P. of the Philadelphia & Reading Railroad, who had previously experimented by removing the side rods from an eight-wheeled engine; by increasing the weight on the single pair of drivers it was his intention to secure sufficient adhesion to start an ordinary train, while the advantages gained by dispensing with the friction of an extra pair of wheels and the side rods would add materially to its speed. This engine has been in service since July 3, 1895, hauling their fast trains between New York and Philadelphia, and she has given such good results that her double was ordered and built and is now sharing honors with the 385. This engine has drivers $84\frac{1}{4}$" in diameter with a trailer $54\frac{1}{4}$" in diameter. The cylinders are of the Vauclain system of compounding and are 13" and 22" in diameter with a 26" stroke. The boiler is of the Wooten type, 56" in diameter at smallest ring; made of $\frac{5}{8}$" steel and carries a working pressure of 200 pounds. The fire box is 114x96" and has 324 tubes $1\frac{1}{2}$" in diameter. Total weight of engine, 115,000 pounds, with 48,000 pounds on the drivers. The driving journals are $8\frac{1}{2}$x12". From the position of the cab the reverse lever is attached direct to the lifting shaft. The tank is equipped with a water scoop. This engine has made several extraordinarily fast runs, which are given on page 506.

PASSENGER ENGINE 101. PENNSYLVANIA RAILROAD.

DIMENSIONS, WEIGHT, ETC.

Cylinders	$18\frac{1}{2}$x26 in
Driving wheels	80 in
Steam pressure	185 lbs. per sq. in
Weight on drivers	93,100 lbs
Total weight of locomotive in working order	134,500 lbs
Weight of tender loaded	82,000 lbs

Passenger Engine No. 15, Buffalo, Rochester & Pittsburg Ry.

PASSENGER ENGINE 15. BUFFALO, ROCHESTER & PITTSBURG RY.

The handsome eight-wheel locomotive shown by the half-tone engraving on page 481 bears eloquent testimony to the advances made within the last year by the builders, the Dickson Locomotive Works, of Scranton, Pa. The engine is one of an order given by the Buffalo, Rochester & Pittsburgh Railway, and is intended for passenger service. The cylinders are 18x24 inches, drivers 68 inches diameter, and boiler 58 inches diameter at smallest ring. There are 256 2-inch tubes. The fire-box is 96x42 inches, and provides 138.32 square feet of heating surface. The total heating surface is 1,784 square feet, and the grate area is 28 square feet. The working boiler pressure is 180 pounds. The weight in working order is 118,200 pounds, 77,300 pounds of that resting on the driving wheels.

At the last convention of the Railway Master Mechanics' Association a very valuable report was submitted on "The proper ratio of heating surface and grate area to cylinder volume," in which rules were given for the proportions of a locomotive that would approximate to the most successful engines in service. This new Dickson locomotive, whether by design or accident, comes very closely to the rules laid down by the committee referred to, except in the size of steam and exhaust ports, an exception which proves the good sense of the designers. The following table gives a comparison of the leading ratios of what we consider the best form of bituminous-burning, eight-wheel passenger engine mentioned by the Master Mechanics' Committee (No. 23 in table G, Master Mechanics' Report) and those of the Dickson locomotive:

	M. M. Locomotive.	Dickson Locomotive.
Weight on drivers	79,000 lbs.	77,300 lbs.
Size of cylinders	18x24 ins.	18x24 ins.
Ratio of grate to cubic feet of cylinder volume	3.39	3.96
Ratio of grate area to area through tubes	6.19	5
Ratio of heating surface to cylinder volume	210	252.7
Ratio of heating surface to grate area	62	63.7
Ratio of firebox heating surface to total heating surface	10	12.9
Ratio of tractive force to adhesive weight	0.21	0.226

These particulars show that the Dickson locomotive has a margin of efficiency in all respects over what may be called the ideal engine described in the Master Mechanics' report.

PASSENGER ENGINE 111. LOUISVILLE, NEW ALBANY & CHICAGO R. R.

DIMENSIONS, WEIGHT, ETC.

Type .... 8-wheeled passenger
Class ....

Passenger Engine No. 111, Louisville, New Albany & Chicago R.R.
Brooks.
General Dimensions Given on Page 482.

Gauge .... 4 ft. 8½ in
Fuel .... Bituminous coal
Diameter and stroke of cylinders................................. 18½x24 in
Diameter of driving wheels... 72 in
Diameter of truck wheels.. 33¼ in
Driving wheel base... 8 ft. 6 in
Engine wheel base.. 23 ft. 5 in
Total wheel base.. 48 ft. 6 in
Diameter of boiler... 62 in
Number, diameter and length of tubes.............. 300—2 in. x 11 ft. 7 in
Length and width of firebox................................. 97 and 41 in
Boiler pressure.. 190 lbs
Weight on drivers... 79,000 lbs
Weight on truck... 42,800 lbs
Total weight... 121,800 lbs
Tank capacity.. 4,000 gallons

PASSENGER ENGINE NO. 250. PITTSBURG & WESTERN RY.

DIMENSIONS, WEIGHT, ETC.

Type .... Ten wheel
Fuel .... Bituminous coal
Gauge of track... 4 ft. 8½ in
Total weight of engine in working order...................... 145,000 lbs
Total weight of engine on drivers................................ 113,000 lbs
Driving wheel base of engine....................................... 13 ft. 8 in
Total wheel base of engine... 24 ft. 6 in
Total wheel base of engine and tender....................... 51 ft. 9½ in
Height from rail to top of stack.................................. 14 ft. 7¼ in
Cylinders, diameter and stroke.................................. 20x26 in
Piston rods... Steel, 3½ in. diameter
Type of boiler.. Extended wagon top
Diameter of boiler at smallest ring............................ 60 in
Diameter of boiler at back head............................... 71⅛ in
Crown sheet supported by radial stays.................... 1⅛ in. diameter
Stay bolts.................. 1 in. diameter, spaced 4 in. from center to center
Number of tubes.. 231
Diameter of tubes... 2¼ in
Length of tubes over tube sheet................................ 14 ft. 6 in
Length of fire box, inside.. 108 in
Width of fire box, inside.. 42 in
Working pressure.. 180 lbs
Kind of grates... Cast iron, rocking
Grate area... 31.5 sq. ft
Heating surface in tubes.. 1,961.3 sq. ft
Heating surface in fire box... 158.0 sq. ft
Total heating surface... 2,119.3 sq. ft
Diameter of driving wheels outside of tires............... 68 in
Diameter and length of journals................................ 8x10 in
Diameter of truck wheels.. 32 in
Diameter and length of journals................................. 5x10 in
Slide valves... American balance
Type of tank... Level top
Water capacity of tank... 4,000 U. S. gallons
Fuel capacity of tank.. 280 cu. ft
Weight of tender with fuel and water........................... 77,000 lbs
Type of brakes.. Westinghouse American
Train Signal.. Westinghouse

Passenger Engine No. 250, Pittsburg & Western Ry.
Pittsburg.

Passenger Engine No. 376. Illinois Central Railroad.
Rogers.
General Dimensions Given on Page 488.

Passenger Engine No. 1818, Southern Pacific of California.
Cocke.
General Dimensions Given on Page 488.

487

PASSENGER ENGINE 376. ILLINOIS CENTRAL RAILROAD.
DIMENSIONS, WEIGHT, ETC.

Gauge of track	4 ft. 8½ in
Total weight of engine	144,500 lbs
Total weight on drivers	109,000 lbs
Wheel base, drivers	13 ft. 3 in
Wheel base, total engine	24 ft. 3½ in
Wheel base, total engine and tender	50 ft. 7½ in
Height of stack	15 ft
Heating surface, firebox	177 sq. ft
Heating surface, total	2,031 sq. ft
Grate area	28 sq. ft
Boiler tubes, o. d. 2 in. 264	13 ft. 5 in
Driving wheels	69 in., cast steel centers
Driving journals	8x10½ in
Cylinders, diameter	19½x26 in
Steam ports	1½x20 in
Exhaust ports	3x20 in
Bridge	1½ in
Valves	American balanced
Valves, maximum travel	5½ in
Valves, outside lap	1 in
Valves, inside clearance	1-16 in
Valves, lead in full gear	1-16 in
Nozzles	Single
Boiler	Belpaire
Boiler, diameter of boiler outside first course	66 in
Tender frame	White oak
Capacity of tender—water	4,200 gallons
Capacity of tender—coal	6 tons

PASSENGER ENGINE 1818. SOUTHERN PACIFIC OF CALIFORNIA.
DIMENSIONS, WEIGHT, ETC.

Diameter of cylinders	19 in
Stroke of cylinders	24 in
C to C	86 in
Ports	18 in
Diameter of boiler	58 in
Boiler tubes	246, 2 in. x 12 ft. 6 in
Fire box	96x42 in
Total engine and tender wheel base	47 ft. 10 in
Total wheel base of engine	22 ft. 6 in
Diameter of driving wheels	54 in
Center driving wheel	48 in
Tire	3 in
Axle	7½ in
Engine truck wheels	Four
Engine truck wheels	26 in
Axle	5½ in
Engine truck center	Swivel

THE NEW PENNSYLVANIA, CLASS H 5, CONSOLIDATION ENGINE.

The Pennsylvania Railroad Company have recently built a very large freight locomotive at the Juniata shops at Altoona. It has been in service only a short while, but is showing evidence of splendid working ability. In appearance it so closely resembles the other large engines that we have illustrated that we shall simply give its general dimensions:

Weight in working order (est.)	218,000 pounds
Weight of tender, loaded	92,600 pounds
Capacity of cistern	6,000 gallons
Number of pairs of driving wheels	Four
Diameter of drivers	56 inches
Size of driving axle journals	9x13 inches
Length of driving wheel base	17 feet 6 inches
Total wheel-base of engine	25 feet 11½ inches
Number of wheels on engine truck	Two
Diameter of wheels on engine truck	30 inches
Spread of cylinders	90 inches
Size of cylinders	23½x28 inches
Travel of valve	6 inches
Type of boiler	Belpaire
Minimum internal diameter of boiler	70½ inches
Number of tubes	369
Outside diameter of tubes	2 inches
Length of tubes between tube sheets	168 inches
Size of firebox, inside	40x120 inches
Fire-grate area	33 1-3 square feet
Total heating surface of boiler	2,977 square feet
Steam pressure	185 pounds
Tractive power per pound of M. E. P.	288

A radical departure from Pennsylvania standards is found in the H5 cylinders. These are bolted to the saddle, which is a single casting. Ball joints between the cylinders and saddle prevent the escape of steam. This method of fastening the cylinders will allow a variety of repairs to be made without shopping the engine.

Another new feature is the position of the injectors, which are placed at the head of the boiler, above the fire-box door, instead of at the back of the cab.

In the preliminary tests made on the Pittsburg division, it was found that the H5 could handle 578 tons light, and 643 tons loaded. The record of the standard Class R freight engine is 350 tons light and 383 tons loaded, and for the Mogul, 375 tons light and 433 tons loaded. It will be seen from this that the class H5 will eclipse any other locomotive on the Pennsylvania system.

The engine is too big to be turned on any of the Altoona turntables, and she therefore is used as a helper above Altoona, backing down the mountain.

CONSOLIDATION ENGINE 700. CLEVELAND, CINCINNATI, CHICAGO & ST. LOUIS RY.

DIMENSIONS, WEIGHT, ETC.

Gauge of track	4 ft. 8½ in
Total weight of engine in working order	150,500 lbs
Weight on drivers	134,650 lbs
Weight of tender loaded	80,000 lbs. (about)

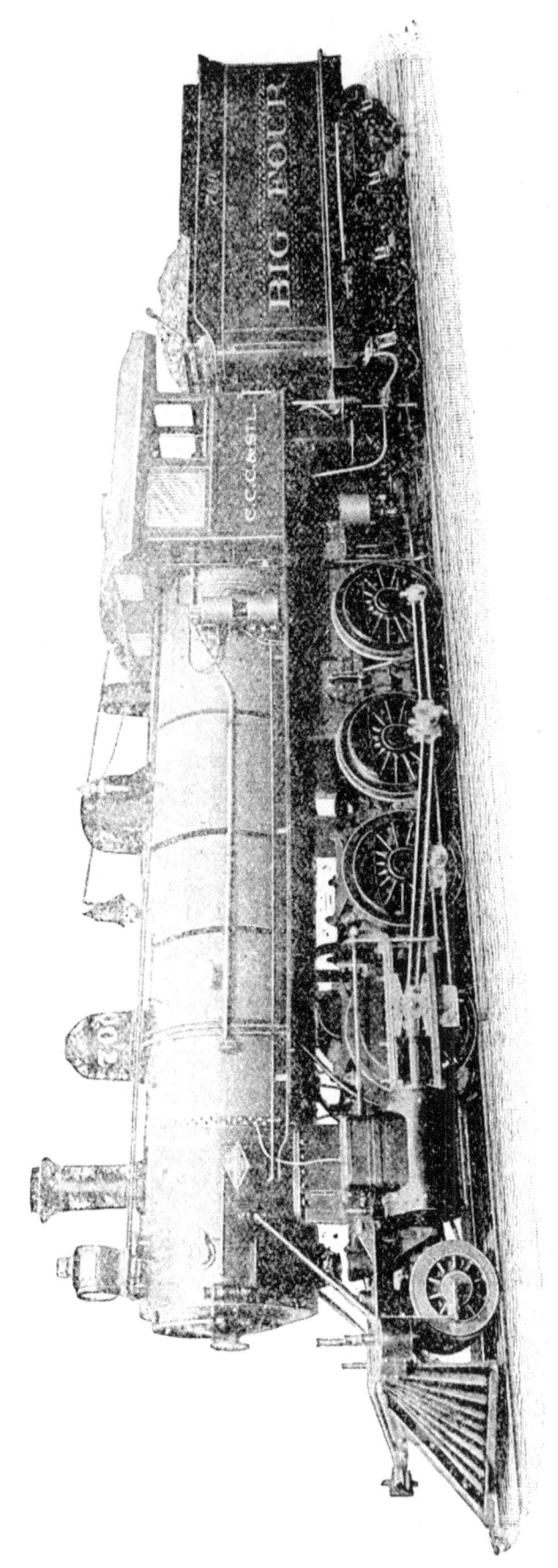

Consolidated Engine No. 700, Cleveland, Cincinnati, Chicago & St. Louis Ry.
Richmond.
General Dimensions Given on Page 429.

Consolidated Engine No. 1278, Baltimore & Ohio Railroad.
Baldwin.
General Dimensions Given on Page 492.

Cylinder diameter and stroke.................20x26 in
Driving wheels diameter........................51 in
Journals.....................................8½x11 in
Diameter boiler, 1st course......................64 in
Total heating surface.......................2,431 sq. ft
Tube heating surface........................2,260 sq. ft
Grate area..................................3,523 sq. ft
Boiler pressure................................190 lbs
Total wheel base engine....................23 ft. 8½ in
Driving wheel base engine......................16 ft
Tank capacity..............................4,500 gallons

CONSOLIDATED ENGINE 1278. BALTIMORE & OHIO RAILROAD.

DIMENSIONS, WEIGHT, ETC.

CYLINDERS.

Diameter ....21 in
Stroke ....26 in
Valve .... Balanced

BOILER.

Diameter ....60 in
Thickness of sheets.....................⅝ and 11-16 in
Working pressure..............................180 lbs
Fuel ....Soft Coal

FIRE BOX.

Material .... Steel
Length ....116 in
Width ....42 in
Depth, front................................68¾ in
Depth, back.................................66¼ in
Thickness of sheets, sides.....................⅜ in
Thickness of sheets, back......................⅜ in
Thickness of sheets, crown...................7-16 in
Thickness of sheets, tube......................½ in

TUBES.

Number 214
Diameter ....2¼ in
Length ....14 ft. 2½ in

HEATING SURFACE.

Fire box..................................179.91 sq. ft
Tubes ....1,780.26 sq. ft
Total ....1,960.17 sq. ft
Grate area................................33.83 sq. ft

DRIVING WHEELS.

Diameter, outside..............................50 in
Diameter of center.............................44 in
Journals...............................7½ in. x 9 in

ENGINE TRUCK WHEELS.

Diameter ....30 in
Journals ....5 in. x 8¾ in

WHEEL BASE.

Driving	15 ft. 2 in.
Total engine	23 ft. 2 in
Total engine and tender	51 ft. 1 in

WEIGHT.

On drivers	about 137,000 lbs
On truck	about 13,000 lbs
Total engine	about 150,000 lbs
Total engine and tender	about 234,000 lbs

TENDER.

Tender wheels, diameter	33 in
Journals	4¼ in. x 8 in
Tank capacity	3,500 gallons
Weight, empty	41,500 lbs

SERVICE.

Heavy freight.

COMPOUND FREIGHT ENGINE 62. NORFOLK & WESTERN RY. CO.

SHOWN ON PAGE 349.

DIMENSIONS, WEIGHT, ETC.

Gauge	4 ft. 9 in
Cylinders	23, 35x32 in
Drivers	56 in
Total wheel base	24 ft. 6 in
Driving wheel base	15 ft. 6 in
Weight, total, about	186,000 lbs
Weight on drivers, about	166,000 lbs
Boiler, diameter	68 in
Number of tubes	306
Diameter of tubes	2¼ in
Length of tubes	14 ft. 6 in
Fire box, length	121 in
Fire box, width	41¾ in
Heating surface, fire box	195 sq. ft
Heating surface, tubes	2,593.96 sq. ft
Heating surface, total	2,788.96 sq. ft
Tank capacity	4,000 gallons

CONSOLIDATED ENGINE 969. ATCHISON, TOPEKA & SANTA FE R. R.

DIMENSIONS, WEIGHT, ETC

Weight on driving wheels	143,500 lbs
Total weight of engine in working order	160,000 lbs
Driving wheel base	15 ft. 2 in
Total wheel base of engine	23 ft. 3 in
Total wheel base of engine and tender	50 ft. 7½ in
Diameter of cylinders	21 in
Stroke of piston	28 in
Diameter of driving wheels	57 in
Diameter of engine truck wheels	30 in
Working boiler pressure	180 lbs
Diameter of boiler at front end	66¾ in
Length of fire box	102 1-16 in

Consolidated Engine No. 969, Atchison, Topeka & Santa Fe Ry.
Dickson.
General Dimensions Given on Page 493.

Consolidated Engine No. 93, Delaware & Hudson Canal Co.
Dickson.
General Dimensions Given on Page 496.

Width of fire box...41¼ in
Number of tubes..233
Length of tubes..13 ft. 10 in
Diameter of tubes..2 in
Tube heating surface.................................1,687.15 sq. ft
Fire box heating surface.............................153.60 sq. ft
Total heating surface................................1,840.75 sq. ft
Grate area...29.23 sq. ft
Tank capacity..5,000 gallons

CONSOLIDATION ENGINE 93. DELAWARE & HUDSON CANAL COMPANY.

DIMENSIONS, WEIGHT, ETC.

Weight on driving wheels...........................111,000 lbs
Total weight of engine in working order........135,000 lbs
Driving wheel base....................................14 ft. 9 in
Total wheel base of engine..........................22 ft. 11 in
Total wheel base of engine and tender..........49 ft. 1 in
Diameter of cylinders....................................20 in
Stroke of piston..24 in
Diameter of driving wheels..............................50 in
Diameter of engine truck wheels.......................32 in
Working boiler pressure..............................160 lbs
Diameter of boiler..60 in
Length of fire box......................................120 in
Width of fire box...96 in
Number of tubes..248
Length of tubes..11 ft. 4 in
Diameter of tubes...2 in
Tube heating surface................................1,685.25 sq. ft
Fire box heating surface..............................201.93 sq. ft
Total heating surface................................1,887.18 sq. ft
Grate area..80 sq. ft
Tank capacity..3,000 gallons

PASSENGER ENGINE 908. CHICAGO & NORTHWESTERN RY.

DIMENSIONS, WEIGHT, ETC.

Gauge ....4 ft. 8½ in
Weight, total..125,600 lbs
Weight on drivers......................................78,000 lbs
Weight, tender (empty).............................39,670 lbs
Wheel base, driving....................................8 ft. 6 in
Wheel base, total.......................................23 ft. 1 in
Drivers, diameter (O. S. of tire).......................75 in
Driving journals.............................8½ in. dia. x 11½ in. long
Cylinders ....19 in. dia., 24 in. stroke
Steam ports..............................1⅝ in. wide x 16 in. long
Exhaust ports...........................3 in. wide x 16 in. long
Bridges ....1⅜ in. wide
Valves, kind.........................Allen—American balance
Valves, greatest travel..................................5¾ in
Valves, lap........................outside, 1¼ in.; inside clearance, ¼ in
Valves, lead in full gear..................................— in
Exhaust nozzles..........................4¾, 5 and 5¼ in. dia
Exhaust, style...........................Company's style, single
Boiler, style.............................Extended wagon top
Boiler sheets.............................9-16, ⅝, 11-16, ½ and 7-16 in
Boiler, O. D. 1st ring....................................62 in

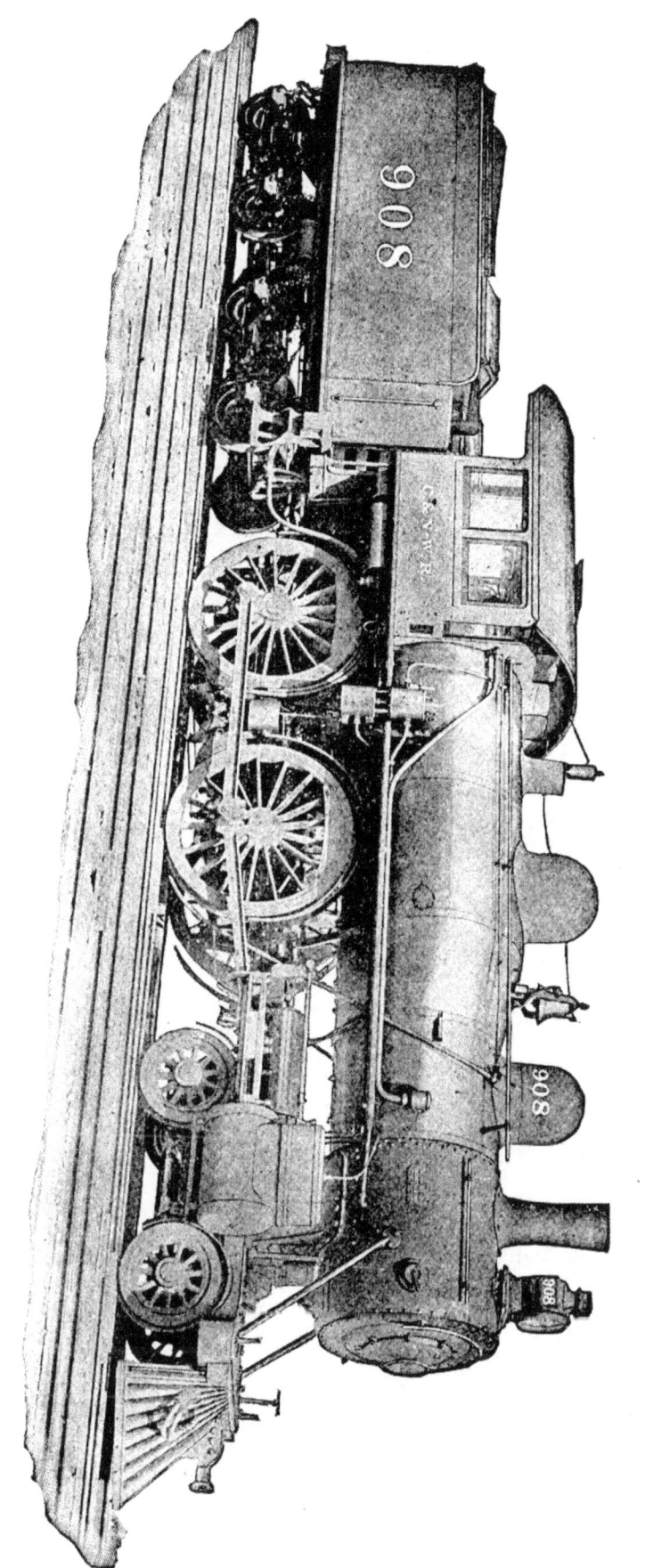

Engine No. 908, Chicago & North-Western Ry.—Passenger.
Schenectady.

Boiler working pressure.....................................190 lbs
Heating surface, tubes..................................1,763.4 sq. ft
Heating surface, water tubes.............................14.4 sq. ft
Heating surface, fire box...............................158.7 sq. ft
Heating surface, total................................1,936.5 sq. ft
Grate surface..26.96 sq. ft
Boiler tubes...No. 295, 2 in. dia
Tender frame.......................................10 in. steel channel
Tender water capacity............................4,500 U. S. gallons
Tender coal capacity...........................8 (2,000 lb.) tons
Brakes...................W. A. A. B. on drivers, tender and for train
Sand blast...Leach D-1 double

MOGUL FREIGHT LOCOMOTIVE NO. 862.—GRAND TRUNK RY.

From designs supervised by Mr. F. W. Morse, Superintendent of Motive Power of the Grand Trunk Railway, the Schenectady Locomotive Works have recently built for that road six mogul freight locomotives. These engines are just entering service, and we understand that their initial performance gives excellent promise of exceptionally satisfactory results. Our excellent photographic view of one of these engines on page 499, together with the appended data, afford a fair idea of the general design and detail followed:

GENERAL DIMENSIONS.

Gauge..4 ft. 8½ in
Fuel..Bituminous coal
Weight in working order............................152,850 lbs
Weight on drivers..................................127,650 lbs
Wheel base, driving.................................15 ft. 8 in
Wheel base, rigid...................................15 ft. 8 in
Wheel base, total...................................24 ft. 1 in

CYLINDERS.

Diameter of cylinders....................................20 in
Stroke of piston...26 in
Horizontal thickness of piston.................5⅛ in. and 5⅝ in
Diameter of piston rod..................................3¾ in
Kind of piston packing........................Cast iron rings
Kind of piston rod packing..............United States Metallic
Size of steam ports.............................20 in. x 1⅝ in
Size of exhaust ports............................20 in. x 3 in
Size of bridges...1⅝ in

VALVES.

Kind of slide valves..................Railway company's style
Greatest travel of slide valves..........................5½ in
Outside lap of slide valves..............................⅞ in
Inside clearance of slide valves........................1-16 in
Lead of valves in full gear....................Line and Line
Kind of valve stem packing.............United States Metallic

WHEELS, ETC.

Diameter of driving wheels outside of tire.................62 in
Material of driving wheel centers..........................
................Main, cast steel; f. and b., steeled cast iron

Mogul Freight Engine No. 909, Grand Trunk Ry.
Schenectady.
General Dimensions Given on Page 498.

Tire held by..................................Shrinkage and retaining rings
Driving box material.....................................Steeled cast iron
Diameter and length of driving journals................9½ in. dia. x 12 in
Diameter and length of main crank pin journals........6½ in. dia. x 6 in
Diameter and length of side rod crank pin journals (Main, side 7½ in. dia x 5¼ in.); f. and b. ....5½ in. dia. x 4 in
Engine truck, style.........................2 Wheel, swing bolster
Engine truck, journals............................6½ in. dia. x 10½ in
Diameter of engine truck wheels...37 in
Kind of engine truck wheels..........Steel tired, cast iron spoke center

BOILER.

Style ....Extended wagon top
Outside diameter of first ring...62 in
Working pressure ....200 lbs
Material of barrel and outside of fire box...................Carbon steel
Thickness of plates in barrel and outside of fire box..............
....21-32 in., ¾ in. and ½ in
Horizontal seams ....Butt joint,
   sextuple riveted, with welt strip inside and outside
Circumferential seams..................................Double riveted
Fire box, length..120 in
Fire box, width..40⅞ in
Fire box, depth..............................F. 73¾ in., b. 65 in
Fire box, material..Carbon steel
Fire box plates, thickness........................
........Sides, 5-16 in.; back, ⅜ in.; crown, ⅜ in.; tube sheet, ½ in
Fire box, water space............Front, 4 in.; sides, 3½ in.; back, 4 in
Fire box, crown staying..................Radial stays, 1⅛ in. diam
Fire box, stay bolts..................Ulster Special iron, 1 in. diam
Tubes, material.......................Charcoal iron, No. 12 W. G.
Tubes, number of..291
Tubes, diameter ....2 in
Tubes, length over tube sheets....................................11 ft. 11 in

PASSENGER ENGINE 1027. ATLANTIC CITY RAILROAD.

ILLUSTRATED ON PAGE 507.

DIMENSIONS, WEIGHT, ETC.

Gauge ....4 ft. 8½ in
Cylinders ....13, 22, 26 in
Drivers ....84¼ in
Total wheel base......................................26 ft. 7 in
Driving wheel base.....................................7 ft. 3 in
Weight, total, about..................................142,900 lbs
Weight on drivers, about..............................78,600 lbs
Boiler, diameter......................................58¾ in
Number of tubes.......................................278
Diameter of tubes.....................................1¾ in
Length of tubes.......................................13 ft
Fire box, length......................................113⅞ in
Fire box, width.......................................96 in
Heating surface, F. B.................................136.4 sq. ft
Heating surface, tubes.....1,644.9 sq. ft.. Combustion chamber 53.8 sq. ft
Heating surface, total................................1,835.9 sq. ft
Tank capacity ....4,000 gallons

COMPOUND FREIGHT ENGINE NO. 4. LAKE SUPERIOR & ISHPEMING RY.

ILLUSTRATED ON PAGE 369.
DIMENSIONS, WEIGHT, ETC.

Fuel	Bituminous coal
Gauge of track	4 ft. 8½ in
Total weight of engine, in working order	147,600 lbs
Total weight of engine on drivers	132,800 lbs
Driving wheel base of engine	15 ft. 6 in
Total wheel base of engine	23 ft. 6 in
Total wheel base of engine and tender	52 ft. 10½ in
Height from rail to top of stack	14 ft. 11¼ in
Cylinders, high pressure, diameter and stroke	20x28 in
Cylinders, low pressure, diameter and stroke	31x28 in
Slide valves	Richardson balance
Piston rods	Steel, 3¾ in. diameter
Type of boiler	Straight
Diameter of boiler at smallest ring	64 in
Diameter of boiler at back head	67 in
Crown sheet supported by radial stays	1⅛ in. diameter
Stay bolts	1 in. diameter, spaced 4 in. from center to center
Number of tubes	240
Diameter of tubes	2¼ in
Length of tubes over tube sheet	14 ft. 7 in
Length of fire box, inside	108 in
Width of fire box, inside	42⅜ in
Working pressure	180 lbs
Kind of grates	Cast iron, rocking
Grate surface	31.78 sq. ft
Heating surface in tubes	2,049.5 sq. ft
Heating surface in fire box	148.6 sq. ft
Total heating surface	2,198.1 sq. ft
Diameter of driving wheel outside of tire	56 in
Diameter and length of journals	8x9 in
Diameter of truck wheels	30 in
Diameter and length of journals	5x9 in
Type of tank	Level top
Water capacity	4,000 U. S. gallons
Fuel capacity	280 cu. ft
Weight of tender with fuel and water	76,200 lbs
Type of brakes	Westinghouse American

PASSENGER ENGINE NO. 1. DELAWARE, LACKAWANNA & WESTERN RY.

SHOWN ON PAGE 389.
DIMENSIONS, WEIGHT, ETC.

Cylinders, diameter of H. P.	21 in
Cylinders, diameter of L. P.	30 in
Piston stroke	26 in
Center to center	83 in
Ports	20 and 22 in
Driving wheels diameter	62 in
Center	56 in
Tire	3 in
Axle	8 in
Boiler diameter	64 in

```
Tubes ............................................................. 242
    Diameter ..................................................... 2 in
    Length ....................................................... 13 ft
Fire box ......................................................... 96x144 in
Engine truck wheels (four) ....................................... 30 in
    Axle ......................................................... 5½ in
    Center ....................................................... Swivel
Total wheel base, engine and tender .............................. 47 ft. 9 in
Total wheel base of engine ....................................... 22 ft. 8 in
Wheel base, rigid engine ......................................... 11 ft. 8 in
Total weight of engine in working order .......................... 145,000 lbs
Weight on drivers ................................................ 109,000 lbs
Weight on truck .................................................. 36,000 lbs
Weight of tender (loaded) ........................................ 80,000 lbs
```

COMPOUND CONSOLIDATED ENGINE NO. 673. CANADIAN PACIFIC RY.

Shown on Page 353.

DIMENSIONS, WEIGHT, ETC.

```
Gauge of track ................................................... 4 ft. 8½ in
Total weight of engine in working order .......................... 142,650 lbs
Weight on drivers ................................................ 126,300 lbs
Weight of tender loaded, about ................................... 75,000 lbs
Cylinders, diameter and stroke, H. P. ............................ 20¼ in. x 26 in
Cylinders, diameter and stroke, L. P. ............................ 32 in. x 26 in
Driving wheels, diameter ......................................... 51 in
Driving journals ................................................. 8½x10 in
Diameter of boiler, first course ................................. 62 in
Total heating surface ............................................ 1,996 sq. ft
Tube heating surface ............................................. 1,845 sq. ft
Grate area ....................................................... 32.7 sq. ft
Boiler pressure .................................................. 200 lbs
Total wheel base of engine ....................................... 22 ft. 6 in
Driving wheel base ............................................... 14 ft. 6 in
Tank capacity .................................................... 3,840 gallons
Total wheel base of engine and tender ............................ 49 ft. 6½ in
```

ELECTRIC LOCOMOTIVE, BALTIMORE & OHIO R. R.

With the view principally of abating the nuisance of smoke and gases arising from steam locomotives in drawing passenger trains through long tunnels, the B. & O. R. R., with the co-operation of the General Electric Co., has brought into service this powerful electric locomotive shown on page 503 to use in their tunnel under the city of Baltimore, Md. It weighs 96 tons, the whole weight resting on the drivers, which are 62 inches in diameter; the wheel base of each truck is 6 feet 10 inches, and the length over all 35 feet; height, 14 feet 3 inches. It has a motor on each axle, each motor being 360 H. P. The trolley support shown is diamond shape, and compressible, contracting and expanding as may be necessary, and leaning to one side or the other of the overhead conductor. The tunnel where this engine is used is 7,339 feet long, with a grade of 8 degrees; the total haulage by electric traction is 3 miles. Passenger trains are taken through the tunnel at a speed of 35 miles per hour. Under a test this engine developed a speed of almost 60 miles per hour; but its power is more wonderful than its speed. A test made October 6, 1895, has added to the list of extraordinary performances of this locomotive, the character of which is heightened by the fact that the train which it moved measured 1,800 feet in length and weighed 1,900 tons, and was started from rest in the tunnel.

The train consisted of 43 loaded cars and 3 locomotives. In starting, it is said, not a sputter, spark or slip of the wheels occurred. The draw bar pull is given at 60,000 pounds; and the train was quickly brought to a speed of 12 miles per hour when the draw-bar stress was but 40,000 pounds. This locomotive has been in use about one year.

FAST RUNS.

The greatest speed ever attained by a locomotive, of which there is any record, was made by Engine 999 of the N. Y. C. & H. R. R. R. on May 9th, 1893, having covered five consecutive miles at the rate of 102.8 miles per hour. This remarkable performance was made with the Empire State Express, a train consisting of four heavy parlor cars, on a descending grade of 20 feet per mile; it is claimed a single mile was covered at the rate of $112\frac{1}{2}$ miles per hour, but there is no official record of such performance. Many other remarkable runs have been made by this engine, and it is without doubt the fastest locomotive in the world to-day.

It was the intention of the designer that this engine should be capable of attaining a speed of 100 miles per hour. It was designed by Mr. Wm. Buchanan, Supt. M. P. of the N. Y. C. & H. R. R. R. and built by that company. It has cylinders 19x24", driving wheels 87 inches in diameter and weighs in working order 102 tons. It was used to haul the "Exposition Flyer" during the World's Fair in 1893, where it became famous for its extraordinary fast running. This engine is equipped with a speed indicator and recorder which is driven from the truck axle in order to eliminate slippage. We have shown a cut of this engine on page 505.

On May 19th, 1893, Engine 903, a double of the 999, covered the same ground with the same train at the rate of 100 miles per hour.

Previous to that time there were five records, varying from 87.8 to 97.3 miles per hour. These runs were made on the Philadelphia & Reading and the Central R. R. of New Jersey, in 1890 and 1891 and 1892. Three of these records, including the highest one, was made by a Baldwin four-cylinder compound engine 385 of the C. R. R. of N. J.

On October 24, 1895, engine No. 564, of the Lake Shore & Michigan Southern Ry. Co., run from Erie, Pa., to Buffalo, N. Y., a distance of 86 miles, at an average speed of 72.91 miles per hour, having covered a single mile at the rate of 92.3 miles per hour.

On November 11, 1895, engine No. 590 of the Chicago, Burlington & Quincy Ry. Co. made a run with six heavy mail cars from Galesburg, Ill., to Mendota, Ill., a distance of 80 miles, at an average speed of 67.6 miles per hour, covering a single mile at the rate of 88 miles per hour.

On April 21, 1895, Pennsylvania Railroad Engine No. 1658,

Passenger Engine No. 999, New York Central & Hudson River R. R. Record, 102.8 Miles per Hour. Description and General Dimensions Given on Page 504.

"Class P," pulled a combination baggage car from Camden, N. J., to Atlantic City, N. J., a distance of 58.3 miles, at an average speed of 76.46 miles per hour; while from Liberty Park to Absecon, a distance of 49.8 miles, the speed averaged 79.7 miles per hour; from Berlin to Absecon, 35.6 miles, it averaged 82.9 miles per hour, and from Winslow Junction to Absecon, a distance of 24.9 miles, the speed was 83 miles per hour; the fastest single mile being made at the rate of 87.8 miles per hour.

On May 7, 1894, engine No. 655 of the Lehigh Valley R. R. Co. covered a single mile at the rate of 82½ miles per hour. This engine has cylinders 20x25", with 5' 9" drivers and the Wooten fire box.

Engine 385 of the Philadelphia & Reading Ry., the single driver Vauclain compound shown on page 333, hauled five cars from Wayne Junction to Bound Brook, a distance of 54.9 miles, at an average speed of 75 miles per hour.

On September 25, 1895, the New York Central & Hudson River Ry. Co. ran a newspaper train, consisting of two cars, from Albany to Syracuse, a distance of 147.84 miles, in 130 minutes, average 68.23 miles per hour.

An engine of the Delaware, Lackawanna & Western R. R. hauled a special train from Binghamton to East Buffalo, a distance of 197 miles, at the rate of 60.64 miles per hour.

The fastest regular trains in this country are run between New York and Philadelphia, and from Philadelphia to Atlantic City over the Philadelphia and Reading, Atlantic City Railway and the Central Railway of New Jersey. These trains average about sixty miles per hour. The schedule time of the Empire State express is 53.33 miles per hour.

ENGINE NO. 1027. OF THE ATLANTIC CITY RAILROAD.

The Baldwin Vauclain compound shown on page 507 pulls the fastest train in the world.

The distance from Camden (which is just across the Delaware river from Philadelphia) to Atlantic City is 55½ miles and the schedule time of this train is 50 minutes, an average speed of 62.2 miles per hour; exclusive of stops, the actual running time is about 70 miles per hour; this exceeds any of the fast trains of the British Isles, while the schedule time of the Empire State Express is only 53.33 miles per hour. On page 509 we have shown the daily performance of this engine, covering a period of two months.

Mr. Angus Sinclair, of *Locomotive Engineering*, was permitted to ride on this engine during one of its recent trips, and we quote the following as his experience:

"Half an hour before starting time the engine was backed up to train, which consisted of seven passenger cars. I happened to be exceptionally fortunate to take notes of an extraordinary feat

Passenger Engine No. 1027, Atlantic City Railroad.
Pulls the Fastest "Schedule" Train in the World.
Baldwin "Vauclain" Compound.
Description and Record, Page 506. General Dimensions Given on Page 509

of fast train running, for it was the first time that seven cars had been hauled on this train, five or six cars having been the usual load last season. Each car averages 75,000 pounds, and the engine, in working order, with tender, weighs about 218,000 pounds, so there were 525,000 pounds of train, making a total of 743,000 pounds, or 371½ tons, to be moved.

"I found a crowd of interested admirers about the engine, watching every move of the engineer and fireman, both of whom were quietly attending to the duties of preparing the engine to do its work without chance of failure. The engineer, Mr. Charles H. Fahl, kept moving about the engine scanning every part, and dropping a little oil on the parts that needed the greatest amount of lubrication. While I remained watching him he oiled the principal bearings twice, and then carried his cans to the cab, apparently satisfied that his full duty had been performed. The fireman, Mr. John Pettit, was engaged throwing a few shovelfuls of coal at brief intervals into the enormous firebox, which has 86 sq. feet of grate area, and watching at intervals to find a thin spot that needed covering up.

"These trains were run for three months last year on the 50 minutes schedule, with the same men on the engine, without a single mishap, or without losing a minute of time. The engine never had a hot pin or bearing, and, in spite of the tremendous work put upon it it was always ready to turn round and take out another train, without a minute's delay. That fine record was due to the care in seeing that everything was in good order before the start was made. In conversing with Vice-President Vorhees I found that he attributed the successful running of this train in a great measure to the care and skill of the engineer and fireman.

"At 3:56 precisely the signal came to start and the engine moved ahead without slip or quiver. A few turns of the great driving wheels forced the train into good speed and away we rushed out through the yards, through the suburban residences, and away past smiling vegetable farms. On reaching the first mile-post to be seen, which was about a mile out, I had my watch in hand and the second one was passed in 68 seconds. An interval of 62 seconds brought us to the following post, and then the succeeding notations were 60, 59, 56, 52, 50, 48, 46, 52, 53, 53, 51, 50, 52, 49, 50, 53, 52, 50, 49, 44, 45, 42, 44 seconds for each succeeding mile. Then I made up my mind that the high speed was authentic and put my watch in my pocket the better to note particulars about the handling of the engine.

"I was sitting on the fireman's side and could not see how the engineer was handling his reverse lever and throttle lever, but I noticed that there was no change in the point of cut-off after the train was going forty miles an hour, and it seemed to me that the steam was permitted to follow the piston a little more than half

THE LOCOMOTIVE UP TO DATE.

JULY.

Weight of 5-car train, including locomotive and tender, 255 tons of 2240 pounds. Weight of 6-car train, including locomotive and tender, 280 tons of 2240 pounds.

STATIONS.	Distance.	Schedule Time.	2d	3d	5th	6th	7th	8th	9th	10th	12th	13th	14th	15th	16th	17th	19th	20th	21st	22d	23d	24th	26th	27th	28th	29th	30th	31st

(Detailed numerical schedule data for stations Camden, W. Collingswood, Haddon Heights, Magnolia, Clementon, Williamstown Junc., Cedar Brook, Winslow Junc., Hammonton, Elwood, Egg Harbor, Brigantine Junc., Pleasantville, Meadow Tower, Atlantic City, Number of cars, Running time, Miles per hour (av'e).)

AUGUST.

STATIONS.	Distance.	Schedule Time.	2d	3d	4th	5th	6th	7th	9th	10th	11th	12th	13th	14th	16th	17th	18th	19th	20th	21st	23d	24th	25th	26th	27th	28th	30th	31st

Engineer Fahl. Conductor Stokes.

stroke. The steam pressure gauge could be easily noted, and the safety valve blew off at 230 pounds per gauge pressure. The fireman appeared to do his best to keep the pressure about five pounds short of the popping point, and he did his work well, but the indications were that he had more difficulty in keeping the steam down to the popping point than in letting it rise. He did not seem to work much on the fire. He watched it very closely, and threw in a few lumps occasionally, but there was no hard work in supplying all the steam needed to do the enormous work of pulling the heavy train at the speed noted. The coal used was small lump, similar to house furnace coal.

"The road is a little undulating, but the rises and descends seemed to make little difference to the speed. Out through stretches of farm lands, away through spreading woods and moor-like regions of scrub oaks the train rushed along, neither curve nor grade seeming to restrain its velocity. The engine rode with astonishing smoothness. When I have ridden on other engines working hard and keeping up speed over 70 miles an hour, there was always a harsh vertical vibration due probably to the jerk of compression, but that disagreeable sensation was entirely absent in this compound. The work done gauged in horse power per hour was enormous, and perhaps unprecedented for a locomotive, but it was performed with remarkable smoothness, and the impression was always present that the engine still had some margin of power in reserve which could be used if necessary.

"About four miles from Atlantic City a signal was against the train and the speed was reduced to about 20 miles an hour before the signal was lowered. That was about three-quarters of a mile from the succeeding mile post. I noted the time from that mile post to the next one and the mile was run in 60 seconds. That will give a good idea of the power of the engine.

"Two minutes were used in running the last two miles through the switches. At least one minute was lost with the signal check. With these deductions I calculate that the average run was made at a speed of over 70 miles an hour."

The general dimensions of this engine are given on page 500.

LONG DISTANCE RUNS.

The fastest long distance run on record was made October 24, 1895, over the Lake Shore & Michigan Southern railway from South Chicago to Buffalo, N. Y., a distance of 510.1 miles, at an average speed of 65.07 miles per hour, excluding stops. This run was made with five Brooks engines, engine No. 564 making the best record, covering 86 miles at an average speed of 72.92 miles per hour. The train consisted of two sleepers and a private car; total weight, 304,500 pounds.

On September 11, 1895, the New York Central & Hudson River Railroad company ran a train from New York to Buffalo, a distance of 436½ miles, in 407 minutes, an average of 64.22 miles per hour, exclusive of stops. Three engines were used, Nos. 999, 903 and 870; the two latter are 100-ton engines, and the 999 weighs 102 tons; the train weighed 361,000 pounds. Engine No. 903 made the best time, having covered 145.60 miles at an average of 65.75 miles per hour; this engine has 19x24 inch cylinders; drivers 78 inches in diameter.

The London & Northwestern railway and the Caledonian railway (West Coast Route) made a run on Aug. 22, 1895, from London to Aberdeen, a distance of 540 miles, at an average speed of 63.93 miles per hour, exclusive of stops. Four engines were used, engine No. 904 making the best record, covering 141.25 miles at an average speed of 67½ miles per hour; this engine has 17x24 inch cylinders, with 78-inch driving wheels; the weight of the train was 150,080 pounds.

The Delaware, Lackawanna & Western Railroad company ran a special train from Hoboken to East Buffalo, a distance of 404 miles, at an average speed of 50½ miles per hour, including stops.

December 29, 1893, a special train was run from Chicago to Atlanta, Ga., a distance of 733 miles, at an average speed of 45 miles per hour. The run was made over the Chicago & Eastern Illinois, Evansville & Terre Haute, Louisville & Nashville, Nashville, Chattanooga & St. Louis, and the Western Atlantic Railway, the best time being made on the Chicago & Eastern Illinois Railway.

During December, 1893, new mail and express trains were put on the Chicago & Northwestern Railway and the Chicago, Burlington & Quincy Railway between Chicago and Omaha; the distance over the Chicago & Northwestern Railway is 490 miles, and over the Chicago, Burlington & Quincy, 499 miles. These trains run in competition, carrying the cars of rival express companies. On December 27, 1893, they reached Omaha, the C. & N. W. Ry. beating the C., B. & Q. two minutes; both trains running into the Union Pacific Railway station at Council Bluffs, opposite Omaha. The time of the former was 11 hours and 23 minutes, or at the rate of 46 miles per hour, deducting stops. The Northwestern train carried one express car and two mail cars, while the Burlington train was composed of six cars, therefore the C. & N. W. had the advantage of both distance and weight.

On August 26, 1894, a special train was run from Jacksonville, Fla., to Washington, D. C., a distance of 780 miles, in 880 minutes, which is an average speed of 53.18 miles per hour; from Ashley Junction to Florence, a distance of 96 miles, the run was made in 99½ minutes, including two stops.

GENERAL MACHINE SHOP WORK.

(ERECTING DEPARTMENT.)

SHOES AND WEDGES.

Shoes and wedges are castings fitted to the pedestal jaws; they are also planed to fit into the grooves on the forward and back faces of the driving boxes and form a bearing between the driving box and the pedestal jaws. The object of these castings is to hold the driving box in its correct position and at the same time permit of its adjustment when desired. The shoe is usually fitted to the forward jaw and the wedge to the back jaw; the wedge should be so adjusted as to take up all lost motion between the forward and rearward jaw, but not so tight as to cause the box to stick; in other words, prevent the driving box from moving up or down in the jaw. The two flanges on the driving box prevent it from working laterally. The importance of having the shoes and wedges properly fitted upon a locomotive cannot be over estimated, and yet there are railroad shops in which this kind of work is considered inferior, or second class work, while in reality there is no job upon the locomotive of more importance, as a set of shoes and wedges improperly fit up may cause the engineer much trouble with rod brasses and driving box brasses, or cause the tires to cut and sometimes break crank-pins or side-rods.

In order to explain how to lay off shoes and wedges properly it is first necessary (like building a house) to see that the foundation is properly laid, or in other words, to see that the frames are set in place correctly. The following method is given for fitting pedestal braces and setting locomotive frames:

FITTING PEDESTAL BRACES AND SETTING FRAMES.

By Ira A. Moore, of Cedar Rapids, Iowa.

When frames are new, or when the back frames have been removed for extensive boiler repairs or other purposes, it is important that when they be put in place again that they be properly lined and squared up. In what follows we shall endeavor to show how this can be accomplished.

The last operation when taking down old frames should be to place them bottom side up on thick blocks or horses; if you do this they will be in convenient position for fitting the pedestal braces.

We will suppose that new pedestal binders are necessary, and the first thing to be done in fitting them is to prepare the frame, or bottom ends of pedestal jaws, for receiving them.

The fit $a\ a'$, $b\ b'$, Fig. 1, should be filed perfectly at right angles to the frame's length, and all to the same level, which should be about 7 degrees from a right angle to the top of frame. b' and a' should be $\frac{1}{8}$ inch below the face of pedestal jaw, as shown in Fig. 1, to allow the jaws to be refaced without destroying the fit. Before the pedestal binder is laid out it should be planed on one side.

Now bolt two straps of iron, about $8 \times 1\frac{1}{2} \times \frac{1}{8}$ inch ($c\ c'$, Fig. 1), to the frame, as shown, and let them extend at right angles to the frame. Lay the binder on these strips, planed side down, and against the side of the pedestal jaw, being careful to have the

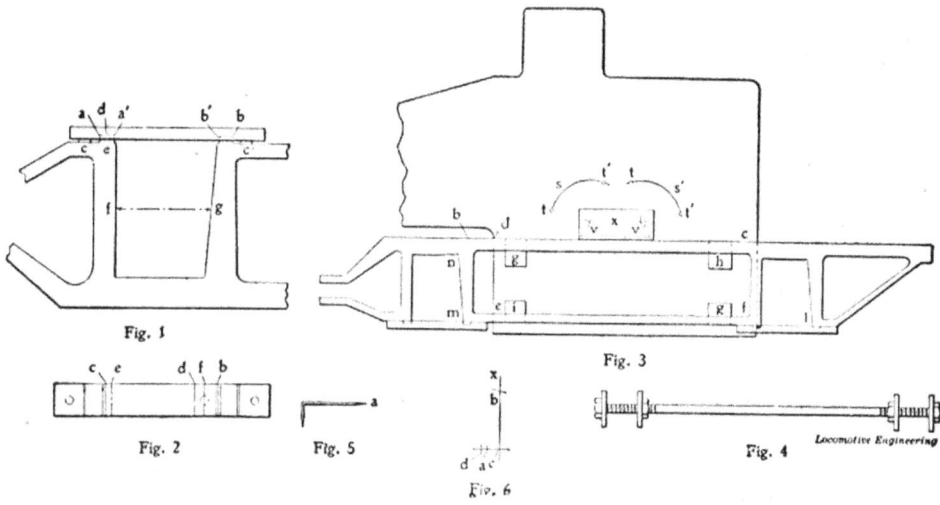

Fig. 1   Fig. 3   Fig. 2   Fig. 5   Fig. 4   Fig. 6

ends extend the same distance back and front of the fit. After setting the planed side of the pedestals at right angles to the side of the pedestal jaw, by bending the straps $c\ c'$ up or down, clamp it in that position. Now with a small straight-edge, held against the fit and against the bottom of the binder, scribe a line on binder. Do the same at the four places. Scribe lines on the edge of binder next to frame to show the right bevel to plane to.

Make the depth of recess 1-16 inch more than the distance $d\ e$, Fig. 1. After planing off the recess to these lines, if the work has been accurately done, the binder will drop to within $\frac{1}{8}$ inch of the bottom of frame. It should now be fitted down 1-16 inch farther by filing when the holes may be laid out. If possible the bolt holes should be laid out and drilled, so that no reaming will be necessary, since reaming the holes weakens the frame. It is important that the hole for the wedge-adjusting bolt be in the proper position. If it is too far from the face of pedestal jaw, it will interfere with the driving box. If the hole is too far the other way, it will come in contact with the face of pedestal jaw.

To find its proper position, proceed as follows: Fig. 2 is the pedestal binder. The line b represents the face of back pedestal jaw, and c the face of front jaw. The line $f\,g$, Fig. 1, is parallel with top of frame and passes through the center of pedestal. If the distance between d and e, Fig. 2, is $12\frac{3}{4}$ inches, and the driving box is $11\frac{1}{4}$ inches, it is evident that the thickness of driving shoe will be $\dfrac{12\frac{3}{4}-11\frac{1}{4}}{2}=\frac{3}{4}$ inch. Hence scribe the line e, Fig. 2, $\frac{3}{4}$ inch from c. The distance between e and d equals the size of driving box $=11\frac{1}{4}$ inches. It is now plain that the space $d\,b$ represents the thickness of the bottom of wedge. Suppose the diameter of that part of the bolt that enters the wedge to be $1\frac{1}{4}$ inches, then the center of bolt hole in binder should be 11-16 inch back of line d, or on the line f. This will allow 1-16 inch clearance between the bolt and driving box.

Having finished this part of the work, we will proceed to put frames in place. But before putting them up, the expansion plate studs should be examined carefully, and if any of them show signs of leaking, they should be replaced with new ones.

Any studs that prevent the frame liners from sliding out and in when the frame is in place should be taken out, and the new ones not put in until after liners are fitted. A die nut should be run over the old studs that are good to straighten up the thread.

Now place some blocks across the pit directly under where the pedestals will come when frames are in place, to support them while fitting the liners and buckles. Set the frame on the blocks, and then raise or lower it to the proper height, which may be determined by using the buckles as a gauge. When the frame is the right height the buckle will slide on the studs.

Now put in the splice bolts, then fasten the deck in place. Set inside calipers to the distance between the frames at the deck. Then by means of rods and plates of iron like Fig. 4, which should be made of at least $\frac{3}{4}$-inch iron, placed one at l, one at m and one at n, Fig. 3, set the frames the same distance apart at these points that it is at the deck, using the calipers as a gauge.

Now run lines through the center of cylinders, letting them extend to the back end of frames. Then measure the distance from the outside of pedestal jaws to the lines. This distance should be the same at all the pedestals, but very likely will not be.

Suppose the distance from the left front pedestal to the line to be $11\frac{1}{2}$ inches, and right front 12 inches. This indicates that the frames are $\frac{1}{4}$ inch too far to the left in front. To draw them over insert an iron wedge between the boiler and frame at d and at e, Fig. 3, on right side and drive them down until the frame has been drawn over the required distance, which in the present case is ¼ inch. As the right frame is drawn out, the left will be drawn in by the rods, previously mentioned, which bind the frames together.

Now insert iron wedges between frame and boiler at d and e on left side of engine, but do not drive them down any, as these are merely to fill up space between frame and boiler to hold the frames in place after they have been drawn over.

We will now go to the back pedestals. Suppose the distance from the left pedestal to the line through cylinders to be $11\frac{5}{8}$ inches, and from right pedestal to line to be $11\frac{7}{8}$ inches; then the proper distance from pedestal to line on both sides is

$$\frac{11\frac{5}{8} + 11\frac{7}{8}}{2} = 11\frac{3}{4} \text{ inches.}$$

Hence the frames are ⅛ inch too far to the right. Draw them to the left by means of wedges at c and f on left side, then insert wedges on right side to hold them firmly in place. Now try the front pedestals again, as setting the back ones will be liable to throw them slightly out of line. If such is the case they can be put in line again by driving the wedges farther down on the side that is farthest from the line, being careful to first raise the wedges on opposite side.

We now have the frames the same distance from the lines at all four pedestals, but something more is necessary. They must be at right angles to a line drawn across their tops. We will try them at the front pedestals first.

Put a straight-edge across the frames at b, Fig. 3, and then place the short side of a 2-foot square against the straight-edge, when the side of pedestal should be parallel with long side of square. Suppose it is found to be ⅛ inch away from the square at bottom end on left side, then the right pedestal will be that distance from square at top end, since the frames are held parallel by the rods. They could be squared up by raising the bottom wedge (in front) on right side and driving the one on left side down, but that would throw both frames about 1-16 inch too far to the left. To square them up, and at the same time keep them in line, proceed as follows:

Raise the bottom wedge on right side enough to allow the bottom of pedestal to go toward the boiler 1-16 inch, and drive bottom wedge on left side down enough to draw bottom of pedestal out 1-16 inch. This will leave them out of square the same as they were, but only half as much, and they have been drawn 1-32 inch too far to the left. Now raise the top wedge on left side enough to allow top of pedestal to go toward boiler 1-16 inch, and drive top wedge on right side the same amount.

We now have the frames square and have drawn them back into line. Proceed in the same manner with the back pedestals. The frames now have the proper position, and in order to determine whether they move or not, and to know when the liners are the right thickness, the position of the frames should be marked in some way, and a very good one is to use a tram similar to the one shown in Fig. 5. Make a center punch mark on the side of boiler, near *c, d, e* and *f,* Fig. 3; then with the point *a,* Fig. 5, in these marks scribe arcs on the frames at the four places.

Do the same on other side of engine. It is plain that these arcs must come to the same position when the liners are fitted and the buckles on. We are now ready to fit the liners *g, h, i, j,* Fig. 3. Generally the old liners can be used again. If they are too thin to fill space between boiler and frame, a piece of boiler plate can be riveted onto the side next the frame. If a piece of the exact thickness cannot be had, rivet one on slightly thicker than is required, then plane it down to the exact thickness.

After the liners are in place, the studs can be screwed in through the holes in them by using a stud nut.

The buckles should be loose enough on frame to allow them to slide on it without binding when the boiler is expanding or contracting.

The lateral, or cross, braces can now be put on. If they are not the right length, have the blacksmith lengthen or shorten them to suit.

To lay out the holes in a new expansion plate when the studs are in the boiler, with any degree of accuracy, is generally not very easily done. The following method has been found to give satisfaction: Make a small center punch mark in the center of each stud that is to pass through the plate. Then set dividers to any convenient radius—say, 10 inches—and with centers of studs as centers scribe the arcs *s s',* Fig. 3, on the side of boiler, and make two center punch marks *t t'* on each arc, about 90 degrees apart if possible; more or less will answer the purpose, but 90 degrees will give the best results. Now lay the expansion plate *x,* Fig. 3, on top of frame, as shown (this is not the position usually occupied by an expansion plate, but will serve our purpose), with the part that goes next to boiler against the ends of studs. If we now use the points *t t'* as centers, and with the same radius used to scribe *s s',* scribe arcs *v v'* on expansion plate, their point of intersection will not lie in a line with the center of stud, but will be to the side of this line nearest the arc *s* or *s'*; or, in other words, the radius used was too short. The correct radius with which to draw the arcs *v v'* can be found thus:

On a board or other plain surface draw two indefinite lines at right angles to each other, Fig. 6; then lay off the distance *a c* equal to the length of the studs, and *a d* equal to the thickness of the expansion plate. With *a* as a center and the same radius that the arcs *s s',* Fig. 3, were drawn with, scribe the arc *b,* Fig.

THE LOCOMOTIVE UP TO DATE.   517

6, across the line $x\ y$. Now set the dividers to the distance $b\ d$, which is the correct radius with which to scribe the arcs $v\ v'$, Fig. 3, on expansion plate, using $t\ t'$ as centers, to have their point of intersection in line with center of stud. Hence this point will be center of hole in expansion plate.

It is hardly necessary to say that an arc must be scribed on side of boiler from center of each stud that passes through the plate.

PORTABLE MILLING MACHINES FOR FRAME JAWS.

A very neat and handy tool for milling locomotive frame jaws while in place is in use at the Big Four shops at Urbana, Ill. It consists of a frame which carries a spiral fluted milling cutter, a worm and worm wheel and suitable gears. It may be clamped to the frame and is capable of accurate adjustment. It may be driven by steam, air, electricity, rope transmission or hand power. One hundred and eighty revolutions gives a feed of one inch per minute. It is the invention of Mr. F. Davidson and has been in use about three years and is said to be giving good results.

SETTING THE CENTER CASTING.

As the center casting leads the engine it is very important that it should be set perfectly central between the frames, which will prevent crowding the engine to one side and thereby cutting the tires; in order to do this place the casting central with the cylinder saddle, front and back, then set it central between the frames, measuring either from the turned surface of the casting or from its dead center. When properly set, ream or rosebit the holes, and use "driving fit" bolts to hold it permanently in this position.

OBJECT OF "SQUARING UP" AN ENGINE.

In order to avoid trouble with driving box brasses and rod brasses, and to prevent cutting the tires, it is necessary that the main driving axle or shaft should be set at perfect right angle with the cylinder centers and held secure in this position, and that the centers in each other pair of driving wheels (on each side) should be an equal distance from the centers in the main wheels, and they also should be held secure (this does not imply that each pair of drivers should be an equal distance from the main pair). Various methods are in use for locating a square line on each frame of the engine from which to lay off the shoes and wedges; some use a fish tail tram from the center casting and locate a center on the inside of each main jaw, while others use a three-pointed tram from a center located midway between the frames on the back end of the cylinder saddle. The centers of the rocker boxes are also used for this purpose; others line one cylinder or both; but whichever method is employed the object is the same, that is to secure a square center upon each main jaw from which to lay off the shoes and wedges. Perhaps the most mechanical way to obtain these centers would be to run a fine line through both cylinders, being positive that the center casting is set central, as this method will keep the main shaft at right angle with the cylinders and avoid any twist in the driving box or rod brass bearing, which would surely cause them to run hot.

LINING THE CYLINDERS.

Use as fine a line as possible that will withstand the strain, passing it through the cylinder and letting it extend back of the forward part of the main jaw; fasten the line at either end in any convenient manner, set the line by the counter-bore at each end of the cylinder after the counter-bore has been scraped clean. If the back cylinder head is in position set the line at the back end of the cylinder by the stuffing box.

HOW TO LOCATE THE SQUARE CENTERS.

When both cylinders have lines extending through their centers and are set properly, measure the distance between the lines just back of the cylinder and near the back end of the lines and see if the two lines are parallel, also see that the pedestal braces (or binders) are properly tightened; then use a six-foot square, or a straight-edge, and a two-foot square; if a straight-edge is used fasten it securely against the front face of each main jaw, or shoe, keeping each side an equal distance from the top of the frames. Now set a two-foot square against the straight-edge and try it with the line on each side of the engine and see if the straight-edge is at perfect right angle with both lines; if it is not, then place liners between the straight-edge and the jaw until the straight-edge is at perfect right angle with both lines; if the two lines are not parallel divide the difference as near as possible. If a large square is used a two-foot square will be unnecessary. Now when the straight-edge or square is properly set, use a pair of hermaphrodites and scribe a short line on each main jaw inside and outside and far enough forward to clear the flange of each main shoe. Then carry four lines an equal distance from the top of each frame to intersect the other lines. The points where these lines cross each other will be the square center indicated by the point Z on both illustrations on page 519. A small circle should be described around these centers to keep them from being confused with other centers. Now that you have secured the square center use a T square or place a face plate on the top of each frame and use a two-foot square and set one edge of the square true with the square center and scribe the line L. M. down the full length of each frame on the outside, as shown on the illustrations. This will be the square line from which to lay off the shoes and wedges.

LOCATING CENTERS IN THE JAWS.

When both jaws taper as shown by Fig. 1 scribe the lines A B above the jaw, as shown on the outside of the frame, and true with the face of each jaw, then use a pair of hermaphrodites

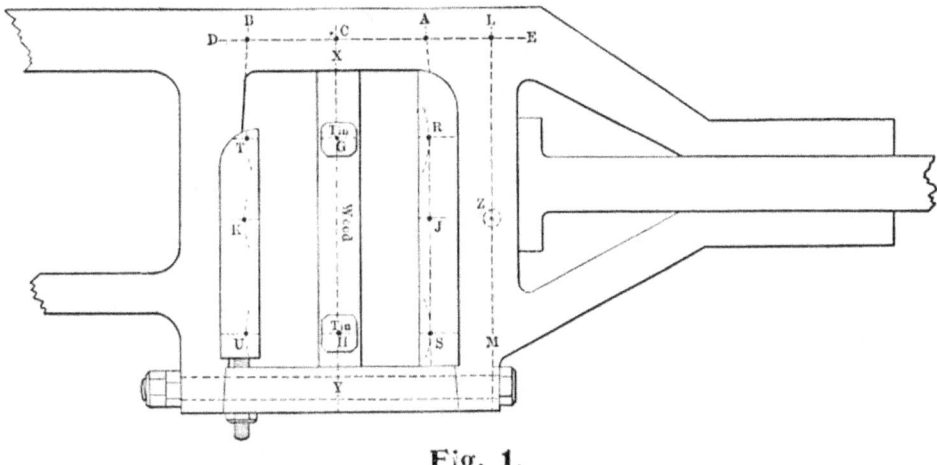

Fig. 1.

and scribe the line D E parallel with the top of the frame, then with a pair of dividers locate the center C midway between the intersection of lines A B and D E on the line D E. This point indicated by the letter C will be the center of the jaw. Now use the dividers again and see if the point C is an equal distance from

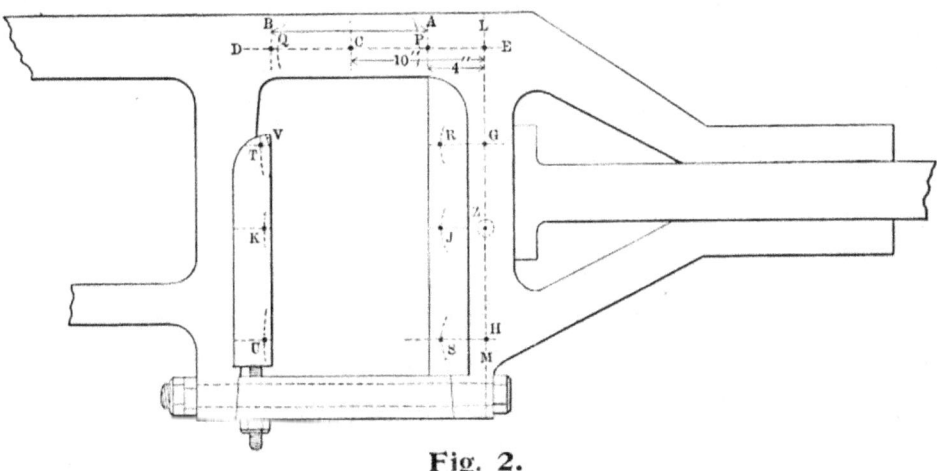

Fig. 2.

the square line L M on both sides of the engine, if so the centers in the main jaws are correct; if not, then move one center forward and the other backward a distance equal to one-half the difference between the point C and the line L M on the two sides of the engine. When the front jaw is at right angle to the top of the frame and the back jaw is tapering, as shown in Fig. 2, a different method must be followed. First ascertain the thickness of the shoe and the thickness of the wedge at its top; they should each be the same thickness; if they are not, you must figure on your liners accordingly, clamp each shoe and wedge securely in its proper position, letting the bottom of the wedge clear the binder or pedestal brace about $\frac{1}{4}''$ to allow for pulling down the wedge, should the driving brass run hot; then use a straight-edge on the face of both the shoe and the wedge and scribe the lines A B on the frame, as shown in Fig. 2. Now locate the center C midway between the lines A and B and see if this center is an equal distance from the line L M on each side of the engine; if not, move each center as previously explained.

HOW TO MAKE THE CENTERS OF THE JAWS TRAM.

Locate a center in each jaw the same as for the main jaws; then use a pair of long trams and see if the centers of the other jaws on each side are an equal distance from the centers in the main jaws, which you know are correct. Different pairs of wheels need not be the same distance from the main wheels, but corresponding pairs should tram perfectly on each side, therefore the jaw centers should correspond. If you find they do not tram perfectly you can move one or both centers the same as you did with the centers in the main jaws, unless solid side rods are used, in which case you must line up the shoes and wedges to accommodate the rods. If the boiler sets down between the frames you should allow $1\text{-}32''$ for expansion of the frames; that is, make the jaw centers (and likewise the driving box centers) tram $1\text{-}32''$ shorter than the length of the side rods. Do not allow for expansion between the main jaw and the forward jaw. When lining up old shoes and wedges for solid rods it is not necessary to locate the center of the jaw on the frame unless you want all the shoes and wedges lined central. You can lay off the main shoes from the square line and the other shoes from the main shoes, using the length of the rod and lay off each wedge from its own shoe.

LAYING OFF NEW SHOES AND WEDGES.

Various methods are employed to perform this kind of work, but we shall explain the two methods most generally used. We shall assume that the square line and jaw and centers are properly located as previously explained. We shall first call the reader's attention to Fig. 1; here we find two pieces of tin tacked onto a long wooden center and fit between the binder and frame, as shown by the cut; the face of these two tins should be flush with the outside flanges of the shoe and wedge. If the fire-box or other parts of the engine will not interfere with the use of a long tram it will be necessary to put these wooden centers in the main jaws only, but if a long tram cannot be used then put wooden centers in all of the jaws and lay off all of them the same way. Be sure the wedge is set up $\frac{1}{4}''$ above the binder or wedge bolt when one is used above the binder. Then use a two-foot square from the top of the frame and scribe the perpendicular line X Y on both pieces of tin, letting the line pass through the jaw center C. Now with a pair of hermaphrodites set to any convenient distance from the top of the frame scribe two more horizontal lines across both shoe and wedge and the tin, as shown by the cut, and where these two lines cross the line X Y locate the centers G and H. Then caliper the driving box; we shall assume it measures 12''; now use a pair of dividers, set to 6'' (which is one-half the thickness of the driving box), from the point G and H scribe the four small arcs shown on the shoe and wedge, and at the intersection of these arcs and the horizontal lines make the four centers R S T U. If a shoe and wedge gauge is used, which should be done, then before describing these arcs, add to the 6'', which is one-half of the box, the distance from the face of the gauge to the point of the gauge, which we will assume is 1''; therefore set the dividers to 7'', and then describe the arcs. The gauge will prove the work of the planer hand and also of the man that lays off the shoes and wedges, and when planed the point of the gauge should enter each center without crowding. Now you have two points on the outside of each shoe and wedge by which to set it on the planer, but it will be necessary to locate another point on the inside of each shoe and wedge in order to have them square across their faces. So scribe a line from the point R to the point S; set a pair of dividers from the square center Z to the line R S and with this length, and from the point Z on the inside of each frame locate the point J on the inside of each shoe. Next set a pair of dividers from the point R to the point T; now with this length go to the inside again and from the point J locate the point K, keeping the points J and K an equal distance from the top of the frame. Now you have three points on both shoe and wedge, so lay off all shoes and wedges the same way. When you have located these points on the main jaws you can sometimes set a pair of long trams to the jaw centers and lay off all other shoes and wedges from them; if you cannot, you must lay off all of the others the same as the main jaws, then you have finished.

ANOTHER METHOD.

Another but slower method is to avoid the use of wooden centers altogether as they are liable to be moved, and to have the shoes laid off, planed and properly fitted up before the wedges are laid off. We will now call the reader's attention to

Fig. 2, and assume that we have the square line and jaw center located, and that the driving box calipers 12″ in thickness. We shall also assume the use of the shoe and wedge gauge, as previously explained, to be used upon the shoe only. First locate the points G and H any convenient distance from the square center Z and through the two points scribe a line on both shoe and wedge, as shown, keeping each line parallel with the top of the frame. Now as the distance from the jaw center C to the line L M measures 10″ and one-half the thickness of the box is 6″ and the gauge is 1″, which makes a total of 7″, the difference is therefore 3″ (see how to measure uneven surfaces, page 559);

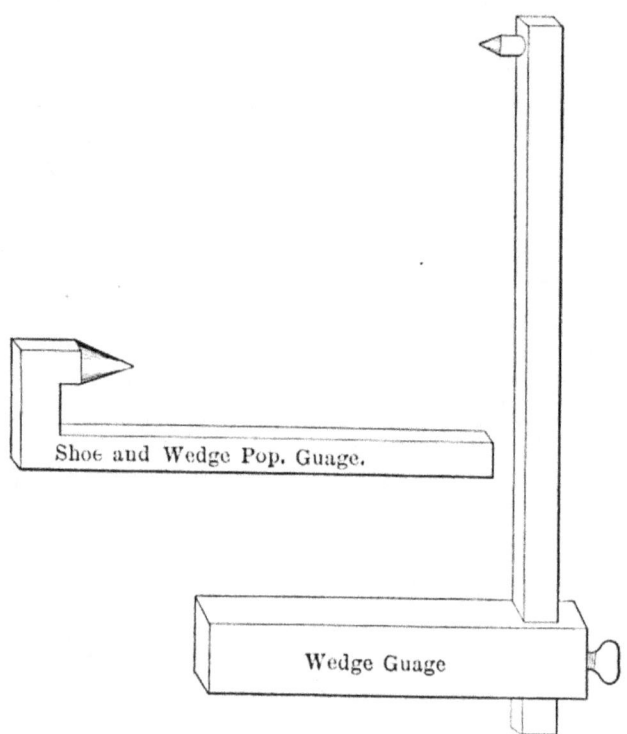

set a pair of dividers to 3″, and from the points G and H locate the two points R and S on the outside flange of the shoe. Now with the same length and from the square center Z on the inside of the frame locate the point J. Now take down the main shoes and have them planed and filed perfectly square across their faces; then if the points R S and J are correct with the shoe gauge them from these points and with a long tram set to the jaw centers (presuming all boxes are of equal thickness), lay off all the other shoes; have them planed, filed square and tram perfectly before you lay off any of the wedge when this is done. Below set the wedge, below the pedestal brace or nut on wedge bolt. Then with a wedge gauge, small tram, or pair of dividers set to any convenient length from the points R S locate the points T and U on the outside of each wedge, and with the same length from the point J locate the point K on the inside of the wedge. This process gives three points by which to set the wedge for planing. Now set a pair of inside calipers to the exact size of the driving box, then clip and file a spot anywhere on the wedge, and when you have located a point that calipers exactly the same between shoe and wedge, then make the small prick-punch mark indicated by the letter V. This mark should be just barely scraped when planing and should be visible when finished.

LINING UP OLD SHOES AND WEDGES WITH BOXES OF DIFFERENT SIZE AND BRASSES BORED OUT OF CENTER.

We will assume that the square center and jaw center are properly located, as previously explained. First clamp all shoes and wedges securely in their proper places, setting the wedge $\frac{1}{4}''$ above the binder or nut on wedge bolt, then try the face of each and see if they are square with the frame; if they are, then scribe the lines A B on the frame above the jaw, as shown in Fig. 2. Now if it can be done conveniently and to avoid mistakes truck all the driving boxes to their respective plates and lay them on the floor with the top of each box turned toward the rail and the outside of each box turned upward. Now use a transfer plate (or gauge for the purpose) and scribe a line on each side of the box on its outside face, and true with its bearing for both shoe and wedge. Then put a center in each driving box. Now set a pair of dividers from the center in the box to the forward line and transfer it to the frame, marking the distance in front of the center C, as shown by the letter P, Fig. 2. Then set the dividers from the center of the box to the line indicating the back face of the box, and with this length from the jaw center C describe the arc Q, Fig. 2; do the same on all jaws. Now the distance between the lines A and P indicate the thickness of the liner required behind the shoe to keep it central, and the distance between the lines B and Q indicates the thickness of the liner that should be riveted to the wedge. To both these thicknesses you should add 1-32" to allow for planing up. If the liners are to be planed inside, add an additional 1-32" to allow for it. When all the shoes and wedges have liners of the proper thickness riveted to them, add to the distance C P the amount of the gauge 1"; then add to the distance C Q the amount of the gauge 1". You can then employ either method, as previously explained, to lay off the shoes and wedges, or with the 1" added to the distance C P and C Q you can carry down square lines from the top of the frame and locate the four points R S T U and from these locate the points J K on the inside of the jaw, as previously explained. If the wedge is laid off with calipers then to protect the man who lays off the wedges a small gauge should be used from the top on the face of the wedge and a small circle described on the outside flange of the wedge; this will be a proof mark.

TRAMMING THE DRIVING BOX CENTERS.

When the shoes and wedges are all planed fasten them up in their respective places. Use a long straight-edge black leaded and see if they are true across their faces and use a pair of inside calipers and see if each wedge is parallel with its shoe, trying four points with the calipers, also try their width as compared with the driving box. File all of them (if necessary) perfectly true both ways, then they are finished. Now, regardless of the method employed to lay off the shoes and wedges, it is considered very good practice to prove your work before the wheels are put under the engine. This is done by placing all the boxes up in the jaws and putting up each shoe, wedge, and binder and setting the wedge; it should be set up moderately tight, but not enough to cause the box to stick; the box should be permitted to fall by its own weight. Now put centers in all the boxes and see that each pair of boxes tram perfectly with the main boxes and try the centers of the main boxes with the square line on the frame; they should each be the same distance from the line. If any alterations are necessary, now is the time to make them, when they are all perfect; then before taking the boxes down scribe a line on the inside and outside of the frame at the top of each wedge. Then when the wheels are put under the engine the wedge may be set to these lines and avoid a jumping of the wheels.

TRAMMING WHEEL CENTERS.

When you tram the wheel centers of an old engine see that the wedges are properly set up before you begin; if two wheel centers are out, say $\frac{1}{8}''$, then a full 3-32'' liner or a scant $\frac{1}{8}''$ line will make the wheel centers tram about right.

HANDLING DRIVING BOXES.

The illustration shown is self-explanatory. The upright shown is in two parts and easily clamped to the axle. A small crane with a chain pulley swings from its top; the clamp shown below the axle is in two parts and it will hold the box without the necessity of an eye-bolt.

When the old methods of fitting driving box brasses are considered this device will be seen to be a great labor-saving device, and with the very large driving boxes now in use something of

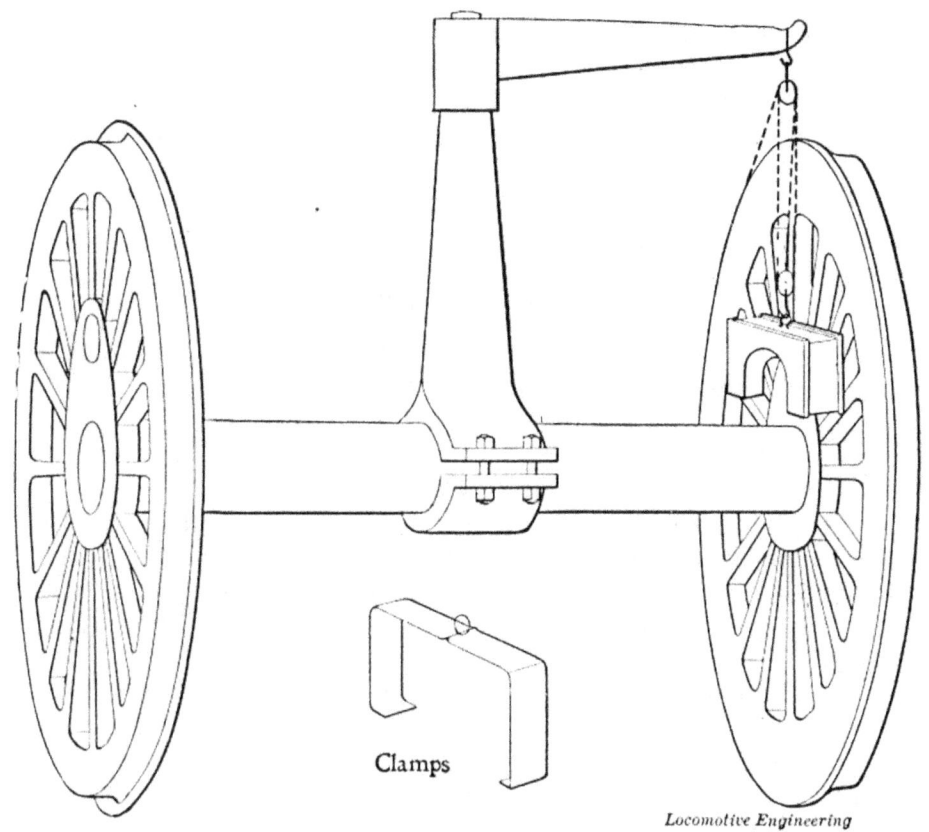

Locomotive Engineering

the kind is rendered necessary. This device is used in the Northern Central shops at Elmira, N. Y.

LINERS.

When liners are needed they should be riveted solid to the shoe or wedge, using five or eight ¼" rivets, and if the liner is very heavy it should be trued up on the planer after it is riveted on.

The habit of placing loose liners behind shoes and wedges is false economy and is not practiced by good workmen.

FITTING DRIVING BRASSES INTO THE BOXES.

In this you must be governed by the strength of your box. If it is a very light box and you are fitting a circle brass, four or five tons' pressure will be sufficient. But if it is a very heavy box, from fifteen to eighteen tons' pressure may be put on. For medium size driving boxes ten tons is the rule.

If gib brasses, three tons' pressure on each gib; be very careful not to file the brass tapering or you may burst the box.

FITTING DRIVING BRASSES ONTO THE JOURNALS.

Always file out crown of brass about 3" wide and scrape to a good bearing on sides. The weight of the engine will soon bring her down on crown. Leave no rough file marks on brass, and clean out oil holes. Don't let it bear on fillet. Let it move freely on journal when finished. A long board is generally used when fitting driving boxes, but in some shops where very heavy boxes are in use a block and tackle with a hook is used to raise and lower box onto journal.

LATERAL BETWEEN HUBS.

On a smooth road bed 1-32" is enough lateral between hubs of drivers, or truck wheels. On a rough roadbed 1-16" should be allowed. Brass hub plates are most generally used to take up side play—although on some roads babbitt is used either on the hub or on the driving boxes and is said to wear well.

If you have too much lateral motion between hubs, face off hubs and turn up and put on new brass hub plates, or put plates on outside of boxes; in either case fasten plates with screw bolts with counter-sunk heads, with top of head 1-16" or $\frac{1}{8}$" below the surface.

AN IMPROVED DRIVING BOX CELLAR.

From the illustrations here shown it will be seen that this form of cellar can be replaced without slacking or removing the binder, whether there is a cellar on the axle or not. This appears to be a great improvement upon the ordinary method of

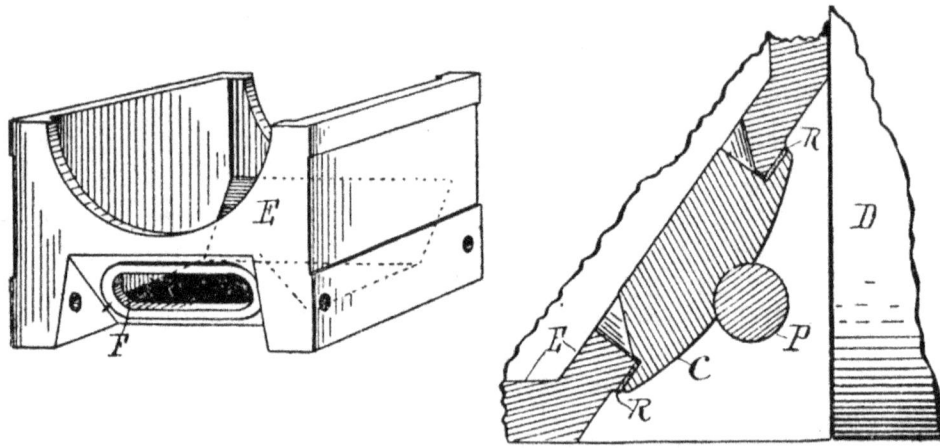

re-packing cellars and will save much time and labor, as may be seen by the drawings. The lid which covers the opening in the cellar is held secure by the cellar bolt. Therefore, it is only necessary to remove one cellar bolt in order to repack the cellar. This device is the invention of Messrs. Frank H. Taylor and Frank Riley, of Joliet, Illinois, and is patented.

THE LOCOMOTIVE UP TO DATE. 527

NEW METHOD OF SPLICING FRAMES.

This invention for splicing locomotive frames was invented by Mr. C. J. O'Reilly, of Manchester, N. H., who describes his invention as follows:

"My invention consists of an improved method of fastening the locomotive main frame and frame front together; and

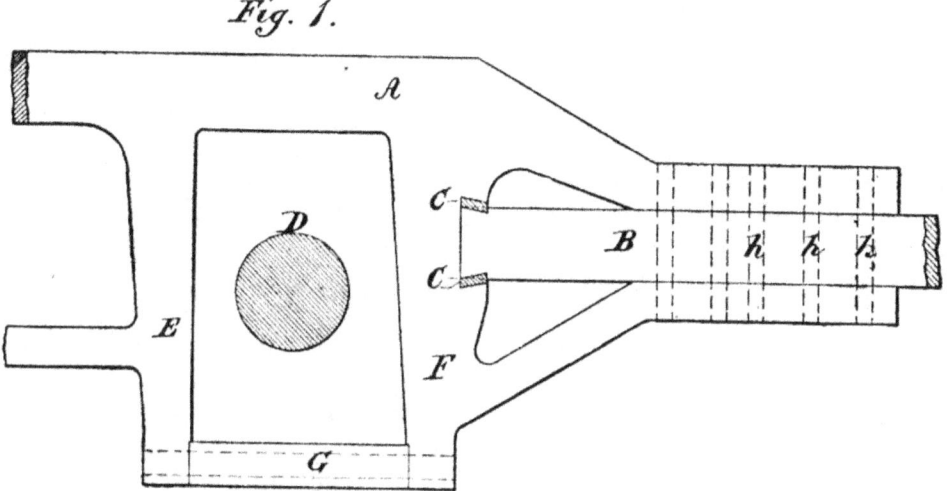

the object of my improvement is the binding together of these two parts in such a manner that they will become to all intents and purposes one piece, and not be liable to separate by vibration and the shearing off of the horizontal bolts, as is the case on ordinary locomotive frames, and to save the expense of

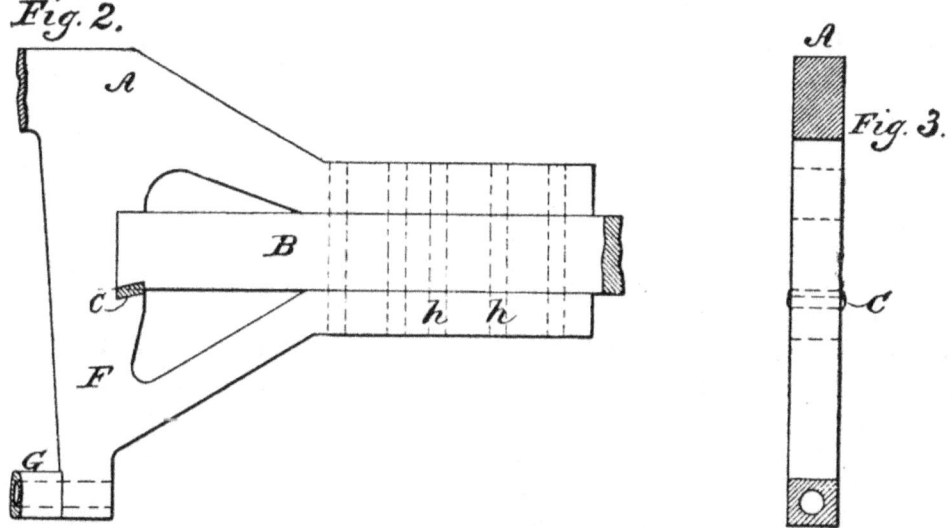

welding feet on end of frame-front, top and bottom, by which frame-front has heretofore been bolted to main frame by two horizontal bolts, in conjunction with vertical bolts h. These objects are attained by me in the manner shown on the accompanying drawings, in which

"Fig. 1 is an elevation of the frame and frame-front where they join. Fig. 2 is the same curtailed; and Fig. 3 is a cross-section of the same, showing key C slightly riveted over.

"A is the main frame and B is the frame-front. C C are keys. D is the forward axle of driving wheels. E and F are the pedestals of frame, and G is the thimble of frame. $h\ h\ h$ are bolts which, in conjunction with keys C C, hold main frame and frame-front together. The method generally employed prior to my invention in binding frame-front B to main frame pedestal F consisted in welding a foot on B, top and bottom, where keys C C are now shown, and binding same to pedestal F by a bolt through each foot and through pedestal horizontally."

This method of securing the frame was used upon thirty locomotives built by the Rhode Island Locomotive Works.

STRESSES ON PEDESTAL JAWS.

It is commonly supposed that the stresses upon each jaw are equal, but such is not the case.

Suppose an engine having 18x24-inch cylinders, 72-inch drivers and carrying 175 pounds of steam. The engine is, by hypothesis, pulling a train, and we will further suppose that it is running slowly, so as to be in full gear. Then the maximum pressure on the pedestal jaws will be approximately when the crank pins are on the upper and lower quarters. The area of an 18-inch diameter piston (neglecting the rod) is about 254 square inches. With 175 pounds boiler pressure we may realize 140 pounds on the piston. Multiplying 254 by 140 gives 35,560 pounds, the thrust on the piston rod. Consider now this as a force, applied in a line parallel to the track to the crank pin on the top quarter.

If the center of the crank pin and the point of contact of the wheel with the rail be considered the extremities of a lever A C 4 feet long, the fulcrum is at the rail C; the application of the force to the pedestal jaw is at the axle B, 1 foot from the crank pin A, and the lever is one in the second order. The lever arms are here 3 to 4. Consider the forces in equilibrium and we have a stress of 35,560 pounds acting at the crank-pin end, 1 foot from the axle, or 4 feet from C. Multiplying 35,560 by 4, the length of the lever arm A C in feet, and dividing by 3, the length of the other lever arm B C, we obtain 47,413 as the number of pounds exerted at the axle against the front pedestal jaw.

The difference of these two forms, or 11,853 pounds, gives the force acting at the end of the lever, which in this case is resisted by the friction between the wheel and rail.

Consider now the fact that not only is the steam pushing the piston forward in the cylinder, but that it is also pushing on the back cylinder head with a force of 35,560 pounds. This is transferred by the frames, and as it is in an opposite direction to the force pushing forward on the front pedestal jaw, the difference of these two, or 11,853 pounds, must be taken as the net pull transferred to the draw-bar.

Let us now proceed to the other position, with the crank on the lower quarter. Here we have a lever B C, 3 feet long, one of whose ends B is the center of the axle, and the other the point of contact of the wheel with the rail C. The fulcrum is at the latter point as before, but the lever is now one of the third order.

With a force of 35,560 pounds on the crank pin A, as before,

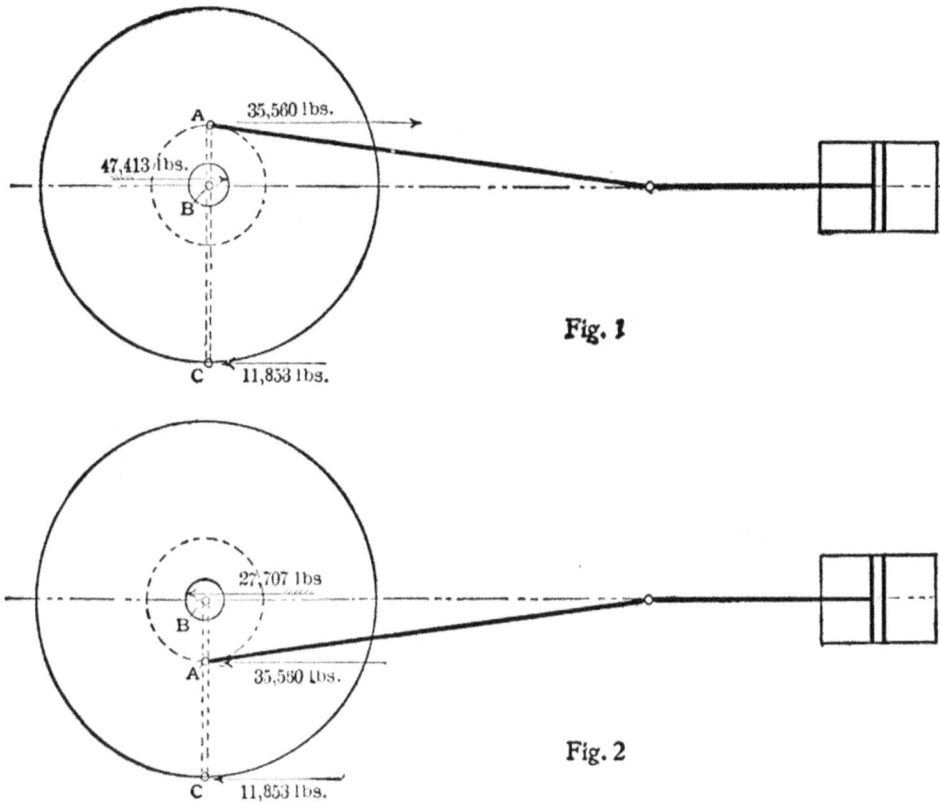

one-third of this will go to the rail and two-thirds to the axle, the forces being divided inversely proportionately to the respective lengths of the lever arms. This gives 11,853 pounds at the rail as before, and 23,707 pounds against the back pedestal jaw. Remember now, however, that the steam is also pushing against the front cylinder head with a force of 35,560 pounds, and this is transferred by the frames as before, so that the difference of these two amounts, or 11,853 pounds, is the net effort applied to the draw-bar.

We therefore see that the stresses are double as much on the front pedestal jaw as they are on the back one.

WHEELS AND AXLES.

A NEW MACHINE FOR TRUEING TIRES WHILE UNDER THE ENGINE.

The illustrations herewith shown present a new device for trueing locomotive driving wheel tires while the wheels are under the engine. It is a product of 1898 and the inventors are Wallace J. Lewis and Peter H. Murphy, of St. Louis, Mo., to whom we are indebted for the following claims and description of the device. Fig. 1 is a side elevational of our improved machine in operative position. Fig. 2 is a plan view of the same, certain parts being omitted. Fig. 3 is an end elevational view of the

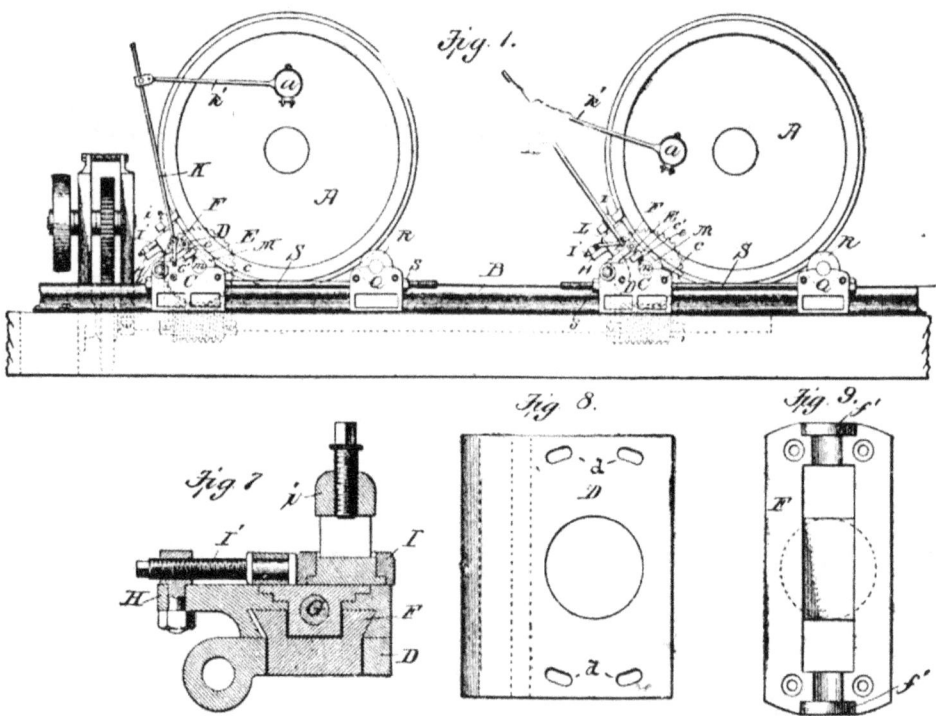

rail-clamp and its carried cutting tool, Fig. 4, is a side elevational view of the same. Fig. 5 is a top plan view of the same. Fig. 6 is a top plan view of the clamp which carries the idle trunnion. Fig. 7 is a sectional view through the tool-holder and its associate parts. Fig. 8 is a plan view of the swinging frame on which the tool-holder guide is mounted. Fig. 9 is a plan view of the tool-holder guide. Fig. 10 is a sectional view through the tool-holder frame and guide, showing the manner of adjusting the latter. Fig. 11 is a side elevational view of the worn gearing for imparting motion to the live trunnion.

The object of our invention is to provide a machine of the character described which is adapted to co-operate with the engine to true the wheels thereof without necessitating the dismantling of the engine.

It is the general practice at the present time when wheels become flat or worn, so as to be irregular, to remove the wheels from the engine and put them in a lathe for the purpose of trueing the tread and flange. It is obvious that such methods are expensive. In our device an engine may be run into a round-house or repair-shop, where power may be obtained, the cutting tool and device clamped to the rail, so as to support the wheels to be trued on rollers, and power being applied to said rollers will cause the cutting-tool to true up the tires, said cutting-tool being fed automatically across the face of the tread.

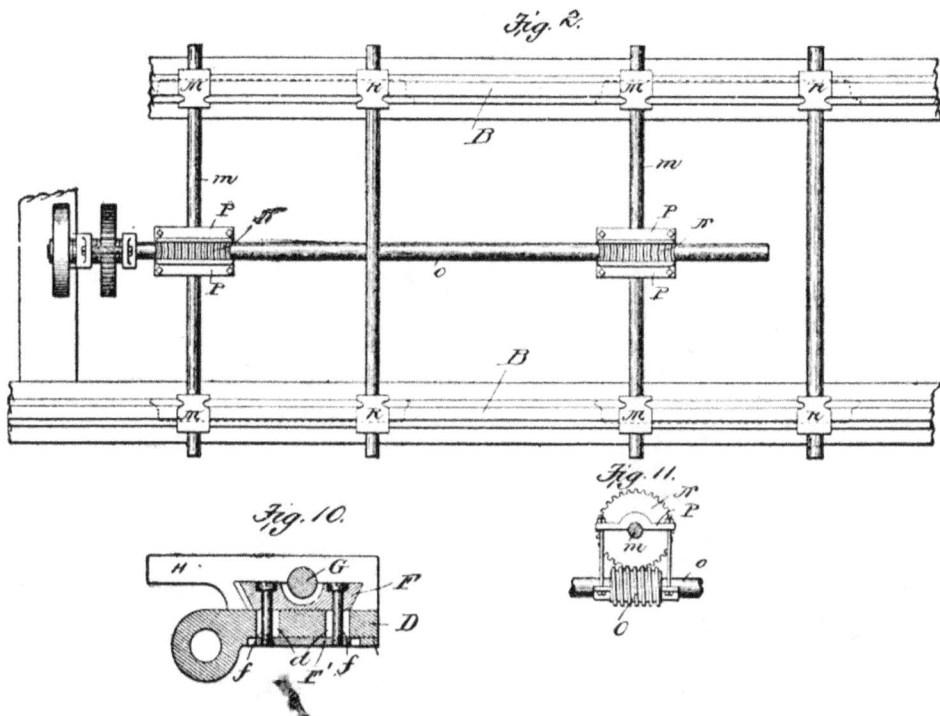

The invention consists in connecting the trueing tool with the wheel being operated upon in such a manner that rotation of the wheel to be trued will feed the trueing device; other features of the device are the novel construction of the rail-clamp; the tool holder, means for adjusting the tool holder, mechanism for moving the tool holder laterally to move the tool along the face of the wheel being operated upon, and the construction and arrangement of the transmitting gearing for imparting motion to the live rollers.

In the drawings A indicates the driving wheels of a locomotive, and *a* the crank-pins thereof; B indicates the rails on which said driving wheels run. We have shown these rails as being arranged alongside of the pit over which the engine whose wheels are to be trued is run.

After the engine is in position our machine is applied to the rails to raise the driving wheels from the track and true them, the engine not being dismantled during the operation of trueing its wheels.

C indicates a clamp adapted to be applied to the rails; the clamp being formed of two like parts for engaging the rail, said parts also supporting the live roller and the cutting-tool. The parts of this clamp are bound to the rail by the cross bolt c.

D indicates a supporting frame for the cutting tool, which frame is pivoted on the cross bolt d. E indicates a bolt having an eye in its upper end, which eye is pivotally mounted to the free end of the frame D. Bolts E pass down through the lug c' on the side of clamp section C and jam nuts e on this bolt impinge against the upper and lower faces of this lug to adjust the free end of frame D. One of these bolts E is arranged on each side of the frame D.

F indicates a guide formed with a circular projection on its under side and about its middle, as shown in Fig. 7, which projection fits in an opening in the frame-plate D, which opening is

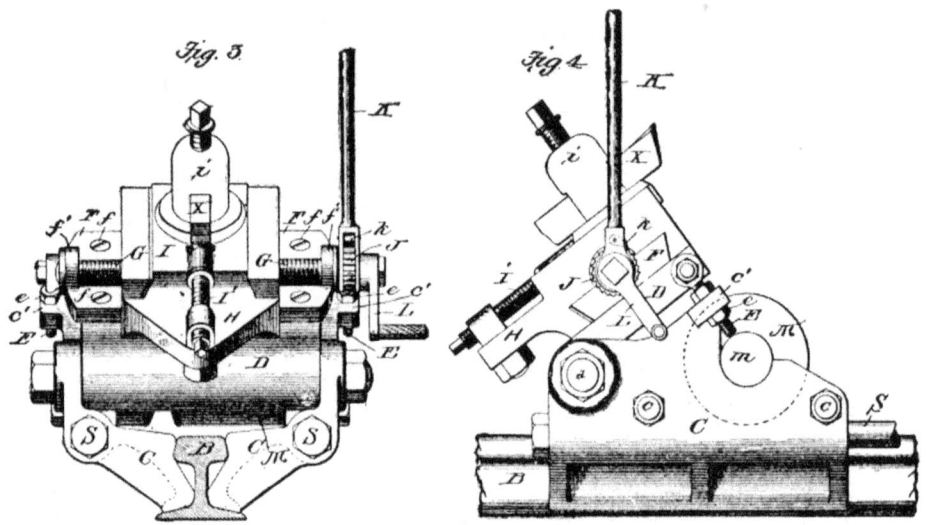

shown in Fig. 8. This guide is provided with bolt-openings at its ends, through which pass screw-bolts f, said screw-bolts passing through slots d, concentrate with the circular opening in the frame-plate D, and into a locking block or nut F', arranged on the under side of the frame-plate D, as shown in Fig. 10. By loosening these screw-bolts in the locking-block the guide F may be adjusted at an angle to the frame-plate and locked in such adjusted position to guide the cutting-tool in an oblique line to form the inclined tread on the wheels. This guide F is formed with ears f' at its ends, in which ears is mounted a feed-screw G.

H indicates a laterally-moving tool-carriage mounted upon guide F and threaded upon the feed-screw G. I prefer to provide a circular bearing-block about the middle of this carriage to act as a nut for the feed-screw.

I indicates a tool-block formed or provided with a dome *i*, in which the tool is clamped. This block is slidingly mounted in longitudinally-disposed guideways and is adjusted to its work by a threaded rod I', which engages therewith at one end, the other end of said rod passing through the nut or projection on the carriage H.

J indicates a rigid wheel fixed to the outer end of the feed-screw G, with which wheel engages a pawl *k*, mounted in a yoke pivoted on an extension of the feed-screw, said yoke being the end of an arm K. I prefer also to provide a crank-handle L on the end of the feed-screw, by which the tool-carriage may be returned after it has traveled across the face of the tread, as is obvious. It will be understood that the pawl *k* is raised to permit this.

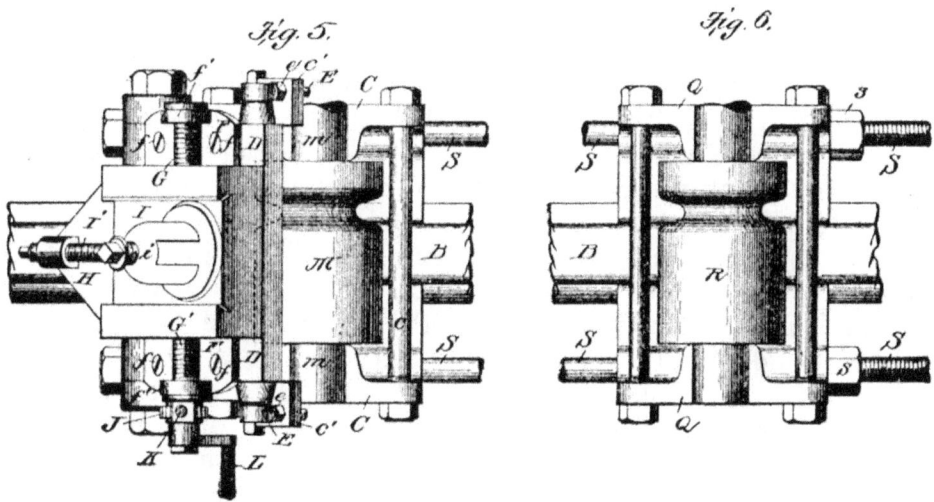

M indicates a roller mounted in suitable bearings in the clamp-sections C. This roller is arranged upon a shaft *m*, extending across the track, said shaft having rollers mounted upon both of its ends.

A worm-wheel N is keyed to shaft *m* and engages a worm O, keyed to a longitudinally-disposed shaft *o*. Suitable bearings may be provided for shaft *o*: but in Fig. 11 I have shown yokes P resting upon the shaft *m* on each side of the worm-wheel, from which yokes are suspended by suitable bolts half-bearings for the shaft *o*. Power is imparted to shaft *o* in any suitable manner, we having shown in Figs. 1 and 2 a power-driven pulley and gearing for driving said shaft. By referring to Fig. 2 it will be seen that when shaft *o* is driven it will operate both screws, shafts *m*, and drive four live-rollers.

It is obvious that while we have shown our device applied to trueing four driving-wheels more can be driven and trued or a single pair can be operated upon, if desired.

Q indicates track-clamps corresponding in design with the

clamps C, with the exception that clamps Q do not carry means for trueing the wheel. In fact, clamps Q act simply as bearings for idle-rollers R. Clamps C and Q are provided with longitudinally-disposed alining openings on each side of the rail for the reception of tie-bolts S. These tie-bolts S are headed at one end, while their other end is threaded and receives a nut s. By tightening the nut s the clamps are turned toward each other on the rail and under the wheel to be trued. By this arrangement the wheel to be trued is lifted above the rail, so that when the live-rollers are driven the wheel to be trued will ride freely on these supporting-rollers.

There are two live-rollers and two idle-rollers mounted on their respective shafts for each pair of wheels, and the rail-clamps forming bearings for these rollers are individually adjustable by means of the tie-rods, whereby it will require but little power to apply the device and get it in operative position. When in operative position, the arm K is connected by a link k' to the crank-pin a of the driving-wheel. When the live-rollers are driven, they cause the wheels to be trued to rotate, thereby revolving the crank-pins, which, connected by a link to the arms K, oscillate said arms and rotate the feed-screw G through the medium of the pawl-and-ratchet connection. The guide F being previously adjusted will cause the cutting-tool X to move obliquely and form an inclined face on the tread. The outer end of link k' is adjustable lengthwise the arm K, so as to impart different throws to said arm and thereby regulate the rotation of the feed-screws. By adjusting the frame D up or down the cutting edge of the tool is placed at a proper angle and in proper position relative to the wheel to be trued and in this way can be adjusted to different-sized wheels.

QUARTERING.

If the wheels and axle are both new the axle should first be quartered in the lathe and the keyways planed or milled out. See how to quarter an axle in a lathe, "Machine Work," page 569. Then proceed as follows: Place the axle on two horses or frame built for the purpose as shown on page 541; drive each wheel on the axle as far as you can with heavy wooden blocks, being sure to have the right crank-pin lead the left pin one-quarter of a turn. Now put centers in all four holes, crank-pin holes (whether bored out or not) and in center of wheels. Clamp a straight-edge true with these centers on each side and use a spirit level and adjust the wheels until one straight-edge is perfectly horizontal and the other vertical; then scribe the keyways in both wheels. If the crank-pin holes are not bored out they can be finished in a quartering machine or lathe afterwards. If an old axle and new wheels proceed the same as before and scribe the keyways in the new wheels. If old wheels and new axle do not quarter the axle in the lathe proceed as previously explained and when both wheels are properly set scribe the keyways on the new axle to correspond with the keyways in the old wheels, but if the old wheels have the crank-pins intact it will be a little more difficult, but may be accomplished in various ways. If you have an offset straight-edge similar to the one shown in our illustration on page 541 then proceed as follows: Locate a small center in the center of the wheel and set a straight-edge to this center, letting one end of the straight-edge rest on the top of the collar of the crank-pin, then at the other end of the straight-edge scribe a short line on the wheel or tire. Now change the straight-edge, letting it touch the bottom of the crank-pin and keep it true with the wheel center and scribe another line on the wheel or tire same as before. Now with a pair of hermaphrodites and dividers locate the exact center between these two short lines; this center must therefore lie in a true line with the wheel center and center of the crank-pin. Therefore set the straight-edge true with this center and also true with the wheel center and clamp it in this position, then use a spirit level on the straight-edge and adjust the wheel until the straight-edge is perfectly level. On the other side you can set another straight-edge the same way, or you can describe a circle from the wheel center equal to the diameter of the collar on the crank-pin and drop a plumb line around the collar and adjust the wheel until both lines are true with the circle. Whatever method is employed the object is the same to set the centers of the crank-pins at perfect right angle. Perhaps a more simple method would be to describe a circle from the center of each wheel equal to the diameter of each crank-pin collar, then set a plain straight-edge, letting it rest on the collar of the crank-pin and setting it true with the circle, clamp and level the straight-edge and then drop a plumb line on the other side as previously explained. When set true then scribe the keyways on the axle.

HOW TO FIT DRIVING WHEEL KEYS.

The key-ways are usually straight. The key should fit on the sides, but should be 1-32" loose top and bottom. Plane up key accordingly, allow 1-64" stock sideways for fitting and drive with small sledge. Keys are usually planed on the end of a forging which has a head or heel on it to back out with when too tight. Set a jack under end of key to prevent bending.

HOW TO TRUE UP BEARINGS ON OLD CRANK-PINS IN HUB.

There is a small machine for this purpose, to turn by hand. A center enters center in pin, and set screws fasten to inside collar; three dogs rest against outside of inside collar.

CAST STEEL WHEEL CENTERS.

A great many of the large driving wheel centers for modern passenger engines are now made of cast steel. With this kind of a wheel center the weight is greatly reduced, while the strength is increased.

A VERY SIMPLE CRANK-PIN GAUGE.

To press a crank-pin out of a driving wheel to determine positively whether it is bent or how badly it is worn out of true is rather an expensive job. The most simple form of a gauge for this purpose that we have yet seen consists of a small light V block which rests on the bearing and an adjustable band encircles the remainder of the bearing. An adjustable needle slips through the V block so the face of the wheel can be tried all around, or with a bent needle a circle may be drawn on the outer collar (or face) of the pin. The band can be adjusted to fit any sized bearing.

CRANK-PIN PRESS.

The first cut shows a plain, simple form of portable press for forcing crank-pins in or out of driving wheels, while the wheels

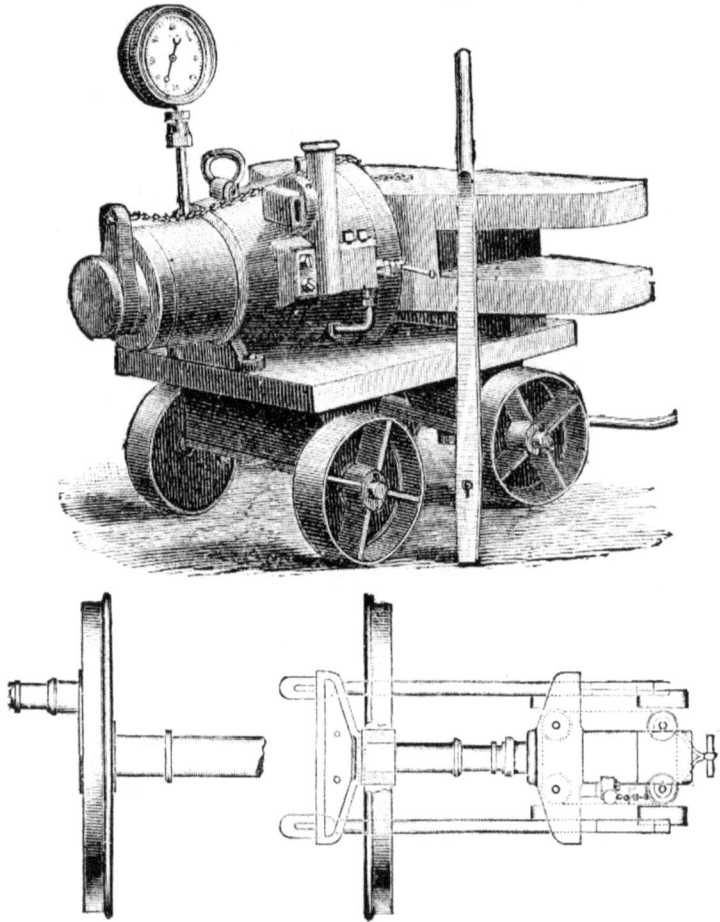

are under the engine. The second cut shows the method of attachment and operation. It is a very useful tool.

MODERN COUNTER-BALANCING.

The adjusting of counter-balance in driving wheels is receiving much more attention than was formerly devoted to it, on account of the agitation of the subject that was forced on the heads of mechanical departments by the increased weights and higher speeds of power. Much has been said and written concerning this subject within the past few years, but railway officials and locomotive builders are indebted to Prof. Wm. F. M. Goss, of Purdue University, more, perhaps, than to any other man for the numerous experiments and tests made covering this particular part of locomotive construction. It is a well established fact that a poorly balanced locomotive is very injurious to the road bed and track, causing what is known as "the hammer blow" upon the rail; such a condition is also very severe on the machinery of the locomotive, causing the engine to jerk and pound. An experiment was made at Purdue University (where a locomotive is mounted on a testing plant) by letting a long piece of wire run under the driving wheel while the engine was running at a high rate of speed. It was clearly demonstrated by the unevenness of the wire after the test, that a poorly counter-balanced driving wheel not only pounded but actually left the rail. It is therefore very essential than an engine should be balanced as nearly correct as possible, but it should be borne in mind that it is impossible to have an engine perfectly balanced at all times, for, if it is balanced perfectly under steam pressure it will not be when steam is shut off. Remember that the force that propels the engine is in the cylinder and not in the crank-pin. It is, however, possible to parts approximately correct so that excessive strain will not be imposed, the object being to balance the wheels almost perfectly while running, but to overcome a part of the stress upon the crank-pins when steam is shut off. There is more or less of an air of mystery surrounding the operation, and the veil is made opaque or transparent in proportion to the facilities in a shop for doing the work.

There are nearly a score of different empirical rules in use for balancing and the results obtained by them have varied all the way from very good to very bad.

The two-thirds proportional weight of the reciprocating parts to balance as given in our formula was carefully computed as a result of recent scientific experiments upon the testing plant. While this proportion is subject to modification under certain circumstances, yet on a general average for all classes of engines it is considered perfect. For engines with very light reciprocating parts 50 per cent or one-half the weight of the reciprocating parts would be sufficient to balance, while upon compound engines with very heavy reciprocating parts 75 per cent or three-quarters of the weight should be balanced.

Formula for Counterbalancing.

Report of Counterbalance in Engine No. _281_ _Greenville_ Shops, _Jan. 18_ 189_3_.

RECIPROCATING PARTS.

	Lbs.
Weight of Piston, Packing Rings and Crosshead, complete,	
Weight of Front End of Main Rod, complete,	90
Total Weight of Reciprocating Parts,	
⅔ Weight of Reciprocating Parts,	
Counterbalance on each Wheel for Reciprocating Parts,	91

WEIGHTS TO BE COUNTERBALANCED.

FRONT WHEEL.	MAIN WHEEL.	BACK WHEEL.
Lbs.	Lbs.	Lbs.
Weight of Front End of Side Rod _10_	Weight of Back End of Main Rod _90_	Weight Back End of Side Rod ____
Proportion of Reciprocating Parts _48_	Weight of Middle Connection ____	Proportion of Reciprocating Parts ____
	Weight of Front End of Side Rod ____	
	Proportion of Reciprocating Parts ____	
Total weight to be balanced _2.3_	Total weight to be balanced ____	Total weight to be balanced ____

CONDITION BEFORE RE-BALANCING AND CORRECTION

	Present Counterbalance, weighed at Crank Pin.	Correct Counterbalance, weighed at Crank Pin.	Present Counterbalance, Light.	Present Counterbalance, Heavy.
Front Wheel,				
Main Wheel,				
Back Wheel,				

Counterbalance corrected as above

_____ Master Mechanic.

NOTE.—Divide two-thirds of the weight of the reciprocating parts on each side by the number of driving wheels on a side, and add quotient to weight of revolving parts on each wheel. Under Weights to be Counterbalanced Main Wheel, the "Weight of Middle Connection" to be used for six coupled engines, "Weight of Front End of Side Rod" for four coupled engines.

For eight-wheeled engines it is considered excellent practice to divide the total weight to be balanced among the four wheels. For engines having more than four driving wheels each wheel should be balanced for all of its own revolving weight and its proportion of three-fourths (or other proportion used) of the reciprocating weight; this three-fourths of the reciprocating weight is to be divided among the wheels in proportion to the revolving weight carried by each wheel. The designing of the wheel and counter-balance is done in the drawing room, but unless the wheels are for duplicate engines the counter-balance is invariably made a little heavy as it is easier and better to turn off a part of its weight than to add weight. The wheels are counter-balanced in the machine shop after each wheel has been finished and pressed onto its axle, but it is not necessary that the crank-pins should be pressed into place, as they can be weighed. The pair of wheels to be balanced should be mounted on a frame, or pair of horses; most every good shop has a frame built especially for this purpose. The one shown on page 541 is in use in the Wabash Railway shops at Springfield, Ill. It is of neat and convenient form and has an adjustable screw to level it by. The horses, or frame, should have an iron top and be leveled up; place one crank-pin on the top quarter and you are ready to begin.

LOCATING THE CENTER OF GRAVITY.

The first thing necessary will be to ascertain the distance from the wheel center to the center of gravity. It is impossible to locate the exact center of gravity in a counter-balance, owing to its peculiar shape, but it can be located near enough for all practical purposes. Various methods are used to determine this point, but the two methods hereafter explained are those generally used. First, use a large piece of pasteboard or thin planed board for a templet and lay it off and have it cut or sawed out the exact shape of the counter-balance, then balance the templet lengthwise on anything with a sharp edge, or on a straight-edge, and mark a line on each side of the straight-edge as shown by Fig. 1; then balance it the other way, or locate a center midway between its two ends and midway between the lines; this point is considered the center of gravity.

Another method, and perhaps a better one, is what is known as the "string method;" after the templet is cut out as previously explained, locate three points on the templet at any convenient place; they are usually located near the corners of the templet as shown by Fig. 2, and from each of these points let the templet suspend, also allowing a chalked line to suspend from each point and with it make a line on the templet each time. You will find these three lines will intersect each other at the same point; this point is considered the center of gravity. It now becomes necessary to determine what distance this point is from the wheel center; to determine this distance clamp the templet to the wheel, keeping it true with the counter-balance, then measure the exact distance between the two centers.

Now clamp a straight-edge to one wheel, true with the wheel center and the center of the crank-pin or crank-pin hole (see page 541). Then go to the opposite wheel, where the crank-pin hole should be at the top quarter; plumb this center with the wheel center; if the crank-pin is intact describe a circle around the wheel the size of the collar on the crank-pin, and drop a plumb line around the collar and set both lines to the circle, then return to the opposite side, using a sharp pointed stick or piece of iron between the straight-edge and a pair of scales. This stick should be kept in a vertical position with its top set a distance from the wheel center equal to the distance from the wheel center to the center of gravity. The weight transmitted to the scale by the stick should then be ascertained and marked down for reference; then weigh the other side and each other wheel the same way, keeping the weight of each separate.

A sort of trial balance account should then be opened with

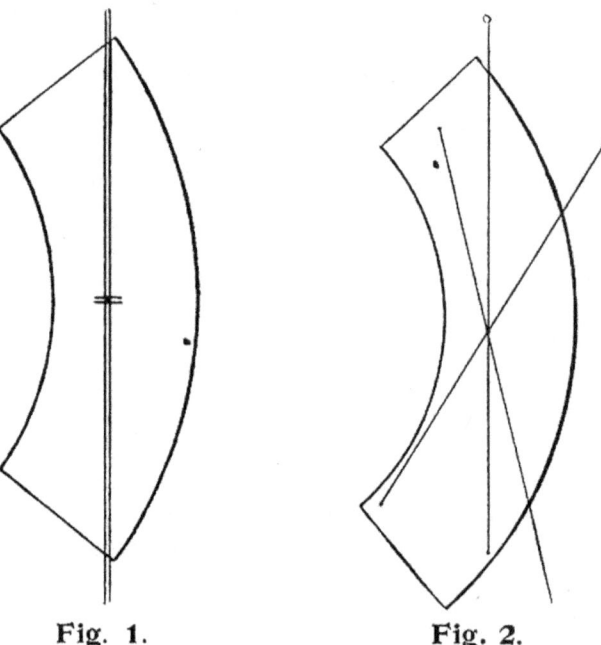

Fig. 1.   Fig. 2.

each wheel, entering the weight of the counter-balance on one side and the weight of the revolving parts and a proportion of the reciprocating parts upon the other.

Now, if an eight-wheel engine, weigh the side rods, straps, brasses, oil cups and all; divide the weight of each rod between the two wheels and if the crank-pins are not yet pressed into place add the weight of each to its respective wheel; then level the main rod with strap, brasses, etc., and weigh it at the back end and add the weight to each main wheel. If there are more than four driving wheels connect all the rods and level them as shown by the illustration on page 541, and weigh it at each point where it fits onto the crank-pin. The weight of the back end of the main rod should then be added to each main wheel. Now in ordinary practice two-thirds the weight of the reciprocating parts should be balanced and this weight should be divided equally among the driving wheels. The reciprocating parts are the piston-head rod, etc., the cross-head and forward half of the

THE LOCOMOTIVE UP TO DATE. 541

main rod, with all bolts, nuts, oil cups, etc. When this is finished see how the account of each wheel will balance and add to or subtract weight from each counter-balance, as the case requires. This may be done by pouring lead into the counter-balance or turning off a part of the balance in the lathe, weighing the iron turned off until the required weight is turned off. Another method of balancing, extensively practiced, is to hang weights onto the crank-pin until the wheel is perfectly balanced, and then compare the weight obtained with the weight of the rods, reciprocating parts, etc.

METHOD OF COUNTER-BALANCING ON THE WABASH RY.

The following method for counter-balancing on the Wabash Railway, as described by Mr. J. B. Barnes, Superintendent of Motive Power, we are pleased to present with drawings:

The wheels are placed on the wooden horses and leveled up in each direction, with the side which is to be worked on in the position as shown, so that a line through the crank-pin center

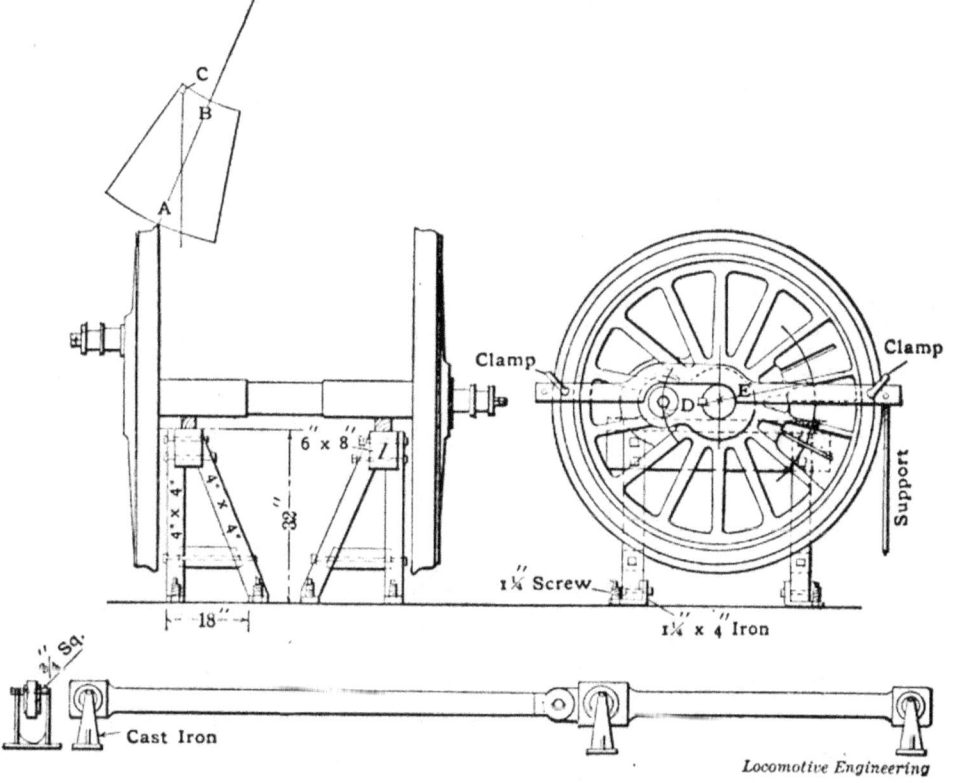

and axle center will be perfectly level. The rods are attached to the crank-pin, the outer ends of the rods being supported so that they will also be level. Weights are then suspended from the pin, such as will represent the proper proportion of reciprocating weight acting upon the wheel. The balance is then completed by adding to, or deducting from the counterbalance.

Each wheel is to be balanced for all of its revolving weight, and its proportion of three-fourths of the reciprocating weight; this three-fourths of the reciprocating weight is to be divided among the wheels in proportion to the revolving weight carried by each wheel.

To find the weight to be added to or deducted from the counterweight, take the surplus or deficient weight as shown at the crank-pin, and multiply it by the distance D, then divide that product by the distance E, and the quotient is the number of pounds to be added to or deducted from the counterweight.

The same results may be obtained without placing the rods on the pin, by clamping a straight-edge to the wheel as shown. At a point equal to three times the distance D from the axle center, place a support under the straight-edge, with the lower end resting on scales, and note the weight. Multiply this weight by three, and note the difference between the product and the weight of the rods and reciprocating parts belonging to the wheel. This difference multiplied by the distance D, and divided by the distance E, will give the amount in pounds to be added to or deducted from the counterweight. In order to retain the center of gravity, the additions or deductions should be made to extend over the whole surface of the counterweight.

To weigh the rods, place them on supports as shown, with all bolts, keys and oil cups in place, and block up under the supports, so that each block may be removed in turn and the weight at that point taken on scales; the main rod to be weighed in the same manner. To find the center of gravity of the counterweight, make a thin wood template of the balance block, showing the center line AB. Suspend the template from any point, as C, and where the plumb line from C intersects the line AB, will be the center of gravity. The revolving weights are the crank-pin, the crank-pin hub, the side rods and back end of main rod. The reciprocating weights are the piston and rod, the crosshead and the front end of main rod.

CYLINDERS.

FITTING TOGETHER.

Use a long straight-edge and set the back faces of the cylinder head joints perfectly true with the straight-edge at all four points and see that the bottom of the saddle is set true where the center casting fits. Use a spirit level and level each valve seat both ways. Put a center in the end of each cylinder and see that each end trams perfectly, and each end the same distance apart. If the cylinders are perfectly true at all the above points, then proceed to rose-bit all the holes and bolt them together with diving fit bolts. If they are not true then use liners between them wherever necessary until they are true; then return one cylinder to the planer, calling the planer hand's attention to the thickness and location of each liner.

FACING OLD CYLINDER HEAD JOINTS.

Ordinarily this job is done with a file and scraper, but in some shops a small machine is used for this purpose; it has a cross-feed and can be clamped into the cylinder head studs. It is a great labor-saving device as it is a very tedious job to file and scrape a cylinder head. Mr. William McKenzie, Master Mechanic of the Southern Pacific Ry. at West Oakland, Cal., in describing the method employed upon their road to secure tight cylinder head joints, says: We shear off a piece of sheet copper the proper thickness and width and drive it into the groove (which is cut into cylinder heads to allow the center of the head to blow out in case of accident to the rod or strap, thereby saving the cylinder). We allow sufficient stock to let the copper extend outside the head a sufficient distance to true it up in the lathe. The joint is turned V shape and will remain tight while the engine remains in service.

GRINDING CYLINDER HEAD JOINTS.

Short work is made of this job in locomotive works; it being done in a large drill press or boring mill, the head being raised by the spindle at intervals to apply emery; but in railroad shops this is seldom done. Proceed as follows: If possible upend the cylinder, fasten a short pole or board onto the head and grind it with oil and emery, using No. 2 emery. When you cannot upend the cylinder use a long rod through the cylinder or a brace and ratchet outside; slack the head frequently to apply emery and thereby avoid cutting grooves on the joint.

FITTING CYLINDER SADDLE WHEN SMOKE BOX IS RIVETED TO BOILER.

Let the arch rest on the saddle of cylinders in proper place; now level boiler with cylinders, with spirit level used on valve seat and on ring for the dome cap, also on shell of boiler. Now line both cylinders and divide fire box central with cylinder lines. Now drop plumb line over wagon top of the boiler and keep fire box plumb with the lines on both sides.

Now, if the saddle will finish so as to hold the boiler proper height up and down, lay off the saddle with hermaphrodites, as follows:

Set hermaphrodites under the center of the smoke arch, with points perpendicular with the center of the smoke arch. Now scribe a line all around at both ends and on both sides, and be careful to hold your hermaphrodites in the same perpendicular position all the way around.

Now separate boiler and saddle, and chip saddle down to a

good bearing; use a straight-edge to chip to lengthwise. When you have finished put some red lead on the saddle and bolt down solid.

If the circle of the saddle is too large for the boiler, and will not true up, put in a sheet iron or boiler plate liner, and figure on the thickness of the same when laying it off.

FITTING CYLINDER SADDLE WHEN SMOKE BOX IS TO BE FIRST BOLTED TO CYLINDER.

Some of the larger shops have a boring bar for this purpose, but you can lay it off as follows: Clamp board up at front and back end of saddle to be used for center of smoke arch; tack a piece of tin on each for a center. Now get the outside diameter of smoke arch and carry a line up on one board from lowest point of saddle, that distance. Now put four centers in your cylinders and set trams for cylinder center to a point where both marks will cross, exactly on the line you have carried up on one board from saddle; mark a small center at that point. Keep your trams set to same length and scribe two lines from center of cylinder at the other end, and make another small center where these lines cross.

Now you have both centers, so set your trams to one-half the outside diameter of the smoke arch and scribe a line on each side of the saddle. Set a straight-edge on the outside of saddle on each side, and set to each line, and scribe a straight line.

Now it is laid off, so take down your wooden centers and chip to your lines. When you have finished bolt your smoke arch solid to saddle. Set boiler perfectly central between frames at back end, and perfectly level both ways. Then have arch riveted to boiler.

BORING OUT OLD CYLINDERS WITH BORING BAR.

Set bar central at each end by counter bore; or if back head is up, use taper gland in stuffing box. If cylinder is worn out very bad, take two cuts; use a coarse feed on first cut and a fine feed on the second cut. Recounter bore each end (if it needs it) with offset tools. If the tool chatters, hang a heavy weight on the back end of the boring bar. The reason a cylinder is counter-bored is to prevent the piston wearing a shoulder in the cylinder; the counter-bore should be made a sufficient length so the outside edge of the cylinder packing ring will just travel over the inner edge of the counter-bore. It should be made about $\frac{1}{4}''$ deep.

Small engines and motors are often used in the round-house to drive the boring bar. Steam is sometimes taken from the locomotive for this purpose.

NEW LOCOMOTIVE CYLINDER BORING MACHINE.

The device here shown is the invention of Mr. James Buchanan, Assistant Superintendent of Motive Power of the N. Y.

C. & H. R. R. Co., at West Albany, N. Y. It consists in the combination of a cylinder borer and valve seat facing machine to be operated in conjunction, or separately, and driven by a small oscillating engine. The operation of the device is clearly shown by the illustration. The principal improvement in this device over methods now in use is the addition of the small os-

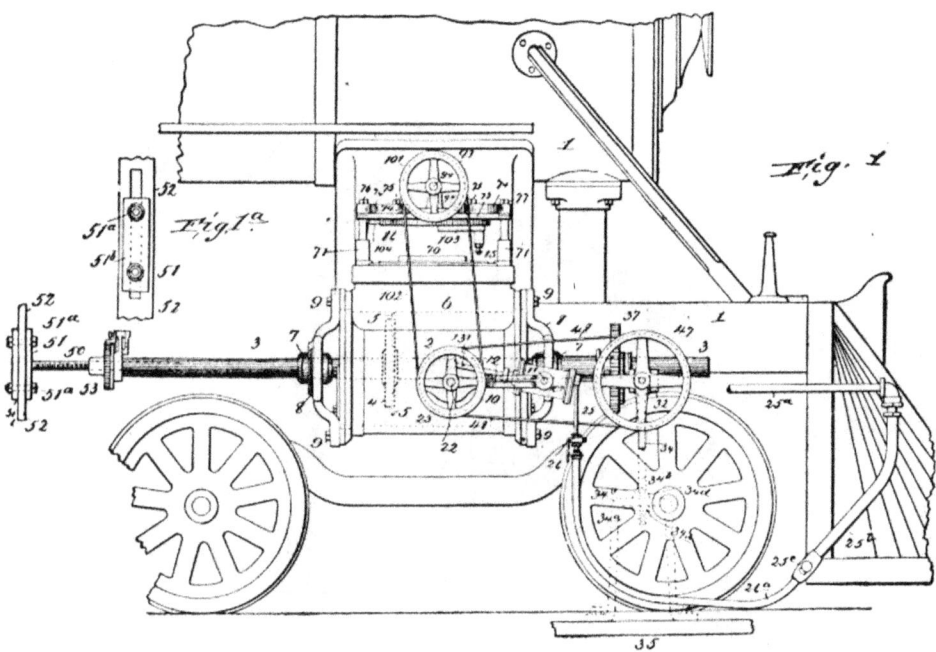

cillating engine, thus permitting the cylinders to be bored out or valve seats faced while the engine is in the round-house or elsewhere. Steam or air may be used and may be furnished by the locomotive upon which the machine operates. This invention also consists in means of giving the boring tool a continuous rotary motion with an intermittent feed, so as to produce the most perfect boring within the cylinder.

FACING VALVE-SEATS, ETC.

ROTARY VALVE FACING MACHINE.

This machine is made to face locomotive valve-seats, and is used in nearly all the principal railroad shops in this country. It is considered one of the greatest labor-saving machines in use. You can face all sizes of locomotive valve-seats; also cylinder and dome joints, these machines being made in two sizes.

This machine is secured to the cylinders for facing valve-seats by placing the four tubular legs A onto the four corners, or any of the steam chest studs, and tightening up the set screws F of said legs, and tightening the nuts on the steam chest studs, and is secured in the same manner to the dome or cylinder head studs. With this machine you can face a valve-seat in about one-fourth

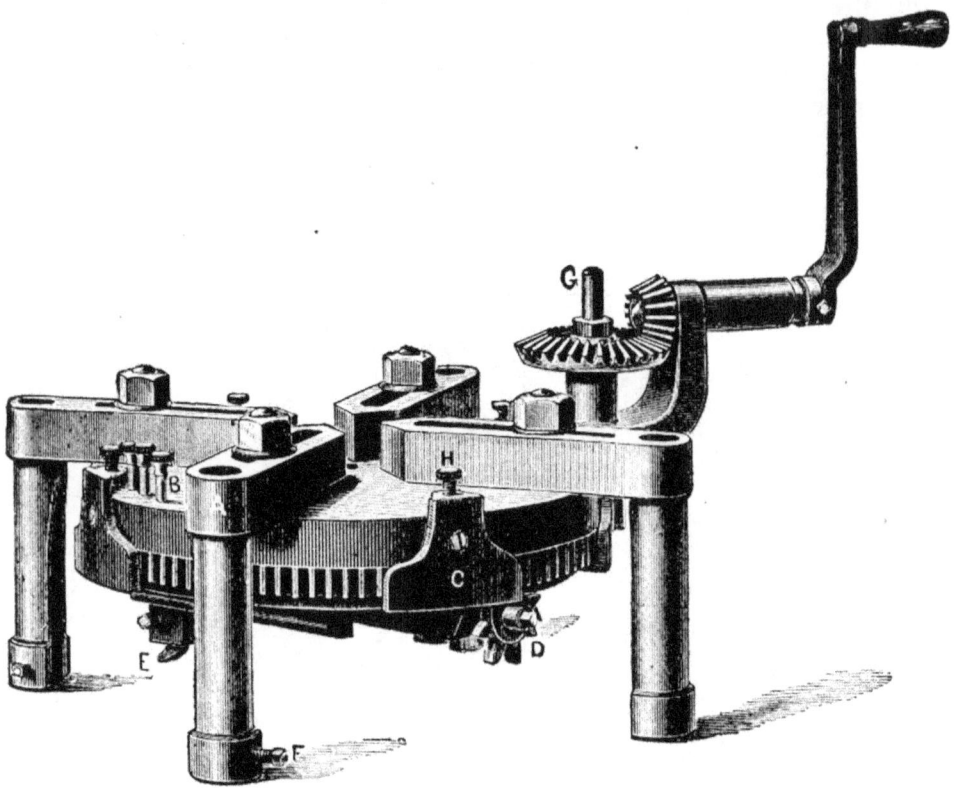

the time required to do it by hand. If the machine is in good order it leaves the surface almost perfectly true, therefore the seat requires very little filing or scraping.

TRUEING UP AN OLD VALVE-SEAT.

If you have a valve facing machine use it to true up the valve-seat; see that the machine is properly cleaned and oiled before using, and that all lost motion in the machine is taken up; scrape the face of the cylinder at the four points where the legs of the machine are to be fastened. Usually the legs will not have a good bearing on the bottom owing to the slight elevation for the steam chest joint; in such cases four thin parallel strips should be used under the legs of the machine (pieces of old cylinder packing rings are good for this purpose), set the machine so that the four corners of the valve-seat will finish at the same time; this will save considerable turning by hand, and when set tighten the nuts on the steam chest studs securely and use a broad nose tool; a proof mark should be left when finished; remove the machine and use a large face plate on the seat, or the valve will do if it is planed, and scrape the seat to a good black-lead bearing.

THE LOCOMOTIVE UP TO DATE.

If you have no valve facing machine and the seat is worn very badly, use a hand hammer and chisel, and chip down the high places, chip true with a straight-edge and keep the seat the same height from the cylinder all around, then use a bastard file until it shows a pretty good bearing with a face plate, then use a fine file or a scraper and bring it down to a good bearing all over the seat.

PUTTING ON NEW FALSE VALVE-SEAT.

First true up the cylinder, or what is left of the old seat, either with a valve facing machine or by hand, bring it to a good bearing, also the bottom of the false seat; if the seat has wings to fit inside the steam chest, first screw both corner steam chest studs into place at each of the four corners, then clamp the false seat true with the edges of the steam ports in the cylinder or old seat; now put on the steam chest and scribe each of the four wings inside of the steam chest; remove the chest and false seat and have the ends of the wings jumped off, or fit and file them to fit the steam chest.

If the false seat has no wings then on the face of the old seat drill, counterbore and tap about 16 holes for screw bolts, placing two or three in each bridge. Screw the bolts down solid, tapping the seat lightly with your hand hammer to see that each bolt is tight; if the screw bolts are brass cut them off flush with the face of the valve seat, but if they are iron cut them off so that the top of each head will be 1-16" below the face of the valve-seat; when all the bolts are in place use a face plate and scrape the seat to a good bearing.

LINING DOWN PRESSURE PLATES.

Regardless of the form of balanced valve used, when the steam chest cover is tightened down the pressure plate should permit the valve to move freely without binding it in any position; in order to line down the pressure plate properly first have the gaskets put on the steam chests, then use two straight-edges and a scale and measure the exact height of the steam chest. We will assume it measures $8\frac{1}{4}"$; now place the valve on the valve-seat, place a straight-edge on top of the valve and measure the distance from it to the seat on the cylinder; we shall assume this distance is $7\frac{1}{2}"$; now you must make allowance for tightening down the gaskets and for clearance between the valve and pressure plate, 1-32" being sufficient for clearance; if the gaskets are new and made of $\frac{1}{4}"$ copper wire the gaskets will let the steam chest cover squeeze down 3-32" before the cover is considered tight, while if the gaskets are made of thin flat copper they will not squeeze to exceed 1-32", but we shall assume that we have $\frac{1}{4}"$ copper wire gaskets, therefore add the 1-32" clearance to the 3-32", which the gaskets will squeeze, which amounts to $\frac{1}{8}"$; now subtract $\frac{1}{8}"$ from $8\frac{1}{4}"$, which leaves $8\frac{1}{8}"$, and as the top of the valve is 7½" high the difference, which is ⅝", is the distance the pressure plate should extend below the joint on the steam chest cover; so line the pressure plate accordingly, allowing 1-32" stock for the plate to true up on the planer; use turned brass washers for liners between the plate and cover; they should all four be turned the same thickness. Let the screw bolts pass through then, and tighten the screw bolts as tight as possible without breaking them.

LINING GUIDES.

It is very important that the guide bars of an engine be lined perfectly true with the piston's travel, otherwise trouble may be expected with the guides and crosshead. In order to keep the crosshead perfectly true with the piston it should be keyed onto the piston rod for planing while the finishing cuts are taken on the crosshead, and the cylinder should be properly lined with a very fine, strong line, and the line should be set perfectly central by the counterbore at the front end and by the stuffing box at the back end.

HOW TO LINE UP FOUR BAR GUIDES.

Upend the crosshead upon your bench and fit a piece of flat copper into the hole for the piston rod, and with a pair of hermaphrodites locate the center of the hole, making a very small prick-punch mark in the exact center. Now place two parallel strips or straight-edges against the two wearing surfaces on the bottom side of the crosshead, letting the end of each extend above the end of the crosshead, and clamp them in this position, or have some one hold them while you set the guide gauge, letting the gauge touch both straight-edges and setting the needle to the center of the hole for the piston rod.

Next bolt up both bottom guides in their respective places, then line the cylinder perfectly true. Now line both guides until the needle of the guide gauge will just touch the cylinder line at both front and back ends. Keep both bottom guides perfectly true across their faces, also lengthwise and perfectly level with the top of the cylinder, and an equal distance from the cylinder line and allow 1-32" for lateral motion. Now place the crosshead on the two guides and move it two and fro and see if there is any rock motion in the crosshead and notice the points where it bears on the guides. Line the bars to take out all twist and see that the crosshead has a good bearing, then put the inside top guide in place and line it down to the crosshead and keep it true with the bottom inside guide by using an adjustable square or straight-edge. It should be lined down as close as possible without binding the crosshead, which should move freely by hand. Now put on the outside top guide and line it the same way; remember you can spring any of the guides up or down in the center (if they are not too thick) by using narrow liners either in front or back of the bolt. When the guides are properly lined remove one bolt at a time, using clamps in its stead and rosebit each hole; fit the bolts and tighten them, then trim off the liners and make a neat job of it.

LINING TWO BAR GUIDES.

MOGUL.

First put a center in the crosshead same as for four bar guides, then clamp both gibs to the crosshead without any liners and measure the exact distance from the face of the bottom gib to the center, also try the center and see if it is central sideways.

Now place both guides in their respective places and bolt them securely, then line the cylinder, and with a guide gauge, or straight-edge and scale, try the two guides and see if they are central and parallel with the cylinder line; line the bottom guide the right distance from the cylinder line at both ends, now caliper the crosshead outside of both gibs, and line the top guide close enough to let the crosshead and gibs between them without any liners under the gibs; see that each guide is perfectly square across its face, and level with the cylinder and the sides perpendicular, then put up the crosshead and slip both gibs into place and put on the outside plate, or plates, and see that the crosshead will move freely, then rosebit the holes for new guide bolts. If on account of the thickness of the guide blocks it becomes necessary to place liners under the top gib, see that the oil holes are cut in the liners before putting them into place.

If the engine is on blocks and there is no support between the fire box and the front end of the engine, it is considered good practice to set the back end of both guides 1-32" low at the back end; when the weight is on the wheels the guides will be central.

When lining old crosshead gibs place the crosshead at the back end of the guides, and caliper between the bottom guide and the piston; if the piston and guide are parallel place the liners under the top gib; if the piston is low at the back end, place enough liners under the crosshead to make the piston parallel with the bottom guide.

BOTH GUIDES ABOVE THE CYLINDER CENTER.

Locate a center in the crosshead as previously explained, find the exact distance from the wearing surface of the top gib to the center of the crosshead, then bolt the top guide in its place and line the cylinder; set the top guide the right distance from the line at both ends, using a twelve-inch scale with an adjustable square head; see that the guide is central with the cylinder line, and that its face is level with the cylinder, then remove the cylinder line and put up the bottom guide; caliper the gib and the bottom guide that distance from the top guide, keep the bottom guide level across its face and true with the top guide; now slip the solid crosshead gib into place, line close and see that the gib will move freely, then put up the crosshead and you have finished.

LINING ONE BAR DIAMOND GUIDE.

Fasten the guide in its place, put up the crosshead and line the cap closely so the crosshead will move freely upon the guide; now line the cylinder, letting the line extend through the hole in the crosshead, then with a pair of inside calipers see if the cylinder is central in the hole in all positions; if not, line the guide until it is.

LAYING OFF NEW GUIDE BLOCKS.

FOUR BAR GUIDES.

Locate a center in the crosshead and find the distance this center is above or below the wearing surface on the bottom of the crosshead, tighten all four guide blocks in their proper places, keeping each perfectly level across its top face, then line the cylinder perfectly; now clamp a short straight-edge across the face of both front guide blocks, keeping it level with the cylinder and the same distance from the cylinder line as the crosshead center is from the bottom wearing faces of the crosshead, and clamp another straight-edge to the faces of the two back guide blocks, setting it the same way; use a spirit level and see that the two straight-edges are perfectly true with each other, and level with the top of the cylinder, and each an equal and correct distance from the cylinder line, then scribe a line across all four guide blocks; next caliper the wings of the crosshead; they should both be the same thickness; so lay off another line across the face of each guide block above the first line and parallel with it, and a distance from it equal to the thickness of the wings of the crosshead; work close and have the machine man split your lines, and when the guides are hung it will not be necessary to use a single liner.

TWO BAR GUIDES.

These guide blocks should be laid off exactly the same way as for four bar guides, but they can be laid off many different ways, such as measuring from center of stuffing box to center of guide block hole, etc., but the method we have explained will be found to give the best results.

STRAIN ON TOP GUIDE.

When the center of the cylinder is between the top and bottom guides, remember most of the wear is on the top guide when in the forward motion. Remember the power is in the cylinder, and not in the crank-pin.

ROD WORK.

TO DETERMINE THE LENGTH OF A MAIN ROD.

If the guides are not yet hung and the engine is still on blocks proceed as follows: If the main jaw is square with the top of the frame place a long straight-edge across both main jaws and measure the distance to the face of the back cylinder head. Now add to this length the thickness of the main shoe and one-half the thickness of the driving box, and from the total length subtract the following: Distance from the center of the crosshead pin to the front face of the crosshead; one-half of the stroke of the crosshead; the clearance and the length of the front guide block and the remainder will be the length of the main rod. If the construction of the engine is such that the cylinder centers are above the wheel centers, then see "How to secure measurements on unequal surfaces," page 559.

If the guides are hung and the main driving boxes are properly set in the jaws, or if the main wheels are under the engine and wedges set up, proceed as follows: Set the crosshead exactly central in the guides and use a square from the center of the main driving box, or wheel center; now use a long stick, letting the forward end of it touch the crosshead pin and mark a line on the back end of the stick true with the wheel center. Add to the length one-half of the diameter of the crosshead pin and you will have the correct length of the main rod.

DIVIDING THE CLEARANCE FOR THE CROSSHEAD.

When adjusting a main rod to the proper length, you should always notice whether the key in the back end is in front or behind the crank-pin. When in front you lengthen the rod as the brass wears and you drive the key down, and when behind you shorten it. Divide your clearance accordingly.

Another mechanical point to be considered is the space occupied by the area of the piston rod; when divided central the back end should have 1-32" more clearance than the front end to equalize the exhaust. In the round-house it is sometimes necessary to disconnect a main rod and put it up again. When the engine cannot be pinched onto center, in this case scribe a line on the guides at either end of crosshead before you disconnect and see that crosshead is right with same line after you have connected.

TO DETERMINE THE LENGTH OF PARALLEL RODS.

If wheels are under engine, get length from wheel centers. If not, take length from center of jaws (if boxes are lined central), and in either case, if engine is cold when length is taken, make the back rods 1-32" long to allow for expansion; providing fire box sets down between two back pair of wheels.

TRAMMING CRANK PINS AND PUTTING UP RODS.

Plug up your wheel centers with lead and find exact center of each wheel, and try them with a tram. If they show out of tram, then try main centers with centers in rocker-arms and line shoes, and wedges wherever needed to bring into tram and also to keep square with rocker boxes (which are supposed to be right), also see that all your wedges are set up moderately tight before you tram main centers, and also notice if any of her tires are beginning to cut; if so, that wheel should be lined forward or opposite wheel back.

Now place pins on top and bottom eight, and jump wheels until they tram. If engine has more than two pairs of driving wheels and "M" pin is longer than others, set tram accordingly. (See page 559.) Now, when pins tram, pinch one side of engine on dead center and put up rods on that side. Then pinch other side on dead center and put up that side. Line snug between pins.

When up, the front rods should work freely on pins, but if boiler is cold line the back rod tight between pins to allow for expansion of boiler. Each brass should have at least 1-64" lateral. See that rods divide right sideways.

TO LOCATE WHICH PAIR OF WHEELS ARE OUT OF TRAM.

If you tram an engine and find wheel centers "out of tram," you should at once locate which pair is out. You may ascertain this by using inside calipers between frame and largest turned face of wheel inside, at front and back of wheel. Each pair of wheels should be square with the frame.

HOW TO FIT ROD BRASSES.

File out the top and bottom and give a good side bearing, but don't let them bear on the fillet. Key up brass to brass in the strap and whirl on the pin, but take up all lost motion, or it may pound.

File the front end of the main rod brass open 1-32", but on all others key up brass to brass. Use soap on hot pin.

FINISH ENDS OF RODS.

For convenience, finish the ends of rods, valve rods, etc., before welding, as they are more convenient to handle.

HOLES FOR OIL CUPS.

Always counter-sink all holes that are to be tapped for oil cups, so you can leave fillet on the stem of cup, which strengthens stem.

FORGED SIDE ROD CUPS.

On many roads where solid end side rods are in use, they forge the oil cup solid with the rod, then there is no danger of losing oil cups, the lid or cover being the only removable part.

ECCENTRIC BRASSES.

Some roads use solid brasses in the side rods, which are turned eccentric shape and are capable of adjustment by simply moving a small lever which is secured to a quadrant, which is fastened to the rod. The short levers are adjusted while the crank-pins are on center, thereby adjusting the length of the rods to suit the wheel centers on each side.

PISTON, PACKING RINGS, ETC.

HOW TO FIND LENGTH OF PISTON ROD.

If your guides and crosshead are up, set crosshead in exact center of guides, and get distance from front face of crosshead to inside of back cylinder head.

Now get exact distance between inside faces of cylinder heads. Now get distance from back face of spider or piston head to extreme front point, whether it is the heads of the follower bolts or nuts on end of piston. Add this to travel of crossheads and see what clearance you have inside of cylinder. Now add one-half this clearance to original length from face of crosshead to inside face of back head and one-half the clearance.

This will give length of piston from shoulder to shoulder, then add on whatever you need on either end.

If back head is not up, add the thickness of back head, front guide block, one-half of the travel, and one-half the clearance on guides, and one-half the clearance in cylinder, and you have the distance from shoulder to shoulder.

DRAW FOR PISTON OR VALVE STEM KEYS.

Find out the length of key and width at each end and how much taper to one inch, then see how much you will have left to drive, and multiply by the amount to the inch, then allow about 1-32" for filing out, if keyway is to be drilled.

EXTENSION PISTON RODS.

Extension piston rods are used upon most all low pressure cylinders on compound engines, and on many other engines with very large, heavy pistons. They hold up the weight of

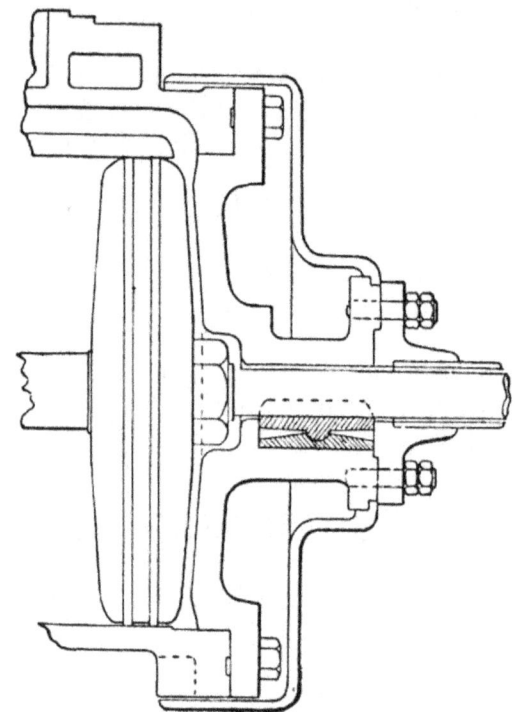

the piston and prevent wearing the cylinders. The cut below shows the method of their application.

CAST STEEL PISTON HEADS.

Cast steel piston heads are also in use in low pressure cylinders upon many compound engines.

A MALLEABLE IRON PISTON HEAD.

An experiment has been made on the S. C. & G. R. R. with a malleable iron piston head. The head is in two sections riveted together, and has two ordinary cylinder packing rings, the advantage being in a reduction of weight. It weighs about one-third less than the ordinary cast iron head with follower plate, and is giving very good results.

HOW TO LAY OFF STEAM PIPES.

See that nigger-head is securely fastened. Haul pipe up into place with small block and tackle; fasten temporary, putting thin board between top of pipe and nigger-head (thickness you want the top ring). Now block up the bottom so top will divide central both ways with nigger-head, and keep the bottom in proper position.

Measure the thickness each ring should be. Now use hermaphrodites and carry a line around top and bottom, true with flange on nigger-head, and face on cylinder joint below. Make the line the right distance from each face. By that we mean the thickness of each ring.

PUT IN THROTTLE BOX AND CONNECT THROTTLE ROD.

Use block and tackle; let the stand pipe and throttle box, or both, down into the dome; fasten solid to dry pipe with key bolts and strap, or flanges with bolts and nuts. Keep the top of it level; then fasten to the sides of the dome and put in the throttle. Jam the nuts on the rod that goes through throttle valve so valve will turn around. Now connect with rod that extends through boiler head. Split all cotter pins inside of boiler; then pack throttle rod at boiler head and tighten gland.

LAYING OFF AND FITTING UP AN ENGINE TRUCK.

When the truck frame is planed and ready to be laid off, first measure the length of the frame over all and in its exact center carry a central line across the frame. Now find out the required distance between the wheel centers and lay off two more square lines across the truck frame an equal distance from your central line and parallel with it and one-half the distance required between the wheel centers. These two lines will indicate the centers of the truck boxes. Now caliper each truck box (we shall assume they are all an equal size, which they should be). Now clamp the four corner jaws onto the frame, keeping the faces of each one-half the thickness of the truck box from the lines last mentioned; also keep each jaw true with each other jaw; that is, true across their faces and sides; use a long straight-edge for this purpose. Also see that each jaw is square with the frame. Now use a pair of hermaphrodites, or a gauge, and locate a center on the top end of the jaw an equal distance from the face and side and each jaw the same; these four centers should tram perfectly lengthwise, crosswise and diagonally when they are right. Place all of the truck boxes and other jaws in their places; place a strip of tin between the boxes and each of these last four jaws, then clamp them to the other jaws and clamp these jaws to the frame. Next try the lateral motion and see if the outside faces of the boxes will go between the hubs of the wheels. They should have at least $1\text{-}32''$ play sideways. If the truck brasses are to be bored, place centers in the boxes and lay them off an equal distance apart and an equal distance from the central line. Now lay off the holes in the jaws, if they are new, or in the truck frame; if it is new and the jaws are old drill and rosebit all holes and have "driving fit" bolts put in; each bolt should have a large cast die planed to their own standards, so the only work necessary to set the truck jaws is to clamp them to these dies, and drill and rosebit the holes. The truck center should be laid off in the exact center of the truck.

MISCELLANEOUS POINTS, TABLES, ETC.

HOW TO DETERMINE THE HORSE POWER OF AN ENGINE.

To get the area of piston in square inches multiply the diameter of piston in inches by itself, then multiply the product by .7854; this will give the area in square inches. To get the speed of piston in feet per minute, multiply the length of the stroke by 2 (because there are two strokes to each revolution), then multiply the product by number of revolutions per minute the engine is making. This will be the speed of the engine in feet per minute. Take the area of the piston in square inches and multiply it by the mean pressure per square inch on piston, then multiply this product by the speed of piston in feet per minute; now divide the last product by 33,000 and you will have the horse power of your engine.

One horse-power implies sufficient power to lift 33,000 pounds one foot high in one minute, or one pound 33,000 feet high in one minute, or equivalent power.

CUTTING A WATER-GLASS.

When a water-glass is too long the usual method employed for cutting off a short piece of it to make it fit is to use a small rat-tail file inside or a glass-cutter outside. If you have neither of these tools it can be accomplished with two matches in the following manner: Wet the sulphur end of one match and mark the inside of the glass all around at whatever point you wish it cut; then ignite the other match, hold the glass over the blaze and keep revolving the glass until well heated, then break the end off with your fingers, and you will find that it will make a nice, smooth cut.

HOW TO TAP HOLES PERFECTLY TRUE.

If you are tapping holes on any level surface, use a square to keep your tap true; but when tapping holes on a circle, or you wish your stud to stand at an incline (like studs for steam gauge bracket on back end of boiler head), lay off your holes, then set dividers to any size and scribe a circle; when you have your tap entered well, try dividers from end of tap to your circle, and keep it central.

MAKING FITS TO THE BOILER.

First level, or square up the piece of work as it should be when finished. If a fit is to be made on top or bottom of boiler, hold the points of your hermaphrodites perpendicular while you mark it all around. If it fits on the side of the boiler, hold the points of your hermaphrodites horizontal while you mark it all around. So many mistakes are made by holding the points of the hermaphrodites on a line with the center of the boiler all around, which makes a larger circle, and therefore will not fit. Many good men make this error every day.

HOW TO LEVEL UP STACK BASE BEFORE LAYING OFF.

Put a center in base and drop a plumb bob to center of the exhaust nozzle, or center of exhaust port openings, and then level both ways with a spirit level, and lay it off.

FITTING UP EXPANSION PLATES.

In fitting up expansion plates and buckles, do not clamp the frame, or fit them close, but leave space enough to slip a thin piece of tin between them and the frame. If any bolts go through expansion plates, or buckles and the frame, file and chip the holes oblong $\frac{1}{4}"$ to allow for expansion. Do likewise on back boiler braces—chip oblong holes.

If new plates, have the holes drilled in the plates first, then with a scriber lay off the holes on the boiler sheet. If the studs are in place lay off the holes with a surface-gauge or pair of hermaphrodites.

ADDITIONAL POINTS.

Never tighten the stem of a globe valve very tight, as the stem will expand and cause the valve to stick.

A globe valve should always be connected with the constant pressure against the bottom of the valve; then if the seat is well ground the valve will not leak around the stem, and may be packed at any time.

A safety boiler cock which has a check valve in its end, which will close in case of accident to the cock, is another improvement.

When drilling, rosebitting, reaming or tapping holes in wrought iron or steel always use oil. For cast-iron or brass do not use oil.

When drilling close work always use a twist drill, draw to center with gouge chisel and face off all holes for head of bolt or nut, and counter-sink all holes for rivets or oil cups.

Never use a washer on an engine when you can avoid it; cut the thread up farther on the bolt or stud, or bore one or two threads out of the nut.

When connecting side rods, eccentric blades and such work, always try them and see how they divide sideways.

Always put a drain cock in the bottom of the boiler check or branch pipe to avoid freezing.

Manure put into a boiler will stop small leaks.

Never use rubber gaskets for steam-tight joints where you can use copper.

Spring hanger gibs are being replaced with round steel pins.

Allow from $\frac{3}{4}"$ to $1\frac{1}{4}"$ draw on all spring hangers; be governed by the dimensions, number of leaves, etc., and weight of the engine.

Many roads use a soft plug (filled with lead or some alloy which fuses at low temperature), in the crown sheet, to avoid burning the engine. They have proven very successful.

The old style expansion plates and buckles are being replaced with two heavy cast-iron plates. One is fastened to the boiler and the other to the frame; a long key holds them intact.

The plunger oil cups are considered the standard for main and parallel rods. Most all oil cups now made use an iron stem, which is much stronger than brass.

When grinding steam-tight joints use coarse emery on iron and flour emery on brass. You should always make at least $\frac{1}{8}''$ of good bearing to secure a good joint.

Cast-iron running board brackets are now used; also sheet-iron running boards.

The channels of solid parallel rods are sometimes filled with rubber to deaden the noise.

A little giant boiler-washer, which will wash out a boiler with hot or cold water, is used extensively.

Hollow crank pins are used on many roads. They will withstand a greater strain than solid pins, with the same amount of metal.

An automatic closing valve for steam fountains is another new device. It acts as a safety valve in case of an accident, and can be closed at any time to repack any of the cocks.

HOW TO MAKE A PERFECT RIGHT ANGLE.

First draw the horizontal line A B and locate the two centers upon it, E and F any distance apart. Then with a pair of dividers

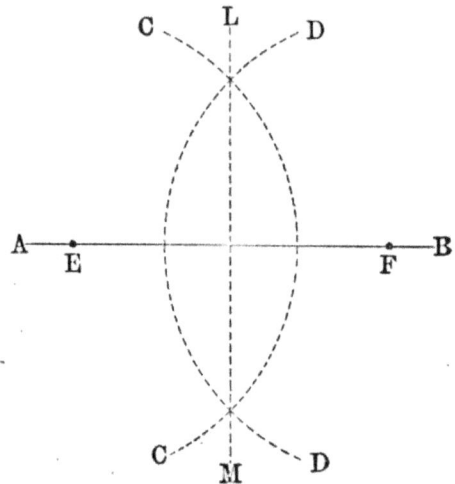

set to any convenient length, and from the point E describe the arc C C and from the point F describe the arc D D as shown. Now erect the perpendicular line L M, letting it pass through the two points where the arcs C C and D D intersect. This will make a perfect right angle.

LOCATING A CENTER FROM ANY PORTION OF A CIRCLE.

Locate three points upon that portion of the circle which is known as A B C in the figure. It is not necessary that these points be an equal distance apart. With a pair of dividers set

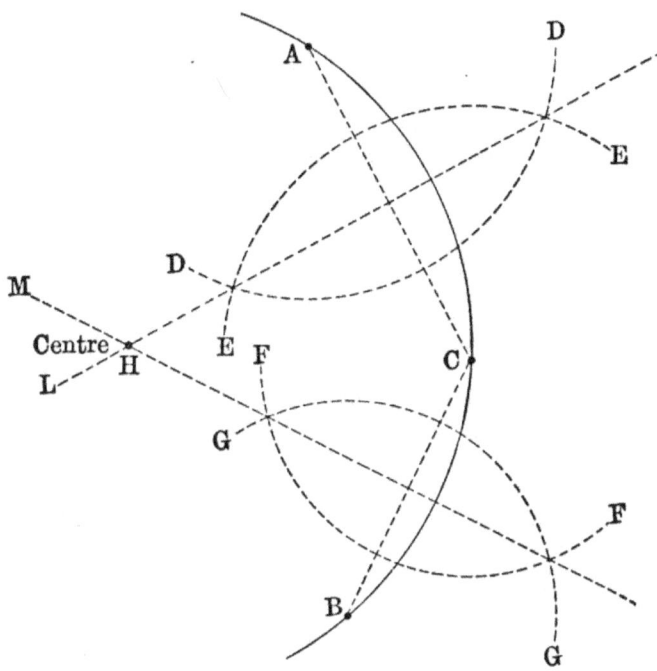

to any convenient length, and from the point A, describe the arc D D and with the same, or any other length, from the point C describe the arcs E E and F F and from the point B describe another arc G G. Now through the two points where the arcs D D and E E intersect draw the straight line L and through the two points where the arcs F F and G G intersect the straight line M. Now the point H, where the lines L and M cross each other, will be the correct center. This knowledge will often be found valuable in a machine shop, in laying off cylinder patches, etc.

MEASUREMENTS ON AN UNEQUAL SURFACE.

When it is necessary to tram crank pins of unequal length, or to lay off holes on a back cylinder head where the center you work from is above, or below, the face to be laid off, the following rule will be found both simple and accurate:

Presuming you wish to lay off a 20" circle on a back cylinder head, on which to locate the holes for the cylinder head studs, and you find the face for the guide blocks (and therefore your center) is 4" above the face where the holes are to be laid off, use a square and scribe the two lines A and B on a face plate as shown in the Figure. Make line A 10" long and line B 4" long;

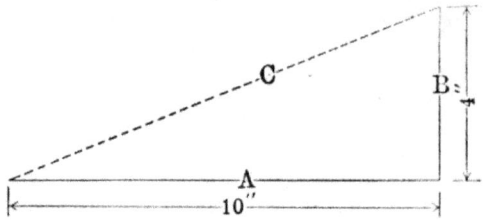

then set a pair of dividers equal to the length of line C, which will be the correct radius with which to scribe the 20" circle.

DECIMAL EQUIVALENTS OF FRACTIONAL PARTS OF AN INCH.

	1-64=.015625		17-64=.265625		33-64=.515625		49-64=.765625
1-32	=.03125	9-32	=.28125	17-32	=.53125	25-32	=.78125
	3-64=.046875		19-64=.296875		35-64=.546875		51-64=.796875
1-16	=.0625	5-16	=.3125	9-16	=.5625	13-16	=.8125
	5-64=.078125		21-64=.328125		37-64=.578125		53-64=.828125
3-32	=.09375	11-32	=.34375	19-32	=.59375	27-32	=.84375
	7-64=.109375		23-64=.359375		39-64=.609375		55-64=.859375
1-8	=.125	3-8	=.375	5-8	=.625	7-8	=.875
	9-64=.140625		25-64=.390625		41-64=.640625		57-64=.890625
5-32	=.15625	13-32	=.40625	21-32	=.65625	29-32	=.90625
	11-64=.171875		27-64=.421875		43-64=.671875		59-64=.921875
3-16	=.1875	7-16	=.4375	11-16	=.6875	15-16	=.9375
	13-64=.203125		29-64=.453125		45-64=.703125		61-64=.953125
7-32	=.21875	15-32	=.46875	23-32	=.71875	31-32	=.96875
	15-64=.234375		31-64=.484375		47-64=.734375		63-64=.984375
1-4	=.250	1-2	=.500	3-4	=.750	1	=1.000

TABLE OF U. S. STANDARD TAPS.

Diameter.	Number of Threads.		Diameter.	Number of Threads.	
	U. S. Standard.	Other Threads used.		U. S. Standard.	Other Threads used.
1-8	48	..	1 1-4	7	..
5-32	40	32	1 5-16	7	..
3-16	24	32	1 3-8	6	..
1-4	20	18-22-24	1 7-16	6	..
5-16	18	16-20	1 1-2	6	..
3-8	16	14-18	1 5-8	5½	5
7-16	14	12-16	1 3-4	5	..
1-2	13	12-14	1 7-8	5	4½
9-16	12	14	2	4½	..
5-8	11	10-12	2 1-8	4½	..
11-16	11	12	2 1-4	4½	..
3-4	10	11-12	2 3-8	4	..
13-16	10	9	2 1-2	4	..
7-8	9	10	2 3-4	4	..
15-16	9	8	3	3½	..
1	8	..	3 1-4	3½	..
1 1-16	8	..	3 1-2	3¼	..
1 1-8	8	7	3 3-4	3	..
1 3-16	8	7	4	3	..

TAP DRILL SIZES FOR U. S. STANDARD THREAD.

Size.	Diameter of Drill.	Size.	Diameter of Drill.	Size.	Diameter of Drill.
1-4 in.	0.189	5/8 in.	0.512	1 3/8 in.	1.167
5-16 in.	0.244	3/4 in.	0.625	1 1/2 in.	1.292
3-8 in.	0.298	7/8 in.	0.737	1 5/8 in.	1.398
7-16 in.	0.349	1 in.	0.844	1 3/4 in.	1.500
1-2 in.	0.405	1 1/8 in.	0.947	1 7/8 in.	1.625
9-16 in.	0.459	1 1/4 in.	1.072	2 in.	1.722

The above table gives the diameter of drills in thousandths of an inch for holes to be tapped U. S. Standard, and is an allowance above actual bottom diameter size of thread of from four thousandths of an inch for a 1/4-inch tap to ten thousandths for a 2-inch tap.

STANDARD DIMENSIONS OF WROUGHT IRON PIPE.
BRIGGS' STANDARD.

Nominal Inside.	Diameter of Tube. Actual Inside.	Actual Outside.	Thickness of Metal.	Screwed Ends. Number of Threads per inch	Length of Perfect Thread at Bottom.	Size to Drill Holes for Tapping.
1/8 in.	0.270 in.	0.405 in.	0.068 in.	27	0.19 in.	5-16 in.
1/4 in.	0.364 in.	0.540 in.	0.088 in.	18	0.29 in.	7-16 in.
3/8 in.	0.494 in.	0.675 in.	0.091 in.	18	0.30 in.	9-16 in.
1/2 in.	0.623 in.	0.840 in.	0.109 in.	14	0.39 in.	5-8 in.
3/4 in.	0.824 in.	1.050 in.	0.113 in.	14	0.40 in.	7-8 in.
1 in.	1.048 in.	1.315 in.	0.134 in.	11 1/2	0.51 in.	1 1-8 in.
1 1/4 in.	1.380 in.	1.660 in.	0.140 in.	11 1/2	0.54 in.	1 7-16 in.
1 1/2 in.	1.610 in.	1.900 in.	0.145 in.	11 1/2	0.55 in.	1 11-16 in.
2 in.	2.067 in.	2.375 in.	0.154 in.	11 1/2	0.58 in.	2 3-16 in.
2 1/2 in.	2.468 in.	2.875 in.	0.204 in.	8	0.89 in.	2 5-8 in.
3 in.	3.067 in.	3.500 in.	0.217 in.	8	0.95 in.	3 1-4 in.
3 1/2 in.	3.548 in.	4.000 in.	0.226 in.	8	1.00 in.	3 3-4 in.
4 in.	4.026 in.	4.500 in.	0.237 in.	8	1.05 in.	4 1-4 in.
4 1/2 in.	4.508 in.	5.000 in.	0.246 in.	8	1.10 in.	....
5 in.	5.045 in.	5.563 in.	0.259 in.	8	1.16 in.	....
6 in.	6.065 in.	6.625 in.	0.280 in.	8	1.26 in.	....
7 in.	7.023 in.	7.625 in.	0.301 in.	8	1.36 in.	....
8 in.	7.982 in.	8.625 in.	0.322 in.	8	1.46 in.	....
9 in.	9.000 in.	9.688 in.	0.344 in.	8	1.57 in.	....
10 in.	10.019 in.	10.750 in.	0.366 in.	8	1.68 in.	....

Taper of conical tube-ends, 1 in 32 to axis of tube (3/4-inch per foot, or 1-16 inch per inch).

SHRINKAGE.

Allow on all tires 1-80 of an inch for each 12" in diameter.

$1\frac{1}{4}"\times1\frac{1}{2}"$ finished cylinder band, allow $\frac{1}{8}"$.

$1\frac{1}{4}"\times1\frac{1}{2}"$ rough cylinder band bored out, allow 3-16".

$1\frac{1}{2}"\times1\frac{1}{2}"$ rough band for hub of wheel and pin hub, allow $\frac{1}{4}"$ to 5-16".

$\frac{1}{2}"\times\frac{3}{4}"$ finished air pump band, allow 1-16".

$\frac{1}{4}\times\frac{1}{2}\times3"$ finished band, allow 1-32".

and other bands in proportion.

GENERAL MACHINE WORK.

GEARING THE LATHE.

SIMPLE GEARED.

Multiply the number of threads (per inch) you desire to cut by any small number, and put that gear on your screw; then multiply the number of threads (per inch) on your screw by the same number and place that gear on the spindle. For example: We will use number 4, although any small number will answer the same; say you wished to cut 10 threads, multiply 10 by 4, which equals 40; put 40 on your screw. Now multiply the number of threads (per inch) on your screw (we will presume it is 6) by 4, which equals 24; and put 24 on the spindle.

Another way is to take any small gear you may have (put it on the spindle) and multiply it by the number of threads desired, and divide the product by the number of threads on your screw (per inch) and put it on the screw. If you haven't those two gears try another one, and so on until you have two that will cut it.

Many small lathes have a stud geared into the spindle, which stud only runs one-half as fast as the spindle. In finding the gears for these lathes first multiply the number of threads to be cut, as before, and then multiply the number of threads on the screw as double the number it is. For example: If you want to cut 10 threads, multiply by 4; so put 40 on the screw; then if your screw is 6, call it 12 and multiply by 4 and it will give you 48; put that on the spindle. Many of these lathes have a pin in end of spindle which changes the gear without the necessity of changing your gears; simply pull out the pin and the lathe will cut double the thread it is geared to cut.

The *American Machinist* gives the following rule for selecting gears to cut threads on lathes, for both single and compound gearing: If the lathe is simply geared and the stud runs at the same speed as the spindle, then select some gear for the screw, and multiply its number of threads per inch by the lead screw, and divide this result by the number of threads per inch you wish to cut. This will give you the number of teeth in the gear for the stud. If this result is a fractional number, or a number which is not among the gears which you possess, then try some other gear for the screw. But if you prefer to select the gear for the stud first, then multiply its fixed compound gears. In the instance given if the lead screw had been $2\frac{1}{2}$ threads per inch, then its pitch being 4-10 inch we have the fractions of 4-10 and

25-32, which, reduced to a common denominator, are 64-160 and 125-160, and the gears will be the same as if the lead screw had 125 threads per inch, and the screw to be cut 64 threads per inch.

COMPOUND.

If the lathe is compound, select at random all the driving gears, multiply the number of their teeth together, and this product by the number of threads you wish to cut. Then select at

LATHE TOOLS.

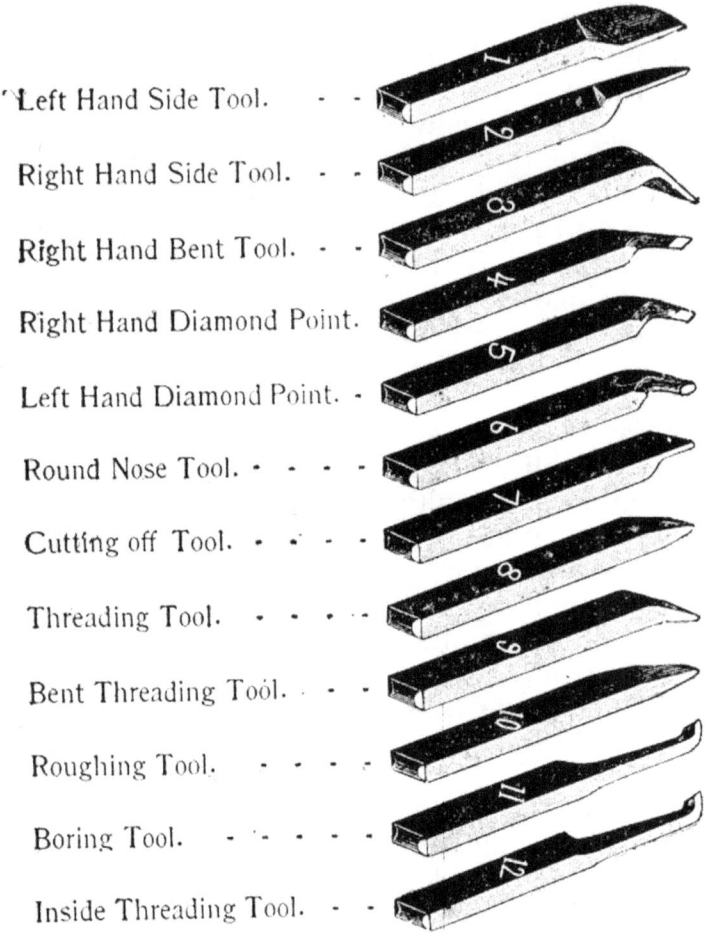

Left Hand Side Tool.
Right Hand Side Tool.
Right Hand Bent Tool.
Right Hand Diamond Point.
Left Hand Diamond Point.
Round Nose Tool.
Cutting off Tool.
Threading Tool.
Bent Threading Tool.
Roughing Tool.
Boring Tool.
Inside Threading Tool.

random all the driven gears except one, multiply the number of their teeth together, and this product by the number of threads on the lead screw. Now divide the first result by the second, and you will have the number of teeth in the remaining driven gear. But if you prefer you can select at random all the driven gears. Multiply the number of their teeth together, and this product by the number of threads per inch in the lead screw. Then select at random all the driven gears except one. Multiply the number of their teeth together, and this result by the number of threads per inch of the screw you wish to cut. Divide the first result by the last, and you will have the number of teeth in the remaining driver. When the gears on the compounding stud are fast together and cannot be changed, then the driven one has usually twice as many teeth as the other, or driver; in which case you can, in the calculations, consider the lead screw to have twice as many threads per inch as it actually has, and then ignore the compounding entirely. Some lathes are so constructed that the stud on which the first driver is placed revolves only half as fast as the spindle. You can ignore this in the calculations by doubling the number of threads of the lead screw. If both the last conditions are present you can ignore them in the calculations by multiplying the number of threads per inch in the lead screw by four. If the thread to be cut is a fractional one, or if the pitch of the lead screw is fractional, or if both are fractional, then reduce the fractions to a common denominator, and use the numerators of these fractions as if they equaled the pitch of the screw to be cut, and of the lead screw respectively. Then use that part of the rule given above which applies to the lathe in question. For instance, suppose it is desired to cut a thread of 25-32 inch pitch, and the lead screw has 4 threads per inch. Then the pitch of the lead screw will be $\frac{1}{4}$ inch, which is equal to 8-32 inch. We now have two fractions, 25-32 and 8-32, and the two screws will be in the ratio of 25 to 8, and the gears can be figured by the rule above, assuming the number of threads to be cut to be 8 per inch, and those on the lead screw to be 25 per inch. But this latter number may be further modified by conditions named above, such as a reduced speed of the stud, or fixed compound gears. In the instance given if the lead screw had been $2\frac{1}{2}$ threads per inch, then its pitch being 4-10 inch we have the fractions of 4-10 and 25-32, which, reduced to a common denominator, are 64-160 and 125-160, and the gears will be the same as if the lead screw had 125 threads per inch, and the screw to be cut 64 threads per inch.

THREAD CUTTING.

UNITED STATES STANDARD SCREW THREADS.

The accompanying sketch shows the three forms of screw threads now in use in this country. The United States standard form has been adopted by the United States Government, the Master Mechanics' and Master Car Builders' Association; Locomotive Works, Machine Bolt Makers, and many manufacturing establishments throughout the country, and is recommended by the Franklin Institute of Philadelphia. The thread has an angle of 60 degrees with flat top and bottom, equal to one-eighth of the pitch. The advantages of this form of thread over the sharp

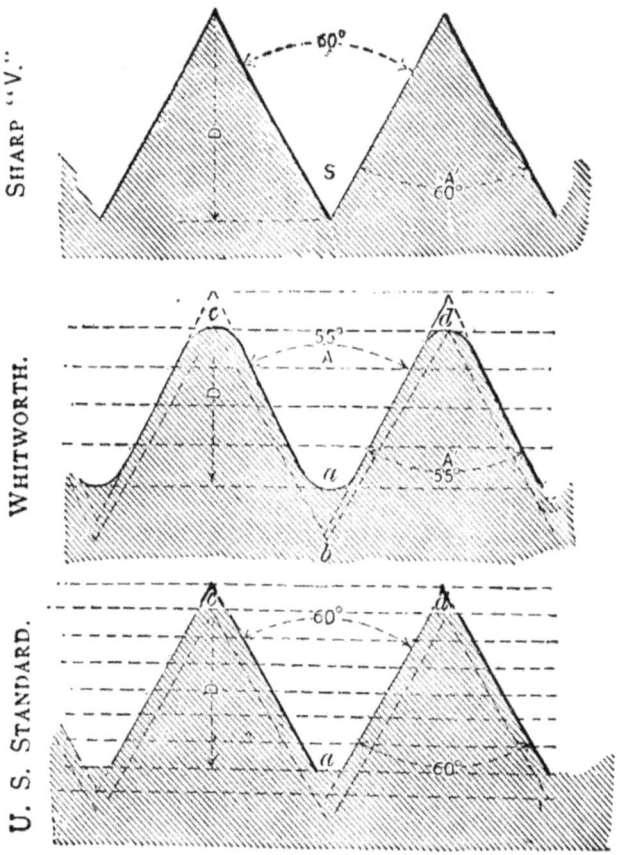

V are that in the tap the edges of the thread are less liable to accidental injury, and will wear and retain their size and form longer, and in the bolt the flat top and bottom give increased strength and an improved appearance.

SCREW OR V-SHAPED THREADS.

When cutting V-shaped threads, set your tool at right angle with lathe centers, and look at thread carefully on both sides, and see that threads do not lean like fish scales. Always grind your tool to a gauge.

SQUARE THREADS.

When cutting square threads it is always necessary to get the depth required with a tool somewhat thinner than one-half the pitch of the thread. After doing this dress another tool exactly one-half the pitch of the tread and use it to finish with, cutting a slight chip on each side of the groove. Then polish with a pine stick and emery. Square threads for strength should be cut one-half the depth of their pitch, while square threads for wear should be cut three-fourths the depth of their pitch.

MONGREL THREADS.

Mongrel, or half V and half square threads, are usually made for great wear, and should be cut the full depth of their pitch, or even more. They are sometimes cut one and one-half the depth of their pitch; the point and the bottom of the grooves should each be in width one-quarter the depth of their pitch.

DOUBLE THREADS.

The face plate on every lathe has at least four slots in it which carry the dog; so after you have cut one thread change your dog to slot directly opposite on face plate.

LEFT-HAND THREADS.

Gear up the lathe the same as for a right-hand thread, and then change the shifter on the end of your lathe. If your lathe has no shifter, put in an extra gear, so as to move the carriage from left to right. Then begin to cut the thread at the left hand and cut to right. The same way for inside or outside threads.

DIFFICULT CHUCKING.

To chuck an odd shaped piece of work neatly and securely often requires much deliberation, considerable skill and mechanical ingenuity, and is next in importance to being able to make a good fit, and an observant foreman can easily tell an experienced lathe hand by the manner he performs this kind of work. When a job of this kind is brought to your lathe you should immediately begin to plan the best and simplest way to chuck the particular piece of work in order to keep all its finished faces (if it has any) perfectly true, and in a manner not to interfere with your tool or tool post. When possible complete all plans in your mind before you begin the job. If by placing the chuck or face plate on the bench until you get it set would be an advantage, do so; if the dogs of your chuck are too short, perhaps you may be able to clamp it to a face plate, or a short tool placed from one dog to another may assist you, or perhaps the use of a nut against one or more of the dogs, or V blocks, or a pipe center, brace or steady rest, might help you, or a block of wood cut out the proper shape may be better. It would be impossible to enumerate the many jobs of this kind, such as goose necks, injectors, etc.; meditate a little and you will most likely think of a good way.

HOW TO SET CALIPERS FOR FITTING.

Some men do all their fitting with the inside calipers, but the best mechanical way is to set the inside calipers to the exact size of the hole, then set the outside calipers to the exact size of the insides, and make all fits with the outside calipers, letting them drag over the work while in the lathe. Experience will teach the amount of drag to allow for different pressures suitable to the stiffness of your own calipers. No book can teach you how to make a fit; experience alone can do this, as it is all done by the sense of feeling, yet many valuable points may be obtained by the recorded experience of other men. With an ordinary pair of 6" calipers 1-18" thick, a 5-16 drag of the outside calipers will give a 30-ton fit, while a ⅜" drag will give a 70-ton fit; much of course will depend upon the softness or otherwise of the metal and the length of the fit. The amounts above mentioned are suitable for crank pins and driving axles. Expansion must be considered when fitting bushings, and bolts should be made as near the exact size of the hole as possible.

HOW TO FIT BUSHINGS.

You must use judgment in doing this work; if your bushing is very heavy fit almost as you would anything solid, for the harding process will make it a little larger; but if a light bushing allow about 1-64" on a 2" one, and 1-32" on a 4" bush and bore out that much larger to allow for closing. A cylinder bushing make the same size as hole or a shade smaller, owing to so much bearing.

HOW TO FIT ALL KINDS OF BOLTS.

First, measure the distance through your hole, allow for nuts, thimble point, etc., and cut off right length under heads. Put in new centers, turn up the end for thread 1-64" under size of tap (to avoid the sharp edge on the thread) and allow plenty of draw. Then cut threads in lathe or have it done in bolt cutter. Now set your calipers to the exact size of the hole, and if an ordinary 1" bolt 6" long, make as near the exact size as possible. The smaller the bolt and the less the bearing, the tighter you can drive it in, and the larger the bolt, and the more bearing, the less you must allow for fitting and driving in.

Always use a little judgment; don't drive a bolt very tight in a light casting or you may crack it.

TAPER BOLTS.

Get the size of the hole at each end, and turn to the required size at each end, if there is no taper attachment on your lathe. Then move back head toward you and use scale between your tool and the finished size of the bolt at each end; when your scale shows the same at each end, tighten back head and turn the bolt down to that size. Boys, learn to use your calipers, don't be afraid of spoiling a bolt. Remember you are not paid a journeyman's pay.

HOW TO SET TOOLS IN LATHE.

FOR IRON.

Set your tool above the center for outside turning and below the center for boring out.

FOR BRASS.

Set your tool below the center for outside turning, and above the center for boring out.

FINISH BRASS WORK THAT CANNOT BE TURNED.

A strop is used over two pulleys, put glue and emery on strop and hold the work against it.

GRINDING TOOLS.

When grinding machine tools always give the cutting edge sufficient clearance and angle, the top face to a degree just suf-

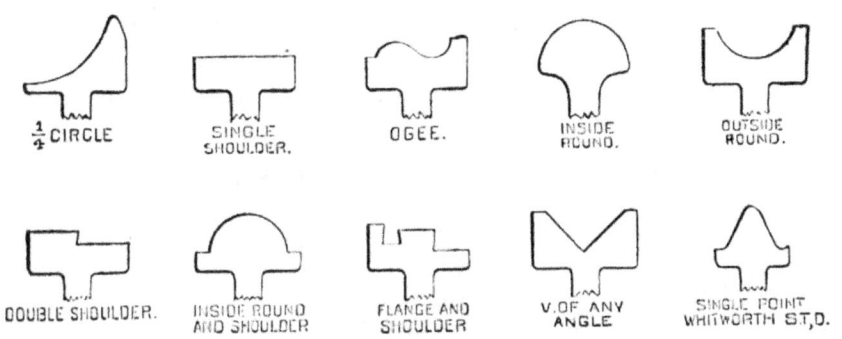

ficient to insure a good clean cut, so as to prevent the tool tearing, which would leave a very rough finish.

TURNING AND FITTING.

HOW TO TURN AND FIT DRIVING AXLES.

Put centers in each end, put in lathe and drill small hole in each center 1" deep. Get length over all and cut off and face each end, and cut small groove on ends. Turn down to largest finished part, space off the fit for journal bearing, and space for eccentrics, and turn each to required size. Now cut down space between eccentrics about 1-16" smaller than size for eccentrics, and fit same as crank pins, only let outside calipers drag about $\frac{5}{8}$" for 70-ton.

HOW TO QUARTER A DRIVING AXLE IN A LATHE.

To do this scribe a large circle on face plate of lathe and divide into four equal parts and prick punch each quarter.

Now use post or solid tram from bed of lathe and set to one of your four centers; put sharp-pointed tool in tool post, run carriage along and scribe a fine line, the full length of the fit on right end of axle. Now turn lathe one-quarter turn ahead and set to next center, then scribe another line on left end of axle.

These two lines indicate the center of your key, so lay off keyways required size.

HOW TO TURN UP AND FIT CRANK PINS.

First see that pin will true up to required size; put a large center in each end of pin and drill a small $\frac{1}{4}$" hole about $\frac{1}{2}$" deep in each center; this is done to retain original center. Now face off outside end of pin perfectly square and cut a small groove around center to find center afterward.

Turn collar required size. Find distance pin should stand out and cut shoulder, then finish your bearings for brasses in proper place and required diameter; reverse pin in lathe and fit to hole. To do this (if your calipers are medium, say 6" by 1-18" thick), set inside caliper to exact size of hole, set outside pair to them and let outside pair drag 5-16" over fit. That will give you about a 30-ton fit. Do not cut off inside center, but counter-bore about ¼" deep and leave collar to rivet inside.

To find the diameter of a crank pin multiply the diameter of the cylinder by .234.

HOW TO TURN UP AND FIT PISTON RODS.

New Rod.—Put two deep centers in rod and put in lathe, use a square center and get each end to run true. Now replace square center with round center, use diamond point tool, and turn rod off to about 1-16" larger than finished size. Now measure distance rod should go into cross-head and cut a shoulder that length from end. Now caliper taper hole in cross-head at each end, and fit, same as to fit taper bolts (see how to fit taper bolts, page 568); allow 1-16" to grind and drive to shoulder. Now turn rod down to finished size, using very fine feed. Now take out tool marks with a lathe file, and polish rod with emery cloth, or emery paper.

Now reverse in lathe, cut other shoulder exact distance rod should be, and fit to piston head; let calipers drag ¼" for 20-ton pressure, then counter-bore end to rivet over. The diameter of a piston rod should be about 1-6 the diameter of the cylinder.

TURN UP VALVE STEM.

Drill a small hole in end of stem, put in lathe, using a bolt and nut or screw jack in yoke to keep from springing. Cut off end right length. Now cut shoulder and fit to valve rod (same as fitting taper bolts), then turn down stem to required size.

TURNING AND FITTING SOLID ROD BRASSES.

Chuck collar and bore out 1-32" larger than the pin, then turn the outside 1-32" larger than the hole. Reverse and turn off collar. If pressed into rods, bore out in boring mill, making 1-64" loose.

TURN UP AND FIT THROTTLE VALVE.

A hole is usually cast through valve. Drive a mandrel through hole, or use lathe centers and face off each end. Turn wings and flanges on each end down to fit throttle box, making 1-32" loose.

Find the exact distance between the two seats in throttle box and make seats on valve, trying with black lead until you get good bearing on each seat. If you have a templet make to it.

TURN AND FIT PISTON HEAD, BULL RING, CYLINDER PACKING RINGS, AND FOLLOWER PLATE.

USE UNIVERSAL CHUCK.

Piston Head.—Chuck, bore out hole for piston and ream with large reamer, turn outside, or largest part 1-32" smaller than cylinder, reverse in chuck, turn off lugs and true up both faces to required size and leave collar for follower plate.

Bull Ring.—Chuck, bore out about $\frac{1}{8}$" larger than outside of lugs on piston head, turn off 1-32" smaller than cylinder, turn

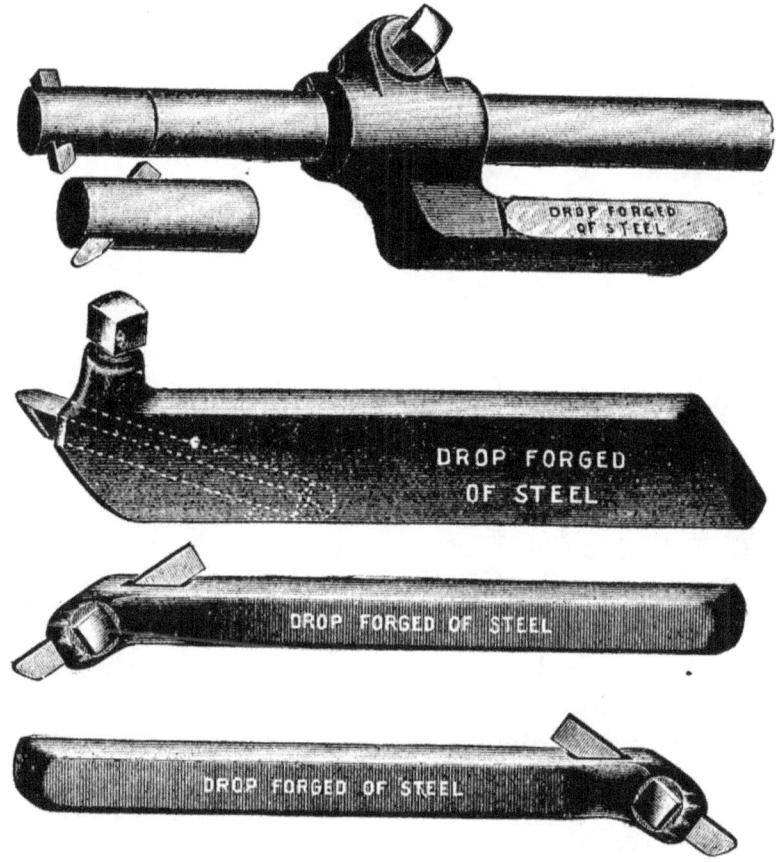

outside flanges 3" smaller than largest part, face off 1-64" wider than distance from one face of piston head to other, just enough for follower plate to clamp bull ring without cracking plate. If solid bull ring, chuck and face off one side, reverse in chuck and face off right width, bore out 1-16" larger than lugs on piston head and cut grooves proper width and 1-16" deeper than size of packing rings.

Packing Rings.—Caliper cylinder. Make outside of packing rings 3-16" larger, if the two diameters of bull ring differ 3", then make packing rings about ¼" less, therefore bore out packing casting 2½" less than outside of same; use square nose tool and cut off rings right width. Dunbar packing, which is one narrow flat ring and an L ring sawed into many pieces and set out against the cylinder with a flat spring, is considered the best but most expensive cylinder packing in use; it will fit a cylinder that is worn out of round better than any other kind. It is used on many roads.

Follower Plate. — Chuck and face off true, turn off outside 1-32" smaller than cylinder, and face off holes for follower bolts.

TO FIT UP BULL RING, PACKING RINGS, ETC.

First—See that follower plate clamps bull ring when tightened, put one dowel pin on each side of bull ring about 4" apart, and when you place in the cylinders put dowels to the bottom.

Saw out of packing rings three and one-seventh times the difference between size of cylinder and outside of packing rings,

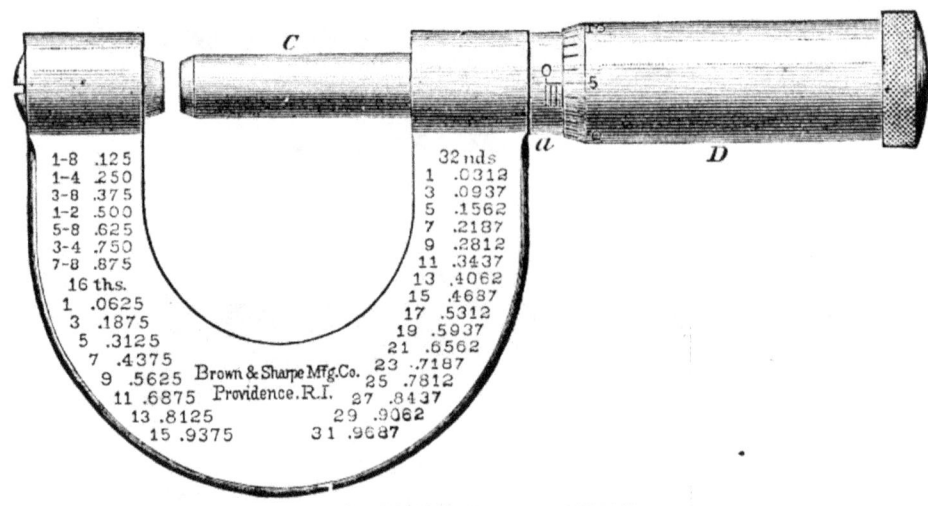

MICROMETER CALIPER.

and see that they clear dowel pins and slip through smallest part of cylinder. If no dowel pins are used saw rings so they will lap. Put in piston head, or spider, bull ring and packing rings and line up piston head central with cylinder, placing liners between bull ring and lugs on piston head, then put on follower plate and tighten up follower bolts.

TURNING A CROSS-HEAD PIN FOR MOGUL ENGINE.

Put center in each end and make it run true. Measure distance through head and add to it the thickness of your washer and two nuts and ½" for the thimble point, and ¼" for draw, and face off both ends to right length over all.

Now you must cut the three shoulders, measure from larger end of pin and make the first two the exact distance to shoulder on cross-head, but allow ¼" draw on the outside shoulder. Now turn pin down to required size, and fit to holes in cross-head;

when tight let shoulders stand off ¼" to allow for draw, cut threads on the stem and fit nuts and washer.

TURNING A SOLID CROSS-HEAD PIN.

In some shops there is a device for this purpose, large and small gear, and it turns by hand. Small gear is in two halves, with a short tool and feed inside.

The old way: Use a tool made to a half circle and put a wooden pole in piston hole and turn and feed by hand in lathe, cutting one-half circle at a time; chip clearance for tool.

HOW TO FINISH EXHAUST NOZZLE.

Chuck in lathe, bore out and turn off to required size, making a close fit in end of nozzle box, with groove for set-screw.

TURNING AND BORING OUT TIRES.

When you have a set of tires to turn, always look for flat spots. Caliper each tire and turn smallest one first, make all others the same size. When boring out tires allow 1-100 of an inch for each foot in diameter for shrinkage. Turn the tread and flanges to a gauge. See new machine for this purpose, page 530.

TURNING, FITTING AND CLAMPING CYLINDER BUSHINGS.

The subject of clamping or holding cylinder bushings in position has received much discussion from practical railway men of late and a great diversity of opinion obtains as to the best method of holding the bushing in place and likewise the length to make it. Some claim that better results can be obtained by making the bushings a length equal to the distance between the front and back counter-bore and to be fastened to the wall of the cylinder by a number of brass screw bolts, but the writer believes the old method of letting the cylinder heads hold it intact to be the best, as the former method increases the clearance space, which is already too great, being very wasteful on steam. A few screw bolts may be placed in the walls of the cylinder when considered advisable and if the bushing be made a shade larger than the cylinder a rib may be left when drilling or milling out the ports in the bushing; this will prevent a leaky joint. If made a length equal to the cylinder or a shade less, no such bridge will be required, making the joint on the cylinder.

If the cylinder is heated when applying the bushing, as is done in many shops, the bushings may be made a trifle larger than the cylinder, about 1-64", but if driven in with a large wooden block, or drawn into place with long rods when the cylinder is cold, it should be no larger than the cylinder, but should be as near the exact size as it can be made; the large and long bearing surface will make it tight enough.

Cylinder bushings are usually bored out and turned up in a large lathe. To perform this job, proceed as follows: Clamp two large blocks onto the carriage of the lathe, and bore out wood the exact size of the outside of the bushing when rough; then clamp your bushing onto those two blocks; use a boring bar and bore out to the required size and counter-bore same. Then set your bushing onto a large mandrel for the purpose, which has set-screws at each end with which to adjust the bushing. Set it to run perfectly true with the inside at each end; then turn off and fit to the cylinder as previously explained, unless otherwise instructed. Make it the exact size of the cylinder, making a good finished face at each end. When you are pressing it into place use your judgment and do not crack the cylinder.

BORING.

HOW TO BORE OUT CRANK PIN HOLES.

Most large shops have a quartering machine for this purpose. The two sides of the bed are at perfect right angles, with a head and boring bar on either side. So it is impossible to bore the holes out of square. The heads are adjustable so the stroke can be changed if desired.

When there is no quartering machine, the large face plate on the wheel lathe is usually quartered with four large holes drilled in it for dowel pin, and small boring bar fastens onto carriage of lathe.

BORING OUT CYLINDERS.

NEW CYLINDER IN BORING MILL.

Set tool by inside of cylinders, and run tool through and see if there are any low or warped spots; then see if the inside face of the cylinder will plane off the required size; if it is all right then bore out to required size; then bore out counter bore at each end about $\frac{3}{8}''$ deep. Get the length between the faces, and face off the ends. If the tool chatters, hang a weight on the end of the bar.

IN LATHE.

Clamp two large blocks onto your carriage and bore them out to the same circle as outside diameter of rough cylinder; then clamp the cylinder onto the blocks and proceed to bore them out the same as in boring mill.

TO BORE OUT THROTTLE BOX.

Clamp on boring mill, set parallel with boring bar and turn joint on bottom end. Now reverse on bed; keep true with boring bar, set central with rough holes for throttle valve; bore out top and bottom to right sizes and face off top. Now use half-round or bevel tool in boring bar and make both seats for valve, making the top one the largest. If you have a templet make it to it.

HOW TO FINISH NOZZLE BOX.

If for double nozzle, put a center in each end and face off each end in lathe, making right length. If for single nozzle, clamp on boring mill, keeping bottom face square with bed, and bore out and face off end for nozzle. Reverse, keeping other end square, and face off bottom joint. Or if done in lathe, put center in bottom end and put that end toward face plate on lathe; tighten the other center against piece of iron across the other end; get it to run true and face off near end, enough for steady rest; then put on steady rest and bore out and face off top end. Reverse in lathe chuck with universal chuck, and tighten back center and face off the bottom joint.

BORING ROD BRASSES.

When finished a rod brass should have a good bearing front and back, but should not bear top and bottom, therefore put a piece of Russia iron between the brasses before boring them out, then key up tight and bore out 1-32" larger than crank pin. Take out Russia iron when bored.

BORING OUT DRIVING BRASSES.

To Bore Out.—Driving brasses should not be bored out until after they are pressed into box. You can bore them out in boring mill or lathe. Set perfectly central sideways, leaving all the stock possible in crown; bore the exact size of smallest part of journal, and keep oil cellar in box when boring out.

PLANER WORK.

HOW TO PLANE UP A NEW CYLINDER.

Set the inside face square with the bed of the planer. Put a center in each end of the cylinder, and set each center perfectly true with the tool. Now plane the inside face to one-half the width the cylinders should be from center to center. Plane the top of cylinder outside of seat; then turn cylinder around, one-quarter turn. Set the inside face square with the bed of the planer up and down, and set two centers in the cylinder perfectly square across the bed of the planer; use square from center to square line on bed. Now finish the valve seat and the outside of it. Now turn the cylinder over and plane the bottom square with the inside face, the same distance from center of cylinder as the other cylinder, mate to it.

When taking a cut down the inside face of a cylinder, or a

side cut on any other piece of work, always adjust the circle head of your planer slightly to one side (whichever side you are planing on), so tool won't drag.

HOW TO PLANE UP SHOES AND WEDGES.

NEW.

Clamp on bed of planer face down and true up edge of flanges. Now turn them over and clamp flanges down, and keep sides true with tool. Now take light cut across face, just enough to true up. Now caliper driving boxes inside and finish both outsides, making shoe or wedge 1-64" smaller than box. Now clamp all shoes down solid in chuck and finish inside; unless otherwise instructed make each flange same thickness, and make wide enough to slip over widest part of jaw. Now chuck all wedges and block up top end of each, so as to set inside faces true with tool, and finish flanges the same as shoes.

OLD OR NEW WHEN LAID OFF.

Set shoe or wedge by the pop marks on the outside, using surface gauge or hermaphrodites, and clamp down solid to face of planer. Line up until pop marks show exactly same height all around. Now plane off face to required size, and make plenty of fillet.

HOW TO PLANE UP DRIVING BOXES.

Clamp on bed of planer and finish on face, turn over, clamp planed face to bed and finish other face, making box the correct width. Now clamp one side against angle plate on bed of planer and plane out for shoe, making each flange required thickness and depth. Now reverse, and plane the other side the same. Now if you wish to bevel or taper flanges place a liner between box and angle plate and finish one end, then change the liner and finish the other end. Boxes are usually slotted for brasses, but if you have a circle feed on planer place all boxes top down on bed, clamp and plane to required circle. Keep all the boxes the same size when possible.

BEVEL INSIDE OF DRIVING BOXES.

On very rough railroads it is well to bevel inside of driving boxes (where shoes and wedges fit) at both ends, leaving about 4" of bearing in the center; this saves flanges of box from breaking.

CIRCLE FEED ON PLANER.

Some shops have planer head with circle feed, which is very handy for planing out driving boxes for brasses, as any number of boxes may be planed at once and all exactly the same size; it saves time and work for the slotter.

PLANE UP VALVES.

Clamp valve on planer face down and true up top to set by. Now turn over and true up face of valve. Turn over again and clamp on parallel strips, find centers at each end, and set true with tool. Now face off top to required height, and plane off sides where yoke fits and outside edges of valve to required size and keep perfectly central, and keep bottom flanges of valve required thickness. Now use a square nose tool and cut two grooves for long strips, proper width and depth, and right distance apart, keeping perfectly central. Now set valve crosswise on bed, still using parallel strips and set perfectly square with bed. Now plane off outside ends right length and keep central, and cut end grooves for strips. Have inside of valve slotted, or lay off, chip clearance at each end and plane out.

PLANE VALVE STRIPS.

Clamp in chuck, top faces down and true up bottom lugs, turn over in chuck and finish top side, making same height as grooves in valve. Now chuck and true up one side of each, then turn over and plane off other side, fitting snug to grooves in valve. You must finish lugs on short strips in jumper, then measure distance between grooves on valve and jump off ends of long strips 1-32" shorter to allow for expansion of long strips.

TRUE UP FACE OF OLD VALVE.

Clamp on bed of planer, top down, and true up. Use broad nose tool with coarse feed.

PLANING CROSS-HEADS.

Clamp on bed of planer and finish the top and bottom and outsides, but leave about $\frac{1}{4}$" stock on the wings that bear on the guides, also on the sides that fit between the guides, to be finished when keyed onto piston. To finish: Place two V blocks in grooves of the planer and clamp piston down, key on crosshead and set perfectly square. Now finish the head to the right size all around, letting the tool run over both wings without changing, and plane perfectly central.

PLANE ROD STRAPS.

Finish one side, turn over and put a piece of tin under the end that goes on the rod, then finish that side. The tin makes a taper in the strap, so brasses will slip on easy. The inside of the straps are usually finished on a milling machine where there is one; if not, then on a slotter or planer.

THE METRIC SYSTEM OF WEIGHTS AND MEASURES.

HISTORY OF THE METER.

To France belongs the honor of making the first systematic attempt to break through the customs of antiquity, and to substitute a new metrology for the old. Until the latter part of the eighteenth century there was the same condition of affairs in that country that prevails to-day in the United States; and in Méchain and Delambre's *Base du Systeme metrique decimal* (Paris, 1806), we read of "le système incohérent de nos mesures," "l'étonnante et scandaleuse diversité de nos mesures," etc. Several ineffectual attempts to reform the French system of weights and measures had been made previous to 1790, but it is from that year that the present "metric system" dates. In May, 1790, M. de Talleyrand proposed to the National Assembly of France that a new system of measures should be devised on strictly scientific principles, and that the units of length and weight in this system should be based on some natural and invariable standard. On the 8th of May the Assembly passed a resolution requesting the king, Louis XVI, to open a correspondence on the subject with the king of England, desiring him to invite the British Parliament to coöperate with the National Assembly of France in fixing the "natural unit" which was to serve as the basis of the proposed system of weights and measures. The work was to be put in the hands of a joint commission of scientific men, half of whom were to be appointed by the French Academy of Sciences, and the other half by the Royal Society of London. In conformity with the resolution, Louis laid the matter before the English king; "but owing to the temper and the public troubles of the times, his overture met with no response. Similar applications to other nations were more successful, and in subsequent proceedings Spain, Italy, the Netherlands, Switzerland, Denmark, and Sweden, participated by sending delegates to an international commission. The system itself was, however, matured by the labors of a committee of the Academy of Sciences, embracing Borda, Lagrange, Laplace, Mongé, and Condorcet, five of the ablest mathematicians of Europe." Lavoisier, though not a member of this committee, contributed largely to its proceedings, and the standards that were afterwards prepared were made under his supervision. Lavoisier, for some reason, has not received the recognition that is due him in this matter, his work being usually credited to Borda.

There were two reasons advanced for basing the new unit of length upon some absolute and invariable quantity in nature. It was said that the selection of an arbitrary unit would violate the most fundamental principle of the proposed reform; for the one distinctive feature of the new system was to be the exclusion of the last vestige of arbitrariness. This argument seems to us to be hardly worth serious consideration, for however satisfactory the unit finally chosen might be, the *selection* of that unit would necessarily be arbitrary, and hence the unit would itself be so, in some sense. The other reason advanced for the selection of a natural unit was much more logical. It was urged that a system of weights and measures based on some permanent and invariable natural quantity could be entirely reconstructed, with any desired degree of precision, even though every standard in existence were utterly destroyed.

M. de Talleyrand proposed to adopt, as the fundamental unit of length, the length of a pendulum vibrating seconds in latitude 45°, "ou toute autre latitude qui pourroit être préférée" (or such other latitude as might be preferred). The committee of the Academy of Sciences considered the advisability of adopting the seconds pendulum as the unit of length, but finally rejected it, principally because it involved the conception of *time*. It seemed to them preferable to base the proposed unit on *the length of some object actually existing in nature:* and after much deliberation it was decided to adopt some one of the dimensions of the earth itself. Of the different dimensions proposed, the two that met with the most favor were the equatorial circumference, and the meridian quadrant; and of these two the meridian quadrant was finally selected, because it could be measured with greater accuracy. (The measurement of a *meridian* involves observations of *latitude*, while the measurement of the *equator* involves observations of *longitude;* and before the invention of the electric telegraph, longitude measures were subject to large errors). In order to guard against possible differences in the lengths of the various meridian quadrants, it was decided to recommend that the new system of measures be based on the particular meridian that passes through Paris. This length (about 6,000 miles) would be entirely out of the question as a practical unit for business purposes. However, a rough calculation showed that a standard having a length equal to one ten-millionth part of this quadrant would be convenient, and the Meter (as the new unit was named) was therefore defined to be the *ten-millionth part of the distance from the equator to the north pole, measured along the sea level, on the meridian passing through Paris.*

The report of the committee, embodying the points explained

above and dated March 19, 1791, was approved by the National Assembly, and the work of measuring the meridian passing through Paris was begun. This operation involved immense labor, and seven years were required to complete it. The meridian through Paris strikes the North Sea at Dunkirk, near the Strait of Dover, in latitude 51° 02' 8.85", and the Mediterranean Sea at Montjouy (a small place in the suburbs of Barcelona, Spain), in latitude 41° 21' 44.96". The northern half of this arc was measured under the direction of Delambre, and the southern half under Mechain. A brief account of the method of measurement employed, and of the subsequent computations, will be found in the first chapter of Clark's *Geodesy;* and the operations and calculations are described in full in Mechain and Delambre's *Base du Systeme metrique decimale*, to which we have already referred. The unit of length used in measuring the meridian was the *toise*, and the final result of the seven years' work was, that the distance from the Barcelona end of the line to the Dunkirk end was 551,584.7 toises. It will be found from the data given above that the difference in latitude between the two ends was 9° 40' 23.89". Now the latitude of the equator is 0°, and the latitude of the pole is 90°; and therefore if the earth were a true sphere the distance from the equator to the pole would be given by the simple proportion 9° 40' 23.89" : 90° :: 551,584.7 toises : distance required. The first term of this proportion, when reduced to degrees and decimals of a degree is 9.6733°; and if we substitute this value of it, and then solve the proportion by the ordinary rule-of-three we find

Distance from equator to pole = 5,131,922 toises.

In the actual calculation a small allowance had to be made for the fact that the earth is not a perfect sphere (the polar diameter being about 26 miles shorter than the equatorial diameter). It was found that to take this into account it would be necessary to subtract 1,181 toises from the distance as calculated above. Hence, it was concluded that the true distance from the equator to the pole, along the meridian of Paris, is 5,130,741 toises. One ten-millionth part of this is 0.5130741 of a toise, and this, therefore, was to be the length of the new unit, or *meter*.

The next task was to prepare a bar of platinum which should have this length. When this had been done, the length of the bar was verified by the international commission referred to above, and the commission then proceeded in a body to the Palace of the Archives, in Paris, and there they formally deposited the bar of platinum which was to be ever afterward the standard meter of the world.

PRESENT STATUS OF THE METRIC SYSTEM.

For many years the advocates of metric system of weights and measures have been laboring unceasingly to have the United

States Government adopt this system and make it the only legal one in this country. As this is a subject of vital importance to every manufacturer and railroad company, and very interesting to mechanics we will endeavor to briefly explain what is meant by the metric system. This system of weights and measures was designed to remove the confusion arising out of the excessive diversity of weights and measures prevailing throughout the world, by substituting in place of the arbitrary and inconsistent systems actually in use, a simple one constructed on scientific principles and resting upon an invariable standard. The system has been successfully adopted by more than half the civilized world, which includes the following countries: Holland, Belgium, Spain, Portugal, Italy, the German Empire, Greece, Roumania, British India, Mexico, New Granada, Ecuador, Peru, Brazil, Uruguay, The Argentine Confederation, and Chili. Switzerland, without adopting the system in full, has given to all her standards metric values and Denmark has done the same for her standards of weights. Austria has adopted the system for custom house purposes; and Turkey has introduced a metric measure of length. In Great Britain the use of metric denominations in business transactions has been made legally permissible, and now a committee of the House of Commons has recommended that within two years the metric system of weights and measures shall be rendered compulsory in the British Isles. In the United States metric weights and measures were legalized by an act of Congress passed July 27, 1866, and now the House Committee on coinage and weights and measures has recommended that the metric system be adopted by the various departments of the government July 1, 1898, and by the Nation at large January 1, 1901. It is therefore fair to presume that within a few years this system will be the only legal one used in this country; so it will be well for every mechanic to study and familiarize himself with the table of metric threads shown on page 582, which were adopted by the German Engineers; we do not think a better table could be given:

    A French meter equals 39.37 inches in length.
    A decimeter equals 3.937 inches in length.
    A centimeter equals 0.3937 inches in length.
    A millimeter equals 0.03937 inches in length.

We would no doubt profit by adopting this system of weights and measures (measures of capacity) as much confusion exists, the various States having different standards, while the metric system measures of capacity, divides by tenths like its linear measurement. There are also objections to our linear measurement because it does not divide evenly like the metric system. For example, our rod contains $16\frac{1}{2}$ feet and our mile 5,280 feet. But when the effort is made to substitute the French meter for our inch the difficulties in the way of carrying out the change will become apparent. No objections exist to the meter as a unit of measurement, except that the parts of our existing system cannot be represented in divisions of the meter without the use of numerous figures, which would cause endless confusion. Our standards have been built up and adopted by Railroads and Manufacturers at a cost of millions of dollars. Perhaps the most important and expensive are our screw thread, and it would be almost impossible to express the number of our threads to any part of the meter without changing the pitch of the threads, which would entail an enormous expense. Manufacturers who

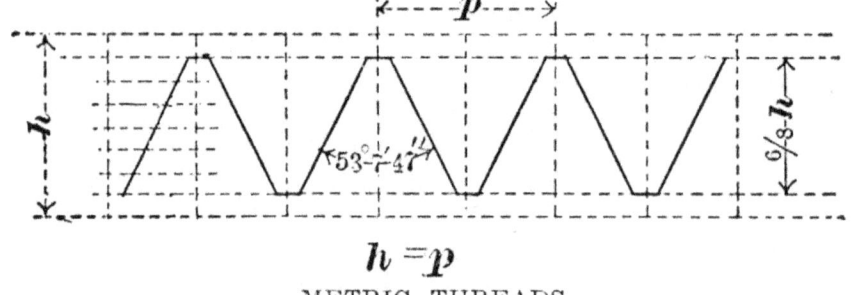

$h = p$

METRIC THREADS.

have a large foreign trade might profit by the change, but an overwhelming majority of the manufacturers and railroads would be forced into an unnecessary expense. The most enthusiastic advocates of this system in this country are theorists, who do not realize at what cost a change of measurement could be effected; however, their efforts have been remarkably successful and the change seems inevitable. Machines are now manufactured and used in this country which are geared to cut both United States standard and Metric threads.

TABLE OF METRIC THREADS.

Diameter of Screw. mm.	Pitch. mm.	Diameter of Screw. mm.	Pitch. mm.
1.	0.25	8.	1.2
1.2	0.25	9.	1.3
1.4	0.3	10.	1.4
1.7	0.35	12.	1.6
2.	0.4	14.	1.8
2.3	0.4	16.	2.
2.6	0.45	18.	2.2
3.	0.5	20.	2.4
3.5	0.6	22.	2.8
4.	0.7	24.	2.8
4.5	0.75	26.	3.2
5.	0.8	28.	3.2
5.5	0.9	30.	3.6
6.	1.	32.	3.6
7.	1.1	36.	4.
		40.	4.4

COMPRESSED AIR.

We are indebted to Dr. Denys Papin, who lived nearly two centuries ago in the town of Blois, France, for the first suggestion of conveying compressed air through pipes as a means of transmitting power. His fertile brain conceived the idea of conveying parcels through a tube by means of compressed air; but, like many other great inventions, its real value was not at that time appreciated. We do not again hear of any atmospheric systems of propulsion until 1810, when George Medhurst took out a patent in England, "For a means of conveying goods, letters, parcels and passengers by means of a tube and a blast of compressed air." From that time on compressed air began to be recognized as a simple and valuable power, and it grew constantly in popularity. When Air Brakes for railroad trains were invented it gave compressed air such an impetus that it has since been used for almost every conceivable purpose; in the last decade, with the advent of improved air compressors, its growth has been phenomenal, until to-day compressed air is used for a greater variety of purposes than any other known power. It lacks the objectionable features of the other forces, such as fire, smoke, heat and electricity. It is cleaner and more desirable for use, especially where combustible material is stored or handled. It has been wisely remarked, "That if a fractional part of the brains and money that have developed electricity had been spent on air, our street cars would be running by air to-day." Compressed air is yet in its infancy. Perhaps the most beneficial use made of air so far is upon the air brakes of railways, although it is used as a means of propulsion on a few railway locomotives, numerous mining locomotives, stationary engines, elevators, street cars, mining coal, drilling rock, pumping water, and for transmitting cash, letters and packages, and in a few of the larger cities it is piped throughout the city and sold by the cubic foot; the city of Paris, France, using 25,000 horse-power daily. We will not attempt to enumerate the many uses made of compressed air, but will confine ourselves to its use as applied to railways, which will be more interesting to our readers.

COMPRESSED AIR LOCOMOTIVES.

Compressed air locomotives are not used in ordinary railway service, but are used by many large manufacturing plants on account of the absence of smoke, which renders it feasible to run them inside of buildings where operations are being carried on to which fire or smoke would be fatal. Many of them are also

used in coal mines in preference to the dangerous electric cars, and where steam could not be used on account of the smoke. They also furnish an additional supply of fresh air in a mine. They are also extensively used upon cotton plantations and around cotton warehouses and yards. In most of these locomotives the air is pressed through a hot water tank before passing into the cylinders, which increases the efficiency of the air. The illustration we have shown is that of a mining locomotive built by the Dickson Locomotive Works at Scranton, Pa. It stands five feet high. The wheels are 26" in diameter and the cylinders are 9x16"; it weighs 16 tons; its storage pressure is 600 lbs., and the working pressure 125 lbs.; the capacity of the storage tanks amounts to 170 cubic feet.

USE OF COMPRESSED AIR UPON RAILWAYS.

In addition to its use in railroad shops, which we shall treat at length, its principal use upon railways is the air brakes, but it is also used for signals and switches, for pumping water, for bridge building, bell ringers, and pneumatic sanders, signal whistle, for operating cylinder cocks, and for heating trains with the exhaust from the air pump, and for many other purposes which are constantly increasing. We shall first describe a few track sanding devices, the bell ringer and the pneumatically operated cylinder cocks, a separate chapter being devoted to the air brake.

THE LEACH SANDER FOR LOCOMOTIVE.

Three or four different track sanders are now in use upon locomotives, but as the Leach sander is the most extensively used, we have illustrated this particular device and explained its advantages and its method of application and operation.

This device is intended to reduce to a minimum the evils due to sanding the track for preventing the slipping of locomotive driving wheels. The ordinary sand lever arrangement is hard to operate, and is also very wasteful of sand, thereby causing unnecessary wear of tires and rails, and also resistance to trains.

A sand trap, or pocket, in which is an air nozzle, one trap being provided for each sand pipe, and attached to the outside of the sand box, that it may be readily inspected, is the main feature of the device. The trap receives its supply of sand through an auxiliary passage which is normally open, and delivers it into the sand pipe as may be required, by means of the air blast. This arrangement does not interfere with the sand lever and valves, which are preferably retained for emergency use.

Suitable arrangements are provided in the trap for receiving the wear of the sand blast, and also for removing stones and other foreign substances.

The compressed air used is taken from the air brake pipe which leads from the main reservoir to the engineer's brake valve, the pipe conveying this air to the sand box being placed, whenever possible, under the boiler jacket, in order to heat the air.

The feed valve which regulates the air pressure and thereby the sand delivery, is a small globe valve having a fine thread on the stem, and with the angle of the valve seat nearly parallel with the stem. This allows of very fine adjustment, as a considerable movement of the valve wheel is necessary in order to

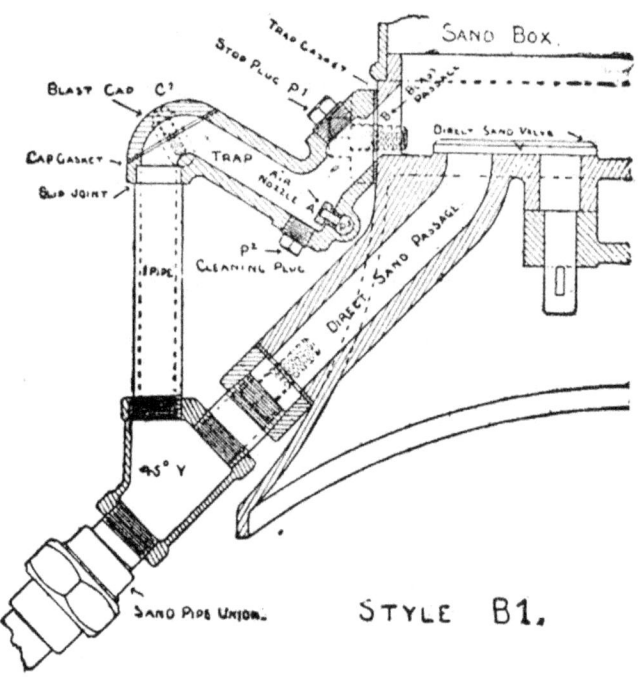

STYLE B1.

secure a perceptible opening of the valve. This fine adjustment of the feed is an important feature, as a pressure of but two to five pounds at the air nozzle is required for light feeding, and a higher pressure would result in a waste of air and sand, which must be avoided.

In order to prevent the feed valve being forgotten and thereby left open longer than needed, a tell-tale attachment is provided in the valve wheel, which sounds a warning whenever the valve is opened for sanding, unless prevented by the pressure of the operator's hand. If the hand is removed for other duties, the feed continues, but the warning sounds until the hand is replaced, or the valve closed.

The first cut illustrates the principle of the device, but this form is only occasionally used. The style in most common use is shown in the second cut, which represents an inverted sand box base with the sand traps and pipes attached, but with the sand lever and valves removed. For new work, the auxiliary passages are cast instead of bolted in.

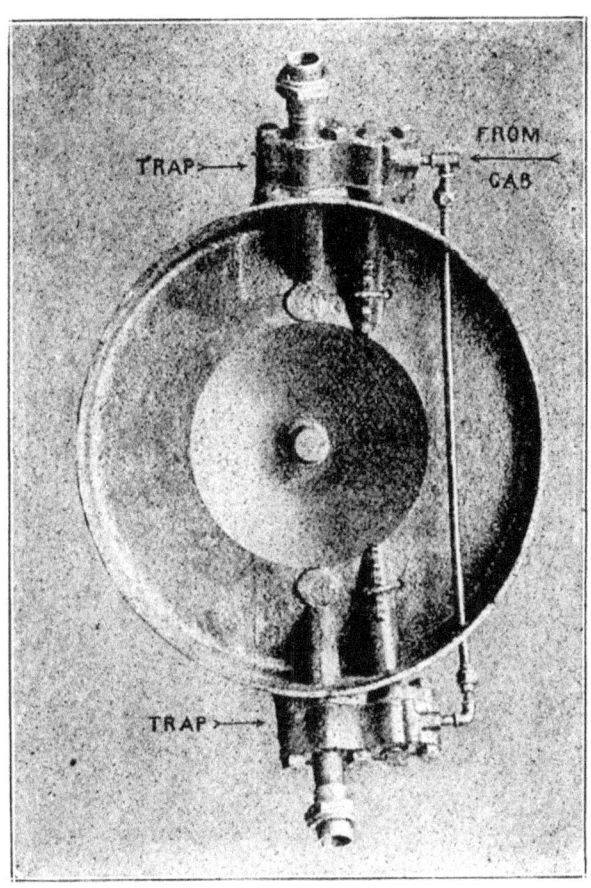

Double traps, designed for two pairs of sand pipes on the same sand box, are made and applied in a similar manner.

LOUISVILLE AND NASHVILLE TRACK SANDER.

We quote the following from *Locomotive Engineering* in regard to this device:

"This is the very latest and only adjustable feed out. It works in all positions of engineer's brake valve, or not, as you like; is automatically closed; can't be forgotten, etc. We have had these in service for the past twelve months on this division of the Louisville & Nashville, and they have never been even looked at. There are also several sets on other divisions of this system giving entire satisfaction. Mr. Leeds, Superintendent of Motive Power, says it is all right, and he will put them on. This arrangement retains old sand valves, to be used in case of a failure of air supply, and also uses the old valves as a cleaner for pneumatic valves; thus keeping both ready for service. It always applies full in emergency. We have just received an order from the N. C. & St. L. for trial set. We put them on any road wishing to try them, free of charge, and let them do their

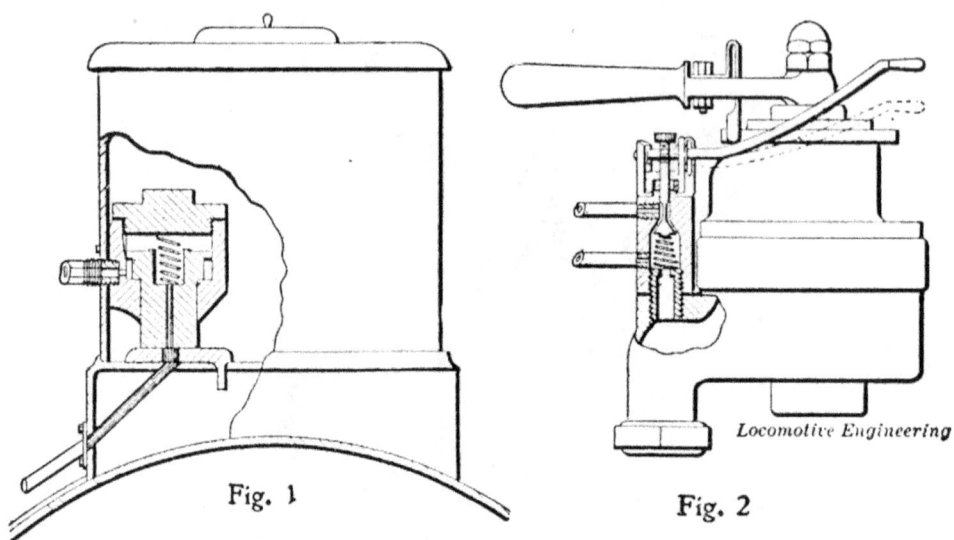

Fig. 1   Fig. 2

own talking. Have yet to find an engineer who is willing to exchange for any other kind.

J. H. WATTERS,
Division M. M., L. & N. R. R.
Anniston, Ala."

The owners of this device, Messrs. Walters, Howden and Jacobs, of Anniston, Ala., make the following claims for their sanding device:

They are adjustable and can be regulated by the engineer while running; it will work in service application of the brakes, or not, as you like, thus making the brakes work as well in bad weather as in good weather.

They always work in emergency application of the brakes adding 25 per cent to the braking power, thereby avoiding accidents.

They never stop up.

They have been used for the past two years and have not cost a cent to maintain.

They are the easiest put on.

They are cheap to manufacture and the most durable.

They do not interfere with old sand valves.

They discharge through the old sand ports in the sand box and air jet strikes on sand and not against any metal surface, and do not cut out as others do.

We guarantee them in every respect, and put them on and let them do their own talking.

THE JONES TRACK SANDER.

This device is the invention of Mr. Evan Jones, of Creston, Iowa, who describes his invention as follows:

In the accompanying drawings Figure 1 is a transverse sectional elevation taken through the sand-box of a locomotive; Fig. 2, a horizontal sectional view taken on the line 2 of Fig. 1;

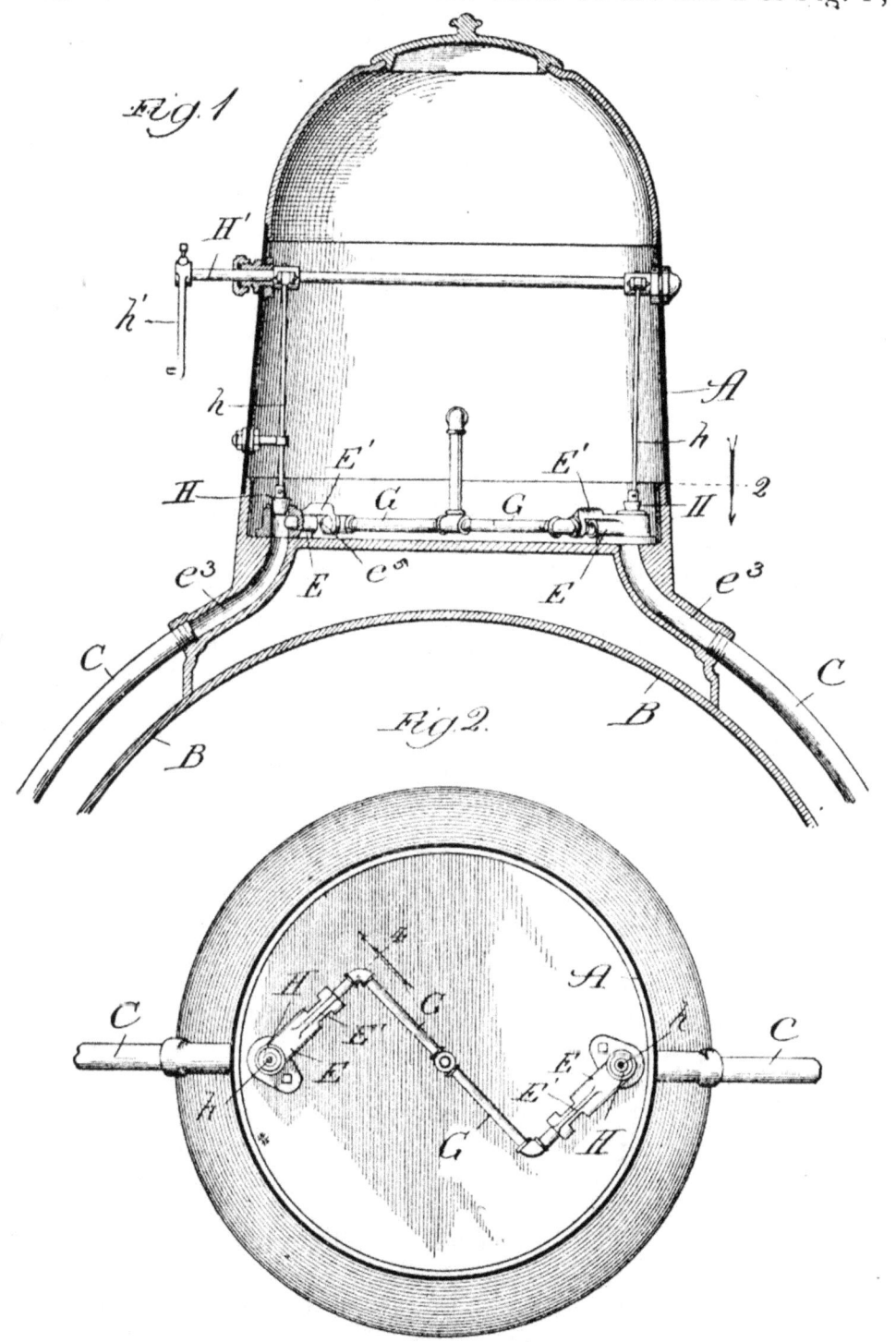

Fig. 3, an enlarged sectional view of the compound valve, taken on line 4 of Fig. 2; and Fig. 4, a transverse section taken on the line 5 of Fig. 3.

In constructing my improvements and fitting them to a locomotive I provide a sand-box A, of the usual type and of any desired size and shape, and attach it to the body of a locomotive-boiler B by riveting it thereto or in any of the ordinary and well-known methods. Adjacent to the lower side portions of the sand-box I provide delivery-pipes C, that lead therefrom to a point directly under the driving wheel or wheels, so that as sand is delivered from the box it strikes the rail or rails adjacent to the under portion of the driving-wheels for the purpose of enabling the same to obtain better traction on the surface of the track.

In order to feed the sand, as desired, in the box to the delivery-pipes, I provide what I term a "compound valve" E, arranged in the bottom of the box and horizontally so as to provide a substantially horizontal passage e through the same and which opens at a point adjacent to the bottom of the sand-box, as at e'. The other end is provided with an opening e^2, which connects directly with the delivery-channel e^3, with which the delivery-pipes are connected. The compound valve is provided with a horizontally-projecting portion E', that practically covers the upper portion of the inlet-opening e', so as to prevent the sand from directly entering such opening, or, in other words, keep the weight of the larger portion of the sand off of such opening, but in such a manner as to permit the sand to enter a slot e^4 between the projection and the main body of the valve. This prevents the sand from being jarred or shaken out of the box while the train is in motion. Arranged at the front part of this projecting portion E' and substantially in line with the passage e is an injecting-perforation e^5, in which is secured a pipe G, that has its opposite end connected with a source of fluid-pressure—compressed air. This pipe is provided, as is shown to the left of Fig. 3, with a valve g, which may be operated to open or close connection with the air-reservoir G' for the purpose of shutting off or admitting the desired quantity of fluid-pressure, the action resulting in injecting the sand that may be in the slot e^4 into the passage e and forcing it down through the delivery-channel to a point underneath the driving-wheel.

The advantages of the structure are that the sand will not spill or run down the channel without the use of the fluid-pressure, and the passages and the arrangement of the injector are such that the valve may be outside of the sand-box and the flow of sand be regulated to any degree required.

It is often desirable, especially in climbing or descending grades, to save all of the fluid-pressure for the purpose of the brake mechanism, and in such conditions it is desirable to have mechanism, operated, preferably, by hand, that will allow or permit of the sand being fed to the track by the force of gravity only. In order to accomplish this result, I provide the compound valve with a second vertical passage E^2, directly in line with the delivery-channel e^3, and in this opening I fit a vertically-movable valve H, having its stem portion l connected with a rock-shaft H', this rock-shaft in turn being provided with a bell-crank lever h', having an actuating-rod with its free end extending into the cab of the locomotive, so that it may readily be

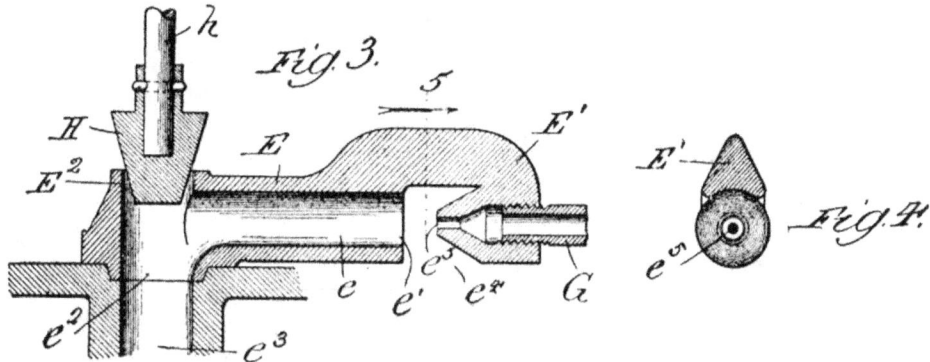

reached by the engineer or fireman for the purpose of operating the valve H and admitting sand into the delivery-pipe or preventing it from entering such pipe.

The advantages of the entire structure are that the engineer at all times is furnished with a mechanism that permits him to control the outflow of sand without the use of pneumatic pressure, and at times, when desired, pneumatic pressure can be used for forcing out desired quantities of sand for the usual purpose.

THE TIRMANN TRACK SANDING DEVICE.

This device is the invention of Mr. Hugo Tirmann, of Cleveland, Ohio, who gives the following description of the device:

The annexed drawings and the following description set forth in detail one mechanical form embodying the invention, such detail construction being but one of various mechanical forms in which the principle of the invention may be used.

In the annexed drawings, Figure I represents a vertical transverse section of the lower portion of the sand-box of a locomotive provided with my improved track-sanding device; Fig. II, a horizontal section of the sand-box and device; Fig. III, a vertical section of the sanding device, said section being taken on the lines III III in Fig. I; and Fig. IV a face view of the inner side of the valve-seat.

The sand-box A is of the usual construction, secured upon the top of the locomotive-boiler shell B. The usual holes a are formed at opposite sides of the sand-box, near the bottom of the same, and casings C are secured over said openings, and are formed with necks c, to which the sand-pipes may be attached. Annular valve-rings D are secured to the inner edges of the casings within the sand-box and said valve-seats are formed with inwardly-bulging hoods d at their upper portions. Cup-shaped valves E have their edges seated against the valve-seats, bearing against the outer faces of the same, so that the convex sides of the valves may face outward within the casings. The cup-shaped valves are formed with openings e, which openings are in the portions of the valves which are normally turned upward. The valves have central pivots e', which are journaled in bearings c' in the casings, and said pivots are surrounded by sockets e^2, within which are fitted spiral springs e^3, which bear against the bottoms of the sockets and the sides of the casings, so

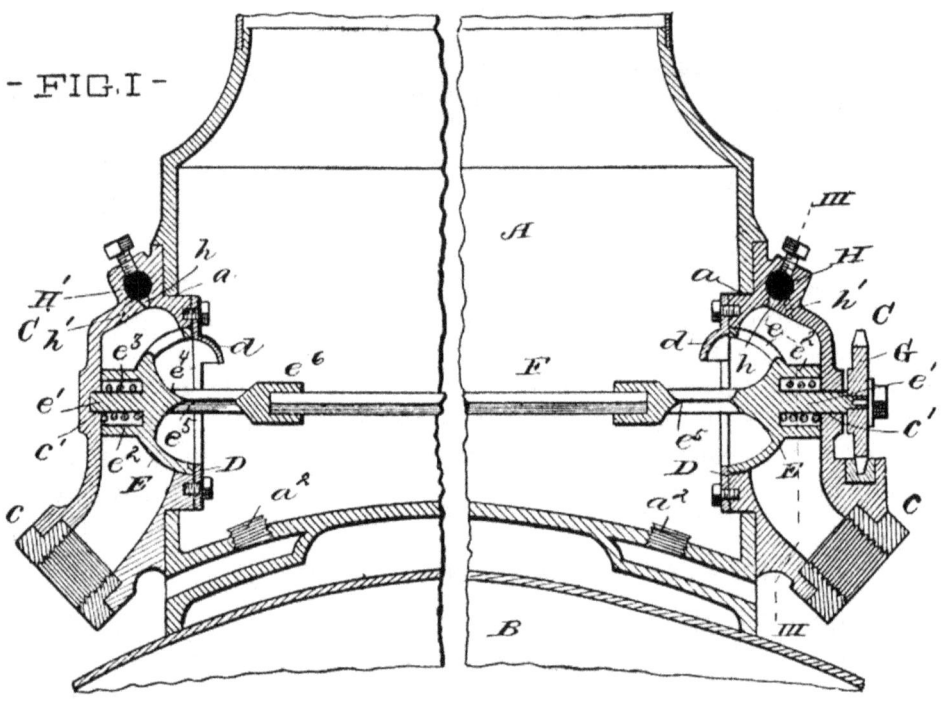

as to force the valves against the seats. Shanks e^4 project from the centers of the inner concave faces of the valves, and said shanks are formed with circular openings e^5, registering with the openings in the valves, and in the line of the air blast, which will be referred to later. Sockets e^6 are formed at the inner ends of the shanks, and the ends of a rod F are fitted in said sockets, so that the two valves are connected by said rod and may be revolved together.

A gear-wheel G is secured upon one of the valve-pivots, outside of the casing, and said gear-wheel meshes with a rack-bar g, guided to slide horizontally upon that casing; a lever g^1 is fulcrumed upon the sand-box and is pivotally connected to one end of the rack-bar, and said lever may be rocked from the locomotive cab by means of a rod g^2 entering the cab at a convenient place exactly in the same manner as the sand-valve-operating rod, now usually employed with the ordinary sanding apparatus.

Air pipes H extend from a suitable source of compressed air into air jet-chambers H' in the upper portions of the casings C. Said chambers are formed with jet openings h, adapted to send air-jets in an inwardly oblique direction through the openings in the valves and the circular openings in the valve shanks down into the sand-box and against the bottom of the same. As the sand blast created by the air-jets would be likely to wear through the bottom of the sand-box, I provide removable plugs a^2 at the points of the bottom of the sand-box, which are directly opposed to the jet openings, and said plugs are preferably made from some soft material, such as wood, vulcanite, or other similar

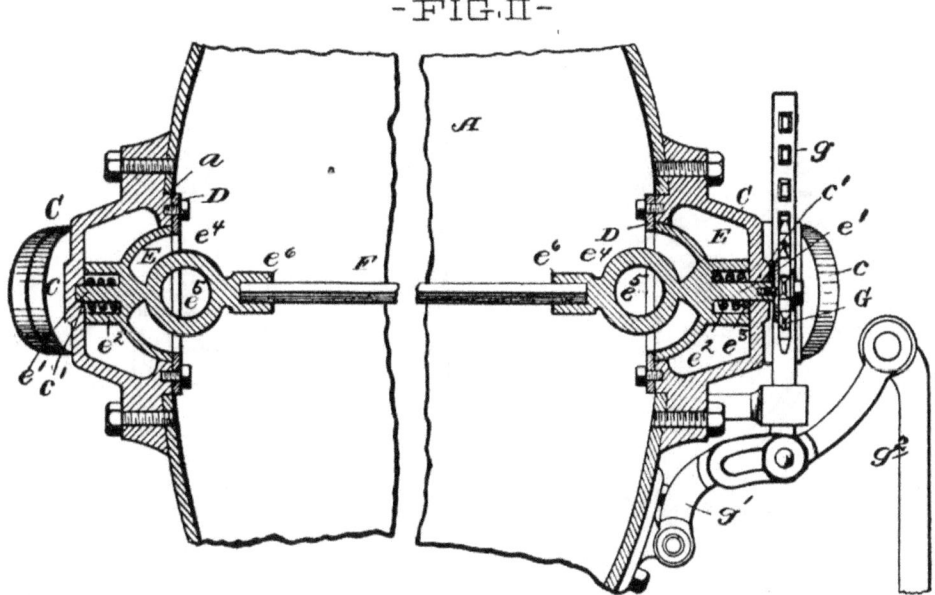
-FIG. II-

material which will not be cut by a sand blast, such as the hard metals. The air-jet chambers are formed with outwardly inclined jet-openings h^1, controlled by needle valves h^2, and said jet openings serve to admit small jets of air to the outer sides of the valves, so that any sand which may become lodged in the valve casings or on the valves may be loosened and removed by means of such blast.

When the device is in operative position upon the locomotive, the actuating rod and the cock which controls the compressed air supply are arranged within convenient reach of the engineer. The valves are normally turned so as to bring the openings in the same to face upward. When the engineer desires to apply sand to the track in the ordinary quantity, he admits compressed air through the pipes to the jet-chambers, whence the air will issue through the jet-openings in inwardly-inclined jets. Such jets will stir up the sand in the sand-box and will cause an upward eddy, which will carry the sand with it up through the openings in the valves, whence it will fall down through the valve-casings and the sand-pipes. The outwardly-inclined jet-openings will prevent sand from lodging in the valve-casings and upon the valves; but the jets issuing from said openings are not required to be of sufficient strength to act in any way to force the sand through the sand-pipes.

When an extraordinary amount of sand is required upon the track, the operating-rod is pulled, which will cause the valves to be revolved so as to bring their openings downward, when the sand may freely flow from the box through the valve openings and casings into the sand-pipes and onto the track. The

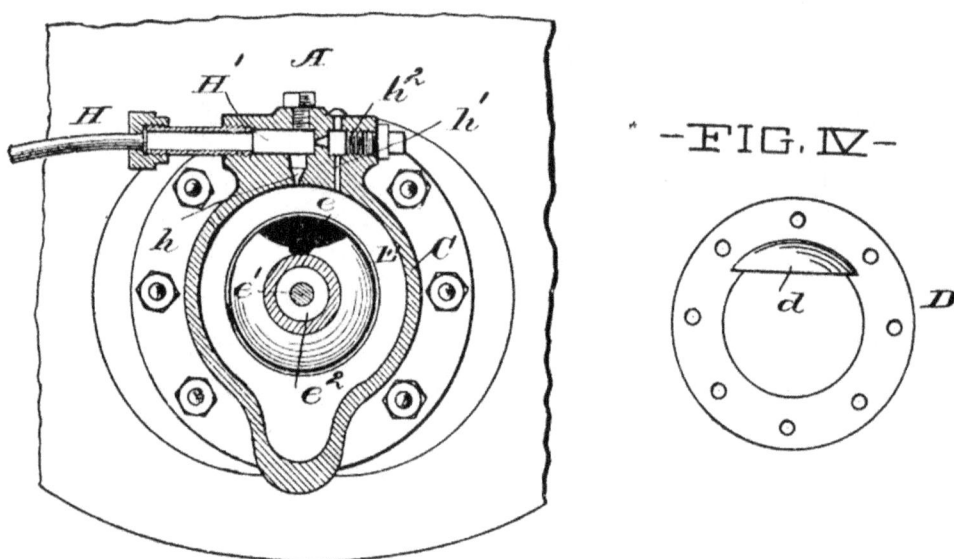

hoods upon the inner sides of the valve-seat rings serve to guide the eddy of sand out through the valve-openings when the air-blast is employed.

This sanding device may be applied to the sand-boxes ordinarily used upon locomotives without any further change than the removal of the valves now in use, the securing of the valve-casings and valves in the openings from which the old valve-casings have been removed, the connection of the sand-pipes to the new valve-casings, and the connection of the valve-operating rod to the lever which actuates the rack-bar. As the air-blast is downward and into the sand in the box, the jet openings cannot be clogged with sand; neither can the valve-casings or pipes be injured by the sand-blast.

As long as the sand-box is filled to a distance of about six inches from the bottom the sand-blast will not injure the bottom, whether plugs are provided to meet the sand-blast or not; but I consider it judicious to provide the plugs or other means for receiving the force of the sand-blast, so as to provide for the possibility of the sand becoming lowered too far in the box and the blast then cutting through the bottom of the same.

The entire device is simple of construction and is not liable to get out of order, as the bearing-surface of the valve upon its seat is so small that pebbles cannot lodge between the valve and the seat, and sand will not be liable to accumulate in sufficient quantities to interfere with the working of the valve.

All parts of the device are easily accessible and may easily be detached for renewal or repair.

DEAN'S PNEUMATIC TRACK SANDER.

The operation of this device is as follows:—A three-eighths inch pipe connects main air reservoir with sanders in sand box, and air is controlled by a three-eighths inch globe valve placed

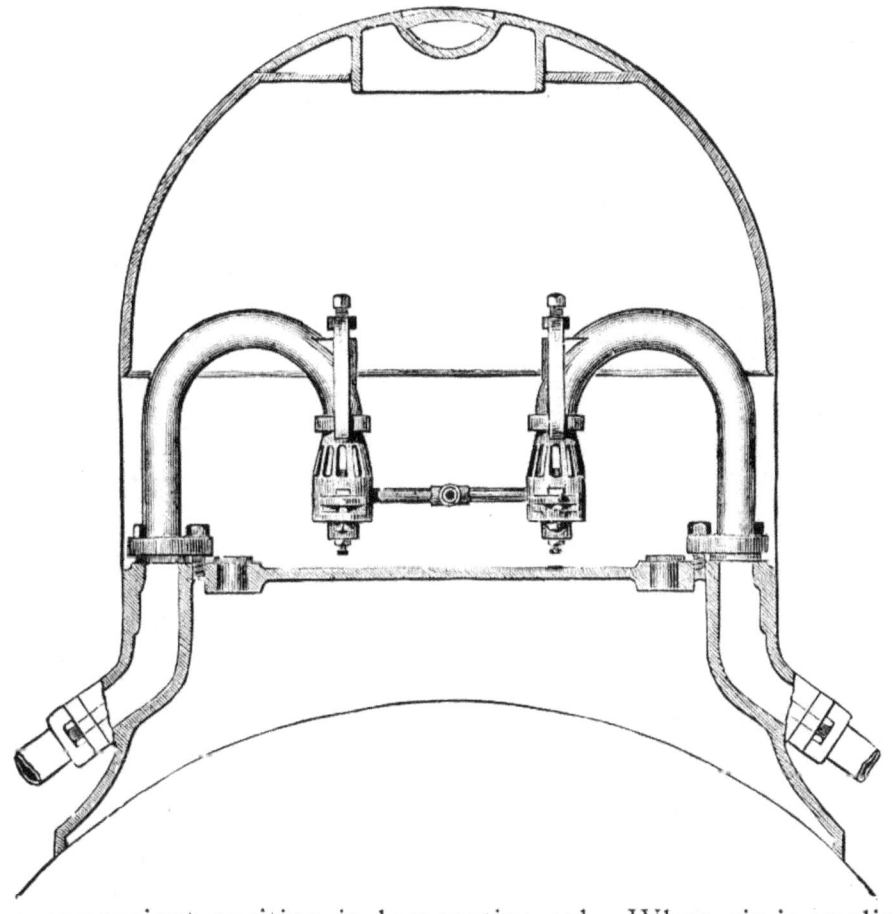

in a convenient position in locomotive cab. When air is applied to sanders it passes into cage above diaphragm through holes in cup nut, beside cleaning needle out of nozzle. The sand filling the cage is then blown to the rail.

The jar of the engine may cause sand to become packed in air nozzle and rest on point of cleaning needle, but when air is applied it forms a pressure in cavity of cage, which forces the diaphragm down, withdrawing point of needle from the sand in air nozzle, and leaves the sand free, so that it is easily blown out, thereby making it positive in its action, and it can never become clogged by sand. When air is shut off from main reservoir, diaphragm and cleaning needle are forced back into position by spring.

DIRECTIONS FOR APPLYING AND OPERATING.

Blow out air pipe leading from main reservoir to sander before coupling up, as scale or dirt might plug up air nozzle.

Main sand pipe should be carried as close to point of contact of wheel and rail as possible. If one and one-quarter inch pipe is used the lower end should be reduced to one inch.

If air comes through sand pipes and no sand comes with it, turn on full pressure of air and hold hand on bottom of sand pipe. This will blow any obstruction away from cage and allow sand to flow.

If when sanders are started they work for a moment and then stop suddenly and continue to do this every time air is applied, it is probable that diaphragm leaks and equalizes the pressure in the cage, closing the cleaning needle.

THE FOLLOWING CLAIMS ARE MADE FOR THIS SANDER.

1. Will give you sand when needed.
2. It is positive in its action.
3. It works as well in poor sand as others do in the finest lake sand.
4. You need sand in emergencies, and an engineer can have it without fail, instantaneously, by using the Dean pneumatic track sander.
5. You can use as little or as much sand as you wish, but cannot "stall" an engine by the sand piling on the rail.
6. Sander can be applied without removing old sand valves if desired.
7. Crooked pipes and sharp bends do not interfere with placing sand where desired.

BELL RINGERS.

During the past few years steam and air bell ringers for locomotives have come into general use; at first they were used only upon yard engines, but at present they are applied to locomotives for all kinds of service. The advantages to be derived from a good bell ringer are clearly apparent, and as a result inventors have devised no less than one hundred different forms of bell ringers within the past five years.

We have illustrated but two of the best forms of these devices which will suffice to give the reader a general knowledge of the construction and operation of bell ringers.

THE BREITENSTEIN BELL RINGER

With a view of providing a simple and positive bell ringing mechanism for use upon locomotives and which will not be liable to get out of order, the device illustrated in the accompanying drawings was designed by Mr. Henry Breitenstein, of Laramie, Wyoming. Referring to the drawings, Fig. 1 shows an eleva-

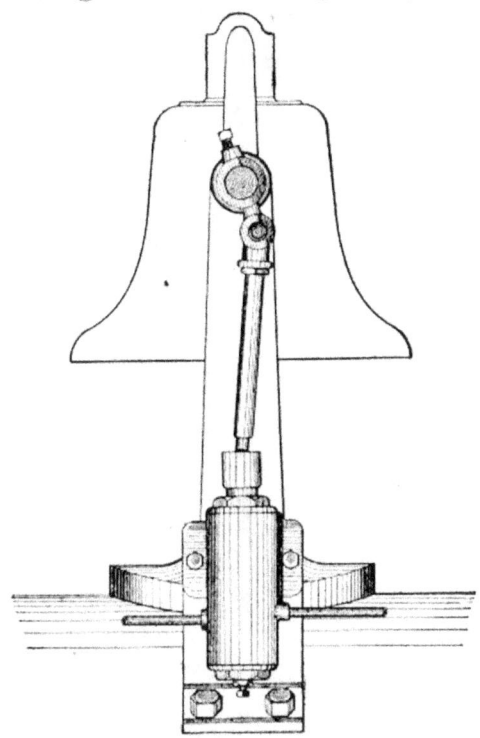

Fig. 1.

tion of the ringer attached to the bell yoke and bell; Fig. 2 is a vertical section and above Fig. 2 is shown a horizontal section taken on the line Z, Z, of Fig. 2; Fig. 3 shows a half elevation and half section through the cylinder. The cylinder A contains the piston B cast together with the piston rod C, which is connected with the main driving shaft of the bell yoke in the usual manner. The valve for operating the piston is contained in the lower part of the casting under the cylinder, the valve being in two parts, the upper part E containing the steam passages and upon the lower end the piston E' is carried. This piston is cast in one piece with the valve. The valve E operates over the inlet port e in the valve chamber D and over the exhaust port f leading to the outside. From the upper end of the cylinder A a port g leads into the exhaust port f to avoid compression in the upper end of the cylinder. In the case of air being used for motive power this port is displaced by a hole through the head of the cylinder. In Fig. 3 a port h is shown leading from a point just below the piston B after it has reached its extreme uppermost position, and passing downward below the piston o the valve in the chamber D'. Through the center of the valve E, Fig. 2, a port i connects the cylinder A with inlet port e, when

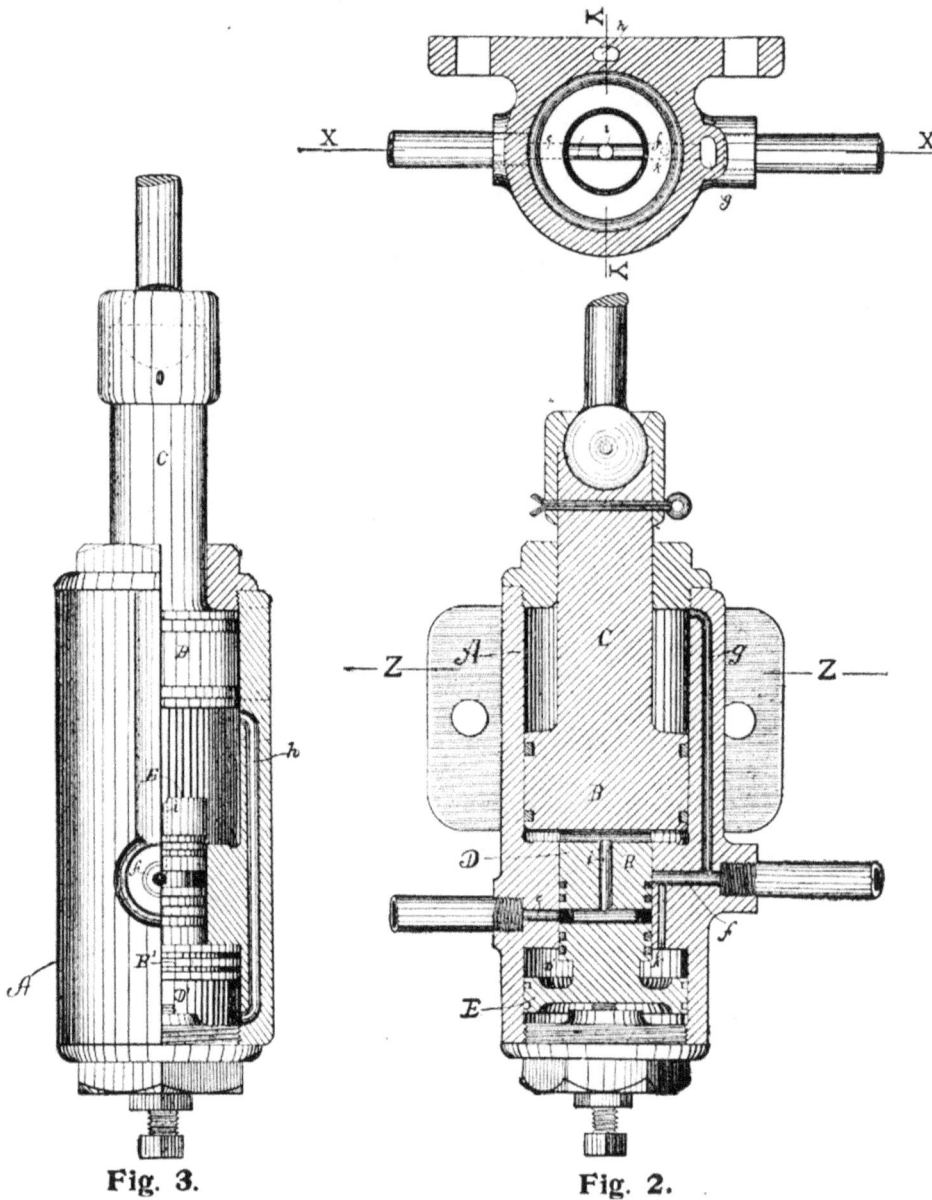

Fig. 3.   Fig. 2.

the valve is forced to its lowermost travel by the piston B striking upon its top, as shown in Fig. 2. In this position the valve admits steam under the piston B, forcing it upward until it uncovers the port h when the live steam is admitted by means of this port to the chamber D'. This forces the valve upward, cuts off the communication with the admission port e and opens communication to the exhaust. The valve owes its upward motion to the fact that the piston E' is larger in area than the main portion of the valve E. A port k opens from the upper side of the valve piston into the exhaust, preventing compression at this point, upon the upward movement of the valve.

Upon the exhausting of the steam below the piston B, the weight of the bell carries the piston downward and in striking upon the top of the valve E it returns the valve to its lower position. Immediately before the piston B reaches the top of the valve E the port h is uncovered by the upper side of the piston B and the pressure below the valve piston E' is discharged into the exhaust. The apparatus is now ready for a repetition of the operation, which becomes continuous when steam or air pressure is once applied. The device has been in use for five years upon locomotives of the Union Pacific Railway and the only change that has been made in the construction during that time has been the addition of the adjusting screw upon which the valve piston rests when in its lower position. The connecting rod is made of half inch pipe, with a blank welded in the top end, threaded for the connection with the bell crank. The ringer is of cast iron, as are also all of its parts with the exception of the bearing at the bell crank, which is of brass. The claim made by the inventor for this device is that the valve motion is positive and not depending upon mechanical connections for its operation, the liability of derangement is greatly reduced. This device should not be expensive to manufacture and the bearing parts are sufficiently large to resist the effect of wear for a long time. The simplicity of its construction is one of its strong points. This device has been patented by Mr. Breitenstein.

THE GOLLMAR BELL RINGER.

The feature of this device is the automatic starter as shown by the illustrations. Every time the whistle is sounded the bell will also ring; it is simple of construction and has been adopted by many of the leading railroads in this country. It works by

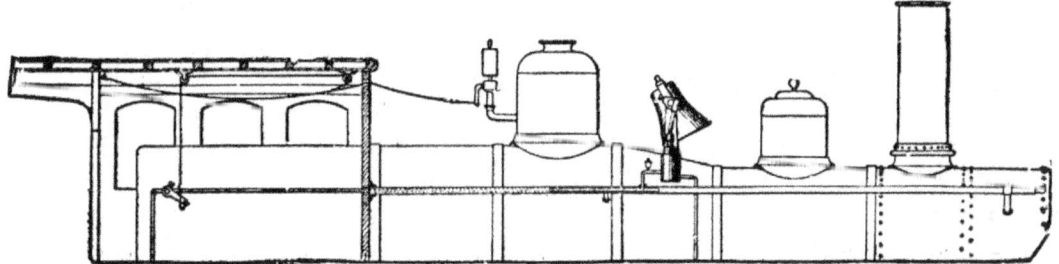

either steam taken from the boiler or air from the air-brake. It entirely obviates the use of bell rope and is a simple, durable and reliable bell ringer. It can be used as an automatic bell ringer by ringing the bell continuously, or with the use of the automatic starter may be rung at intervals when required. The automatic starter causes the bell to ring immediately upon sounding the whistle. It has also the desirable feature of restarting the bell to ringing when it is stopped from any cause and forgotten on roads that require the bell to ring continuously. The first crossing signal does the work, and it does not forget to ring the bell when it should be rung, as often occurs when the engineer and fireman are depending on each other to start the bell ringing. It meets with the general approval of railroad officials when their attention is called to this feature. Besides affording protection to life it has many advantages of security, which render such a device almost indispensable. It has means of adjustment and is economical in the use of air or steam.

ITS CONSTRUCTION.

The construction and its action are as follows: There are two openings for pipes; the upper one is the inlet, the lower is the exhaust. Pressure is admitted through upper opening, opposite an annular groove in valve 18, through which four holes are drilled, admitting the pressure under the single acting piston 10; this causes piston 10 to rise, forcing the bell to swing. Piston 10 has a stroke of 1¼ inches when at its extreme travel; crank 2 has a stroke of 4 inches. The connecting rod is in two sections, 6 and 7, which allows the crank 2 to make a complete revolution without causing piston 10 to move. When the ringer is started to work the piston 10 will be driven upward, causing the bell to swing, and valve stem 17 will raise valve 18, closing inlet port and use pressure expansively by traveling the length of the lap before it will open the exhaust port. The bell having received an impulse will continue its motion after the piston 10 has reached the upper end of its stroke, the crank box 6 sliding on rod 7. The impetus which the bell receives being expended it will fall; the set bolt 4 will strike the end of rod 7, the piston 10 will be forced downward, coming in direct contact with valve 18, closing exhaust port and opening inlet port after cushioning on the pressure remaining under piston 10 after exhaust is closed. It will be seen that valve 18 is only operated at the terminations of the piston 10 stroke. Packing rings 15 on piston and on main valve are packing rings D 6-No. 8 Westinghouse Air-Brake Co., Catalogue 1890, and are duplicates of those used in air-pumps. As the rings are kept in stock by all railroads using air-brakes no extra supply need be carried by them. This is the only bell ringer that can be adjusted to use pressure in proportion to the power required. This is accomplished by means of valve stem 17, which is secured in its adjusted position by jam nut 16. No change in length of connecting rod is required in making this adjustment. These bell ringers are in use cutting off pressure

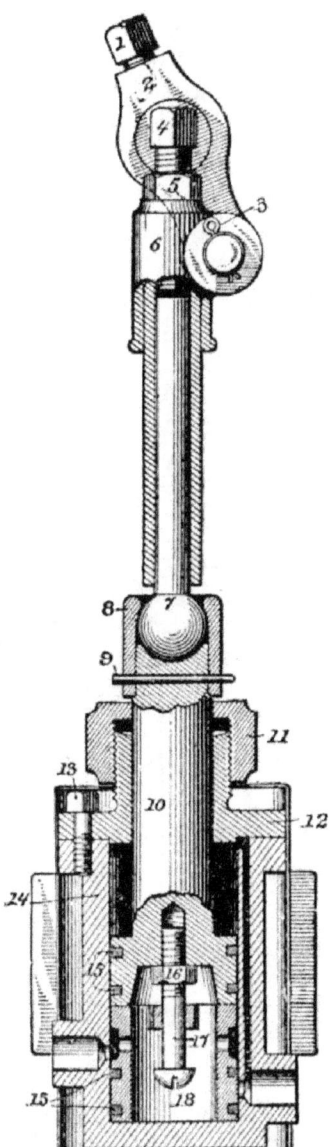

after piston has been moved ¾ inch of its stroke. This arrangement makes it so economical in use of pressure that air is always used in preference to steam, and it has never caused any trouble with train brakes.

PNEUMATICALLY OPERATED CYLINDER COCKS.

This device is the invention of Messrs. J. W. Thomas, Jr., and Otto Best, of the N. C. & St. L. shops, at Nashville, Tenn.

The following is a description of the device furnished by Mr. Thomas:

The air cylinder A is placed underneath the running board, attached to back cylinder head casing, or may be located at any other convenient point. If placed underneath the running board, it may be connected to vertical cylinder cock lever B, as shown on print. If secured to back cylinder head casing, piston C is coupled direct to cylinder cock rod E. A quarter-inch pipe is run from cylinder A to an ordinary quarter-inch operating cock, such as is used on auxiliary reservoirs, this cock being located in the cab within easy reach of the engineman. The spring G operates to hold cylinder cocks open, and as it requires a pressure of say 65 pounds to compress spring, cylinder cocks will open should pressure in main reservoir fall

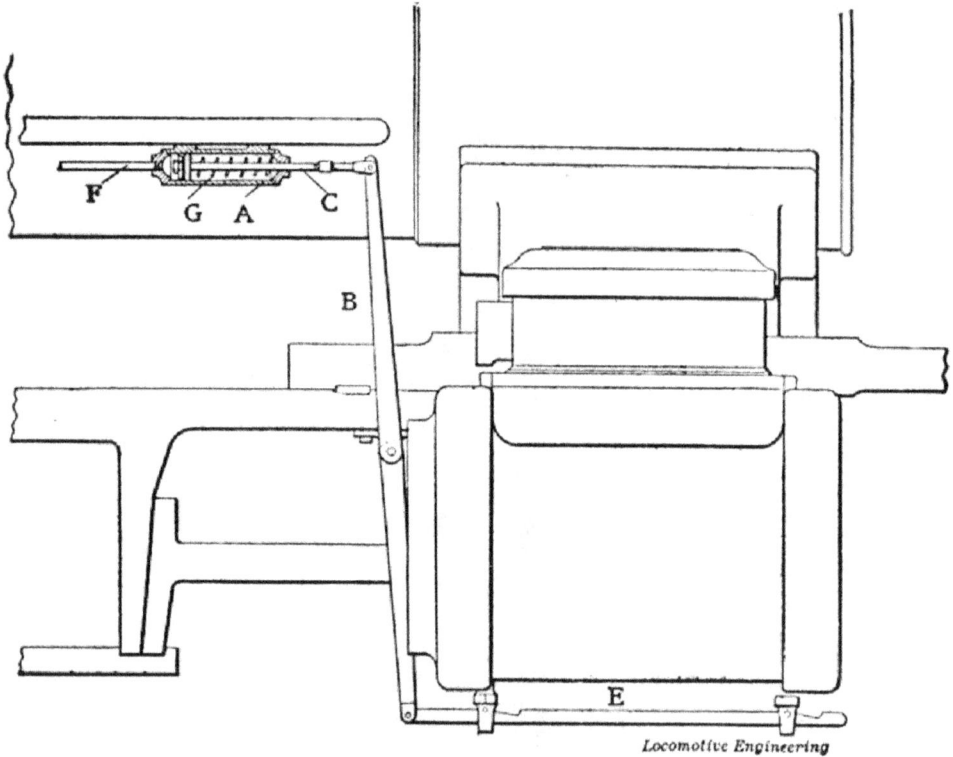

Locomotive Engineering

below 65 pounds, thus notifying the engineman that the pressure in brake system is getting low, the reduction of pressure having taken place on account of pump stopping, leaks, or too frequent application of the brakes.

As the cylinder cocks open automatically when the pressure falls below 65 pounds, the risk of cylinders bursting during freezing weather is entirely obviated. It should also be noted that as the air pressure must be, say 65 pounds, before cylinder cock can be shut, and as most locomotive throttles leak more or less, hostlers cannot close the cylinder cocks and rush the engines out of the round-house with cylinders full of water, and even if they should have the necessary air pressure to shut the cocks, the water in the steam cylinders will have practically been blown out while the engine is generating enough steam to pump the air pressure up to 65 pounds. One hundred and forty engines on the N. C. & St. L. Ry. are equipped with this device.

HEIDELBERG'S IMPROVED PILOT COUPLING.

The object of this device is clearly shown by the illustration; it relates to an improved pilot coupling for locomotives. The small oscillating cylinder No. 13 being located inside of the pilot and resting upon the small casting No. 7; a small air drum (26) being located in the back of the pilot and the valve (35) to be operated by hand, or it can be operated direct from the cab. There is also a connecting rod attached to the pilot bar at 40,

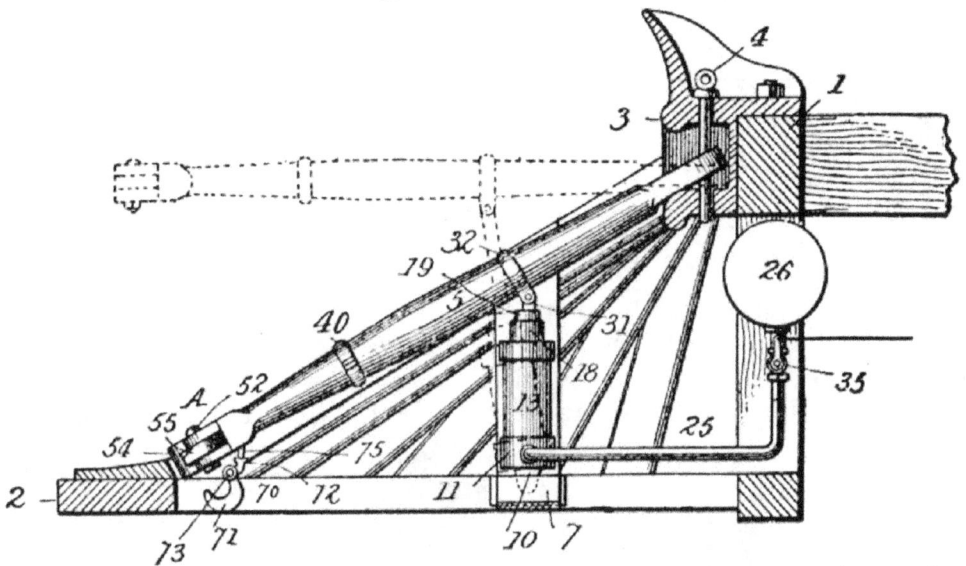

the other end of which is attached to a crank which is located near one end of the bumper casting and permits the pilot bar to be moved laterally; a small release hole in the valve (35) permits the bar to be lowered when desired. The pilot bar also has a peculiar form of coupler section designed to operate in connection with any of the ordinary sections of the Jenney type coupler. This device was invented by Mr. Samuel R. Heidelberg of Palestine, Texas.

AIR IN LOCOMOTIVE SHOPS.

The perfection of economical air compressors is, perhaps, the direct cause of the abnormal growth in the use of compressed air during the last few years. The use of compressed air has developed and become established more rapidly and more fully in the railroad shops of the country than anywhere else, simply because it has been easy to produce a supply of the compressed air and to augment the means of compressing it as the demand of its service increased; this, of course, was the direct outgrowth of the air brake. Most every small shop used one air pump to supply air for the shop, while the larger shops were obliged to use from ten to fifteen pumps to supply the demand and the growing needs and uses made of the air. The question of economy soon arose, and it was found that air could be supplied by a good air compressor much cheaper than by using air pumps. Many experiments were made, and at last an official test was made under the auspices of the Railway Engineering and Me-

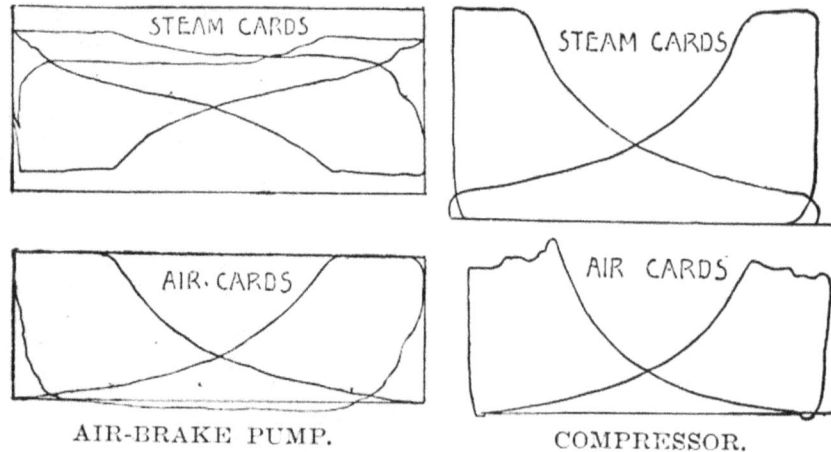

AIR-BRAKE PUMP. COMPRESSOR.

chanics of Chicago, in October, 1894. The result of the test was remarkable and is embodied in the following words: "We find that air brake pumps use 5¼ times as much steam as the compressor." The reason for this result will be apparent to anyone

who will examine the diagrams taken at that time, which speak for themselves. Regarding the brake-pump cards, the steam distribution is so abnormal as to need a word of explanation: First, we should say that the steam cards from the brake pump are not wrong-side up, as might be imagined; on the contrary, they represent the real cards, the lower line of which shows the enormous back pressure in the steam cylinder, this back pressure being introduced for the purpose of offering a resistance to the motion of the piston in the early part of the stroke, to make up for the fact that during this portion of the stroke the resistance offered by the air is very small. It will be seen that as the air pressure increases the back pressure diminishes until at the end of the stroke, when the expulsion of the air takes place, the back pressure becomes very small. Examination of the air cards will show that another serious loss is introduced in consequence of the absence of a crank, which causes the piston to stop before the end of the cylinder is reached, which action prevents the free expulsion of the air from the cylinder. Compressors are made in all sizes and for all purposes, and there are about as many designs as there are of the stationary engines. We illustrate a very simple belt compressor, especially adapted for use in railroad shops. It regulates itself automatically, maintaining the pressure at any desired point. It is compounded and water jacketed. It runs only so long as air is being used and delivers 44 cubic feet of free air per minute.

THE AIR HOIST.

Air hoists are an invaluable addition and improvement to railroad shops, where so much heavy material is constantly handled, and they have superseded the old rope and chain pulleys wherever air is in use. These hoists are made both vertical and horizontal, so they can be used under low ceilings or located to clear shaftings, cross beams, or other supports. They are attached to cranes of all kinds, hung from rods, rafters, beams, or other supports, and are moved to all parts of the shop by means of transverse pulleys and guide bars. They are in use everywhere in large shops and used for all purposes. The vertical hoist which we have illustrated is the kind most generally used in machine shops; they have an automatic stop and retaining valve, only admitting enough air to raise and keep the load at the desired height, giving the operator the free use of both hands to adjust his work in the lathe or other machine. The horizontal hoists are usually attached to cranes. They operate a chain pulley attachment and can be locked and sustain their load indefinitely, air or no air. The following are a few of the uses made of the air hoist in the machine shop: Lifting heavy material for planers, boring mills, lathes, slotters, drill presses, and various other machines; handling air pumps, stacks, front ends, steam chest covers, and other such work in the erecting department, and transferring material to different parts of the shop. In boiler shops and blacksmith shops large sheets and heavy forgings are handled. In the round-house, stacks, air pumps and other work is handled, while outside, tires, castings, and such things are moved around and cars loaded and unloaded.

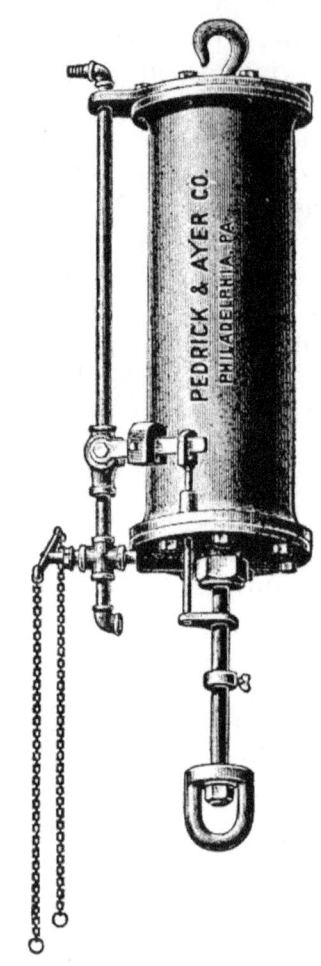

We give a table of the lifting capacities of direct acting hoists with a maximum pressure of 80 lbs. and the cubic feet of air used for each full lift:

Diameter.	Capacity.	Lift.	Cubic feet.
3 inch	400 lbs.	4 feet	1.04
4 inch	825 lbs.	5 feet	2.32
5 inch	1,450 lbs.	5 feet	3.63
6 inch	2,100 lbs.	5 feet	5.33
7 inch	2,600 lbs.	5 feet	7.12
8 inch	3,750 lbs.	5 feet	9.55
9 inch	4,300 lbs.	5 feet	11.78
10 inch	6,000 lbs.	5 feet	14.54
12 inch	8,500 lbs.	5 feet	20.94
14 inch	12,000 lbs.	5 feet	28.05
16 inch	15,800 lbs.	5 feet	37.23

In many shops they make their own air hoists out of wrought iron pipe. A simple method for cleaning out the inside of the pipe for this purpose is to force a steel die through the pipe. This method is successfully used in one of our large railroad shops.

AIR MOTORS.

The air motor is another valuable tool in railway shops. The illustration shows the general form of these motors. They are made in all sizes and adapted to almost every use, in some cases replacing steam power, but more often, perhaps, performing work which was formerly done by manual labor, doing the same work much more rapidly. The following is a partial list of the uses made of the air motor in railroad shops: Propelling separate machines when working overtime; driving grindstones, emery wheels, fans, or machines in small shops or round-houses;

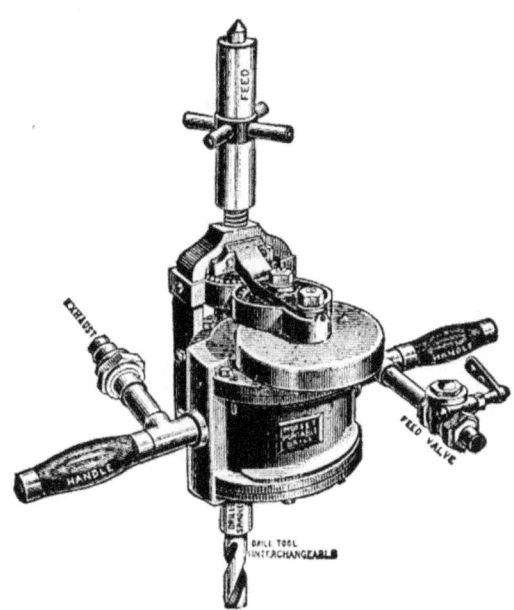

drilling, tapping, reaming, and screwing, all kinds of work. In the erecting department of machine shops, driving cylinder boring bars, valve facers and flexible shafts and turning the rollers when setting valves. In the boiler shop it is used for drilling, tapping, and screwing in stay bolts and other work; running portable drills, rolling flues, and supplying blast for forges. In the blacksmith shop and copper shops they are used for blast purposes. In some places they propel the transfer table. Sandpapering cars is another use to which it is put.

JACKS AND PRESSES.

Many shops use air jacks in their drop pits, where it is laborious and sometimes dangerous to use hydraulic jacks. Air presses are used for various purposes. They are used for bending air brake and other pipes, wooden dies being used for forming the pipe, also for punching and shearing and driving out old bolts. For applying couplings on air brake hose and for pressing and forming tinware. Long presses similar to air hoists are also used in boiler shops for holding sheets while chipping, and other kinds are used for flanging and forming. In the blacksmith shop and yard they are used for straightening old iron rods.

PNEUMATIC HAMMERS.

Pneumatic hammers are used for beading flues, riveting tanks, calking, and chipping castings, and all kinds of extensive hammering.

OTHER USES MADE OF AIR IN RAILROAD SHOPS.

Running engines out of the shop, kindling fires, cleaning flues, cleaning out lubricators, oil cups, eccentric strap oil cellars, blowing out pipes, sifting sand, fire alarm whistles, breaking old castings, sanding car roofs, whitewashing, cleaning cars, cushions, etc., and pumping water. The above list is incomplete, but is sufficient to give the reader a fair knowledge of the uses and advantages of a supply of compressed air in any shop.

AIR BRAKE.*

THE AUTOMATIC BRAKE.

Examination questions and answers for enginemen, and the Revised List of the Air Brake and Signal Instructions.

We have prefixed an introductory chapter to this chapter, which will help those who do not have the advantages of a regular air brake instructor with his instruction car. This subject of self instruction is so large in all its details that only a small part of it can be taken up, but the principal parts are brought to notice.

The Air Brake and Signal Instructions approved by the Master Car Builders' and Master Mechanics' Associations in 1892 were revised by a Joint Committee of these Associations in 1897. This revised form is a very complete and short form of instructions. Every employé who has any interest in the air brake should carefully study it. For that purpose it is included in this book.

The questions and answers in this chapter are intended as additional information to engine and train men by taking up points not treated on before. It has been the aim to ask other questions not found in the M. C. B. instructions, which have a practical application to every day operations.

The list of examination questions is intended to call attention to matters connected with these operations which you should know, if you wish to handle the brake with confidence. They can be used in examination for promotion to engineer and cover a pretty fair range of air brake practice. Bear in mind that good judgment is the first requisite for a successful air brakeman, and that the addition of knowledge and correct methods of operating to good judgment will make a skillful one under all conditions.

INTRODUCTION.

There is a demand for a short form of instruction in air brake practice, not so much to instruct the beginner on all the points as to put him in the way of learning them himself, and this introductory chapter is intended to help those who set out to learn the theory and have a chance to operate the brake or see it operated. This can be best done by learning the foundation principles first, studying the action of the important primary parts of the machine; the secondary parts will then work their way in, so you understand the whole properly. Much time may be wasted by beginning at the wrong end to unravel air brake

---

*This chapter was especially prepared for this book by the well-known authority, Mr. Clinton B. Conger.

operations. If you are too hasty and jump at conclusions you may be wrong; better not know anything about it than know it wrong. Therefore, take time enough at first to learn it right; you will never regret it.

When you see the air brake working every day, sometimes making a good stop, at others not controlling the speed as you think it should, the operation may seem mysterious, but it is governed by fixed laws of mechanics and forces. If you take pains to learn these laws and about the forces and examine each part of the mechanism it will be clear to you.

There is nothing mysterious about the operation of the air brake. Each part has its own duty to perform. Take each part by itself and study it up, then get an idea of its relation to the other parts and you will find out that it is easy. You cannot learn it all at once, or by once reading over an instruction book. In studying the construction and principle on which it operates it is an advantage to have help from some one who can instruct you.

When you come to operate it, the machinery in actual operation is the best instructor.

Attention is called to explanations of some of these operations in the succeeding pages.

In the first place, all the parts of the brake (you will find them enumerated on page 621) are charged with compressed air; the foundation principle on which it operates is changing the relations of these pressures in different parts of the equipment; the effort compressed air makes to equalize, by the high pressure air pushing against the low pressure air, moves the different parts of the air brake that can be moved in this manner away from the high pressures.

When it is once fixed in your mind what pressure you have in each place and that any change of pressure will cause the movable parts of the valves to change their positions, closing some of the openings through which the air can pass and opening others, it is plain that the next step is to find out just what openings the air must pass through at each operation, whether setting, holding set or releasing the brake.

We will take up the plain triple valve first, as the process of equalization is best explained with it.

When the brake is ready to operate the pressure is equal in the train pipe, in the triple valve on both sides of the triple piston and in the auxiliary reservoir.

You will notice (see cut on next page) that the triple piston 5 is the dividing line when the pressures are unequal; that the train pipe pressure is against the lower side of this piston and auxiliary pressure on top. There is a small passage cut in the cylinder around the piston, called a feed port, at *m*, through which air can pass from the train pipe to auxiliary when triple piston is clear up in release position; this is the opening through which air can equalize in train pipe and auxiliary. This port *m* is very small, and equalization takes place very slowly through it. A brief explanation of this is found on page 646. As soon as any reduction of pressure is made in the train pipe the auxiliary pressure will be greater and force the triple piston down, following the decrease of pressure in train pipe end of triple; the first movement of the piston 5 closes feed port *m*.

This movement of the triple piston, in itself, does not set the brake, but when it comes down it brings the main air valve 6 down with it and opens a port which admits compressed air

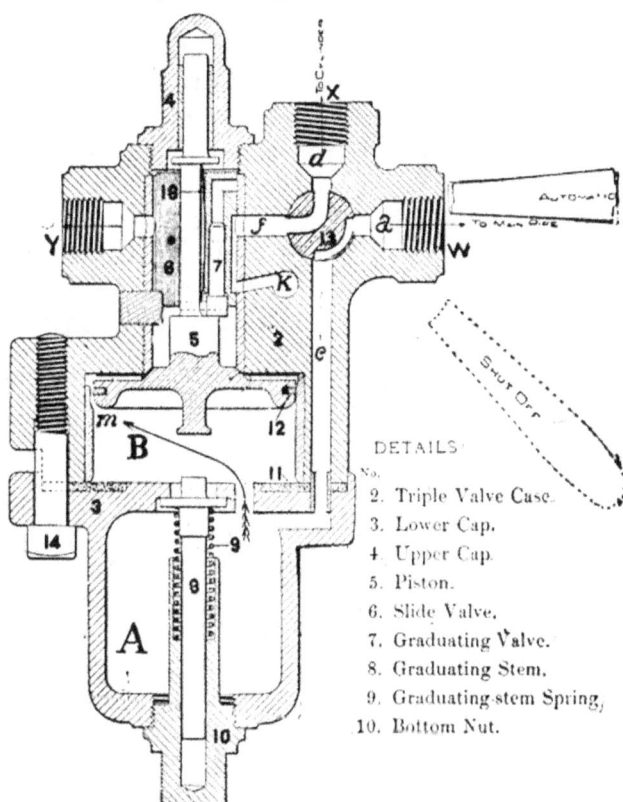

DETAILS:
No.
2. Triple Valve Case.
3. Lower Cap.
4. Upper Cap.
5. Piston.
6. Slide Valve.
7. Graduating Valve.
8. Graduating Stem.
9. Graduating-stem Spring.
10. Bottom Nut.

from the auxiliary to the brake cylinder, which also has a movable piston in it that pushes against levers that are so coupled up that the brake shoes are forced against the wheels. The operation of this triple piston with a moderate reduction of train pipe pressure, say from 70 pounds down to 63, will show the exactness of this equalization principle.

When the triple piston 5 comes down with a reduction of 7 pounds in the train pipe, and slide valve 6 and graduating valve 7 open, the air in the auxiliary at 70 pounds begins to expand into the brake cylinder; as soon as enough has gone into the cylinder to reduce the auxiliary pressure a little below 63 pounds, the train pipe pressure is then greatest; so triple piston moves up, also moving the graduating valve 7; closing it. This cuts off the flow of air into brake cylinder, setting the brake with a partial application and holding it set; as the piston does not move the slide valve it does not move up and open the exhaust port k to air port f. For a more detailed statement of the operation of the graduating valve see page 635. Thus you see the triple piston moves between the train pipe and auxiliary pressures, always towards the lesser one.

If the relations between the pressures on either side of the triple piston are changed, it will move toward the lower pressure till the limit of its travel is reached, or till the relation between the pressures is changed the other way, which will stop its movement if pressures are equalized, or move it the other way if pressure is increased. Increasing the pressure in train pipe side of triple over the auxiliary moves the triple piston clear up, moves the slide valve 6, opens exhaust port, allows the air to escape from brake cylinder and releases the brake, so you see charging up the train pipe to standard pressure releases the brake. As the feed port m is also opened when the piston 5 is clear up, the air flows into auxiliary, equalizing its pressure with the train pipe. To change the relation between the pressures in any other way is done by letting out some of the air—the train man releases the brake by bleeding out the auxiliary pressure till it is lower than train pipe pressure.

After thoroughly posting yourself on the way in which the plain triple operates by the reduction or equalization of pressures, you can then take up the engine equipment. It is a good plan to know just how the pump operates, its care and management, but that can be left till later. The pump generally goes ahead with its work from beginning to end of the trip without much attention from the engineer; the rest of the equipment depends on the skill and knowledge of the engineer for its successful operation, as it responds directly to his manipulation.

The brake valve controls the passage of air between the main reservoir and train pipe, or train pipe and the atmosphere, and the hardest task when learning the brake valve is to trace the course of the air through the equalizing discharge valve. The main reservoir pressure is always in the top of the valve (see page 613), holding the rotary valve 13 on its seat; the train pipe air is under the equalizing piston 17 at all times. Train pipe air is also on top of piston 17 when valve is set on running position. This is further explained in detail on pages 625 and 626. This equalizing piston 17 is moved up, opening the discharge or train line exhaust port n, or down, closing the port, by a change of pressure on either side, just the same as the triple piston, and the pressures are changed by opening or closing the various ports in the valve. The rotary 13 is moved by the engineer, its office is to put in communication the various openings or ports that will let the air pass through. Locate these ports, next find out just what they are for and in what positions of the rotary they are open and shut. The best way to do this, if you have not got a sectional valve to study on, is to get a complete valve for a few hours and dissect it, using a piece of fine copper wire to run through the ports, which will show the course of the air; this

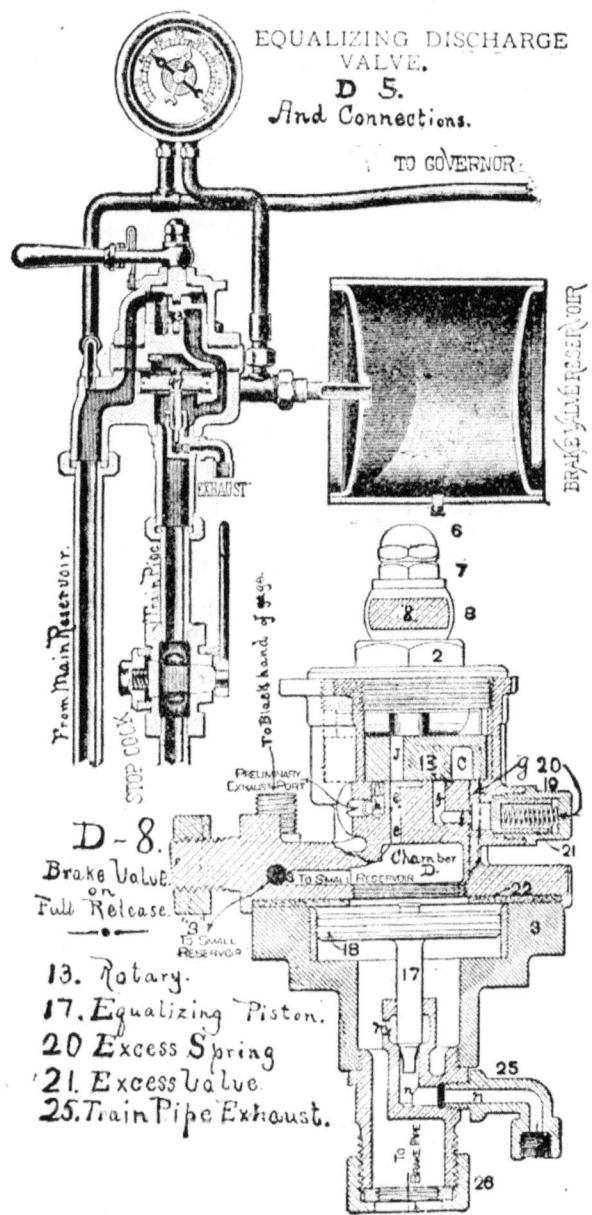

wire can be bent in any direction and its use is not likely to scratch the seats or valve. After locating all the ports through which main reservoir air can flow into train pipe or the chamber D over piston 17, then see in what position of the rotary all ports are covered, and figure out which ports are covered first and why it is necessary to stop the flow of air through them.

You will probably notice that before any ports are opened to allow air to escape from the train line the main reservoir air is cut off from any escape, so it can not supply the train pipe while you are reducing that, or triple valve would not feel any reduction of pressure and brake would not be set. Then when you come to locate the ports that are opened to reduce the train pipe pressure and actuate the triple, it will be necessary to know just exactly the principle of operation of the equalizing discharge valve. On a long train the reduction of train pipe pressure must be the same at each triple if we expect each brake to be set at the same time and with the same relative power. To make this reduction of pressure alike for all the triples, or what is the same, for all the cars in the train, we must allow the air to escape from the train pipe gradually, so the reduction will not be any more violent from the first car than from the last one, nor should the escape of air be closed till the same reduction has been made in each car. The discharge should not be stopped suddenly before the pressure in the last cars has equalized with the first ones, or the momentum of air flowing from rear cars, as well as equalizing pressures in all cars, will raise the train pipe pressure in the cars nearest the brake valve and tend to release their brakes. This gradual closing of the train pipe exhaust the brake valve is intended to do automatically. Its principle of operation is, the engineer makes the proper reduction of pressure in the brake valve over the equalizing piston 17, and the action of the piston 17 reduces the train pipe pressure to an equal amount in all the cars, whether few or many.

Before you move the rotary far enough to open the preliminary exhaust port h, the equalizing port g, which allows train pipe air to pass from train pipe to chamber D, over piston 17, is closed; this cuts off chamber D from any other pressure, you can then make a reduction on top of piston 17, so train pipe pressure will raise it up and hold discharge n open till the pressure below is a little less than it is above, when piston moves down, closing train pipe discharge.

With the equalizing discharge valve, the black hand of the double gauge is connected with chamber D at all times; if the rotary is in either full release or running position the equalizing port g connects it to the train pipe pressure, so it shows that also. When the brake is set with a service application the pressures equalize so nearly on each side of piston at the instant train line exhaust closes, that black hand is expected to show train pipe pressure then also. In the emergency position the black hand does not at once show the amount of the reduction. On page 629 you will find this further explained.

If the packing ring in piston 17 leaks very much, the black hand will show train pipe pressure when rotary is on lap, as the air pressures can equalize past this leaky packing ring; all of them leak a little. Look out for this defect when operating the brake. The brake valve reservoir is connected to the valve for the purpose of giving a larger volume of air to chamber D, to insure a gradual reduction of pressure there. See pages 625 and 631.

Up to this point we will assume that the student has followed the action of the brake in a service or graduated application. There is what we call the emergency or quick action, which is produced by a different set of operations and is peculiar to the

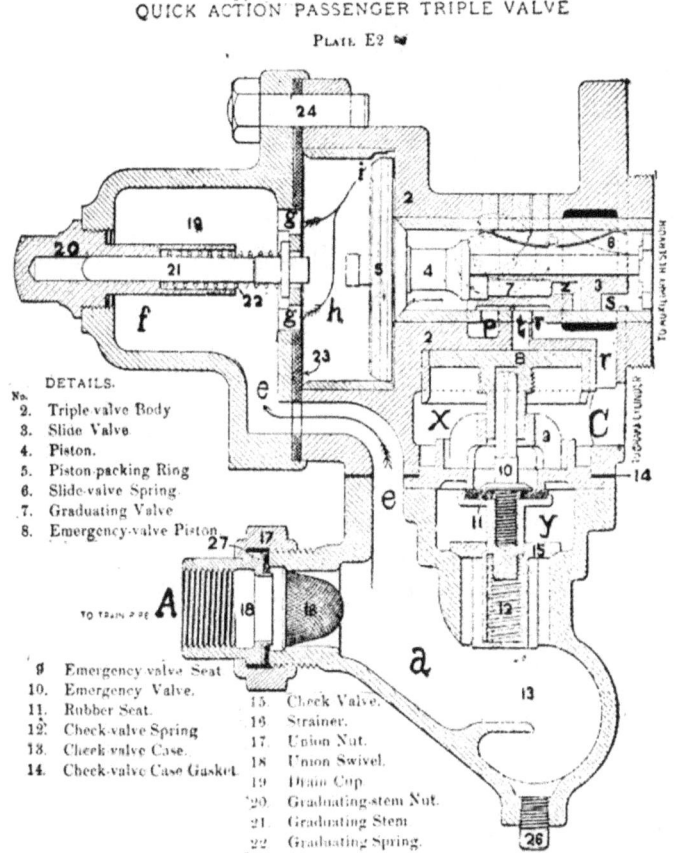

QUICK ACTION PASSENGER TRIPLE VALVE
PLATE E2

DETAILS.
No.
2. Triple-valve Body.
3. Slide Valve.
4. Piston.
5. Piston-packing Ring.
6. Slide-valve Spring.
7. Graduating Valve.
8. Emergency-valve Piston.
9. Emergency-valve Seat.
10. Emergency Valve.
11. Rubber Seat.
12. Check-valve Spring.
13. Check-valve Case.
14. Check-valve Case Gasket.
15. Check Valve.
16. Strainer.
17. Union Nut.
18. Union Swivel.
19. Drain Cup.
20. Graduating-stem Nut.
21. Graduating Stem.
22. Graduating Spring.

quick action only. We will go to the beginning and inquire why this action is necessary.

On a long train of air brake cars, to avoid a severe shock to the rear part of train when brakes are applied from the head end of train very suddenly, as bursting a hose, or breaking apart of the train, or in case of danger when it is necessary to set the brake from the engine very hard so as to stop as quickly as possible, the brakes should set on the rear cars quickly enough so the slack will not run up against head cars and damage cars or draft gear. Then, in case of danger, every second after the brake is applied at the engine, before it begins to set on last cars and hold them, the train is getting nearer the danger, so a brake that can be set instantly on the whole train will stop the train quicker than one which sets slowly from car to car. With the plain triples the air in train pipe will be exhausted at one place only, wherever the opening is made in train pipe or at brake valve; this takes several seconds to affect the farthest car. If an opening can be made to exhaust this air at each car and reduce the train pipe pressure, the action of the brakes on a long train can be made nearly simultaneous, so nearly so that the brakes are all set before the slack can run out. The quick action triple is designed to exhaust a portion of the train pipe air at each triple, so as to set the next brake suddenly. To thoroughly understand how the quick action triple can exhaust some of the air from the train pipe suddenly and reduce the pressure so as to affect the next triple in the same manner, it will be necessary to study the construction of the quick action valve. See cut on page 615. This triple has the same openings to admit air from the train pipe to the auxiliary that we find in the plain triple. In the quick action they are shown at e through g and feed port i. But there is another channel for air to pass from the train pipe into the brake cylinder in this valve, which is from the check valve case 13 into y, then into x and into brake cylinder at c. These openings are ordinarily kept closed to the passage of the air in either direction, the rubber seated emergency valve 10 keeps the train pipe air from getting into the cylinder, and the check valve 15 keeps the air in brake cylinder from getting back into train pipe. These two valves are held on their seats by a spring of very light tension, shown at 12, as well as by the air pressures. Now it follows that if you wish this triple to make an opening to let the air out of train pipe suddenly, that this valve 10 must be moved away from its seat against the train pipe pressure and the strain of the spring 12. For this purpose piston 8 is used. A port t, which is shown just over the figure 8, can be opened, letting the air from the auxiliary on top of piston 8; with auxiliary reservoir pressure over this piston and no pressure at all or a very low one below it, piston 8 goes down instantly, forcing valve 10 away from its seat. The train pipe air then holds check 15 up and flows through c into brake cylinder till pressures equalize.

When these valves 10 and 15 are opened in this manner, the train pipe air goes past them like a flash through the large ports into the empty brake cylinder, setting the brake with the pressure at which the train pipe and brake cylinder can equalize, which is somewhere near 20 pounds. At the same time port s, in the end of slide valve 3, is open, air from auxiliary flows through r and piles in on top of train pipe air in cylinder, raising the cylinder pressure at full equalization to 60 pounds. The train pipe air equalizes first with cylinder, through large ports, and auxiliary pressure last, through small ports.

Now for the means employed to let the auxiliary reservoir pressure on top of piston 8 at one time to produce "quick action" and keep it out at another, to preserve the graduated application. As long as this triple is used with a graduated application the slide valve 3 does not move over far enough to uncover emergency port l, as with a gradual reduction of train pipe pressure the auxiliary pressure will be reduced equally with the train pipe pressure through the graduating valve 7 and its port Z. But if the train pipe pressure is reduced so suddenly and to such an amount that the graduating valve can not reduce auxiliary pressure equally with train pipe reduction, the greater auxiliary pressure will move piston 4 and slide valve 3 far enough so port l will be uncovered, piston 8 and valves 10 and 15 will move at once. On page 637 is an explanation differently worded. Any defect in valve 7 that will prevent air getting past it will sometimes cause the quick action operation with a moderate service application if you have a short train. The equalization of the pressures is the foundation principle to look for in the operation of the quick action triple.

It is necessary to restrict the flow of air through some of the openings in triples and brake valve in order to be sure to handle a long train with safety. This refers more particularly to the "feed ports" in the triples for recharging auxiliaries, and the "preliminary exhaust port" h, train pipe exhaust n and "equalizing port" g in brake valve.

The proper size of these ports has been determined by the experience of many years.

Perhaps it would be well to study on the matter of equalization of different pressures of air in the equipment, as if this is well understood you can solve other problems in air brake operations more easily. It is the law that where compressed air in a certain sized vessel expands into an empty one, that the pressure is reduced in the full one in proportion to the increased volume the air has to occupy. From this you can see that in the case of the brake cylinder and auxiliary that the auxiliary pressure will be reduced more if the brake cylinder is large in proportion to the auxiliary, than if it is small. Apply this law to the cylinders having different piston travels, you see that a cylinder having a long piston travel holds a greater volume of air than the one with short travel, so we can expect the one with long travel to reduce the auxiliary pressure to a lower point than the short travel, and it is found that a travel of 11 inches of the freight brake piston gives a final equalized pressure of close to 45 pounds, while a short travel of 4 inches gives a final pressure of 57 pounds; a travel of 8 inches, which is between 4 and 11 inches, will give a final equalization of about 50 pounds. This will give different brake power on different cars in the train, where it should be equally proportioned to the weights of the cars; unequal brake power makes some cars hold less than others, so the strain is not equally distributed throughout the train, a point in equalization worth studying on.

The final point at which auxiliaries and their cylinders equalize cuts quite a figure in operating the triples to release the brakes. As was stated in explaining the operation of the triple piston, the train pipe pressure must be greater than the auxiliary pressure to move the triple piston up so slide valve will open exhaust port and let air in brake cylinder equalize with the atmosphere, when brake piston will have no air pressure on either side of it; this relieves the strain on brake levers and shoes.

If auxiliary pressures are unequal and train pipe pressure is not raised at once higher than the highest auxiliary pressure all the brakes will not release at once. This leads us to consider the question of equalization of the train pipe and main reservoir pressures when you desire to release the brakes.

If the train pipe is long it will take more air from main reservoir to equalize at a certain stated pressure than if it is short, for a long train pipe holds more air than a short one. Then, again, if the train pipe has considerable pressure left in it after setting the brake, it will equalize at a higher point than if it is empty. This emphasizes the fact that it will be hard work to release the brakes on a long train if you exhaust all the air in making an application.

Another place where equalization is important is on the second application shortly after releasing brakes. It takes time for the train pipe and auxiliary pressures to equalize. If you do not wait this proper time the auxiliary will not have charged to standard pressure, and, of course, when brake is set it will not reach as high a final pressure on brake piston, which reduces the braking power. Equalization between train pipe and auxiliary on making the reduction for a second application is very important, because if train pipe has a high pressure which the auxiliary has not reached, the triple piston cannot move till the train pipe pressure has been drawn down a trifle lower than the auxiliary. If one is 80 and the other 60 it means a reduction of 20 pounds before brake begins to set, and about 20 more to set all brakes tight. This effective handling of the brake is called "overcharging the train pipe," and is further explained on page 642.

And lastly, we will define some of the terms which are used to shorten the explanations you may hear.

A partial application means that the brake is set with part of its full force; that the brake cylinder pressure has not equalized with the auxiliary reservoir pressure.

A full application means that the brake cylinder pressure has equalized with the auxiliary pressure and has, therefore, got the full pressure that can be obtained from the air stored in the auxiliary. No matter how many reductions of train pipe pressure you make it is only one application till it releases. You can reduce the train pipe pressure a few pounds at a time and make eight or ten successive reductions, but it is only one application if it has not been released.

A graduated or service application means a gradual reduction of train pipe pressure which sets the brake slowly.

An emergency application means a very sudden and heavy enough reduction of train pipe pressure to set the brake with full force at the first movement of the triple valve. With this application the quick action triple works the quick action feature of the valve, so air from the train pipe passes into brake cylinder as well as equalizing the cylinder and reservoir pressures.

The plain triple does not let any air pass from the train pipe to the cylinder, but it opens the air port f full width instead of all the air going past the graduating valve.

Excess pressure is the difference between the main reservoir and train pipe pressures when the brake valve is in running position so that the excess valve 21, see page 613, or the feed valve 63, on page 655, can maintain a difference between the two pressures. In full release position these valves are "cut out" by closing the air ports leading to them. Air can pass through an open port from the main reservoir to the train pipe and equalize, so in this position there is no excess. If you carry excess you aim to prevent this equalization and thus have a greater amount of air in main reservoir to equalize into train pipe when necessary to release brakes.

Brakes are "cut in" when the cock in cross over pipe or the plug 13 in the plain triple is in proper position for air to pass through it from the train pipe to the triple piston and auxiliary. If they are "cut out" this passage is closed to the passage of air in any direction. See pages 645 and 648.

When operating the brake valve you should listen to the sound of the air discharging from it, because that sound tells how many cars you have in your train—see page 632—and how the valve is doing its work, just as the exhaust of the locomotive tells whether the valve motion is in order; any unusual sound notifies you something is wrong.

Have your air gauge placed so it can be seen without taking your eyes too far off the track or signals. Consult it often till you learn the air brake business.

It pays to inspect and test your engine equipment carefully before leaving the engine house; it may save a failure on the road.

Draining the main reservoir does not take out the water that is in the triples and other parts of the equipment where it may collect; the tender triple should be drained regularly in cold weather.

Too much oil used in the air end of the pump does more

damage than not enough, as it chokes up all the small openings in the engine equipment. The piston rod packing needs more oil than the air piston; the air valves do not need any.

When you make a test of the train brake before starting out, make the same kind of an application as when stopping at a station, then you will know how the brake will work when making station stops. It should be a full application to get the full piston travel.

If the brake leverage on your train is adjusted for 70 pounds train pressure it is not safe to carry either more or less. If you carry less you cannot stop quickly when you have to; if you carry more and skid the wheels you certainly cannot, and will spoil a lot of wheels.

WHEN OPERATING THE AUTOMATIC BRAKE REMEMBER

That the compressed air stored in the main reservoir is used to charge up the train line and auxiliary reservoirs, and that it is used to release the brake. Do not have any water in any reservoir, as it takes up the room needed for air.

That the compressed air stored in auxiliary reservoir is used to set the brake. There is an independent supply for each brake. Keep a full supply in each auxiliary.

That the brake is set by any reduction of pressure in the train line, no matter how it is made, if it is sufficient to move the triple piston and valve.

That the train pipe pressure must be reduced 5 to 7 pounds at first application, or brake pistons will not travel over leakage grooves, allowing brake to leak off.

That the train line pressure must be raised above the auxiliary pressure, or the auxiliary pressure reduced by bleeding, before the brake will release.

That you cannot recharge an auxiliary reservoir until the exhaust port in triple is wide open, unless air leaks past triple piston, as the feed port does not open until after the exhaust port is open.

That a second application after release does not set the brake as tight as the first full application, unless the auxiliaries have had time to recharge to standard pressure. This takes from 25 to 45 seconds.

That the small reservoir attached to brake valve is put there to give a larger supply of air for the preliminary exhaust of brake valve so you can make a gradual reduction.

That if your driver brake does not work quickly and hold well with service application, in 99 times out of 100 it is on account of a leak.

That in all these questions and answers it is understood—unless otherwise stated—that 70 pounds is the standard train-line and auxiliary pressure; 90 pounds main reservoir pressure; and 8 inches, the standard piston travel for all passenger, freight and tender brake pistons. The brake piston travels an inch farther when train is running than with a standing test, so travel should be adjusted to less than 8 inches.

Answers apply to Westinghouse brake.

1. Q. What are the essential parts of the automatic brake and what service does each part perform?

A. The air pump, the main reservoir, the engineer's brake valve, the train pipe with its hose and couplings, the auxiliary reservoir, the triple valve, the brake cylinder, the gauge and the pump governor. The air pump compresses the air for setting and releasing the brake; the main reservoir is used to store a supply of air for charging the train pipe and auxiliary reservoirs when empty, as well as to hold the supply for increasing the train pipe pressure when the brake is to be released and charge the train pipe and auxiliaries ready for the next application; the brake valve governs the passage of the air from the main reservoir to the train line, from the train line to the atmosphere, or stops the flow of air through it in any manner. The brake can be set, gradually or "full on," held set or released, when this valve is properly handled by the engineer. The train pipe, with its hose and couplings, extends from the brake valve and supplies each auxiliary reservoir with air for setting the brake. Each brake has an auxiliary reservoir, in which air is stored for setting it. The triple valve is connected to the train pipe, auxiliary and brake cylinder; it is used to control the charging of the auxiliary with air and regulate the time in which this is done, to open a valve to admit air from auxiliary to brake cylinder to set the brake, or by another movement to close this valve and open the exhaust port so air can get out of brake cylinder to the atmosphere and release the brake. Thus the functions of the triple valve are three-fold, to charge the auxiliary, set the brake and release it. The triple valve is operated by a variation of pressures between the train pipe and auxiliary; this variation is controlled by the brake valve. The brake cylinder, with its piston connected to the brake levers, sets the brake when the triple valve lets air into it. The gauge shows with the red hand the main reservoir pressure, with the black hand the pressure in the brake valve above the equalizing piston and in the brake valve reservoir; when brake valve is in full release or running position it also shows the train pipe pressure. The single-hand gauge used with the old style brake valves shows the train pipe pressure direct. The pump governor is located in the steam pipe to the pump; it is operated by the air pressure and shuts off steam from the pump when the air pressure reaches the standard amount carried.

2. Q. Explain the difference between the operation of the automatic and straight air brakes.

A. The automatic brake is set by a reduction of pressure in train pipe operating the triple valve, which admits air from the auxiliary to the brake cylinder. With straight air the air goes from the main reservoir through engineer's brake valve and train pipe direct to brake cylinder.

The automatic brake has its supply of air to set the brake stored under each car. With straight air it is stored in main reservoir only, the train pipe is empty before the brake is applied. This pipe and hose, as well as the brake cylinders, must be filled with air at 50 pounds pressure to make a full application; on a long train this cannot be done; when brake is released all this air is wasted.

The automatic uses only 20 to 25 pounds out of entire train pipe, and only so much from auxiliary as fills the brake cylinder, so automatic uses less air at each application of the brakes.

The automatic sets the brake with same piston pressure each time, and can thus be adjusted to get the full braking power on the car, and no more. With straight air, the piston pressure may be 90 pounds at one application, and with a longer train it may be only 30 for same pressure in main reservoir; leverage would have to be adjusted for the high pressure to avoid sliding wheels; with the low pressure you could not hold the train.

The automatic can be set from any car, straight air from the engine only. The automatic applies itself when hose bursts, or is pulled apart in case train breaks in two; in this case the rear portion of an all air-brake train usually stops first, diminishing the risk of parts running together. Straight air cannot be used in case of an accident that breaks train pipe or hose; that is just where the automatic does its best at once.

Straight air cannot be used with quick-action triples, as air is cut out from brake cylinder when cock in cross over is shut.

3. Q. What are the duties of an engineer as to his air brake equipment when leaving the roundhouse?

A. To start his pump slowly and increase its speed after 15 or 20 pounds of air have picked up; to be sure that pump is in good order and will pump a full supply of air promptly; to know that governor shuts off the pump when the proper pressure is reached and allows it to start promptly; to see that lubricator has oil enough in it for the trip; to know that there is no water in the main reservoir, drain cup, triple valves or auxiliary reservoirs; to test all the joints in piping, also brake valve and triple valves for leaks, and have leaks made tight; to see that tender and driver brake pistons have the proper travel and do not leak off when set; to test the air signal if one is used.

4. Q. Why must the pump be started slowly, oil used cau-

tiously, triple valves reservoirs and strainers be drained, and how often?

A. The pump must be started slowly to allow the condensed water to get out of steam end, and run slowly till the air pressure rises, or the piston will strike the heads of air cylinder. The triple valves, reservoirs and strainers, or drain cups, should be drained every day in cold weather, once a week in warm weather. Oil should be used sparingly in air end of pump. It should never be put in through the air inlets of 8-inch pump, as it soon collects dirt and chokes up the air passages, which helps to make the pump run hot.

5. Q. How do you test for leaks in the engine equipment?

A. When full pressure is obtained—70 in train line, 90 in main reservoir—shut off pump, place valve on lap; if red hand drops and black hand is stationary, it is a sign of a leak somewhere in main reservoir line, which begins at valves in pump and ends at brake valve. It may be in joints of piping, in main reservoir drip plug, in the air signal line, in valves of pump or brake valve. If main reservoir pressure falls rapidly when you are sure it is not going into train line under rotary, examine each of the places mentioned. With the use of the "cut out cock" under brake valve a leak under rotary is soon detected. Set the brake full on, place the valve on lap, shut the cut out cock; if rotary leaks into train line the black hand will soon show same pressure red one does. With a leak in train line and cut out cock shut or valve on lap, the brake will set; when cut out cock is opened and valve on release, brake will let go and main reservoir pressure fall. There are many other ways of testing for leaks—using a torch is one.

6. Q. Why must there be no leaks in your train pipes, or any other part of your air brake supply?

A. If train pipe leaks, brake will continue to set tighter when brake valve is put on lap, and stop the train before you want it to, so that it is necessary to let it off and make another application for an ordinary stop. If cars are cut off from engine, they must be bled at once if their train pipe or angle cocks leak. Train pipes sometimes get worn through where they rest on or rub against something, so they are tight when standing still and leak when moving or shaken around. This sets the brake when train is in motion, and no leak can be heard when standing still.

7. Q. Why must all hose couplings be hung up properly when not in use? Why should they always be blown out at rear of tender before uniting to other couplings? What is the difference between an air brake and an air signal coupling?

A. So no dirt or foreign matter will get into the open coupling and work into the triple or brake valve or stop up strainers. So couplings and gaskets will not get injured or broken dragging over rails and crossings. If blown out each time, any water, sand or dirt in the tender piping will be blown out. Air brake and air signal couplings are of different sizes—made so purposely—so the brake line cannot be coupled to the signal line. The opening and lip of the lock in brake coupling is much wider than the signal coupling, so the brake coupling will not go into it. It is the practice to paint the signal couplings red so they are more easily distinguished when taking hold of them to couple up.

8. Q. If main reservoir has water in it, how will it affect the operation of the brake?

A. The water in main reservoir reduces the supply of air stored there in proportion to the amount of water contained. The brake will set the same, but on a long train will not release as readily, as there will not be enough air stored to recharge the train line quickly, and you must wait to have it pumped. The main reservoir should be entirely clear of water, if it is necessary to drain it each trip, so as to get a prompt release and recharging of train.

9. Q. Of what use is the extra main reservoir pressure, and does the size of the reservoir have anything to do with the amount of excess pressure you carry?

A. It recharges the train pipe and forces the triple pistons up into exhaust position quicker and surer, so that all brakes release about the same instant; recharges the auxiliary to full pressure in less time, ready for the next application. With a large main reservoir there is a greater volume of compressed air stored to draw from, so a less number of pounds of excess pressure will do the work than with a small reservoir. With a short train, good work can be done with less excess than on a long train. The main reservoir should always be drained of water so it will be full sized.

10. Q. Could it release the brakes with an empty train pipe as readily as when the pressure in the train pipe had been reduced only 20 or 25 pounds? Why?

A. No. When the air from main reservoir expands into an empty train pipe, it will not fill it up and equalize at as high a pressure as when the train pipe has some pressure left in it. For instance, the train pipe line of 25 freight cars holds about as many cubic inches of air as an ordinary main reservoir. If this train pipe is entirely empty and the main reservoir has 90 pounds, it will equalize into twice the space, and show half the pressure, or 45 pounds in each. The brake would be set at 50 pounds; with that pressure above triple piston, brakes could not release until the pump had raised the pressure over five pounds. Now, if the train line has been reduced 25 pounds, having 45 pounds still left in it, 90 in main reservoir and 45 in train line would equalize at a little over 65, which would raise triple pistons so brakes would release promptly.

11. Q. Would you run your pump as fast to recharge an

empty train pipe as one with 45 or 50 pounds in it? Is there any economy in retaining as much air as possible and keeping the pump cool?

A. The pump would have to run faster to recharge an empty train pipe than one with 45 or 50 pounds in it. When you empty the train pipe of 25 cars it wastes as much air as when you empty a small main reservoir; smaller trains in proportion. This would make some pumps hot to supply. Always save your air and keep the pump cool, no matter what length of train you handle.

12. Q. Do you understand what is meant by the equalizing discharge brake valve? Have we more than one pattern?

A. Yes, we have two kinds of them in service, called D-5 or E-6, and D-8, from the number of the plate on which each is illustrated in the Westinghouse catalogue.

13. Q. Describe the principle on which it operates and what difference there is between the two patterns.

A. This brake valve has a piston in it called the equalizing piston, with a train pipe exhaust valve on the bottom side of it which is designed to automatically reduce the train pipe pressure. When brake valve is not being operated this piston has an equal pressure on both sides of it, so it remains stationary, holding train pipe exhaust valve closed. When it is used to set the brake, the reduction of air pressure is not made by the engineer direct from the train pipe, as with the "three way cock" or the small brass valve named B-11. If the engineer wishes to reduce the train pipe pressure any specified amount—say seven pounds—he moves the rotary to "service application" position. As the rotary passes "lap" position, the ports which allow the air to pass from one part of the brake valve to another are all closed. The main reservoir air is held on top of the rotary as it is not used when the brake is set, only when releasing and charging the train pipe and auxiliary reservoirs. The air above the equalizing piston and the brake valve reservoir being cut off from all other air may be called "brake valve air," it is what operates the automatic part of the brake valve to equalize the discharge of the train pipe air which is below the piston. When the pressure of the "brake valve" air is reduced by allowing some of it to escape from the preliminary exhaust port—it does not reduce the train pipe air through the same opening; so the equalizing piston having less pressure above it, raises up, opening the train pipe exhaust valve on the bottom of this piston and air flows out of the train pipe. As soon as the brake valve air is reduced the amount the engineer wishes (and the amount of the reduction is shown with the black hand of gauge) he closes the preliminary exhaust port by a movement of the rotary to "lap." The pressure of the brake valve air then remains stationary; while the train pipe air flows out through the train pipe exhaust till it is reduced a little lower than the brake valve air, which then moves the piston down gradually and closes the train pipe exhaust. It takes longer to reduce pressure in a long train pipe than in a short one through the small train pipe exhaust port because of the greater volume of air in the long pipe, so the train pipe exhaust is held open by the train pipe air till the pressure is reduced the whole length, then closed automatically by the pressure of the brake valve air.

Each of these valves uses a double hand gauge and has a small reservoir about 12 inches long connected to it by a small pipe; this equalizing reservoir is used to supply the cavity over equalizing piston with a larger volume of air, so a more gradual reduction of pressure can be made through the preliminary exhaust port from this cavity.

EQUALIZING DISCHARGE VALVE
WITH FEED VALVE ATTACHMENT.
1892 MODEL.

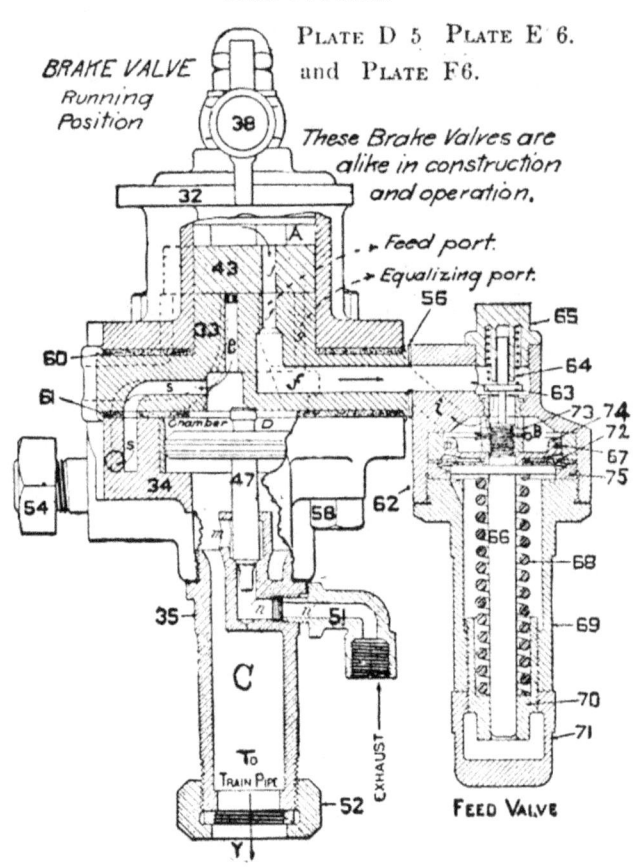

In the above illustration the main reservoir air is above the rotary 43 in A and feed port f. The train pipe air is in B over the feed valve piston 74 and in C under the equalizing piston 47. The brake valve air is in chamber D over piston 47 and in the equalizing reservoir connected to brake valve at port S.

The brake valve air is connected with train pipe air through equalizing port *g*, which is drilled through the seat under rotary shown by dotted lines and more plainly on this page.

Either of these valves when placed on "emergency position" opens a large port which lets the air from the train pipe direct to the atmosphere, making a sudden reduction, which causes the brake to go on suddenly and with full force.

The older pattern D-8 has the excess valve in one side where

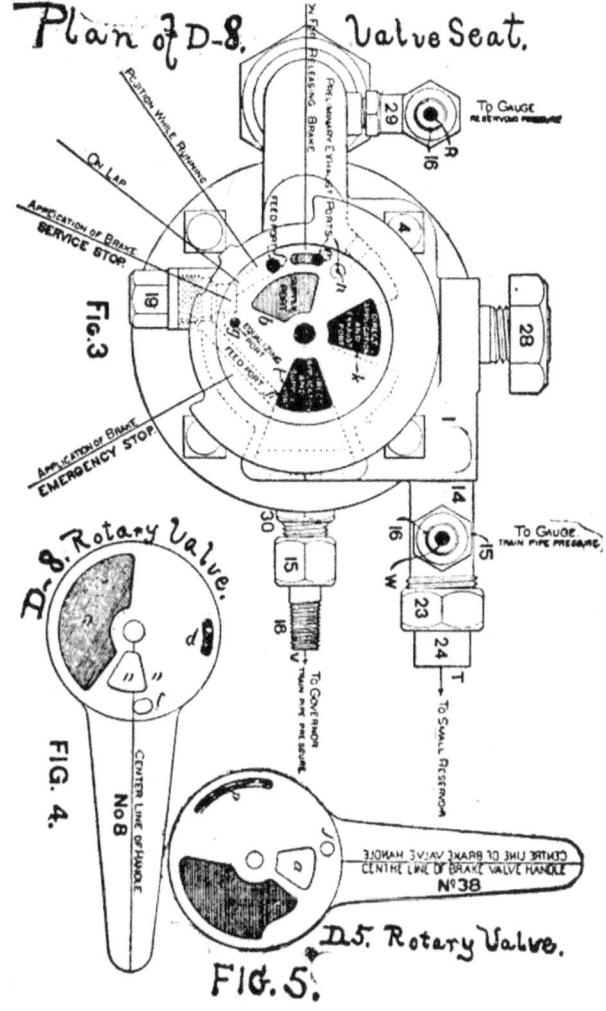

it can be taken out to clean or regulate, when air is let out of train pipe. When the brake valve is on "running position" there is no other passage for the air to go from main reservoir to train pipe except by this valve, and it is held on its seat by a spring, which is stiff enough to maintain about 20 pounds difference between the pressure in train line and main reservoir and hold the excess in main reservoir. The pump governor is piped to train pipe and set at 70 pounds.

The D-5 has a reducing or feed valve attached in the place of the excess valve, which is set to regulate the train line pressure at not over 70 pounds, at which pressure it closes and no more air can pass to train pipe from main reservoir till the train pipe pressure falls below what the feed valve is set at, when it opens again; with this valve the governor is piped to main reservoir and set at 90 pounds.

There are some other differences in the construction of these valves which make their operation a little different. The preliminary exhaust cavity "p" in the rotary of D-8 valve is short, so that as soon as the valve is moved from "service application" to emergency position the preliminary exhaust port is lapped or covered, and the pressure over equalizing piston 17 can not be reduced while in this position, unless air leaks out past piston packing ring or through joints.

With D-5 the preliminary exhaust cavity in rotary is a long groove, so preliminary exhaust port is open from "service application" to "full emergency." This can let all the air out of cavity over equalizing piston—which is named "chamber D"—and equalizing reservoir, so black hand will drop back to nothing as soon as air can escape.

With D-8 the running position feed port "j" through rotary comes over equalizing port "g" when at a certain place on emergency, so main reservoir pressure can get in "chamber D" and black hand will show the same pressure as red one. In this position there is no air in train pipe as direct application port is open. These ports do not connect with each other in the D-5, so main reservoir air can not get into "chamber D" in this position.

With D-8 set on the ledge between "running position" and "lap," the main reservoir pressure is shut off by port "j" closing at "f," so no air can pass through to train pipe and equalizing port "g" remains partly open so train pipe pressure is still connected to black hand of gauge. The D-5 does not show this way, as both ports "g" and "f" are closed at same point of movement of rotary.

14. Q. What is the red hand of the gauge coupled to?

A. The main reservoir pipe in the brake valve, and it always shows the main reservoir pressure.

15. Q. What is the black hand of the double gauge coupled to?

A. It is coupled to the pipe leading from the brake valve to the brake valve reservoir, and it always shows the pressure in "chamber D" above the equalizing piston and in brake valve reservoir.

16. Q. Why is it coupled in this manner?

A. Because when setting the brake the engineer must know exactly how much he reduces the pressure; as he makes the reduction above the equalizing piston, therefore the black hand must show the exact pressure there and nowhere else, while making a service application.

If brake is set with direct application, as the gauge does not show how much the train pipe pressure is reduced, you have no means of knowing it exactly, unless you let the air all out.

17. Q. In what position of brake valve does it also show train pipe pressure?

A. When brake valve is in full release or running position or anywhere between full release and lap. In these positions the "equalizing port," which is the communication between the train pipe and the cavity D, is open. In any other position it is shut to the train pipe pressure so it does not show on black hand.

18. Q. Then the black hand does not show the exact train pipe pressure when on lap, or past lap towards emergency?

A. No, not immediately, and you can easily prove this by placing the valve on lap and opening the angle cock at rear end of tender; the train pipe pressure will drop to nothing at once, which the black hand will not do. Usually the equalizing piston packing ring leaks a little, and the black hand will drop back slowly as the air leaks out into the empty train pipe; if there are no leaks in the brake valve, or connections to gauge or brake valve reservoir, it will not drop any. Unless the packing ring leaks considerable it does little harm.

19. Q. When the D-8 valve is on running position, do the red and black hands show the same pressure?

A. Not always. The red hand should show the most. This is because the air from main reservoir cannot get into the train line without going by the excess pressure valve, and if the spring in this valve is set at 20 pounds, there will be 20 pounds more in the main reservoir than in the train line, and the gauge will show it.

20. Q. What is the advantage of carrying this excess pressure?

A. When releasing brake, it supplies the train pipe with a higher pressure than brake was first set at; this makes the movement of all triples to release position much quicker and surer. With a long train it is absolutely necessary for this purpose. It recharges the auxiliaries quicker, ready for the next application of the brake. It charges empty cars quicker that are taken on the train. When brakes "creep on," they can be released at once by placing the brake valve on full release for a second or two, just long enough to raise the triple to exhaust position and not long enough to charge the reservoirs to a higher pressure, then returning it to running position.

21. Q. What difference between the D-8 and D-5 valves in regard to carrying this excess pressure?

A. With D-8 valve placed in "running position," you get the excess pressure before the train line begins to raise any, no matter at what pressure you start, as the main reservoir pressure must raise enough first to get by the excess valve before going into train line. With D-5, both pressures raise together till train line stands at 70 pounds, then the feed valve shuts and excess begins to pick up in main reservoir. So you have excess first, with D-8 valve, and hold it while valve is in running position; with D-5, you get train line pressure first up to 70 pounds, then excess afterwards.

22. Q. When the D-8 valve has been left on release position till train line and main reservoir have equalized at 70 pounds, and is then placed on running position, are the brakes apt to creep on at once? Why is this?

A. When the D-8 valve is placed on running position, it shuts off the air from train line till the excess pressure is picked up in the main reservoir. If train line leaks, the brake will set. In such a case, run your pump a little faster for a few minutes —not over five—so as to get the excess quicker. If train is under motion and you feel a brake dragging, put the brake valve in full release for a second only, then place it in running position; this may have to be done a second or third time until air begins to go through excess pressure valve, when it will hold brakes off. A short rule for this is: Keep your excess all the time by not using the full release position, except at the time of releasing the brakes, then running position will hold them off.

23. Q. Please define the different positions of the brake valve and the course the air takes passing through in each position.

A. The handle of brake valve has a spring pin in it. This pin is clear over against left side full stop when in "full release," against right side full stop when in "full emergency," against left side of middle stop when on "running position," against right side of middle stop when on "lap," or all ports covered. Between "lap" and first emergency comes the "service application." When on "full release" the air from main reservoir goes through the direct supply port to train pipe, also from main reservoir through feed port in rotary into supply port for the preliminary exhaust "e" and into cavity D, at the same time charging up the equalizing reservoir. Air can also pass to cavity D from train pipe through equalizing port "g." The warning port is open. Both preliminary and emergency exhaust or direct application ports are covered; also the feed port leading to excess valve or feed valve is closed.

When on running position the direct supply port is covered so main reservoir air cannot get direct into train pipe, and the supply port to preliminary exhaust is also covered, so main reservoir air cannot get into cavity D over equalizing piston. The feed port "f" to excess valve is opened and air must pass through there to get into train pipe and from train pipe into cavity in the rotary valve and thence through equalizing port to get into cavity D and the equalizing reservoir.

"On lap," all ports are closed so no air can pass under or through the rotary.

On "service application," the preliminary exhaust port is the only one opened directly by a movement of the rotary; the equalizing piston raising opens the train pipe exhaust.

On "emergency," the direct application port is opened, allowing the air in the train pipe to escape to the atmosphere very suddenly. With D-5 valve the preliminary exhaust is also left open. With D-8 the preliminary exhaust is closed when leaving service application.

24. Q. Do leaks in the brake valve interfere with its work?

A. Yes; if there is a leak under the rotary valve from the main reservoir to train-line, the brake will release when valve is placed on lap. A leak from train-line under rotary valve, or through train-line discharge valve to atmosphere, or a leak between equalizing reservoir and brake valve will set the brake tighter than you want it. If it leaks through gasket from main reservoir to cavity over equalizing piston 47 in D-5 valve, brake cannot be set in service application. Using the brake valve on emergency habitually will tend to cut the rotary and seat quicker, as it brings sand and scales of iron rust up from the train pipe on the seat, which the service application will not do. If the brake valve is fastened close to the boiler head so it gets very hot, the leather gaskets get burned and crack so they leak badly. A bad leak past the equalizing piston will cause engine brakes to release when set with a light "direct application." This is because air leaks from equalizing reservoir past piston and raises train pipe pressure in the short train pipe on engine and tender.

25. Q. What is the effect if equalizing reservoir pipe is broken so a blind joint has to be made?

A. The brake cannot be set with a gradual application; there is so little air above the equalizing piston, it escapes out of preliminary exhaust so quickly that brake works with full application; sometimes emergency with a very short train.

26. Q. What should you do in such a case?

A. If joints cannot be made so as to use equalizing reservoir again, a blind joint should be made at its connection with brake valve; the elbow in train-pipe exhaust should be plugged and valve used with direct application port, taking care to make a gradual reduction so brake will not go on with emergency, and closing valve slowly so the brakes on head end will not be "kicked off." The elbow has a thread cut in it for plugging.

27. Q. With the equalizing discharge valve, why does the air blow out of the train line exhaust when brake is released, if working brake on engine and tender only?

A. Because the train line is charged up through a large hole in rotary valve; the cavity over equalizing piston and brake valve reservoir is charged from the main reservoir through the small supply port *e*, for preliminary exhaust and by equalizing port *g*. If the train line is short, it will charge up to a full pressure quicker than the space above piston; train pipe pressure will then raise piston and discharge valve, allowing air to blow out of train line exhaust elbow for a second or two.

28. Q. Can this action of the valve be of advantage to you?

A. Yes; if you hear this escape of air from train line exhaust when releasing brake on a train, it is a sign of a short train line; and is a notice to the engineer that an angle cock at the head end of train is closed, or something has got into the train pipe and stopped it up. You should see at once if an angle cock is not shut by some mistake or malicious intent. Check chains swinging against the handle will close it.

29. Q. Does the amount of air which blows out of train line exhaust when setting the brake give you any idea of the number of cars in your train working air?

A. Yes; with engine and tender only, the train line exhaust does not blow much, if any, longer than preliminary exhaust. With a long train it takes some seconds for the train line pressure to be reduced and equalize its whole length. You can, after some practice, tell whether you have a long or short train working air by listening to the amount of air escaping from train line exhaust. This test shows the length of train line cut in and filled with air, not the number of brakes that set.

30. Q. What is the stop cock under brake valve for? Will it assist you in locating leaks? How?

A. To cut out the train pipe from brake valve when "double heading," so only one engineer can handle all the brakes. For this purpose it is absolutely necessary. Yes, it will assist in locating leaks. When shut, after charging train pipe and auxiliaries, if there is a leak in train pipe, brake will set at once; if the rotary leaks either into or out of the train line, it will show it very soon, as there is so short a train line to leak into or out of. A little observation will teach you many ways of using this cut-out cock in testing for leaks.

31. Q. If you had D-5 valve and the brake would not go on in service application, nor the black hand fall, nor the service exhaust open, while air came readily from preliminary exhaust, what would be the matter?

A. I would look for a leak at the joint on lower gasket where a leak would allow air to get from main reservoir direct to cavity over equalizing piston No. 47. This would give main reservoir pressure to black hand. A brake valve with this leak would show no excess pressure. No air could come out of service exhaust, as the pressure could not be reduced over the piston so valve could be raised. To set the brake use direct application port, opening and closing it slowly.

32. Q. If you had a continual blow at the train pipe exhaust port of the brake valve and could make no air, where would the difficulty be apt to be found?

A. Stuck or leaky piston D-17, dirt on its valve seat or bad leak in pipe to brake valve reservoir. Would put valve on lap, then on emergency for a moment and see if that would stop it.

33. Q. What is the difference in the operation of the equalizing discharge brake valve and the old style small brass valve named B-11?

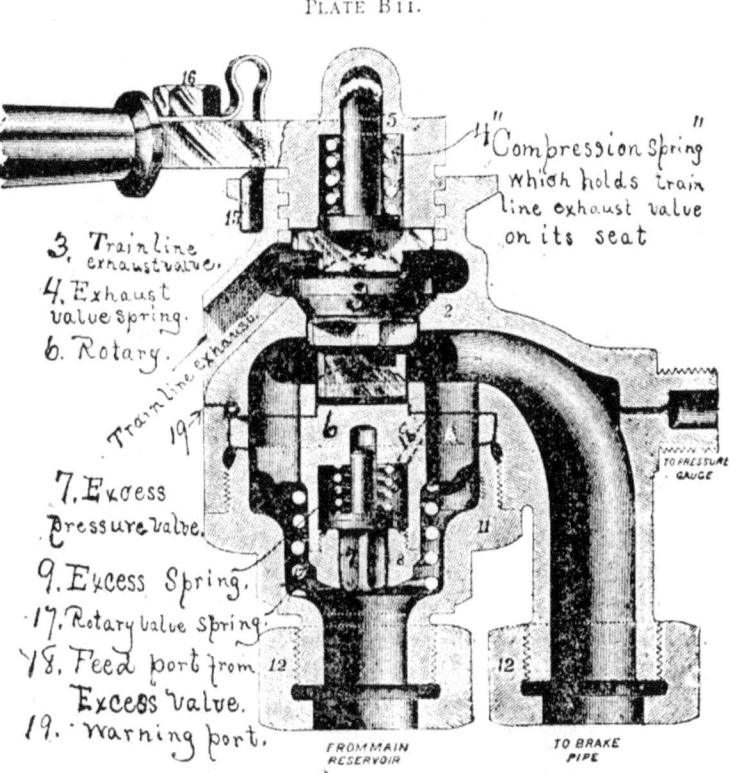

PLATE B 11.

A. In the small engineer's brake valve used before the equalizing discharge valve was perfected, the train line discharge valve "3" is held on its seat against the train pipe pressure by a short, stiff compression spring "4," located in the handle of the brake valve. This handle has a quick thread screw on it; when turned to set the brake, it relieves the spring of its strain so train pipe pressure will lift the discharge valve and allow the air to escape. By the same movement of the handle the rotary valve is turned and closes the supply port "a," so no more air can pass from main reservoir to train pipe. When train pipe pressure is reduced so the spring has power enough to hold the valve on its seat, it closes and shuts off the escape of air. A light reduction on the tension of the spring will make a light reduction in train pipe pressure; this is controlled by the distance you unscrew the handle. If this spring is too weak, the air will escape from pipe before supply ports are "lapped" and it cannot be used on "following" engines in "double heading," as valve will not hold the train line pressure when forward engine releases the brake. In this case, if there is no cut-out cock in train pipe under the small brake valve, it is necessary to make a blind joint there. The excess pressure valve in this brake valve is located in the rotary; a port "18" is shown by dotted lines, that takes air from excess valve to train pipe. To clean or adjust the excess valve, the brake valve must be taken apart.

The later pattern of this valve has an exhaust cavity in the rotary valve which connects the train pipe direct to atmosphere when in emergency position. This valve has the same positions as the D-8, viz., full release—running position—lap—service application and emergency. All these positions come at the same place each time valve is used, except lap and service, which positions depend on the tension of compression spring.

This brake valve cannot be used successfully, except on very short trains. The air ports are too small to handle a long train, especially the excess valve port 18, and very few of them equalize the discharge of air from the train pipe, which makes it a back number at the present day.

34. Q. How should the brake valve handle be placed when running or standing with brake released, unless auxiliaries are being charged? Why?

A. Always in running position, because this is the only position in which you can carry excess pressure, which is needed to release brakes promptly. With D-5 valve on full release the train line pressure will run up as high as pump governor will allow; this high pressure is apt to slide the wheels. When on running position the opening through brake valve from main reservoir to train pipe is a smaller one than on full release. If the train breaks in two or conductor's valve is opened to stop the train in case of accident, the brakes operate instantly. If valve is on full release the brakes will not all set tight till main reservoir pressure is reduced also. A small blowhole is put in the rotary valve to warn engineer that valve has been left in full release. All D-5 valves should have this warning port; if it gets stopped up, it is a sign that there is dirt on top of rotary valve, which should be taken out and cleaned at once.

35. Q. Can a gradual application of the brake be made, that is with only part of its full force?

A. Yes, by reducing the train line pressure only a few pounds, say 5 to 7 pounds at first application; this reduction is necessary to make brake piston move over leakage groove; a lighter reduction than 5 pounds will not always do this; 2 to 3 pounds at each of the succeeding applications.

36. Q. Why does this reduction of only a few pounds in the train line pressure make a light application of the brake?

A. With a light reduction the triple piston moves down slowly, opening the air valve slowly; the air from the auxiliary reservoir passes into brake cylinder through a small port and graduating valve; as soon as the auxiliary pressure is a very little lower than train line pressure, the triple piston raises and closes the graduating valve so no more air can go into brake cylinder, thus setting the brake lightly. To illustrate this, suppose we let out 7 pounds of air, reducing train line pressure from 70 to 63 pounds—that leaves 70 pounds above the triple piston (auxiliary pressure) and 63 pounds below it—the piston moves down towards the lower pressure, opening the graduating valve; when enough air has gone into the brake cylinder to reduce the auxiliary pressure a little below 63 pounds, the piston rises towards the lower pressure and closes the valve; another reduction produces the same effect, each time setting the brake tighter.

The piston moves the main air valve at the first reduction, but only closes and opens the graduating valve at the following reductions till a full application is made.

37. Q. How much do you reduce the train pipe pressure to make a full service application of the brake?

A. About 20 pounds, or from 70 pounds down to 50, or until the auxiliary pressure has equalized with brake cylinder.

38. Q. Why does a reduction of 20 pounds set the brake "full on?"

A. If the brake is in good order, with a piston travel of 8 inches, a reservoir pressure of 70 pounds will fill the brake cylinder and equalize in both at 50 pounds, that will leave 50 pounds on top of triple piston. If the pressure on the train pipe side or under the triple piston is any less than 50 pounds, the piston will stay down and hold the air valve open. One pound less will hold it down as well as any amount. When it has equalized, no more air will pass from auxiliary to brake cylinder, pressure on brake piston will not rise above 50 pounds, and brake cannot be set tighter. Any reduction of train line pressure that leaves it lower than auxiliary pressure will set the brake tight. If a reduction of 20 pounds opens the air valve and holds it open, any further reduction will not produce any effect on it, and as far as that brake is concerned is only a waste of air which must be supplied from main reservoir when you want to release brake. If any check valves in quick action triples leak, a reduction of train line pressure below brake cylinder pressure will let the brake leak off through this check into train pipe.

39. Q. What is necessary to have brakes set alike, with same reduction of train line pressure and release at same time, with same increase of train line pressure?

A. 1st. The auxiliary pressures must all be the same, so

triples will move down towards the same reduced train line pressure. For example, if one auxiliary has 70 pounds, another 60, a reduction of train line pressure below 70 will set the first brake, but it takes a reduction of below 60 to set the second one. 2d. All piston travels must be the same, for a short piston travel will equalize at a less reduction and with a higher piston pressure than a long piston travel. When train line pressure is increased, brake with long piston travel will release first, as the auxiliary pressure is lower. Thus, brake with long piston travel

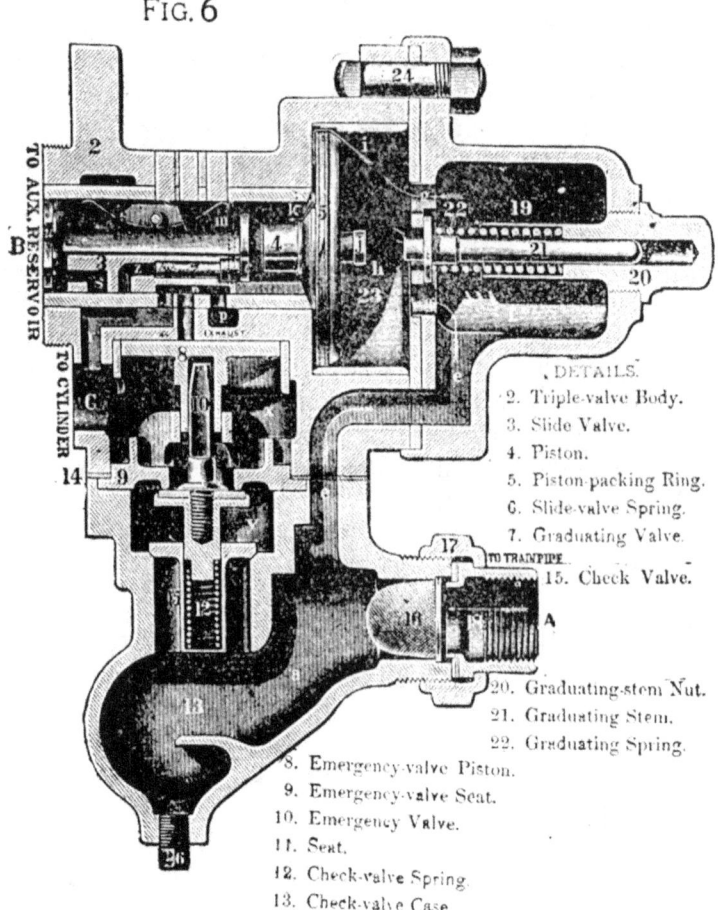

Fig. 6

DETAILS.
2. Triple-valve Body.
3. Slide Valve.
4. Piston.
5. Piston-packing Ring.
6. Slide-valve Spring.
7. Graduating Valve.
15. Check Valve.
20. Graduating-stem Nut.
21. Graduating Stem.
22. Graduating Spring.
8. Emergency-valve Piston.
9. Emergency-valve Seat.
10. Emergency Valve.
11. Seat.
12. Check-valve Spring.
13. Check-valve Case.

sets tight last, with less piston pressure, and lets go first. 3d. That all triples and brake pistons are in good working order and no leaks anywhere.

40. Q. What is the emergency or direct application?

A. If the train pipe pressure is suddenly reduced at the first application 10 pounds or more, with the quick-action triple, the emergency part of the triple valve is brought into action, opening a large port in the triple so the air goes from the train pipe direct into the brake cylinder, not only setting the brake quick-action, but also reducing the train pipe pressure suddenly at that point, instead of all the air going clear to brake valve to escape and reduce pressure. This sudden reduction sets the next triple in the same manner, which sets the next one, and so on to the last car; its action from one car to another is so quick that even on a long train it seems to catch all at once. When a quick-action triple takes air from the train pipe and sets the next triple quick-action, it also takes air from its auxiliary through a small port "s," after the train pipe has equalized, so the full application is made at 60 pounds, about 10 pounds more than the piston pressure in full service application.

A sudden reduction of train pipe pressure which will pull the triple piston down hard enough to compress the graduating spring and let piston make a full travel will open the large air port in plain triple valve, so brake will set somewhat quicker, but does not set with any higher piston pressure.

41. Q. Explain the operation of the quick-action triple when used on the "emergency."

A. A sudden reduction in the train pipe pressure right at the triple must be made, so the triple piston will make a full stroke, compress the graduating spring and open the emergency port in seat under slide valve, which will admit the auxiliary pressure over the emergency piston 8. This in turn pushes the emergency or rubber-seated valve 10 off its seat, and the train pipe air can then go direct to brake cylinder through large port C, raising the train line check valve 15 to do this. As soon as the train pipe and brake cylinder pressures have equalized the check 15 seats itself, and a spring in this check valve pushes the rubber-seated valve 10 up against its seat. At the same time that train pipe air is passing into the cylinder, the air from the auxiliary is also going through a small port in the end of slide valve 3. This port is made very small to give the train pipe air a chance to equalize first, then the auxiliary pressure equalizes with the brake cylinder afterward, at about 60 pounds. The emergency valve 10 is used to hold the train pipe air out of the brake cylinder, therefore in quick action it must be moved off its seat against the train pipe pressure; this is done only when triple piston makes a full stroke suddenly. If it moves down slowly, the graduating valve will allow air to pass from auxiliary into brake cylinder before emergency port is opened, and reduce the auxiliary pressure as fast as the train pipe pressure falls, so graduating spring will have power to prevent a full stroke of piston.

42. Q. Can you get the emergency action of the quick-action triple while brakes are set with a service application?

A. Not unless they are set with a light application. The pressure in auxiliary and train pipe must be considerably higher than in the brake cylinder, or the emergency piston will not move valve 10 off its seat, nor will check 15 raise to allow air to pass through. Then, if a partial service application has been made, the graduating valve can open first, which reduces the auxiliary pressure some and retards the full stroke of the piston a little in opening the emergency port. Even if all these emergency valves operate after a moderate service application, only a very little air will pass through them, not enough to affect the triples behind it, or raise the pressure in cylinder very much.

43. Q. Is it practicable to attempt to get the emergency action of the brake by suddenly recharging the train line for one or two seconds and then opening the direct application port wide?

A. No. The triple piston will not move till you have reduced the train pipe pressure a little lower than the auxiliary pressure, and no air can pass into the brake cylinder from either train pipe or the auxiliary till piston moves and opens the valves. By this movement you will partially release some of the brakes and get a lighter service application the second time than you had at first. Don't try it. Unless you have time to recharge auxiliaries to 70 pounds, hang on to what you have.

44. Q. When is it necessary to use the emergency application?

A. Only in case of accident, or sudden danger to train or persons.

45. Q. Is it safe to try and retain air in a train pipe in the emergency application, and why not?

A. It is not safe as a general rule. In an emergency, when life or property are in danger, you must act quickly. The point is to get stopped dead as soon as possible, and see about getting started afterwards. An emergency application is the last resort and you must get it when you need it. If you do not let nearly all the air out of a long train pipe, some of the triples will not act quick enough. If three or four triples are cut out, or there are three or four plain triples close together at the head end of the train, the "quick action" will not catch behind them and all the air must be let out at head end of train to reduce the pressure as quickly as possible. With a "double header" it is generally necessary to let the air out at brake valve of rear engine to catch the quick action on the train. With a full train of quick action triples a sudden reduction of 25 or 30 pounds at the engine will catch them all and leave considerable air in the train pipe, so you can release and back up out of the other train's way if the brake stops you in time. This is the only special exception to the general rule. It is easy to hold part of the air when making tests or in the instruction car; but when you think some one is going to get killed it is not quite as easy as "clear over" to "full emergency." Opinions vary on this question.

46. Q. How does the quick action triple operate if graduating pin is broken?

A. With the emergency on a light service application. If the graduating pin is broken the graduating valve will be held on its seat by auxiliary pressure, and the emergency port is the first one to open.

If the graduating valve is gummed up or dirty so the air can not flow past it properly, the triple will work with emergency when you make a moderate service application.

47. Q. If while making a moderate service application your brakes would "fly on" and at the same time the air would stop running for a moment from train pipe exhaust and then begin again, where would you look for the trouble?

A. In one of the quick action triples. This action of the brake valve shows that one of the triples is working quick action only, in advance of the rest, even with a service application. Probably the graduating pin is broken. Sometimes a triple with graduating spring gone will do this. It takes some skill to locate this defective triple. Watch the action of brakes in different parts of the train. When you find it cut it out.

48. Q. If a quick action triple should refuse to release, but kept blowing from the exhaust port with the brake on, what would be the matter and what would you do?

A. The emergency valve No. 10 was probably off its seat or was worn out and leaked badly. If out on the road and valve would not quit leaking after a few emergency applications, would cut out that brake.

49. Q. What is the difference between the plain engine and tender triple valve and the car or quick-action triple valve? And why should not the plain triple do as well on a long train?

A. When a long train of plain triples is used it takes some seconds for the last triple to operate, the pressure in train pipe is reduced slowly on last cars, which makes them set gradually; the train pipe reduction is made at the engine only.

In an emergency application the quick action triple allows some of the train pipe air to escape at the triple so that the train pipe pressure is suddenly reduced at this triple; this also operates the next triple "quick action," which reduces the train pipe pressure still more, so that all the triples act quicker than when the reduction is made at the brake valve, and all the brakes are set at nearly the same instant. Thus there is less shock to the rear cars of a full air brake train, as the action of the triples travels from one to the other faster than the slack can run up from car to car.

The quick action triple has two separate actions; in one a service application only operates the plain part of it; in the other, plain and quick action parts are operated at once. A quick action triple is not needed on an engine or tender, as they are so close to the brake valve that they operate quickly enough. None of the quick action triples can be set to work straight air; some of the plain triples can.

A plain triple gets all its supply of air to set the brake from the auxiliary reservoir only; the quick action triple gets it all from the auxiliary on a service application; when used with emergency it gets air from both train pipe and auxiliary.

50. Q. How do you locate a leak that lets off the brake?

A. If it leaks off through piston packing leather the air will blow out of the hole in spring case or lower head in push down brake. With a pull up brake, around piston rod or through the vent hole in top head. With a leak that lets air leak slowly out of brake cylinder a leaky graduating valve will let the brake release with a full application, although not so quickly as with a partial one.

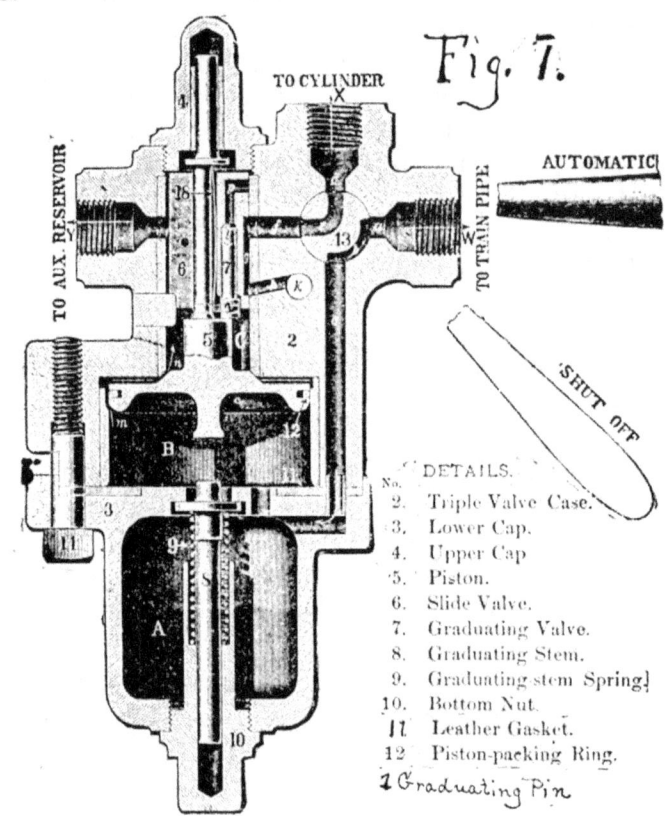

When a leaky graduating valve lets off the brake with a light application, it is because the air from the auxiliary leaks past the seat of valve 7 into the brake cylinder, until the auxiliary pressure is enough lower than train pipe pressure so triple piston moves air slide valve up into exhaust position, releasing air from brake cylinder through the exhaust port. This it cannot do with a full application, as in this case the air pressure has equalized between the auxiliary and cylinder so a leaky valve cuts no figure; air will not pass through after pressures are equal.

To locate a leaky graduating valve, set the brake with a light application, it will release through exhaust port of triple about as quick as you lap the brake valve. Then, after recharging the auxiliary, set with a full application and brake should stay set.

brakes equalize at, and train line pressure must be raised correspondingly higher to release tender brake. Then the tender triple gets more sand and dirt in it than any other triple, which causes it to wear and get defective. A leaky triple piston packing ring will allow any brake to stick unless very high excess is used, as it will allow air to equalize past the triple piston with auxiliary without moving piston up to exhaust position.

54. Q. If the train line is charged up with a high pressure from main reservoir when brake is released, will the brake set again at once with a small reduction of train line pressure?

A. It will not set again until the train line pressure is reduced below the auxiliary pressure. For example: If the brake has been set tight, the auxiliary pressure will be about 50 pounds for the first application; if you turn 90 pounds into the train line you must let 40 pounds out again, to draw train line pressure below 50, before the triple piston will move; all this time your train is getting nearer the stopping point. This is one of the reasons why you run by when trying to make a stop this way; it takes so long to draw your train line pressure down where it was before. In case you expect to apply the brake at once after releasing it wholly, or partly, put the brake valve on full release for an instant, just long enough to charge up the train pipe, and then put it on lap. This movement will hold your train pipe pressure so near the auxiliary pressure that the triple is ready to act instantly with light service application.

55. Q. Is it possible to let off part of the brakes and leave part of them set?

A. Yes. In the case of a full application, by moving the brake valve handle slowly over on to "running position" the train line will gradually recharge; as soon as the train line pressure is a little higher than the lowest pressure in any auxiliary, that triple will move up into exhaust position, letting off that brake. Then auxiliary will commence to recharge through feed port and hold train line pressure down till that auxiliary is charged up high enough, when another brake will let off, and so on, till all are let off. The brake that holds least lets off first; it generally has longest piston travel, consequently lowest auxiliary pressure, and the operation here described takes place when piston travels are unequal. If the triples all work alike with a light service application, when auxiliaries all equalize with the train line pressure, the effect of a very gradual rise of train line pressure through excess pressure valve is to send a small wave of air back through the train pipe; this can be so handled as to affect the engine triples first, and cause them to release first, others afterward. In case of a defective triple with a leaky packing ring, it will not let off at all, but will keep brake set (at train line pressure, sometimes), as air will get past leaky packing ring instead of forcing triple piston up into exhaust position.

56. Q. Why do some of the brakes "creep on" when the train is running?

A. Because there is a leak in the train line, triple valves or auxiliary reservoirs; or the auxiliaries have not all equalized after releasing the brake.

57. Q. How can these brakes be released the quickest and surest way?

A. By moving the brake valve handle from "running position" to "full release" just long enough so the rush of air will charge up the train pipe, and putting it back to running position before any of the auxiliaries are charged any higher. This forces the triple valves of the sticking brakes up into release position, so air from brake cylinder exhausts and does not give time to raise the pressure in any reservoir. Sometimes this must be done a second and third time to release all of them. If brake valve is held on full release long enough to charge a reservoir higher than before, that brake will be sure to set as soon as brake valve is returned to running position.

58. Q. If governor is set at 70 pounds with D-8 valve, and train line is charged from main reservoir higher than that pressure, is the brake apt to creep on?

A. Yes; the pump is stopped and will not start again till train line pressure is lowered to 70 pounds. During this time brake is pretty sure to go on.

59. Q. How can this be avoided?

A. By not allowing main reservoir to charge train line and auxiliaries at over 70 pounds. When standing at a water tank, or any stop, with brakes set, the main reservoir pressure is apt to run very high. If all of this is turned into train line and allowed to equalize at over seventy pounds, with brake valve carried in full release regularly, there is no way to prevent the brake setting. In this case, set it a little and at once release it; this will reduce the train line and auxiliaries below 70 pounds, so pump will go to work and you can hold brake off.

60. Q. In making a stop how should you release the brakes on a freight train? On a passenger train?

A. On a freight train not till it has entirely stopped, or you run the risk of train breaking in two. The train pipe pressure on a long train is increased next the engine first; so brakes let go there first, even if it is only a few seconds sooner. Part of the shock is from unequal piston travel, which gives unequal piston pressure; brakes with long piston travel let go first after a full application. With a "part air" train the slack of entire train runs up against the head cars; releasing brake while train is moving is liable to part the train; working steam before slack is all evened up in train is sure to break it in two.

With a passenger train, just a few feet before the train stops, so there will be just enough power to stop the train and avoid tilting the coach truck forward at the instant the train stops. If the brake beams are hung from the body of the car the truck will not tilt forward.

61. Q. Why should a brake on a passenger train be left off just before coming to a full stop?

A. Because the brake shoes pulling down on the forward end of the truck, and pushing up on the back part of the truck, tilt the truck; and if brake is not let off until after the train stops, when the truck rights itself it rolls the wheels back a little and throws the body of the coach back, annoying the passengers, even if it is not severe enough to throw them against the seats. This trouble is not felt so plainly by the engineer when he has a good driver and tender brake, as the brake on the coach is what jerks the coach. Then less power is required to stop a train going very slow, as at the instant of stopping, than when running at full speed; if power enough is left on to hold a train at full speed it must stop very forcibly at a slow speed. The brakes should begin to release about half a rail length from where the train finally stops; a little farther if going very fast, a little less if a very slow stop is being made. Practice will teach you the distance.

There is an exception to this rule in the case of a very long passenger train, say over twelve coaches, especially if it is not vestibuled and the buffer spring slack all taken up solid between the cars. Experience will teach you that in stopping a train of this length that less shock will be given the front end of train if brake is held on moderately tight just at the instant of stopping till train stops, i. e., handle a very long passenger train about the same as a freight train of the same length. Try this plan for yourself, that is the sure way to find out. If the couplings between coaches are slack there is more shock at the instant of stopping.

62. Q. If you are stopping at a water tank with part air train with air brake set and with hand brakes set on rear cars, would you release the air brake and work steam in case you were stopping too quick? If you did what would happen? Why?

A. No. If air brakes were released on head end while hand brake is still set on rear end, the shock is sure to be severe. Air brake should be held on until train stops or hand brakes let off. If you worked steam you would risk a break in two. Don't try it.

63. Q. Can you describe the position of the handles to all valves and cocks in the air brake and signal equipment, whether open or shut? Do so.

A. All the handles, except to angle cocks, stand at right angles or crosswise of the pipes when they are open; parallel to pipe when "cut out;" plain triples and pressure retainers follow the same rule, their handles are horizontal or crosswise when "cut in." The crooked handle of angle cock is parallel with pipe when cut in. This is so the hose will protect handle from being struck by anything flying under the car and getting shut off as the old style straight handled cock is liable to. A small groove square across the end of plug shows whether cock is open or shut, as the groove runs same way with hole in plug.

64. Q. Do you understand that all air cars in a train should be connected and train pipes charged with air, whether brakes are cut in or not? Why?

A. Yes. All train pipes should be coupled up and air working through them, so that if the train breaks in two anywhere in the line, all brakes will be set that are working.

65. Q. 1. After coupling to train why should you not immediately try to apply the brakes for inspection? 2. How long should you wait? 3. Can an auxiliary reservoir be recharged without releasing the brake?

A. First. Because you must wait till a full pressure of 70 pounds is stored in auxiliaries so a full application of brakes can be obtained to get the piston travel. Second. The time you should wait depends on the pressure maintained in the train line from the moment of coupling on; if 70 pounds is held steadily, two and one-half minutes is the shortest time, but it must equalize in auxiliaries and train pipe at this pressure—even if it takes longer—before testing. When the governor stops the pump, with standard pressures shown with both hands, is generally time enough. Third. The ports are so located in the triple valve that the "feed port" through which auxiliary is charged does not open till after exhaust port is open, which releases the brake first, recharges the auxiliary afterward. By the use of a pressure retaining valve the auxiliary can be recharged without releasing the brake altogether.

66. Q. Should the train brakes be inspected? How? When? Why?

A. Yes, by applying them with full service application, examining each car to see that the piston travel is the proper length and that there are no leaks that will let brakes off; then releasing them and examining each car to see that all release, and that there are no leaks through exhaust port. They should be inspected at all terminals and tested whenever train breaks in two, or cars are taken on or set off, as the wrong angle cocks may be closed or left closed at such points. This is necessary because it is not safe to depend on a brake till it is shown that it will set and release properly. Hand brake should always be let off before testing.

67. Q. Would you consider a train safe to leave with if the brakes had been tested by opening angle cock at rear of train, and how would this affect your main reservoir pressure?

A. No, sir. Not unless some other test has been made. This would not set all the brakes unless the brake valve was on lap. It would draw down main reservoir pressure and waste air without doing any good. This test is only good to show that air hose are coupled, angle cocks open and train pipe charged from engine to last car.

68. Q. If you release the brake and apply it again immediately, would you expect to obtain the same power you had before? How *long* would it take to regain the original pressure?

A. No, sir! never. About 40 seconds, if main reservoir had 35 or 40 pounds excess over auxiliaries, sometimes less time. The feed ports in triple valves which regulate the time of charging are not always the proper size for the reservoirs they supply. A short train and light application would reduce this time to 20 or 25 seconds. Generally it takes longer than the tests show it with everything in good working order, the feed ports are not always clean and strainers free. The pressure at which auxiliary equalized after first application is what you begin with on second application after first release, generally it is 50 after first full application; with full release of brake and immediate application you get 35 and a little more on second full application; the third time you will have less than 30 pounds piston pressure.

69. Q. Why does it take so long to regain the original pressure in the auxiliaries after releasing brakes?

A. Because the feed port in the triple through which the air passes from train pipe to auxiliary is small. This feed port is shown at "m" in the plain triple in Fig 7 and at "i" in the quick-action triple in Fig. 6. It is necessary to have this port small for two reasons; first—when setting the brake, the feed port must be small; or when train pipe pressure is reduced at brake valve for a light service application, the auxiliary air could flow around the triple piston through the feed port "i" as fast as it is taken out of train pipe; so triple piston would not move. Second—if feed ports were larger, when brakes are to be released, it would be impossible to charge up a long train pipe from the engine and hold the pressure up, quick enough to release all the brakes at as nearly the same instant as possible, as the first few ports to open would take some of the train pipe air and hold the pressure down; if they were large enough a few of them would do this. These feed ports are so located in triple valve cylinder that they are open when triple piston is in exhaust position and must be the proper size for the auxiliaries they supply, so different sized auxiliaries will charge to the same pressure in the same time from the same train pipe. The auxiliary reservoir for a 10 inch coach brake holds about 3,200 cubic inches, that for an 8 inch freight brake holds about 1,650 inches; therefore a feed port for a 10 inch brake reservoir must be the right size to pass twice as much air through in the same given time as for an 8 inch brake. This is the reason for using only the proper triple for each reservoir. Then the reservoirs are a certain size for the brake cylinders they supply, so an auxiliary pressure of 70 pounds will equalize with brake cylinder of 8 inches piston travel at 50 pounds. This in turn gives a standard piston pressure to arrange the brake leverage on each car or engine for, so as to get the full effective braking power. The plain triples used for driver and tender brakes with either 10x24 or 12x33 auxiliaries have feed ports the proper size for the 12x33 auxiliary used with a 10 inch coach brake. This gives quicker recharging to standard pressure, so you have better service from engine brakes in switching or rapid work. If engine brakes creep on from this cause when coupled to a train, they are easily released, as main reservoir and brake valve are close to them.

70. Q. What are "leakage grooves" for? Where are they? Is it necessary to allow for them when setting the brake? How do you do it?

A. Leakage grooves are small grooves cut in the inside of passenger, freight and some tender brake cylinders, at the top of cylinder, so that when piston is clear back in full release this groove is uncovered, so that a small amount of air leaking by the triple, or from the joints, or from a *very* light application of the brake will get past the piston and escape without moving the piston. They are long enough so a movement of the piston of three inches is necessary to cover it, after which no air can get by the piston. It is necessary to allow for them at the *first* application, by making it strong enough so piston will go far enough the first move to cover the "leakage groove." Five to seven pounds reduction will do this, if everything is in good order; after that two to three pounds reduction will set this brake tighter. It is of vital importance to be sure leakage grooves are covered at first application or the air from the auxiliary will be wasted. The second application will waste it again and thus a number of small applications will waste all the air so train cannot be stopped. This is a very common fault in the operating of the brake.

If the hand brake is set on a passenger car, or the piston travel shortened to less than 3 inches, that brake will not hold, as air will get by the piston through leakage groove. Driver brakes do not have leakage grooves; none are needed, as you can release the driver brake from the main reservoir pressure if they creep on or stick.

71. Q. Does the difference in travel of pistons in brake cylinders increase or decrease your braking power? Why?

A. Long piston travel decreases the braking power because it gives less air pressure on piston, short piston travel gives higher piston pressure. With 8 inch piston travel, 70 pounds auxiliary pressure gives 50 pounds on piston per square inch.

An inch difference in the travel makes close to 2 pounds in pressure, thus 7 inches will give nearly 52 pounds, 9 inches a little over 48 pounds. The piston travel can be correct with a heavy car and high leverage, and the shoes will not clear the wheel much when released. If levers and brake beams spring much with 8 inch travel, the shoes will not have much slack when let off. Brake levers may catch on something so piston travel is correct and shoes not touch the wheels.

72. Q. How do you cut out the brake on engine and tender without disturbing train pipe?

A. By turning the four-way cock in top of plain triple so the handle is at an angle of 45 degrees; this will lap all ports and allow no air to pass from train pipe or auxiliary to brake cylinder; see that brake is entirely released first, and open bleeder in auxiliary.

73. Q. What is the difference between cutting the air out from a car and cutting it out from a brake?

A. Shutting the angle cock at the end next engine cuts out that car and all behind it; shutting the cock between train pipe and triple cuts out that brake only and allows all the rest to operate.

74. Q. If one brake beam under a car was broken how would it affect that brake? How would you cut out the brake on that car and allow air to pass to other cars?

A. If one brake beam or rod is broken, the brake on that car is useless and it must be cut out by shutting the cock in the crossover from train pipe to triple, or by turning the fourway cock in plain triple. This will allow air to pass through train pipe to other cars without operating disabled brake. Be sure the brake with plain triple on either engine, tender or coach is released before it is cut out, as no air can get out of brake cylinder after cock is turned. The plain triple used on freight equipment before the quick action triple was perfected, which is still in service on a great many freight cars, bleeds the brake cylinder when the handle of plug cock in triple is turned to the cut-out position, but does not bleed the auxiliaries, so the brake is likely to set when handle of cock is turned to automatic again. There is no bleeder in the cast iron auxiliary for this triple, the air escapes from brake cylinder through a bleed hole in the plug of cut-out cock. If this hole gets stopped up, cut out the car from the others, open the stop cock at hose, turn the cock in triple for straight air and air will escape from brake cylinder through train pipe, after which the cock in triple can be set for cut-out position. All quick action brakes can be bled by opening the bleeder in auxiliary reservoir and allowing all air to escape, as the cut-out cock does not close the communication between brake cylinder and the bleed cock in auxiliary.

75. Q. What should be done with a car on which the train pipe is broken?

A. If it cannot be plugged at leak and allow air to pass freely to cars behind it, it must be switched behind all other air cars, have air in hose that is coupled to next car in front, brakeman should look after that car and all behind it. If you have two ¾ inch air brake hose, the signal hose can be taken off signal line, brake hose put on, and signal line used for brake line through that car to get air back to other cars.

76. Q. If the pipe at one end of the car should come loose would you consider it dangerous? Why?

A. Yes. If the pipe at end of car gets loose so cock will bounce up and down and strike the handle end of plug against the dead wood or any part of car, it is likely to work shut gradually. This is caused by the spring, which holds the plug in its seat, turning plug a little each time it strikes. If the spring is wound one way, it works open; if the other way, it works shut.

77. Q. On a freight train can you tell about how many cars are working air by the exhaust from train pipe in service application? Could you tell the difference if some of the cars were cut out? Is this important? Why?

A. It takes considerable practice to tell how many cars are coupled on. By this test it gives the number of car lengths of train pipe in use; if the triple is cut out on any car it gives you no notice. When some of the cars are cut out by closing angle cocks, a less amount of air will come out than with all of them. Yes, it is important to know this, as some of the angle cocks may be closed, thus cutting off all the cars behind the closed one.

78. Q. In going down a long, steep grade, which would be the best practice, to make a number of applications, or as few as possible, and keep your brakes on until compelled to recharge them? How would you control your train while brakes were released and recharging?

A. Air braking on a long hill is a business to be learned on that particular hill; no set rule can be laid down. You must know just where the steep and the easy parts of the grade are located so as to have a good pressure for the steep part of the grade and be ready to recharge on the easy part of grade. Generally you should make a moderately heavy application at first and go as far as possible before making another one or a third one. On a short hill this should hold the train until the bottom is reached. If on a long hill make as few applications as possible between releases, use the "retainers," try to pick out a place where the grade is easy to recharge auxiliaries. "Hill work" is the most exacting of any in operating the brake.

79. Q. What is the difference between handling a train having part air in front and one entirely of air?

A. A great difference. It requires more skill and practice

to make a good stop with a part air train than with a full air train. With part air you must be careful to "bunch the train" so slack will run up easily against the air brake cars before setting the brake very tight; this takes some seconds; if you make a second reduction before the rear end feels the effects of the first one, the two light applications make one heavy one, as far as the shock to the rear cars is concerned. When backing up, extra care must be taken, or train will break in two and merchandise be damaged in cars.

80. Q. If in going around a curve you apply your brake to steady the train, when should you do it—on straight line or curve?

A. On straight line nearing point of curve *every time*.

81. Q. In applying your brakes do you understand that the force with which you are retarding the train is exerted on the rail; that it is better to bring these strains lengthwise of the rail, as on a straight line, than crosswise as on a curve?

A. Yes, it is better for the track, also for the train, as it may crowd some of the cars against the outside rail, hard enough to derail them. This may happen with air brakes ahead, empties in the middle and loads behind. Where brake beams are hung from the body of the car it interferes with the curving of the truck if brake is set very hard on a curve—just how much no one can say positively.

82. Q. If you had a freight train with "part air" cars in operation and you used the emergency application, would it make any difference whether the slack was out or not? In case there was a shock, on what part of the train would it fall?

A. Using the emergency brake with part air train always sets the head end hard and solid; if slack is all run up against the engine the shock is not as great. In any case the rear end gets all the damage; the weakest cars and draft gear behind air cars suffer.

83. Q. What would you do if an air hose burst? How would you know it? Should you have extra hose? Of what kinds?

A. Put brake valve on lap; whistle out a flag. If in a dangerous place to wait, or when a train is close behind, shut the first cock ahead of bursted hose; let off brakes on head end from engine; bleed the cars behind bursted hose; get to a safe place and replace the bursted hose with a new one. If with bad grades or all air train, put in a new hose any way, if possible. It would be known at once because brake would set; main reservoir pressure would run down quickly; black hand would drop way down. Put brake valve on lap to save your air. To locate the bursted hose, put brake valve on running position just so you will keep a little pressure in the hose and trainmen can hear the air blowing out of bursted hose and find it.

Extra hose should be carried on engine, one of each kind used. Trainmen should have a 1 inch, a 1¼ inch, a signal hose and one double end or splice coupling to use in case drawheads or coupling of cars are so long the regular hose and couplings will not meet each other.

84. Q. What course would you take should your train break in two and set the brakes?

A. Put brake valve on lap, shut off steam, whistle out a flag, shut the open angle cock on rear end of last car connected to engine, let off brakes on head section from the engine. When they are let off and you get a signal to do so, back up to rear section; after coupling up to it, if brakes can not be let off from engine, bleed a few of the sticking ones at back end of train until train can be started. Be very careful to shut the bleeder as soon as air begins to escape from triple exhaust port or you will set some of the others, and that will hold the train longer than necessary. All air bled out is wasted; it is done only to save time, which is valuable in a case of breaking in two.

85. Q. Do you know what the "pressure retaining valve" does? And how? If the pipe leading to this valve should break off would you plug it? If you did how would you affect the brake?

A. The pressure retaining valve holds some of the compressed air in the brake cylinder after the triple valve has moved to exhaust position. It is attached to exhaust port of triple valve by a piece of pipe and placed where it can be conveniently reached when train is in motion. When set to operate, its handle is turned up to a horizontal position, which closes the direct opening, so the air goes out slowly under a weighted valve; when pressure falls to 15 pounds per square inch in brake cylinder, this valve shuts off the escape altogether and holds the air in there, keeping the brake set at 15 pounds; this allows the auxiliary reservoir to be recharged to full pressure again. It is used on long steep grades. If the pipe was broken off and end plugged, brake would not let off at all, as there would be no way for air to get out of triple exhaust. A steady leak at the pressure retainer comes from the triple and turning the handle of retainer will not stop the leak at the triple; that brake will creep on.

86. Q. If the air signal on the engine whistled each time you released the brakes, what would be the trouble? If the whistle blows frequently when not in use, what is the matter? If it blows one long blast? If the whistle is weak on engine will it usually help it to blow out the signal hose on the rear of tender?

A. If an air signal whistles each time brake is released with standard braking pressure, it is a sign the reducing valve is dirty and stuck open, so air goes back into main reservoir from signal line each time main reservoir pressure is reduced in recharging train. In this case signal line has main reservoir pressure. Clean the reducing valve before the air signal hose bursts. If the spring over diaphragm in reducing valve is too stiff it will do also this. The reducing valves are set at about 45 now; the old valves were set at 25 pounds. This is so as to carry a lower pressure in signal line than is used to operate the brake. If the signal whistle blows frequently when not in use, there is a leak somewhere, which the jar of the engine may open for an instant, or the reducing valve may be out of order. If it sticks a little in its seat as in cold weather a very small leak will cause the whistle to give a strong blast—or a jar may unseat signal valve. Shut the cut out cock at reducing valve; if signal line leaks it will whistle as soon as the leak can reduce the signal line pressure. When it blows one long whistle some of the valves on engine are stuck, or the car discharge valve is opened a second and third time before the whistle stops blowing the first blast; the pressure in signal line must equalize each time between the blasts to make it work accurately. If the whistle bell works loose so it does not make a clear sound, or is located near partly opened windows so a strong draft of air blows across it, when train is running fast, the sound will be very weak. Blowing out the signal hose at rear of tender gives all the valves a chance to make a full opening and clean out the dirt.

87. Q. Which engineer should handle the brake in "double heading," and what should the other engineer do?

A. The head engineer should handle the brakes in double heading. Second engineer should shut the cut-out cock under his brake valve, keep his pump running and a full supply of air; brake valve on running position; if no cut-out cock put brake valve on lap, and plug the train line exhaust elbow so head engineer can release brake, without losing air through your valve. To help the head man release sticking brakes, when signaled to do so by first engineer, open cut-out cock and place valve in full release, shutting it again as soon as train is moving, so first engineer can stop the train at once if necessary. With old style small brass brake valve it is sometimes necessary to make a blind joint in train pipe of second engine at brake valve if there is no cut-out cock and the compression spring is weak.

88. Q. How should brake valve be carried when backing up the train, or when expecting the trainmen to set brake from rear end?

A. On running position; so the brake will be applied as soon as train pipe pressure is reduced, when brake valve should be placed on lap at once.

89. Q. Does it require the same skill and judgment to stop a train properly as it does to get it under full headway by manipulating an engine?

A. Same kind of skill acquired from practice. More judgment.

90. Q. Are you familiar with the air brake and signal instructions as approved by the M. C. B. and Master Mechanics' Association? What other books on air brake instruction have you studied?

A. Yes, sir. Also Westinghouse Instruction Book and Phelan's Air Brake Practice.

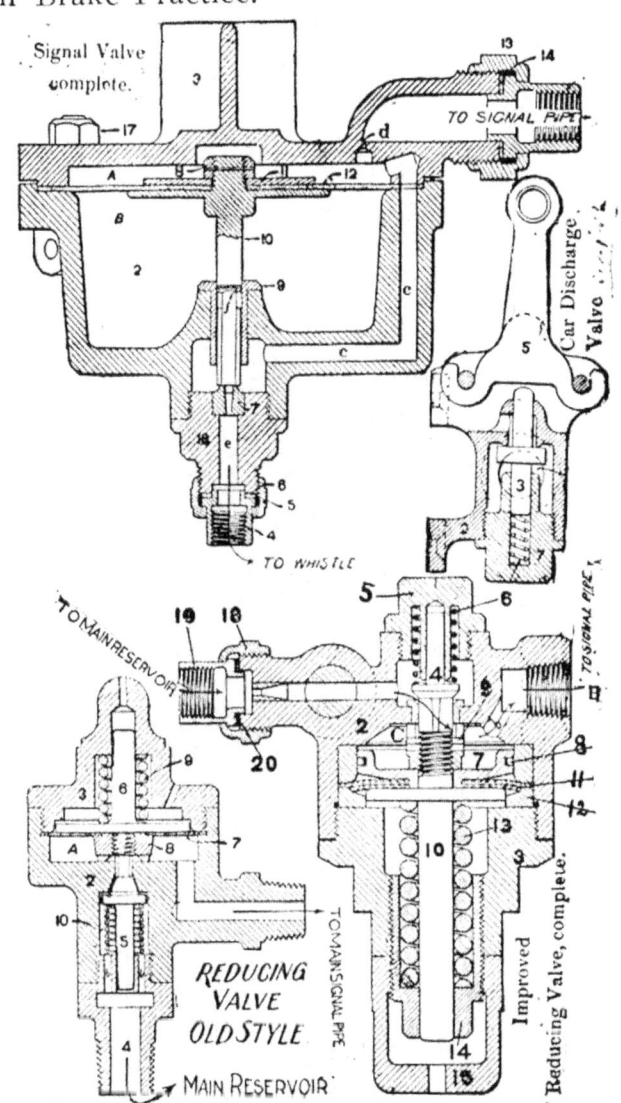

91. Q. How does the pump governor operate?

A. It is located in the steam pipe close to the pump; at union connection No. 70 it is coupled to the train pipe with D-8 and B-11 brake valves; to the main reservoir pressure with D-5 brake valve. When the air pressure raises to the standard amount, which with D-8 we will assume to be seventy pounds, it raises the diaphragm No. 67 against the tension of the regulating spring No. 66 so the air valve is raised off its seat. This allows the air, at seventy pounds pressure, to flow in over governor piston No. 53, and as this piston has several times more area than the steam valve No. 51, it will force the steam valve shut, thus cutting off the steam from the pump. As soon as the air pressure falls below seventy pounds the regulating spring moves the diaphragm down, closing the air valve; air leaks out from over governor piston which allows it to raise and steam goes to pump again.

92. Q. What might prevent the governor from shutting off the steam and stopping pump when maximum pressure is obtained?

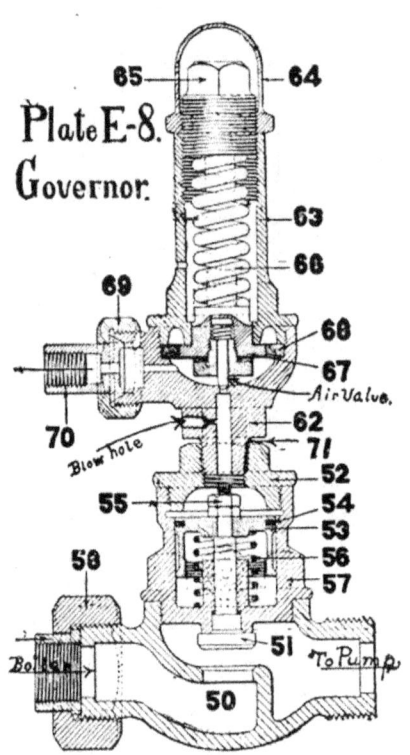

A. If the regulating spring is screwed down too tight it will not allow the diaphragm to raise at 70 pounds and lift the air valve off its seat. If too much oil is used in air end of pump the air valve gets gummed up where it rests on its seat, so air can not get through after air valve raises. This is the most common cause of the pressure getting higher than the governor is set for. To cure this trouble take out diaphragm and clean off air valve and its seat so air can get through freely when air valve raises. If the air leaks past piston as fast as it comes through air valve, the piston will not be moved down as there will be no pressure above piston. This is a very common fault with the old governor D-9. Putting in a tight packing ring cures this unless the cylinder is worn out of true. If the governor piston sticks so air pressure will not force it down, steam will not be shut off. If the waste pipe in side of steam end of governor is stopped up so steam or air is confined below piston, the governor will not shut off at any pressure. If anything gets in over diaphragm so it cannot raise, that will hold air valve shut so air cannot get on piston to shut off steam valve. If valves and seats are kept clean, and all parts allowed to move as they should, governor will work accurately.

93. Q. Where should you look for the trouble if governor shuts off pump at less than standard pressure?

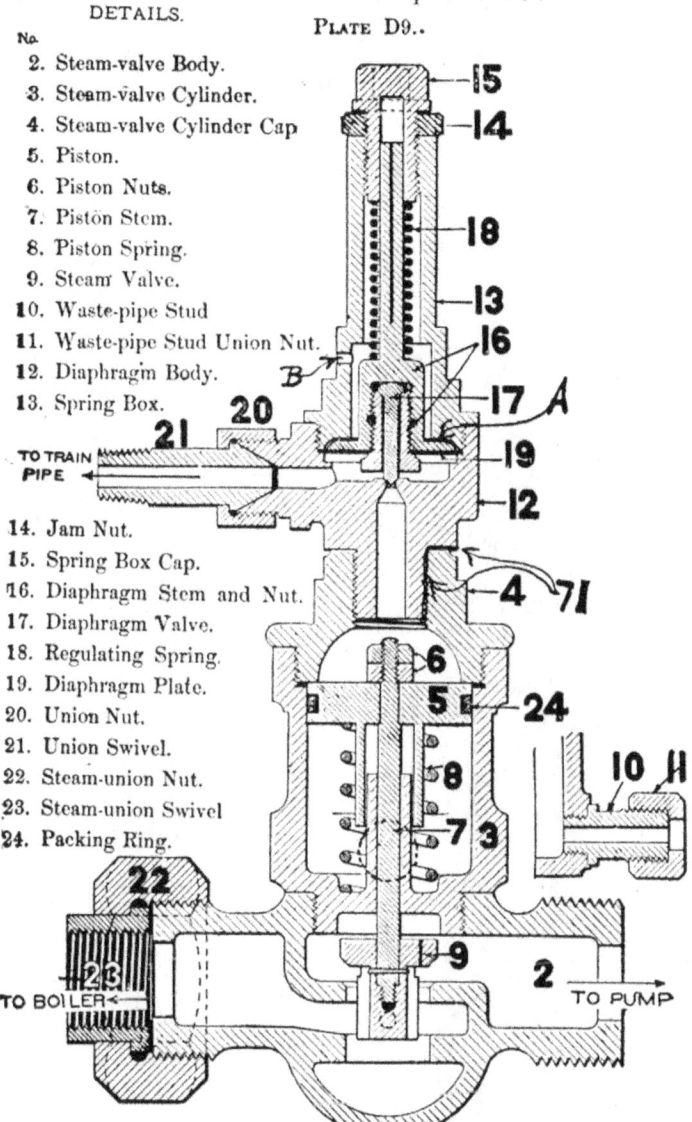

DETAILS.   PLATE D9.

No.
2. Steam-valve Body.
3. Steam-valve Cylinder.
4. Steam-valve Cylinder Cap
5. Piston.
6. Piston Nuts.
7. Piston Stem.
8. Piston Spring.
9. Steam Valve.
10. Waste-pipe Stud
11. Waste-pipe Stud Union Nut.
12. Diaphragm Body.
13. Spring Box.
14. Jam Nut.
15. Spring Box Cap.
16. Diaphragm Stem and Nut.
17. Diaphragm Valve.
18. Regulating Spring.
19. Diaphragm Plate.
20. Union Nut.
21. Union Swivel.
22. Steam-union Nut.
23. Steam-union Swivel
24. Packing Ring.

A. Generally there is dirt or a scale holding the air valve off its seat so air can get through on top of piston steadily, in which case the governor will shut off steam as soon as air pressure on the top of governor piston will more than balance steam pressure on steam valve. If this air valve seat is injured so it leaks, or a new valve has been put in that is too short to make a good joint, a very low air pressure, less than forty pounds, will shut off the steam. A broken regulating spring will also do this.

94. Q. If the pump will not start up soon after the air pressure has reduced below that at which the governor is set, what should be done to governor?

A. When the air valve closes, the air is shut up in cylinder

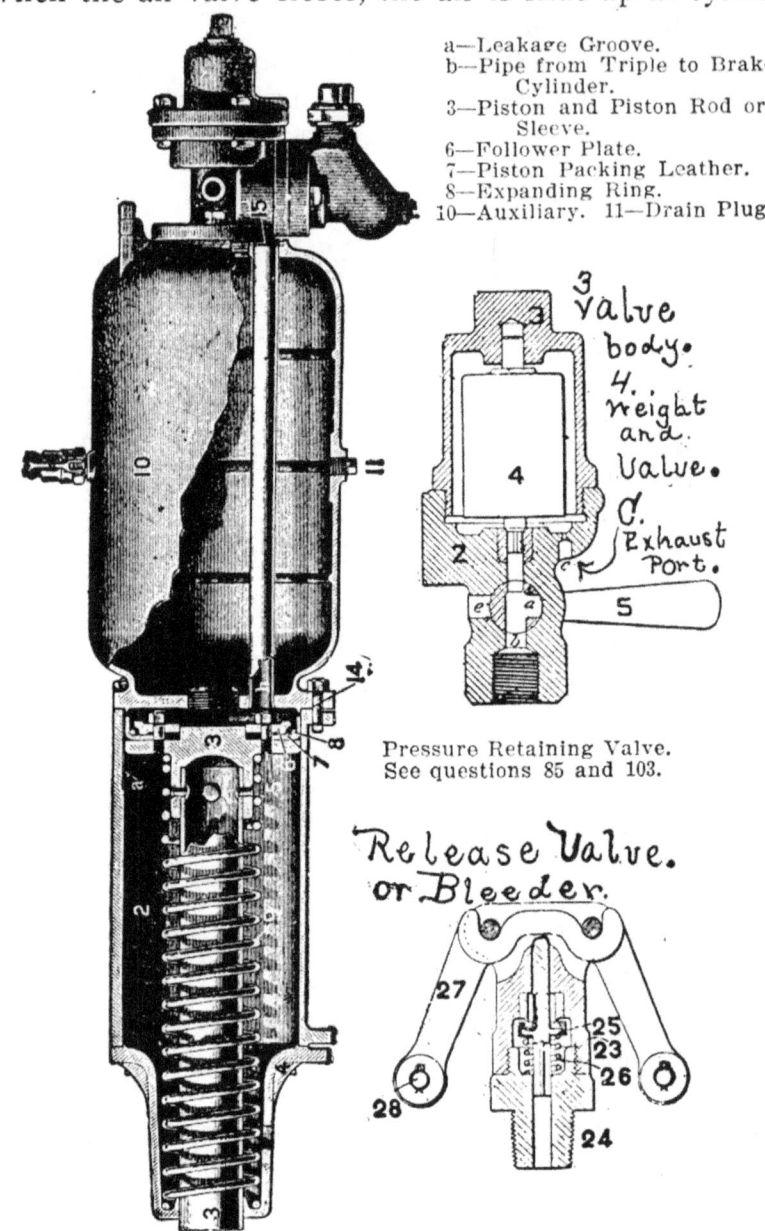

a—Leakage Groove.
b—Pipe from Triple to Brake Cylinder.
3—Piston and Piston Rod or Sleeve.
6—Follower Plate.
7—Piston Packing Leather.
8—Expanding Ring.
10—Auxiliary. 11—Drain Plug.

Pressure Retaining Valve. See questions 85 and 103.

over governor piston and must leak out before piston can raise and open steam valve. The new style of governor E-8 has a small blow hole drilled in the side of No. 62, below air valve seat, which lets enough air escape after standard pressure is reached to keep pump running steadily. Another way is to cut a small crease through the threads, where No. 62 is screwed into No. 52; so the air over piston will leak out in about two seconds and let pump go to work. If this crease gets stopped up with gum it should be cleaned out; after which governor should let pump go to work promptly. This crease is shown at "71" in cut of D-9 governor.

FOR TRAINMEN.

95. Q. When coupling the engine to an air brake train, equipped with quick action triples and already charged with air, which angle cock should be opened first?

A. The one on engine always, so as to fill the hose from engine. If cock on car is opened first, the train brake is liable to set with emergency action. Get in the habit of opening the cock on engine first, whether train is charged or empty.

96. Q. When coupling an empty car to other cars already charged and working, how should the angle cocks be opened?

A. Open the one on empty cars first, so the empty train pipe and empty hose will be connected. Then open the angle cock on the charged car slowly so the pressure in train pipe will not be reduced any faster than the engine can supply it. This will prevent the brakes setting on head end of train, which they will do with emergency action if angle cock is opened suddenly. A little practice will teach you the advantage of this.

97. Q. If an angle cock at head end of train is only partly opened or there is an obstruction in the train pipe, how will it affect the operation of the brake?

A. The brake can be set with service application, but it releases very slowly as the air does not get back fast enough to move all the triple valves to release promptly. With angle cock on tender partly open, you cannot always get the emergency action of the brake. When passing over the top of the train, angle cocks can be inspected, as they are generally far enough outside the end of car so the handles are visible from top of car. When cocks are wide open the handles are exactly over the hose. The old style plug shut-off cocks come in the straight pipe just under the end of car and cannot be seen when passing over the cars.

98. Q. Can an air brake train be made up so it will be impossible to get the emergency action of the brake from the engineer's brake valve?

A. If there are four cars with the brakes cut out at cross over near triples, or four cars with train pipe only or with plain triples, next to the engine, the reduction of air pressure in train pipe will be so gradual on the fifth car that you cannot get the emergency application of the quick action triples. It takes a sudden reduction at the first quick action triple to get the emergency. Switch the plain triples among the quick actions; you may need them to make a sudden stop in an emergency.

Note.—Questions 95, 96, 97 and 98 refer to quick action equipment. With the plain triple valves, such as are still used on many roads on their passenger equipment, it does not make very much difference which angle cock is opened first. It is, however, better to fill the hose in all cases from the engine or head end of train.

99. Q. How can the piston travel on a freight car be tested and then taken up the proper length when car is not charged with air and brake operated?

A. See that the push rod going from piston to brake cylinder lever is clear in against the bottom of piston sleeve. Make a mark on the push rod even with end of the sleeve. Set the

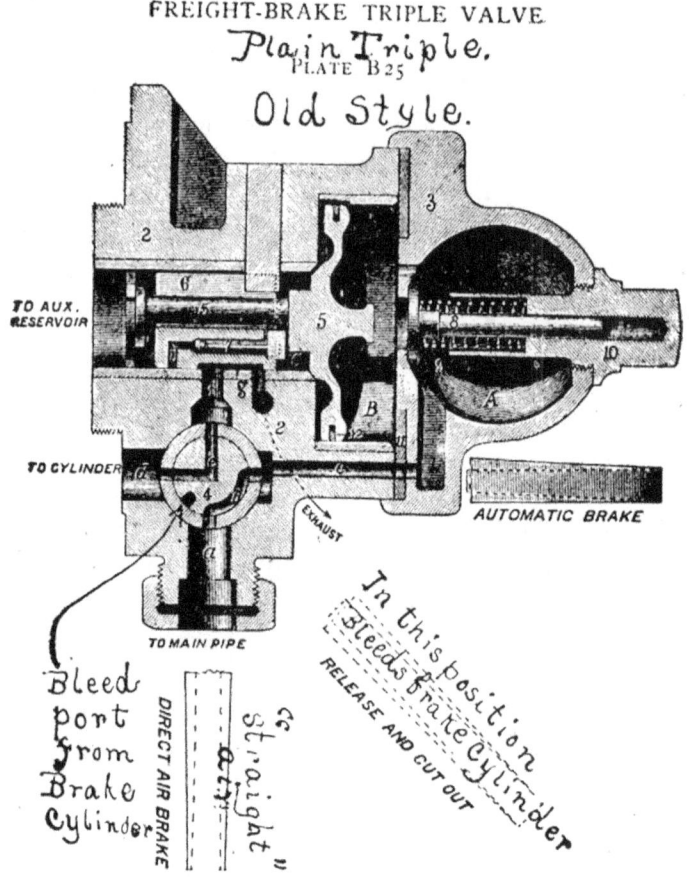

FREIGHT-BRAKE TRIPLE VALVE
Plain Triple.
PLATE B 25
Old Style.

brake by hand as tight as possible, with a club if necessary; the distance push rod is pulled out of sleeve is the piston travel. There is generally over an inch more piston travel when car is moving than when standing. The piston travel on an empty car may be very short, say 4 inches, and when loaded the same car may have 9 inches. When testing from the engine, have the brakes set with full service application, so you will get full piston travel.

100. Q. If the brake sets tight when you are charging the auxiliary reservoir with air when first coupling the hose to another car, should you cut out that brake?

A. If it is a quick action triple it is a sign that air leaks through some of the joints or valves in the triple into the brake cylinder. Have the engineer set and release the brake suddenly, once or twice; if there is dirt on the rubber seat of the emergency valve which causes the trouble, it will sometimes blow it off; if it does not make the brake work all right, cut it out and bleed it. With the freight brake there may be a leak in the pipe from the triple valve to the brake cylinder which passes through the auxiliary reservoir, nothing can be done on the road for a leak of this kind but cut out the brake. Most always in these cases the air blows out of exhaust port or at the pressure retaining valve. With the plain triple the plug cock in triple may be turned for "straight air." This will allow the air to go direct from train pipe to brake cylinder; none of it will come out of exhaust port, as the triple is cut out from train pipe and cylinder. In this case cut it in for automatic. If the handle is gone, examine the marks on the end of plug which show which way the openings are. If this plug cock leaks, the air can get past it from train pipe to brake cylinder. If brake will not work after one or two applications, cut it out. With all plain triples the brake should be released first, although the plain triple used on freight equipment is built to bleed the brake cylinder when brake is cut out. Sometimes this bleed hole, which is in one side of plug cock in freight triple valve, gets stopped up, in which case it may be necessary to let all the air out of the train pipe—set this triple for straight air which will bleed brake cylinder, after which cock in triple can be placed in cut out position.

101. Q. If the piston travel is too long or too short what affect does it have on the brake as to its holding power?

A. If it is too short it will not cover the leakage groove, and air will leak out of the cylinder; it must travel three inches to cover this groove. If it is too long it will strike the cylinder head, which will get the force instead of the brake shoes; it must travel twelve inches to do this. All brake pistons on coach, freight and tender equipment of standard gauge have twelve inch piston travel.

102. Q. If air blows past piston packing so freight brake leaks off, can it be fixed on the road?

A. Sometimes this is from want of oil in the cylinder; if the oiling plug near back cylinder head where it makes a joint with cast iron auxiliary reservoir is taken out and 4 or 5 table-spoonsful of black oil put in, it will soften the packing so it will be tight. This should be attended to by car inspectors, but is not always done. In no case should oil or water be put in the hose and be blown back into the triple with the air. It will carry the dust and sand in the pipe back towards the triple; this stops up the strainers, and if any gets by the strainers it spoils the rubber seat of the emergency valve, and cuts the triple to pieces very fast. Putting oil in the hose will destroy the efficiency of the brake in very short time.

103. Q. When air blows out steadily from the pressure retaining valve, should it be closed or left open?

A. Left open by all means. The air that blows out there comes from a leak in the triple valve; shutting the pressure retainer only stops air coming out there and sets the brake, or if leak is a small one, makes it go out through the leakage groove in the brake cylinder. If "pressure retainer is turned up," even if the brake does not set right off, it will stay set when engineer sets it and tries to let it off. Never turn up retainers unless you want to hold the brake set the next time engineer releases it. If the pressure retainer is broken off or the pipe leading to it from triple is broken or leaking bad that does not affect the operation of the brake in any way, except that the retainer cannot be used on that car; on nearly level roads they are not needed; you will notice that very few coaches have them, only those running on mountain roads. If retainer is broken off and pipe plugged, the brake cannot be released at all from the engine, as there is no way for air to escape from triple valve exhaust. If there is a leak in pipe from triple valve to retainer the retainer is of no use, as air will escape from the pipe at leak when retainer is set to work. Sometimes the pipe to pressure retainer gets stopped up so air cannot get through it, in which case the brake will set once and not release till bled off. It is not unusual to find nests of insects in the pipe right at retainer.

104. Q. How can the air signal whistle be operated from the cars most successfully?

A. By allowing just enough air to escape at the car discharge valve to reduce the air signal line pressure clear to signal valve on engine, so that it will operate promptly, then allowing car discharge valve to close and remain closed till signal line is recharged to standard pressure; this sometimes takes two seconds. The whistle will give only one blast if the car discharge valve is opened a second and third time before the whistle stops blowing the first time.

If you make a second and third reduction before the reducing valve on engine has had time to charge signal line to standard pressure, the second and third blasts of whistle will be very weak; in cold weather the reducing valves do not always work perfectly. Sometimes when a car discharge valve is opened, a sufficient amount of air will seem to blow out there, but on account of an obstruction near the train pipe Tee under car it does not reduce the pressure enough at the engine to operate the signal valve, and the whistle cannot be operated from that car when it works from other cars. If the whistle blows when engine is coupled to train and cannot be sounded afterward, look for a bad leak near rear end of train.

THE LOCOMOTIVE UP TO DATE. 661

If the whistle cannot be sounded from any cars back of a certain car, the cock in back end of that car is shut, or train pipe is stopped up so you cannot make a sudden enough reduction there to affect the signal valve on the engine.

If one blast of the whistle is used to start the train without using any additional signal, remember that one blast of the whistle can be given (without opening car discharge valve) when you do not want the train started. For instance, if the signal hose has been uncoupled (without the knowledge of the engineer) for any purpose; when the cock is opened enough air goes into empty hose to sound the whistle, giving signal to start the train before the man coupling hose can get out from between coach platforms. Other causes may cause the whistle to give one blast when not intended, therefore it is not always safe to use one blast of the whistle when standing still, to start the train.

Note.—Questions 2, 6, 7, 46, 48, 56, 63, 64, 65, 66, 67, 68, 69, 70, 71, 73, 74, 75, 76, 83, 84, 85, 86 of interest to trainmen also.

In addition to the questions for trainmen—95 to 104 inclusive—they should know about the subjects treated of in the following questions and be able to explain why.

Do you understand that all the air brake cars in a train should be connected and the train pipe charged with air whether brakes are cut in or not?

If the train pipe is broken on any car, what should be done?

After coupling an engine to an air brake car or train, that is not charged with air, how long do you wait before testing the brakes?

Why is it necessary to wait?

Where is it necessary to test the brake? Name all places.

How is this done? Give full description of your duties.

What is the proper piston travel?

How do you adjust it for that length?

If a brake is broken or disabled on a car, what should you do?

Must a brake be released before cutting it out?

Would you consider a train safe to leave with if the brake had been tested by opening the tail hose on the last car, if no other test had been made?

What effect does a leak in train pipe or hose have on the operation of the brake? When set from the engine? When car is cut off from engine?

What should be done in case of a bursted hose?

How is it located?

What should be done in case the train breaks in two?

How does the pressure retainer operate?

If the pressure retainer is broken off can the brake be operated on that car?

When and how should the conductor's valve be operated?

How is the air signal operated?

On an air braked passenger train, in case the engineer whistles for brakes, what is your duty?

If hand brakes are used on a part air train, on which cars should they be used?

Under what circumstances would you set the hand brakes on the rear cars of a part air train?

AIR BRAKE EXAMINATION QUESTIONS.

1. What style of power brakes is in general use on this road?
2. What are the essential parts of the automatic brake? What does each part do?
3. How should the pump be started and lubricated?
4. Should oil be put in through air inlets of 8 inch pump? Why?
5. How often should main reservoir and pipes be drained? Why?
6. If main reservoir is partly filled with water can the brake be set as readily as when it is drained of all water?
7. Can the brake be released as readily? Why?
8. Should you test the engine and tender equipment for leaks?
9. How do you test for leaks in main reservoir and pipes from pump to brake valve?
10. How for a leak in train line? In signal line?
11. How do you locate a leak that lets off the brake?
12. What pressure should you have before testing?
13. What controls the air pressure in the train line?
14. Is pump governor connected to train pipe with all kinds of brake valves?
15. What might prevent governor shutting off the steam and stopping pump when maximum pressure is obtained?
16. Where should you look for trouble if governor shuts off the pump at much less than standard pressure?
17. If the governor does not allow pump to start up at once when the pressure is reduced below that at which the governor is set, what should be done to the governor?
18. If too much oil is used in air end of pump, is it likely to gum up governor so it will not work accurately?
19. When an engine is left standing alone and the pump running, why must the brake valve not be left on lap?
20. If necessary to hold air brake set, how will you keep the air pressure from getting too high in main reservoir?
21. Is there any difference with the D-5 valve about these positions?
22. How many kinds of engineer's brake valves have we in service on this road?

51. Do you hear this blow when brake valve is used on direct application?

52. How do you tell if the brake valve is working properly?

53. When setting the brake can you tell about how many cars are coupled up with air by the amount of air escaping from the train pipe exhaust?

54. What is meant by "straight air"?

55. Can "straight air" be worked with quick action triple valve? Why?

56. How much do you reduce the train line pressure from 70 pounds to set the brake as tight as possible?

57. Why will this reduction do that?

58. What are the functions or uses of the triple valve?

59. How many forms in use on this road?

60. Where does all the air come from that enters the brake cylinder through the plain triple when setting the brake?

61. Does all the air that goes into the brake cylinder of a quick acting brake come from the auxiliary reservoir?

62. Please explain the action of the quick action triple when used on the "emergency."

63. Can you get the emergency action after a service application?

64. Does it take a sudden reduction of train pipe pressure right at the triple to work the "emergency," or will a slow heavy reduction do this? Why?

65. What is the function of the graduating valve?

66. Where is it located and how does it operate?

67. If the graduating valve leaks will the brake release as soon as the handle of brake valve is on lap with light application? Why?

68. Will it do the same with a full application? Why is this?

69. How does the quick action triple work if graduating pin is broken? Why?

70. How does the air get from the train pipe into the auxiliary?

71. Why is this port so small?

72. In what position of the triple is this port open?

73. How rapidly does an empty auxiliary charge up to 70 pounds?

74. How rapidly from 50 pounds up to 70 pounds?

75. What regulates the time of charging each sized auxiliary?

76. In making a service stop why should the brake valve handle not be moved past the positions for service application?

77. Is this movement of brake valve liable to kick off brakes on head end of train? Why?

78. What is the proper position to place brake valve handle in after releasing brake, if the brake is to be set again immediately? Why?

23. Explain the principle the engineer's equalizing discharge brake valve operates on.

24. Describe its operation to set the brake with service, or emergency applications, and in releasing the brake.

25. Where is the feed valve, and excess spring and valve located in the equalizing discharge valve?

26. Where in the old style brake valve named B-11?

27. Why is it necessary to keep these valves clean?

28. Why is excess pressure necessary?

29. What amount is necessary?

30. How do you regulate the amount of excess pressure with D-8 valve? With D-5 valve?

31. Is more excess needed to let off all the brakes promptly on a long train than on a short train?

32. Do you consider a cut rotary valve or seat dangerous? Why?

33. Will using the valve in emergency instead of service application cause the seat to cut any quicker? Why?

34. Name the different positions of the equalizing discharge valve.

35. What ports are open and what ports are closed in each position.

36. Is equalizing port open in all positions of the brake valve? Why?

37. Is this port the connection from train pipe to black hand?

38. Is it ever open to main reservoir pressure?

39. With D-8 valve why does the black hand show the same pressure as the red hand with valve on emergency?

40. Does the D-5 valve show the pressure in this way? Why?

41. What are the positions of the old brake valve?

42. Why is the equalizing discharge valve better than the old brake valve?

43. If the equalizing piston becomes gummed or sticks, how will it affect its operation, and setting the brakes?

44. What is the purpose of the small reservoir that is connected to the equalizing discharge valve?

45. If the pipe leading from valve to small reservoir is broken off or leaking badly, what will you do?

46. Where is the first air taken from in making a service stop? What port does it blow out of?

47. Where next does it come from? Where next?

48. Why does the air blow out of the train pipe exhaust of the equalizing valve just after releasing the brake?

49. Do you always hear it when releasing the brake on engine and tender only?

50. Do you hear it when releasing the brake on a train of over two cars?

THE LOCOMOTIVE UP TO DATE. 665

79. Why is it dangerous to apply and release the brake repeatedly in making one station stop?

80. Do you understand that brake cylinders have leakage grooves?

81. Where are they located and how long are they?

82. What are leakage grooves provided for?

83. Do you allow for them when setting the brake? How?

84. As a rule how much reduction in train pipe pressure is necessary to force piston past leakage groove?

85. What should be done after coupling to a train and before pulling out?

86. What pressure should you have in train pipe and auxiliaries before testing the brake?

87. How do you know you have 70 pounds in the auxiliaries?

88. What tests of air brakes are called for by our rules?

89. What is the only reliable method of testing a train? Do you call this a terminal test?

90. Why is a terminal test absolutely necessary?

91. At what other times should this test be made?

92. If a brake is broken or disabled, how will you prevent it from operating on that car and let it work on other cars?

93. How do you cut out the brake on engine or tender?

94. Is it necessary to release the brake before cutting it out?

95. What is your understanding about the piston travel; if it is too long or too short?

96. What is the proper piston travel?

97. How is the slack taken up to secure this adjustment?

98. At what travel should the driver brake piston be adjusted?

99. What danger is there if cam driver brake piston has too much travel?

100. What precautions are necessary when taking up slack of cam brake?

101. How is slack of a six wheel brake taken up?

102. What is necessary in order to have all brakes work alike?

103. When brakes go on suddenly and are not operated by the engineer, what should you do?

104. To what causes would you assign this?

105. If air brake train broke in two, how do you proceed to get train ready to go ahead again?

106. How do you proceed in case of a bursted hose, and how can you help trainmen to locate it?

107. Would it be necessary in these cases to make a terminal test?

108. If there was a steady leak from exhaust port of triple valve what is likely to be the trouble?

109. What is the pressure retaining valve and what is its use? How is it operated?

110. How many pounds of air is the retainer intended to close up on and hold in brake cylinder?

111. Does the brake let off any slower till it gets down to this pressure? How is this done?

112. In descending a grade how can you best keep a train under control?

113. What precautions must be observed in making a stop with a "part air" brake freight train?

114. In making a stop with a freight train when would you let off the brake to make a smooth stop? Why?

115. When with a passenger train? Why?

116. When two or more engines are coupled together which one should do the braking?

117. Can both engineers use the brake at the same time safely?

118. How will you proceed to give the leading engineer complete control of the train?

119. What should the other engineer do?

120. If there is no cut out cock under brake valve, what should you do?

THE AIR BRAKE AND SIGNAL INSTRUCTIONS.

Latest Revised List Submitted by the Joint Committee of the Master Car Builders' Association and the American Railway Master Mechanics' Association.

GENERAL INSTRUCTIONS.

The following rules and instructions are issued for the government of all employés of this railroad whose duties bring them in contact with the maintenance or operation of the automatic air brake and train air signal. They must be obeyed in all respects, as employés will be held responsible for the observance of the same, as strictly as for the performance of any other duty.

Every employé, whose duties are connected in any way with the operation of the air brake, will be examined from time to time, as to his qualification for such duties, by the Inspector of Air Brakes or other person appointed by the proper authority, and a record will be kept of such examination.

INSTRUCTIONS TO ENGINEMEN.

General.—Engineers, when taking their locomotives, must see that the air brake apparatus, on locomotive and tender, is in good working order; that the air pump and lubricator work properly; that the regulator prevents the train pipe pressure exceeding a maximum of 70 pounds; that an excess pressure of not less than 20 pounds can be maintained in the main reservoir when the handle of the engineer's brake valve is placed in position 2 (Running Position); that the engineer's brake valve works properly in all different positions of the handle; and that, when the brakes are fully applied, the driver brake pistons do not travel less than one-third nor more than two-thirds of their stroke, and the tender brake piston does not travel less than five nor more than eight inches.

Engineers must report to round house foreman, at the end of the run, any defect in the air brake or signal apparatus.

Making Up Trains and Testing Brakes.—The train pipe under the tender must always be blown out thoroughly before connecting to the train. Be sure to have 70 pounds train pipe pressure on the engine, with the handle of the engineer's valve standing in position 2, before connecting to the train.

When the locomotive has been coupled to the train and it has been charged with an air pressure of 70 pounds, the engineer shall, at a signal from the inspector or trainman, apply the brakes with full service application of not less than 20 pounds reduction, and leave them so applied until the brakes on the entire train have been inspected and the signal is given to release. He shall then release the brakes, and shall not leave the station until it has been ascertained that all brakes are released and he has been informed by the inspector or trainmen that the brakes operate all right. This test must be made after each change in the make-up of the train, and before starting down such grades as may be designated by special instructions. Where the train air signal is used, the signal to release the brakes, in testing, will be given from the rear car of the train, to show that the signal connections have been properly made.

Service Application.—In applying the brakes to steady the train upon descending grades, or for reducing the speed for any purpose, be very careful not to make too great a reduction of pressure in the outset, as the speed of the train will be too quickly or too much checked, and it will be necessary to release the brakes and apply them again later, perhaps repeating the operation. *Apply the brakes lightly at a sufficient distance from the stopping point, and increase the braking force gradually, as is found necessary, so as to make the stop with one application, or at most two applications of the brakes.*

With freight trains first allow the slack to run up against the engine. Great care must then be taken to apply the brakes with five to seven pounds reduction and not make a second reduction until the effect of the first reduction is felt on entire train, in order to prevent shocks which otherwise may be serious.

In making a service stop with a passenger train, *always release the brakes a short distance before coming to a dead stop,* except on heavy grades, to prevent shocks at the instant of stopping. Even on moderate grades, it is best to do this, and then, after release, to apply brakes lightly, to prevent the train starting, so that when ready to start, the release will take place quickly. This does not apply to freight trains, upon which the brakes must not be released until the train has stopped.

Emergency Applications.—The emergency application of the brakes must not be used, except in actual emergencies.

Brakes Applied from an Unknown Cause.—If it is found that the train is dragging at any time without a rapid fall of the black pointer, move the handle of the engineer's valve into the full release position for a few seconds, and then return it to the running position.

If, however, the brakes go on suddenly, with a fall of the black pointer, it is evidence that (*a*) a conductor's valve has been opened, (*b*) a hose has burst or other serious leak has occurred, or (*c*) the train has parted.

In such an event, place the handle immediately in position 3, lap position, to prevent the escape of air from the main reservoir, and leave it there until the train has stopped, the brake apparatus has been examined and a signal to release is given.

Braking by Hand.—*Never use the air brake* when it is known that the trainmen are operating the brakes of the air brake cars by hand, as there is danger of injury to the trainmen by so doing.

Cutting Out Brakes.—*The driver and tender brakes must always be used automatically at every application of the train brakes,* unless defective—except upon such grades as shall be designated by special instructions.

When necessary to cut out either driver or tender brake, on account of defects, it shall be done by turning the handle of the four-way cock in the triple valve down to a position midway between a horizontal and a vertical position, first releasing the brake and leaving the bleed cock open. With the special driver brake triple valve, close the cut out cock in the branch pipe.

Double Headers.—When two or more engines are coupled in the same train the brakes must be connected through to and operated from the head engine. For this purpose a cock is placed in the train pipe just below the engineer's valve. The engineer of each engine, except the head one, must close this cock and place the handle of the engineer's valve in position 2. He will start his air pump and let it run, as though he were going to use the brake, for the purpose of maintaining air pressure on his engine and enabling him to assume charge of the train brakes should occasion require it.

An Extra Air Brake Hose and Coupling must always be carried on the engine for repairs in case of a burst hose. Upon engines having the air signal a signal hose and coupling must also be carried for the same purpose.

INSTRUCTIONS TO TRAINMEN.

Making Up Trains and Testing Brakes.—When the engine has been coupled to the train, or when two sections have been coupled together, the brake and signal couplings must be united, the cocks in the train pipes—both brake and signal—must all be open, except those at the rear end of the last car, which must be closed, and the hose hung up properly in the dummy coupling, when cars are so equipped.

After the engineer has charged the train with air he must then be signaled to apply the brakes, as provided for in the train rules. When he has done so the brakes of each car must be examined to see if they are properly applied. When it is ascertained that each brake is applied the engineer must be signaled to release the brakes. When the train air signal is to be used the signal to the engineer to release the brakes must be given by means of the air signal from the rear car of the train. The brakes of each car must then be examined to see that each is released.

If any defect is discovered it must be remedied and the brakes tested again—the operation being repeated until it is ascertained that everything is right. The conductor and engineer must then be notified that the brakes are all right. This examination must be made every time any change is made in the make-up of the train and before starting down such grades as may be designated by special instructions. At points where there are no inspectors trainmen must carry out these instructions. No passenger train must be started out from an inspection point with the brakes upon any car cut out or in a defective condition without special orders from the proper officers. The air brakes must not be alone relied upon to control any freight train with a smaller proportion of cars with the air brake in service than the division time card specifies. When hand brakes are also used they must be applied upon those cars next behind the air-braked cars.

Detaching Engine or Cars.—First close the cocks in the train pipes at the point of separation, and then part the couplings, invariably by hand. If the brakes have been applied do not close the cocks until the engineer has released the brakes upon the whole train.

Couplings Frozen.—If the couplings are found to be frozen together or covered with an accumulation of ice, the ice must first be removed and then the couplings thawed out by a torch to prevent injury to the gaskets.

Brakes Sticking.—If brakes are found sticking, the engineer must be signaled "brakes sticking," as provided for in train rules. If the engineer cannot release the brakes, or if the brakes are applied to detached cars, the release may be effected by opening the bleed cock in the auxiliary reservoir until the air begins to release through the triple valve, when the reservoir cock must immediately be closed.

Train Breaking into Two or More Parts.—First close the cock in the train pipe at the rear of the first section and signal the engineer to release the brakes. Having coupled to the second section, observe the rule for making up trains—first being sure that the cock in the train pipe at the rear of the second section has been closed, if the train has broken into more than two sections. When the engineer has released the brakes on the second section the same method must be employed with reference to the third section, and so on. When the train has been once more entirely united the brakes must be inspected on each car to see that each is released before proceeding.

Cutting Out the Brake on a Car.—If, through any defect of the brake apparatus while on the road, it becomes necessary to cut out the brake upon any car, it may be done by closing the cock in the cross-over pipe near the center of the car where the quick-acting brake is used, or by turning the handle of the cock in the triple valve to a position midway between a horizontal and vertical where the plain automatic brake is used, first releasing the brake. When the brake has been thus cut out, the cock in the auxiliary reservoir must be opened and left open upon passenger cars, or held open until all the air has escaped from the reservoir upon freight cars. *The brake must never be cut out upon any car unless the apparatus is defective,* and when it is necessary to cut out a brake the conductor must notify the engineer and also send in a report stating the reason for so doing.

Conductor's Valve.—Should it become necessary to apply the brakes from the train, it may be done by opening the conductor's valve, placed in each passenger equipment car. *The valve must be held open until the train comes to a full stop, and then must be closed again.*

This method of stopping the train must not be used except in case of emergency.

Burst Hose.—In the event of the bursting of a brake hose, it must be replaced and the brakes tested before proceeding, provided the train be in a safe place. If it is not, the train pipe cock immediately in front of the burst hose must be closed, and the engineer signaled to release. All the brakes to the rear of the burst hose must then be released by hand, and the train must then proceed to a safe place where the burst hose must be replaced and the brakes again connected and tested as in making up a train.

Brakes Not in Use.—When the air brakes are not in use, either upon the road or in switching, the hose must be kept coupled between the cars or hung up properly to the dummy couplings, when cars are so equipped.

Pressure Retaining Valve.—When this valve is to be used,

the trainmen must, at the top of the grade, test the brakes upon the whole train, and must then pass over the train and turn the handles of the pressure retaining valves horizontally (position 2) upon all or a part of the cars, as may be directed. At the foot of the grade, the handles must all be turned downward again. (Position 1.) Special instructions will be issued as to the grades upon which these valves are to be used.

Train Air Signal.—In making up trains, all couplings and car discharge valves on the cars must be examined to see if they are tight. Should the car discharge valve upon any car be found to be defective while on the road, it may be cut out of use upon that car by closing the cock in the branch pipe leading to the valve. The conductor must always be immediately notified when the signal has been cut out upon any car, and he must report the same for repairs.

In using the signal, pull directly down upon the cord during one full second, for each intended blast of the signal whistle, and allow two seconds to elapse between the pulls.

Reporting Defects to Inspectors.—Any defect in either the air brake or air signal apparatus discovered upon the road must be reported to the inspector at the end of the run; or, if the defect be a serious one in passenger service, it must be reported to the nearest inspector, and it must be remedied before the car is again placed in service.

INSTRUCTIONS TO ENGINE HOUSE FOREMEN.

General.—It is the duty of engine-house foremen to see that the air brake and signal equipment is properly inspected upon each engine after each run. It must be ascertained that all pipe joints, connections and all other parts of the apparatus are air tight, and that the apparatus is in good working order.

Air Pump.—The air pump must be tested under pressure, and if found to be working imperfectly in any respect, it must be put into thoroughly serviceable condition.

Pump Governor.—The pump governor should cut off the steam supply to the pump, when the train pipe pressure has reached 7 pounds with D-8 brake valve, and at 90 pounds main reservoir pressure with D-5, E-6 or F-6 valve. If it does not, it must be regulated to do so.

Engineer's Brake Valve.—This valve must be kept clean and in perfect order. With the handle in position 2, the main reservoir pressure must not be less than 20 pounds greater than train pipe pressure. The valve must be tested with the handle in positions 4, "service application," and 3, "lap position," to note that the equalizing piston responds promptly, and that there are no leaks from port to port under the rotary disc valve.

Adjustment of Brakes.—The driver brakes must be so adjusted that the pistons travel not less than one-third nor more than two-thirds of their stroke. When the cam brake is used care must be taken to adjust both cams alike, so that the point of contact of the cams shall be in line with the piston rod. The tender brake must be adjusted by means of the dead truck levers, so that the piston travels not less than five nor more than six inches when the air brake is applied and the hand brake is released. This adjustment must be made whenever the piston travel is found to exceed eight inches.

Brake Cylinders and Triple Valves.—These must be examined and cleaned once every six months, and the cylinders oiled once in three months. If the driver brake cylinders are in a position to be affected by the heat of the boiler, they must be oiled more frequently. A record must be kept of the dates of last cleaning and oiling for each engine.

Draining.—The main reservoir, and also the drain cup in the train pipe under the tender must be drained of any accumulation after each trip. The auxiliary reservoirs and triple valves must also be drained frequently, and daily in cold weather, and the train pipe under the tender blown out.

Air Signal.—The train air signal apparatus must be examined and tested by opening and closing the cock in the signal pipe, at the rear of the tender, to see that the whistle responds properly. A pressure gauge must be applied to the air signal pipe, once each month, to ascertain that the reducing valve maintains the proper pressure of 40 pounds per square inch in the train signal pipe.

INSTRUCTIONS TO INSPECTORS.

General.—It is the duty of all inspectors to see that the couplings, the pipe joints, the conductor's valves, the air signal valves, and all other parts of the brake and signal apparatus are in good order and free from leaks. For this purpose they must be tested under the full air pressure as used in service. No passenger train must be allowed to leave a terminal station with the brake upon any car cut out, or in a defective condition, without special orders from the proper officer.

If a defect is discovered in a brake apparatus of a freight car, which cannot be held long enough to give time to correct such defect, the brake must be cut out and the car properly carded, to call the attention of the next inspector to the repairs required.

The division time card rules specify the smallest proportion of freight cars, with the air brakes in good condition, which may be used in operating the train as an air brake train.

Making Up Trains and Testing Brakes.—In making up trains, the couplings must be united and the cocks at the ends of the cars all opened, except at the rear end of the last car, where the cocks must be closed and the couplings properly hung up to the dummy couplings. After the train is charged, the engineer must be signaled to apply the brakes. When the brakes have been applied, they must be examined upon each car to see that they are applied with proper piston travel. This having been ascertained, the inspector must signal the engineer to release the brakes, using the train air signal from the rear car, upon passenger trains. He must then again examine the brakes upon each car to note that each is released. If any defect is discovered, it must be corrected and the testing of the brakes repeated, until they are found to work properly. The inspector must then inform both the engineer and conductor of the number of cars with brakes in good order. This examination must be repeated if any change is made in the make-up of the train before starting.

Cleaning Cylinders and Triple Valves.—The brake cylinders and triple valves must be kept clean and free from gum. They must be cleaned and oiled as often as once in six months, upon passenger cars, and once in twelve months upon freight cars. The dates of last cleaning and oiling must be marked with white paint upon the cylinder in the places left for such dates opposite the words, which will be stenciled with white paint, in one inch letters, upon the cylinder or reservoir, as follows:

Cylinder Cleaned and Oiled.............................

Triple Cleaned and Oiled.............................

The triple valve and auxiliary reservoir must be frequently drained, especially in cold weather, by removing the plug in the bottom of the triple valve and opening the small cock in the reservoir.

Adjustment of Brakes.—The slack of the brake shoes must be taken up by means of the dead truck levers.

In taking up such slack, it must be first ascertained that the hand brakes are off, and the slack is all taken out of the upper connections, so that the live truck-levers do not go back within $1\frac{1}{2}$ inches of the truck timber or other stop, when the piston of the brake cylinder is fully back at the release position. When, under a full application, the brake piston travel is found to exceed eight inches upon a passenger car or nine inches upon a freight car, the brake shoe slack must be taken up and the adjustment so made that the piston shall travel not less than five inches nor more than six inches.

Braking Power.—Where the cylinder lever has more than one hole at the outer end, the different holes are for use upon cars of different weights.

It must be carefully ascertained that the rods are connected to the proper holes, so that the correct braking power shall be exerted upon each car.

Repair Parts.—Inspectors must keep constantly on hand for repairs a supply of all parts of the brake and signal equipment that are liable to get out of order.

Hanging Up Hose.—Inspectors must see that, when cars are being switched or standing in the yard, the hose is coupled between the cars or properly secured in the dummy coupling, when cars are so equipped.

Responsibility of Inspectors—Inspectors will be held strictly responsible for the good condition of all the brake and signal apparatus upon cars placed in trains at their stations; they will also make any examination of brake apparatus or repairs to the same, which they may be called upon to do by trainmen.

GENERAL QUESTIONS
Regarding the use of the
AIR BRAKE AND TRAIN SIGNAL.

(All parties who have to do with the use, adjustment, care or repairs of air brakes should be thoroughly examined on these questions, in addition to the special questions for each class of men following them.)

1. Question. What is an air brake?
Answer. It is a brake applied by compressed air.
2. Q. How is the air compressed?
A. By an air pump on the locomotive.
3. Q. How does the compressed air apply the brakes?
A. It is admitted into a brake cylinder on each car, and it pushes out a piston in that cylinder which pulls the brake on.
4. Q. How does the piston get back when the brakes are released?
A. There is a spring around the piston rod which is compressed when the brakes are applied, and when the air is allowed to escape to release the brakes, this spring reacts and pushes the piston in again.
5. Q. Where is the compressed air kept ready for use in the automatic air brake?
A. In the main reservoir on the locomotive, in the smaller or auxiliary reservoir on each car, and in the train pipe.
6. Q. Where does the compressed air come from directly that enters the brake cylinder when the automatic brake is applied?
A. It comes from the auxiliary reservoir on each car in service application and from the auxiliary reservoir and train pipe in emergency application.
7. Q. How does it get into the auxiliary reservoir?
A. It is furnished from the main reservoir on the locomotive through the train pipe and triple valve when the brakes are released.
8. Q. How is the automatic brake applied and released?
A. The automatic brake is applied by reducing the air pressure in the train pipe below that in the auxiliary reservoir, and

is released by raising the train pipe pressure above that remaining in the auxiliary reservoir.

9. Q. Why does the compressed air not enter directly into the brake cylinder from the train pipe?

A. Because the triple valve used with the automatic brake prevents the air from entering directly from the train pipe to the brake cylinder when the pressure in the train pipe is maintained or increased.

10. Q. What other uses has the triple valve?

A. It causes the brake cylinder to be opened to the atmosphere under each car, and releases the brakes when the pressure in the train pipe is made greater than that in the auxiliary reservoir, and it opens communication from the train pipe to the auxiliary reservoir by the same movement; when the pressure in the train pipe is reduced, it closes the openings from the train pipe to the auxiliary reservoir and from the brake cylinder to the atmosphere, and then opens the passage between the auxiliary reservoir and the brake cylinder by the same movement, so as to admit the air and apply the brakes.

11. Q. How many forms of triple valves are there in use, and what are they called?

A. Two; the plain triple and the quick-acting triple.

12. Q. How can you tell the plain triple from the quick-acting triple?

A. The plain triple has a four-way cock in it, with a handle for operating the cock; the quick-acting triple has no such cock in it, but there is a plug cock in the cross-over pipe leading from the train pipe to the triple, when the quick-acting triple is used.

13. Q. What are these cocks for in both cases?

A. They are to be used to cut out brakes on one car, without interfering with other brakes on the train, if the brake on that car has become disabled.

14. Q. How does the cock handle stand in the plain triple valve when the pipe is open for automatic action?

A. It stands in a horizontal position.

15. Q. In what position does the same handle stand when the brakes are cut out by closing the cock?

A. It stands at an inclined position midway between horizontal and vertical.

16. Q. How does the handle of the plug cock in the cross-over pipe, used with the quick-acting triple, stand for automatic action?

A. It stands with the handle cross-wise with the pipe, and the cock is then open.

17. Q. How does it stand when the cock is closed and the brake cut out of action?

A. It stands with handle lengthwise of cross-over pipe.

18. Q. How is the train pipe coupled up between the cars?

A. By means of a rubber hose on each end of the train pipe, fitted with a coupling at the loose end.

19. Q. How is the train pipe closed at the rear end of the train?

A. By closing the cock in the train pipe at the rear end of the last car.

20. Q. How many such train pipe cocks are there to a car, on the air brake train pipe and on the air signal train pipe, and why?

A. Two for each pipe on each car, because either end of any car may sometimes be at the rear end of the train.

21. Q. How many kinds of train pipe cocks are there in use at the ends of the cars?

A. Two.

22. Q. Describe each and give the position of the handles for opened and closed in each case?

A. The older form of train pipe cock is a straight plug cock in the train pipe, not far from the hose connection; the handle stands crosswise with the pipe when it is open, and lengthwise with the pipe when closed; it is now found principally on the air signal pipe. The other form of train pipe cock now used on the air brake pipe is an angle cock placed at the end of the train pipe and close to the hose. The handle of the angle cock stands lengthwise with the pipe when open, and crosswise with the pipe when closed.

23. Q. What uses have these train pipe cocks besides to close the pipe at the rear end of the train?

A. They are to be used to close the train pipe at both sides of any hose coupling which is to be parted, as when the train is cut in two.

24. Q. Why is it necessary to close the train pipe on both sides of the hose coupling before it is parted?

A. To prevent the escape of air from the train pipe which would apply the brakes.

25. Q. How must the hose coupling be parted when it is necessary to do so, and why?

A. The air brake must first be released on the train from the engine, then the adjacent train pipe cocks must both be closed and the coupling must be parted by hand, to prevent the possibility of injury to the rubber gasket in the coupling.

26. Q. Why must the brakes be fully released before uncoupling the hose between the cars?

A. Because if the brakes are applied upon a detached car, they cannot be released without bleeding the auxiliary reservoir, and thus wasting air.

27. Q. In coupling or uncoupling the hose between cars, what must be done if there is ice on the couplings?

A. The ice must first be removed and the couplings thawed

out, so as to prevent injury to the rubber gaskets in uncoupling, and to insure tight joints in coupling the hose.

28. Q. What must be done with a hose coupling which is not coupled up, such as the rear hose of a train, or any hose on a car which is standing or running, but not in use?

A. It must be placed in the dummy coupling if provided for, in such manner that the flat pad on the dummy will close the opening in the coupling.

29. Q. What pressure should be carried in the train pipe and auxiliary reservoir?

A. Seventy pounds pressure to the square inch.

30. Q. Why should the pressure be 70 pounds?

A. Because this pressure is necessary, to get the full braking force which each car is capable of using, and, if it be exceeded, there will be danger of sliding the wheels.

31. Q. How much pressure can be obtained in the brake cylinder by the service application of the brakes with 70 pounds in the auxiliary reservoir?

A. About 50 pounds pressure to the square inch, with an eight-inch piston travel.

32. Q. Why can only 50 pounds pressure be obtained under these circumstances?

A. Because the air, at 70 pounds pressure in the auxiliary reservoir expands into an additional space when the auxiliary reservoir is opened to the brake cylinder, and when the pressure has become equalized, it is thus reduced to 50 pounds.

33. Q. How much must the train pipe pressure be reduced, in order to get 50 pounds pressure in the brake cylinder, in ordinary service?

A. Twenty pounds; or from 70 pounds down to 50 pounds in the train pipe also.

34. Q. Can the brakes be applied so as to get only a portion of this 50 pounds pressure in the brake cylinder, and how?

A. They can be so applied by reducing the train pipe pressure less than 20 pounds.

35. Q. If the train pipe pressure be reduced 10 pounds what will the pressure be in the brake cylinder?

A. About 25 pounds.

36. Q. How is this graduated action obtained?

A. By means of the graduating valve in the triple valve.

37. Q. Is it important to keep all the air brake apparatus tight and free from leaks?

A. Yes.

38. Q. Why is this important?

A. In order to get full service from the air brakes, and to prevent the waste of air, and also to prevent the brake applying automatically by reason of leak in the train pipe.

39. Q. Is it important to know that the train pipe is open

throughout the train and closed at the rear end before starting out?

A. Yes, this is very important.

40. Q. Why is this very important?

A. Because if any cock in the train pipe were closed, all the brakes back of the cock which is closed would be prevented from working.

41. Q. How can you know that the train pipe cocks are all open when the train is made up?

A. By testing the brakes; that is, by applying and releasing them, and observing whether they all operate.

42. Q. Do you understand then no excuse will be acceptable for starting out the train without first testing the air brakes?

A. Yes.

43. Q. Why is this rule absolute?

A. Because the safety of passengers and of property depends upon the brakes being properly coupled up and in an operating condition before the train is started.

44. Q. At what other times should the brakes be tested, and how?

A. After each change in the make-up of the train and before starting the train down certain designated grades, and the test should be made with a full service application of the brakes.

45. Q. How much air pressure should be carried in the air signal train pipe?

A. Forty pounds pressure.

46. Q. Is it important that this train pipe and its connections be also kept tight?

A. Yes.

47. Q. After taking up the slack of the brake shoes, how far should the brake piston travel in the cylinders on cars and tenders with a full application of the brake?

A. Not less than five inches nor more than six inches.

48. Q. What would happen if the piston traveled less than five inches when brakes are fully applied?

A. A partial application of the brakes might not close the leakage groove in the brake cylinder provided for the escape of small amounts of air.

49. Q. Why should the piston travel not be permitted to exceed eight inches on passenger cars and tenders, or nine inches on freight cars?

A. Because, if it travels further than this when sent out, a little wear of the brake shoes will cause the piston to travel far enough to rest against the back cylinder head when the brakes are applied, and this cylinder head would then take the pressure instead of its being brought upon the brake shoes.

50. Q. How far should the driver brake piston travel with a full application of the brakes, and why?

A. Not less than one-third of the full stroke of the piston nor more than two-thirds of its full stroke, for reason similar to those given for cars and tenders.

51. Q. If the brakes stick upon any car so that the engineman cannot release them at any time, how should they be released?

A. By opening the release cock in the auxiliary reservoir and holding it open until the air begins to escape from the triple valve and then closing it again.

52. Q. What is the pressure retaining valve, and what is its use?

A. The pressure retaining valve is a small valve placed at the end of a pipe from the triple valve, through which the exhaust takes place from the brake cylinder. It is used to retard the brake release on heavy grades and hold the brakes partially applied, so as to allow more time for the engineman to recharge the auxiliary reservoir.

53. Q. What precautions are necessary on every train in regard to hose couplings?

A. Every train must carry at least two extra hose and couplings complete, for use in replacing any hose couplings which may fail or become disabled. These extra hose and couplings to be carried on such part of the train as is required by the rules and regulations.

SPECIAL FOR ENGINEMEN.

54. Q. How should the air pump be started?
A. It should be started slowly, so as to allow the condensation to escape from the steam cylinder and prevent pounding, which is more likely to occur when the air pressure is low.

55. Q. Why should the piston rod on the air pump be kept thoroughly packed?
A. To prevent the waste of air and steam.

56. Q. How should the steam cylinder of the air pump be oiled, and what kind of oil should be used?
A. It should be oiled as little as necessary through a sight feed lubricator, and cylinder oil should be used.

57. Q. How should the air cylinder of the air pump be oiled; what kind of oil, and why?
A. It should be oiled very little by once filling the oil-cup with West Virginia well oil daily. Cylinder oil, lard oil and other animal or vegetable oils must not be used, as their use causes the engineer's brake valve and the triple valves to gum up. The oil must never be introduced through the air inlet ports, as the practice would cause the pump valves to gum up.

58. Q. What regulates the train pipe pressure?
A. The pump governor with D-8 valve and the feed valve attachment with the D-5, E-6 or F-6 valve.

59. Q. Why should the train pipe pressure not exceed 70 pounds?

A. Because 70 pounds train pipe pressure produces the strongest safe pressure of the brake shoes upon the wheels. A higher train pipe pressure is liable to cause the wheels to slide.

60. Q. Why is the equalizing engineer's valve better than the older forms?

A. Because it enables the engineer to apply the brakes more uniformly throughout the train, and with less shock to the train, especially when the quick-acting triple valves are used. It also prevents the brakes from being kicked off on the forward end of the train when the engineer closes the valve after applying the brakes.

61. Q. Why does the equalizing engineer's valve produce these results in ordinary service stops?

A. Because the engineer does not, in such cases, open the train pipe to the atmosphere direct, but he only reduces the air pressure above the piston in the engineer's valve, which causes that piston to open the train pipe to the atmosphere, and to close the opening gradually when the train pipe pressure has been correspondingly reduced.

62. Q. What does the excess pressure valve in the D-8 brake valve accomplish, and do you regard it important to have it working properly?

A. It maintains an excess pressure of about 20 pounds in the main reservoir above the pressure in the train pipe, and it is important that it be kept clean and in working order so as to have this excess pressure to insure release, and for use in recharging the train quickly after the brakes are released.

63. Q. What does the feed valve attachment of the D-5, E-6 or F-6 engineer's valve accomplish?

A. When properly adjusted it restricts the train pipe pressure to a maximum of 70 pounds with the engineer's valve in running position. When this valve is used the pump governor is attached to main reservoir pressure and may be set to carry whatever pressure is desired therein.

64. Q. How often should the brake valve be thoroughly cleaned and oiled?

A. At least once every two months.

65. Q. If the rotary disk valve in the engineer's valve is unseated by dirt or by wear, what may be the result, and what should be done?

A. It may be impossible to get the excess pressure; when the brakes have been applied they may keep applying harder until full on, or when they have been applied they may go off. The rotary disk valve should be thoroughly cleaned, and if worn it should be faced and ground to a seat.

66. Q. If the piston in the engineer's valve becomes

gummed up or corroded from neglect to clean it, what will be the result?

A. It will be necessary to make a large reduction of pressure through the preliminary exhaust port before the brakes will apply at all, and then the brakes will go on too hard, and will have to be released.

67. Q. When the engine is standing alone and the pump is running, why must the D-8 engineer's valve not be left standing in the lap position (No. 3)?

A. Because the main reservoir pressure may become so high that, when the handle of the engineer's valve is again placed in the release position, it will cause the train pipe and tender auxiliary reservoir to be charged with too high pressure, which might injure the adjustment of the pump governor as well as cause the tender wheels to be slid with the first application of the brakes.

68. Q. How and why should the train pipe under the tender always be blown out thoroughly before connecting up to the train?

A. By opening the angle cock at the rear end of the tender and allowing the air from the main reservoir to blow through. This blows out the oil, water, scale, etc., which may accumulate in the pipe, and which would be blown back into the train pipe and triple valves if not removed before coupling to the train.

69. Q. When the engine is coupled to the train, why is it necessary to have the full train pipe pressure and the excess pressure on the main reservoir?

A. So that the brakes will all be released and the train quickly charged when the engineer's valve is placed in the release position.

70. Q. Why should the driver brakes always be operated automatically with the train brake?

A. Because it adds greatly to the braking force of the train, and the brakes can be applied alike to all the wheels for ordinary stops, and in an emergency the greatest possible braking force is at once obtained by one movement of the handle.

71. Q. In making a service application of the brakes, how much reduction of the train pipe pressure from 70 pounds does it require to get the brakes full on?

A. About 25 pounds reduction.

72. Q. What should the first reduction be in such an application, and why?

A. From five to seven pounds, so as to insure moving the pistons in the brake cylinders past the leakage groove, yet not apply the brakes too hard, until the slack in drawbars and drawsprings is first taken up.

73. Q. What is the result of making a greater reduction of pressure than 25 pounds?

A. A waste of air in the train pipe, without getting any

more braking force, and therefore requiring more air to release the brakes.

74. Q. How many applications of the brakes are necessary in making a stop?

A. Generally only one; by applying them lightly at first with five or seven pounds reduction of air in the train pipe, and afterward gradually increasing the force of the application. Two applications are as many as should ever be required.

75. Q. Why is it dangerous to apply and release the brakes repeatedly in making stops?

A. Because every time the brakes are released the air in the brake cylinders is thrown away, and, if it is necessary to apply them again before sufficient time has elapsed to recharge the auxiliary reservoirs, the application of the brakes will be weak, and after a few such applications the brakes are almost useless on account of the air having been exhausted from the auxiliary reservoirs.

76. Q. In releasing and recharging the train, how long should the handle of the engineer's valve be left in the release position?

A. Until the train pipe pressure has risen nearly to 70 pounds again.

77. Q. In making service stops with passenger trains, why should you release the brakes a little before coming to a full stop?

A. So as to prevent stopping with a lurch; it also requires less time for the full release of the brakes after stopping.

78. Q. In making stops with freight trains, why should the brakes not be released until after the train has come to a full stop?

A. Because long freight trains are apt to be parted by releasing the brakes at low speed.

79. Q. In making service stops, why must the handle of the engineer's valve not be moved past the position for service applications?

A. So as to prevent unnecessary jerks to the train, and the emergency action of the triple valve when not necessary.

80. Q. If you find the train dragging from the failure of the brakes to release, how can you release them?

A. By placing the handle of the engineer's valve in the lap position until an excess of pressure is attained, and then throwing it into full release position.

81. Q. When the brakes go on suddenly when not operated by the engineer's valve, and the gauge pointer falls back, what is the cause, and what should you do?

A. Either a hose has burst, or a conductor's valve has been opened, or the train has parted. In any event, the handle of the engineer's valve must be immediately placed in the lap position to prevent the escape of air from the reservoir.

82. Q. Are the brakes liable to stick on after an emergency application, and why?

A. The brakes are harder to release after a severe application, because they are on with full force, and it requires higher pressure than usual in the train pipe to release them again. In this case it is necessary always to have in reserve the excess pressure on the main reservoir to aid in releasing the brakes. With the quick-acting triple valve this is especially necessary, because air from the train pipe as well as from the auxiliary reservoir is forced into the brake cylinder when a quick application of the brake is made, thus increasing the pressure in the brake cylinder without the usual reduction of pressure in the auxiliary reservoir, and requiring a high pressure in the train pipe afterward to cause the brakes to be released.

83. Q. In using the brakes to steady the train while descending grades, why should the air pump throttle be kept well open?

A. So that the pump may quickly accumulate a full pressure in the main reservoir for use in recharging the train when the brakes have been released again.

84. Q. In descending a grade how can you best keep the train under control?

A. First, by commencing the application of the brakes early, so as to prevent too high a speed being reached. Second, by applying the brakes lightly at first, then increasing the brake pressure as needed, and by slowing the train down just before it is necessary to release the brakes for recharging, so as to give time enough to refill the auxiliary reservoirs before much speed is again attained.

85. Q. If the train is being drawn by two or more engines, upon which engine should the brakes be controlled, and what must the enginemen of the other engines do?

A. The brakes must be controlled by the leading engine, and the enginemen of the following engines must close the cock in the train pipe just below the engineer's valves. The latter must always keep his pump running and in order, and main reservoir charged, with the engineer's valve in the running position, so that he may quickly operate the brakes if called upon to do so.

86. Q. If the air signal whistle only gives a weak blast, what is the probable cause?

A. Either the reducing valve is out of order so that the pressure is less than 40 pounds or the whistle itself is filled with dirt or not properly adjusted, or the port under the end of signal valve is partially closed by gum or dirt.

87. Q. If the reducing valve for the air signal is allowed to become clogged up with dirt, what will the result probably be?

A. The signal pipe might get the full main reservoir pressure, and the whistle will blow when the brakes are released.

88. Q. If you discover any defect in the air brake or signal apparatus while on the road, what must be done?

A. If it is something that cannot be readily remedied at once, it must be reported to the engine-house foreman as soon as the run is completed.

89. Q. What is the result if water be allowed to collect in the main reservoir of the brake apparatus?

A. The room taken up by the water reduces the capacity for holding air, and the brakes are more liable to stick. In cold weather also the water may freeze and prevent the brakes from working properly.

SPECIAL FOR ENGINE REPAIRMEN.

90. Q. How often must the air brake and signal apparatus on locomotives be examined?

A. After each trip.

91. Q. Under what pressure must it be examined?

A. Under full pressure, i. e., 70 pounds on the air brake train pipe, 20 pounds excess in the main reservoir, and 40 pounds pressure upon the air signal train pipe.

92. Q. How will you be sure that proper pressures are upon the two train pipes?

A. By regulating, and, if necessary, cleaning the pump governor so that it will shut off steam from the pump when 70 pounds train pipe pressure is reached, and by examining, and, if necessary, cleaning the pressure reducing valve for the signal train pipe, so that it maintains 40 pounds pressure in the train pipe.

93. Q. If you do not obtain 20 pounds excess pressure in the main reservoir when the handle of the D-8 engineer's valve is in the running position, what is the cause?

A. Either the excess pressure valve needs cleaning, or the rotary disk valve in the engineer's valve is unseated and allows air to leak from one port to another.

94. Q. Why must the air-pump piston rod be kept well packed?

A. To prevent leakage of steam and air.

95. Q. How often must the main reservoir and the drain cup under the tender be drained?

A. After each trip.

96. Q. How often must the triple valves and the cylinders of the driver and tender brakes be cleaned and oiled?

A. They must be thoroughly cleaned and oiled with a small amount of mineral oil once every six months, and the cylinders must be oiled every three months. If the driving brake cylinders are so located that they become hot from the boiler, they may require oiling more frequently.

97. Q. If there are any leaks in the pipe joints or anywhere in the apparatus, what must you do?

A. Repair them before the engine goes out.

98. Q. How is the brake shoe slack of the cam driver brake taken up, and what precautions are necessary?

A. By means of the cam screws, and it is necessary to lengthen both alike, so that when the brake is applied the point of contact of the cams will be in a line with the piston rod.

99. Q. How is the brake shoe slack of driver brakes on a locomotive with more than two pairs of driving wheels taken up?

A. By means of a turn buckle or screw in the connecting rods.

100. Q. How is the slack of the tender brake shoes taken up?

A. By means of the dead truck levers; if they will not take it up enough, it must be taken up in the underneath connection, and then adjusted by the dead lever.

101. Q. How far should the driver brake piston travel in applying the brakes?

A. Not less than one-third nor more than two-thirds of the full stroke of the piston.

102. Q. What travel of piston should the tender brakes be adjusted for?

A. Not less than five inches nor more than six inches, and such adjustment must be made whenever the piston travel is found to exceed eight inches.

SPECIAL FOR TRAINMEN.

103. Q. How should you proceed to test the air brakes before starting out, after the change in the make-up of the train, or before descending certain specially designated grades?

A. After the train has been fully charged with air, the engineman must be signaled to apply the brakes; when he has done so, the brakes must be examined upon each car to see that the air is applied and that the piston travel is not less than five inches nor more than eight inches on a passenger car, nor more than nine on a freight car. The engineman must then be signaled to release the brakes, and this signal must be given by the train air signal from the rear car, if it is used upon the train; after he has done so, each brake must be examined again to see that all are released. The engineman and conductor must then be notified that the brakes are all right, if they are found so.

104. Q. In starting out a passenger train from an inspection point, how many cars must have the brakes in service?

A. Every car upon the train.

105. Q. When might you cut out a brake upon a passenger car?

A. Never; unless it gets out of order while on the run, in which case it must be reported to the inspector at the end of the run, or upon the first opportunity which may give sufficient time to repair it.

106. Q. If a hose bursts upon the run what must be done, if the train is in a safe place?

A. The hose must first be replaced by a good one, and the engineman then signaled to release the brakes. The train must not proceed until the brakes have been reconnected and tested upon the train to see that all are working properly.

107. Q. If the train is not in a safe place when the hose bursts, what must be done?

A. The train pipe cock immediately ahead of the burst hose must be closed and the engineer signaled to release the brakes. The brakes at the rear of the burst hose must then be released by bleeding the auxiliary reservoirs, and the train must then proceed to a safe place to replace the hose and connect up the brakes, after which the brakes must be tested.

108. Q. If the train breaks in two, what must be done?

A. The cock in the train pipe at the rear end of the first section must be closed, and the engineman signaled to release the brakes. The two parts of the train must then be coupled, the hose connected and the brakes again released by the engineman. When it is ascertained that the brakes are all released, the train may proceed.

109. Q. Explain how the pressure retaining valves are thrown into action or thrown out of action, and when this must be done.

A. The pressure retaining valve is thrown into action by turning the handle of the valve to a horizontal position, and it is thrown out of action again by placing this handle in a vertical position pointing downward. This handle should be placed in a horizontal position at the top of a heavy grade, and it should always be returned to a vertical position at the foot of the grade, as otherwise the brakes will drag on any cars which still have the handle of the pressure retaining valve in the horizontal position.

110. Q. If the brake of any car is found to be defective on the run, how should you proceed to cut it out?

A. By closing the cock in the cross-over pipe of the quick-acting brake, or in the triple valve of the plain automatic brake, and then opening the release cock in the auxiliary reservoir upon that car, leaving it open, if a passenger car, or holding it open until all the air has escaped from it, if a freight car.

111. Q. When it is necessary to cut out a defective brake upon a car, why should it always be cut out at the triple valve and

never by the train pipe cock at the end of the car, even if it is the last car of the train?

A. The train pipe should always be open from the engine to the rear end of the last car, so that if the train breaks in two the brakes will be automatically applied before the parts of the train have separated sufficiently to permit damage to be done by their coming together again, and so that the brakes may be applied with the conductor's valve upon any car.

112. Q. Should the train pipe burst under any car, what must be done?

A. The train must proceed to the nearest switching point, using the brakes upon the cars ahead of the one with the burst pipe, where the car with the burst pipe must be switched to the rear of the train; the hose must then be coupled up to the rear car and the cock at the rear end of the next to the last car opened and the cock at the forward end of the last car closed, so that if the train should part between the last two cars the brakes will be applied.

113. Q. What is the conductor's valve, and what is its use?

A. It is a valve at the end of the branch pipe leading from the train brake pipe upon each passenger car; it is to be opened from the car in any emergency when it is necessary to stop the train quickly, and only then. When used it should be held open until the train is stopped, and then it should be closed.

114. Q. What is the air signal for, and how is it operated?

A. It is to signal the engineman, in place of the old gong signal, and it is operated by pulling directly downward on the cord one second for each signal given and releasing immediately, allowing two seconds to elapse between pulls.

115. Q. If the car discharge valve on the air signal system is out of order or leaking on any car, how can you cut it out?

A. By closing the cock in the branch pipe leading from the train signal pipe to the discharge valve; to do so the handle of this cock should be placed lengthwise with the pipe.

116. Q. How is the slack taken up so as to secure the proper adjustment of piston travel?

A. By means of the dead truck lever, and if that is not sufficient, one or more holes must be taken up in the underneath connection and the adjustment then made by the dead truck lever.

SPECIAL FOR INSPECTORS.

117. Q. Do you understand that no passenger train may be started out with any of the brakes cut out of service?

A. I do.

118. Q. Why is it important that no leaks should exist in the air brake service?

A. Because they would interfere with the proper working of the brakes and might cause serious damage.

119. Q. What must be done with the air brake or air signal couplings when not united to other couplings on cars equipped with dummy couplings?

A. They must be secured in the dummy coupling, so that the face of the dummy coupling will cover the opening of the hose coupling so as to prevent dust and dirt from entering the hose.

120. Q. If air issues from the release port of the quick-acting triple valve when the brakes are off, what is the cause?

A. It is probably due to dirt on the rubber-seated emergency valve.

121. Q. How often must the cylinder and triple valves be examined, cleaned and oiled?

A. As often as once every six months on passenger cars and once in twelve months on freight cars, and the cylinders

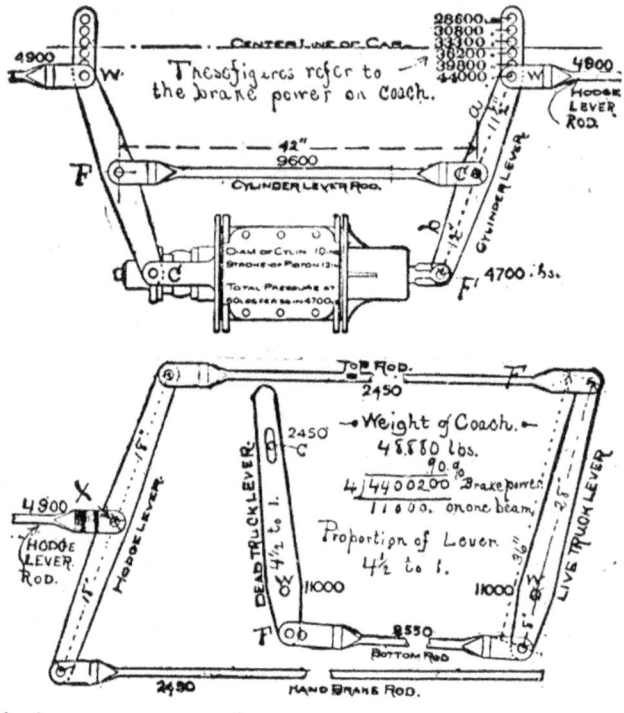

must be oiled once every three months with a small quantity of mineral oil. The dates of the last cleaning and oiling must be marked with white paint on the cylinders.

122. Q. To what travel of piston must the brakes be adjusted?

A. Not less than five inches nor more than six inches, and this adjustment must be made whenever the piston travel is found to exceed eight inches on a passenger car or nine inches on a freight car.

123. Q. How is the slack taken up so as to secure this adjustment?

A. By means of the dead truck lever, and if that is not

sufficient, one or more holes must be taken up in the underneath connection and the adjustment then made by the dead truck lever.

124. Q. What are the different holes in the outer end of the cylinder levers for, and why must the connections be pinned to the proper hole for each car?

A. These holes are to enable the adjustment of the brake pressure to be made according to the weights of different cars. The connection must be made to the proper hole in each case, according to the weight of the car, so as to give proper braking power, otherwise the brake will be inefficient, or the wheels may be slid under the cars.

BRAKE LEVERAGE.

A knowledge of the method of calculating brake leverage is not always essential to operators of the air brake, but it comes very handy to know something about it. The rules used are not complicated. Certain formulas help to shorten the calculations. You must first learn how the several classes of levers operate and the difference between those of the first, second and third kind.

On page 688 is a small cut of the arrangement of a coach brake according to the Hodge system, which we will use to illustrate this explanation. The Hodge lever rod shown there is supposed to connect the cylinder lever to the Hodge or floating lever at the figures 4900, or x. The pressure at F, where the piston is attached to cylinder lever is 4700 lbs. for a 10-inch cylinder with a quick action triple. This multiplied by 12, the distance from power F to the fulcrum c, and divided by $11\frac{1}{2}$, the distance from fulcrum c to W, gives 4900, which is the strain on the rod to the middle of Hodge lever at x. As this lever is equally divided, each end gets half of this power, or 2450 lbs., which now becomes the force F at top end of the live truck lever. We next multiply 2450 by the distance from F to the fulcrum c, which is 36 inches; divide product by distance from c to W, 8 inches, and we have 11000 lbs., the strain on that brake beam. Now, if the force at F is 2450 lbs., and at W is 11000 lbs., the resistance at c must be 8550, as the sum of the strains at the ends of the live lever must balance the strain in the middle. Therefore, the strain on the bottom rod, 8550 lbs., is what goes to the dead lever at F, and must be multiplied by the distance to c and divided by the distance from c to W. If the dead lever is the same proportion as the live one, it will give 11000 lbs. To get the proportion of a lever, divide the whole length of the lever between outside holes by the distance from c to W in the case of the live truck lever shown, which is 8 into 36, or $4\frac{1}{2}$ to 1. Look at the cylinder lever tie rod and you will see its strain is 9600 lbs., which is the sum of 4700 and 4900, the strains on both ends of the live cylinder lever. This strain in turn is exerted at F on the other lever so that we get a braking power on both ends of the coach amounting to double what the piston has, but we get it because the piston travels twice as far as it would if the fulcrum at c was fixed stationary. Both cylinder levers need not be the same length, but they must be the same proportion if the same strain is to go to each end of the car. Coaches have cylinder levers exactly alike, freight cars do not always, although they are the same proportion. In the computations for this example the odd pounds are not shown.

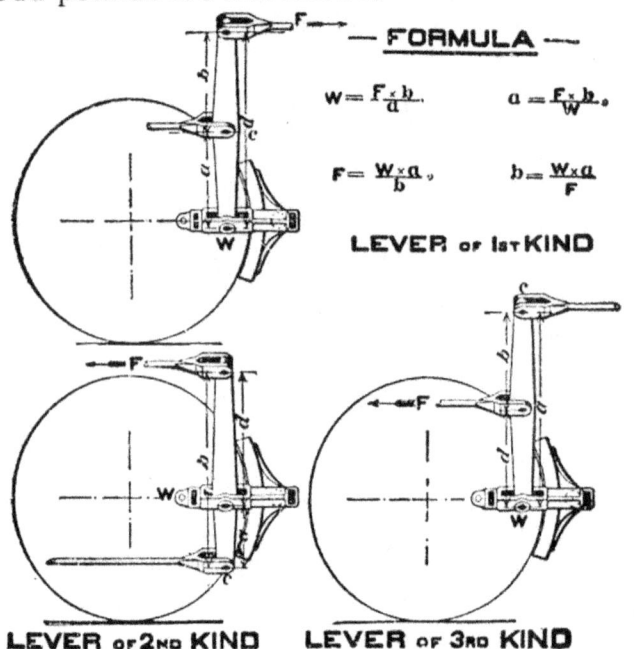

Possibly this explanation will make the formula shown in the cut plain to you. The same formula is used for each of the three kinds of levers, but the letters used to represent pounds or inches are changed in their positions on the levers. In each case F represents the pull which the piston transmits to the brake lever, c is the fulcrum, and W the brake beam. b is the distance from F to c, and a the distance from c to W; d can be found when you have a and b. If you have three of these quantities you can always find the fourth by the use of the formula provided for that case.

CAM DRIVER BRAKE LEVERAGE.

The limited space in this book will not allow a full description of how the cams and levers are designed, but some information on calculating their brake power will come handy to the men operating them. The illustration of the cam brake shows its various parts.

These cams are really segments of wheels with x-x for the centers. If they are properly laid out no matter how far they roll down, the point of contact at the edges of the wheels will always be on the line between the centers x-x.

The cam as used with the brake is a bell crank with the long arm from g to x and the short arm from x to a. A true bell crank requires a fixed fulcrum at x to act as a brace to transmit the power at g to a, but in the case of the cams no fulcrum is needed there, for the faces of the cams rolling against each other act as fulcrums.

To calculate the brake power, set the brake full on and measure the distance between the cam screw pins at a-a. Also measure the distance between the cam link pins g-g and subtract this distance from the distance a-a, one-half of this remainder will be the long arm of the bell crank included in the design of each cam, which distance we will call X in the formula.

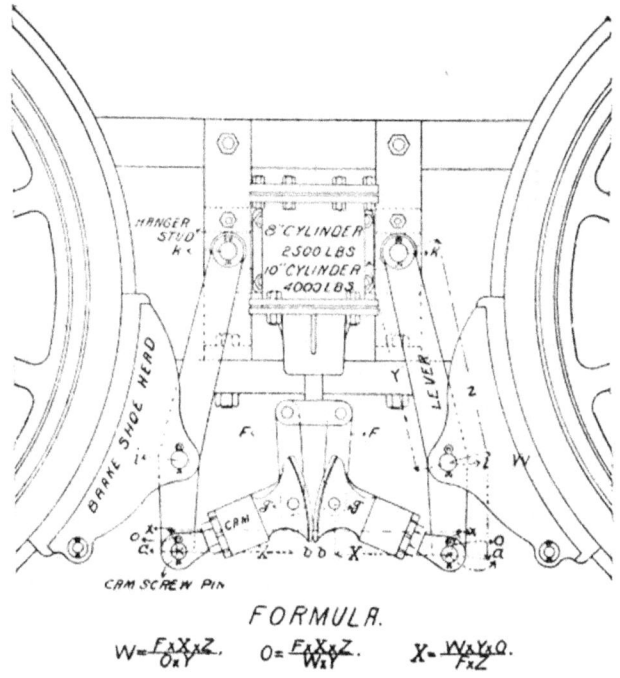

FORMULA.

$$W = \frac{F \times X \times Z}{O \times Y}. \quad O = \frac{F \times X \times Z}{W \times Y}. \quad X = \frac{W \times Y \times O}{F \times Z}.$$

We do not measure clear to the face of the cam, because the power is applied at g-g, one-half of the power exerted by the piston going to each cam.

As the cams roll down against each other when the brake is set their faces touch at one point only, which we will call the point of "rolling contact." Place a straight-edge from one of the cam screw pins at a to the other on a line with their centers and measure from the straight-edge up to the point of rolling contact; this distance is the other arm of the bell crank, it is called the "offset," and is the distance from a to x also; this is named O in the formula. This last distance divided into the length of the line from b to a—the long arm of the bell crank—gives the leverage of the cam.

Multiply this leverage by 1250 for an 8 inch cylinder or by 2000 for a 10 inch cylinder, which will give the power delivered at the bottom end of the lever at a. Multiply this power by the whole length of the lever from a to k called Z in the formula, and divide the product by the distance from the pin k to the pin i in the brake shoe head, which is distance Y, this quotient is the brake power delivered at that shoe; four times the power for one shoe will be the brake power for all shoes, which should be 75 per cent of the weight on drivers. In all these calculations we use 50 pounds as the air pressure per inch on the brake piston.

To calculate the other way take 75 per cent of the weight on the rail at the drivers, one-fourth of that will be the power required at each shoe. Multiply this amount by the length in inches of the lever from i to k and divide the product by the length from k to a; this last amount will be the power required at a delivered by the cam.

Divide this by 1250 for an 8 inch cylinder or by 2000 for a 10 inch cylinder, the quotient will be the "leverage" of the cam, and should correspond exactly with the cam in use. To get the leverage of the cam divide the length at X by the offset.

The cams are designed to give the full brake power of 75 per cent of the weight on drivers when the shoes and tire are worn down to their limit. The brake power increases as the length of the cam X is increased by wear of shoes and tires, but this does not affect the "offset" O. Therefore, with thick new tire and new shoes you will not get the full brake power, because the long lever of the bell crank in the cam is not the full length as laid out for a thin tire. To avoid the difficulty of having too much leverage with thin tire the radius of the face of the cam is struck from a point one and one-fourth inches further out than at x. A thick or thin shoe does not change the power as much where the cams are long with long wheel base as with very short cams.

If you find that with a short piston travel, say two inches, the cams do not roll down so that their faces separate at the lower corners, as shown in the illustration, the cam links are too short. It is not unusual to find these links put up too short, and this defect reduces the brake power very materially. Changing the brake heads and putting on a wide one in the place of a narrow head also reduces the brake power, as it shortens the length of the cam. A thick shoe shortens the cam in the same way.

In any brake the proportion between the piston travel and the brake shoe travel is the leverage. For instance, if the piston travel is four inches and the brake shoes travel one-half inch the proportion is eight to one, so the power from the piston is multiplied eight times. If you can get the exact brake shoe travel of a brake and divide it into the piston travel you can easily find the brake power.

THE HIGH SPEED BRAKE.

A short description of this improvement on the quick-action automatic brake for passenger equipment may be valuable to those who have not seen it. For ordinary speed, below 20 miles an hour, the automatic brake is able to control the speed in the ordinary manner, but when the speed is much higher, more power is required in proportion as the speed is higher. It is the friction of the brake shoes on the wheels that arrests the speed of the train and finally brings it to a stop. In addition to arresting the momentum of the train at speed this friction must also arrest the rotary motion of the wheels turning around. This friction up to a certain point is proportioned to the amount of force applied to hold the shoes against the wheels and this force is called brake power. There is a difference in the amount of the friction of the same shoes and wheels at different speeds, it being greater at a low speed than at a high one.

What is called the co-efficient of friction is about .074 at 60 miles an hour and increasing to .241 as the speed is reduced to 10 miles an hour, so you see the brake shoes really hold less at a high speed than at a low one, and more brake power can be applied at the high speed than could be safely used at a low one.

Now it follows that if the full brake power was the same for all speeds, if it was the proper power for a moderate speed it would be much too low for a high speed. If a high speed was the standard the full brake power would be too great for the low speed, the wheels would "skid" on the rail and a loss of nearly all the brake power would result.

Therefore, an attachment to the brake that would give a very high brake power when first applied and gradually reduce this brake power at about the same rate the speed was reduced, would be proper for all speeds.

This brake power for moderate speeds has been fixed at 90 per cent of the weight of the coach when all the wheels have brake shoes applied to them, and is about all that can be used without sliding the wheels.

For emergency action the high speed brake is intended to apply the brakes at first with a brake power of 125 per cent of the weight of the coach and gradually reduce the brake cylinder pressure as the speed is reduced, till it reaches the standard amount of 60 pounds, which gives a 90 per cent brake power, at which point the reduction ceases.

This is accomplished by an automatic reducing valve, which is here illustrated.

This reducing valve, shown in Fig. 1, is fastened by the bracket at to the coach frame (see Fig. 6) and connected to the brake cylinder by suitable piping at z (see Fig. 2). When the air enters the cylinder at the time the brakes are applied, it also comes in on top of piston 4. This piston is held up by the spring 11 against a pressure of 60 pounds, so that if no more than 60 pounds comes into the cylinder the reducing valve remains stationary in the position shown in Fig. 3.

But if more than 60 pounds pressure per inch comes on piston 4, as is the case with the emergency application with this brake, piston 4 is forced down to the lower limit of its travel

THE AUTOMATIC REDUCING VALVE

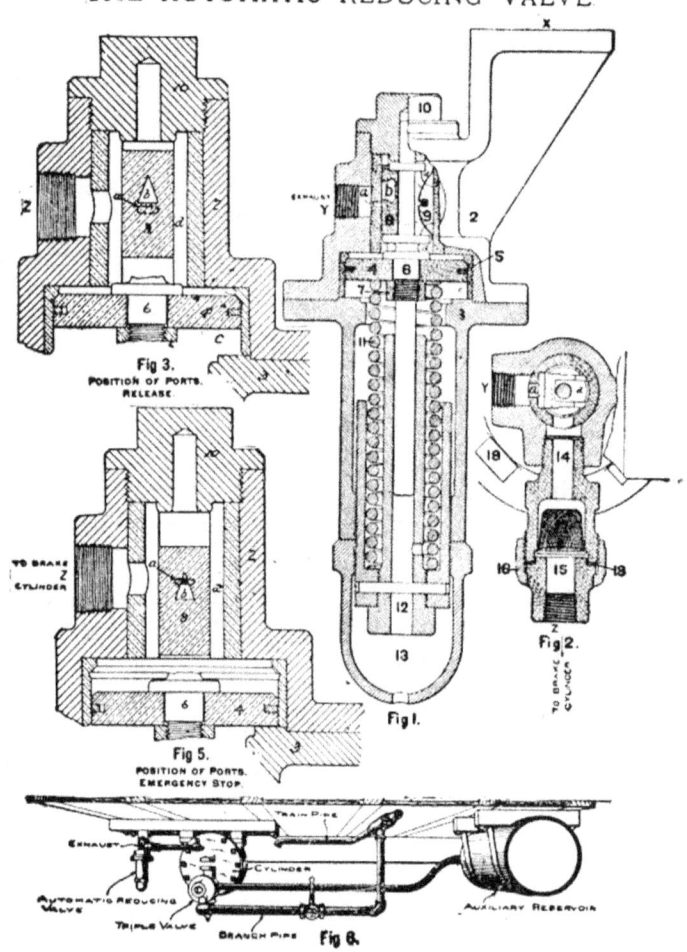

against the tension of spring 11 into the position shown in Fig. 5. This moves the slide valve 8 attached to piston 4 down also and opens an exhaust port *b* through the slide valve and allows air to escape slowly from the cylinder to the atmosphere through a port *a* in the seat and the opening Y. This port *b* is triangular in shape, the narrow part at the top being open at the highest pressure and opening wider through the action of the spring 11 pushing up the piston as the pressure decreases. When the pressure is reduced to 60 pounds, the spring 11 has moved the slide valve up to lap ports *a* and *b*, and the piston to the top limit of its travel so no more air can escape from the cylinder. This position is shown in Fig. 3.

The size of this exhaust port *b* has been determined by experiment so as to reduce the pressure in the brake cylinder proportionate to the reduction of speed.

When the brake is first set with emergency at a high speed the pressure is about 85 pounds in the cylinder; as the speed of the train is reduced by the action of the brakes, the pressure is also reduced by the reducing valve at about the same rate, till it reaches 60 pounds, where it remains till the brake is released in the regular way.

This gives a very high brake power ready to use if found necessary at high speeds and still leaves the service application feature unchanged, ready for use in ordinary stops. With the service application the reducing valve remains in position as shown in Fig. 3. A reduction of 20 pounds applies the brake fully and any further reduction of train pipe pressure is a waste of air, as this reduction will fill the brake cylinders at 60 pounds, the full pressure for a service application; as well as leaving a high auxiliary reservoir pressure ready for another full service application of the brake if found necessary before recharging; during these moderate service applications the reducing valve does not move.

The train pipe and auxiliary pressure is set at 110 pounds with this type of brake—it may be more if the conditions seem to call for it. As the engines equipped for drawing these high speed braked trains may be used on coaches without the high speed attachments, some arrangement is needed for changing the standard train pipe pressures from 70 pounds to the higher pressure, and vice versa.

There are two "feed attachments" on the engine coupled to the brake valve with suitable piping. One of them is set at 70 pounds, the other at 110, and there is a cock between them which can be turned to "cut in" either one for service as is desired. Likewise there is a duplex governor for the air pump, one side of which is set for 90 pounds main reservoir pressure, the other side set at the higher pressure required, and a suitable cock to cut out the 90 pound side when using the higher pressure. The tender is equipped with a quick-action triple and reducing valve the same as a coach. An engine truck brake is a necessary part of this equipment, which is supplied with air from the driver brake triple; a reducing valve similar to the coaches is used. Any extra coaches placed on these high speed braked trains require a reducing valve, although a safety valve set to blow off at 60 pounds through a restricted opening can be used temporarily by screwing it into the oiling plug hole in the cylinder head.

INDEX.

A

Accidents to locomotives	401
Admission commences	22, 49
Effect of early	24
late	27
Equalizing	179
line	217
Point of	29, 30, 36
Advice to young runners	325
Air Brake, Adjusting	658, 671, 673, 678, 688
auxiliary reservoir	620, 621, 635, 646
Charging	620, 645, 646
breaking in two	651, 670, 686
Cam brake leverage	691
creeping on	642, 643
cutting out	648, 649, 658, 659, 668, 670, 686
Defective	641, 658, 659, 671, 686
double heading	632, 634, 638, 652, 668, 686
emergency application	615, 619, 636, 657, 668
equalizing brake valve	613, 614, 625, 679
reservoir	620, 625, 631
examination questions	661, 662
excess pressure	619, 624, 629, 641, 680, 684
feed port	610, 620, 646
graduating valve	617, 634, 637, 638, 640, 641
gauges	285
Utica	285
high speed	693
hose, Bursted	650, 670, 682, 686
Kicking off	632, 641, 642, 680
leaks	620, 631, 632, 640, 641, 660, 688
leakage groove	634, 647, 678
leaking off	620, 631, 640, 641, 660
leverage	673, 688, 689, 690
piston travel	617, 647, 658, 659, 667, 678, 685
pressure on brake piston	635, 637, 642, 646, 677
pump governor	643, 653, 654, 655, 656, 671, 679
questions	621
signal	651, 660, 671, 672, 683, 687
sticking	641, 642, 658, 669, 683
train pipe, Overcharging	618, 642, 643
valve, D-5	626, 627, 628, 629, 632, 680
D-8	613, 627, 628, 629, 630, 643
Plain triple	610, 611, 639, 648, 649, 658, 675
Pressure retaining	651, 656, 660, 670, 679, 686
Position of	629, 630, 634, 652
Quick action triple	615, 616, 636, 637, 638, 639
Releasing	641, 642, 643, 644, 645, 667, 682
service application	610, 619, 634, 635, 667, 681
Testing	645, 646, 667, 669, 672, 685
for leaks	623, 632, 641, 642
train line pressure	620, 621, 625, 642, 667, 677
train men's questions	657, 661
Air Cards	218, 604
compressed	583
in shops	603
on railways	585
compressor	604
gauges	280, 285
hoist	605
Table of lifting capacities of	606
locomotive	584
motors	607
Albin boiler check	267
Allan Valve Gear	192

INDEX.

Allen Valves 51
              Lead of 52, 175
Allen-American Valve 77
Allen-Richardson Valve 75
Alterations, Proving 180
              Main rod 185
American Balanced Valves 76
Angle, How to make a right 558
Angularity of connecting rod 117
                Effect of 33, 117, 119
                Error due to 135
Angular advance 119, 174, 190
Ashcroft Steam Gauges 283
Atmospheric line 218
Axles, Broken Driving 415, 420
       Truck 419
       tender 420
       How to turn and fit 569
       Quartering driving 569

B

Balanced valves .... (See Valves)
Baldwin locomotives 333, 349, 491, 508
       system of compounding—two cylinder 347
              Vauclain 332
              Accidents with 346, 351
Barnes' Balance Slide Valve 83
Bell Ringer, Breitenstein's 597
       Gollmar's 599
Bell's Front End 434
Best's Pneumatic Cylinder Cocks 601
Bischoff on the injector, Article by 239
Blow-off Cock, McIntosh's 447, 448
       Reed's 449, 450
       Surface, Hornish's 455
              McIntosh's 446
Blows and Pounds, Locating 313
Boiler checks, Albin's 267
       De Sanno's 266
       Foster's 262
       Heintzelman's 263
       Linstrom's 264
       McLeod's 267
       Repairing 261
       Cleaner, McIntosh's 446
              The Hornish Mechanical 450
       expansion, Effect of 189
       Fitting parts to 556
       foaming 403, 450
       lagging 309
Bolt, Broken elic 415
       Fitting 568
            taper 568
Boring 574
Boxes .... (See Driving Boxes)
Brass-Finishing 569
Brasses, Driving .... (See Driving Brasses)
       Rod .... (See Rod Brasses)
Break Downs 401
Breitenstein's Bell Ringer 597
Bridges, Valve Seat 25
Briggs' Balanced Slide Valve 85
       standard pipe sizes 561
Brooks locomotives 2, 473, 483
       system of compounding—Tandem 377
              Two cylinder 372
              Accidents with 376, 380
Brown's Balanced Slide Valve 86
Brownley injector 259
Bull Rings, Fitting up 571
Bushings, Fitting 568
       Cylinder 568, 573

C

Cab Window, Tinker's 298
Canby's Draft Regulator 442

INDEX.

Cause of valves sounding "out" 316
Celler, Improved driving box 526
Center casting, Broken truck 415
                  Fitting up 519
                  Lining shoes and wedges by the 519
                  Setting the 517
           Finding a square 519
           How to locate a dead 154, 178
           Meaning of dead 154, 177
           line of motion 119, 190
                  and angular advance 190
           of gravity, Locating 538
Chime Whistle 310
Chucking, Difficult 567
Checks ....(See Boiler Checks)
Chicago, Rock Island & Pacific Ry. Engine 1101 479
Circle, Finding center of part of a 559
Clariometer 312
Clearance 23
         inside 21, 44, 188
                amount used 27
                effect of 27
              Meaning of 23
Cock, Drain 239
Combustion 421
Compressed air 583
            locomotives 584
Compression 22, 216
            begins 30
            Effect of 28
            line 217
            Meaning of 22
            Point of 29, 30
            Valve for reducing, Farrer's 64
                  Fay's 62
Compound Gearing 564
          locomotives ....(See Locomotives)
Compressor, Air 604
Cooke locomotives 389, 429, 487
        system of compounding 387
                  Accidents to 392
Corey's Force Feed Lubricator 276
Counter balancing, Formula for 538
              Locating centre of gravity for 539
              Modern 537, 541
Crank pin, Broken 406
           gauge 536
           holes, Boring out 574
           Hollow 557
           press 536
           Tramming 552
           Trueing up old bearings on 535
           Turning and fitting 569
Crosby's Indicator 222
           Care of 226
        Pop Safety Valve 288
        pressure gauge tester 286
        Steam gauge 283
                Thermostatic 283
Cross-Head, Broken 406
           Extreme travel of 179, 551
           Lining up 549
           pins, Turning solid 573
               Fitting Mogul 572
           Planing 577
Cut-off 21
       Effect of 36, 38
            change of lead on 180
       How to equalize the 176, 177, 183, 184, 185
            running 189
       Point of 29
           how to determine 38
           travel will effect 38
           trying 160
Cups, Forged rod 553
      Holes for oil 552
      Plunger oil 557
Currents, Induced 231

INDEX.

Cylinders 542
    axis above wheel centers 118
    Boring out 544, 574
    cocks, Pneumatic 601
    Fitting bushings into 573
        together 542
    heads, Broken 407
    relief 82
    joints, Facing 543
    joints grinding 543
    New boring machine for 545
    Planing 575
    packing 316, 571
    properly, How to line 518
    saddles, Fitting 543, 544

D

Dean's track sander 595
Decimal parts of an inch 560
DeLancey's Exhaust Nozzle 431
DeSanno's check valve 266
Detroit Metallic Packing 297
    Lubricator 272, 273
        Tippet attachment 270, 271
DeWallace Train order signal 393
Diaphragm, Adjusting 422
Dickson locomotives 481, 494, 495
    system of compounding 392
Direct motion 109, 191
    To distinguish between indirect and 188, 189
Disconnecting one side 402
    both sides 403
Distribution of steam, Events in the 45
    Equalizing the 161, 176, 177, 183, 185, 189, 190
Dodge injector 257
Drain cock 239
Draft Regulator, Bell's 434
    Canby's 442
    DeLancey's 431
    Lord's 438
    Wallace & Kellogg's 425
    Warren's 435
Drawbar, Broken 420
Driving boxes, Broken 420
    Beveled flanges 576
    celler improved 526
    Fitting 526
    Handling 524
    New style 320
    Planing 576
    Tramming 524
    brasses, Broken 420
    Boring out 575
    Fitting 525
    Hot 322
Dynamometer, Traction 311

E

Eccentric, Angular advance of 119, 174, 190
    blade adjuster 150
    blades, Connecting 182
        Changing length of 179
        Error due to the angular vibration of 137
        Length of 104, 124
        or bolts, Broken 409
        pins back of link arc, Error due to locating 139
        Slipped 409
        Whether to lengthen or shorten 179
    brasses for side rods 553
    slipped 408
    strap, Breakage of 410
        Broken 409
        bolts, Broken 410
        Hot 322
Eight, Meaning of 177
Electric locomotive 503
    headlights 310, 481
Elic bolt, Broken 415

INDEX.

Engine .... (See Locomotives)
Engineer's duties 323, 401
Errors of the link motion 131
Equalizing the cut-off 176, 183, 184, 189
Equalizers, Broken stands for 414, 415
Events in the distribution of steam 45
Examination questions on air 609
Exhaust 22
  Equalizing the 163
  line 217
  Meaning of 22
  Muffling the 424
  nozzle, Blown out 313, 314
   box, Leak at bottom of 313, 314
   DeLancey's variable 431
   How to finish 573
   New South Wales 433, 434
   Wallace & Kellogg's variable 425
   Size to use 421
   Smith's triple expansion 427
Expansion, Duration of 30
  Effect of boiler 189
   steam 215
  Initial 215
  line 217
  Meaning of 22
  plates, Fitting 557
Extension rod, Length of 107, 110
  Broken 411
  piston rods 553

F

Farrington's Valve Setting Machine 147
Fast Runs 504
Farrer's valve for reducing compression 64
Fay's valve for reducing compression 62
Feed Water Heater, Wallace & Kellogg's 299
Firemen, Hints to 423
Fitting parts to boiler 556
Flue burst 404
Foaming boiler 403
Follower bolts, Loose 318
  plate, Turning 571
Formula for balancing valves 83
  counterbalancing 538
  valve setting 165
Foster's boiler check 262
Frame, Broken engine 420
  truck 420
  Laying off engine truck 555
  Setting 512
  Splicing 527
Friction 95
Front end, Bell's 434
  Broken 412
Full gear, Meaning of 177

G

Garfield injector 257
Gearing, Compound 564
  Simple 563
General machine shop work 512
Glands, Broken piston or valve stem 406
Globe Valve, Improved 308
Gollmar Bell Ringer 599
Gooch valve gear 190, 191, 192
Goodspeed's cutoff valve 53
Gould balanced valves 65
Gravity, Locating center of 538
Gauges, Steam and Air 280
  Crank pin 536
  Thermostatic 283
  Tester for 286
  Testing 285
Guide blocks, Laying off 547, 550
  bolts or yoke, Broken 407
  Broken 407
  Lining 547, 550
  Why strain is on top 550

H

Haley's slide valve	57
Hancock inspirator	254
Hand holt plate blown out	413
Hangers, Broken spring	413, 415
Draw on spring	557
Link	106
Harris Metallic Packing	295, 296
Headlight, Electric	310
Heat, Latent	216
Sensible	216
Unit of	216
Heidelberg's pilot coupling	603
Heintzelman's boiler check	265
Hints, to Firemen	423
History of the locomotive	13
Holt slide valve	83
Hornish Mechanical boiler cleaner	450
Horse power	556
Indicated and net	216
Hose strainer	260
Hot bearings	322
Howe's link	98

I

Improved tools	571
Inch, Fractional parts of an	560
Incrustation	445
Indicator diagrams	218
Reading	217
Taking	216
Crosby's, Description of	221
Sectional view of	222
Construction of the	213
Definition of technical terms used with the	215
Invention of the	212
Method of applying	214
planimeter, Lippincott's	229
reducing wheel, Victor	228
Steam	212
Use of the	213
Indirect valve motion	100, 188, 189
To distinguish between direct and	188
Induced currents	231
Information, General	298
Injector action, Theory of	231
Bischoff, J. A., Article on	239
Brownley	259
Care of	237
Check	(See Boiler Checks)
Colvin, F. H., Article on	231
Construction of	252
Dodge	257
Essentially active parts of	240
Garfield	257
Giffard	234
Invention of the	230
Location of	244
Little Giant (Rue)	254
Luckenheimer	258
Meaning of numbers on	239
Metropolitan 1898	250
Monitor, Improved 1888	247
P. R. R. standard	248
Standard	249
Ohio	255
Parts of	240
Repairing	259
Restarting	236
Seller's improved 1893	245
1876	246
strainer	243
Won't work	413
Inspirator, Hancock	254
Intercepting valve	331

J

Jacks, Air	607
James' link	97
Jaws, Finding center of pedestal	519
Milling machine for	517
Stresses on pedestal	528
Tramming pedestal	520
Jerome Metallic Packing	295
Jones' Track Sander	589
Joy Valve Gear	203

K

Keying up rods	321
Keys, Draw on piston and valve stem	553
How to fit driving wheel	535
Keyways, Laying off eccentric	167
Kunkle Pop Safety Valve	288

L

Lagging, Magnesia sectional	309
Lap	21
Determining amount of	174
distinguished from lead	179
Exhaust	21, 22, 26, 27
Inside	44, 188
Effect of	23, 26
Meaning of	20, 21
Outside	21
Determining amount of	35
Effects of	23, 35
on point of cut-off	26
Meaning of	20
Steam	21
Valve without	23
Lateral between hubs	526
Lathe gearing, Compound	563, 564
Simple	563, 564
thread cutting	565
tools	564, 571
Grinding	569
Setting	568
work	569
Leach track sander	585
Lead, Amount used	175
with Allen valves	52, 175
Changing	172, 183
effect of	180
Constant	98, 192
How to distinguish from lap	179
Effect of	24, 27, 176
engine, Right and Left	188
Equalizing	176
given, Why	24, 175
when boiler is cold	189
Inside	44
Effect of	27
Meaning of	21
Meaning of	22
opening	29, 30, 153
Point of	29, 30
Stationary link gives constant	189
Trying the	157
Lewis valve gear	196
Link	99, 113
Allan	192
Broken	411
Connecting eccentric blades to	182
Finding the radius of a	112, 190
gives constant lead, Stationary	189
hanger, Length of	106
Broken	411
pins of	411
Invention of the	98
Laying off	112, 113
Length of	114
lining up or down	187

INDEX.

Link motion		97
Errors of		131
radius, To find		190
saddle, Adjustable		185
Adjustment of the		185
backset, Amount of	113,	142
Effect of		185
Why		115
Broken		411
Cut-off equalized by changing		185
suspension, Point of		126
Slip of		115
Shifting	97,	100
Stationary	97,	190
Warren's improved		143
Line of motion, Center		190
Linstrom's check valve		264
Lippincott's planimeter		229
Little Giant injector		254
Locomotive, Accident to		401
Baldwin No 1278, B. & O. Ry		491
system of compounding two cylinder		347
accidents to.		351
"Vauclain"		332
accidents to		346
two cylinder compound, No. 62, N. & W. Ry		349
"Vauclain" compound, No. 1027, A. C. R. R.		507
385, P. & R. R. R.		333
Brooks, No. 100, Great Northern Ry	2,	473
111, L., N. A. & C. Ry		483
system of compounding "Tandem"		377
accidents to		380
two cylinder		372
accidents to		376
Chicago, Rock Island & Pacific, No. 1101		479
Cooke, No. 1, N. Y., P. & O. Ry	427,	429
1818, S. P. of Cal		487
Compound, No. 1, D., L. & W. Ry		389
system of compounding		387
accidents to		392
Dickson, No. 969, A., T. & S. F. Ry		494
15, B., R. & P. Ry		481
93, D. & H. C. Co		495
system of compounding		392
accidents to		398
Electric, No 1, B. & O. Ry. Co		503
History of the		13
N. Y. C. & H. R. R. R., No. 999		505
P. R. R., No. 101		3
Class H5		488
Pittsburg, No. 250, P. & W. Ry		485
95, U. R. R.	1,	470
Compound, No. 4, L. S. & I. Ry		369
system of compounding		367
accidents to		372
Pneumatic		584
Rhode Island system of compounding		399
accidents to		399
Richmond, No. 700, Big Four Ry		490
Compound, No. 673, C. P. Ry		353
system of compounding		352
accidents to		358
Rogers, No. 376, Illinois Central R. R.		486
system of compounding		381
accidents to		386
Schenectady, No. 908, C. & N. W. Ry		497
Compound, No. 3, N. P. Ry		359
No. 909, Grand Trunk Ry		499
system of compounding		358
accidents to		366
Three cylinder		400
Lord's draft regulating device		438
Lost motion adjusting machine		149
in valve gear, Effect of		176
of main rod	157,	178
Louisville & Nashville track sander		587

INDEX.

Lubricators 269
    action, Mode of 270
    Air pump 274
    Attaching 270
    Corey's force feed 276
    Detroit 272, 273
    Filling 270
    Nathan 269
    Operating 270, 274
    Repairing 275
    Tippett attachment for 270
    Won't work 413
    Wallace & Kellogg's cylinder 299
Lunkenheimer's Injector 258

M

Machine, Cylinder boring 544
    Farrington's valve setting 147
    for facing valve seats 545
      turning tyres 530
    Lost motion adjusting 149
    Portable milling 517
    work, General 563
Main rod, Alteration of 185
    Angularity of 117
    brasses, Fitting 552
    Broken 406
    Error due to angularity of the 135
    length, Changing 551
      Finding 551
      upon the cut-off, Effect of 135
    Lost motion in 157, 178
    Position for keying up 321
    strap, Broken 406
Malleable iron piston head 554
Malone's balance slide valve 88
Margo valve 95
McDonald's balance valve 84
McIntosh's blow-off cock 447
    surface blow off 446
McLeod's turbine check 267
Meady Muffled Safety Pop 288
Measurements of unequal surfaces 559
Metallic packing ....(See Packings)
Meter, Length of 581
Metric system 578
    History of the 578
    Present status of 580
    threads, Table of 582
Metropolitan injector "1898" 250, 257
Midgear, Meaning of 177
    Travel of valve in 174
Modern locomotives 470
    counterbalancing 537, 541
Monitor injector of 1888 247
    Standard 249
    P. R. R. Standard 248
Motor, Air 607
Miscellaneous points 556

N

Nathan Lubricator 269
Nicholson's balance valve 90
Nozzle Box, How to finish 575
    Correct size of exhaust 421
    DeLancey's variable exhaust 431
    How to finish 573
    Lord's 438
    Reamers 422
    size on draft, Effect of 421
    Wallace & Kellogg's variable exhaust 425

O

Ohio injector 255
Oil cups, Holes for 552

Oiling 322
O'Reilly's Frame Splice 527
Overtravel .... (See Travel)

P

Packing, Air pump 294
        Cylinder blowing 313
            Fitting up 553
            Repairing 316
            Turning 571
        Metallic 292
            Detroit 297
            Double 297
            Harris 295
            Jerome 295
            piston rod 292, 295, 296, 297
            United States 291, 292, 293, 294
            valve stem 294
Parallel rods .... (See Side Rods)
Pedestal braces, Fitting 512
        jaws, Stresses on 528
Petticot pipe, Adjusting 422
Pilot coupling, Heidelberg's 603
Pipe sizes, Brigg's standard 561
Piston rod, Broken 405
        Extension 553
        gland, Broken 406
        keys, Draw on 553
        length, Finding 553
        Turning and fitting 570
        displacement 216
        head, Turning 571
            Cast steel 554
            Malleable iron 554
        travel 216
        valve 72
Pitkin's double throttle valve 302
Pittsburg locomotives 1, 485, 501
        system of compounding 367
            accidents to 372
Planimeter, Lippincott's 229
Planer, Circle feed for 576
        work 575
Plug blown out of boiler 413
Pneumatic cylinder cocks 601
        Hammers 608
        locomotives 584
        Track sanders .... (See Track Sanders)
Points, Additional 557
        Technical 116
Pop valve, Blown out 413
        Crosby safety 288
        Kunkle safety 288, 290
        Meady muffled 288, 290
        Setting 291
Port, Exhaust 26
        Marking 152
        Steam 174
        marks from tram marks, How to distinguish 178
            to work from, Which 179
            Trying 178
Pounds, Blows and 313
        Locating 318
Power required to move a valve 95
Press, Crank pin 536
Pressure, Absolute 215
        Back 215
        Boiler 215
        gauge tester 286
        Initial 215
        plate, Lining down 547
Problems relating to lap of the slide valve 35
Pump won't work 412

Q

Quadrant 111
        Laying of a new 166

INDEX.

Quarter, Meaning of 177
Quartering driving wheels 534

R

Radius of a link, Correct 190
               Finding 112
Reach rod, Broken 411
        length, Finding 106
               Trying 187, 188
Reazor's eccentric blade adjuster 150
Reducing wheel, The Victor 228
Reed's blow-off cock 449
Relation between motion of crank-pin and piston 31
Relative positions of crank-pin and eccentric at full and half stroke 122
         eccentric 172, 173, 176
         piston 31
         tumbling shaft and rocker 129, 111
Release, Meaning of 22
      Point of 29, 30, 39
         trying 163
Reverse lever, Broken 411
Rhode Island system of compounding 399
                Accidents to 399
Richardson balance valve 74
          Allen- 75
Richmond locomotives 353, 490
         system of compounding 352
               Accidents to 358
Right angle, How to make a 558
Right lead engines 188
Rocker arms, Finding backset of standard 107, 108
             direct motion 107, 109
        length of 111
        Laying off, turning and fitting 110
       box, Broken 407
        Effect of lining 184
Rod Brasses, Boring out 575
        Eccentric 553
        Fitting 552, 570
        Hot 322
        Keying 321
Rogers locomotive 486
     system of compounding 381
               Accidents to 386
Rollers for valve setting 147
Rules for valve setting 172
Runs, Extraordinary fast and long distance 504
Running cut-off, How to equalize 189
       board brackets 557
Runners, A word to young 325

S

Saddle .... (See Link Saddle)
Safety pop valve .... (See Pop Valves)
Sander, Track .... (See Track Sanders)
Schenectady locomotive 497, 499
     system of compounding 358
               Accidents to 366
Schwab & Sercomb's valve motion model 151
Seal, Meaning of 23
Seller's injectors 245, 246, 248
Set screws broken in rods 406
Setting up wedges 320
     valves .... (See Valves)
Shelb valve gear 209, 210, 211
Shifting link 97, 100
Shoes and wedges 512
           Finding a square center 518
           Laying off 520, 521
           Liners for 525
           Lining old 523
           Planing 576
Shop practice 102
Shrinkage 562

Side Rod brasses, Boring out	575
Eccentric	553
Fitting	552, 570
cups, Forged	553
Broken	406
Finding length of	551
Finish ends of	552
Keying	321
Putting up	552
set screws, Broken	406
strap, Broken	406
Planing	577
Signal, DeWallace train order	303
Slide valve	(See Valves)
Slip of the link	20, 41
Slipped eccentric	408
blade	409
Smith's roller balance valve	92
triple expansion exhaust pipe	426, 427
Smoke box, Bell's	434
door broken	412
Soft plug	557
Spark arrester, Northern Pacific Ry	440
Speed recorder	312
Spring hanger, Broken driving	413
truck	415
Draw to allow for	557
Indicator	224
Squaring up an engine	517, 518
Stack base, Laying off a	557
Starting a train	323
Stationary link	97, 98
gives constant lead	189, 192
Stay bolts	463
Steam chest or cover, Broken	405
Lining the pressure plate	546
gauges	280
pipes, Laying off	554
ports	25, 174
Saturated	216
Superheated	216
Strainer, Improved hose	260
The wicked	243
Stresses on guides	550
pedestal jaws	528

T

Table Briggs' standard pipe sizes	561
Metric threads	582
United States standard taps and threads	560
drill sizes	561
Tandem Compound, Brooks	377
Taps, United States standard	560, 561
Tapping holes true	556
Technical points	116
Templet, Link	123
Testing plant for locomotives	311
for blows and pounds	314, 318
Tester, Crosby's pressure gauge	286
Thermostatic Gauge, Crosby's	283
Threads, Cutting	565
Metric	582
Sharp V; Square; Mongrel; Double and Left Hand	566, 567
Throttle box, Boring out	574
Throttle valve, Balanced	303
Double	302, 303
Fitting	570
rod, Broken or disconnected	412
Connecting a	555
Throw of eccentric	102, 174
Tinker's cab window	298
Tippett attachment for lubricators	270
Tools, Lathe	564
Grinding	569
Improved	571
Setting	568

INDEX.

Track sanding device, Dean's 595
                    Leach's 585
                    Louisville & Nashville Ry 587
                    Jones' 589
                    Tirman's 591
Tramming Crank Pins 552
Tram marks distinguished from port marks 178
           Equalizing 180
           Meaning of 152
           Scribing 153
Train order signal, The DeWallace 303
Travel of Cross-head, Marking extreme 179
          valve 27, 102
              Finding 102
              in mid-gear 174
              Meaning of 23
              Over 23
              will affect point of cutoff 38
          Piston 216
              Marking extreme 179
Travel marks 156
Truck frame, Laying off engine 555
          Broken 420
Tumbling shaft arms, Broken 410
            bent, Effect of 183
            Length of 130
            Leveling up 184
            Lining 184
            Position of 11, 129
            stands, Broken 410
Turning, Lathe work 569
Turbine check, McLeod's 267
Tyres, Broken 415
    Cutting 419
    Turning 573
       while under engine 530

U

Unit of heat 216
    work 216
United States metallic packing 291, 292, 293, 294
       standard threads 566
       taps 560
       screw threads 565
Utica gauges, Air 285
       Steam 282

V

Valve, Allen ported 50, 51
       Lead for 52, 175
    Balanced slide 64
          Allen-American 77
          -Richardson 75
          American 76
          Barnes' 83
          Briggs' 85
          Brown's 86
          Formula for balancing 83
          Gould's 69
             quick action 65
          Hole in top of 95
          Malone's 88
          Margo's 95
          McDonald's 81
          Nicholson's roller 90
          Piston 72
          Power required to move a 95
          Richardson's 74
          Smith's roller 92
          strips blowing 313
          Vandeventer's roller 94
       throttle 302, 303
    Cam 99
    check .... (See Boiler Checks)
    gear, Allan 97, 98, 192, 193, 194
       Direct motion 109, 191
       Gooch 98, 190, 191, 192

Valve gear, Indirect motion..100, 188, 189
         Joy ....98, 203, to 209
         Lewis ....98, 196 to 203
         Lost motion in 176
         Pennsylvania Railroad 193
         out of square 316
         Shelb 209, 210, 211
         Standard shifting link 99, 100
                     Model of 151
         Walschaert's 97, 98, 194, 196
         Warren's 143
Improved globe 308
Intercepting 331
Planing up 577
rod, Adjusting 183
    Broken 406
    Length of 105
Safety pop ....(See Pops)
seat bridges 25
        Broken 412
        Construction of 24
        exhaust port 26
        False 547
            blowing 313
        Facing 546
            machine for 546
        Longitudinal width of 26
    steam port 25, 174
setting 146
        Alterations in 158
                to equalize the cut-off.................................. 161
                proved by trial 180
        Eccentric blade adjuster for 150
        Farrington's machine for 147
        Finding a dead center 154
        Formula for 165
        implies 147
        Introduction to 146
        Lost motion adjusting machine for.. 149
        Marking port openings when 152
        on a new engine 189
        Preparation for 151
        Reazor's eccentric blade adjuster for... 150
        Recent practice in 163
        Rollers for 148
        Rules for 172
        Trying the lead 157
                cut-off 160
Slide 19, 20
    blowing 313
    broken 412
    clearance ....(See Clearance)
    Construction of 20
    cut-off, Goodspeed's 53
             Haley's 57
             Metzger's 60
    Elementary principles of the... 20
    Farrer's 64
    Fay's 62
    Holt 53
    Invention 20
    Lap of ....(See Lap)
    Lead of ....(See Lead)
    Merits of the 28
    Planing 577
    reducing compression, Farrer's... 64
                          Fay's 62
    Travel of ....(See Travel)
sounding out 316
stem, Broken 412
    glands, Broken 406
    keys, Draw on 553
    Length of 105
    Turning 570
Vacuum relief 96
yoke, Broken 404
    Cracked 404

Valve yoke, Fitting.. 106
            Laying off 106
Victor reducing wheel 228

W

Wallace & Kellogg's draft regulating device............................ 425
                   feed water heater 299
Walschaert's valve gear..97, 98, 194, 195, 196
Warren's draft regulating device 435
        improved link 145
Water, Amount to carry 424
      foaming in boiler 403
      glass, Broken 413
          Cutting 556
Wedge bolt, Broken 420
          Laying off and fitting up...................................... 520
          Planing 576
          Setting 320
Wheels and Axles 530
        Driving, Broken 415
          Cast steel centers 536
          Fitting axle to 569
              keys into 535
          Lateral between hubs 526
          main, Effect of changing end for end........ 169
          Quartering 534
          Tramming 524
              which pair are out 552
          Turning tyres on 530, 573
        Tender, Broken 420
        Truck, Broken 419
Wheel, Victor reducing 228
Whistle, Chime 310
      blown out 413
Work, Unit of 216
Wire drawing 215

Y

Yoke, Broken guide 407
      valve 404
      Laying off a valve 106

ADVERTISEMENTS.

If you seek for information
   You may find it in this wise,
By looking through the pages
   Of those who advertise.

And remember as you thus search
   Did these pages not appear,
The book's price would be double
   What we ask you for it here.

And so we give our blessing
   To the men who advertise,
And urge that every reader
   These firms now patronize.

FULLER'S "WALKEASY" Artificial Legs

Established 1857

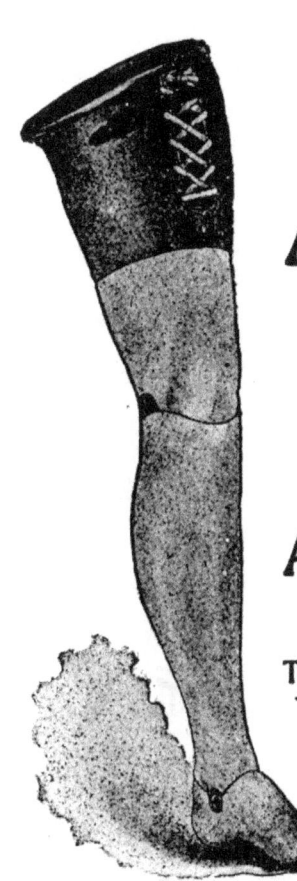
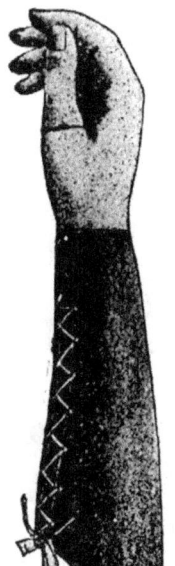

With Sponge Rubber Feet and BALL-BEARING JOINTS represent the latest progress in Artificial Limb Construction. Wood, Leather or "Neverchafe" Sockets.

Arms With Ball and Socket Wrists, Automatic Elbow, etc. : : : :

SEND FOR BOOK.

Trusses, Crutches, Elastic Stockings, etc.

GEORGE R. FULLER CO.

15 S. St. Paul Street, Rochester, N. Y.

Branch, 306 Main Street, Buffalo, N. Y.

Air Brake Catechism

BY C. B. CONGER

ON HANDLING THE AUTOMATIC BRAKE, WITH A LIST OF EXAMINATION QUESTIONS FOR ENGINEMEN AND TRAINMEN : : : : : : :

Containing the Revised List of the Air Brake and Signal Instructions of the Joint Committee of the Master Car Builders and Master Mechanics' Associations.

This book on the Automatic Brake is by an up-to-date air-brake man whose explanations are full and clear so that they are easily understood.

PRICE: 25c a Copy or 10 Copies for $2.00.

APPLY TO

LOCOMOTIVE ENGINEERING

256 BROADWAY, NEW YORK.

Baldwin Locomotive Works
Builders of Single Expansion and Compound Locomotives

BURNHAM, WILLIAMS & CO.
PHILADELPHIA, PA., U. S. A.

ADAPTED TO EVERY VARIETY OF SERVICE.

ANNUAL CAPACITY 1,000.

Richmond Locomotive and Machine Works
Richmond, Va. U.S.A.

BUILDERS OF SIMPLE AND COMPOUND ::: **LOCOMOTIVES** FOR ALL CLASSES OF SERVICE

Brooks Locomotive Works, Dunkirk, N.Y.

BUILDERS OF LOCOMOTIVE ENGINES FOR ANY REQUIRED SERVICE....

COMPOUND LOCOMOTIVES FOR PASSENGER AND FREIGHT SERVICE ::::

ANNUAL CAPACITY 400

ROGERS LOCOMOTIVE COMPANY

PATERSON, NEW JERSEY

Address Paterson, N. J., or 44 Exchange Place, N. Y.

BUILDERS OF

Locomotive Engines

OF EVERY DESCRIPTION

R. S. HUGHES, President.
G. E. HANNAH, Treasurer.

G. H. LONGBOTTOM, Secretary.
REUBEN WELLS, Superintendent.

JOHN S. COOKE, PRESIDENT.   F. W. COOKE, VICE-PRES'T AND GEN'L MGR.   C. D. COOKE, SEC'Y AND TREAS.

COOKE LOCOMOTIVE AND MACHINE CO.
PATERSON, N. J., U. S. A.

NEW YORK OFFICE,
ROOM 506, 141 BROADWAY.

WESTERN OFFICE,
1439 MONADNOCK BLOCK, CHICAGO.

Designers and Builders
of all Classes of....

LOCOMOTIVES

New Works completely equipped with all modern facilities for doing high grade work economically.

Sole Builders of the "ROTARY" SNOW PLOW

THE INSTRUCTOR IS A MODEL DESIGNED TO TEACH THE SETTING OF THE VALVE OF AN ENGINE

FULL NICKEL $15.00. JAPANNED AND NICKELED $10.00.

29½ inches in length and 10 inches high. 13½ pounds.

SCHWAB & SERCOMB

PATENTEES AND SOLE
MANUFACTURERS ::::::

271 Clinton Street...Milwaukee, Wis.

Correspondence Solicited.

The Instructor is one-eighth the size of a 17 x 24 locomotive engine, showing a section through the cylinder and steam-chest, together with the piston and valve.

The connections that require changing for the purpose of setting the valve are adjustable.

The driver (balance wheel) is turned, edges faced, axle (crank shaft) is centered and fits the shaft boxes, which are bored.

By means of a spring a report is given, which indicates the exhaust and can be regulated.

It is a mechanical design of the common slide valve with link, piston, rocker arm and all necessary connections, and is an invaluable aid in the study of link and valve motion.

Low Fuel Records

Can always be made on engines that are equipped with ::::

The Detroit Lubricators

With Tippett Attachment

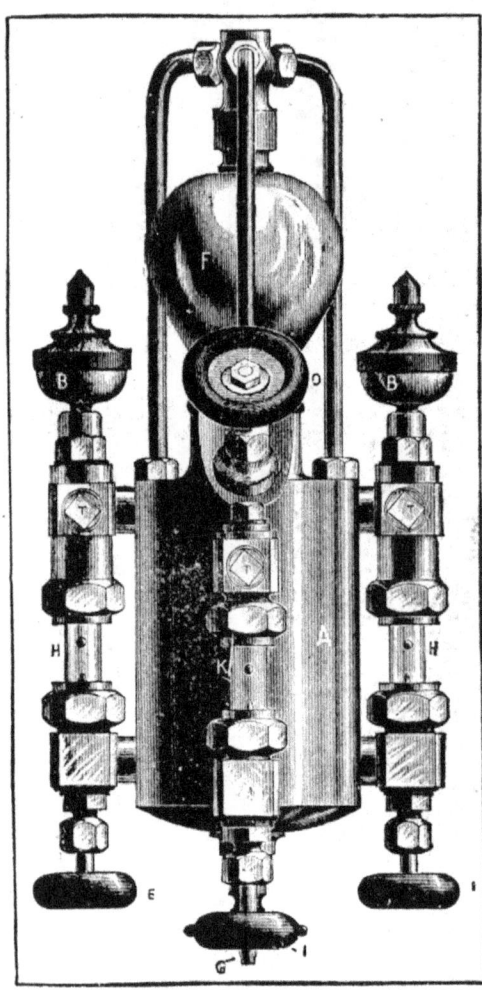

They overcome the back pressure absolutely, insure regular and perfect lubrication of the valves and cylinders, and thus prevent waste of power from valve friction. When no power is wasted in this way, less steam and less fuel are necessary to do the legitimate work of the locomotive.

Our New Catalogue

showing our lubricators in section, and blue print of Tippett Attachment, will be sent on application.

Detroit Lubricator Co., Detroit, Mich.

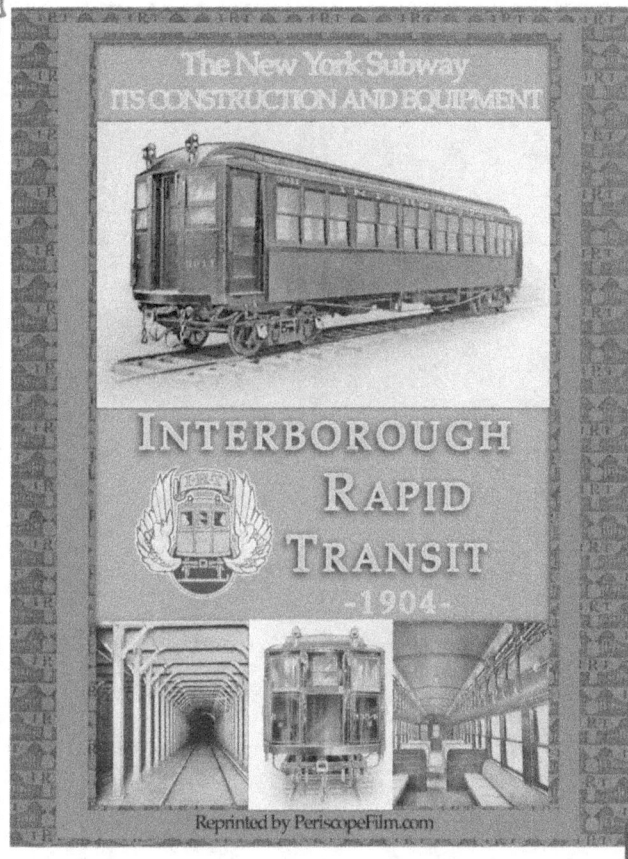

On October 27, 1904, the Interborough Rapid Transit Company opened the first subway in New York City. Running between City Hall and 145th Street at Broadway, the line was greeted with enthusiasm and, in some circles, trepidation. Created under the supervision of Chief Engineer S.L.F. Deyo, the arrival of the IRT foreshadowed the end of the "elevated" transit era on the island of Manhattan. The subway proved such a success that the IRT Co. soon achieved a monopoly on New York public transit. In 1940 the IRT and its rival the BMT were taken over by the City of New York. Today, the IRT subway lines still exist, primarily in Manhattan where they are operated as the "A Division" of the subway. Reprinted here is a special book created by the IRT, recounting the design and construction of the fledgling subway system. Originally created in 1904, it presents the IRT story with a flourish, and with numerous fascinating illustrations and rare photographs.

Originally written in the late 1900's and then periodically revised, A History of the Baldwin Locomotive Works chronicles the origins and growth of one of America's greatest industrial-era corporations. Founded in the early 1830's by Philadelphia jeweler Matthais Baldwin, the company built a huge number of steam locomotives before ceasing production in 1949. These included the 4-4-0 American type, 2-8-2 Mikado and 2-8-0 Consolidation. Hit hard by the loss of the steam engine market, Baldwin soldiered on for a brief while, producing electric and diesel engines. General Electric's dominance of the market proved too much, and Baldwin finally closed its doors in 1956. By that time over 70,500 Baldwin locomotives had been produced. This high quality reprint of the official company history dates from 1920. The book has been slightly reformatted, but care has been taken to preserve the integrity of the text.

NOW AVAILABLE AT
WWW.PERISCOPEFILM.COM

THE CLASSIC 1911 TROLLEY CAR BUILDER'S REFERENCE BOOK

ELECTRIC RAILWAY DICTIONARY

By Rodney Hitt
Associate Editor, Electric Railway Journal

REPRINTED BY PERISCOPEFILM.COM

THE CLASSIC 1915 TROLLEY CAR AND INTERURBAN RAILWAY BOOK

ELECTRIC RAILWAY ENGINEERING

By Francis H. Doane, A.M.B.

REPRINTED BY PERISCOPEFILM.COM

©2008-2010 Periscope Film LLC
All Rights Reserved
ISBN #978-1-935700-21-0

www.PeriscopeFilm.com

www.ingramcontent.com/pod-product-compliance
Lightning Source LLC
Chambersburg PA
CBHW080527300426
44111CB00017B/2631